Molecular Wires and Nanoscale Conductors

University of Manchester
31 August 2005–02 September 2005

FARADAY DISCUSSIONS

Volume 131, 2005

RSC Publishing

The Faraday Division of the Royal Society of Chemistry, previously the Faraday Society, founded in 1903 to promote the study of sciences lying between Chemistry, Physics and Biology.
http://www.rsc.org/science/physical.htm

Faraday Discussions (Print ISSN 1359-6640, Online ISSN 1364-5498) is published 3 times a year by the Royal Society of Chemistry, Thomas Graham House, Science Park, Milton Road, Cambridge, UK CB4 0WF.
Volume 131 ISBN 0-85404-988-6
2006 annual subscription price: Print+Online £470, US $861; online £423, US $775. Customers in Canada will be subject to a surcharge to cover GST. Customers in the EU subscribing to the electronic version only will be charged VAT. All orders accompanied with payment should be sent directly to RSC Distribution Services, Portland Press Ltd., Commerce Way, Whitehall Industrial Estate, Colchester, UK CO2 8HP. If you take a full price institutional subscription to any RSC primary or review journal you are entitled to free site-wide electronic access to that journal. You can arrange access *via* Internet Protocol (IP) address by registering your site's IP address at www.rsc.org/ip. Customers should make payments by cheque in sterling payable on a UK clearing bank or in US dollars payable on a US clearing bank. Periodicals postage paid at Rahway, NJ. Airfreight and mailing in the USA by Mercury Airfreight International Ltd. Inc., 365 Blair Road, Avenel, NJ 07001, USA and at additional mailing offices.

USA Postmaster: send address changes to *Faraday Discussions*, c/o Mercury Airfreight International Ltd. Inc., 365 Blair Road, Avenel, NJ 07001. All despatches outside the UK by Consolidated Airfreight.

PRINTED IN THE UK.

Faraday Discussions documents a long-established series of *Faraday Discussion* meetin which provide a unique international forum for the exchange of views and newly acquired results in developing areas of physical chemist biophysical chemistry and chemical physics.

Typeset by Macmillan India Ltd., Bangalore, India
Printed by Cambrian Printers, Aberystwyth, UK

Molecular Wires and Nanoscale Conductors

Faraday Discussions

www.rsc.org/faraday_d

A General Discussion on Molecular Wires and Nanoscale Conductors was held at the University of Manchester, UK on 31st August, 1st and 2nd September 2005.

CONTENTS

ISSN 1359-6640; ISBN 0-85404-988-6

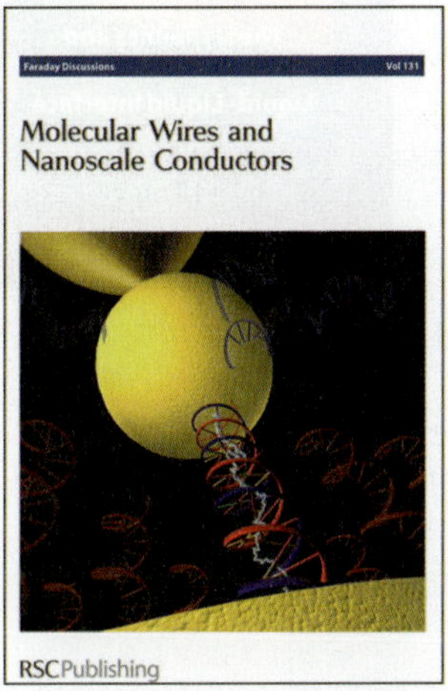

Cover
See Hezy Cohen, Claude Nogues, Daniela Ullien, Shirley Daube, Ron Naaman and Danny Porath, *Faraday Discuss.*, 2006, **131**, 367–376.
Electrical transport through a double-stranded DNA molecule attached on opposite ends by thiols to a gold nanoparticle and a metal substrate is measured using a conductive AFM tip. Cover design by Noam Hamou.

Image reproduced by permission of Dr Danny Porath and Noam Hamou from *Faraday Discuss.*, 2006, **131**, 367.

INTRODUCTORY LECTURE

PAPERS AND DISCUSSIONS

ADDITIONAL INFORMATION

Electronic supplementary information (ESI) is available *via* the online article (see http://www.rsc.org/esi for general information about ESI).

Spiers Memorial Lecture

Molecular mechanics and molecular electronics

Robert Beckman,[a] Kris Beverly,[a] Akram Boukai,[a] Yuri Bunimovich,[a] Jang Wook Choi,[a] Erica DeIonno,[a] Johnny Green,[a] Ezekiel Johnston-Halperin,[a] Yi Luo,[a] Bonnie Sheriff,[a] J. Fraser Stoddart[b] and James R. Heath*[a]

Received 15th September 2005, Accepted 20th September 2005
First published as an Advance Article on the web 5th October 2005
DOI: 10.1039/b513148k

We describe our research into building integrated molecular electronics circuitry for a diverse set of functions, and with a focus on the fundamental scientific issues that surround this project. In particular, we discuss experiments aimed at understanding the function of bistable [2]rotaxane molecular electronic switches by correlating the switching kinetics and ground state thermodynamic properties of those switches in various environments, ranging from the solution phase to a Langmuir monolayer of the switching molecules sandwiched between two electrodes. We discuss various devices, low bit-density memory circuits, and ultra-high density memory circuits that utilize the electrochemical switching characteristics of these molecules in conjunction with novel patterning methods. We also discuss interconnect schemes that are capable of bridging the micrometre to submicrometre length scales of conventional patterning approaches to the near-molecular length scales of the ultra-dense memory circuits. Finally, we discuss some of the challenges associated with fabricated ultra-dense molecular electronic integrated circuits.

I. Introduction

Several years ago we began a research program that was focused around the concept of building a molecular electronic computing machine. Our goals were several-fold, including minimization of energy consumption per bit operation, development of computational and memory architectures that are both defect-tolerant and amenable to chemical assembly,[1] and the development of parallel manufacturing approaches that scale to molecular dimensions. The longest term goal is to construct integrated nanoelectronic circuits with multiple functions. These functions include the traditional electronic tasks of routing, logic and memory, but can also include non-traditional tasks such as biomolecular or chemical sensing, photonics, actuation, *etc.*, that are enabled through molecular scale fabrication. Our 'vision' for such an integrated circuit is shown in Fig. 1. The crossbar architecture that is the dominant circuit theme shown in the figure is loosely based upon the Teramac supercomputer—a machine that we demonstrated could provide an excellent model system for nanoscale and molecular electronics.[1]

The challenges towards constructing such a multifunctional computational machine are non-trivial. One challenge surrounds the development of patterning techniques that can approach molecular dimensions in terms of both feature size and feature pitch. A second requires the development of architectures that allow for the bridging of length scales between the micrometre and sub-micrometre world of lithographic patterning to the nanometre scale world of self-assembly (the multiplexers of Fig. 1). A third challenge revolves around the development of molecular

[a] *Caltech Chemistry, MC 127-72 1200 East California Boulevard, Pasadena CA 91125, USA. E-mail: heath@caltech.edu*
[b] *The California NanoSystems Institute and Department of Chemistry and Biochemistry, UCLA 405 Hilgard Avenue, Los Angeles CA 90095-1569, USA. E-mail: stoddart@chem.ucla.edu*

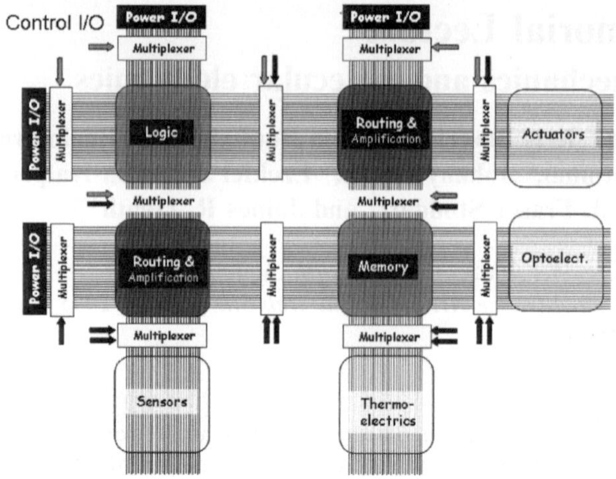

Fig. 1 A molecular and nanoelectronic computational platform based upon a patchwork of crossbar structures. Various traditional and non-traditional functions are illustrated, with multiplexers (and demultiplexers) controlling communication between the various functions and input/output (I/O) with the outside world. This particular circuit architecture is defect tolerant and amenable to both traditional and non-traditional patterning methods.

electronic switches. However, the fundamental scientific issues that must be addressed to meet each of these challenges are very rich. In this paper, we will discuss some of our work in these various areas, with a focus on both the fundamental scientific problems that have been addressed, and the outstanding questions that must still be answered.

Note that although many applications are labeled in the Fig. 1 circuit diagram, there is only one architecture—that of the crossbar—for all functions. The major hardware components that differentiate these various functions are the nanowire materials and the molecules that are coated on top of the nanowires or sandwiched in between the junctions. For the memory applications, the molecular systems that we have explored have been bistable molecular mechanical systems known as electrochemically switchable donor–acceptor [2]rotaxanes and [2]catenanes. It is towards these molecules, and the associated molecular electronic devices, that we first turn our attention.

II. Molecular mechanical switches

Over the past several years two previously distinct areas of chemical research—the synthesis of molecular mechanical systems, and the fabrication of solid-state molecular electronics devices[2] and circuits—have been conjoined. The resulting union has led to the scientific demonstration of a number of interesting devices, including molecular electronic switches,[3] light harvesting devices,[4] molecular electronic based random access memory and configurable logic circuits,[5] molecular mechanical biosensors,[6] actuated molecular valves, ion channel mimics, molecular muscles,[7] and novel electrochromic devices.[8] Equally interesting has been the emerging picture of large-amplitude molecular actuation and how it is influenced and modified by the surrounding physical and chemical environments.[9]

In Fig. 2 we present the molecular structure of a bistable [2]rotaxane ($RATTF^{4+}$) molecular switch, along with a potential energy diagram describing the bistable nature of the switch. This bistable [2]rotaxane is composed of π-electron-accepting cyclobis(paraquat-*p*-phenylene)($CBPQT^{4+}$) ring (blue) that encircles either a tetrathiafulvalene (TTF) unit (green) or a 1,5-dioxynapthalene (DNP) unit (red), both π-electron-donating systems. These mechanically-interlocked molecular compounds and other closely related bistable rotaxanes[10] as well as rotaxanes constructed from different donor–acceptor units[11] and from hydrogen-bonded systems[12] and transition metal templates,[13] have been investigated in depth previously. Under ambient conditions in acetonitrile solution, the $CBPQT^{4+}$ ring in $RATTF^{4+}$ encircles the TTF unit preferentially

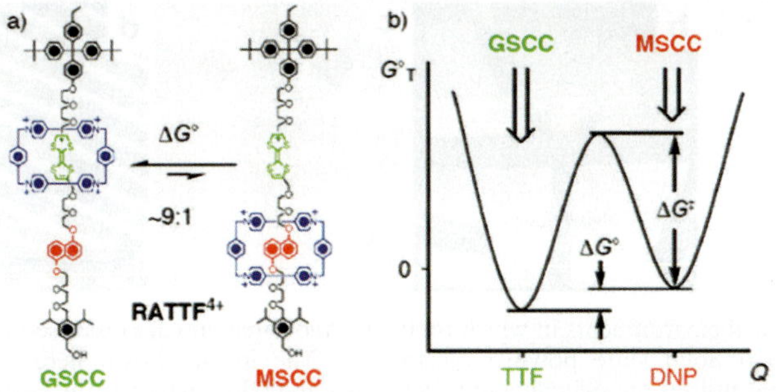

Fig. 2 (a) Structural formulas of the two translational isomers of the bistable rotaxane $RATTF^{4+}$ corresponding to the ground state co-conformation (GSCC) and the metastable state co-conformation (MSCC). (b) Potential energy surface for the bistable $RATTF^{4+}$ where the energy wells correspond to the GSCC and MSCC. The free energy difference ΔG°, between the wells and the free energy barrier to relaxation, $\Delta G^\ddagger$, from the MSCC to the GSCC are defined against a normal coordinate, Q, representing translation of the ring along the dumbbell component of the [2]rotaxane.

(>90%) with respect to the DNP unit. This equilibrium is described by the $\Delta G_{298}{}^\circ$ change shown in Fig. 2b where $\Delta G^\circ = +1.6$ kcal mol^{-1} when the $CBPQT^{4+}$ ring moves from the TTF to the DNP unit. Hence, the co-conformation (CC) with the $CBPQT^{4+}$ ring encircling the TTF is unit is referred to as the ground state co-conformation (GSCC). The first two oxidation states of $RATTF^{4+}$ correspond to the $TTF^0 \rightarrow TTF^{\bullet+} \rightarrow TTF^{2+}$ processes. Upon formation of $TTF^{\bullet+}$ cation radical, the Coulombic repulsion between the $CBPQT^{4+}$ ring and the $TTF^{\bullet+}$ results [14,15] in translation of the ring to the DNP unit. This process is fast (on the millisecond time scale) and is believed to convert all of the GSCC into the MSCC. When the $TTF^{\bullet+}$ cation radical is reduced back to TTF^0, the $CBPQT^{4+}$ ring remains around the DNP unit for a period of time. This translational isomer of the GSCC is the metastable state co-conformation (MSCC). Recovery of the MSCC/GSCC equilibrium distribution (~1 : 9) is an activated process. This switching cycle can be detected by a number of experimental observations. First, the lowest oxidation potential (corresponding to $TTF^0 \rightarrow TTF^{\bullet+}$) of the GSCC is +490 mV, while that for the MSCC is +310 mV. (All potentials referenced to an Ag/AgCl electrode.) Second, the colors of GSCC- and MSCC-dominated solutions are green and red, respectively. Thus, electrochemistry and spectroscopy can be employed to quantify the MSCC/GSCC ratio in such a bistable rotaxane at any given time. Third, the (activated) relaxation of an MSCC- back to a GSCC-dominated distribution is temperature-dependent and so the kinetic parameters may be quantified through time- and temperature-dependent measurements. For example, the $\Delta G_{298}{}^\ddagger$ for this process in the case[16] of $RATTF^{4+}$ is 16.2 (±0.3) kcal mol^{-1}.

Consider, for example, a molecule that, when incorporated into an electronic device, acts as a solid-state molecular resistor[17,18] or rectifier,[19–23] or perhaps generates some unusual device characteristic, such as negative differential resistance.[24,25] Such properties are uniquely studied in the device setting and may arise from a distinctly molecular property, or from the result of the molecule/electrode interaction. When these molecules are studied in the liquid solution phase, those same properties are no longer directly revealed in experiments, although donor–acceptor charge transfer spectroscopy can provide an indirect probe of molecular rectification. In other words, certain properties arise through a combination of the molecular design and the physical (device) environment in which that molecule is incorporated. Molecular mechanical motion, however, has certain key characteristics that are experimentally observable and that are independent of environment. This is especially so for the case of the synthetically flexible catenanes and rotaxanes highlighted in this Discussion.[26] There are not only multiple ways to design and control molecular mechanical motion in these systems, but there are also a variety of experimental methods that can be employed to interrogate it, regardless of environment.

Our approach towards understanding switching within molecular electronic device settings has been to quantify how the environment of the molecular switch impacts the kinetic and thermodynamic parameters associated with electrochemically driven, bistable molecular mechanical switching. We have established two criteria that define a good molecular switch. First, the thermodynamics and kinetics of the molecular mechanical switching processes should correlate

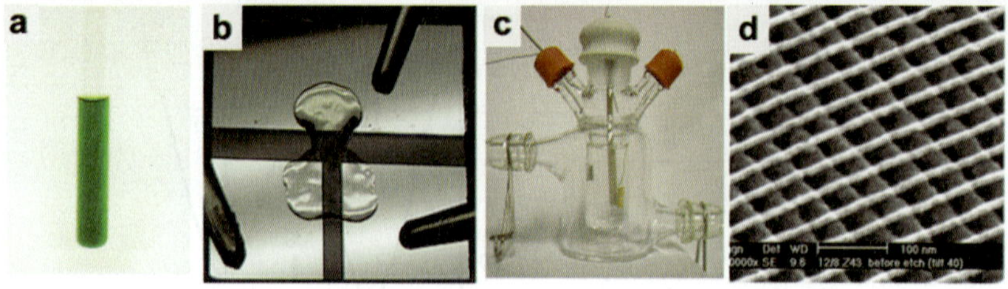

Fig. 3 The four physical environments in which rotaxanes and catenanes are discussed within this Lecture. (a) In organic solvent. (b) Solid state polymer electrolytes. The image shown here is of a three-terminal electrochemical cell that utilizes the switching of [2]rotaxanes and [2]catenanes as the basis of an electrochromic device. (c) Self-assembled monolayers on the surface of a working electrode. The example shown here is of a [2]rotaxane assembled on the surface an a working Au electrode within an electrochemical cell. (d) A molecular switch tunnel junction. Shown is a crosspoint memory in which a Langmuir monolayer of [2]rotaxanes is sandwiched between a top Ti/Al electrode and a bottom poly-Si electrode.

to those same properties when measured in the solution phase. Second, control molecules shouldn't switch. The change between the solution phase and the sterically restricted environment of a molecular electronic device is likely to significantly modify at least some subset of the thermodynamic and kinetic parameters for switching. Thus, our approach has been to investigate bistable switching in [2]rotaxanes and [2]catenanes in a host of environments,[27] including the solution phase, solid-state polymer electrolyte gels, self-assembled molecular monolayers tethered to the surface of working electrodes within electrochemical cells,[15] and molecular switch tunnel junction devices. Representative images of each of these environments are shown in Fig. 3.

Much of our work on quantitating the switching kinetics and the equilibrium thermodynamics in these various environments has been published elsewhere. From those measurements, the mechanistic picture illustrated in Fig. 4 has emerged as being universal for both [2]catenanes and [2]rotaxanes across all physical environments. We initially discuss the switching kinetics (*e.g.* measurements of $\Delta G^{\ddagger}$ and other associated parameters from Fig. 2), starting with a self-assembled monolayer of bistable [2]rotaxanes on the surface of a working Au electrode.[15] The objectives of these measurements are to quantify the recovery of the ground state equilibrium distribution (described by ΔG° of Fig. 2 with an equilibrium constant of $K_{(D/T)}{}^{4+}$ as shown in Fig. 4) following electrochemical switching of a GSCC-dominated population (ground state) to an MSCC-dominated population (metastable state) of [2]rotaxanes.

Cyclic voltammetry (CV) scans, and accompanying descriptions of the redox processes, are shown in Fig. 5 for the surface-tethered [2]rotaxane and its dumbbell component.

In Fig. 6 we present two CV scans, taken in rapid succession, of just the reversible oxidization of the TTF groups on the same surface-tethered bistable [2]rotaxane.

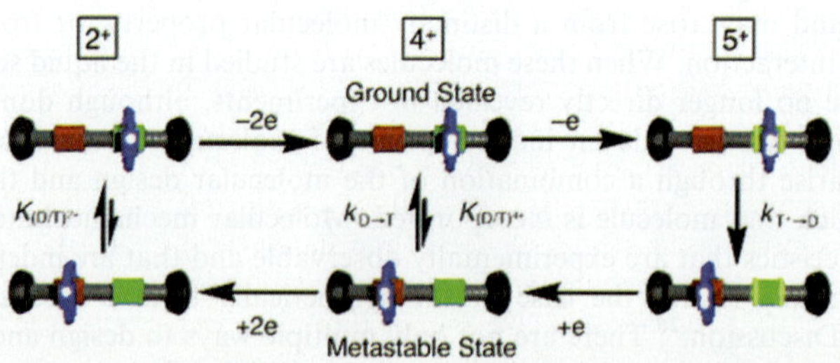

Fig. 4 The switching cycle of a bistable [2]rotaxane, starting with the ground state conformer. The green and red sites refer to the TTF and DN recognition units. When the TTF is oxidized, it is drawn as a highlighted green unit. The $CBPQT^{4+}$ ring is modeled as the blue ring encircling the green site, with positive charges indicated as white spots. The metastable and ground state co-conformers are isoelectronic. The various rate constants indicated are discussed in the text. The charge states labeled mid-line in the drawing refer to the rotaxane structures below the labels. An exactly analogous mechanism fits the switching of bistable [2]catenanes that are constructed from similar host/guest components.

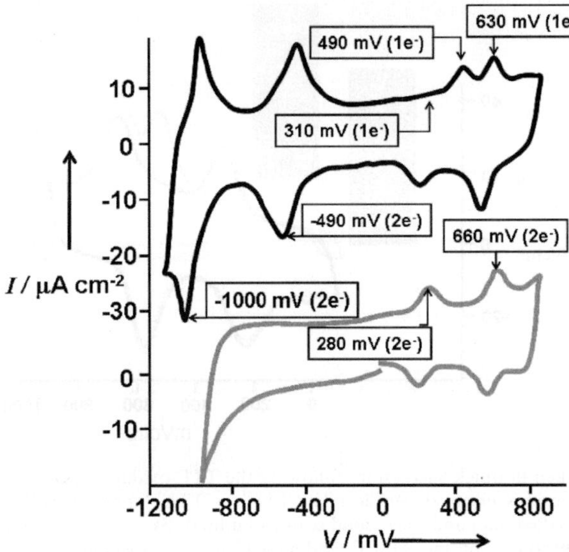

Fig. 5 Cyclic voltammograms (CVs) of a bistable [2]rotaxane tethered (through a disulfide linkage) to the surface of an Au working electrode within an electrochemical cell. All voltages are referenced to Ag/AgCl. The top CV trace is of a bistable [2]rotaxane containing TTF and DN recognition groups. In the ground state conformation, a $CBQT^{4+}$ ring preferentially encircles the TTF group. The two oxidation peaks at 490 and 630 mV correspond to $TTF^0 \rightarrow TTF\cdot^{+\bullet} \rightarrow TTF^{2+}$ oxidation steps, which are reversed on the return sweep. The two reduction peaks at −490 and −1000 mV each correspond to two-electron processes, and are the $CBQT^{4+} \rightarrow CBQT^{2+} \rightarrow CBQT^0$ reduction processes. The bottom trace is a CV of the dumbbell component of the tethered [2]rotaxane (*e.g.* no $CBQT^{4+}$ ring). Note that the $TTF^0 \rightarrow TTF^{\bullet}+$ process is shifted by −210 mV, although the $TTF^{\bullet+} \rightarrow TTF^{2+}$ process is relatively unchanged. This indicates that the $CBQT^{4+}$ ring translates to the DN site after the TTF group is singly oxidized.

Note that the first oxidation peak is observed to shift by ∼−200 mV between the first and second scans. This hysteresis implies that the $CBQT^{4+}$ ring encircles the TTF-site during the first scan, but it has not yet returned to the TTF site when the second cv scan was carried out—*i.e.* it is still on the DN site. If a long time period existed between the first and second scans, then the two scans are observed to overlay (no hysteresis is observed in the first oxidation peak). This is because the ground state equilibrium distribution would have been re-established. Thus, this type of CV data can be measured as a function of temperature and time, and the kinetic parameters associated with the recovery of the GSCC-dominated equilibrium from a MSCC-dominated monolayer can be established. In a similar way, solution-phase data and data within polymer electrolytes can be collected. For all of these environments, the relative GSCC and MSCC confirmations can be established quantitatively. For molecular switch tunnel junction devices, the corresponding experiment is to electronically switch the device from the low- to the high-conducting state, and then follow the time-dependent decay back to the low-conducting state.

Such measurements can be carried out as a function of temperature. In this way, Eyring plots can be constructed for similar molecules in all four environments, and such plots and a table of the derived kinetic parameters are presented in Fig. 7. These type of measurements have been done for several [2]rotaxanes and [2]catenanes, and the picture that has emerged is that the switching (relaxation) kinetics are strongly dependent upon both the molecular structure of the switch, and the physical environment that the switch has been placed in. In fact, both effects can be harnessed to yield approximately equal kinetics control. For example, relaxation times in polymer-based electrochromic devices can be controlled over a dynamic range of approximately 10^4 through control over the molecular structure of the switch.[8] Similarly, taking an identical switch (with adaptations to the stopper groups that allow for it to be incorporated into different environments), the relaxation times are reduced by approximately 10^4 or 10^5 when moving from the solution phase to the polymer to the self-assembled monolayer to the molecular switch tunnel junction device.

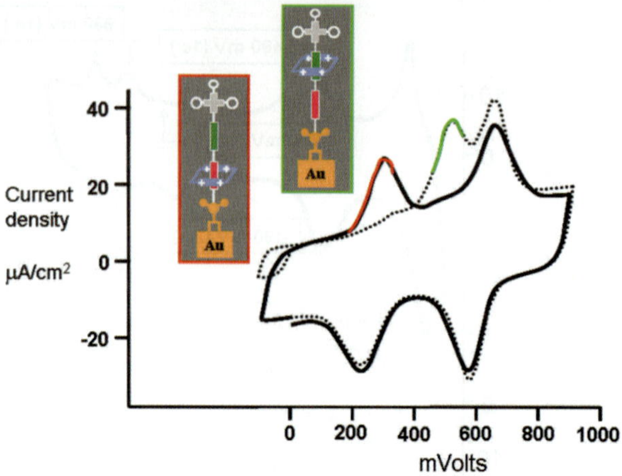

Fig. 6 Two cv scans taken in quick succession, following the TTF-oxidation process of the tethered bistable [2]rotaxane on the Au working electrode surface. The $TTF^0 \rightarrow TTF^{\bullet+}$ process exhibits a hysteretic response between the first scan (dashed line) and the second scan (solid line). By integrating these two peaks (green and red) the relative concentrations of the MSCC and GSCC may be evaluated as a function of time and temperature, thereby allowing for a quantitative analysis of $\Delta G^{\ddagger}$ and associated kinetic parameters.

We have also measured the equilibrium thermodynamics within a wide variety of environments, and that work will appear in press soon.[9b] Briefly, we designed three molecules based upon differential scanning calorimetry measurements of the various host–guest complexes. One molecule ($RTTF^{4+}$) utilized the same [2]rotaxane switching motif shown in Fig. 2, but the hydrophilic stopper was replaced with a much bulkier group that contained three polyethylene glycol oligomers. For the second [2]rotaxane ($RBPTTF^{4+}$), we replaced the TTF with a bispyrrolotetrathiafulvalene (BPTTF) unit. A third control [2]rotaxane ($RBLOCK^{4+}$) was designed with an ethyl group at the base of the TTF group, and this served to lock the $CBQT^{4+}$ ring so that it could not translate away from the TTF site, whether or not the TTF site was oxidized. $RTTF^{4+}$ was designed to exhibit a temperature-independent ground state equilibrium that is dominated (90%) by the $CBQT^{4+}$ ring encircling the TTF site. $RBPTTF^{4+}$ exhibits a strong temperature dependent equilibrium that is

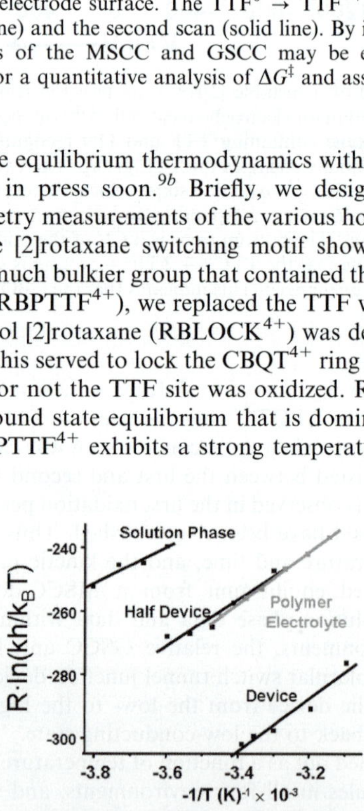

DATA	E_a (kcal mol⁻¹)	ΔG‡(kcal mol⁻¹)	ΔH‡ (kcal mol⁻¹)	ΔS‡ (cal mol⁻¹ K⁻¹)
Solution:	14.1 (± 1.2)	15.9 (± 1.2)	13.5 (± 1.2)	-8.9 (± 5)
SAM ½ device:	17.7 (± 0.6)	18.0 (± 0.7)	17.6 (± 0.6)	-1.4 (± 3)
Polymer:	19.6 (± 0.2)	18.1 (± 0.2)	19.0 (± 0.2)	3.0 (± 1.5)
Tunnel Junction:	19.8 (± 0.3)	21.4 (± 0.3)	19.8 (± 0.3)	5.5 (± 2)

Fig. 7 Eyring plots for bistable [2]rotaxane molecular switches entrained in four different environments as labeled on the graph. Kinetic parameters extracted from this and similar data are presented in the table. The [2]rotaxane contained TTF and DN recognition units and a $CBQT^{4+}$ ring, but each molecule differed slighted in the terminal stoppers on the dumbbell component. The stoppers were modified to allow for adapting the molecule into the different physical environments.

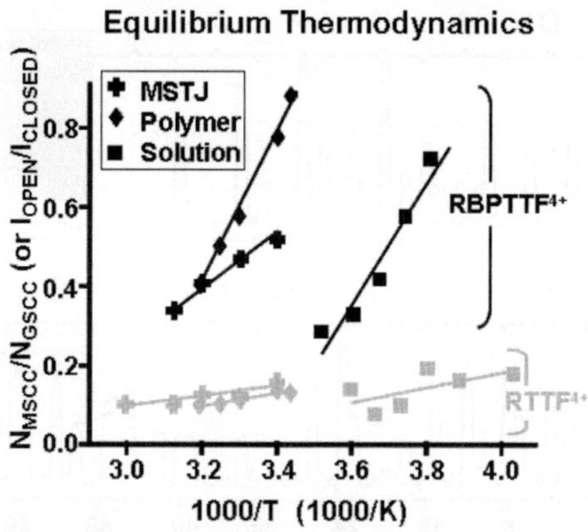

Fig. 8 The temperature-dependent GSCC/MSCC equilibria for three environments are presented for $RBPTTF^{4+}$ and $RTTF^{4+}$. Note that the large (entropically-driven) temperature dependence for $RBPTTF^{4+}$, and the relative temperature independence of $RTTF^{4+}$, is reflected in all three environments.

dominated by very different entropies when the $CBQT^{4+}$ ring encircles the RBPTTF site (−3.6 cal K^{-1} mol^{-1}) or the DN site (−30.8 cal K^{-1} mol^{-1}). The entropy associated with the $CBQT^{4+}$ ring encircling the TTF site was (−22.1 cal K^{-1} mol^{-1}).

The high entropy values are ascribed to configurational entropy that arises from interactions between the $CBQT^{4+}$ ring and the ethylene glycol bridging units. For these measurements, we quantified the amount of GSCC *versus* MSCC for the solution and polymeric environments using cyclic voltammetry. For the molecular switch tunnel junction environment, we evaluated the temperature dependent switching amplitude. Since the high conductance state is associated with the MSCC, a large ground state equilibrium population of the MSCC will reduce the switching amplitude.

$RBLOCK^{4+}$ was observed to be a poor switch in all environments, as expected. Data for $RBPTTF^{4+}$ and $RATTF^{4+}$ are presented together in Fig. 8. This data provides strong evidence that the ground state equilibrium thermodynamics of these molecules is dominated by molecular structure, with the physical environment of the molecular switch playing only a small role.

III. Molecular electronic circuits

One of our motivations for developing molecular switches such as the $RTTF^{4+}$ molecule described above was to be able to apply them as the active (switching) devices within molecular electronic memory circuits. These switches hold a number of advantages (and disadvantages) over more traditional two-terminal switches, such as ferroelectric materials. First, since the switching property is distinctly molecular, devices based upon these switches should scale to near molecular dimensions without a significant change in the switching characteristics. This is distinct from solid state materials such as ferroelectrics. In those materials, a crystallographic domain is 'poled' using an electric field applied between the two junctions. For very small junction sizes, and thus small domains, the hysteretic response of the material vanishes. For magnetic materials, this is referred to as superparamagnetism. The second advantage is that the voltage at which the switches are opened or closed is quite stable, since the switching mechanism is electrochemical in nature (*e.g.* current has to flow in and out of molecular orbitals). For field poled devices, the actual poling process is driven by nucleation, and so is intrinsically statistical. This means that the voltage at which a ferroelectric bit switches can vary from junction to junction, or within a single junction, it varies from cycle to cycle. The net result is that, within a crosspoint memory architecture, it is difficult to get reliable performance from solid-state materials. Our molecular electronic switches, however, work well within a crosspoint environment. A more detailed discussion of this important issue can be found elsewhere.[5]

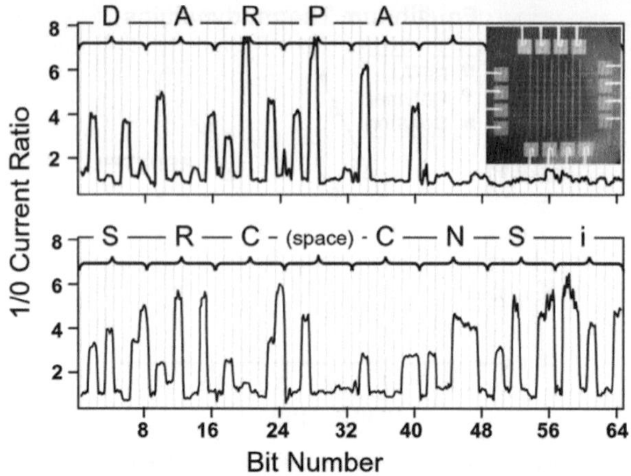

Fig. 9 The performance of a 64-bit molecular electronic crosspoint memory circuit (circuit shown in the upper right optical micrograph). The switching elements were bistable [2]rotaxanes, and the ASCII characters of the text shown above the two traces were stored as 1's and 0's, and then read out.

These molecular mechanical switches are not without their disadvantages. The full switching mechanism contains thermally activated processes, which means that the devices will exhibit temperature-dependent variations in their operation.[9b] Second, molecular switch tunnel junction devices (MSTJs) made from [2]rotaxanes and [2]catenanes have, to date, not been particularly robust performers. For example, a switch that may be cycled 1000 times is, at present, a pretty good switch. Finally, the switching mechanism relies upon high-amplitude molecular mechanical motion, and so it is fairly slow. For different bistable [2]rotaxanes, Willner's group has quantified some of the solid-state kinetic processes and has found that the step that corresponds to $kT^+ \rightarrow D$ occurs on the millisecond time scale.[28] Such slow switching speeds, however, are not a particularly serious limitation when compared to the other two issues, since parallel architectures (of which the crossbar is one) generate speed through circuit design, rather than through device switching speed.

A few years ago we reported on a 64-bit molecular electronic memory circuit (Fig. 10).[5] That circuit utilized bistable [2]rotaxanes, formed as a Langmuir–Blodgett (LB) monolayer. The LB film was sandwiched between a bottom n-type poly-silicon electrode and a top aluminium electrode. Between the Al electrode and the molecular film was deposited a thin (5 nm) Ti adhesion layer. Using this memory circuit, small data strings could be stored, read out, erased, and then new information could be rewritten into the memory. This memory circuit, considering it was the first of its kind, performed reasonably well.

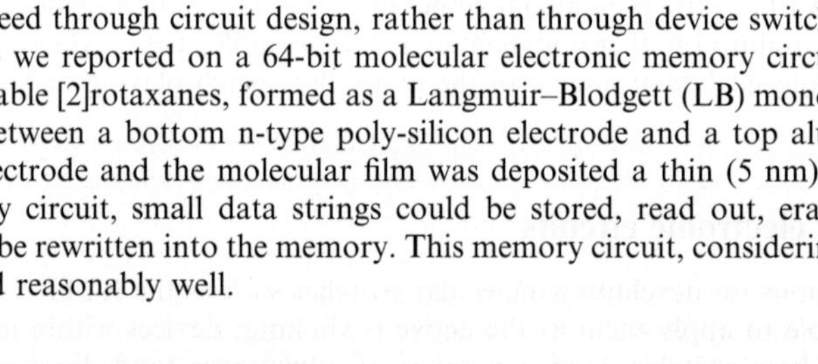

Fig. 10 A molecular electronic memory circuit in which diodes are constructed at each of the crosspoint junctions by establishing a p to n doping gradient through the junction. The data on the right illustrates the rectifying effect of each junction, while the data on the left illustrates the improved memory performance. Note that absolute current values are utilized to read the status of the memory, and that the switching amplitude is at least a factor of 3 for all bits.

A significant advantage of the crosspoint architecture is that, for a given patterning resolution, it will scale to higher bit densities than any other type of electronically addressable memory, a point which we will utilize below. However, crosspoint architectures also have challenges. Note that for the data of Fig. 9, it is the relative, rather than the absolute current values of the individual bits, that were utilized to determine the state of an individual junction. One of the key issues of a crosspoint architecture is that every junction is electronically coupled to every other junction, and thus changing the status of a few bits actually can affect the absolute current values through all of the other bits. One solution is to construct diodes at each of the junctions. Fig. 10 represents a second generation memory circuit in which we constructed a diode in series with each crosspoint junction by establishing a p–n doping gradient through the underlying poly-Si wire. By thinning the poly-Si wire (using reactive ion etching processes) in the regions between the crossed wires, it was possible to establish current pathways through the junction that passed through a p–n diode prior to passing through the molecular junction. Characteristic data from that memory circuit, are shown in Fig. 10. Note that this memory has a significantly improved performance—the signal level between a '1' and a '0' is recorded as an absolute current value, and the ability to discriminate between a '1' and a '0' is significantly improved. This diode concept also decreases the energy requirements of such a memory, and enables somewhat more flexible operation. In fact, this particular circuit (and the data string shown in Fig. 10) was configured to act as a complementary input XOR wired-logic gate. The details of wired-logic gates may be found elsewhere.[29]

Molecular electronic memory and (wired) logic circuits, based upon bistable molecular mechanical systems, can clearly be constructed using conventional lithographic patterning methods. For example, the data presented in Figs. 9 and 10 was from memories that were patterned using a combination of optical lithography and metals deposition through evaporation shadow masks. However, one of the key advantages to molecular electronics is that it should scale to near molecular dimensions. The challenge of constructing circuits at such dimensions is not unique to the field of molecular electronics. In fact, pushing the limits of patterning methods continues to be a key challenge for much of the electronics industry. The highest resolution patterning method is electron beam lithography (EBL), which can pattern silicon structures with dimensions of just a few nanometers (although with considerable structure-to-structure variability). However, EBL is severely limited in terms of the density of structures that can be patterned. For example, high aspect ratio silicon wires can be patterned at a center-to-center separation distance (pitch) of about 60–70 nm.[30] Thus, the development of methods for fabricating electronic circuits with molecular scale feature sizes and pitches constitute a significant challenge. We turn now to a discussion of ultra-high density electronic circuitry.

A couple of years ago we reported on the superlattice nanowire pattern transfer, or SNAP, method for constructing ultra-high density nanowire and crosspoint circuits.[31] This patterning method, which translates the atomic-level control over layer thicknesses that can be achieved within $GaAs/Al_xGa_{(1-x)}As$ superlattices into atomic-level control over widths and pitches of nanowires, has been demonstrated to produce nanowire circuits at a density and with a level of fidelity that far exceeds any competing approach. For example, 8 nm wide nanowires patterned at a pitch of 16 nm and with lengths of hundreds of micrometres to a few millimetres have been fabricated.[31] A further advantage of the SNAP method is that it can be utilized to fabricate nanowires of virtually any metal or semiconducting material, including complex structures, such as nanowires constructed of multiple layers of material. We subsequently developed non-destructive doping and reaction ion etching processes for silicon nanowire fabrication. By coupling those processes with the SNAP patterning method, we were able to fabricate silicon nanowires with bulk-like conductivity characteristics.[32] This is significant because of the extreme sensitivity of the electronic properties of one-dimensional conductors towards even very small numbers of defects. Finally, working with colleagues at Hewlett Packard, we have recently reported upon methods for replicating these structures in high throughput.[33] This provides us with the key step towards having a realistic manufacturing method that can approach molecular dimensions in scale.

In Fig. 12 we present scanning electron micrographs of a 4.5 kbit memory. This memory was wired up so that many of the wires were connected to one another to make the testing simpler. Only 64 effective bits, with each effective bit containing between 4 and 200 actual junctions, were individually probed. In the inset of Fig. 12 we present data measured from a similarly fabricated circuit. The data represents the switching responses from several of the smallest of those bits (four

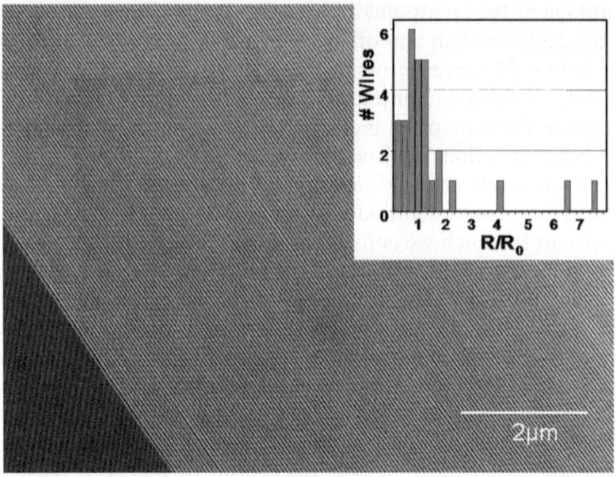

Fig. 11 SNAP-patterned silicon nanowires. Each nanowire is 15 nm wide by 30 nm tall, patterned at 34 nm pitch, and approximately 3 mm long. These wires were produced on a Si(111) on insulator (SOI) substrate. The electrical characteristics of a statistical number of these nanowires are shown in the top inset. R_0 is the calculated resistivity of a nanowire based upon knowledge of the doping level of the SOI wafer. R is the experimentally measured value. Thus these wires are essentially defect-free as evidenced by the fact that their electrical properties are as good as they can possibly be.

or five junctions each). About half of these devices worked, but the device performance is not too dissimilar to what was observed for the data presented in Fig. 11. Each junction in this circuit is about 10 nm × 40 nm wide, and so contains on the order of 300 molecules. The memory circuits shown in Figs. 9 and 10 each contained on the order of 10^8 molecules or so.

Thus, the circuit of Fig. 12 provides a compelling demonstration that these bits will scale well to reasonably small numbers of molecules. We are currently fabricating and testing 160,000 bit memory circuits fabricated using SNAP (Si) nanowires as the bottom electrodes and SNAP metallic wires as the top electrodes. Those circuits are at a density of 10^{11} bits cm^{-2}, and represent a bit density that would correlate to around the year 2020 according to standard Moore's law protocols.

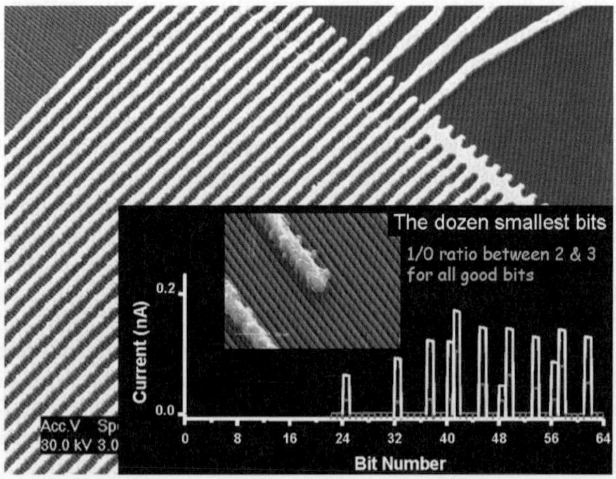

Fig. 12 (Background image) Electron micrograph of a 4.5 kbit molecular electronic memory circuit consisting of 10 nm wide Si bottom electrodes, 40 nm wide Ti/Al top electrodes, and a layer of bistable (RTTF^{4+}) rotaxanes sandwiched in between. Each junction contains ~300 molecular switches. The inset shows electrical contacts to the Si nanowires, as patterned by electron-beam lithography. Note that each contact touches three to four nanowires. Also shown is the performance of the smallest 13 bits (out of ~24 such bits).

Interconnecting ultra-high density molecular electronics circuits

One final issue that we have addressed involves interconnect architectures that can fully take advantage of ultra-dense electronic circuitry.[34] Note that in Fig. 12, EBL is not sufficiently high-resolution to enable the addressing of individual nanowires. Architectural concepts for meeting the challenge of electrically addressing (demultiplexing) individual nanowires that are patterned at sub-lithographic densities should satisfy three criteria. First, the demultiplexer architecture must bridge from the micrometre or submicrometre dimensions achievable through lithography to the few-nanometre dimensions achievable through alternative patterning methods.

Second, the architecture should allow for the addressing of many nanowires with a few large wires. Third, the manufacture of the multiplexer should be tolerant of fabrication defects.

Proposed demultiplexer architectures[35,36] have been based upon combining the concepts of crossbars[37] (lithographically patterned demultiplexing address wires crossing the nanowires) with multi-input binary tree demultiplexers.[38] Binary trees, by their very nature, exhibit order $2\log_2(N)$ scaling, where N is the number of nanowires, and $2\log_2(N)$ is the number of the large demultiplexing wires utilized to address the nanowires. An example of a binary tree demultiplexer is shown in Fig. 13.

Nanowire assemblies are often characterized by some randomness in organization, as well as defects, such as broken or non-conducting nanowires. To compensate, the proposed demultiplexing schemes have contained a certain amount of redundancy (extra wires). The major cost of this redundancy is that the use of additional address wires implies that certain nanowire addresses will be redundant or non-active, and circuit testing must be carried out to determine the good addresses, and memory must be devoted to storing those addresses. Kuekes and Williams[35] described a diode- or resistor-based decoder that utilizes $5 \times \log_2(N)$ large microscale address wires crossing an array of N nanowires. DeHon, Lincoln and Savage[36] described an architecture that utilizes no more than $[2.2 \times \log_2(N)] + 11$ address wires. Their scheme is based upon the field-effect gating of nanowires by the demultiplexer, and requires control over the doping profile along the axial dimension of the nanowires. Such nanowires have been realized experimentally,[39–41] and Lieber's group has used them to demonstrate a demultiplexer that bridges fabrication methods (*i.e.* self-assembly *versus* lithographic patterning), but not length scales.[42] Both of these architectures are based upon placing controllable regions on the surface of the nanowires. An individual nanowire, which is initially in the non-conducting state, will conduct only when all of the control regions are field- or voltage-addressed—*i.e.* it is the logical equivalent of a multi-input AND gate.

Our approach takes full advantage of the perfect fabrication characteristics of the SNAP patterning method, and it takes full consideration of both the full possibilities and limitations of conventional patterning methods. Fig. 14 illustrates our concept. We have demonstrated this approach, which is described in significant detail elsewhere,[34] does work for selectively addressing

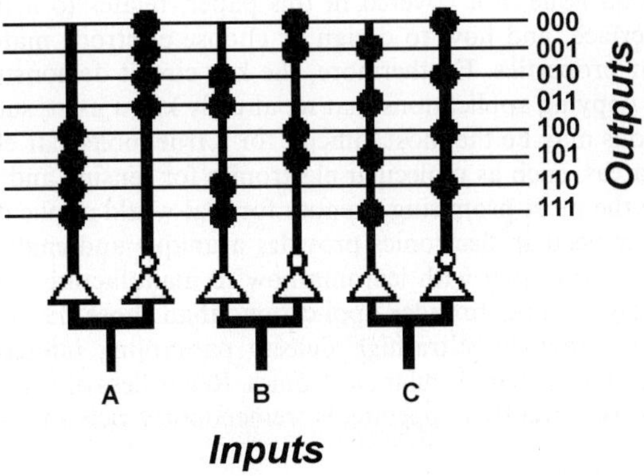

Fig. 13 A binary tree demultiplexer. The operation of this circuit may be understood by considering each of the eight horizontal wires as having an output that has the truth table of a three-input AND gate, with inputs defined by A, B and C. There are eight possible input addresses, each defining one of the eight horizontal wires. Note that 2^3 outputs are defined by three input wire pairs, so that the scaling is $2\log_2(N)$ for N outputs.

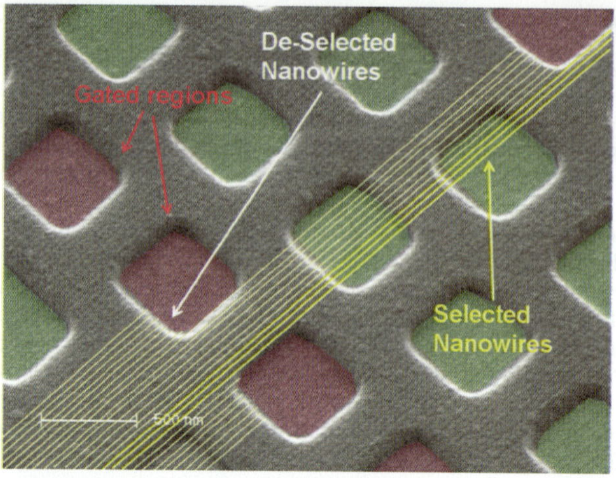

Fig. 14 The nanowire demultiplexer concept, shown as a drawing on top of an electron micrograph. The outline of SNAP-fabricated Si nanowires can be seen as the diagonal strips proceeding from bottom left to top right. The trenches represent regions in which a high-*k* dielectric is deposited on the nanowire surfaces. Application of a high voltage in certain regions (shaded red) decreases the conductivity of the underlying nanowires, through a field effect. A few nanowires are highlighted in yellow. Only three nanowires pass through this part of the demultiplexer circuit in the high-conductance state, and so are selected. Note that the EBL pattern demux structure is patterned at a much larger feature size and pitch than are the underlying nanowires.

nanowires, although significant challenges remain before it can be fully exploited for demultiplexing large and ultra-dense Si nanowire arrays. Many of those challenges relate to achieving molecular or atomic control over the nanowire/dielectric interfaces within the demultiplexer structure.

Conclusions

In this Discussion we have tried to present some of the key aspects that are required for building a science and technology that is molecular electronics. However we have also just touched at the surface, with only a few of the key issues described, and even those descriptions are not in depth.

One such issue relates to the ability, through thermodynamic and kinetics measurements, to establish a feedback loop between molecular synthesis and the performance of switching molecular electronic devices. Other groups are taking different routes towards establishing similar relationships. However, a second issue, not covered in this paper, relates to achieving control over the molecule–electrode interface, and how to design or choose electrode materials that can take full advantage of molecular properties. Furthermore, the key circuit demonstrations discussed herein are the result of simply copying applications that we already know exist, such as memory and logic. Such existing applications may be the most difficult for a true molecular electronics technology to compete in, but other areas, such as molecular electronics for sensing and responding, mechanical actuation, *etc.*, may be the most promising avenues for real world applications. This is because in certain of those areas, molecular electronics provides a unique and enabling pathway. However, certain of the challenges associated with learning how to manufacture with high precision at the molecular scale are likely to have broader applications than those discussed here. For example, issues of defect tolerant circuitry, ultra-high density patterning, interconnects, *etc.*, are much broader than their applications to molecular electronics. Regardless of what applications eventually emerge, the science that is currently happening is tremendously rich and fun.

Acknowledgements

This work was funded by the Department of Energy, the DARPA Moletronics Program, and the MARCO Center for Advanced Materials and Devices. We would also like to acknowledge the many current and former members of the Stoddart and Heath groups who have contributed to the science described in this article.

References

1 J. R. Heath, P. J. Kuekes, G. Snider and R. S. Williams, *Science*, 1998, **280**, 1716.
2 R. L. McCreery, *Chem. Mater.*, 2004, **16**, 4477–4496.
3 C. P. Collier, G. Mattersteig, Y. Li, E. W. Wong, K. Beverly, J. Sampaio, F. Raymo, J. F. Stoddart and J. R. Heath, *Science*, 2000, **289**, 1172–75.
4 L. Sheeney-Haj-Ichia and I. Willner, *J. Phys. Chem. B*, 2002, **106**, 13094–13097.
5 Y. Luo, C. P. Collier, K. Nielsen, J. Jeppesen, J. Perkins, E. DeIonno, A. Pease, J. F. Stoddart and James R. Heath, *ChemPhysChem*, 2002, **2002**(3), 519.
6 E. Katz, L. Sheeney-Haj-Ichia and I. Willner, *Angew. Chem., Int. Ed.*, 2004, **43**, 3292–3300.
7 (*a*) M. D. Jimenez, C. Dietrich-Buchecker and J.-P. Sauvage, *Angew. Chem.*, 2000, **112**, 3422–3425; (*b*) M. D. Jimenez, C. Dietrich-Buchecker and J.-P. Sauvage, *Chem. Commun.*, 2003, 1613–1616; (*c*) M. J. Marsella and R. J. Reid, *Macromolecules*, 1999, **32**, 5982–5984; (*d*) E. W. Taylor, *Science*, 1993, **261**, 35–36.
8 D. W. Steuerman, H.-R. Tseng, A. J. Peters, A. H. Flood, J. O. Jeppesen, K. A. Nielsen, J. F. Stoddart and J. R. Heath, *Angew. Chem.*, 2004, **43**, 2–7.
9 (*a*) A. H. Flood, A. J. Peters, S. A. Vignon, D. W. Steuerman, H.-R. Tseng, S. Kang, J. R. Heath and J. F. Stoddart, *Chem. Eur. J.*, 2004, **10**, 6558; (*b*) J. W. Choi *et al.*, Ground state equilibrium thermodynamics and switching kinetics of bistable [2]rotaxane switches in solution, polymer gels, and molecular electronic devices, *Chem. Eur. J.*, 2005, in press.
10 (*a*) J. O. Jeppesen, J. Perkins, J. Becher and J. F. Stoddart, *Org. Lett.*, 2000, **2**, 2631–2634; (*b*) J. O. Jeppesen, J. Perkins, J. Becher and J. F. Stoddart, *Angew. Chem., Int. Ed.*, 2001, **40**, 1216–1221; (*c*) J. O. Jeppesen *et al.*, *Chem. Eur. J.*, 2003, **9**, 2982–3007; (*d*) T. Yamamoto, H. R. Tseng, J. F. Stoddart, V. Balzani, A. Credi, F. Marchioni and M. Venturi, *Collect. Czech. Chem. Commun.*, 2003, **68**, 1488–1514; (*e*) H.-R. Tseng *et al.*, *Chem. Eur. J.*, 2004, **10**, 155–172; (*f*) S. Kang, S. A. Vignon, H.-R. Tseng and J. F. Stoddart, *Chem. Eur. J.*, 2004, **10**, 2555–2564; (*g*) I. C. Lee, C. W. Frank, T. Yamamoto, H.-R. Tseng, A. H. Flood, J. F. Stoddart and J. O. Jeppesen, *Langmuir*, 2004, **20**, 5809–5828; (*h*) J. O. Jeppesen, S. Nygaard, S. A. Vignon and J. F. Stoddart, *Eur. J. Org. Chem.*, 2005, 196–220.
11 (*a*) E. Katz, O. Lioubashevsky and I. Willner, *J. Am. Chem. Soc.*, 2004, **126**, 15520–15532; (*b*) D. Ryan, S. N. Rao, H. Rensmo, D. Fitzmaurice, J. A. Preece, S. Wenger, J. F. Stoddart and N. Zaccheroni, *J. Am. Chem. Soc.*, 2000, **122**, 6252–6257; (*c*) B. Long, K. Nikitin and D. Fitzmaurice, *J. Am. Chem. Soc.*, 2003, **125**, 15490–15498.
12 (*a*) A. M. Brouwer, C. Frochot, F. G. Gatti, D. A. Leigh, L. Mottier, F. Paolucci, S. Roffia and G. W. H. Wurpel, *Science*, 2001, **291**, 2124–2128; (*b*) A. Altieri, G. Bottari, F. Dehez, D. A. Leigh, J. K. Y. Wong and F. Zerbetto, *Angew. Chem., Int. Ed.*, 2003, **42**, 2296–2300; (*c*) A. Alteri, F. G. Gatti, E. R. Kay, D. A. Leigh, D. Martel, F. Paolucci, A. M. Z. Slawin and J. K. Y. Wong, *J. Am. Chem. Soc.*, 2003, **125**, 8644–8654; (*d*) F. Cecchet, P. Rudolf, S. Rapino, M. Margotti, F. Paolucci, J. Baggerman, A. M. Brouwer, E. R. Kay, J. K. Y. Wong and D. A. Leigh, *J. Phys. Chem. B*, 2004, **108**, 15192–15199.
13 (*a*) A. Livoreil, C. O. Dietrichbuchecker and J. P. Sauvage, *J. Am. Chem. Soc.*, 1994, **116**, 9399–9400; (*b*) J. P. Collin, P. Gavina and J. P. Sauvage, *New J. Chem.*, 1997, **21**, 525–528; (*c*) L. Raehm, J. M. Kern and J. P. Sauvage, *Chem. Eur. J.*, 1999, **5**, 3310–3317; (*d*) I. Poleschak, J. M. Kern and J. P. Sauvage, *Chem. Commun.*, 2004, 474–476.
14 D. W. Steuerman, H.-R. Tseng, A. J. Peters, A. H. Flood, J. O. Jeppesen, K. A. Nielsen, J. F. Stoddart and J. R. Heath, *Angew. Chem., Int. Ed.*, 2004, **43**, 6486–6491.
15 H.-R. Tseng, D. Wu, N. X. Fang, X. Zhang and J. F. Stoddart, *ChemPhysChem*, 2004, **5**, 111–116.
16 A. H. Flood, A. J. Peters, S. A. Vignon, D. W. Steuerman, H.-R. Tseng, S. Kang, J. R. Heath and J. F. Stoddart, *Chem. Eur. J.*, 2004, **10**, 6558–6564.
17 E. E. Polymeropoulos and J. Sagiv, *J. Chem. Phys.*, 1978, **69**, 1836.
18 T. Lee, W. Wang, J. F. Klemic, J. L. Zhang, J. Su and M. A. Reed, *J. Phys. Chem. B*, 2004, **108**, 8742.
19 R. M. Metzger, *Chem. Rev.*, 2003, **103**, 3803–3834.
20 R. L. Carroll and C. B. Gorman, *Angew. Chem.*, 2002, **114**, 4556–4579.
21 G. Maruccio, R. Cingolani and R. Rinaldi, *J. Mater. Chem.*, 2004, **14**, 542–554.
22 (*a*) G. J. Ashwell, J. R. Sambles, A. S. Martin and W. G. Parker, *J. Chem. Soc., Chem. Commun.*, 1990, 1374–1376; (*b*) A. S. Martin, J. R. Sambles and G. J. Ashwell, *Phys. Rev. Lett.*, 1993, **70**, 218–221.
23 G. Ho, J. R. Heath, M. Kondratenko, D. F. Perepichka, K. Arsenault, M. Pézolet and M. R. Bryce, *Chem.Eur. J.*, 2005, **11**, 2914–2922.
24 J. Chen, M. A. Reed, A. M. Rawlett and J. M. Tour, *Science*, 1999, **286**, 1550.
25 N. P. Guisinger, M. E. Greene, R. Basu, A. S. Baluch and M. C. Hersam, *Nano Lett.*, 2004, **4**(1), 55–59.
26 A. R. Pease, J. Jeppesen, J. F. Stoddart, Yi Luo, C. P. Collier and J. R. Heath, *Acc. Chem. Res.*, 2001, **34**, 433.
27 A. H. Flood, J. F. Stoddart, D. W. Steuerman and J. R. Heath, *Science*, 2004, **306**, 2055–2056.
28 E. Katz, L. O. Shenney-Haj-Ichia and I. Willner, *Angew. Chem., Int. Ed.*, 2004, **43**, 3292–3300.

29 C. P. Collier, E. Wong, M. Belohradsky, F. Raymo, J. F. Stoddart, P. J. Kuekes, R. S. Williams and J. R. Heath, *Science*, 1999, **285**, 391.
30 C. Vieu *et al.*, *Appl. Surf. Sci.*, 2000, **164**, 111.
31 N. Melosh, A. Boukai, F. Diana, B. Geradot, A. Badolato, P. Petroff and J. R. Heath, *Science*, 2003, **300**, 112.
32 R. Beckman, E. Johnston-Halperin, Y. Luo, N. Melosh, J. Green and J. R. Heath, *J. Appl. Phys.*, 2004, **96**(10), 5921–5923.
33 G. Y. Jung, W. Wu, Z. Yu, S. Y. Wang, Z. Li, R. S. Williams, E. Johnston-Halperin, J. E. Green, A. Boukai, Y. Bunimovich and J. R. Heath, 2005, submitted.
34 R. Beckman, E. Johnston-Halperin, Y. Luo, J. E. Green and J. R. Heath, *Science*, 2005, DOI: 10.1126/science.1114757.
35 P. J. Kuekes and R. S. Williams *US Pat.* 6,256,767 (3 July 2001).
36 A. DeHon, P. Lincoln and J. Savage, *IEEE Trans. Nanotechnol.*, 2003, **2**, 165.
37 J. Heath, P. Kuekes, G. Snider and R. S. Williams, *Science*, 1998, **280**, 1716.
38 R. P. Feynman, *Lectures in Computation*, ed. A. J. G. Hey and R. W. Allen, Addison-Wesley, Menlo Park, CA, 1996.
39 M. S. Gudiksen *et al.*, *Nature*, 2002, **415**, 617.
40 Y. Wu, R. Fan and P. Yang, *Nano Lett.*, 2002, **2**, 83–86.
41 M. T. Björk *et al.*, *Nano Lett.*, 2002, **2**, 87–89.
42 Z. Zhong, *Science*, 2003, **302**, 1377.

Molecular rectification: stabilised alignment of chevron-shaped dyes in hybrid SAM/LB structures in which the self-assembled monolayer is anionic and the Langmuir–Blodgett layer is cationic

Geoffrey J. Ashwell,* Marta Sujka and Andrew Green

Received 25th April 2005, Accepted 16th June 2005
First published as an Advance Article on the web 30th September 2005
DOI: 10.1039/b505785j

The chevron-shaped dye, *N*-butyl-2,6-bis-[2-{4-(2-(4-dibutylaminophenyl)-vinyl)-phenyl}-vinyl]-pyridinium iodide, has a hydrophilic apex that is electron-accepting and hydrophobic π-bridged limbs that are electron-donating. It forms non-centrosymmetric monolayers at the air–water interface and when deposited, its LB films exhibit second-harmonic generation and asymmetric current-voltage (*I–V*) characteristics. However, the behaviour is short lived: molecular reorganisation causes the absorption maximum to shift from *ca.* 400 to 700 nm with suppression of the SHG and rectification. Long-term stability results when the cationic dye is LB deposited on an anionic surface, for example, a SAM formed *via* chemisorption of sodium 3-mercapto-1-propanesulfonate on a gold-coated substrate. The hybrid Au/SAM/LB device exhibits asymmetric *I–V* curves with rectification ratios of *ca.* 25 at ±1 V when investigated by scanning tunnelling spectroscopy and contacted by PtIr probes.

Introduction

The nonlinear optical and electrical properties of LB films are limited by the hydrophobic alkyl tail, which reduces the concentration of donor–(π-bridge)–acceptor chromophores for SHG[1,2] and provides an insulating barrier between one of the electrodes and the active moieties in molecular rectifying devices.[3–5] An elongated alkyl tail, typically dodecyl to docosyl, is usually required for alignment of molecules at the air–water interface[6] but there have been attempts, albeit only three to date, to eliminate the substituent whilst retaining a non-centrosymmetric arrangement: (a) by depositing D–π–A molecules with an extended π-electron bridge which acts as a hydrophobic component[7,8] and (b) by suppressing the tendency of polar molecules to align centrosymmetrically by utilising a chevron-shaped D–π–A–π–D dye, *N*-butyl-2,6-bis[4-(2-(4-dibutylaminophenyl) vinyl)]pyridinium iodide.[9] In this work, we now report the LB alignment of a larger five-ring chevron-shaped molecule in which the apex is hydrophilic and the extended π-bridged limbs are hydrophobic (Fig. 1). SHG and asymmetric *I–V* curves confirm the non-centrosymmetric alignment and the films exhibit long-term stability when fabricated as a hybrid Au/SAM/LB structure in which the SAM is anionic and the LB layer is the cationic dye. It is only the ninth donor-acceptor chromophore to exhibit molecular rectification when aligned as a SAM[10–14] or LB film[3–5,9,15] and contacted by non-oxidisable electrodes although others exhibit asymmetric *I–V* characteristics when the contact is protected by a surface coating: *e.g.* Pt/(LB film)/Mg/Ag (refs. 16–18) and Au/(LB film)/Hg–S–C_nH_{2n+1} (ref. 19).

The first report to be linked with molecular rectification was published in 1964: it focused upon adjacent layers of n-type and p-type materials,[20] which were contacted by lead with the rectification

The Nanomaterials Group, Cranfield University, Cranfield, UK, MK43 0AL.
E-mail: . E-mail: g.j.ashwell@cranfield.ac.uk

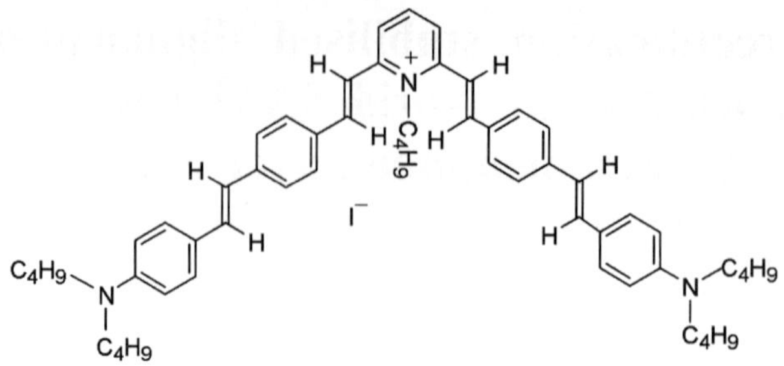

Fig. 1 Molecular structure of the chevron-shaped dye.

probably dominated by oxide-induced Schottky barriers. Electrical asymmetry has since been obtained for Au/(bulk donor)/(bulk acceptor)/Au devices[21,22] but these should be designated as rectifying junctions rather than rectifying molecules. However, molecular rectification has been assigned to a bilayer structure in which the electron-accepting layer is a self-assembled monolayer of a bipyridinium cation (BP^{2+}) and the electron-donating layer is the counterion obtained by metathesis of the SAM with the tetrasodium salt of copper phthalocyanine-3,4′,4″,4‴-tetrasulfonic acid [$CuPc(SO_3^-Na^+)_4$]. The bilayer exhibits a high rectification ratio of 60–100 at ±1 V whereas the trilayer, $Au/BP^{2+}/CuPc(SO_3^-)_4/BP^{2+}/Au$, shows symmetrical *I–V* characteristics,[23] this modified behaviour being expected for a symmetrical acceptor/donor/acceptor arrangement. It confirms that the molecule rather than the contacts induces rectification.

Interest in molecular rectification has primarily focused upon connected donor-acceptor units and, as long ago as 1974, Aviram and Ratner[24] proposed a σ-bridged sequence as the organic counterpart of the p–n junction where the direction of electron flow at forward bias is from cathode to acceptor and from donor to anode. There are three reported examples,[17–19] each studied using protected (coated) oxidisable electrodes, as well as eight sterically hindered π-bridged examples that exhibit rectifying behaviour when contacted by non-oxidisable electrodes.[3–5,9–15] The steric hindrance is necessary to induce a nonplanar D–π–A structure, which in turn suppress through-molecule orbital overlap of the electroactive ends. There is a clear structure/property relationship with the highest rectification ratio to date, albeit unpublished, obtained *via* the chemisorption of *N*-(10-acetylthiodecyl)-4-{2-(4-dimethylaminonaphthalen-1-yl)-vinyl}-quinolinium iodide on gold substrates, the molecule having bulky substituents on either side of the CH=CH bridge.[25] The current ratio of 110 at ±1 V is reduced to *ca.* 18 for SAMs of the less sterically hindered dimethylaminophenyl analogue to *ca.* 1 for the effectively planar 4-{2-(4-dimethylaminophenyl)vinyl}pyridinium derivative.[12] Similar trends have been noted for other series of dyes.[11,13]

Orientation-induced rectification has also been reported for hybrid SAM/LB structures of squaraine dyes in which the molecular donor-acceptor-donor moiety is vertically aligned in the SAM and horizontally aligned in the LB film.[26] The long axis of the D–A–D chromophore contacts the upper electrode whereas only the donor is projected towards the substrate, which provides the basic requirement for electrical asymmetry. Homomolecular films of squaraine dyes with self-assembling groups located at opposite ends of the D–A–D chromophore also show rectifying behaviour, as self-assembly *via* one and two anchoring groups provides vertical and horizontal alignments at the substrate electrode but only vertical alignments for molecules in contact with the upper electrode.[27] Orientations relative to the electrodes are the opposite of those described for the hybrid SAM/LB structure and significantly, the observed rectification is in the opposite quadrant of the *I–V* plot.

Effective control measures have been used to confirm the molecular origin of the electrical asymmetry: (i) to exclude electrode-induced effects, rectification has been verified by electrometer based methods for evaporated Au pads[3b,3c,4,5,9b] and scanning tunnelling spectroscopy using alkanethiolate-coated electrodes[10,12a,13] and PtIr and Au probes;[11–14] (ii) to confirm the rectifying nature of the D–π–A chromophore, the electrical asymmetry has been reversibly switched, first off and then back on, by protonating and then deprotonating either the dialkylamino donor[12] or dicyanomethylene acceptor;[3a,11] (iii) the polarity for rectification has been reversed by inverting the D–π–A orientations, for example, in self-assembled and LB films with respect to each other.[5,12]

Furthermore, as the polarity is dependent upon dipole orientation, self-assembled molecules connected *via* their donors (Au–S–C_nH_{2n}–D–π–A)[12,13,14a] and acceptors (Au–S–C_nH_{2n}–A–π–D)[3a,10,11,14a] exhibit rectification in opposite quadrants of the *I–V* plots.

Confirmation of molecular rectification, as outlined above, is limited to organic structures contacted by non-oxidisable electrodes. There are also several claims for films contacted by oxidisable electrodes (Al/Al_2O_3, Pb/PbO_2 and Ti/TiO_2), which cannot be substantiated as oxide-induced Schottky barrier effects tend to dominate the electrical asymmetry. In relation to this, it is noted that Au/MgPc/Al structures, where MgPc is magnesium phthalocyanine, exhibit symmetrical *I–V* characteristics when fabricated and studied before breaking the vacuum but then have rectification ratios of *ca.* 10^3 at ±1 V when exposed to air and re-evacuated.[28] Films contacted by titanium show similar rectifying behaviour when trace gases are present during deposition[29] and thus, it is vital to limit the study to molecular diodes contacted by non-oxidisable electrodes.

In this work we now focus upon the LB alignment of an unconventional chevron-shaped D–π–A–π–D dye that is cationic but contrary to conventional wisdom, a molecule that has truncated alkyl substituents rather than a long aliphatic tail. Its freshly deposited LB films exhibit SHG and molecular rectification but are inherently unstable, exhibiting suppression of the nonlinear optical and electrical behaviour within hours. However, long-term stability has been achieved by depositing the cationic dye on an anionic surface, Au–S–$(CH_2)_3$–SO_3^-, the resultant structure exhibiting improved electrical asymmetry with a rectification ratio of *ca.* 25 at ±1 V.

Results and discussion

LB film

The chevron-shaped dye, *N*-butyl-2,6-bis-[2-{4-(2-(4-dibutylaminophenyl)vinyl)phenyl}vinyl]pyridinium iodide, was spread from chloroform (0.1 mg cm^{-3}) onto the ultra-pure water subphase of a Nima Technology LB trough and, after allowing 10 min for the solvent to evaporate, the floating layer was compressed at 50 mm^2 s^{-1}. The resultant pressure-area isotherm is featureless: it exhibits film collapse at *ca.* 28 mN m^{-1} and has a limiting area of 1.2 nm^2 $molecule^{-1}$ as the surface pressure approaches zero (Fig. 2). The floating monolayer was maintained at *ca.* 24 mN m^{-1} and deposition was achieved by passing the following vertically through the interface on the upstroke at 0.1 mm s^{-1}: hydrophilically treated glass; gold-coated 10 MHz quartz crystals; gold-coated BK7 glass; gold-coated highly oriented pyrolytic graphite (HOPG). The quartz crystals, which were gold-coated

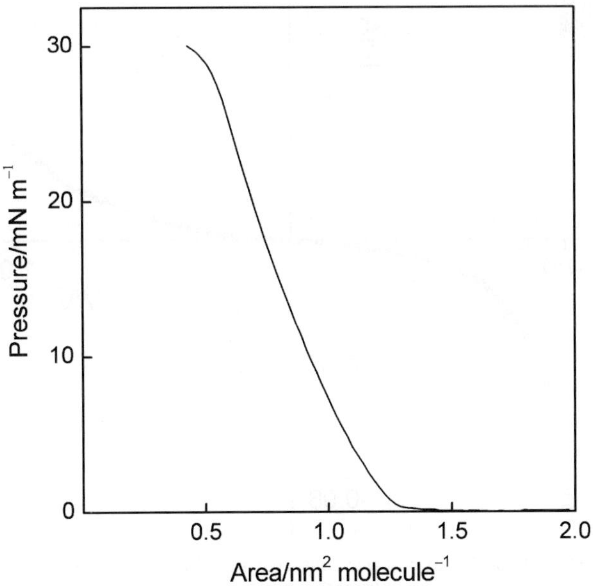

Fig. 2 Pressure–area isotherm.

prior to purchase, were washed sequentially with chloroform, acetone and water and then primed by plasma cleaning using a PlasmaPrep 2 (Gala Instrumente). The BK7 glass and HOPG substrates were gold-coated just prior to use and the dyes deposited without further processing.

A monolayer thickness of 1.9 ± 0.2 nm and real and imaginary components of the dielectric permittivity of $\varepsilon_r = 2.4 \pm 0.2$ and $\varepsilon_i = 0.6 \pm 0.2$ at 532 nm were obtained from a two-layer analysis of surface plasmon resonance data, which were obtained using a Kretschmann configuration[30] for LB films deposited on gold-coated BK7 glass. A mean occupation area of 0.95 ± 0.15 nm^2 molecule^{-1} in the LB film was obtained from a Sauerbrey analysis[31] of the frequency change following deposition on gold-coated 10 MHz quartz crystals. The thickness approximates to the height of the chevron-shaped molecule with its apex pointing up or down, although probably the latter as the pyridinium apex is hydrophilic, and the molecular area approximates to the chromophore footprint.

SHG studies were performed with a Nd:YAG laser beam (1.064 µm, p-polarised) incident at 45° to the LB film and the effective second-order susceptibility obtained by comparing the intensity with the Maker fringe envelope of a Y-cut quartz reference ($d_{11} = 0.5$ pm V^{-1}). Non-centrosymmetric alignment is verified by the second-harmonic intensity, which is as strong as signals obtained from LB films of conventional materials, for example, hemicyanine dyes with long alkyl tails.[1,2] Furthermore, the fact that the susceptibility is not compromised by a thick alkyl layer, as in conventional films, results in an unusually high value of $\chi^{(2)}_{\text{eff}} = 100$ pm V^{-1} at 1.064 µm but this is resonantly enhanced by a broad visible range absorption band. Occurrence of SHG with a high second-order nonlinear optical coefficient demonstrates that the molecular structure may be manipulated to provide the required alignment. We also note that the chevron-shaped dye is only the fourth atypical LB film-forming molecule to date to pack non-centrosymmetrically.[7–9]

Electrical characterisation of the chevron-shaped dye now forms the focus of this work and it adds to the recent reports on molecular rectification, which is now a topic of intense experimental[3–5,9–15] and theoretical[32–34] interest. LB monolayers were deposited on gold-coated HOPG, the gold overlay being thermally evaporated to a thickness of *ca.* 50 nm using an Edwards 306 coating unit. Following a cursory study of the surface structure, using a Nanoscope IV scanning tunnelling microscope, the *I–V* characteristics of freshly deposited films were investigated by scanning tunnelling spectroscopy using a PtIr tip as the contacting electrode. The LB monolayer exhibits weak rectification (Fig. 3) with the higher current in the negative quadrant of the *I–V* plot where

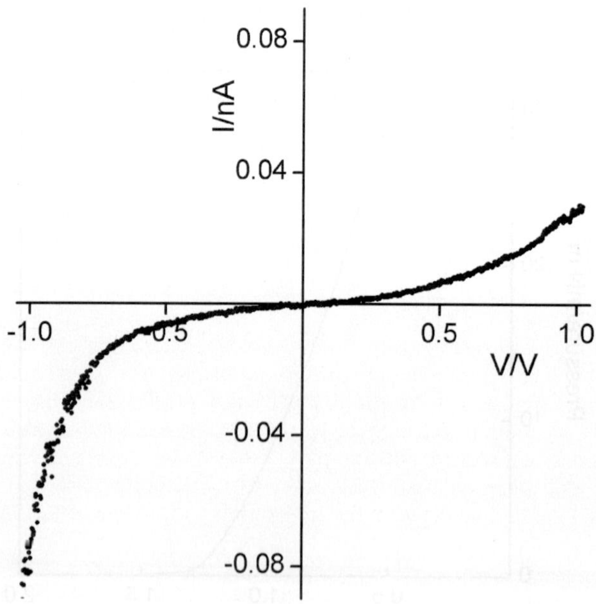

Fig. 3 *I–V* characteristics of the LB monolayer contacted by a PtIr tip: the polarity relates to the substrate electrode and the higher forward bias current (negative quadrant) corresponds to electron flow *via* the molecule in a direction of substrate to tip.

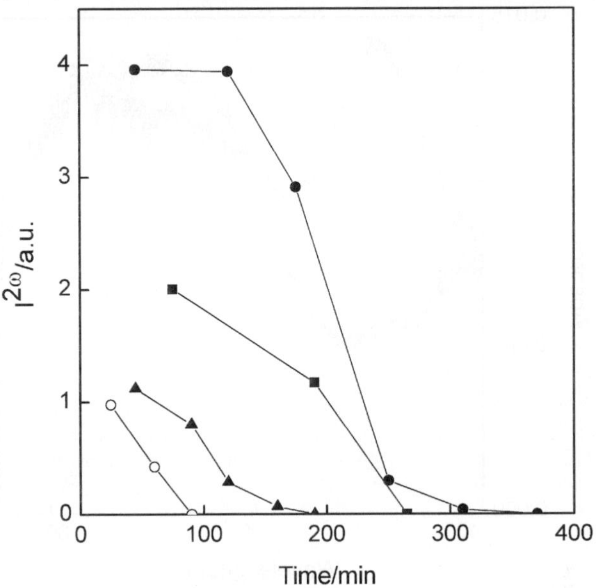

Fig. 4 Time dependence of the suppression of the second-harmonic intensity of four LB films following deposition on glass substrates. The different initial intensities are consistent with the film structure being inherently unstable.

polarity is defined by the substrate electrode. Electrons flow at forward bias in the direction of substrate to tip and, consistent with the Aviram and Ratner model,[24] they tunnel from the cathode to the highest occupied molecular orbital of the acceptor at the substrate interface and from the lowest unoccupied molecular orbital of the donor to the anode at the interface with the tip. Intramolecular tunnelling then follows to restore the ground state. In the opposite direction, there is an energy mismatch of the relevant orbitals and the Fermi levels. The polarity suggests that the pyridinium acceptor is adjacent to the substrate and the dibutylamino donors are directed upwards, this being consistent with the fact that the donors are hydrophobic and, when the molecules are aligned at the air–water interface, point away from the surface. The *I–V* plot suggests that such alignment is maintained during deposition and it is noted that the substrate surface would need to be strongly hydrophobic to induce the opposite orientation.

The electrical properties alter within hours of deposition with the *I–V* characteristics becoming symmetrical. This is attributed to molecular realignment and is accompanied by corresponding changes to the second-order nonlinear optical response and the visible range absorption spectrum. The second-harmonic intensity is initially strong, albeit sample dependent with $I^{2\omega}$(LB)/$I^{2\omega}$(quartz) = 1–4 × 10^{-4}, but the monolayer becomes SHG-inactive within *ca.* 2–6 h (Fig. 4). Concurrent with this, freshly deposited films exhibit a broad visible-range absorption with maxima at *ca.* 400 and 750 nm but the transition at 400 nm diminishes with enhancement of a low-energy band centred about 700 nm (Fig. 5). The loss of SHG confirms that the altered structure is centrosymmetric and, consistent with spectra of other LB film-forming materials, the bathochromic shift suggests that the molecules realign in an antiparallel manner to provide overlap of the neighbouring donor and acceptor groups. The loss of rectification and these associated features are interesting: they provide evidence that the initially observed electrical asymmetry is inherent to the aligned molecules and not induced by asymmetric coupling between the LB monolayer and the contacting electrodes.

Stabilised alignment in a hybrid SAM/LB structure

The nonlinear optical and electrical properties are potentially interesting but are spoiled because the non-centrosymmetric structure is unstable, as might be expected for an LB film obtained from unconventional molecules that are chevron-shaped rather than linear and have truncated butyl

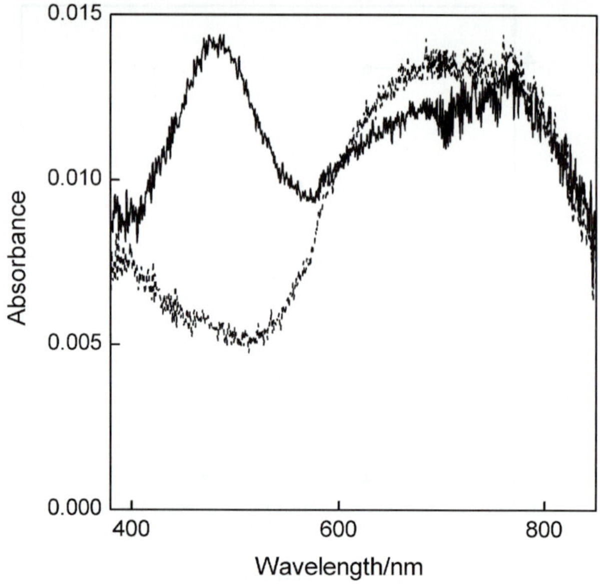

Fig. 5 Visible spectra of the LB film when freshly deposited ($\lambda_{max} = 400$ nm) and after *ca.* 24 h ($\lambda_{max} = 700$ nm).

substituents instead of a long aliphatic tail. However, the instability can be overcome by depositing the monolayer on an anionic surface, which acts as the counterion. In this case, the surface was prepared by immersing gold-coated substrates in a methanol solution of the sodium salt of 3-mercapto-1-propanesulfonic acid (0.4 mg cm^{-3}) followed by rinsing with chloroform and ethanol to remove physisorbed material. Optimum conditions for chemisorption were found by monitoring the frequency change while depositing onto the gold electrodes of 10 MHz quartz crystals. The mean area, determined from a Sauerbrey[31] analysis of the frequency data, saturates to *ca.* 0.3 nm^2

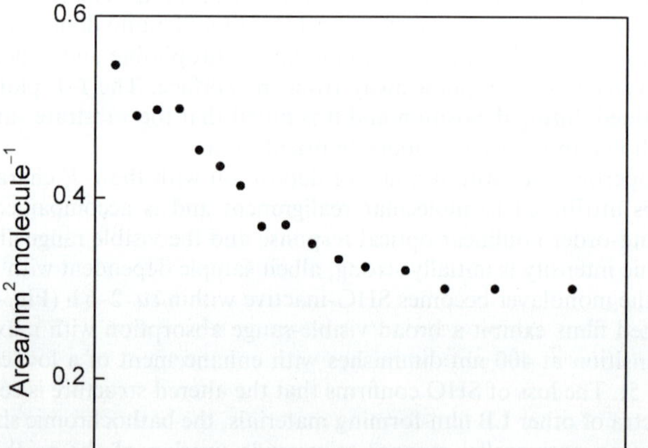

Fig. 6 Mean area occupied by self-assembled Au–S–$(CH_2)_3SO_3^-Na^+$ molecules on gold-coated 10 MHz quartz crystals *versus* the period of immersion of the substrates in a dilute methanol solution of the sodium salt of 3-mercapto-1-propanesulfonic acid (0.4 mg cm^{-3}).

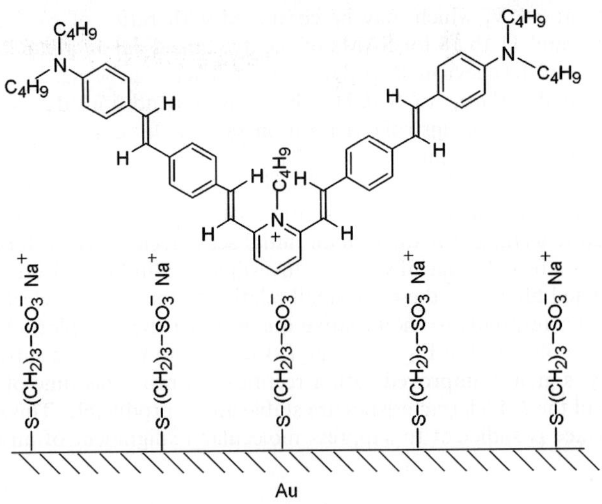

Fig. 7 Hybrid Au/SAMLB structure.

molecule^{-1} for Au–S–$(CH_2)_3$–$SO_3^-Na^+$ after *ca.* 4 h (Fig. 6). The area approximates to the molecular cross-section indicating a closely packed arrangement.

Attempts were made to align the chevron-shaped dye by simply immersing the anionically coated 10 MHz quartz crystals in a dilute chloroform solution of *N*-butyl-2,6-bis-[2-{4-(2-(4-dibutylaminophenyl)vinyl)phenyl}vinyl]pyridinium iodide. The frequency change was consistent with the replacement of Na^+ by the cationic dye at the surface but *I–V* studies of such films on gold-coated HOPG were disappointing and suggest disordered alignments. However, subsequent to this study, linear cationic D–π–A dyes have been aligned by this method and exhibit rectification. For example, solution aligned films of *N*-methyl-4-(2-(4-dibutylaminophenyl)vinyl))quinolinium exhibit rectifica-

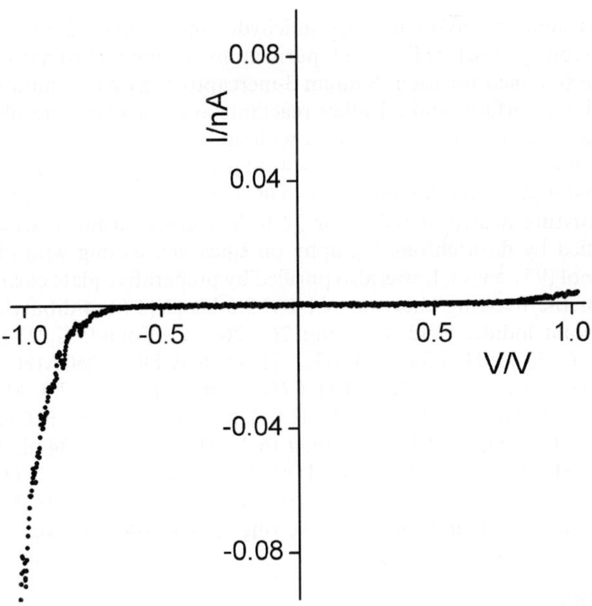

Fig. 8 *I–V* characteristics of the hybrid Au/SAM/LB structure contacted by a PtIr tip: the polarity relates to the substrate electrode and the higher forward bias current (negative quadrant) corresponds to electron flow *via* the molecule in a direction of substrate to tip.

tion ratios of *ca.* 14 at ±1 V, which may be compared with ratios of *ca.* 8 for LB films of the octadecyl derivatives[5] and 11 to 18 for SAMs of the dye linked *via* an alkanethiolate bridge.[12]

Stabilised alignment of the chevron-shaped dye reported here has been achieved by LB deposition on the anionically coated surface, Au–S–$(CH_2)_3SO_3^-$, using conditions described above: a surface pressure of 20 mN m^{-1} and an upstroke deposition rate of 0.1 mm s^{-1}. It is assumed that the cationic molecules replace the sodium ions, with Na^+ and I^- remaining in the water subphase, and that interaction between the cationic dye and the anionic surface stabilises the alignment (Fig. 7). The electrical properties of films deposited on gold-coated HOPG were investigated by scanning tunnelling spectroscopy using a PtIr tip with an initial set current of *ca.* 100 pA at 0.2 V (Fig. 8). This was varied across a broad range of values to investigate tip-induced electrical asymmetries but, as for other investigated films,[10–13] these are small relative to the inherent electrical asymmetry. At forward bias, which corresponds to the negative quadrant of the *I–V* plot, electrons tunnel from substrate to tip as also observed in Fig. 3 for the LB film on an untreated substrate. However, the electrical asymmetry is greatly improved with a rectification ratio this time of *ca.* 25 at ±1 V (*cf.* RR = 2 in Fig. 3) and the *I–V* characteristics are stable and reproducible. This simple modification of the substrate surface is sufficient to suppress molecular realignment of an unconventional LB film-forming dye.

Conclusion

We have described the properties of a new rectifying dye, one of few reported to date,[3–5,9–15] and show that the alignment in LB films may be achieved without resorting to a long aliphatic tail, which limits the second-order susceptibility by increasing the thickness and electrical properties by introducing an insulating barrier between the electrode and the conjugated part of the molecular film. LB films of the dye on Au–S–$(CH_2)_3$–SO_3^- are stable and exhibit high rectification ratios of *ca.* 25 at ±1 V. The method may be adapted for the alignment of other dyes and preliminary work at Cranfield, as yet unreported, has shown improved rectification for cationic D–π–A dyes where the positive charge is in a terminal position.

Experimental

Trans 4-[2-(4-dibutylaminophenyl)-vinyl]-benzaldehyde and *N*-butyl-2,6-lutidinium iodide were synthesized as previously described[8,9a] and purified by column chromatography. Satisfactory analytical data were obtained for each. Sodium 3-mercapto-1-propane-sulfonate, used in the self-assembly of the anionic surface, and all other reactants and solvents were obtained from Sigma-Aldrich (Gillingham, UK) and used as supplied without purification.

To an ethanol solution (70 cm^3) of *trans* 4-[2-(4-dibutylaminophenyl)-vinyl]-benzaldehyde (0.15 g, 0.45 mmol) and *N*-butyl-2,6-lutidinium iodide (0.06 g, 0.23 mmol) was added piperidine (0.1 cm^3) and the resultant mixture heated at reflux for 72 h. The crude product, which was obtained by filtration, was purified by flash chromatography on silica gel, eluting with chloroform and then chloroform : methanol (95 : 5 v/v). It was also purified by preparative plate chromatography to yield red microcrystals of the five-ring chevron, *N*-butyl-2,6-bis-[2-{4-(2-(4-dibutylaminophenyl)vinyl)-phenyl}vinyl]pyridinium iodide: yield 30%; mp 265–268 °C. Found: C, 73.5; H, 7.7; N, 4.3%. $C_{57}H_{72}N_3I$ requires: C, 73.92; H, 7.83; N, 4.53%. ^1H NMR ($CDCl_3$, 250 MHz, *J*/Hz): δ_H 0.98 (t, *J* 7.0, 15H, CH_3); 1.38 (sextet, *J* 7.3, 10H, CH_3CH_2); 1.60 (quintet, *J* 7.0, 8H, CH_2CH_2N); 1.85 (quintet, *J* 7.1, 2H, $CH_2CH_2N^+$); 3.30 (t, *J* 7.4, 8H CH_2N); 4.88 (t, *J* 7.5, 2H CH_2N^+); 6.62 (d, *J* 8.3, 4H, Ar-H); 6.80 (d, *J* 16.2, 2H, C=CH); 7.09 (d, *J* 16.2, 2H, C=CH); 7.34 (d, *J* 16.5, 2H, C=CH); 7.36 (d, *J* 7.8, 4H Ar-H); 7.43 (d, *J* 7.8, 4H, Ar-H); 7.53 (d, *J* 15.2, 2H, C=CH); 7.65 (d, *J* 7.8, 4H, Ar-H); 8.06 (d, *J* 7.8, 2H, Py-H); 8.24 (t, *J* 8.1, 1H, Py-H). *m/z* (FAB): 798 $[M - I^-]^+$. The NMR data confirm an all-*trans* configuration of the five-ring chevron-shaped dye.

Acknowledgements

We acknowledge the EPSRC for financial support, the EPSRC National Mass Spectrometry Service for providing the MS data and Anne Whittam for technical assistance.

References

1 G. J. Ashwell, *J. Mater. Chem.*, 1999, **9**, 1991.
2 C. Bosshard, K. Sutter, P. Prêtre, J. Hulliger, M. Flörsheimer, P. Kaatz and P. Günter, in *Organic Nonlinear Optical Materials, Advances in Optics*, ed. A. F. Garito and F. Kajzar, Gordon and Breach, Basel, 1995, **vol. 1**.
3 (*a*) G. J. Ashwell, A. Chwialkowska and L. R. H. High, *J. Mater. Chem.*, 2004, **14**, 2389; (*b*) N. Okazaki, J. R. Sambles, M. J. Jory and G. J. Ashwell, *Appl. Phys. Lett.*, 2002, **81**, 2300; (*c*) T. Xu, I. R. Peterson, M. V. Lakshmikantham and R. M. Metzger, *Angew. Chem., Int. Ed.*, 2001, **40**, 1749.
4 (*a*) G. J. Ashwell and D. S. Gandolfo, *J. Mater. Chem.*, 2001, **11**, 246; (*b*) G. J. Ashwell and D. S. Gandolfo, *J. Mater. Chem.*, 2002, **12**, 411; (*c*) G. J. Ashwell, B. J. Robinson, M. A. Amiri, D. Locatelli, S. Quici and D. Roberto, *J. Mater. Chem.*, 2005, **15**, 4203.
5 G. J. Ashwell, D. S. Gandolfo and R. Hamilton, *J. Mater. Chem.*, 2002, **12**, 416.
6 M. C. Petty, *Langmuir-Blodgett Films: an Introduction*, Cambridge University Press, Cambridge, 1996.
7 A. Watakabe, H. Okada and T. Kunitake, *Langmuir*, 1994, **10**, 2722.
8 G. J. Ashwell, A. A. Maxwell and A. Green, *J. Mater. Chem.*, 2002, **12**, 2192.
9 (*a*) G. J. Ashwell and M. A. Amiri, *J. Mater. Chem.*, 2002, **12**, 2181; (*b*) J. W. Baldwin, R. R. Amaresh, I. R. Peterson, W. J. Shumate, M. P. Cava, M. A. Amiri, R. Hamilton, G. J. Ashwell and R. M. Metzger, *J. Phys. Chem. B*, 2002, **106**, 12158.
10 G. J. Ashwell, R. Hamilton and L. R. H. High, *J. Mater. Chem.*, 2003, **13**, 1501.
11 G. J. Ashwell, A. Chwialkowska and L. R. H. High, *J. Mater. Chem.*, 2004, **14**, 2848.
12 (*a*) G. J. Ashwell, W. D. Tyrrell and A. J. Whittam, *J. Am. Chem. Soc.*, 2004, **126**, 7102; (*b*) G. J. Ashwell, W. D. Tyrrell and A. J. Whittam, *J. Mater. Chem.*, 2003, **13**, 2855.
13 G. J. Ashwell, A. Mohib and J. R. Miller, *J. Mater. Chem.*, 2005, **15**, 1160.
14 (*a*) P. Jiang, G. M. Morales, W. You and L. Yu, *Angew. Chem., Int. Ed.*, 2004, **43**, 4471; (*b*) M.-K. Ng and L. Yu, *Angew. Chem., Int. Ed.*, 2002, **41**, 3598; (*c*) M.-K. Ng, D.-C. Lee and L. Yu, *J. Am. Chem. Soc.*, 2002, **124**, 11862.
15 A. Honciuk, A. Jaiswal, A. Gong, H. Ashworth, C. W. Spangler, I. R. Peterson, L. R. Dalton and R. M. Metzger, *J. Phys. Chem. B*, 2005, **109**, 857.
16 (*a*) G. J. Ashwell, J. R. Sambles, A. S. Martin, W. G. Parker and M. Szablewski, *J. Chem. Soc., Chem. Commun.*, 1990, 1374; (*b*) A. S. Martin, J. R. Sambles and G. J. Ashwell, *Phys. Rev. Lett.*, 1993, **70**, 218.
17 (*a*) N. J. Geddes, J. R. Sambles, D. J. Jarvis, W. G. Parker and D. J. Sandman, *Appl. Phys. Lett.*, 1990, **56**, 1916; (*b*) N. J. Geddes, J. R. Sambles, D. J. Jarvis, W. G. Parker and D. J. Sandman, *Appl. Phys. Lett.*, 1992, **71**, 756.
18 C. Brady, B. Hodder, A. S. Martin, J. R. Sambles, C. P. Ewels, R. Jones, P. R. Bridden, A. M. Musa, C. A. Panetta and D. L. Mattern, *J. Mater. Chem.*, 1999, **9**, 2271.
19 G. Ho, J. R. Heath, M. Kondratenko, D. F. Perepichta, K. Arsenault, M. Pézolet and M. R. Bryce, *Chem. Eur. J.*, 2005, **11**, 2914.
20 J. E. Meinhard, *J. Appl. Phys.*, 1964, **35**, 3059.
21 C. M. Fischer, M. Burghard, S. Roth and K. von Klitzing, *Europhys. Lett.*, 1994, **28**, 129.
22 T. Reda, A. F. Collings, C. Barton and P. Lukins, *J. Phys. Chem. B*, 2003, **107**, 13774.
23 G. J. Ashwell and J. Ewington, unpublished data.
24 A. Aviram and M. A. Ratner, *Chem. Phys. Lett.*, 1974, **29**, 277.
25 G. J. Ashwell and A. Mohib, unpublished data.
26 G. J. Ashwell and M. Berry, *J. Mater. Chem.*, 2005, **15**, 108.
27 G. J. Ashwell, K. Moczko and J. Ewington, *J. Mater. Chem.*, 2005, **15**, 1154.
28 G. J. Ashwell, J. S. Bonham and L. E. Lyons, *Aust. J. Chem.*, 1980, **33**, 1619.
29 (*a*) R. L. Mcreery, *Chem. Mater.*, 2004, **16**, 4477; (*b*) R. L. McCreery, J. Dieringer, A. O. Solak, B. Snyder, A. M. Nowak, W. R. McGovern and S. DuVall, *J. Am. Chem. Soc.*, 2004, **126**, 6200.
30 E. Kretschmann, *Z. Phys.*, 1971, **241**, 313.
31 G. Sauerbrey, *Z. Phys.*, 1959, **155**, 206.
32 K. Stokbro, J. Taylor and M. Brandbyge, *J. Am. Chem. Soc.*, 2003, **125**, 3674.
33 H. Mizuseki, K. Niimura, C. Majumder and Y. Kawazoe, *Comput. Mater. Sci.*, 2003, **27**, 161.
34 S. Sitha, K. Bhanuprakash and B. H. Choudary, *Synth. Met.*, 2005, **148**, 227.

Carbon/molecule/metal molecular electronic junctions: the importance of "contacts"

Richard L. McCreery, Umamaheswari Viswanathan, Rajendra Prasad Kalakodimi and Aletha M. Nowak

Received 22nd April 2005, Accepted 24th June 2005
First published as an Advance Article on the web 23rd September 2005
DOI: 10.1039/b505684p

Molecular electronic junctions fabricated by covalent bonding onto a graphitic carbon substrate were examined with Raman spectroscopy and characterized electronically. The molecular layer was a 4.5 nm thick multilayer of nitroazobenzene (NAB), and the top contact material was varied to investigate its effect on junction behavior. A 3.0 nm thick layer of copper, TiO_2, or Al(III) oxide (AlO_x) was deposited on top of the NAB layer, followed by a 7.0 nm thick layer of gold. Copper "contacts" yielded molecular junctions with low resistance and showed a strong dependence on molecular structure. Carbon/NAB/AlO_x/Au junctions exhibited high resistance, with current densities three orders of magnitude less than those for analogous Cu junctions. However, Raman spectroscopy revealed that the NAB layer was reduced when the carbon substrate was biased negative, to a product resembling that resulting from electrochemical reduction of NAB. Carbon/NAB/TiO_2/Au junctions showed rectifying J/V behavior, with high conductivity to electrons able to enter the TiO_2 conduction band. Substitution of azobenzene for nitroazobenzene yielded junctions with similar spectroscopic and electronic behavior to NAB, indicating that the nitro group is not essential for rectification. The results are interpreted in terms of the energy levels of the molecule relative to those of TiO_2. The combination of a covalently bonded molecular layer and a semiconducting oxide yields unusual electronic properties in a carbon/molecule/semiconductor/Au molecular junction.

Introduction

A central goal of molecular electronics is the determination of how structure controls electron transport (ET).[1–5] In the case of molecular junctions of the type metal/molecule/metal, it is generally accepted that ET depends on both the structure of the molecule and the nature of the metal/molecule interface at both "contacts".[2–4,6–8] For example, the conductance across a gold/thiol linkage is higher than across a noncovalent alkane/Au bond,[8–11] and varies significantly for Au–O, Au–S, and Au–Se linkages.[12–14] For an Au/thiobiphenyl/Ti molecular junction, the biphenyl/Ti contact showed Schottky behavior with temperature and bias, while the ET through the Au–S bond appeared to be controlled by a thermally activated hopping mechanism.[15] The temperature dependence of carbon/biphenyl/Hg junctions indicated both a tunneling process at low temperature and activated ET at higher temperature, possibly related to ring rotation.[16] Molecular junction paradigms based on self assembled monolayer[15,17–21] and Langmuir Blodgett[22–27] structures have quite different chemistry at the interface between the molecule and metal or metal oxide. In order to understand ET in molecular electronic junctions, it will be important to determine the contributions of the structures of both the contacts and the molecule in controlling ET, and then to integrate these contributions into rational design of molecular electronic devices.

Department of Chemistry, Ohio State University, Columbus OH 43210, USA. E-mail: Mccreery.2@osu.edu

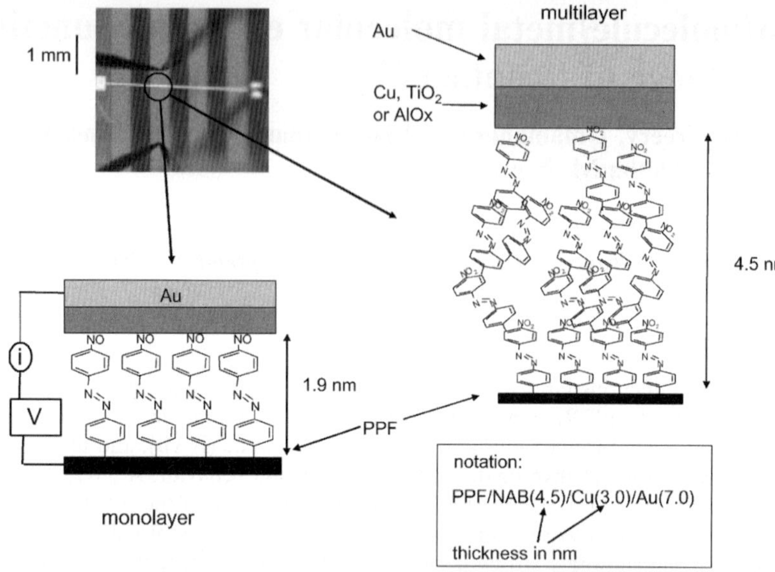

Fig. 1 Structure of carbon/NAB/metal molecular junctions. Photograph shows the crossed-wire geometry of a metal wire on top of a carbon strip. Thicknesses of molecular and metallic layers are shown in parentheses, in nm. Metallic phase between the molecule and the gold top contact is Cu, TiO_2, or Al(III) oxide.

Our laboratory has pursued a unique approach to studying molecular junctions based on a graphitic carbon substrate,[16,28–34] shown in Fig. 1. The primary motive for using the carbon substrate is the strong C–C bond between the molecular layer and the substrate ($\sim$100 kcal mol^{-1}). In addition, the phenyl ring of the molecule is bonded to a phenyl ring in the graphitic "contact", resulting in a nearly symmetric linkage with a possibly low injection barrier. The substrate is a pyrolyzed photo resist film (PPF)[35,36] with a structure and resistivity (0.005 Ω cm) similar to glassy carbon and very flat surfaces (<5 Å rms). Molecules are bonded to PPF by electrochemical reduction of diazonium reagents to yield a molecular layer, which has been characterized by Raman, XPS, FTIR, SIMS and voltammetry.[33,37–45] As shown in Fig. 1, diazonium modification can yield mono or multilayers, and in all cases reported here, the molecular layer thickness was verified with AFM "scratching".[46] Carbon/molecule/metal junctions can be fabricated with a high device yield (>80%) and 5–30% reproducibility of conductance at both low (±0.1) and high (±2 V) applied bias.[34]

Given the novelty of the carbon/molecule/metal junction design, we have used a semi-empirical approach to determine what phenomena control junction conductance. We vary one aspect of the junction structure, such as molecular structure, while keeping all others constant. This approach is conceptually similar to the many structure–reactivity relationships investigated by chemists for centuries. An example is shown in Fig. 2, for PPF/molecule/Cu junctions.[34] The cases shown have identical PPF–molecule bonds, nearly the same molecular layer thickness (1.5–1.7 nm) and the same top contact (Cu). However, the current/voltage behavior of the three junctions varies dramatically, at both low and high bias. The ET mechanisms underlying these variations are currently under investigation, but Fig. 2 clearly establishes that molecular structure is important to junction conductance.

The current report considers variation of the top "contact" and its effect on ET through PPF/nitroazobenzene (NAB) molecular junctions. The junctions consisted of a 4.5 nm thick NAB layer on PPF, and a top contact of 3.0 nm of Cu, Ti, or Al and 7.0 nm of Au. For example, a NAB/Cu junction is designated PPF/NAB(4.5)/Cu(3.0)/Au(7.0), with the numbers in parentheses indicating layer thickness in nanometers. By varying the identity and deposition conditions for the metal or metal oxide layer between the NAB and Au, dramatic changes in junction electronic behavior were observed.

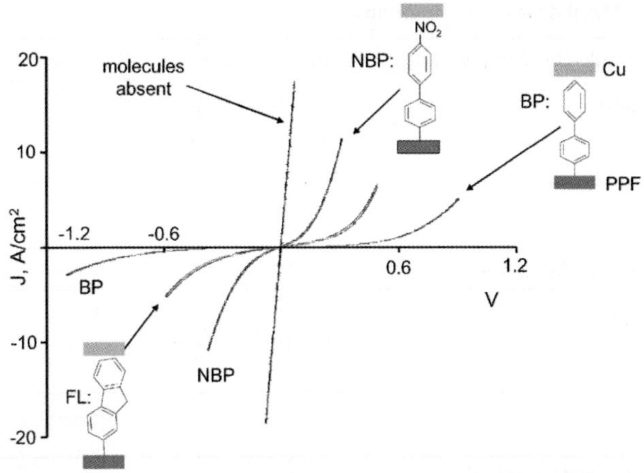

Fig. 2 Current density *vs.* bias curves for three types of PPF/molecule/Cu/Au junctions, plus a "control" junction lacking the molecular layer. In all cases, the bias axis is PPF relative to Au. Scan rate was 1000 V s^{-1}. Adapted from ref. 34.

Experimental

Molecular junctions were of the "crossed wire" type[34] shown in Fig. 1. The preparation of PPF and modification with NAB are described in detail elsewhere.[31,34] The NAB modification yields a multilayer with a thickness of 4.51 ± 0.68 nm, as determined with AFM "scratching".[46] Briefly, a series of line profiles through a trench intentionally cut in the NAB layer were averaged to determine the NAB layer depth. As shown by FTIR,[37,38] SIMS,[39] and AFM,[44,46,47] multilayer formation is possible during diazonium reduction by continued generation of NAB radicals and attack of the first monolayer. Although the multilayer is disordered, it is very low in pinholes and is conjugated throughout its length between PPF and the top contact. Modification of PPF occurred on 1 mm wide strips of PPF ~1 μm thick) on Si_3N_4, with thicknesses of ~1 μm. The PPF/NAB(4.5) samples were rinsed in purified CH_3CN, then the top metal strip was deposited by electron-beam deposition with one of four programs, listed in Table 1. Ti metal was deposited with two sets of conditions, which differ significantly in their oxide composition. The Ti oxide layer formed with lower deposition pressure is referred to herein as "TiO_x", and that from the higher pressure with exposure of Ti to air as "TiO_2". XPS depth profiling showed the TiO_x deposit to be a mixture of Ti(II), Ti(III), and Ti(IV) oxyhydroxides,[48] while "TiO_2" shows only Ti(IV) oxide with no detectable hydroxide.[49] In all cases, the Au top layer was used to protect the metal/metal oxide layer from air and to permit good electrical contact. The metal strip was 0.5 mm wide, yielding a finished junction area of 0.005 cm^2. The notation for describing junction types includes the thickness of the molecular and metal/oxide layers in nanometers, *e.g.* PPF/NAB(4.5)/AlO_x(3.3)/Au refers to a 4.5 nm thick NAB layer, a 3.3 nm thick AlO_x layer, and a top contact of 7.0 nm of Au.

Current–voltage (*J*/*V*) curves were obtained conventionally using a 3-wire configuration which compensated for PPF resistance, and are plotted as current density *vs.* voltage (PPF relative to Au). The metal wire resistance was less than 20 Ω (determined independently) and the ohmic potential error from the metal was ignored. The ~10 nm thick metal contacts are sufficiently transparent to obtain Raman spectra of the junction with a bias applied. As described elsewhere,[32] Raman spectra were obtained from junctions, which were identical to those studied electronically, except for a larger area (1 × 7 mm, or 0.07 cm^2). An Ar^+ laser (514.5 nm, 30 mw at sample) with a line focus (50 μm × 5 mm) was the source, and an *f*/2 spectrograph, back thinned CCD and holographic laser rejection filter detected Raman scattered light.[50]

Table 1 Metal deposition conditions

PPF/Molecule/Cu (3.0)/Au (7.0)	Pressure/Torr	Rate/nm s^{-1}
Cu (3.0 nm)	3×10^{-7}	0.1
Au (7.0 nm)	3×10^{-7}	0.1
PPF/Molecule/AlO_x/Au		
Al (3.3 nm)	5×10^{-6}	0.03
Au (7.0 nm)	5×10^{-6}	0.1
PPF/Molecule/TiO_x/Au		
TiO_x (3.1 nm)	$<5 \times 10^{-6}$	0.03
Au (7.0 nm)	4×10^{-6}	0.1
PPF/Molecule/TiO_2/Au		
TiO_x (3.1 nm)	7×10^{-6}	0.03
Vent with air	~760	(45 min)
Au (7.0 nm)	4×10^{-6}	0.1

Results

When comparing molecular junctions with varying composition, keep in mind that all junctions have the PPF/NAB(4.5) substrate and molecular layer plus the 7.0 nm Au top layer. J/V curves for PPF/NAB(4.5) junctions prepared with different top contacts are shown in Fig. 3. The Cu and TiO_2 junctions are relatively stable with time, with a slow increase in resistance over a period of several months. However, TiO_x junctions become significantly less conductive on a timescale of hours or days. Immediately after metal deposition at low backpressure, the TiO_x junctions are similar to those made with copper. As the Ti is further oxidized during or after metal deposition, the J/V curve becomes asymmetric, and rectification is observed when the Ti is present as TiO_2.

Fig. 4 shows J/V curves for the Cu and TiO_2 junctions with a more sensitive current density scale, as well as a J/V curve for AlO_x (upper panel), and a magnified AlO_x curve (lower panel). Notice that the AlO_x junction exhibits a symmetric J/V response with current densities more than three orders of magnitude lower than the Cu junctions. Even for voltages above +2 V, the current density for the AlO_x junctions remains much lower than either the Cu or TiO_2 junctions.

At first glance, the reason for the dramatic difference between Cu and AlO_x junctions may seem obvious. Transport in PPF/NAB(4.5)/Cu(3.0)/Au junctions relies on ET through two good conductors (PPF & Cu) and a thin, conjugated molecular layer. While there are many interesting questions about what factors control ET through the molecular layer, it is much faster for Cu junctions than for those containing AlO_x or TiO_2, all else being equal. AlO_x is an insulator (band gap of ~8.7 eV), so it is not surprising that conductivity is blocked by a 30 Å AlO_x layer. The low conductance observed in Fig. 4 for $|V| > 2$ V may be due to field emission through the AlO_x layer, or to defects.

However, the Raman spectroscopy results in Fig. 5 show that AlO_x is not acting as a simple insulator in PPF/NAB(4.5)/AlO_x(3.0)/Au junctions. Deposition of AlO_x causes a decrease of the 1340 and 1108 cm^{-1} Raman band intensities and an increase of the 1401/1448 cm^{-1} intensity ratio,

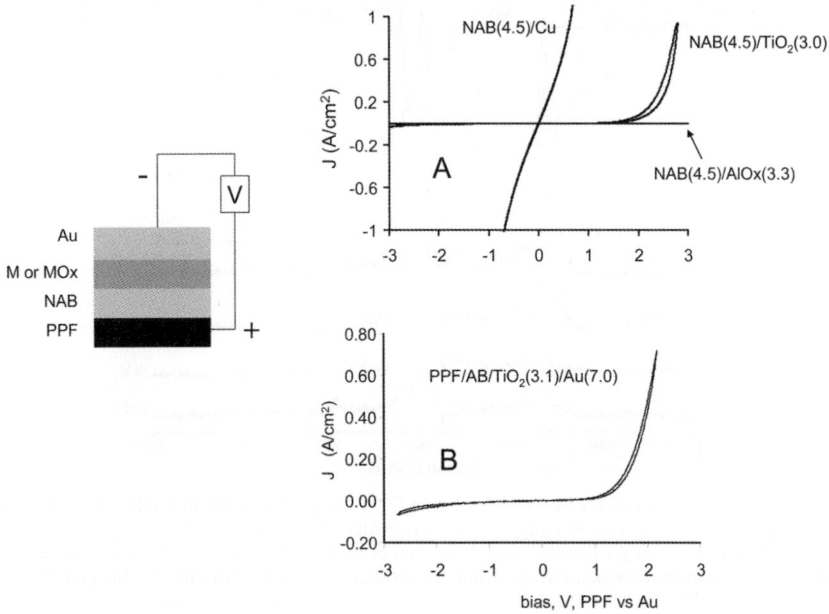

Fig. 3 A. J/V curves for PPF/NAB(4.5) junctions which are identical except for the metal phase in contact with the NAB. Scan rate of 100 V s^{-1}. B. J/V curve for a azobenzene/TiO_2/Au junction, also at 100 V s^{-1}. AB thickness was not determined independently, but was approximately 4 nm.

as shown in the top two spectra of Fig. 5. A study of NAB in solution and NAB bonded to PPF immersed in electrolyte revealed that increases in the 1400/1450 peak ratio and loss of the 1340 cm^{-1} band are associated with NAB reduction to an anion or "methide" species.[33,51] When a negative potential is applied to a completed PPF/NAB(4.5)/AlO_x(3.3)/Au(7.0) junction, the spectra (c) to (f) shown in Fig. 5 result. For progressively negative bias (PPF negative), the 1401/1448 intensity ratio continues to increase. Fig. 6 compares the spectrum obtained at −2 V to that resulting from electrochemical reduction of NAB bonded to PPF and immersed in acetonitrile electrolyte. The

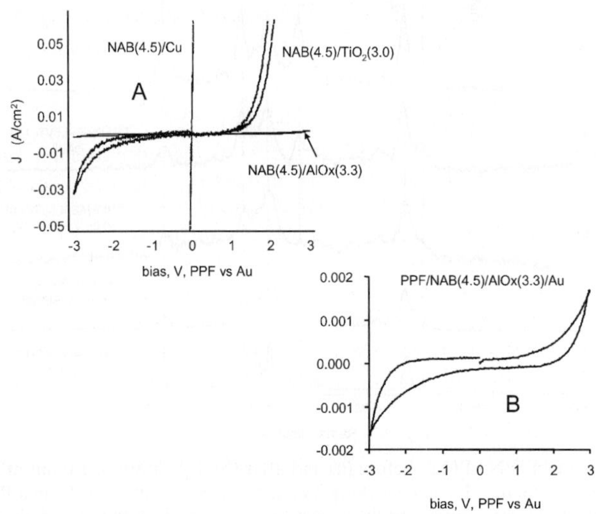

Fig. 4 A. J/V curve of Fig. 3A with expanded current scale. B. J/V curve for PPF/NAB(4.5)/AlO_x(3.3)/Au junction after further expansion of current scale, 100 V s^{-1}.

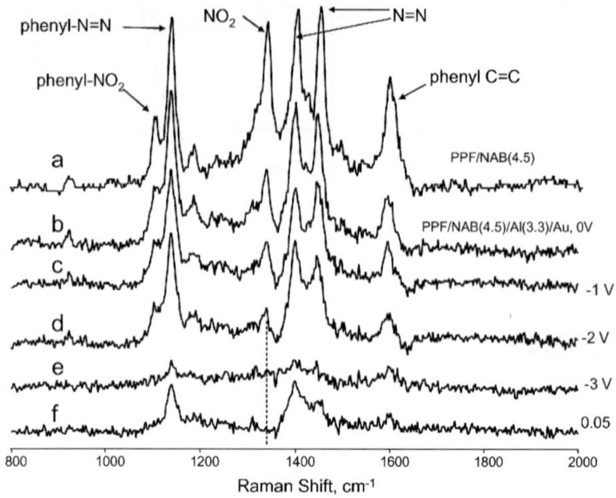

Fig. 5 *In situ* Raman spectra of a PPF/NAB(4.5)/AlO_x(3.3)/Au junction with an applied bias. Spectrum (a) is PPF/NAB(4.5) before AlO_x was deposited, and (b) is after AlO_x deposition without an applied bias. Remaining spectra show changes occurring when the indicated voltages were applied, progressing from top to bottom. All spectra have the same intensity scale, but are displaced vertically for clarity. Spectra (b) through (f) are from the same sample.

spectra are very similar, and the azo stretches at 1401 and 1448 cm^{-1} resemble those of an amino substituted azobenzene also shown in Fig. 6. In fact, the Raman spectrum at −2 V in Fig. 5 is very similar to that of the electrochemically reduced NAB/PPF in which the nitro group had become an amino or hydroxylamine group. Nitro group reduction of aromatic nitro compounds is well understood, and leads to aromatic amines or hydroxylamines.[52,53] The details of the reaction of NAB for negative bias are under investigation, but the Raman results of Fig. 5 provide unequivocal evidence for structural changes under bias, similar to those associated with NAB reduction. Note that the current required for reduction is very small (a few μA cm^{-2}) and much smaller than the

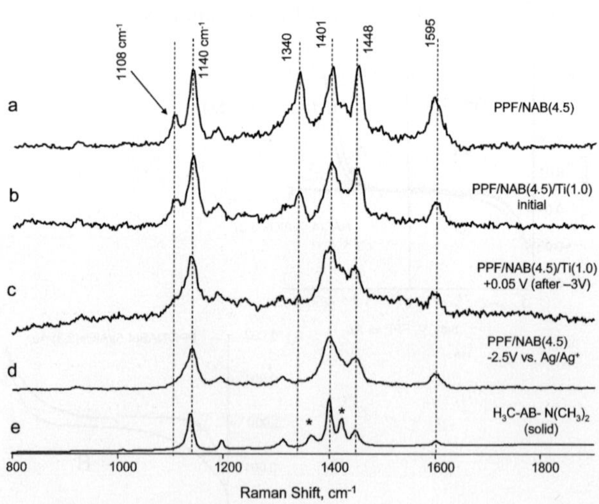

Fig. 6 Raman spectra of PPF/NAB(4.5) before (a) and after (b) deposition of 1.0 nm of Ti and 9.0 nm of Au, and after a potential excursion to −3 V in a working junction. (c). Spectrum (d) is from a PPF/NAB(4.5) surface immersed in 0.1 M tetrabutylammonium tetrafluoroborate in acetonitrile and reduced electrochemically at −2.5 V *vs.* Ag/Ag^+. Spectrum (e) is a Raman spectrum of *N*,*N*-dimethylamino-4′-methyl azobenzene, as a powdered solid. * indicate modes attributable to the methyl groups.

capacitive or tunneling currents apparent in Fig. 4. The disappearance of the spectrum at −3 V is likely due to a shift in the NAB absorption band, with accompanying decrease in resonance enhancement of the NAB bonded to PPF.[49]

Examination of PPF/NAB(4.5)/Cu(3.0)/Au junctions did not reveal changes in the Raman spectrum with an applied bias, although the bias range was limited due to the high currents in Cu junctions. Nevertheless, comparison of the Cu and AlO_x behavior results leads to a working hypothesis. For Cu junctions, ET is rapid through the molecular layer, leading to the high observed current density. As noted elsewhere, the J/V curves for Cu junctions are weakly temperature dependent (10–20% decrease from $T = 313$ K to 221 K).[34] So neither the spectroscopy nor the temperature dependence provides evidence for a structural rearrangement in Cu junctions, although that possibility is not ruled out. With AlO_x present, however, ET through the junction is prevented and a large electric field can develop across the molecule. Even in the absence of dc current flow, the Fermi level of the PPF can be high enough under negative bias to reduce the NAB. An alternative possibility involves a very thin electrochemical cell, in which the AlO_x layer acts as an electrolyte. Under the latter hypothesis, oxidation must occur at the gold electrode, either of some junction component or of gold itself. Under either hypothesis, the AlO_x slows down ET through the junction by three orders of magnitude or more compared to Cu, and changes in molecular structure result.

TiO_2 is a semiconductor with a band gap of 3.0–3.4 eV, depending on phase (anatase or rutile) and disorder.[54,55] We might predict that a PPF/NAB(4.5)/TiO_2(3.0)/Au junction would conduct electrons with sufficient energies to enter the TiO_2 conduction band. The suboxides of Ti(II) and Ti(III) are nearly as conductive as Ti metal, and about 8 orders of magnitude more conductive than TiO_2. We have shown previously Ti deposition also partially reduces NAB on PPF, with a thicker Ti layer resulting in greater reduction.[32,33] Fig. 7, along with our previous report,[32] show that the PPF/NAB/TiO_2/Au junctions exhibit spectroscopic changes under bias, qualitatively similar to those of PPF/NAB/AlO_x/Au junctions. Negative bias causes further increases in the 1401/1448 intensity ratio, and the spectrum following a negative bias excursion is similar to that of electrochemically reduced NAB. As discussed elsewhere, the spectroscopic results indicate an irreversible reduction of the nitro group to an amino or hydroxylamine which is itself electroactive.[32] For the present purposes, it is sufficient to conclude that NAB reduction occurs under negative bias for both PPF/NAB(4.5)/AlO_x(3.3)/Au and PPF/NAB(4.5)/TiO_2(3.0)/Au junctions.

Finally, the role of the nitro group in NAB junctions was investigated by making several junctions from azobenzene (AB). Although AB junctions have not yet been characterized extensively, PPF/AB/TiO_2(3.0)/Au junction show rectification similar to that of the analogous

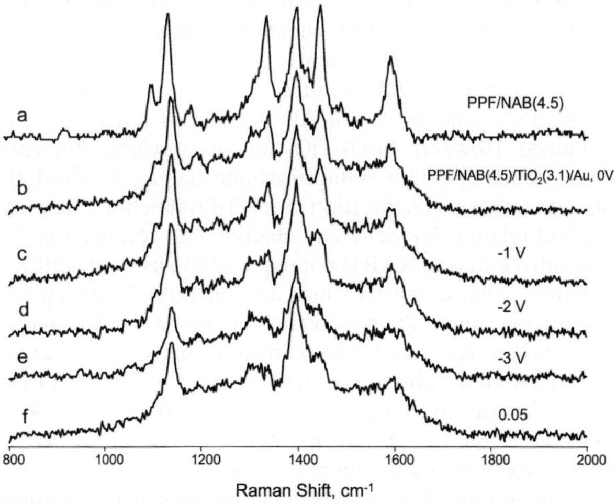

Fig. 7 *In situ* Raman spectra of a PPF/NAB(4.5)/TiO_2(3.1)/Au junction with an applied bias. Spectrum (a) is PPF/NAB(4.5) before TiO_2 was deposited, and (b) is after TiO_2 deposition without an applied bias. Remaining spectra show changes occurring when the indicated voltages were applied. Spectra b–f were obtained in the sequence shown from a single sample, and all spectra have the same intensity scale.

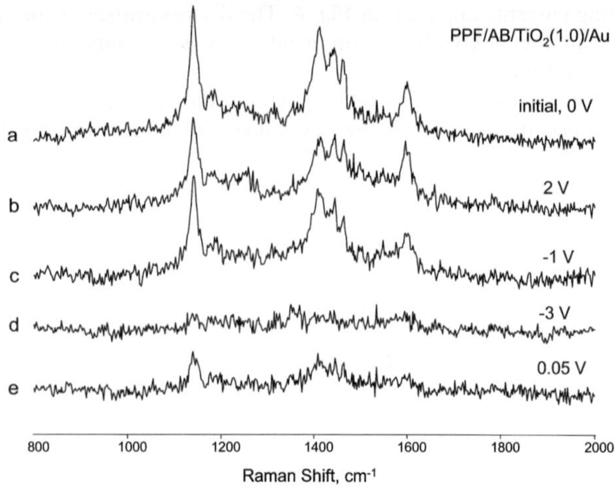

Fig. 8 *In situ* Raman spectra of a PPF/AB/TiO_2(1.0)/Au junction with an applied bias. Spectra progress from top to bottom, with the applied bias as indicated. The spectra are weaker than those of NAB, resulting in lower signal-to-noise ratio. All spectra have the same intensity scale, but are displaced vertically for clarity. Spectra c–e are from a different sample than a–b.

NAB/TiO_2 case, as shown in Fig. 3. AB is a weaker Raman scatterer than NAB, so a more transparent 1.0 nm TiO_2 layer was used instead of 3.0 nm. Spectra of a PPF/AB/TiO_2(1.0)/Au junction are shown in Fig. 8. As expected, the bands for the NO_2 stretch (1340 cm^{-1}) and phenyl–NO_2 stretch (1108 cm^{-1}) are absent. However, AB exhibits a partially reversible loss of intensity under negative bias, similar to that observed for PPF/NAB(4.5)/TiO_2(1.0)/Au and PPF/NAB(4.5)/TiO_2(3.0)/Au junctions. The Raman intensity increases further over the bottom spectrum of Fig. 8 if the potential is returned to +3 V. We conclude that a NO_2 group is not a requirement for the bias induced reduction of either NAB or AB, nor is it required for the junction to exhibit rectification.

Discussion

The principal experimental observations which bear on the mechanism of junction behavior are as follows: First, the NAB/AlO_x and NAB/TiO_2 junctions exhibit bias-induced structural changes, with the Raman spectra following a sufficiently negative voltage excursion (PPF negative) corresponding to those of electrochemically reduced NAB. Second, AlO_x and TiO_2 junctions are much less conductive than Cu junctions, with AlO_x junctions having low conductance over the entire ±3 V range examined. However, NAB/TiO_2 junctions exhibit differential conductance (dI/dV) approximately equal to that of Cu for voltages greater than 2.5 V. Third, the reduction of NAB or AB for negative bias accompanies rectification in the TiO_2 junctions. These two examples do not establish a cause and effect relationship between reduction and rectification, but they do provide a correlation. Finally, the nitro group of NAB is not required to observe rectification, and azobenzene shows spectral and electronic characteristics which are qualitatively similar to those of NAB.

A possible rationale for explaining the observations is based on the vacuum referenced energy levels shown in Figs. 9 and 10. AlO_x and TiO_2 constitute barriers to electron transfer, but with energetically different heights. In the absence of an oxide, ET through the PPF/NAB/Cu junction is rapid, leading to the high current intensities for Cu junctions (Figs. 3 and 4). As noted elsewhere, this ET is due to coherent or diffusive tunneling, and its weak temperature dependence implies negligible structural rearrangement and accompanying nuclear motion.[34,48] Furthermore, although the current through Cu junctions is strongly dependent on molecular structure (Fig. 2), the current densities are all much higher than those observed with oxide junctions. With an AlO_x layer present, electrons cannot pass through the junction, and are presumably blocked by the oxide layer. Although the dc current through NAB/AlO_x junctions is more than three orders of magnitude smaller than that for Cu junctions, the Raman results still indicate that NAB is reduced under

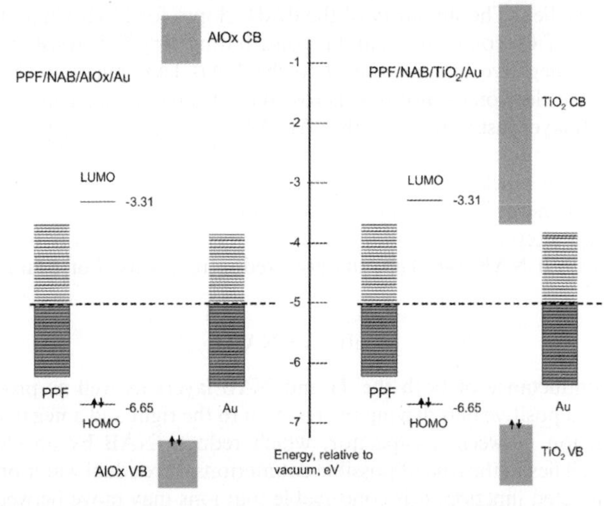

Fig. 9 Vacuum referenced energy levels for NAB, TiO_2, and AlO_x. NAB orbital energies calculated from Gaussian 98. Levels shown are for isolated materials, before junction formation.

negative bias in the AlO_x junction. Apparently, the high electric field across the NAB layer is acting like the ionic double layer in electrochemical reduction, by providing the driving force for injection of an electron from the PPF to the NAB. Electrical double layers in electrochemical cells are typically 10^5–10^6 V cm^{-1}, while the field calculated for a PPF/NAB(4.5)/AlO_x(3.3)/Au(7.0) junction is 4×10^6 V cm^{-1} at -3 V. This estimate assumes a linear potential profile between the PPF and Au, but even higher electric fields are possible if the potential profile is nonlinear.

The behavior of the NAB/TiO_2 junction is similar to that of NAB/AlO_x for ($-1 < V < 1$ V), yielding similar, featureless J/V curves in this region (Fig. 4). Positive bias across a PPF/NAB/TiO_2/Au junction raises the Fermi level of the Au, as shown in the right side of Fig. 10. For low positive bias, the electrons in the Au have insufficient energy to reach the TiO_2 conduction band, and the junction current is small, similar to the NAB/AlO_x case. For higher positive bias, the electrons in Au can pass through the conduction band and the NAB LUMO, resulting in a large increase in current. Under this postulate, the conductance should approach that for the Cu junction once the TiO_2

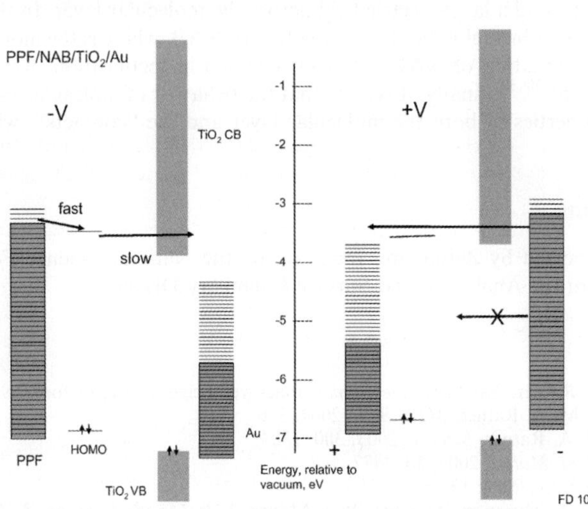

Fig. 10 Postulated effects of applied bias on energy levels and electron transport in PPF/NAB(4.5)/TiO_2/Au junctions. Arrows indicate electron transport between phases for negative (left) and positive (right) bias.

conduction band is reached. The similarity of the dI/dV slopes for NAB/Cu and NAB/TiO_2 above +2 V indicates that the TiO_2 conduction band is contributing very little resistance in the bias range above +2 V. When a negative bias is applied to the NAB/TiO_2 junction, the NAB is reduced, presumably because the electron cannot pass across the TiO_2 band gap. The electron is likely to be injected into the NAB layer just as it was with NAB/AlO_x, but apparently it spends long enough in the NAB layer to permit nuclear rearrangement and NAB reduction. Rectification may result from formation of a Coulomb blockade by the anionic NAB. It is also possible that reduction causes changes in the LUMO energy, which inhibit conduction.

As noted above, an alternative explanation for conductance changes is based on a completed redox cell, in which both NAB and TiO_2 undergo redox reactions. For example, a cell reaction such as:

$$NAB^- + Ti(IV) \rightleftarrows NAB + Ti(III)$$

could change the conductance of both the Ti and NAB layers as well as produce the observed Raman changes, with a positive bias driving the reaction to the right and a negative bias driving it to the left. The distinction between a capacitor, which reduces NAB by an electric field, and a conventional redox cell lies in the role of possible counterions. If residual water or reaction products are present in a completed junction, it is conceivable that ions may move between the organic and oxide layers. In addition, a counter reaction at the Au electrode is necessary for charge balance in a conventional redox cell. Such ions and counter reactions are currently undetermined, if present at all, and would require Au oxidation in the case of PPF/NAB/AlO_x/Au junctions. If adventitious counterions are absent, then the NAB reduction is caused by a large imposed electric field, and the "counterion" is actually the image charge formed on the Au electrode. If counterion motion is involved, it is possible that reduced NAB is stabilized by an unknown cation, possibly H^+. Stated differently, the molecular junctions containing Ti or Al oxide may be acting as small capacitors or small batteries, with possible counterion motion determining the electric field distribution through the junction. In either case, the spectroscopic results unequivocally establish that structural changes occur in the molecular layer in response to an applied electric field.

Two phenomena discussed here may have general significance to molecular junctions which do not involve a carbon substrate. First, metal oxides may play a role in junction behavior, either by undergoing redox chemistry or by forming an insulating phase. UHV conditions are required to deposit Al or Ti without forming oxides, and AlO_x has been identified previously as a factor in metal/molecule/metal tunnel junctions.[56,57] Furthermore, Ti has two oxidation states between Ti(0) and Ti(IV), which can undergo redox reactions and/or participate in the semiconducting properties of TiO_2. Second, the metal oxides can retard or completely prevent electron transport through the junction, thereby imposing a large electric field across the molecular layer. In the present example, the AlO_x and TiO_2 layers brought about the reduction of NAB, whereas the more highly conducting Cu did not. The combination of NAB and TiO_2 resulted in rectification, which did not occur if either phase was absent.[30–32] Finally, it is clear that the behavior of molecular junctions depends on the identity and properties of both the molecular layer and the "contacts", whether they involve metal oxides or not.

Acknowledgements

This work was supported by ZettaCore, Inc., and by the National Science Foundation through project 0211693 from the Analytical and Surface Chemistry Division.

References

1 J. Jortner and M. Ratner, *Molecular Electronics*, Blackwell Science Ltd., Oxford, 1997.
2 C. R. Kagan and M. A. Ratner, *MRS Bull.*, 2004, 376.
3 A. Nitzan and M. A. Ratner, *Science*, 2003, **300**, 1384.
4 R. McCreery, *Chem. Mater.*, 2004, **16**, 4477.
5 R. McCreery, *Interface*, 2004, **13**, 46.
6 G. Leatherman, E. N. Durantini, D. Gust, T. A. Moore, A. L. Moore, S. Stone, Z. Zhou, P. Rez, Y. Z. Li and S. M. Lindsay, *J. Phys. Chem. B*, 1999, **103**, 4006.
7 S. N. Yaliraki, A. E. Roitberg, C. Gonzalez, V. Mujica and M. A. Ratner, *J. Phys. Chem.*, 1999, **111**, 6997.

8 S. M. Lindsay, *Interface*, 2004, **13**.
9 A. M. Rawlett, T. J. Hopson, L. A. Nagahara, R. K. Tsui, G. K. Ramachandran and S. M. Lindsay, *Appl. Phys. Lett.*, 2002, **81**, 3043.
10 G. K. Ramachandran, J. K. Tomfohr, O. F. Sankey, J. Li, X. Zarate, A. Primak, Y. Terazano, T. A. Moore, A. L. Moore, D. Gust, L. A. Nagahara and S. M. Lindsay, *J. Phys. Chem. B*, 2003, **107**, 6162.
11 J. He, F. Chen, J. Li, O. Sankey, Y. Terazono, C. Herroro, D. Gust, T. Moore, A. L. Moore and S. M. Lindsay, *J. Am. Chem. Soc.*, 2005, **127**, 1384.
12 R. J. Hamers, S. K. Coulter, M. D. Ellison, J. S. Hovis, D. F. Padowitz, M. P. Schwartz, C. M. Greenlief and J. N. Russell, Jr., *Acc. Chem. Res.*, 2000, **33**, 617.
13 S. K. Coulter, M. P. Schwartz and R. J. Hamers, *J. Phys. Chem. B*, 2001, **105**, 3079.
14 L. Fang, J. Liu, S. Coulter, X. Cao, M. P. Schwartz, C. Hacker and R. J. Hamers, *Surf. Sci.*, 2002, **514**, 362.
15 C. Zhou, M. R. Deshpande, M. A. Reed, L. Jones and J. M. Tour, *Appl. Phys. Lett.*, 1997, **71**, 611.
16 F. Anariba and R. L. McCreery, *J. Phys. Chem. B*, 2002, **106**, 10355.
17 P. Lewis, C. Inman, Y. Yao, J. Tour, J. Hutchinson and P. Weiss, *J Am. Chem. Soc.*, 2004, **126**, 12214.
18 L. T. Cai, H. Skulason, J. G. Kushmerick, S. K. Pollack, J. Naciri, R. Shashidhar, D. L. Allara, T. E. Mallouk and T. S. Mayer, *J. Phys. Chem B*, 2004, **108**, 2827.
19 J. J. Stapleton, P. Harder, T. A. Daniel, M. D. Reinard, Y. Yao, D. W. Price, J. M. Tour and D. L. Allara, *Langmuir*, 2003, **19**, 8245.
20 J. Chen, W. Wang, M. A. Reed, A. M. Rawlett, D. W. Price and J. M. Tour, *Appl. Phys. Lett.*, 2000, **77**, 1224.
21 J. M. Seminario, A. G. Zacarias and J. M. Tour, *J. Am. Chem. Soc.*, 1999, **121**, 411.
22 J. R. Heath and M. A. Ratner, *Phys. Today*, 2003, **56**, 43.
23 C. P. Collier, J. O. Jeppesen, Y. Luo, J. Perkins, E. W. Wong, J. R. Heath and J. F. Stoddart, *J. Am. Chem. Soc.*, 2001, **123**, 12632.
24 C. P. Collier, G. Mattersteig, E. W. Wong, Y. Luo, K. Beverly, J. Sampaio, F. M. Raymo, J. F. Stoddart and J. R. Heath, *Science*, 2000, **289**, 1172.
25 R. M. Metzger, J. W. Baldwin, W. J. Shumate, I. R. Peterson, P. Mani, G. J. Mankey, T. Morris, G. Szulczewski, S. Bosi, M. Prato, A. Comito and Y. Rubin, *J. Phys. Chem. B*, 2003, **107**, 1021.
26 R. M. Metzger, *Chem. Rev.*, 2003, **103**, 3803.
27 R. M. Metzger, T. Xu and I. R. Peterson, *J. Phys. Chem. B*, 2001, **105**, 7280.
28 S. Ranganathan, I. Steidel, F. Anariba and R. L. McCreery, *Nano Lett.*, 2001, **1**, 491.
29 A. O. Solak, S. Ranganathan, T. Itoh and R. L. McCreery, *Electrochem. Solid State Lett.*, 2002, **5**, E43.
30 R. L. McCreery, J. Dieringer, A. O. Solak, B. Snyder, A. Nowak, W. R. McGovern and S. DuVall, *J. Am. Chem. Soc.*, 2003, **125**, 10748.
31 R. McCreery, J. Dieringer, A. O. Solak, B. Snyder, A. M. Nowak, W. R. McGovern and S. DuVall, *J. Am. Chem. Soc.*, 2004, **126**, 6200.
32 A. Nowak and R. McCreery, *J. Am. Chem. Soc.*, 2004, **126**, 16621.
33 A. M. Nowak and R. L. McCreery, *Anal. Chem.*, 2004, **76**, 1089.
34 F. Anariba, J. Steach and R. McCreery, *J. Phys. Chem. B*, 2005, **109**, 11163.
35 S. Ranganathan, *Preparation Modification and Characterization of a Novel Carbon Electrode Material for Applications in Electrochemistry and Molecular Electronics*, PhD Thesis, Ohio State University, 2001.
36 S. Ranganathan and R. L. McCreery, *Anal. Chem.*, 2001, **73**, 893.
37 J. K. Kariuki and M. T. McDermott, *Langmuir*, 2001, **17**, 5947.
38 J. K. Kariuki and M. T. McDermott, *Langmuir*, 1999, **15**, 6534.
39 C. Combellas, F. Kanoufi, J. Pinson and F. Podvorica, *Langmuir*, 2005, **21**, 280.
40 K. Boukerma, M. M. Chehimi, J. Pinson and C. Blomfield, *Langmuir*, 2003, **19**, 6333.
41 E. Coulon, J. Pinson, J.-D. Bourzat, A. Commercon and J.-P. Pulicani, *Langmuir*, 2002, **17**, 7102.
42 M. Delamar, G. Desarmot, O. Fagebaume, R. Hitmi, J. Pinson and J. Saveant, *Carbon*, 1997, **35**, 801.
43 P. Allongue, M. Delamar, B. Desbat, O. Fagebaume, R. Hitmi, J. Pinson and J. M. Saveant, *J. Am. Chem. Soc.*, 1997, **119**, 201.
44 P. A. Brooksby and A. J. Downard, *Langmuir*, 2004, **20**, 5038.
45 B. Ortiz, C. Saby, G. Y. Champagne and D. Belanger, *J. Electroanal. Chem.*, 1998, **455**, 75.
46 F. Anariba, S. H. DuVall and R. L. McCreery, *Anal. Chem.*, 2003, **75**, 3837.
47 A. J. Downard, *Langmuir*, 2000, **16**, 9680.
48 W. R. McGovern, F. Anariba and R. McCreery, *J. Electrochem. Soc.*, 2005, **152**, E176.
49 R. J. Kalakodimi, A. M. Nowak and R. McCreery, *Chem. Mater.*, 2005, DOI: 10.1012/cm050689c.
50 J. D. Ramsey, S. Ranganathan, J. Zhao and R. L. McCreery, *Appl. Spectrosc.*, 2001, **55**, 767.
51 T. Itoh and R. L. McCreery, *J. Am. Chem. Soc.*, 2002, **124**, 10894.
52 A. Cyr, P. Huot, G. Belot and J. Lessard, *Electrochim. Acta*, 1990, **35**, 147.
53 G. Seshadri and J. A. Kelber, *J. Electrochem. Soc.*, 1999, **146**, 3762.
54 T. Umebayashi, T. Yamaki, H. Itoh and K. Asai, *Appl. Phys. Lett.*, 2002, **81**.
55 J. Portier, G. Campet, C. Kwon, J. Etourneau and M. Subramanian, *Int. J. Inorg. Mater.*, 2001, **3**, 1091.
56 J. J. Hoagland, X. D. Wang and K. W. Hipps, *Chem. Mater.*, 1993, **5**, 54.
57 J. J. Hoagland, J. Dowdy and K. W. Hipps, *J. Phys. Chem.*, 1991, **95**, 2246.

Electrical transport in saturated and conjugated molecular wires

F. Remacle†*[ab] and R. D. Levine[bc]

Received 25th April 2005, Accepted 20th June 2005
First published as an Advance Article on the web 28th September 2005
DOI: 10.1039/b505696a

The mechanism for charge transport in dithio molecular wires tethered between two gold electrodes is investigated, using both a steady state and a time-dependent quantum mechanical approach. The interface with the electrodes is modeled by two gold clusters and the electronic structure of the entire Au_n–S–bridge–S–Au_n system is computed *ab initio* at the DFT level and semi-empirically, with the extended Hückel theory. Current *vs.* applied bias, *I–V*, curves are computed using a scattering Landauer-type formalism in a steady state picture. The applied source–drain and gate voltages are included at the *ab initio* level in the electronic Hamiltonian and found to influence strongly the *I–V* characteristics. The time evolution of a non stationary electronic wave packet initially localized on a gold atom at one end of the extended system shows that charge transfer proceeds sequentially, by a hopping mechanism, to the opposite end. Analysis of the effective one electron Hamiltonian matrix shows that the sulfur atom endows a resistive character to the Au–C–S junctions. The S atoms are however rather well coupled to both the gold and carbon atoms so that typically the super exchange limit for electron transfer is not reached unless the molecular bridge is saturated and the Fermi window function is narrow.

1. Introduction

Understanding and controlling charge transfer in molecular junctions is essential for molecular electronics[1–4] and electrical addressing in molecular logic.[5,6] In a molecular junction, charge transfer can be controlled at different levels, by the intrinsic chemical nature of the electrodes and of the molecular frame, by the substituents on the molecular backbone, and also by the application of external electrical (and magnetic) fields.

In a three terminal device, two external electric fields are applied to the molecule: the source–drain bias, V_{sd}, applied along the molecular backbone between the source and the drain electrodes, and the gate bias, V_g, applied in the perpendicular direction. Charge transfer and current characteristics can be controlled by these two external voltages. The source–drain bias, V_{sd}, defines the energy of a Fermi window, eV_{sd}, (e is the electron charge). The window is that energy range for which there is an excess of electronic population of one electrode with respect to the other, Fig. 1. Current flows when a discrete molecular level is within the Fermi window. Increasing V_{sd} widens the Fermi window, thereby allowing more molecular states to take part in the charge transport. The effect of the gate voltage V_g is to move the energy of the molecular levels with respect to the position of the Fermi window. Varying V_g for a fixed value of V_{sd} brings different levels within the Fermi

[a] *Département de Chimie, B6c, Université de Liège, B4000, Liège, Belgium. E-mail: fremacle@ulg.ac.be; Fax: +32 4 3663413*
[b] *The Fritz Haber Research Center for Molecular Dynamics, The Hebrew University of Jerusalem, Jerusalem 91904, Israel. E-mail: rafi@fh.huji.ac.il; Fax: +972 2 6513742*
[c] *Department of Chemistry and Biochemistry, University of California, Los Angeles, Los Angeles CA 90095-1569, USA*

† Directeur de Recherches, FNRS, Belgium.

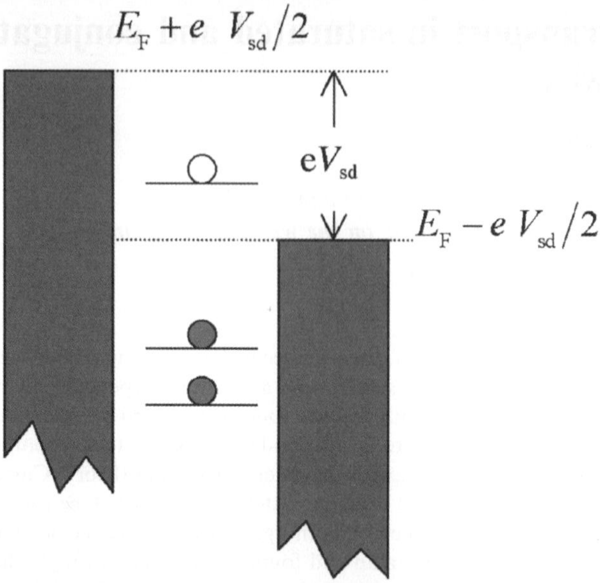

Fig. 1 Schematic representation of the discrete conducting levels of the extended system and the Fermi window of the electrodes. A current flows when the energy of a level of the extended system falls within the Fermi window. The energy shift of the Fermi level of the electrodes induced by the source–drain bias is shown for a symmetric junction, in the case where no electrode is grounded.

window, leading to conductance spectroscopy.[7–19] In addition, especially for molecular systems, the very presence of an external field can modify the electronic structure beyond a simple Stark shift in the position of the level energy and the field can distort the charge density.[20–24] The source–drain voltage is typically low but the gap between the two electrodes is of molecular dimension so that the electric field experienced by the molecule is high. Therefore it is important to include both V_{sd} and V_g in the Hamiltonian used to compute the electronic structure at the *ab initio* level. We discuss below how different orientations of the molecule with respect to V_{sd} and V_g can lead to different *I–V* curve characteristics, depending on the σ or the π character of the conducting molecular level.

Varying the chemical nature of the electrodes can also be used to control the electron transfer. Firstly, because electrodes of a different chemical nature have different Fermi energies, the chemical nature of the electrodes defines the position of the Fermi window with respect to the discrete molecular levels of the system. Beyond this, changing the electrode alters the strength of the coupling between the electrode and the molecule.

When the coupling, β, is weak with respect to the energy difference, ΔE, between the orbitals of the Au cluster and those that are directly coupled to them, that is, $\beta/\Delta E \ll 1$, the super exchange limit[4,25–29] is reached. In this limit the charge is transferred from one interface to the opposite one without there being significant population on the molecular bridge. The reason is that when $\beta/\Delta E$ is very small, it is more favorable for the interface to be coupled, by higher order perturbation theory, to a nearly isoenergetic interface at the other end. The super exchange model has been extensively applied to electron transfer in molecular donor-acceptor complexes, see for example refs. 30–34. We have shown[35,36] that the superexchange coupling limit can also be reached in self-assembled arrays of metallic colloidal quantum dots (QD) in the intermediate compression regime, where the exchange coupling, β, is too small to bridge the energy mismatch between the valence orbitals of the individual dots that are possible when there are wide fluctuations in the size of the dots. We argue in this paper, on the basis of extended Hückel[37–39] (EH) and higher level *ab initio* electronic structure computations, that for the symmetric molecular bridges investigated, with the same Au–S–C junction at both ends, the super exchange limit is typically surpassed because on the average over the valence orbitals the ratios $(\beta/\Delta E)_{Au–S}$ and $(\beta/\Delta E)_{S–C}$ are not small enough. They are however smaller than the ratios $(\beta/\Delta E)_{C–C}$ within the bridge, so that the Au–S–C junction has a

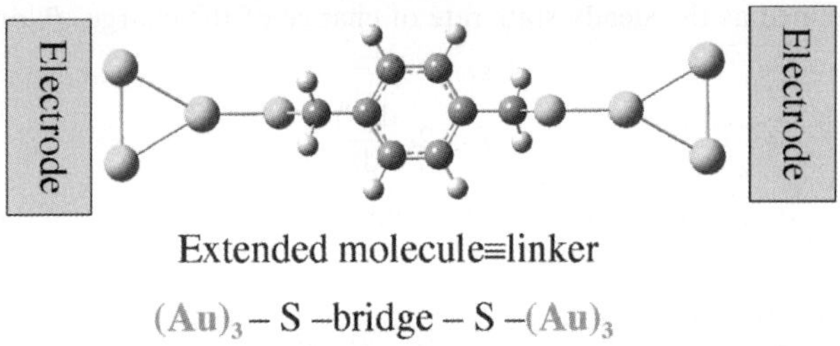

Fig. 2 Schematic representation of the extended system, Au_3–CH_2–C_6H_4–CH_2–Au_3, tethered between two gold electrodes.

resistive character. This conclusion is supported by a time dependent analysis of the electron migration, which at short times, shows sequential hopping of the electron along the chain on a fs time scale, with a finite population on the bridge. At longer times, the population on the bridge and on the two sulfur atoms reaches a constant value, with a beating between the populations on the two end Au clusters.

While the superexchange coupling limit is, on the average, exceeded, there are fluctuations about the average. If the relevant orbitals can be accessed within the Fermi window function then either a strong limit or a superexchange limit can be realized. In particular, the weak coupling, superexchange limit will be relevant for those MO's of the extended system that are significantly localized at the interface with the electrodes. The interface here is the two Au–S–C junctions. For saturated bridges these orbitals will typically be the HOMO and HOMO − 1.[23,24] In terms of the concept of a local ionization potential, IP,[40–42] this happens when the interface has a lower local IP than the center of the bridge. The possibility of designing a weakly coupled interface that favors the superexchange limit or, on the contrary, a strong coupling, suggests that we should explore other manners of linking a linear organic molecule to two metallic electrodes. Specifically we looked at Au–P–C, Au–O–C and Au–N–C interfaces.

The paper is organized as follows. In Section 2, we summarize the formalism used to compute the current. The molecular junctions studied here are symmetric, with the same interface, Au–S–C, on the left and on the right, and neither of the electrodes is grounded. The device is therefore in a short circuit configuration, where the edges of the Fermi window defined by the source–drain bias are given by $E_F + eV_{sd}/2$ and $E_F - eV_{sd}/2$. In our model as in others,[20,43–47] the Au atoms that are perturbed by the bonding to the sulfur atoms of the molecular bridge are included in an extended system, Au_n–S–bridge–S–Au_n. This system is then coupled to the metallic gold electrodes (Fig. 2). The electronic structure of the extended system is computed using a molecular approach with the Gaussian 03 package[48] including external fields in the Hamiltonian. In this approach,[20,43,44,47,49] the external field is applied over a finite region much larger than the molecule so that the boundary conditions remain that of an isolated molecule. The current is obtained using time independent scattering theory formalism[50] which computes the steady state current for a system displaced from equilibrium by the coupling between the interface and the bulk electrode. The scattering theory result is similar to the Landauer approach[51,52] as used for example by Ratner *et al.*, Nitzan, Hush *et al.*, and Datta *et al.*[29,45,53–55] In Section 3, we discuss the mechanism for conduction using a time dependent approach and an analysis of the electronic Hamiltonian matrix. Field effects, and in particular gating effects on the *I*–*V* characteristics are discussed in Section 4. Concluding remarks are given in Section 5.

2. Computation of the current

A scattering formalism based on the generalized Ehrenfest theorem[56] is used to derive the expression for the current under non equilibrium but steady state conditions. The final expression is the same as that used in our work on the conductivity of arrays of metallic quantum dots.[50] We summarize here the main steps of the derivation, in order to relate the expression of the current to the time dependent approach discussed in Section 3.

The current is defined as the steady state rate of change of the charge, P, on the right electrode (see Fig. 2):

$$I = 2e\frac{\mathrm{d}\langle P\rangle}{\mathrm{d}t} \tag{1}$$

where the factor 2 accounts for the two spin directions and e is the charge of the electron. For a 1D system, the rate of change of $\langle P\rangle$ is given by:[56]

$$\frac{\mathrm{d}\langle P\rangle}{\mathrm{d}t} = \frac{1}{h}\sum_{\mathrm{I,F}}(p_{\mathrm{F}} - p_{\mathrm{I}})\langle \mathrm{F}|T(E^-)|\mathrm{I}\rangle\delta(E_{\mathrm{I}} - E_{\mathrm{F}})\langle \mathrm{I}|T(E^+)|\mathrm{F}\rangle \tag{2}$$

The first term of the sum in eqn. (2) is the charge that enters the right electrode and the second term the charge that is leaving it. $T(E^+)$ is the transition operator:

$$T(E^+) = V + VG(E^+)V \tag{3}$$

and $G(E^+)$ the Green's function of the full Hamiltonian, H,

$$G(E^+) = (E^+ - H)^{-1},\ H = H_0 + V \tag{4}$$

The mean value of P can only vary because of the coupling, V, between the bulk electrodes and the interface. H_0 is the Hamiltonian of the full system (electrode–extended system–electrode) in the absence of interaction, $H_0 = H_{\text{left-electrode}} + H_{\text{extended-system}} + H_{\text{right-electrode}}$. $T(E^+)$ induces transitions between the initial and final state $|\mathrm{I}\rangle$ and $|\mathrm{F}\rangle$ that are the eigenstates of the zero order Hamiltonian, H_0, for which the extended system is not coupled to the electrodes. $|\mathrm{I}\rangle$ and $|\mathrm{F}\rangle$ can therefore be written as product states of the one electron states of the electrodes, $|\mathrm{i}\rangle$ and $|\mathrm{f}\rangle$ and of the state of the neutral extended system, $|\Phi_N\rangle$, with N being its number of electrons. We emphasize that the extended system includes the interfaces at both ends, *i.e.*, the Au–S–C junctions, so that the question of the superexchange limit can be discussed in terms of the orbitals of $|\Phi_N\rangle$ or of $|\Phi_{N\pm 1}\rangle$ where N or $N \pm 1$ is the number of electrons.

The effect of the coupling, V, to the electrodes is to transfer an electron from the electrode to the extended system or vice versa. In the initial state, $|\mathrm{I}\rangle$, the state $|\mathrm{i}\rangle$ of the left electrode is occupied by one electron and the state $|\mathrm{f}\rangle$ of the right electrode is empty. The final state, $|\mathrm{F}\rangle$, corresponds to the complementary situation: the state $|\mathrm{i}\rangle$ is empty and the state $|\mathrm{f}\rangle$ is occupied. For an elastic charge transfer, that is, a transfer that leaves the state $|\Phi_N\rangle$ of the extended system unchanged,

$$\begin{aligned} |\mathrm{I}\rangle &= |\mathrm{i}\rangle^1|\Phi_N\rangle|\mathrm{f}\rangle^0 \\ |\mathrm{F}\rangle &= |\mathrm{i}\rangle^0|\Phi_N\rangle|\mathrm{f}\rangle^1 \end{aligned} \tag{5}$$

where the superscript is the occupancy of the state.

In eqn. (2), p_{I} and p_{F} are the thermal weights of $|\mathrm{I}\rangle$ and $|\mathrm{F}\rangle$. The one electron state of the electrodes have a Fermi weight, $f_{\mathrm{i,f}}(\varepsilon - E_{\mathrm{Fermi}} \pm eV_{\mathrm{sd}}/2)$, where $f(x)$ is the Fermi function, that in the very low T limit is a step. The many electron states of the molecule should be weighted by its Boltzmann factor. However, while this factor is important for the transport properties of QD arrays, where the density of electronic states is very high around the Fermi level of the electrodes,[50,57] it does not play a significant role here, since the energy difference between the ground state and the lowest excited state of the molecules investigated is >0.5 eV. Excited electronic states of the N electron molecule are unlikely to contribute as initial or final states and we will therefore consider in the following that in the initial and final states $|\mathrm{I}\rangle$ and $|\mathrm{F}\rangle$ the molecule is in its ground electronic state. As discussed in Section 4, in the compounds investigated below, the Fermi level of the gold electrodes is close in energy to the frontier orbitals[58] of the extended system and they are the ones taking part in the conduction process.

From eqn. (5), the energy of state $|\mathrm{I}\rangle$ is $E_{\mathrm{I}} = \varepsilon_{\mathrm{i}} + E_N$ and that of state $|\mathrm{F}\rangle$, $E_{\mathrm{F}} = \varepsilon_{\mathrm{f}} + E_N$. Making use of the delta function in eqn. (2), one gets

$$\frac{\mathrm{d}\langle P\rangle}{\mathrm{d}t} = \frac{1}{h}\int \mathrm{d}\varepsilon\, g(\varepsilon, V_{\mathrm{sd}}, T)\langle \mathrm{F}|T(E^-)|\mathrm{I}\rangle\delta(E_{\mathrm{I}} - E_{\mathrm{F}})\langle \mathrm{I}|T(E^+)|\mathrm{F}\rangle \tag{6}$$

where the Fermi window, $g(\varepsilon,V_{sd},T)$ depends on the applied V_{sd} and on the temperature, T, and is given by

$$g(\varepsilon, V_{sd}, T) = f_f\left(\varepsilon - E_{Fermi} + \frac{eV_{sd}}{2}\right) - f_i\left(\varepsilon - E_{Fermi} - \frac{eV_{sd}}{2}\right) \tag{7}$$

The matrix elements of the transition operator, $\langle F|T(E^+)|I\rangle$, are computed for a one electron description of the ground electronic state of the molecule:

$$|\Phi_N\rangle = \prod (\phi_\lambda)^{f_\lambda} \tag{8}$$

where f_λ is the occupation number of the molecular orbital (MO), ϕ_λ. Within the approximation of a weak coupling, V, to the electrodes, and taking into account that V cannot induce direct transitions between the states of the electrodes on the left and on the right, eqn. (3) becomes

$$T(E^+) = VG_0(E^+)V. \tag{9}$$

We emphasize that eqn. (9) assumes that the coupling V between the electrode and the extended system is weak. $G_0(E^+) = (E^+ - H_0)^{-1}$ where H_0 is the zero order Hamiltonian of the full system where the extended system and the two electrodes are not coupled, $H = H_0 + V$ (Recall that the extended system includes a gold cluster at each end). V couples that cluster to the bulk electrode. When the coupling induced by V is not weak, one must use the theory of scattering by two potentials to replace V in eqn. (9) by a scattering operator and this will broaden the electronic states of the extended system. It is our experience that at ordinary temperatures this broadening is masked by the finite breadth of the cutoff of the Fermi window at its two edges.

Using eqns. (5), (8) and (9) the expression of the matrix element of the transition operation, $\langle I|T(E^+)|F\rangle$ is then[50]

$$\langle I|T(E^+)|F\rangle = \sum_{\lambda=1}^{b} \frac{\langle i|V|\lambda\rangle\langle\lambda|V|f\rangle}{E^+ - E_\lambda} \tag{10}$$

where E_λ is the energy of the MO $|\lambda\rangle$. The sum in eqn. (10) is over all the orbitals of the extended system. Conduction can either occur through empty MO's, in which case an electron is first transferred to the extended system, leading to an intermediate state where there is no electron in the states $|i\rangle$ and $|f\rangle$ of the electrode, and $N + 1$ electrons in the extended molecule, or through occupied MO's, when an electron is first transferred from the extended system to the state $|f\rangle$ of the right electrode, leading to a $N - 1$ electron extended intermediate, followed by a transfer of an electron from the left electrode to the molecule. This discussion assumes that the coupling between the interfaces and the bulk electrodes is weak but nevertheless strong enough that there is no more than one extra hole or one extra electron at a time on the extended system. If the MO λ is primarily localized at the two interfaces then conduction is by superexchange. The other key assumption in deriving eqn. (10) is that the occupied MOs of the extended system are unchanged when an electron is added or removed. This assumption is necessary to be able to write in eqn. (10) a sum over MO's rather than a sum over many particle states. Below we discuss the computational checks of this assumption.

When the MO's are expressed as linear combination of atomic orbitals eqn. (10) becomes

$$\langle I|T(E^+)|F\rangle = V^2 \sum_{\lambda=1}^{b} \frac{\sum_l^{n_l} c_l^\lambda \sum_r^{n_r} c_r^\lambda}{E^+ - E_\lambda} \tag{11}$$

where the sums over l and over r are over the atomic orbitals (AO) of the Au atoms on the left and on the right of the extended system, Au_n–S–bridge–S–Au_n, that are directly coupled to the electrodes. Note that when we do not use a minimal basis set there are more than just the minimal occupied orbitals on each atom. We further make the resonant approximation for the Green's function so that eqn. (11) becomes:

$$\langle \mathrm{I}|T(E^+)|\mathrm{F}\rangle = -\mathrm{i}\pi V^2 \sum_{\lambda=1}^{b} \left(\sum_{l}^{n_l} c_l^\lambda\right)\left(\sum_{r}^{n_r} c_r^\lambda\right) \delta(E^+ - E_\lambda) \tag{12}$$

If the coupling to the electrode is included in eqn. (11) as an imaginary part in the energies of the MO's then the delta function in eqn. (12) is replaced by a Lorentzian function.[56] This is sometimes called a self-energy term.[54,55] As discussed in ref. 50 for the weak coupling limit, except at very low voltages and very low temperatures, the broadening due to the Fermi window, $g(\varepsilon, V_{\mathrm{sd}}, T)$, is wider than the broadening of the MO energies induced by the coupling to the bulk electrodes.

Using eqns. (12) and (6) in eqn. (2), the final expression for the current becomes

$$I = 2e\frac{\mathrm{d}\langle P\rangle}{\mathrm{d}t} = -2e\pi^2 V^4 \frac{1}{h}\sum_{\lambda=1}^{b} g(E_\lambda, V_{\mathrm{sd}}, T) T_{\mathrm{if}}(\lambda) \tag{13}$$

where $T_{\mathrm{if}}(\lambda)$is the transmission function of the MO λ.

$$T_{\mathrm{if}}(\lambda) = \left(\sum_{l}^{n_l}\sum_{r}^{n_r} c_l^\lambda c_r^\lambda\right)^2 \delta(E^+ - E_\lambda) \tag{14}$$

In the computations reported below, there is the same number of AO's for the Au atoms on the left and on the right, so that $n_l = n_r$. In addition, when computing the transmission function, we can typically neglect the interference between AO localized in the same atom. This further simplifies the expression for the transmission function to

$$T_{\mathrm{if}}(\lambda) = \sum_{l}^{n_l} (c_l^\lambda)^2 (c_r^\lambda)^2 \delta(E^+ - E_\lambda) \tag{15}$$

The electronic levels of the extended systems are obtained from *ab initio* computations. The geometry of the extended system is optimized at the DFT level, with the B3LYP functional. The basis set is the split-valence localized Gaussian AO for the C, H and S atoms and the Los Alamos RECP double zeta (LANL2DZ) pseudo-potential[59] for the gold atoms. This allows for a one-electron description while including part of the electron correlation.

As discussed above, we assume in the final expression for the transmission function (eqn. (10)) that the geometry of the extended system is not modified by adding or removing an electron to it and that when it is charged, the anion or cation keeps the same set of MO's as the neutral. This amounts to assuming that the Koopmans theorem[60] is valid. The most important breakdown of the theorem comes from a change in the equilibrium geometry when the system is charged. The deformation upon charge transfer leads to a breakdown of the Born–Oppenheimer separation and the formalism discussed above no longer applies. We performed computations for the charged (anion and cation) saturated, Au_3–S–$(CH_2)_4$–S–Au_3 and conjugated Au_3–S–$(CH)_4$–S–Au_3 bridges where the geometry is re-optimized. For the *trans* conformers studied here, the geometry changes were not important. There is therefore in this case no significant charge-induced deformation of the extended system. This can be understood from the high mass of the gold trimers with respect to that of the molecular bridges and from its rather rigid nature. We have also recomputed the electronic structure for the cation and the anion of the saturated and conjugated C_4, C_6 and C_8 extended systems at the geometry of the neutral. From the values of the ionization potential (IP = $E(N+1) - E(N)$) and of the electron affinity (EA = $E(N) - E(N-1)$) we can compute the *absolute* hardness,[60–62] $\eta = (\mathrm{IP} - \mathrm{EA})/2 \equiv (e^2/2C(N))$, which gives the Coulomb blockade energy of the entire extended system including the gold trimers. This charging energy is the minimum energy cost to add one extra electron to the N-electron molecule. As expected the Coulomb blocking is higher ($\approx$2 eV) for the saturated molecules (as opposed to $\approx$1.3 eV for the unsaturated bridge) and it decreases monotonically with the length of the linear molecule.

When compared to the energies of the HOMO and of the LUMO of the neutral, we find that the IP is about 10–15% larger than E_{HOMO} while the EA is 10–15% smaller than E_{LUMO}. It is well known that DFT computations underestimate the HOMO–LUMO gap. These results affect the computed value of the current in two ways. The formalism that we use to compute the current is written in a many electron description for the electronic states (see eqns. (5) and (8)) and as long as the many electron initial $|\Phi_N\rangle$, intermediate $|\Phi_{N\pm1}\rangle$and final $|\Phi_N\rangle$ electronic states of the extended systems are computed with the same basis set, it is possible to use different sets of MO's for the neutral $|\Phi_N\rangle$ and the charged $|\Phi_{N\pm1}\rangle$ states. When doing so, we do not expect the changes in the transmission function to be significant because the self-consistent wave function includes first order correction (this is the basis for the validity of the Koopmans theorem) and therefore the changes in the transmission function will only appear in second order. The other consequence is on the energetics. The predictions that we make regarding the strengths of the source–drain and gate voltages required to get a current rely on how well the energies of the HOMO and of the LUMO give the value of the IP and of the EA. The computations reported below are therefore only semi-quantitative, their main objective being to provide understanding and a physical picture for controlling the electric properties of the wire by applying external electric fields.

The applied source–drain and the gate bias are incorporated as an external electrical (dipolar) field [44,47,49] in the electronic Hamiltonian of the quantum chemistry code using boundary conditions for an isolated molecule. The MO's are recomputed at the equilibrium geometry in presence of the external electrical field. An electric field of 1 V nm^{-1} corresponds to 1.93×10^{-3} atomic units of field. Since the lengths of the molecular bridges studied are of the order of a few nm, the charge density of the MO's can be significantly distorted by the external bias. For example, a field applied along the molecular backbone shifts the energies and distorts strongly the charge distribution of σ orbitals whose densities are parallel to its direction. By contrast, it takes higher strengths of the gating field to affect the energies and charge distribution of π orbitals, whose densities are usually perpendicular to the direction of the source–drain field.[22,23,49,63] We show in section 4 below that a gating field applied in the direction perpendicular to the molecular backbone affects the energies of both the π and the σ MO's which move up or down depending on the sign of the applied field. The charge density of the π orbitals remains almost unaffected by the gating field. Including the external field in the computation of the electronic structure is also important because quasi-degenerate MO's in the absence of external field, as is the case for the HOMO and HOMO − 1 in saturated alkane bridges, can be coupled when an electrical field is applied.[24,47]

3. Mechanism for charge transfer

The signature of the superexchange coupling limit is that charge is transferred from one end, (interface), of the system to the other end without any significant population on the intermediate atoms and bonds.[25,26] The bridge in the sense of conventional superexchange theory is the actual molecule from one C end to the other and the superexchange limit could apply if the sulfur atoms in the Au–S–C junction are resistive enough, so that the weak coupling limit, $\beta/\Delta E < 0.01$, is reached for the interface.

The ratio $\beta/\Delta E$ can be evaluated by the analysis of the matrix elements of the one electron Hamiltonian matrix used to compute the electronic structure. In order to get a qualitative insight we have first performed this analysis at the extended Hückel (EH) level[37–39] where the basis set is limited to the AO's of the valence electrons. The parametrized semi-empirical Hamiltonian matrix has a simple structure and can be analyzed to get insight on the strength of the coupling. The geometry used in the EH computations are the equilibrium geometries obtained at the DFT/B3LYP (6-31++G(d,p) for H, C, S and RECP LANLDZ2 for Au). The same analysis is the carried out at the DFT/B3LYP (6-31++G(d,p) level.

From the quantum chemical computations we extract the following physical picture. The Au–S–C molecular junction is made of two chemical covalent bonds, the Au–S bond and the S–C bond. We find that the ratios $(\beta/\Delta E)_{Au–S}$ and $(\beta/\Delta E)_{S–C}$ for the valence AO's are of the order 0.1–0.5 and 1–10 respectively for both the hexane and the hexatriene bridges. Our previous work on the super-exchange limit in arrays of metallic QD's,[35] as well as simulation on a chain of sites show that for the super exchange limit to be reached, $\beta/\Delta E$ must be smaller than 0.01. The ratios obtained for the Au–S–C junctions therefore indicate that the coupling is too strong for the super exchange limit to be realized. However, the effective coupling, $\beta/\Delta E$, for the junction is smaller than the effective couplings between atoms of the backbone of the molecule so that the junction has a resistive character. The effective coupling Au–S is smaller than the coupling S–C to the saturated or conjugated molecular bridge, indicating that the more resistive part of the junction is the Au–S bond. The analysis of the extended Hückel Hamiltonian matrix shows that there is also a non-negligible next-nearest-neighbor coupling, Au–C, with a $(\beta/\Delta E)_{Au-C}$ ratio of the order of the direct coupling $(\beta/\Delta E)_{Au-S}$.

In the EH matrix, the largest matrix elements are between near neighbors and next-nearest neighbors and there is no direct significant effective coupling, $\beta/\Delta E > 0.1$, beyond.

We have performed the same analysis on the matrix elements of the self-consistent Kohn–Sham matrix using increasing sizes for the basis sets: the minimal basis set STO-3G for C, H and S, with the single zeta RECP LANL2MB on Au, and two split valence basis sets, the 6-31G basis and the 6-31++G(d,p), both with the LANL2DZ RECP for Au. The values of the effective couplings $(\beta/\Delta E)_{Au-S}$, $(\beta/\Delta E)_{S-C}$ and $(\beta/\Delta E)_{Au-C}$ for the valence AO's remain basically unchanged when compared to those obtained at the extended Hückel level. However, for the 6-31++G(d,p) basis set which includes diffuse and polarization functions, a few significant values of the off diagonal matrix elements appear between diffuse functions (Rydberg orbitals) centered on more distant atoms. Nevertheless even in the 6-31++G(d,p) basis set, there is no direct coupling between to the two junctions.

Since the Gaussian basis sets are not orthogonal, we have also analyzed the off-diagonal elements of the Kohn–Sham matrix in natural atomic orbitals (NAO).[64–67] These orbitals, unlike the Gaussian basis sets discussed above, are orthogonal by construction because they are defined to diagonalize the one electron density matrix. They are *maximum occupancy* orbitals and basis set independent.[65,66] As a result, their populations provide an unambiguous basis for computing the charge localized on the atoms. We find values of the effective couplings $(\beta/\Delta E)_{Au-S}$, $(\beta/\Delta E)_{S-C}$ and $(\beta/\Delta E)_{Au-C}$ in the same range as obtained at the EH level and at the *ab initio* level with Gaussian AO's. Moreover, in the NAO basis, the off diagonal elements of the Kohn–Sham matrix between atoms more distant than next near neighbors that lead to an effective coupling $\beta/\Delta E > 0.1$ only occur between Rydberg s and p atomic natural orbitals and those orbitals have an occupancy of $<10^{-2}$.

That the effective couplings $(\beta/\Delta E)$ for the Au–S–C junction are too large to be within the superexchange limit is illustrated in Fig. 3. We show in Fig. 3 the values of the energies of the

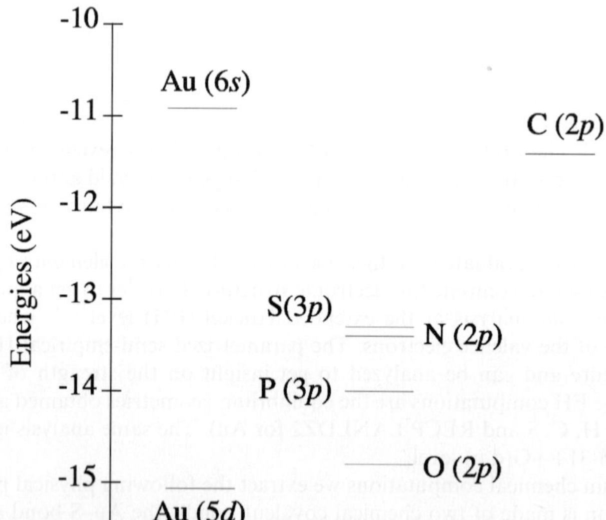

Fig. 3 Energies of the valence AO of different possible molecular junctions with N, P, S, O atoms coupled to Au and C. The energies shown are as parametrized in the extended Hückel theory. The energy differences, ΔE, between the valence AO of the atoms of the junction are of the order of 2 eV or less. We shown in the text that for the Au–S–C and Au–O–C junctions that were analyzed in detail, the ΔE's can be bridged by the off-diagonal elements, β, of the electronic Hamiltonian, thereby leading to values of effective coupling, $\beta/\Delta E$, large enough to induce a sequential migration of the charge along the molecular backbone.

valence AO involved in different kinds of junctions: Au–S–C, Au–O–C, Au–N–C and Au–P–C, as they are parametrized in EH theory. One can see that the highest valence pAO of O, P, N and S fall in between the 5d and 6s AO of Au and the 2p AO of C, with ΔE's of the order of 1 eV or less. Since several off diagonal elements of the electronic Hamiltonian are larger than 1 eV, this explains why the values of the effective couplings for valence AO can be high and of the order of unity. The small energy difference between the AO 6s of Au and 2p of C also explains why the next-nearest neighbor coupling, Au–C, is found to be effective. This conclusion remain true for larger basis sets too: in the case of junctions with N, C, O, P and S, there are always valence orbitals close enough in energy to lead $(\beta/\Delta E)$ of the order 0.1–1. For example, the junction Au–O–C in Au_3-O-$(CH_2)_6$-O-Au_3 was analyzed at the EH and DFT/6-31++G(d,p) levels. At both computational levels, the effective couplings $(\beta/\Delta E)_{O-C}$ are found to be slightly more effective than for S–C, and the Au–O remains the resistive part of the junction, but less so than in the case of the Au–S–C. For the Au–O–C junction, there is also a non negligible direct Au–C coupling with a ratio $(\beta/\Delta E)_{Au-C}$, which is of the order of the direct Au–O coupling.

Having assessed that there are no direct coupling in the Kohn–Sham matrix between the two junctions and that the effective couplings $(\beta/\Delta E)$ for the junction are too large for the superexchange limit to be reached, we examined the charge transfer in the time domain. For doing so, we compute the population in the different subspaces defined by the (CH_2), S and Au_3 units for the alkanes and (CH), S and Au_3 units for the alkenes, that result from the transfer of an electron from the electrode to the extended system. As a basis set for the projectors on the chain units, we use the orthogonalized Gaussian AO[68–70] or the NAO basis. We checked that both atomic basis sets give basically the same results for the time evolution of the populations. As reported above, for the *trans* conformers investigated here, there are no significant geometry changes when an electron is added to or removed from the extended system. Since the time scales reported below for charge migration are of the order of a few femtoseconds, it is reasonable to expect that on this short time scale, there is no significant breakdown of the Born–Oppenheimer approximation.

We compute the population on the different units of the molecular bridge using a time-dependent approach that is similar to that used to get a time-dependent picture of the resonant Raman cross section.[71–74]

Within the weak coupling approximation for V and making the resonant approximation for the Green's function in eqn. (10), $\langle \mathrm{I}|T(E^+)|\mathrm{F}\rangle$ can be written

$$\begin{aligned}\langle \Phi_N \mathrm{f}|T(E^+)|\Phi_N \mathrm{i}\rangle &= -\frac{i}{\hbar}\int_{-\infty}^{+\infty} \mathrm{d}t\, \exp(\mathrm{i}Et/\hbar)\langle \Phi_N \mathrm{f} V|\exp(-iH_0 t/\hbar)|V\Phi_N i\rangle \\ &= -\frac{i}{\hbar}\int_{-\infty}^{+\infty} \mathrm{d}t\, \exp(\mathrm{i}Et/\hbar) C_{\mathrm{FI}}(t)\end{aligned} \tag{16}$$

where $C_{\mathrm{IF}}(t)$ is the cross-correlation function

$$C_{\mathrm{IF}}(t) = \langle \Phi_N \mathrm{f} V|\exp(-\mathrm{i}H_0 t/\hbar)|V\Phi_N \mathrm{i}\rangle \tag{17}$$

$|V\Phi_N \mathrm{i}\rangle$ can be viewed as the non-stationary electronic wave packet $|V\Phi_N \mathrm{i}\rangle \equiv |\Psi_1(0)\rangle$, created on the left inside the extended system by the charge transfer from the electrode. Its time-evolved projection on the final state localized at the other end of the extended system, $|\Psi_r\rangle \equiv |V\Phi_N \mathrm{f}\rangle$, is the cross-correlation function $C_{\mathrm{IF}}(t)$. The square modulus of $C_{\mathrm{IF}}(t)$ corresponds to the probability for an electron initially located on the left $|\Psi_r(t)\rangle \equiv \exp(-\mathrm{i}H_0 t/\hbar)|\Psi_r(0)\rangle$, to be in the final state on the right, $|\Psi_r\rangle$, where H_0 is the Hamiltonian of the extended system uncoupled to the bulk electrodes.

In the time dependent results discussed below, we take the initial electronic wave packet created on the left of the extended system at time $t = 0$ to be a pure state with equal amplitudes on the AO's of the gold atoms on the left:

$$|\Psi_l(0)\rangle = \frac{1}{\sqrt{n_{\mathrm{Au}}}}\sum_{l}^{n_{\mathrm{Au}}} |\mathrm{AO}^l_{\mathrm{Au}}\rangle \tag{18}$$

where n_{Au} is the number of AO's of the gold atoms on the left that are initially excited. We have checked numerically that taking the initial state to be a mixture does not change our conclusions regarding the mechanism for charge transfer. In the basis of the MO's of the extended system, we obtain

$$|\Psi_1(t)\rangle = \exp(-iH_0t/\hbar)|\Psi_l(0)\rangle = \frac{1}{\sqrt{n_{Au}}}\sum_l^{n_{Au}}\sum_\lambda \exp(-iE_\lambda t/\hbar)c_l^\lambda|\lambda\rangle \tag{19}$$

where $C_l^\lambda = \langle\lambda|AO_{Au}^l\rangle$. The population on a given unit of the extended system is given by the projection of $|\Psi_1(t)\rangle$ on the subset of n_A orthogonal AO's, $|a\rangle$, localized on that unit:

$$P_A(t) = \sum_{n_A}|\langle a|\Psi_l(t)\rangle|^2 = \frac{1}{n_{Au}}\sum_{n_A}\left|\sum_l^{n_{Au}}\sum_\lambda \exp(-iE_\lambda t/\hbar)c_l^\lambda c_a^\lambda\right|^2 \tag{20}$$

The populations on the gold trimer on the left, on the S atom on the left, on the bridge C_6H_{12}, on the S atom on the right and on the gold trimer on the right are plotted in Fig. 4 as a function of time for an initial state localized on the left Au_3. In the top panel, the electronic structure is computed at the EH level and at the 6-31++G(d,p)+RECP LANL2DZ in the bottom one. The overall picture is the same for the two computations: at short times, the exit out of the left gold trimer is first to the neighboring S atom, followed by a population of the C_6H_{12}, which is followed by the population of the right S atom and the right gold trimer. At both levels, the charge transfer occurs with a significant population on the alkane bridge, which in agreement with the analysis of the matrix elements of the Hamiltonian matrix above, confirms that the superexchange limit is not reached. The main difference between the two computations is in the population of the S atoms. Except for the initial rise on the left one, it remains small for the 6-31++G(d,p)+RECP LANL2DZ DFT/B3LYP level while it competes for that of the C_6H_{12} bridge at the EH level. The small amount of population on the S atoms is also found at the *ab initio* level with the smaller basis sets STO3G+RECP LANL2MB and 6-31G+RECP LANL2DZ (not shown). It may be related to a more efficient next near neighbor coupling, Au–C, in the junction.

That the charge transfer along the molecular backbone is sequential, from one skeleton atom to the next, is more clearly seen in Fig. 5, where the populations on the S atoms and on the successive (CH_2) groups are plotted at short time on a log scale for an octane bridge. At the EH level, the charge transfer exhibits a completely sequential character, while at the 6-31++G(d,p) level, the charge transfer is sequential up to the last (CH_2) group, after which the population on the right Au_3 trimer rises without significant population on the right S atom. This again may be due to the relatively efficient next near neighbor Au–C coupling identified in the analysis of the Hamiltonian matrix. The same sequential character of the charge migration is also recovered when the propagation at the 6-31++G(d,p) level is carried out in NAOs as well as in the smaller orthogonalized basis sets, STO3G and 6-31G (not shown).

In Fig. 6, we compared the rise of the population on the right gold trimer for alkane bridges of increasing length, $(CH_2)_4$, $(CH_2)_8$, $(CH_2)_{10}$ computed at the DFT 6-31++G(d,p) level. The initial rise for the butane bridge is clearly faster than for the octane and the decane ones, showing that the rate of charge transfer to from the left to the right end is distance dependent. In terms of the *I–V* curves, this dependence is manifested as an exponential fall of the current as a function of the length of the bridge.[49,75–84]

The time dependence of the population for the conjugated bridges is shown in Figs. 7 and 8. Fig. 7 shows the short time behavior of the population for the octatetraene bridge computed at the DFT/6-31++G(d,p) level. While the EH level behavior is completely sequential and very similar to that shown in Fig. 5a for a saturated bridge, all the conjugated bridges (butadiene, hexatriene, octatetraene) studied at the DFT/6-31++G(d,p) level exhibit the pattern shown in Fig. 7. The charge transfer is sequential up to the fourth carbon, after which the population is directly transferred to the right gold trimer, before the end CH_2 units and S atoms are accessed. For the four carbon atom chain, the short time rise on the right gold trimer for the saturated and conjugated bridges are similar (Fig. 8 top) but when the bridge length increases, the not fully sequential electron transfer of the conjugated bridges leads to a faster rise at short time (Fig. 8, bottom).

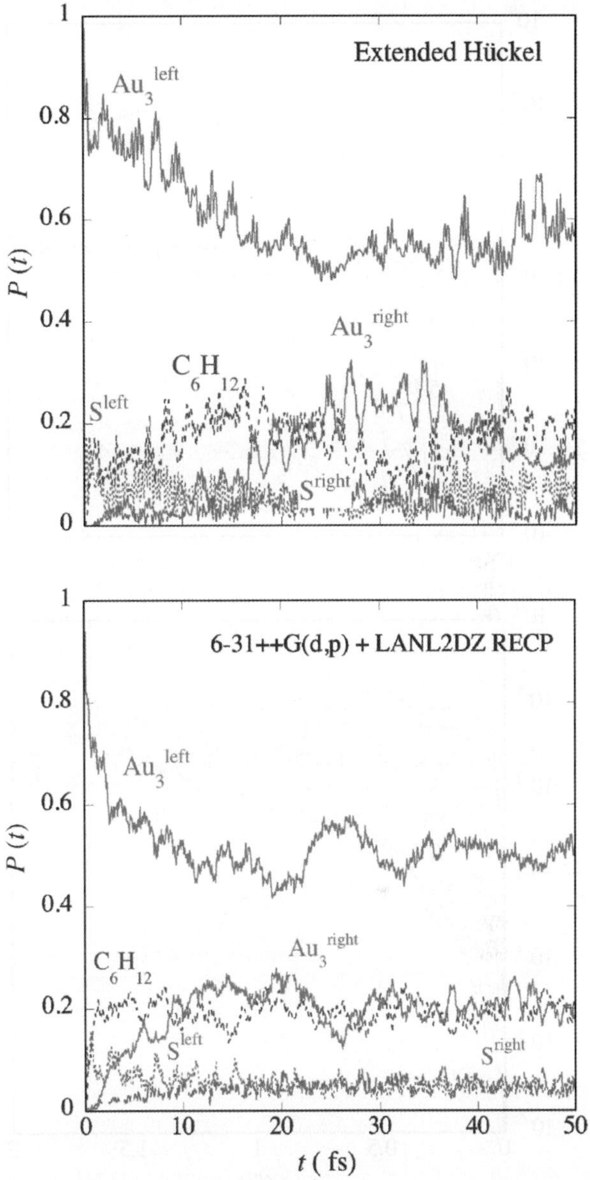

Fig. 4 Populations on the different units of the Au_3–S–$(CH_2)_6$–S–Au_3 extended system for an initial state initial localized on the left Au_3 trimer. The geometry of Au_3–S–$(CH_2)_6$–S–Au_3 is optimized at the DFT/B3LYP 6-31++G(d,p)+LANL2DZ level. The *trans* conformer is the lowest in energy. In the top panel, the populations (eqn. (20)) are computed for the electronic level structure obtained at the EH level, using the optimized geometry. In the lower panel, the populations are computed for the orthogonalized AO's orbitals of the DFT/B3LYP 6-31++G(d,p)+LANL2DZ level. Both levels of computation for the electronic structure lead to a ≈20% population of the $(CH_2)_6$ bridge, and a much smaller population on the S atoms at both ends.

4. Field effects on the *I*–*V* characteristics

While the time dependence of the charge transfer in saturated and conjugated bridges does not exhibit marked differences, their *I*–*V* characteristics do because they have a different MO energy

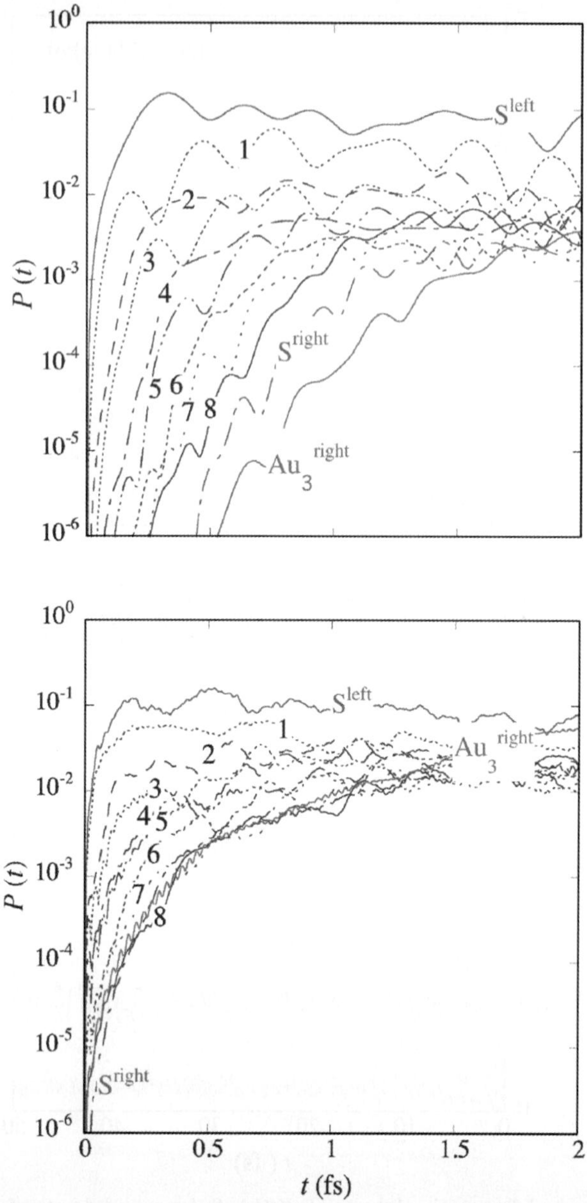

Fig. 5 Computed populations on the different units of a dithio-octane bridge, shown for short times on a log scale, to emphasize the sequential nature of the charge migration. Top panel: The populations are computed from an EH electronic structure, with the geometry optimized at the DFT/B3LYP 6-31++G(d,p)+LANL2DZ level. Lower panel: The populations are computed at the DFT/B3LYP 6-31++G(d,p)+LANL2DZ level. Both computational levels lead to a sequential mechanism for the charge migration. The charge migration to the final Au_3 trimer occurs on a femtosecond time scale, and is shorter at the EH level.

level structure and also because of the different nature, σ or π, of the higher energy MO's that are involved in conduction.

Among the different isomers, the one where Au_3 clusters are in the *trans* configuration with respect to the molecular backbone (see Fig. 9) are the most stable. In that case, the extended system belongs to the C_{2h} group. Of course, when external electric fields are imposed, the symmetry is

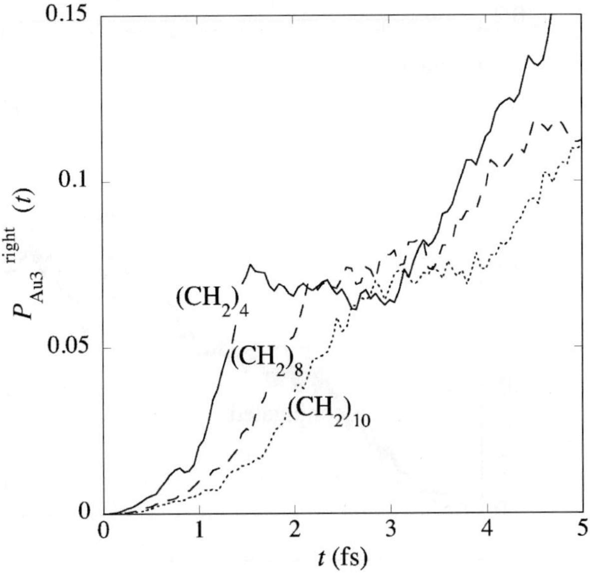

Fig. 6 Population on the right Au_3 trimer for an initial state localized on the left Au_3, computed for increasing lengths, $(CH_2)_4$, $(CH_2)_8$ and $(CH_2)_{10}$, of the dithioalkane chain. The time required to reach the final Au_3 trimer increases with the length of the chain.

broken. In the *trans* configuration, the two highest occupied MO's (HOMO and HOMO − 1) of the alkane saturated bridges are degenerate at zero applied field. These two MO's are localized symmetrically on the two Au–S–C junctions on the left and on the right with no population on

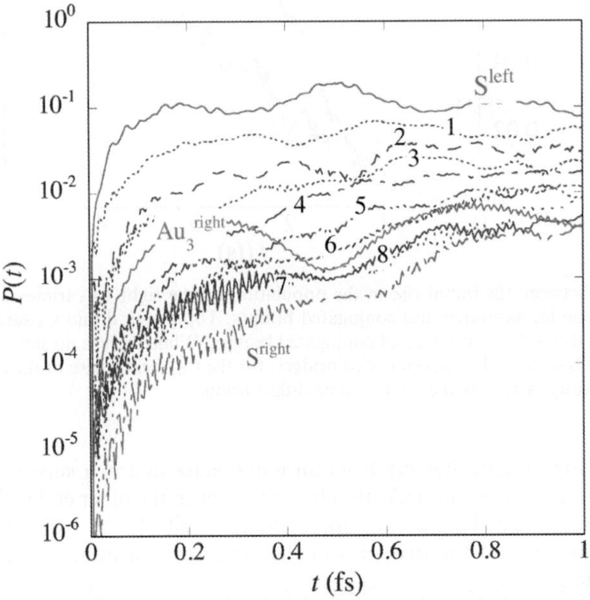

Fig. 7 Populations on the different units of a conjugated dithiooctatetraene bridge, plotted at short times on a log scale, computed at the DFT/B3LYP 6-31++G(d,p)+LANL2DZ level. Unlike for the saturated bridges, the final Au_3 trimer is populated before all the intermediate (CH) units of the chain have been accessed by the initial excitation.

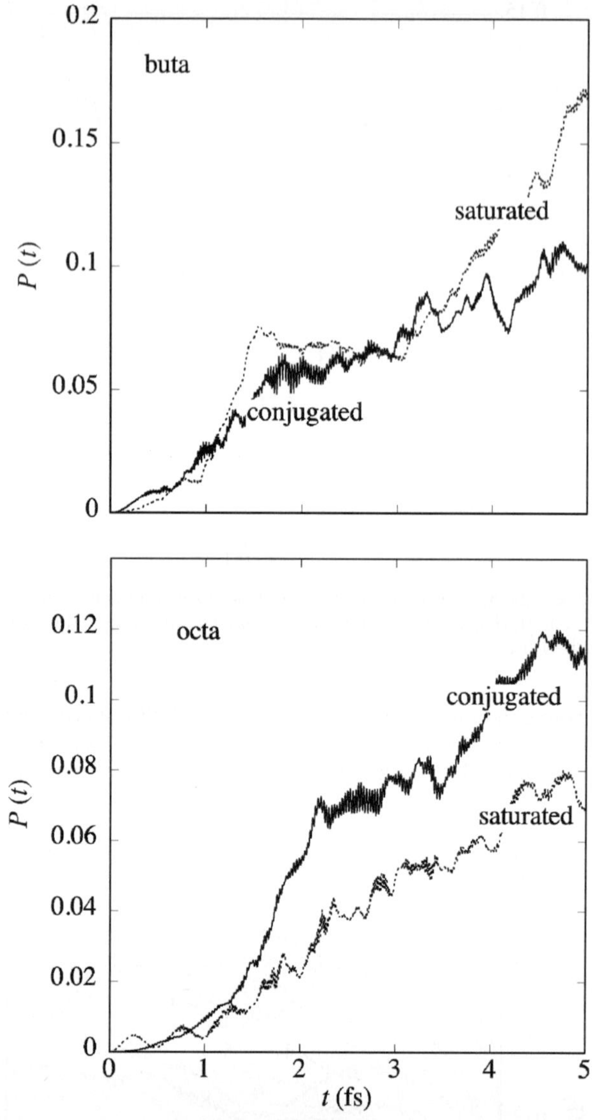

Fig. 8 Comparison between the initial rise in the population on the right Au_3 trimer when the initial state is localized on the left one for saturated and conjugated bridges. Top panel: dithio C_4 saturated and conjugated bridges. Lower panel: dithio C_8 saturated and conjugated bridges. While there is no significant difference for the short time rise in the C_4 saturated and conjugated bridges, for the C_8 one, the rise of the population on the final Au_3 trimer is significantly faster for the octatetraene dithio bridge.

the alkane bridge, Fig. 9, left. Starting from an initial state that is a superposition of these two orbitals and so localized at one interface, the charge will reach the other end with hardly any charge on the molecular backbone. This is the superexchange limit. To be realized it needs molecular orbitals localized primarily at the two ends of the extended system and an initial state of fairly narrow width in energy.

An electrical field applied along the direction of the molecular backbone split the degeneracy of these two levels and localizes the corresponding charge densities at a given end, Fig. 9, right. These two effects of the external field affect the *I–V* curves in different ways. From eqn. (13) it can be seen that for a level to conduct, two conditions must be fulfilled: its energy must fall within the Fermi

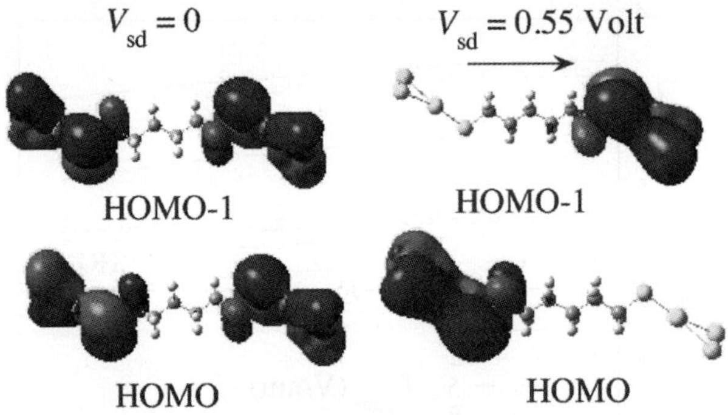

Fig. 9 Charge densities of the HOMO and HOMO − 1 of the saturated Au_3–S–$(CH_2)_6$–S–Au_3 extended system at zero (left) and at $V_{sd} = 0.55$ V (right). The most stable conformer is *trans* and of C_{2h} symmetry, so that for $V_{sd} = 0$, the two highest occupied MO's are degenerate, with charge densities localized on the left and on the right side. As soon as a finite source–drain field is applied, this degeneracy is lifted (see Fig. 13) and for each MO, the charge density localizes at the single end of the molecular backbone.

window (resonance condition) and its transmission function (eqn. (15)), that is the product of the square of its coefficients on the AO's on the left and on right ends, must be non-zero. The shifting in energy can bring the levels into the Fermi window or move them out of it, so that the current is either increasing or decreasing. On the other hand, the fact that because of the applied field, the charge density gets localized on a given end of the extended system leads to a decrease of the transmission function of the level. When the junction is not fully symmetric, for example for conformers which do not have a *cis* or *trans* configuration of the Au_3 trimers with respect to the molecular backbone, the highest occupied quasi-degenerate MO's in the alkane bridges can cross at finite field. We have shown that when this crossing occurs within the Fermi window, it is accompanied by a sharp rise in the current over a small range of bias, thereby providing a mechanism for negative differential conductance.[24,47]

In this paper, we report on gating effects that are included in the Hamiltonian at the *ab initio* level. An electrical field is applied along the molecular backbone to reproduce the effect of the source–drain bias, V_{sd} on the electronic structure and a second one is applied in the direction perpendicular to describe the effect of the gate voltage, V_g. From our previous work[23,49]and the work of others[20,63,85] we expect that the saturated and conjugated bridges will respond differently to the gating field, because the charge density of a σ orbital is oriented parallel to the molecular backbone and therefore sensitive to V_{sd} while because if their symmetry, the density of the π orbitals, are more resilient to the effect of applied electric fields.

In conjugated alkene bridges, the effect of the gating field is to shift linearly the energies of the MO's, with almost no distortion of the charge densities. The energies of the highest occupied and lowest unoccupied MO's of the *trans* conformer of Au_3–S–$(CH)_4$–S–Au_3 bridge are plotted as a function of V_{sd} for different values of the gating field in Fig. 10. The value of V_{sd} is obtained by multiplying the value of the applied field by the projection of the S–S distance on its direction. We report values of the gating field (in V nm^{-1}) as seen by the molecule. The reason is that in an actual device a dielectric layer separates the gate electrode from the molecule so that the actual value of the gate voltage experienced by the molecule very much depends on the details of the device. Because they are of π symmetry, the energies of the MO's shown in Fig. 10 vary very little with the strength of V_{sd}. However applying a gating field that is strong enough shifts the levels down in energy almost linearly except at higher V_{sd} strength. The strength of the applied gating field F_{gate} (along the *z*-direction, perpendicular to the molecular backbone) varies between 0 and 10 V nm^{-1} and it is only from a value of 5 V nm^{-1} and beyond that the shifts in energy are significant. Because of symmetry, reversing the direction of the applied gating field leads to the same energy shifts. In Fig. 11, the values of the transmission function of the HOMO is plotted as a function of the energy of the HOMO (that is shifted by applying V_{sd}, see Fig. 10) for no gating field (bottom abscissa) and for $F_{gate} = 10$ V nm^{-1} (top abscissa). As expected because of the absence of significant distortion of the charge density of the π levels by external fields, the intensity of the transmission function does not

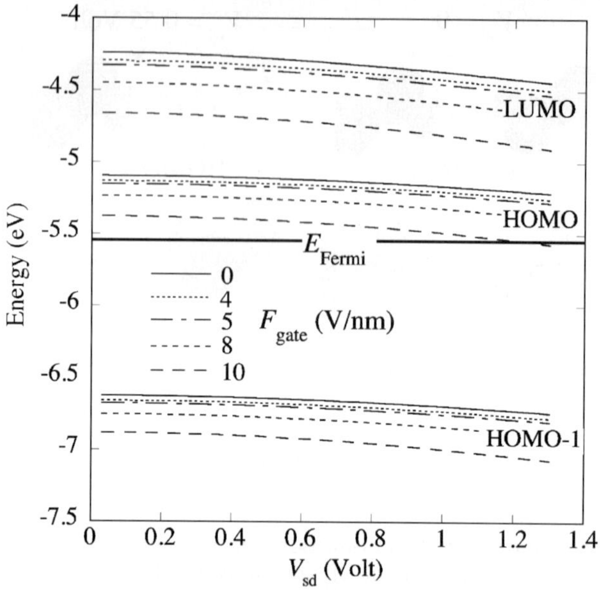

Fig. 10 Energies of the LUMO, HOMO and HOMO − 1 π MO's of the conjugated Au_3–S–$(CH)_4$–S–Au_3 extended system, computed at the DFT/B3LYP 6-31++G(d,p)+LANL2DZ level, in the presence of a source–drain bias and a gating field. The position of the Fermi level, $E_{Fermi} = -5.53$ eV, of the gold electrodes is indicated. The energies are plotted as a function of V_{sd} for different values of F_{gate} as indicated. The energies of the MO's do not vary significantly as a function of V_{sd} and are shifted down by the gating field.

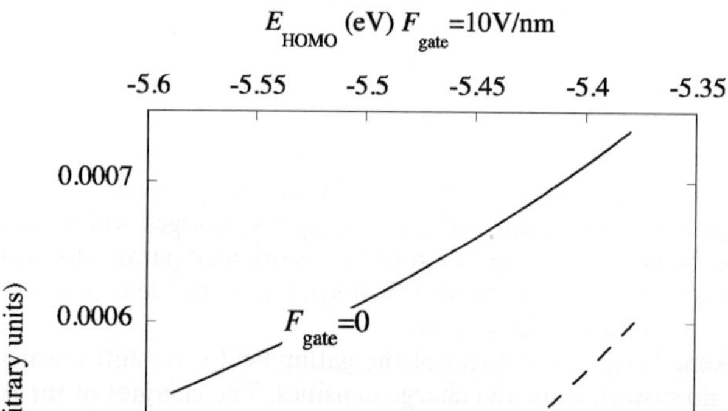

Fig. 11 Value of the transmission function of the HOMO of the conjugated Au_3–S–$(CH)_4$–S–Au_3 extended system, plotted as a function of the energy of the HOMO for the same range of V_{sd} as in Fig. 10, for a 0 V nm^{-1} (bottom abscissa) and 10 V nm^{-1} gating field (top abscissa). The electronic structure is computed at the DFT/B3LYP 6-31++G(d,p)+LANL2DZ level. Because the HOMO is of π symmetry, it is not distorted significantly by the applied external fields, so that its transmission function varies very little both as a function of V_{sd} and F_{gate}.

Charge transfer in molecular bridges can be controlled by applying external fields and this is the basis for conductance spectroscopy. We provided different limiting cases of how the field effects arise in saturated and conjugated bridges. Because in the conjugated bridges, the HOMO and the LUMO are of π symmetry, their charge distribution is not significantly distorted by external field. Their energies are shifted by the gating field, which leads to rather regular and simple effects.

Field effects are more complex for saturated and mixed bridges because for these bridges, the external fields can induce avoided crossing between quasi-degenerate MO's at zero field and lead to highly non-linear behavior of the current both in the gating and in the source–drain directions.

Acknowledgements

This work is supported by the EC-STREP project MOLDYNLOGIC and by the EC NoE FAME. The work of FR is supported by F.R.F.C. 9.4545.03 and 1.5.187.05 (FNRS, Belgium).

References

1 A. Nitzan and M. A. Ratner, *Science*, 2003, **300**, 1384.
2 J. R. Heath and M. A. Ratner, *Phys. Today*, 2003, **56**, 43.
3 J. M. Tour, *Molecular Electronics*, World Scientific, Singapore, 2003.
4 C. Joachim and M. A. Ratner, *Nanotechnology*, 2004, **15**, 1065.
5 S. Ami, M. Hliwa and C. Joachim, *Chem. Phys. Lett.*, 2003, **367**, 662.
6 F. Remacle, J. R. Heath and R. D. Levine, *Proc. Natl. Acad. Sci. USA*, 2005, **102**, 5653.
7 D. L. Klein, R. Roth, A. K. L. Lim, A. P. Alivisatos and P. L. McEuen, *Nature*, 1997, **389**, 699.
8 D. Porath, Y. Levi, M. Tarabiah and O. Millo, *Phys. Rev. B*, 1997, **56**, 9829.
9 B. Alperson, I. Rubinstein, G. Hodes, D. Porath and O. Millo, *Appl. Phys. Lett.*, 1999, **75**, 1751.
10 E. Bakkers, Z. Hens, A. Zunger, A. Franceschetti, L. P. Kouwenhoven, L. Gurevich and D. Vanmaekelbergh, *Nano Lett.*, 2001, **1**, 551.
11 U. Banin, Y. W. Cao, D. Katz and O. Millo, *Nature*, 1999, **400**, 542.
12 L. P. Kouwenhoven and P. L. McEuen, in *Nanotechnology*, ed. G. Timp, Springer Verlag, New York, 1999, p. 4.
13 L. P. Kouwenhoven, D. G. Austing and S. Tarucha, *Rep. Prog. Phys.*, 2001, **64**, 701.
14 O. Millo, D. Katz, Y. W. Cao and U. Banin, *Phys. Rev. Lett.*, 2001, **86**, 5751.
15 W. Liang, M. P. Shores, J. L. Long, M. Bockrath and H. Park, *Nature*, 2002, **417**, 725.
16 Y. Luo, C. P. Collier, J. O. Jeppesen, K. A. Nielsen, E. DeIonno, G. Ho, J. Perkins, H.-R. Tseng, T. Yamamoto, J. F. Stoddart and J. R. Heath, *ChemPhysChem*, 2002, **3**, 519.
17 J. Park, A. N. Pasupathy, J. I. Goldsmith, C. Chang, Y. Yaish, J. R. Petta, M. Rinkoski, J. P. Sthena, H. D. Abruna, P. L. McEuen and D. C. Ralph, *Nature*, 2002, **417**, 722.
18 S. Kubatkin, A. Danilov, M. Hjort, J. Cornil, J. L. Bredas, N. Stuhr-Hansen, P. Hedegard and T. Bjornholm, *Nature*, 2003, **425**, 698.
19 P. Jarillo-Herrero, S. Sapmaz, C. Dekker, L. P. Kouwenhoven and H. S. J. vanderZant, *Nature*, 2004, **429**, 389.
20 Y. Karzari, J. Cornil and J. L. Bredas, *J. Am. Chem. Soc.*, 2001, **123**, 10076.
21 H. Basch and M. A. Ratner, *J. Chem. Phys.*, 2004, **120**, 5761.
22 Y. Karzazi, J. Cornil and J. L. Bredas, *Nanotechnology*, 2003, **14**, 165.
23 F. Remacle and R. D. Levine, *Chem. Phys. Lett.*, 2004, **383**, 537.
24 F. Remacle and R. D. Levine, *Appl. Phys. Lett.*, 2004, **85**, 1725.
25 H. M. McConnell, *J. Chem. Phys.*, 1961, **35**, 508.
26 M. D. Todd, A. Nitzan and M. A. Ratner, *J. Phys. Chem.*, 1993, **97**, 29.
27 C. P. Hsu and R. A. Marcus, *J. Chem. Phys.*, 1997, **106**, 584.
28 L. E. Hall, J. R. Reimers, N. S. Hush and K. Silverbrook, *J. Chem. Phys.*, 2000, **112**, 1511.
29 A. Nitzan, *Annu. Rev. Phys. Chem.*, 2001, **52**, 681.
30 G. Iversen, E. P. Friis, Y. I. Kharkats, A. M. Kuznetsov and J. Ulstrup, *JBIC, J. Biol. Inorg. Chem.*, 1998, **3**, 229.
31 M. Bixon, B. Giese, S. Wessely, T. Langenbacher, M. E. Michel-Beyerle and J. Jortner, *Proc. Natl. Acad. Sci. USA*, 1999, **96**, 11713.
32 H. B. Gray and J. R. Winkler, *Q. Rev. Biophys.*, 2003, **36**, 341.
33 R. H. Goldsmith, L. E. Sinks, R. F. Kelley, L. J. Betzen, W. Liu, E. A. Weiss, M. A. Ratner and M. R. Wasielewski, *Proc. Natl. Acad. Sci. USA*, 2005, **102**, 3540.
34 H. B. Gray and J. R. Winkler, *Proc. Natl. Acad. Sci. USA*, 2005, **102**, 3534.
35 F. Remacle and R. D. Levine, *J. Phys. Chem. B*, 2001, **105**, 2153.
36 J. L. Sample, K. C. Berverly, P. R. Chaudhari, F. Remacle, J. R. Heath and R. D. Levine, *Adv. Mater.*, 2002, **14**, 124.

37 R. Hoffmann, *J. Chem. Phys.*, 1963, **39**, 1397.
38 S. Komiya, T. A. Abright, R. Hoffmann and J. K. Kochi, *J. Am. Chem. Soc.*, 1977, **99**, 8440.
39 M. Tachibana, K. Yoshizawa, A. Ogawa, H. Fujimoto and R. Hoffmann, *J. Phys. Chem. B*, 2002, **106**, 12727.
40 R. Weinkauf, P. Schanen, A. Metsala, E. W. Schlag, M. Buergle and H. Kessler, *J. Phys. Chem.*, 1996, **100**, 18567.
41 F. Remacle, R. D. Levine, E. W. Schlag and R. Weinkauf, *J. Phys. Chem. A*, 1999, **103**, 10149.
42 F. Remacle, R. D. Levine and M. A. Ratner, *Chem. Phys. Lett.*, 1998, **285**, 25.
43 P. A. Derosa and J. M. Seminario, *J. Phys. Chem. B*, 2001, **105**, 471.
44 H. B. Weber, J. Reichert, F. Weigend, R. Ochs, D. Beckmann, M. Mayor, R. Ahlrichs and H. v. Lhoneysen, *Chem. Phys.*, 2002, **281**, 113.
45 Y. Q. Xue, S. Datta and M. A. Ratner, *Chem. Phys.*, 2002, **281**, 151.
46 J. Taylor, M. Brandbyge and K. Stokbro, *Phys. Rev. Lett.*, 2002, **89**, 138301.
47 P. Delaney, M. Nolan and J. C. Greer, *J. Chem. Phys.*, 2005, **122**, 44710.
48 M. J. Frisch, G. W. Trucks, H. B. Schlegel, G. E. Scuseria, M. A. Robb, J. R. Cheeseman, J. J. A. Montgomery, T. Vreven, K. N. Kudin, J. C. Burant, J. M. Millam, S. S. Iyengar, J. Tomasi, V. Barone, B. Mennucci, M. Cossi, G. Scalmani, N. Rega, G. A. Petersson, H. Nakatsuji, M. Hada, M. Ehara, K. Toyota, R. Fukuda, J. Hasegawa, M. Ishida, T. Nakajima, Y. Honda, O. Kitao, H. Nakai, M. Klene, X. Li, J. E. Knox, H. P. Hratchian, J. B. Cross, C. Adamo, J. Jaramillo, R. Gomperts, R. E. Stratmann, O. Yazyev, A. J. Austin, R. Cammi, C. Pomelli, J. W. Ochterski, P. Y. Ayala, K. Morokuma, G. A. Voth, P. Salvador, J. J. Dannenberg, V. G. Zakrzewski, S. Dapprich, A. D. Daniels, M. C. Strain, O. Farkas, D. K. Malick, A. D. Rabuck, K. Raghavachari, J. B. Foresman, J. V. Ortiz, Q. Cui, A. G. Baboul, S. Clifford, J. Cioslowski, B. B. Stefanov, G. Liu, A. Liashenko, P. Piskorz, I. Komaromi, R. L. Martin, D. J. Fox, T. Keith, M. A. Al-Laham, C. Y. Peng, A. Nanayakkara, M. Challacombe, P. M. W. Gill, B. Johnson, W. Chen, M. W. Wong, C. Gonzalez and J. A. Pople, *GAUSSIAN 03, (Revision B.04)*, 2003, Gaussian, Inc., Pittsburgh, PA.
49 F. Remacle, I. Willner and R. D. Levine, *J. Phys. Chem. B*, 2004, **108**, 18129.
50 F. Remacle and R. D. Levine, *Isr. J. Chem.*, 2003, **42**, 269.
51 R. Landauer, *IBM J. Res. Dev.*, 1957, **1**, 223.
52 Y. Imry and R. Landauer, *Rev. Mod. Phys.*, 1999, **71**, S306.
53 Y. Imry, *Introduction to Mesoscopic Physics*, Oxford University Press, Oxford, 2002.
54 S. Datta, *Electronic Transport in Mesoscopic Systems*, Cambridge University Press, Cambridge, 1995.
55 V. Mujica, M. Kemp and M. A. Ratner, *J. Chem. Phys.*, 1994, **101**, 6849.
56 R. D. Levine, *Quantum Mechanics of Molecular Rate Processes*, Dover, New York, 1999.
57 F. Remacle, K. C. Beverly, J. R. Heath and R. D. Levine, *J. Phys. Chem. B*, 2003, **107**, 13892.
58 R. Hoffmann, *Solids and Surfaces: A Chemist's View of Bonding in Extended Structures*, VCH Publishers, New York, 1988.
59 P. J. Hay and W. R. Wadt, *J. Chem. Phys.*, 1985, **82**, 299.
60 R. G. Parr and W. Yang, *Density-Functional Theory of Atoms and Molecules*, Oxford University Press, New York, 1989.
61 R. G. Pearson, *J. Chem. Educ.*, 1987, **64**, 561.
62 G. J. Iafrate, K. Hess, J. B. Krieger and M. Maccuci, *Phys. Rev. B*, 1995, **52**, 10737.
63 S. N. Rashkeev, M. DiVentra and S. T. Pantelides, *Phys. Rev. B*, 2002, **66**, 33301.
64 P.-O. Lowdin, *Phys. Rev.*, 1955, **101**, 1730.
65 A. E. Reed, R. B. Weinstock and F. Weinhold, *J. Chem. Phys.*, 1985, **83**, 736.
66 A. E. Reed, L. A. Curtiss and F. A. Weinhold, *Chem. Rev.*, 1988, **88**, 899.
67 E. D. Glendening, A. E. Reed, J. E. Carpenter and F. Weinhold, *NBO Version 3.1.*
68 P. O. Lowdin, *J. Chem. Phys.*, 1950, **18**, 365.
69 F. L. Pilar, *Elementary Quantum Chemistry*, McGraw-Hill, New York, 1968.
70 A. Szabo and N. Ostlund, *Modern Quantum Chemistry*, Dover, New York, 1996.
71 S. Y. Lee and E. J. Heller, *J. Chem. Phys.*, 1979, **71**, 4777.
72 R. L. Sundberg and E. J. Heller, *Chem. Phys. Lett.*, 1982, **93**, 586.
73 F. Remacle, U. Even and R. D. Levine, *J. Phys. Chem.*, 1996, **100**, 19735.
74 F. Remacle and R. D. Levine, *J. Chem. Phys.*, 1997, **107**, 3382.
75 C. E. D. Chidsey, *Science*, 1991, **251**, 919.
76 J. F. Smalley, S. W. Feldberg, C. E. D. Chidsey, M. R. Linford, M. D. Newton and Y. P. Liu, *J. Phys. Chem.*, 1995, **99**, 13141.
77 S. B. Sachs, S. P. Dudek, R. P. Hsung, L. R. Sita, J. F. Smalley, M. D. Newton, S. W. Feldberg and C. E. D. Chidsey, *J. Am. Chem. Soc.*, 1997, **119**, 10563.
78 R. H. Terrill, T. A. Postlethwaite, C. H. Chen, C. D. Poon, A. Terzis, A. D. Chen, J. E. Hutchison, M. R. Clark, G. Wignall, J. D. Londono, R. Superfine, M. Falvo, C. S. Johnson, E. T. Samulski and R. W. Murray, *J. Am. Chem. Soc.*, 1995, **117**, 12537.
79 J. F. Hicks, A. C. Templeton, S. Chen, K. M. Sheran, R. Jasti and R. W. Murray, *Anal. Chem.*, 1999, **71**, 3703.
80 W. P. Wuelfing, S. J. Green, J. J. Pietron, D. E. Cliffel and R. W. Murray, *J. Am. Chem. Soc.*, 2000, **122**, 11465.

81 X. D. Cui, A. Primak, X. Zarate, J. Tomfohr, O. F. Sankey, A. L. Moore, T. A. Moore, D. Gust, L. A. Nagahara and S. M. Lindsay, *J. Phys. Chem. B*, 2002, **106**, 8609.
82 J. F. Smalley, H. O. Finklea, C. E. D. Chidsey, M. R. Linford, S. E. Creager, J. P. Ferraris, K. Chalfant, T. Zawodzinsk, S. W. Feldberg and M. D. Newton, *J. Am. Chem. Soc.*, 2003, **125**, 2004.
83 A. Salomon, D. Cahen, S. Lindsay, J. Tomfohr, V. B. Engelkes and C. D. Frisbie, *Adv. Mater.*, 2003, **15**, 1881.
84 W. Y. Wang, T. Lee and M. A. Reed, *Phys. Rev. B*, 2003, **68**, 35416.
85 A. W. Ghosh, T. Rakshit and S. Datta, *Nano Lett.*, 2004, **4**, 565.

Electronic and vibrational structure of one-dimensional conductors and superconductors

Sven Larsson

Received 10th May 2005, Accepted 15th June 2005
First published as an Advance Article on the web 26th September 2005
DOI: 10.1039/b506642p

An attempt is made to treat molecular wires by quantum chemical methods. What is the electronic difference between systems (molecules, stacks of molecules, or polymers) that are conducting and almost identical systems that are insulating? At 50% band filling a one-dimensional crystal undergoes a Peierls transition and becomes an insulator (without doping). In the case of 75% band filling, on the other hand, two phases are possible: charge density wave (CDW) and spin-density wave (SDW). The transition between these two insulating phases should be connected to a high conductivity. If CDW and SDW are energetically possible at zero T, vibrational coupling leads to stabilization of a superconducting (SC) ground state and the formation of an energy gap. This hypothesis is exemplified on $(SN)_x$ and $(TMTSF)_2X$ ($X = ClO_4$, PF_6).

1. Introduction

A molecular wire in its simplest form consists of a bridge (B) in a D–B–A system (D = donor; A = acceptor), where the electron can be thermalised only on donor and acceptor. D and A may exchange electrons by tunnelling (=electron transfer; ET). In between wires that just permit tunnelling and true metallic wires there are several possibilities. The electron may be localized on bridge sites and move between them by ET. If the bridge is ordered, various structural effects may lead to non-metallic behaviour. For example in many 1D systems the electrons form spin density waves (SDW) and charge density waves (CDW).[1,2] Another type of 'instability' is the Peierls transition.[3] The 'giant Kohn anomaly'[4,5] is a softening of modes indicating strong electron–phonon interactions.[6] The connection between structural distortions and metallic conductivity (MC) or superconductivity (SC) is unclear at the present. Both CDW and SDW have been connected to MC and SC but both CDW and SDW are insulating.[7,8]

In the case of cuprates and other high-T_c superconductors (SC) the mixed valence model[9,10] has been considered as a viable chemical approach.[11,12] The wave function (or more correctly the electronic density) of a mixed-valence state is a CDW. Such a wave function is not common in any Cu^{2+} oxide or salt, however. An example of mixed valence appears in related $KAuCl_3$ where gold alternates between the valence states +1 and +3.[13] In copper oxide, CuO, the wave function is of SDW type (antiferromagnetic), with alternating spins on the copper sites. In a number of other cases SC appears connected to CDW or SDW or both at the same time.

Recently we have calculated very accurately energy surfaces (PES) of CDW as well as SDW type in a four-center representation of a cuprate.[14] The PES of the CDW state and SDW state have minima for different geometries. The minima have about the same depth and are so close in space that the zero-point vibrational wave functions are considerably overlapping. The CDW state intersects and is lower than the SDW-PES for nuclear geometries that correspond to mixed valence $[Cu^+/Cu^{3+}]$. The motion of the electron pair between two adjacent sites is activationless. In fact the electron pairs are delocalized. This hints to a mechanism for superconductivity. In the present paper we will discuss the connection between CDW, SDW, MC and SC in molecular wires.

Chalmers University of Technology, S-41296, Göteborg, Sweden

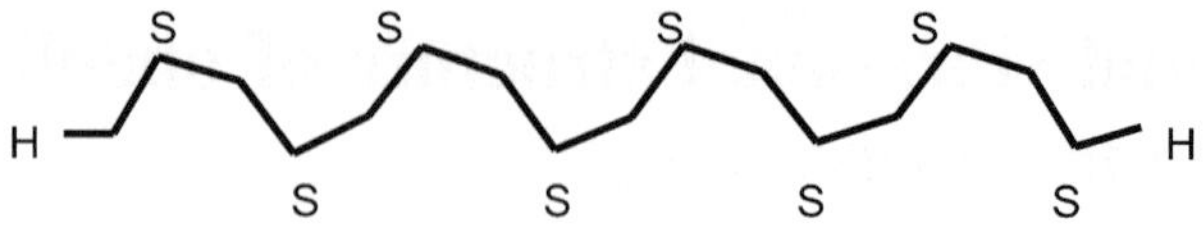

Fig. 1 $HN(SN)_8H$ calculated using the PM3 method.[32]

$(SN)_x$ and a number of other one-dimensional (1D) conductors and superconductors (SC) have been known for some time.[15,16] $(SN)_x$, a polymer (Fig. 1) that may adopt a high degree of crystallinity due to (weak) interchain bonds, is superconducting (SC) below 0.26 K. There is hardly any accepted explanation why just this polymer is superconducting, except that its crystallinity gives it a chance to be at least a metal.

A natural question is whether there is any connection in the conduction mechanisms between different conducting, chain-like polymers and organic crystals where the conductivity is mainly along a stack of aromatic molecules. The difference is that the coupling is very large between the sites in a chain, but usually rather small between the molecules in a stack. In the stack each site consists of a molecule, with associated vibrational motions. In spite of these differences it is believed that the degree of filling of the valence band–conduction band plays the same major role for the conduction properties.

TTF-TCNQ consists of separate stacks of TTF or TCNQ. The stack becomes conducting by transfer of electrons from the TTF to the TCNQ stack. The conductivity increases for decreasing temperature down to a peak value at 58 K. Below this temperature, TTF-TCNQ becomes an insulator.[18] The phase transition has been connected to a Peierls transition, but it appears unlikely that it is the traditional Peierls transition at the half-filled band.

$(CH)_x$, discovered in 1978, was the first conducting, organic polymer (by doping).[19,20] A simple description employed an improvement of the well-known Hückel theory.[21–25] This π-MO-theory is a theory for coupling of electronic and nuclear coordinates (see below in Section 1). The alternating bond-length pattern predicted by Peierls[3] appears. However, since there is no charge alternation, this type of ground state should not be called a CDW state. Conduction is hindered by the Peierls effect and takes place by ET between shallow sites.

$(TMTSF)_2PF_6$ is an organic charge-transfer salt[26] which is super-conducting below $T_c = 0.9$ K at a pressure of 12 kbar.[27] $(TMTSF)_2ClO_4$ becomes superconducting below $T_c = 1.2$ K at ambient pressure.[2,26,27] In this case one electron is moved from every second molecule in the stack to the inorganic molecule that forms an anion. The HOMO band is three-quarter-filled.

A certain amount of confusion exists in the field. Clearly the normal Peierls transition at half-filling opens up a band gap at the Fermi level, so that the compound becomes a semi-conductor. Nevertheless polyacetylene (PA) is conducting after doping. The explanation was first given in terms of 'solitons'. Infinitely extended soliton levels in the middle of the gap provide conductivity. This concept is hardly used nowadays. In reality the band model cannot be used, since there is localization around the alkali ions[28,29] or other dopant. Conductivity is therefore caused by ET between localized sites in the polymer. This is consistent with the fact that alkali doped PA does not show MC but increases its conductivity as the temperature is increased.

The Hückel model is further described in Section 2 and applied to π systems with half-band filling. In Section 3 the method is applied to systems with three-quarter band filling. In the latter case, contrary to the case of neutral $(CH)_x$, we find CDW and SDW states. In Section 5 we will apply this theory to conducting organic 1D crystals.

2. 1D systems with 50% band filling

In planar organic molecules the sp^2-hybridized σ orbitals form inactive CC and CH single bonds in the xy-plane. The π orbitals are expressed as linear combinations of the C $2p_z$ orbitals.[21] The latter form an orthonormal set of N atomic orbitals. The diagonal matrix elements are called α and assumed to be the same for all carbon atoms. The off-diagonal matrix elements (couplings) are called β and assumed to be the same and non-zero only for neighbouring, bonded carbon atoms and independent of bond length in the first approximation. Solution of the matrix eigenvalue problem leads to orbital energies ε_i and eigenvectors where $c_{\mu i}$ is the coefficient for MO i on C atom μ.

π bonding arises since the occupied π MO's are mainly bonding. The almost free rotation around the CC bonds disappears. The bond order, $P_{\mu\nu}$, for $\mu \neq \nu$ and charge, $P_{\mu\mu}$, depend on the occupation numbers n_i according to:

$$P_{\mu\nu} = \sum_{i=1}^{N} n_i c^*_{\mu i} c_{\nu i} \tag{1}$$

The changes in bond length $R_{\mu\nu}$, from the σ bond, assumed to be 1.534 Å, are proportional to the bond order $P_{\mu\nu}$. The proportionality factor is equal to −0.2 Å, meaning that a full π bond ($P_{\mu\nu} = 1$) decreases the σ-bond by 0.2 Å:

$$R_{\mu\nu} = -0.2P_{\mu\nu} + 1.534\ (\text{Å}) \tag{2}$$

Eqn. (2) provides surprisingly good CC bond lengths.

In a cyclic system the eigenvalues and eigenvectors are exact expressions:

$$\varepsilon_\ell = \alpha + 2\beta\cos\frac{2\pi\ell}{N};\ \ell = \pm0, \pm1, \pm2, ...N/2 \tag{3}$$

$$\phi_\ell = \sum_{\mu=0}^{N-1} \chi_\mu c_{\mu\ell} = \sum_{\mu=0}^{N-1} \chi_\mu N^{-1/2} \exp\left(\mu\frac{2\pi\ell}{N}\mathrm{i}\right); \ell = 0, \pm1, \pm2, \ldots, N/2 \tag{4}$$

Real coefficients may be chosen by adding and subtracting functions with different sign of ℓ in eqn. (4). In a system with $N = 4n + 2$ electrons, where n is a natural number, the degenerate highest occupied MO (HOMO) is fully occupied. The bond orders are predicted to be the same, due to mathematical relations among the coefficients. In reality this holds true for benzene, with $n = 1$ and $N = 6$, but for a very large cyclic system bond alternation occurs.

A simple extension of the Hückel model, accounting for the Peierls distortion, is obtained if β depends on the bond length, R, in the following way ($R_0 = 1.4$ Å):[23,25,30]

$$\beta = \beta_0 \exp[-(R - R_0)/a] \tag{5}$$

($a = 0.3106$). An iterative scheme suggests itself. This model was rediscovered in connection with conducting PA and is in that context referred to as the Su–Schrieffer–Heeger (SSH) model.[24] In each iteration, bond lengths are calculated according to eqn. (2). β for individual bonds are calculated and recalculated according to eqn. (5). The Peierls effect for 1/2 band filling appears. In the case of $4n + 2$ systems this may be seen as a second order Jahn–Teller (SOJT) distortion where a low energy excited state rather than a degenerate state (in the ordinary Jahn–Teller effect) mixes into the ground state.[25] For $4n + 2$ systems the SOJT effect becomes relatively larger the longer the chain and as a result the CC bond lengths become alternating, converging to a difference of 0.08 Å. Similar conclusions can be reached for linear oligomers $H(CH)_{2n}H$, which differ from the cyclic systems by rather small end effects (the first and last bonds tend to be short). In cyclic $N = 4n$ systems the bond length alternation decreases from two pure double bonds and two pure single bonds in C_4H_4 to the alternation difference of 0.08 Å, the same as for $4n + 2$ systems.

SOJT introduces a gap in the energy spectrum, consistent with experiments. If $|\beta_0|$ is given the value 3 eV, the band gap is 1.57 eV. The charges remaining after the dopant cause localizations. The conductivity depends on ET between these localized sites, which include several carbon atoms.

The site occupation of π electrons for the $(CH)_x$ system, calculated using $\mu = \nu$ in eqn. (1), is exactly equal to unity. The latter property, common to all alternant systems, appears to agree well with experiments. Consequently there is no CDW. The only contribution to a CDW is the almost negligible difference in charge between short and long bonds.

The ground state is stabilized by SOJT and there is no electronic state within, say, 1 eV. We conclude that a one-dimensional system like $(CH)_x$ with 50% band filling, is a semi-conductor (which can become conducting by doping).

3. $(SN)_x$ a case of three-quarter band filling

$(SN)_n$ can be said to be formed from PA by replacing every second CH by *N* and every other CH by S.[15,16,31] Each S and each N has got two σ electrons. Each S and each N has two lone pair electrons. Assuming that $(SN)_x$ is quasi planar, N has one and S two π electrons. In $(SN)_n$ the number of π electrons is larger by *n* compared to the number in $(CH)_{2n}$. $(SN)_x$ is thus a case of three-quarter π band filling. If semi-empirical methods are used to optimise the geometry[32,33] for large oligomers of the type $(SN)_nS$, almost planar structures are obtained and most changes when the oligomer size is increased are due to end effects.

If the SSH model is applied to the case of three-quarter band filling [using the same parameters as for $(CH)_x$] there are two different types of solution. For example for cyclic C_{104}, with occupancies $n_1 \rightarrow n_{78} = 2$, a band gap opens at the Fermi level, equal to 0.12β. The density of states is increased just below and above the gap. The bond lengths alternate with two long bonds (1.459 Å) followed by two short bonds (1.430 Å). The π-charges alternate as 1.5*e*–1.387*e*–1.5*e*–1.613*e*–1.5*e*– . . . . This solution may be described as a kind of CDW.

If the occupancy is changed to $n_1 \rightarrow n_{77} = 2$; $n_{78} = 1$; $n_{79} = 1$, the charge and bond pattern is completely different. All bonds are now the same (1,444 Å) and all sites have the same π-charge (1.5*e*). There is no change of the density of states compared to eqn. (4). This occupancy corresponds to either a triplet state or a singlet state. The degenerate, un-normalized, singly occupied MO's have the coefficients $(\ldots,1,-\sqrt{2},1,0,-1,\sqrt{2},-1,0,\ldots)$ and $(\ldots,-1,0,1,-\sqrt{2},1,0,-1,\sqrt{2},\ldots)$, respectively. If these two MO's have different spin a SDW is formed.

The crystallographic data do not support a CDW as ground state for $(SN)_x$ since the $(SN)_x$ bonds have alternating bond length [1.593 (outer); 1.628 (inner)].[31] Hence either the ground state is another type of state; a SDW state or some mixed state.

There exist a great number of quantum chemical studies of $(SN)_n$ and the polymer $(SN)_x$.[34–40] A number of SCF calculations on a spin singlet state support the CDW result. RHF and UHF, in the semi-empirical PM3 version, give different results. The RHF solution is a typical CDW solution with two short bonds (S–N–S) followed by two long bonds (S–N–S).[34]

Clearly semi-empirical calculations have to be considered as more accurate than SSH calculations. There are no parameters other than those ordinarily present. It is reasonable to carry out PM3 calculations on as large chains as possible. We expect the nodal properties to be correct. The greatest problem is very likely the lack of electronic correlation.

The lowest states is a SDW state. The triplet state is 0.033 eV above the singlet state. The spin density is alternating on the nitrogen atoms. The bond lengths are almost the same: 1.682 Å (inner) and 1.659 Å (outer). The alternation is 0.023 Å, compared to the experimental value 0.035 Å.

The RHF calculations give a considerably higher energy and a less planar molecule. Two short bonds follow two long bonds. *Ab initio* calculations[34,36] of RHF type on the oligomers $(SN)_2$ and $(SN)_4$ end saturated by hydrogen atoms, give two short bonds around every second sulfur, consistent with the corresponding PM3 results.

How can we understand that three-quarter band filling results in two different electronic states? The SOJT effects are very different from the case of half band filling. In the latter case HOMO covers every two atoms, but has a nodal plane to the next two atoms. This leads to alternating bond lengths.

The MO's occupied from half to three-quarter band filling may be thought of as composed of π orbitals, covering four atoms with two nodal planes perpendicular to the bond axis. The first MO after the above mentioned HOMO at half filling may be thought of as a linear combination of all these orbitals over groups of four atoms, with the same sign. The last orbital to be filled at three-quarter filling has sign variation everywhere between adjacent groups of four atoms. HOMO at three-quarter filling will therefore look like this: If atom 1 is located in the centre of a positive region of an orbital, the atoms with numbers 9, 17 , . . . $(8n + 1)$ will also be in the centre of a positive region, while atoms with numbers 5, 13, 21, . . . , $(8n + 5)$ will be covered by the negative region of the same MO. The atoms in between (3, 7, 11, 15, 19, . . .) will be in a nodal plane.

We may now occupy the component orbitals covering four atoms each by two electrons each, up to three-quarter filling. This will give us the CDW total wave function. Alternatively we may fill all these orbitals with only spin-up electrons. A new set of component orbitals is constructed which are moved two steps to the right. The atom in the nodal plane will now be in a region with a large absolute value. The new set of equivalent MO's is occupied by spin-down electrons. This leads to the SDW total wave function.

The behaviour of the infinite system is thus completely different in the cases of half and three-quarter band filling. The standard Peierls distortion for half filling leads to bond length alternation and a single stable state as ground state, with excited states at a much higher energy. In the three-quarter case of infinite $(SN)_x$ there are two possible states for the ground state, the SDW and CDW singlet states. Both states are formed from the set of MO's with the lowest possible energy. Therefore they may have a close energies. In that case they interact *via* vibrational zero-point vibrations.[41,42] However, since a single Slater determinant solution (DFT or Hartree–Fock), cannot separate the two states, some compromise is reached. Methods that include correlation effects are necessary.

4. Superconductivity

It is interesting that bonding in $(SN)_x$ can be written in (at least) two different ways: (1) with oxidation state on sulfur varying between +2 and +4 and (2) with average oxidation state:

(1) –S–N=S=N–S–N=S=N–S–N=S=N–S–N=S=

(2) ∸S∸N∸S∸N∸S∸N∸S∸N∸S∸N∸S∸N∸S∸

This is not just an exercise in writing chemical formulas. If bonding can be expressed in different ways, different wave functions are usually possible. We know from the previous section that at least two possible wave functions exist, CDW and SDW. The CDW wave function allows electron pairs to form on alternate sites. SDW has a companion in ferromagnetism or triplet states for single molecules. SDW interacts with CDW *via* the nuclear coordinates. For cuprates there are similarly a CDW wave function with alternating Cu^{+} and Cu^{3+} sites and a SDW solution with Cu^{2+} sites.[12] The fact that there are two different types of minima in PES, with different wave functions makes it possible to couple electrons to phonons and stabilize the ground state. However, for conductivity to appear the difference between the two types of state should not be too large since in that case the systems would remain localized in their respective minima. $BaBiO_3$ is a good example.[43] Bi's valence is alternating between +3 and +5. Every second site has a filled 6s orbital and the alternating one an empty 6s orbital. The difference in valence is consistent with the alternating BiO bond length. This is a typical 3D CDW state. Since $BaBiO_3$ is 3D there is no Peierls effect in this case. As is well-known two possible dopings have been found which lead to SC in $BaBiO_3$.[43]

As we have seen above the corresponding 1D system (one electron per site) is a Peierls distorded system described by a unique wave function where SC cannot appear. The two wave functions appearing in the case of three-quarter filling, CDW and SDW may form an SC state. In their relaxed geometries the two states may be referred to as disproportionated ($S^{-1} + S^{+1}$) and the compro-portionated (S + S) states, respectively. S^{-1} and S^{+1} are different in two electrons, so charge transport appears if two electrons move from an S^{-1} site to an S^{+1} site. In the mentioned model for 3-D superconductivity[41] the equilibrium between the two states:

$$S^{-1} + S^{+1} \leftrightarrow 2S, \tag{6}$$

and the vibrational zero point motions may lead to a new state provided the energies are about the same and the zero-point vibrations for different equilibrium points are overlapping. In the latter state the bond lengths are no longer typical for CDW or SDW states. The activation barrier for transition between SDW and CDW may also disappear, which leads to mode softening. This state should be the SC state.

In the one-dimensional case the elementary unit consists of two adjacent sites. For four sites the MO for the first unit is called a and for the next unit b. The SDW may be written as $(ab + ba)(\alpha\beta - \beta\alpha)$ and the CDW as $(aa - bb)(\alpha\beta - \beta\alpha)$. The conducting unit is $(a + ib)(a + ib)(\alpha\beta - \beta\alpha)$, which is the sum of SDW and CDW state wave functions. If the reorganization energy is small for transfer between CDW and SDW, the minima are sufficiently close to interact *via* zero-point motion. It is also necessary that the coupling is very likely sufficiently large to eliminate the activation barrier.[41] The elimination of the activation barrier also explains the mode softening and giant Kohn effect.

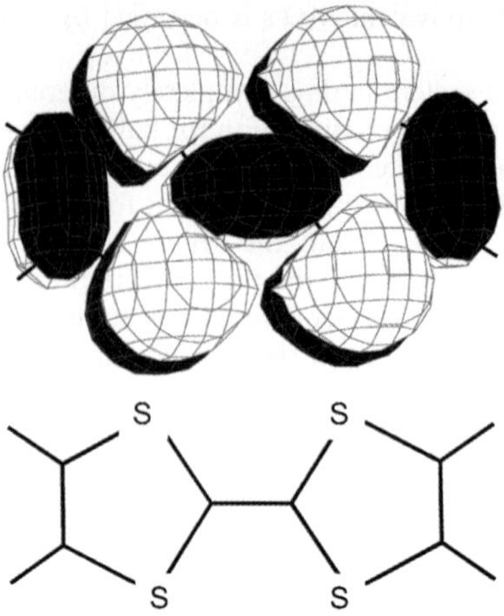

Fig. 2 TTF-HOMO.

5. $(TMTSF)_2X$ has a three-quarter filled band

TMTSF (tetramethyltetraselenafulvalenium) forms stacks of almost planar molecules interacting mainly *via* π-systems between parallel planes. X (=PF_6, ClO_4, *etc.*) is a molecule with a high electron affinity that accepts an electron from every second TMTSF to fill its electronic shell. TMTSF has the same orbital structure near the Fermi level as the simpler molecule TTF (tetrathiofulvalene). TTF-HOMO is shown in Fig. 2. In a TTF sandwich dimer, HOMO forms a bonding and an antibonding MO. If an electron is removed from the latter, bonding in the stack is increased. In Fig. 3 we show the singly occupied HOMO for the dimer obtained in a PM3 UHF calculation. Removal of one half electron from this orbital also weakens the double bonds and makes them slightly longer.

Since every second HOMO is missing one electron, the HOMO band for the stack is filled to three-quarter. The HOMO's of two adjacent TTF interact and form a lower bonding and an upper antibonding, delocalised HOMO. Since the latter orbital is singly occupied there is a net bonding between two neighbouring TTF molecules in the stack. Each dimer has spin 1/2 with alternating spin. In the SDW function one has to assume the spin singlet function $\alpha\beta - \beta\alpha$ for every four TTF molecules in the stack.

Moving the hole on each second dimer to the next one in the stack forms the CDW wave function. Each dimer is now either doubly charged or neutral. In the former case there is more bonding compared to SDW, but also a higher electrostatic repulsion between the monomers. The latter repulsion is to some extent compensated by the presence of the counter ions. The neutral dimer has a much-reduced bonding between the monomers. Unfortunately no calculation has yet been performed which shows the PES for SDW and CDW. In such a calculation it is necessary to relax the bonds within each molecule and also the position of the negative counter ions.

This scenario may be compared to half filling, with one hole on every monomer (and one counter ion per monomer). This clearly introduces a greater bonding in each dimer, since the repulsion between them is offset by the presence of the counter ions. The distance between two monomers is decreased in each dimer. The interaction between two adjacent monomers on different dimers has to be anti-bonding because of the half filling. The monomer HOMO's interact in the same way as the π orbitals of a polyene and the highest occupied stack orbital formed from them has to change sign between every two monomers.

The half filling thus can only lead to a single ground state wave function, the one corresponding to the Hückel solution for stack HOMO's, but no MC or SC. Three-quarter filling leads to SDW and CDW, which are insulating, or SC. Why is then only $(TMTSF)_2X$ with X = ClO_4 superconducting? This may simply be a size effect. The coupling between the CDW and SDW states is essentially due to the overlap between the π orbitals on adjacent sites. The larger the anion, the less

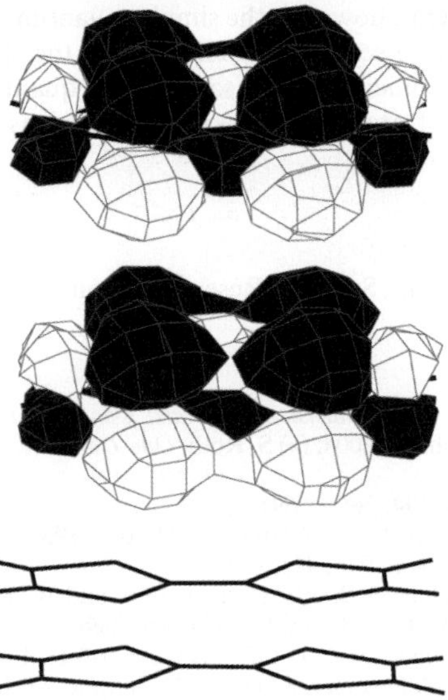

Fig. 3 $(TTF)_2^{+}$, singly occupied MO.

is the overlap. Pressure effects lead to larger overlap and therefore an increased chance for superconductivity.

The SC properties of molecules of a similar structure as TTF may thus be understood using the model of the previous section. If the equilibrium geometry differences are large between the SDW and CDW states, the SC phase cannot appear. There may still be a phase transition at elevated temperatures when the free energies of the comproportionated SDW state and the disproportionated CDW states are equal in the sense:

$$\Delta G = \Delta H - T\Delta S = 0 \tag{7}$$

A transition between two states as in eqn. (7) involves easy ET in the temperature interval of the phase transition. The high conductivity in TTF-TCNQ in a narrow temperature interval may be explained this way. The onset when the temperature is raised towards 59 K and the decrease for $T > 59$ K appears too steep to be explained as simple activated behaviour.

6. Discussion and conclusion

It is clear that in repetitive molecular wires the degree of band filling is a very important property. Covalent wires such as $(SN)_x$ and molecular stacks such as $(TMTSF)_2X$ share conduction properties since they are both three-quarter filled. SDW and CDW are both insulating, but if the energy is about the same they may interact *via* zero-point vibrations and form a new state that may be identified with the SC state (for the case of square-planar grid, see ref. 41).

Recently the SDW and CDW phases in $(TMTSF)_2X$ at quarter and three-quarter filling have been extensively studied experimentally[44] and theoretically.[45–47] In the present paper it was not attempted explain under which conditions the different phases are stable. Instead we have focused on the interaction between the CDW and SDW states *via* phonon coupling. The conclusion is that the CDW–SDW interaction may lead to a high conductivity, for example in the organic crystal TTF–TCNQ. This conclusion is consistent with the recent results of Chollet *et al.*[48] Alternatively we may obtain a SC phase. The mechanism for SC in molecular wires advanced here is the same as used to explain SC in three-dimensional systems.[41]

It is clear that one-determinant methods or the SSH model cannot satisfactorily explain the electronic structure of a molecular wire with three-quarter filling. The SDW state needs a well-known correlated wave function. The CDW state needs a very good description of on-site correlation effects. To some extent, however, the simpler quantum chemical methods give a hint that two states are possible as we have seen above. Historically the correct description of two-electron Coulomb repulsion has been sufficient for a satisfactory quantum description of a system (for example the multiplet theory of atoms). In the case of SC it is clear that correlation effects are important in general.

Acknowledgements

I am grateful for support from the Swedish Research Council (Vetenskapsrådet, VR).

References

1 (*a*) S. Roth and D. Carroll, *One-Dimensional Metals: Conjugated Polymers, Organic Crystals, Carbon Nanotubes*, Wiley-VCH, Weinheim, 2004; (*b*) S. Roth, *One-Dimensional Metals*, VCH, Weinheim, 1995, and further references therein.
2 D. Jérome, *Chem. Rev.*, 2004, **104**, 5665–5691.
3 R. E. Peierls, *Quantum Theory of Solids*, Clarendon, Oxford, 1955, p. 108.
4 S. Kagoshima, *J. Phys. Soc. Jpn.*, 1976, **41**, 2061.
5 (*a*) W. Kohn, *Phys. Rev. Lett.*, 1959, **2**, 393; (*b*) E. J. Woll and W. Kohn, *Phys. Rev.*, 1962, **126**, 1693.
6 (*a*) H. Fröhlich, *Phys. Rev.*, 1950, **79**, 845–856; (*b*) H. Fröhlich, *Proc. Phys. Soc., Sect. A*, 1950, **63**, 778; (*c*) H. Fröhlich, *Proc. R. Soc. London, Ser. A*, 1954, **223**, 296; (*d*) H. Fröhlich, *Adv. Phys.*, 1954, **3**, 325–361.
7 T. Vuletić, P. Auban-Senzier, C. Pasquier, S. Tomić, D. Jérome, M. Héritier and K. Bechgaard, *Eur. Phys. J. B*, 2002, **25**, 319–331.
8 R. H. Friend and D. Jérome, *J. Phys. C*, 1979, **12**, 1441.
9 K. Prassides and P. Day, *J. Chem. Soc., Faraday Trans. 2*, 1984, **80**, 85–95.
10 K. Prassides, P. N. Schatz, K. Y. Wong and P. Day, *J. Phys. Chem.*, 1986, **90**, 5588–5597.
11 W. I. F. David, W. T. A. Harrison, J. M. F. Gunn, O. Moze, A. K. Soper, P. Day, J. D. Jorgensen, D. G. Hinks, M. A. Beno, L. Soderholm, D. W. Capone II, I. K. Schuller, C. U. Segre, K. Zhang and J. D. Grace, *Nature*, 1987, **327**, 310–312.
12 S. Larsson, *Chem. Phys. Lett.*, 1989, **157**, 403–408.
13 (*a*) P. Day, C. Vettier and G. Parisot, *Inorg. Chem.*, 1978, **17**, 2319; (*b*) N. Kojima, H. Kitagawa, T. Ban, F. Amita and M. Nakahara, *Solid State Commun.*, 1990, **73**, 743; (*c*) N. Kojima, M. Hasegawa, H. T. Kikegawa and O. Shimomura, *J. Am. Chem. Soc.*, 1994, **116**, 11368.
14 A. Klimkāns, in Thesis, Chalmers University of Technology, 2001.
15 (*a*) V. V. Walatka, Jr., M. M. Labes and J. H. Pearlstein, *Phys. Rev. Lett.*, 1973, **31**, 1139; (*b*) J. Ferraris, D. Cowan, W. Walatka and J. Pearlstein, *J. Am. Chem. Soc.*, 1973, **95**, 948.
16 R. L. Greene, G. B. Street and L. J. Suter, *Phys. Rev. Lett.*, 1975, **34**, 577.
17 L. B. Coleman, M. J. Cohen, D. J. Sandman, F. G. Yamagishi, A. F. Garito and A. J. Heeger, *Solid State Commun.*, 1973, **12**, 1125.
18 (*a*) M. J. Cohen, L. B. Coleman, A. F. Garito and A. J. Heeger, *Phys. Rev. B*, 1974, **10**, 1298; (*b*) G. Thomas, *Phys. Rev. B*, 1978, **13**, 5105.
19 H. Shirakawa, E. J. Louis, A.G. MacDiarmid, C. K. Chiang and A. J. Heeger, *J. Chem. Soc., Chem. Commun.*, 1978, 578.
20 (*a*) C. K. Chiang, C. R. Fincher, Jr., Y. W. Park, A. J. Heeger, H. Shirakawa, E. J. Louis, S. C. Gau and A. G. MacDiarmid, *Phys. Rev. Lett.*, 1977, **39**, 1098; (*b*) C. K. Chiang, Y. W. Park, A. J. Heeger, H. Shirakawa, E. J. Louis and A. G. MacDiarmid, *J. Chem. Phys.*, 1978, **69**, 5098.
21 (*a*) E. Hückel, *Z. Phys.*, 1930, **60**, 423; (*b*) E. Hückel, *Z. Phys.*, 1931, **70**, 204; (*c*) E. Hückel, *Z. Phys.*, 1931, **72**, 310; (*d*) E. Hückel, *Z. Phys.*, 1932, **76**, 628.
22 Y. Ooshika, *J. Phys. Soc. Jpn.*, 1957, **12**, 1246.
23 H. C. Longuet-Higgins and L. Salem, *Proc. R. Soc. London, Ser. A*, 1959, **251**, 172.
24 (*a*) W.-P. Su, J. R. Schrieffer and A. J. Heeger, *Phys. Rev. Lett.*, 1979, **42**, 1698; (*b*) A. J. Heeger, S. Kivelson, J. R. Schrieffer and W.-P. Su, *Rev. Mod. Phys.*, 1988, **60**, 781–850.
25 L. Salem, *The Molecular Orbital Theory of Conjugated Molecules*, W. A. Benjamin, London, 1966.
26 D. Jérome, A. Mazaud, M. Ribault and K. Bechgaard, *J. Phys. Lett.*, 1980, **41**, L95.
27 (*a*) K. Bechgaard, C. S. Jacobsen, K. Mortensen, H. J. Pedersen and N. Thorup, *Solid State Commun.*, 1980, **33**, 1119; (*b*) K. Bechgaard, K. Carneiro, F. B. Rasmussen, M. Olsen, G. Rindorf, C. S. Jacobsen, H. J. Pedersen and J. C. Scott, *J. Am. Chem. Soc.*, 1981, **103**, 2440–2442.
28 (*a*) S. Larsson and L. Rodríguez-Monge, *Int. J. Quantum Chem.*, 1996, **58**, 517–532; (*b*) L. Rodríguez-Monge and S. Larsson, *J. Chem. Phys.*, 1995, **102**, 7106; (*c*) L. Rodríguez-Monge and S. Larsson, *Int. J. Quantum Chem.*, 1997, **63**, 655–665.
29 S. Larsson, *Int. J. Quantum Chem.*, 1986, **30**, 31–49.
30 S. Larsson and L. Rodríguez-Monge, *Int. J. Quantum Chem.*, 1998, **67**, 107–113.
31 C. M. Mikulski, P. J. Russo, M. S. Saran, A. G. MacDiarmid, A. F. Garito and A. J. Heeger, *J. Am. Chem. Soc.*, 1975, **97**, 6358–6363.

32 J. J. J. Stewart, *J. Comput. Chem.*, 1989, **10**, 209.
33 *HyperChem*, Molecular Visualization and Simulation, Hypercube, Inc.
34 R. C. Haddon, S. R. Wasserman, F. Wudl and G. R. J. Williams, *J. Am. Chem. Soc.*, 1980, **102**, 6887–6993.
35 M. H. Palmer and R. H. Findlay, *J. Mol. Struct.*, 1983, **92**, 373–383.
36 J. Gerratt, S. J. McNicholas, P. B. Karadakov, M. Sironi, M. Raimondi and D. L. Cooper, *J. Am. Chem. Soc.*, 1996, **118**, 6472–6476.
37 C. J. Mei, H. J. Monkhorst and R. J. Bartlett, *Z. Phys. B*, 1996, **101**, 73–78.
38 (*a*) M. Springborg, *Phys. Rev. B*, 1989, **40**, 7928–7938; (*b*) J. Voit, *Phys. Rev. Lett.*, 1990, **64**, 323–326.
39 E. Z. Kurmaev, A. I. Poteryaev, V. I. Anisimov, I. Karla, A. Moewes, B. Schneider, M. Neumann, D. L. Ederer and R. N. Lyubovskaya, *Physica C*, 1999, **321**, 191–198.
40 J. P. Pouget, S. K. Khanna, F. Denoyer, R. Comès, A. F. Garito and A. J. Heeger, *Phys. Rev. Lett.*, 1976, **37**, 437.
41 S. Larsson, *Chem. Phys.*, 1998, **236**, 133–150.
42 A. Klimkāns and S. Larsson, *Int. J. Quantum Chem.*, 2000, **77**, 211–220.
43 (*a*) A. W. Sleight, J. L. Gillson and P. E. Bierstedt, *Solid State Commun.*, 1975, **17**, 27; (*b*) R. J. Cava, B. Batlogg, J. J. Krajewski, R. Farrow, L. W. Rupp, A. E. White, K. Short, W. F. Peck and T. Kometani, *Nature*, 1998, **332**, 814–816; (*c*) D. E. Cox and A. W. Sleight, *Solid State Commun.*, 1976, **19**, 969–973; (*d*) B. Batlogg, R. J. Cava, L. F. Schneemeyer and G. P. Espinosa, *IBM J. Res. Dev.*, 1989, **33**, 208–214.
44 T. Nakamura, *J. Phys. Soc. Jpn.*, 2003, **72**, 213–216.
45 (*a*) H. Kino and H. Fukuyama, *J. Phys. Soc. Jpn.*, 1995, **64**, 1877, 2726, 4523; (*b*) H. Kino and H. Fukuyama, *J. Phys. Soc. Jpn.*, 1996, **65**, 2158.
46 H. Seo, *J. Phys. Soc. Jpn.*, 2000, **69**, 805–820.
47 N. Kobayashi and M. Ogata, *J. Phys. Soc. Jpn.*, 1997, **66**, 3356–3359.
48 M. Chollet *et al.*, *Science*, 2005, **307**, 86–89.
49 (*a*) S. Larsson and A. Klimkāns, *Int. J. Quantum Chem.*, 2000, **80**, 713–720; (*b*) S. Larsson and A. Klimkāns, *Mol. Liq. Cryst.*, 2001, **355**, 217–229.

Effect of internal conversion of vibrational quanta in electron tunneling: a scheme for direct experimental observation and current/voltage relationships

Alexander M. Kuznetsov†*[ab]

Received 11th April 2005, Accepted 10th June 2005
First published as an Advance Article on the web 8th September 2005
DOI: 10.1039/b504746c

Electron tunneling through bridge molecular groups with a strong coupling to a local quantum vibrational mode is considered. A scheme is suggested for direct experimental observation of the effect of internal conversion of vibrational quanta in inelastic electron tunneling. The effect consists of excitation and re-absorption of vibrational quanta in the bridge by tunneling electrons. The tunnel current produced by the absorption of vibrational quanta by the same or another electron can in principle be detected in an experimental setup a scheme of which is suggested. Current/voltage dependences have in general spectroscopic features. Possible effects that can take place in the case of additional strong interaction with classical vibrational modes are discussed.

1. Introduction

The phenomenon of inelastic electron tunneling has been known and in the focus of researchers for a long time (see *e.g.* refs. 1–9). These studies allowed to explain a number of features in the current/voltage characteristics of tunnel contacts in vacuum. The progress in scanning tunneling microscopy (STM) made possible investigations of local effects in electron tunneling, in particular, of electron transfer through a single molecule.[10] Manipulations of the state of single atoms or molecules on the metal surface involving the displacement of atoms and dissociation of molecules also became possible.[11,12] The latter is entirely due to inelastic effects in electron tunneling. The excitation of vibrational quanta of a quantum local mode in a vacuum tunnel contact may result, in particular, in steps on the current/voltage dependence[4] or its derivative.[13] There have been many studies of inelastic electron tunneling in vacuum contacts.[1,2,7,8,11,12,14–16] A number of molecules such as alkanethiol and –dithiol, functionalized organic molecular systems and redox systems were probed as bridge groups. In particular it has been shown that the tunnel current may result in a non-equilibrium stationary average number of vibrational quanta of the local mode.[17,18] This number is determined in general by a balance of the processes of excitation of vibrational quanta by tunneling electrons ("emissive" tunneling) and their damping. The latter involves in general both damping due to interaction with environment and re-absorption of vibrational quanta by the tunneling electrons ("absorptive" tunneling). The "absorptive" electron tunneling contributes to the total current along with elastic and "emissive" tunneling. However in view of the small probability of the former it is usually masked by the latter. The objectives of the present study is to suggest a scheme for direct experimental observation of "absorptive" tunneling. If experiments of this type are successful, one may speak about the phenomenon of *internal conversion of vibrational quanta in electron tunneling*.

[a] *Department of Chemistry, Imperial College London, London, UK, SW7 2AZ*
[b] *A. N. Frumkin Institute of Electrochemistry, Leninskii prospect 31, Bldg. 5, 119071, Moscow, Russia. E-mail: theor@elchem.ac.ru*

† Present address: A. N. Frumkin Institute of Electrochemistry, Leninskii prospect 31, Bldg. 5, 119071, Moscow, Russia.

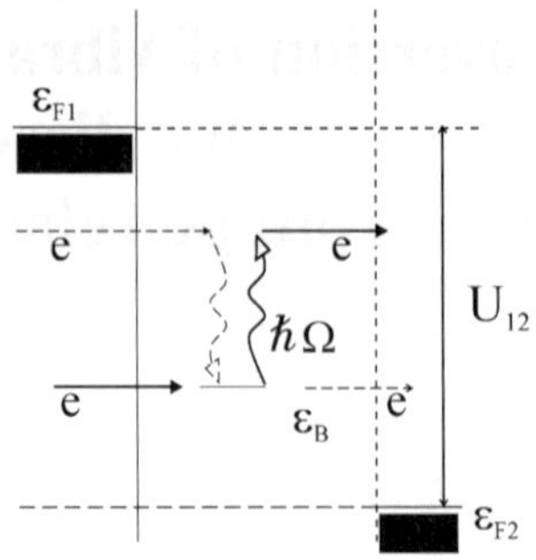

Fig. 1 Electron energy diagram for electron tunneling between two metals *via* a bridge electron energy level ε_B. Dashed arrows show the path with the excitation of a vibrational quantum. Solid arrows correspond to the tunneling with absorption of a vibrational quantum. The combination of dashed and solid arrows shows the path for the excitation/re-absorption process. At a given bias voltage U_{12} tunneling takes place within the energy window between the two Fermi levels of the metals ε_{F1} and ε_{F2}.

2. A scheme of experimental setup

An electron energy scheme for traditional systems used in studies of inelastic tunneling with a single molecule confined between a plane metal substrate and an STM-tip is shown in Fig. 1. If the step-wise approximation for the Fermi distribution functions of the metals is accepted, one may state that the excitation of the vibrational quantum $\hbar\Omega$ is possible only if the difference between the energies of the Fermi levels of the STM tip ε_{F2} and the substrate ε_{F1} is larger than $\hbar\Omega$. Electron transfer occurs in this case from energy levels ε_1 of the substrate to lower energy levels of the STM-tip $\varepsilon_2 = \varepsilon_1 - \hbar\Omega$. In a similar way the "absorptive" tunneling occurs from energy levels of the substrate ε_1' to higher energy levels of the STM-tip $\varepsilon_2' = \varepsilon_1' + \hbar\Omega$. The process of excitation and re-absorption of a vibrational quantum by the same electron ("re-absorptive" tunneling) is also possible (Fig. 1). This does not affect the average stationary number of vibrational quanta. All the processes contribute to the tunneling current between the substrate and the STM-tip along with the elastic tunneling through the energy level ε_B of the bridge molecule B.

The main idea of the present work is to direct the currents which are due to "absorptive" tunneling into another electric circuit. This may be done in principle using the arrangement in Fig. 2. The corresponding electron energy diagram is shown in Fig. 3. The schemes with gate electrode in solid-state contacts or with reference electrode in electrochemical systems may be used to construct the experimental setup corresponding to the scheme of Fig. 2. Fig. 3a gives relative positions of the electron energy levels of the metal substrate M_1, bridge molecular group B and the metal M_2. Relative positions of the energy levels for the electron transfer from the metal M_1 to the metal M_3 *via* the molecular groups B and C are shown in Fig. 3b. It is assumed that the electronic coupling

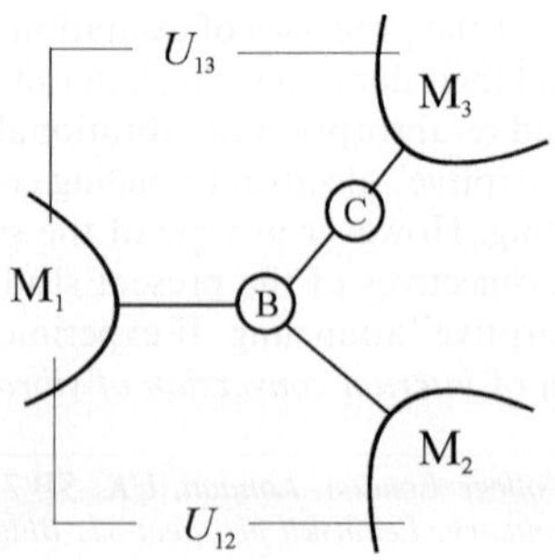

Fig. 2 A scheme for direct experimental observation of internal conversion of vibrational quanta. The excitation of vibrational quanta occurs due to inelastic tunneling between the metals M_1 and M_2 *via* the bridge group B. The current due to the absorption of vibrational quanta passes along the path M_1BCM_3.

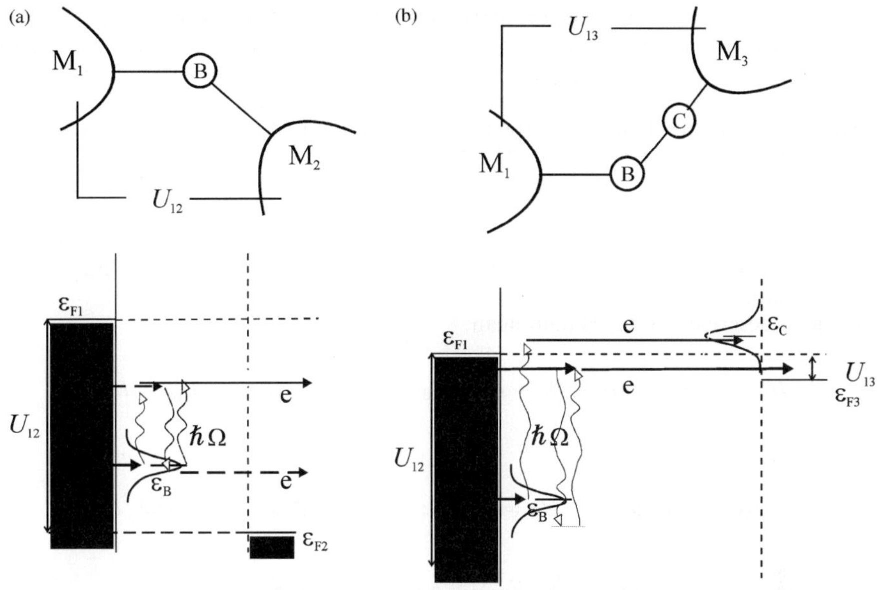

Fig. 3 (a). A scheme and electron energy diagram for tunneling along the path M_1BM_2. It is assumed that direct tunneling between the metals is negligibly small. (b). A scheme and electron energy diagram for tunneling along the path M_1BCM_3. It is assumed that the electron energy level of the bridge group B is located below the Fermi level ε_{F3} whereas the energy level of the group C is above the Fermi level ε_{F3}. A small current due to off-resonant elastic tunneling is possible within the energy window between the Fermi levels ε_{F1} and ε_{F3}.

between the groups B and C is stronger than the coupling of the group B and the metal M_2

$$V_{BC} \gg \Delta_B^{(2)} \tag{1}$$

where V_{BC} is the electronic matrix element and $\Delta_B^{(2)}$ is the partial width of the energy level ε_B due to the interaction of the bridge group B with the metal M_2

$$\Delta_B^{(2)} = \pi \sum_{\varepsilon_2} V_{B\varepsilon_2}^2 \rho(\varepsilon - \varepsilon_2) \tag{2}$$

Here $\rho(\varepsilon - \varepsilon_2)$ is the density of electron states of the metal M_2 and $V_{B\varepsilon_2}$ is the electron matrix element. It is assumed below that the couplings $V_{B\varepsilon_3}$, $V_{C\varepsilon_1}$ and $V_{C\varepsilon_2}$ are equal to zero, *i.e.* only the path M_1BCM_3 is available for the electron transfer to the metal M_3.

3. Mechanisms and basic model

The following mechanisms of electron transfer are feasible, in principle, in the system under consideration:

(i) Resonance elastic electron tunneling along the path M_1BM_2;

(ii) Electron transfer with the excitation of vibrational quantum along the path $M_1(\varepsilon_1)BM_2$ $(\varepsilon_2 = \varepsilon_1 - \hbar\Omega)$

(iii) Electron transfer with absorption of vibrational quantum along the path $M_1(\varepsilon_1)BM_2$ $(\varepsilon_2 = \varepsilon_1 + \hbar\Omega)$

(iv) Electron transfer with absorption of vibrational quantum along the path $M_1(\varepsilon_1)BM_1$ $(\varepsilon_1{}' = \varepsilon_1 + \hbar\Omega)$

(v) Elastic electron tunneling along the path $M_1(\varepsilon_1)BCM_3(\varepsilon_3 = \varepsilon_1)$

(vi) Electron transfer with absorption of vibrational quantum along the path $M_1(\varepsilon_1)BCM_3$ $(\varepsilon_3 = \varepsilon_1 + \hbar\Omega)$

In principle electron tunneling from the metal M_1 to the metal M_3 *via* B and C with the excitation of a vibrational quantum or with the excitation and re-absorption of a vibrational quantum is possible when the energy level of the group C is close to or lower than the energy level of the group B. However this mechanism is not effective for the energy diagrams shown in Figs. 3a and 3b.

Mechanisms (i)–(iii) result in the current between M_1 and M_2, mechanisms (v) and (vi) are responsible for the current from M_1 to M_3. Note that in view of the condition in eqn. (1) the probability of the transition for mechanism (iii) is much smaller than that for mechanism (vi). Therefore practically all the current related with the absorption of vibrational quanta follows the circuit M_1BCM_3. Mechanism (iv) does not result in any current. In particular, it does not operate when $\varepsilon_{F1} - \varepsilon_B > \hbar\Omega$. However in general it affects the value of the stationary average number of vibrational quanta.

The model is described by the Hamiltonian

$$H = \sum_{j=1}^{3} H_{M_j} + H_B + H_C + H_v + H_{BM_1} + H_{BM_2} + H_{CM_3} + V_{BC} + H_{Bv} \qquad (3)$$

where

$$H_{M_j} = \sum_k \varepsilon_k^{(j)} a_{jk}^+ a_{jk}; \quad H_B = \varepsilon_B a_B^+ a_B; \quad H_C = \varepsilon_C a_C^+ a_C; \quad H_v = \hbar\Omega b^+ b \qquad (4)$$

$$\begin{aligned} H_{BM_n} &= \sum_k V_{B\varepsilon_{nk}} (a_B^+ a_{nk} + a_{nk}^+ a_B), n = 1, 2; \\ H_{CM_3} &= \sum_k V_{C\varepsilon_{3k}} (a_C^+ a_{3k} + a_{3k}^+ a_C); \\ H_{Bv} &= \gamma_{Bv} (b^+ + b) a_B^+ a_B \end{aligned} \qquad (5)$$

ε are energies of electron states and a^+ and a are the corresponding creation and annihilation operators, $\hbar\Omega$ is the energy of the vibrational quantum ($\hbar\Omega \gg k_B T$), b^+ and b are the creation and annihilation operators for the quantum vibrational mode, and the constant γ_{Bv} determines coupling of the electron state of the group B with the local vibrational mode.

4. Transition probabilities and electric currents

General expressions for the transition probabilities and electric currents are given in the Appendices A and B. Simplified limiting cases are considered in this section. It will be assumed that the electron energy levels ε_B and ε_C are located respectively far below and far above the Fermi levels $\varepsilon_{F1}$and ε_{F_3} respectively. Total electric current in the circuit M_1BCM_3, i_{13}, is the sum of the current due to elastic tunneling, i_{el}, and the current due to "absorptive" tunneling, i_{abs}:

$$i_{13} = i_{el} + i_{abs} \qquad (6)$$

where

$$i_{el} = ew_{el}; \; i_{abs} = ew_{abs} n \qquad (7)$$

Here w_{el} and w_{abs} are corresponding transition probabilities per unit time and n is stationary average number of vibrational quanta. The latter is determined by the balance of the transitions with the excitation and absorption of vibrational quanta and at $\hbar\Omega \gg k_B T$ this reduces to eqn. (8)

$$n \approx \frac{w_{exc}/\gamma_D + \exp(-\hbar\Omega/k_B T)}{1 + w_{abs}/\gamma_D + w_{iv}/\gamma_D} \qquad (8)$$

where w_{exc} is the transition probability for the inelastic tunneling from M_1 to M_2 with the excitation of the vibrational quantum, w_{iv} is the probability of the transition for the mechanism (iv) and γ_D is vibrational damping constant due to the interaction with the environment. The latter determines the inverse life-time of the excited vibrational state of the quantum mode due to various sorts of friction, *e.g.* due to the excitation of electron-hole pairs in the metal electrodes.

With due account of eqn. (8) the expression for the total current takes the form

$$i_{13} = ew_{\rm el} + ew_{\rm abs}\frac{w_{\rm exc}/\gamma_{\rm D} + \exp(-\hbar\Omega/k_{\rm B}T)}{1 + w_{\rm abs}/\gamma_{\rm D} + w_{\rm iv}/\gamma_{\rm D}} \tag{9}$$

Since the main goal is to detect the current due to the effect of internal conversion of vibrational quanta, the value of the current due to elastic tunneling should be small. This can be reached, *e.g.* by setting up a small value of the bias voltage U_{13} in the configuration shown in Fig. 3b. The problem under consideration is formally rather similar to electron tunneling in systems involving several quantum dots.[19–22] The expression for the "elastic" current can then be written using eqns. (1)–(3) of ref. 22 (Appendix A). Assuming that the couplings $V_{\rm B\varepsilon_3}$, $V_{\rm C\varepsilon_1}$ and $V_{\rm C\varepsilon_2}$ are small we obtain in the limit of small U_{13}

$$i_{\rm el} \approx \frac{4e^2}{\pi\hbar}U_{13}\frac{\Delta_{\rm B}^{(1)}\Delta_{\rm C}^{(3)}V_{\rm BC}^2}{[(\varepsilon_{\rm C}(U_{13}) - \varepsilon_{\rm F1})^2 + \Delta_{\rm C}^2][(\varepsilon_{\rm B}(U_{12}) - \varepsilon_{\rm F1})^2 + \Delta_{\rm B}^2]} \tag{10}$$

where

$$\Delta_{\rm C} = \Delta_{\rm C}{}^{(3)};\ \Delta_{\rm B} = \Delta_{\rm B}{}^{(1)} + \Delta_{\rm B}^{(2)} \tag{11}$$

are the widths of the energy levels.

If the second terms in the nominator and denominator in eqn. (9) are small (due to the condition $\hbar\Omega \gg k_{\rm B}T$ and a small coupling with vibrational mode), it simplifies

$$i_{13} = ew_{\rm el} + ew_{\rm abs}\frac{w_{\rm exc}/\gamma_{\rm D}}{1 + w_{\rm iv}/\gamma_{\rm D}} \tag{12}$$

In this limit the current related with the effect of internal conversion of vibrational quanta is proportional to the product of the transition probabilities for the "emissive" and "absorptive" tunneling. The probability $w_{\rm iv}$ in general depends on the bias voltages. However this dependence is rather weak (linear) and will be neglected. Rather general expressions can be obtained for $w_{\rm exc}$, in particular, taking into account the processes with the excitation of several vibrational quanta (see ref. 4,7 and Appendix B). However here we restrict ourselves to a simple limiting case where only the processes with single-quantum emission are taken into account (as most probable) and the energy level $\varepsilon_{\rm B}$ is located in a broad energy window between the Fermi levels $\varepsilon_{\rm F1}$ and $\varepsilon_{\rm F2}$, *i.e.* the bias voltage U_{12} is assumed to be sufficiently large. The transition probability $w_{\rm exc}$ is then independent of the bias voltages U_{12} and U_{13}, and is roughly equal to (Appendix B)

$$w_{\rm exc} \approx \frac{4}{\hbar}\frac{\Delta_{\rm B}^{(1)}\Delta_{\rm B}^{(2)}}{\Delta_{\rm B}}\left(\frac{\gamma_{\rm B\nu}}{\hbar\Omega}\right)^2\theta(eU_{12} - \hbar\Omega - (\varepsilon_{\rm F1} - \varepsilon_{\rm B}) - \Delta_{\rm B}) \tag{13}$$

The dependence of the current related with the effect of internal conversion of vibrational quanta on U_{13} is determined in this limit by that of the transition probability $w_{\rm abs}$. In the limit where the width of the electron energy level $\varepsilon_{\rm B}$ is much smaller than that of the electron energy level $\varepsilon_{\rm C}$ an approximate expression for $w_{\rm abs}$ can be written as follows (Appendix B)

$$w_{\rm abs} \approx T_q\frac{\Delta_{\rm C}}{[\varepsilon_{\rm C}(U_{13}) - \varepsilon_{\rm B}(U_{12}, U_{13}) - \hbar\Omega]^2 + \Delta_{\rm C}^2} \tag{14}$$

where T_q characterizes the amplitude of the transition probability with the absorption of a vibrational quantum and may be approximately considered to be independent of the bias voltages (Appendix B).

5. Results and discussion

Eqns. (10), (12)–(14) give the dependence of the current in the circuit M_1BCM_3 on the bias voltages U_{12} and U_{13} in the simple limiting case discussed above. As seen from eqn. (13) the contribution due to the effect of internal conversion of vibrational quanta has a threshold character as a function of U_{12} and should manifest itself on the background of decreasing current related with elastic tunneling as a change of the current derivative. The dependence of $i_{\rm el}$ on U_{12} is related with that

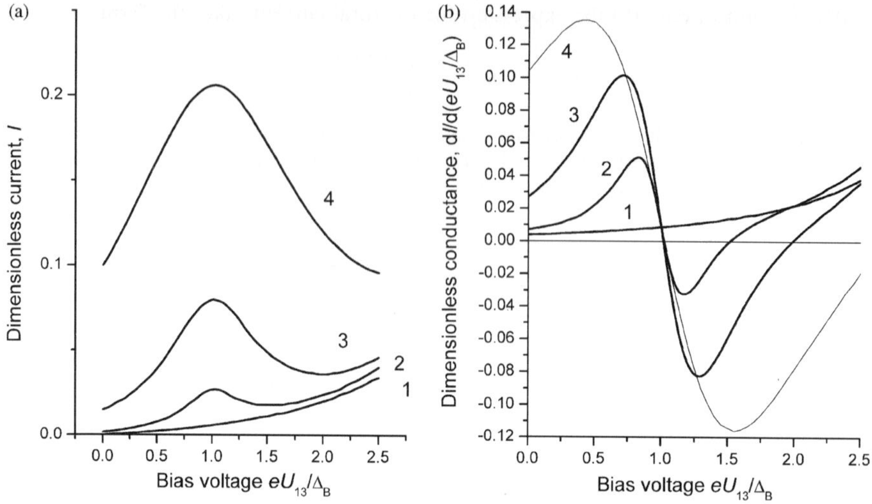

Fig. 4 (a) Dependence of the dimensionless current along the path M_1BCM_3 on the bias voltage U_{13} according to eqns. (10)–(14). Curve 1 corresponds to "elastic" current. $I = i\pi\hbar\Delta_B^3/4e\Delta_C\Delta_B^{(1)}V_{BC}^2$; dimensionless parameter $\tau = \pi T_q\Delta_B/(\gamma_D + w_{IV})V_{BC}^2$; all energies are measured in the units of Δ_B: $\varepsilon_B^0 - \varepsilon_{F1} = -3$; $\varepsilon_C^0 - \varepsilon_{F1} = 5$; $\hbar\Omega = 7$; $\gamma_{13}^{(C)} = 1$; $\gamma_{12}^{(B)} = \gamma_{13}^{(B)} = 0$; 1: $eU_{12} = 5$; $\gamma_{Bv} = 7$; $\Delta_B^{(2)} = 0.2$; $\Delta_C = 1$; $\tau = 1$; 2:10; 3; 0.1; 0.3; 0.1; 3:10; 3; 0.1; 0.5; 1; 3:10; 7; 0.2; 1; 1. (b) Dependence of the conductance along the path M_1BCM_3 on the bias voltage U_{13}. Parameters are the same as in Fig. 4a.

of the electron energy level ε_B. For linear potential drop between M_1 and M_2 the latter can be written as

$$\varepsilon_B = \varepsilon_B{}^0 - \gamma_{12}^{(B)}eU_{12} - \gamma_{13}^{(B)}eU_{13} \tag{15}$$

In a similar way for ε_C we have

$$\varepsilon_C = \varepsilon_C^0 - \gamma_{13}^{(C)}eU_{13} \tag{16}$$

where $\gamma_{12}^{(B)}$, $\gamma_{13}^{(B)}$ and $\gamma_{13}^{(C)}$ are the fractions of the total potential drops U_{12} and U_{13} at the site of the groups B and C respectively. If $\gamma_{12}^{(B)} = 0$, i_{el} is independent of U_{12} (except the threshold condition imposed by the step-wise function in eqn. (13)). The current/voltage dependence in this case is due to that for i_{abs}. Typical curves for the current and conductance for this case are shown in Fig. 4 as functions of U_{13} at two values of U_{12}.

Curve 1 corresponds to the situation where

$$eU_{12} < \hbar\Omega \tag{17}$$

Only i_{el} then exists. If U_{12} exceeds the threshold value, the current has a maximum as a function of U_{13}.

The dependences of the current and its derivative on the position of the electron energy level of the group B at fixed value of U_{13} are shown in Fig. 5. In the step-wise approximation of eqn. (13) the dependences have a threshold character reflected both in the current and in the current derivative. More exact equations result in smoother dependences near the "threshold". As in the case above, the current displays a maximum as the function of the position of the energy level ε_B, *i.e.* of the bias voltage U_{12}.

6. Effect of the interaction with a classical vibrational mode

In some contacts, in particular in the *in situ* electrochemical systems the electron energy levels strongly interact with classical environmental modes. In general these interactions will smear out the inelastic effects under discussion. A particular situation is considered below where it is assumed that only the energy level ε_B interacts with both quantum, Q, and classical, q, vibrational modes.

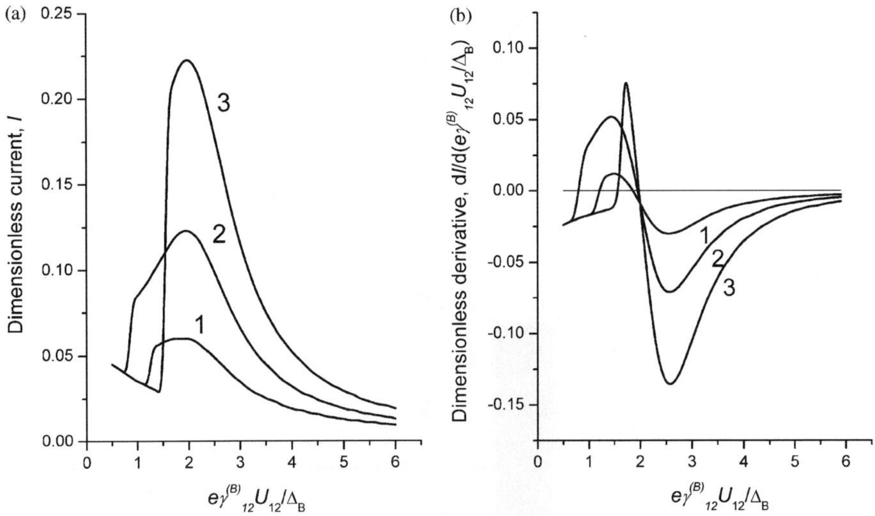

Fig. 5 (a) Dependence of the dimensionless current along the path M_1BCM_3 on the position of the electron energy level ε_B according to eqns. (10)–(14). 1: $\gamma_{B\nu} = 7$; $\Delta_B^{(2)} = 0.2$; $\gamma_{12}^{(B)} = 0.2$; 2: 7; 0.1; 0.1; 3: 3; 0.2; 0.15. $eU_{13} = 3$; $\Delta_C = 1$; $\tau = 1$; other parameters are the same as in Fig. 4a. (b) The same as in (a) for the derivative $dI/d(e\gamma_{12}^{(B)}\,U_{12}/\Delta_B)$.

Commonly accepted linear dependence of the energy level of the group B on the coordinate of the classical mode will be used below

$$\varepsilon_B = \varepsilon_B^0 - \hbar\omega g q \tag{18}$$

where ω is the vibrational frequency, g coupling constant and q the normal coordinate for the classical mode.

A typical situation corresponds to the arrangement of the energy levels shown in Fig. 6 where the vacant (oxidized) energy level ε_B^{ox} is located above the Fermi level ε_{F1} and the occupied (reduced) position of this level ε_B^{red} is located below the Fermi level ε_{F2}. These two configurations correspond to two different equilibrium values of the classical mode coordinate q, $q_{0i} = 0$ and $q_{0f} = g$. Both equilibrium configurations are separated by a potential barrier that depends on the electrode potential and reorganization energy of the classical mode

$$E_r = \frac{1}{2}\hbar\omega(q_{0i} - q_{0f})^2 = \frac{1}{2}\hbar\omega g^2 \tag{19}$$

This barrier determines the activation energies for the redox transformations of the group B.[23,24] The reduction and oxidation of the group B result in electron transfer mainly between the electrodes 1 and 2. These processes determine the average stationary "concentrations" of the oxidized and reduced forms of the group B, n_{ox} and n_{red}

$$n_{ox} = \frac{k^{r/o}}{k^{r/o} + k^{o/r}}; \quad n_{red} = \frac{k^{o/r}}{k^{r/o} + k^{o/r}} \tag{20}$$

where $k^{o/r}$ and $k^{r/o}$ are the rate constants for the reduction and oxidation.

Unlike the situations considered in the foregoing sections the electron energy level ε_B does not remains permanently in the energy window between the Fermi levels of M_1 and M_2. Its appearance in the energy window is the result of fluctuations. Due to thermal fluctuations of the classical vibrational mode the position of the energy levels ε_B also fluctuates. Therefore an additional averaging should be performed in the calculation of the probabilities w_{el}, w_{exc} and w_{abs}

$$w_\mu = \int d\varepsilon_B w_\mu(\varepsilon_B)[n_{ox}(\varepsilon_B^{ox},\varepsilon_B^{red})D_{ox}(\varepsilon_B) + n_{red}(\varepsilon_B^{ox},\varepsilon_B^{red})D_{red}(\varepsilon_B)]; \ \mu = \text{el, exc, abs} \tag{21}$$

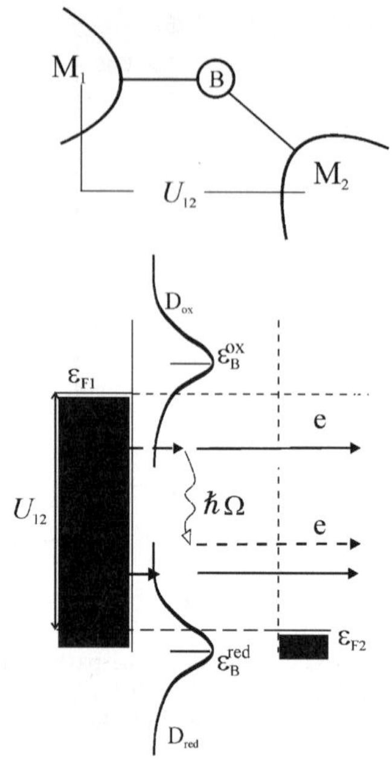

Fig. 6 A scheme and electron energy diagram for electron tunneling along the path M_1BM_2 in the presence of coupling with a classical vibrational mode. The vacant (oxidized) and occupied (reduced) electron energy levels of the bridge group B are located above the Fermi level ε_{F1} and below the Fermi level ε_{F2} respectively. D_{ox} and D_{red} are the distributions of the positions of the energy level ε_B due to fluctuations along the coordinate of classical vibrational mode.

where

$$D_{ox}(\varepsilon_B) = \sqrt{\frac{1}{4\pi E_r k_B T}} \exp\left[-\frac{\left(\varepsilon_B - \varepsilon_B^{ox} - e\gamma^{(12)}U_{12} - e\gamma^{(13)}U_{13}\right)^2}{4E_r k_B T}\right] \quad (22)$$

and

$$D_{red}(\varepsilon_B) = \sqrt{\frac{1}{4\pi E_r k_B T}} \exp\left[-\frac{\left(\varepsilon_B - \varepsilon_B^{red} - e\gamma^{(12)}U_{12} - e\gamma^{(13)}U_{13}\right)^2}{4E_r k_B T}\right] \quad (23)$$

The probabilities $w_{exc}(\varepsilon_B)$ and $w_{abs}(\varepsilon_B)$ are described by eqns. (B.6), and (B.5), (B.7)–(B.9). The expression for $w_{el}(\varepsilon_B)$ is obtained by comparison of eqns. (9), (10) and (A.4). For the absorption process eqn. (21) is valid if the characteristic survival time for an excited vibrational quantum state is larger than the time required for establishing the vibrational equilibrium in the classical mode in reduced and oxidized states. If the survival time is shorter than the latter, the interaction with classical mode totally destroys the effect of internal conversion.

In general the position of the energy levels ε_B^{ox} and ε_B^{red} may be varied by a gate potential (or overpotential in electrochemical systems). In this way the equilibrium between the oxidized and reduced forms can be shifted.

7. Concluding remarks

The scheme of experimental set up suggested provides in principle the possibility for direct experimental observation of the current caused by inelastic electron tunneling with absorption of vibrational quanta which are excited in the same multi-quantum-dot system. This may be called *the effect of internal conversion of vibrational quanta in electron tunneling*. The notion "internal conversion" is used here to emphasize that the excited vibrational quanta being re-absorbed by the electrons *within the same tunnel contact* provide a new path-way for electron tunneling.

The key idea is to direct the current related with the effect of internal conversion to a different electric circuit formed by an additional electrode and bridge group C (Fig. 2). The key factors are the positions of the electron energy levels of the bridge groups, the Fermi levels of the metal electrodes (Fig. 3) and the interactions between them. This current as the function of various bias voltages has spectroscopic features that can help in its detection and separation from the current due to elastic tunneling. If the experiments of this sort will be successful, one may speak about the effect of internal conversion of vibrational quanta in electron tunneling.

Appendix A

Elastic current in a two-quantum-dot system

The general expression for elastic current along the path M_1BCM_3 has the form

$$i_{\mathrm{el}}(\varepsilon_{\mathrm{B}}) = \frac{e^2}{\pi\hbar}\int \mathrm{d}\varepsilon\mathrm{d}\varepsilon' T_{\mathrm{el}}^{(13)}(\varepsilon,\varepsilon';\varepsilon_{\mathrm{B}})f_1(\varepsilon)[1-f_3(\varepsilon')] \tag{A.1}$$

where $T_{\mathrm{el}}^{(13)}$ may be written as[22]

$$T_{\mathrm{el}}^{(13)}(\varepsilon,\varepsilon';\varepsilon_{\mathrm{B}}) = (4\Delta_{\mathrm{B}}^{(1)}\Delta_{\mathrm{C}}V_{\mathrm{BC}}^2/|D(\varepsilon;\varepsilon_{\mathrm{B}})|^2)\delta(\varepsilon-\varepsilon') \tag{A.2}$$

with

$$D(\varepsilon,\varepsilon_{\mathrm{B}}) = (\varepsilon - \varepsilon_{\mathrm{B}} + \mathrm{i}\Delta_{\mathrm{B}})(\varepsilon - \varepsilon_{\mathrm{C}} + \mathrm{i}\Delta_{\mathrm{C}}) - V_{\mathrm{BC}}^2 \tag{A.3}$$

Substitution of eqns. (A.2) and (A.3) into eqn. (A.1) gives

$$i_{\mathrm{el}}(\varepsilon_B) \approx \frac{4e^2\Delta_{\mathrm{B}}^{(1)}\Delta_{\mathrm{C}}V_{\mathrm{BC}}^2}{\pi\hbar}\int_{\varepsilon_{\mathrm{F1}}-eV_{13}}^{\varepsilon_{\mathrm{F1}}} \mathrm{d}\varepsilon([(\varepsilon-\varepsilon_{\mathrm{B}})^2+\Delta_{\mathrm{B}}^2][(\varepsilon-\varepsilon_{\mathrm{C}})^2+\Delta_{\mathrm{C}}^2]+V_{\mathrm{BC}}^4-2V_{\mathrm{BC}}^2 [(\varepsilon-\varepsilon_{\mathrm{B}})(\varepsilon-\varepsilon_{\mathrm{C}})-\Delta_{\mathrm{C}}^2])^{-1} \tag{A.4}$$

If the bias voltage U_{13} is small and the energy levels ε_{B} and ε_{C} are located far from the Fermi level $\varepsilon_{\mathrm{F1}}$, approximate integration in eqn. (A.1) results in eqn. (10) where V_{BC}^2 was neglected in eqn. (A.3).

Appendix B

The transmission function for inelastic transitions

The transmission function $T(\varepsilon,\varepsilon';\varepsilon_{\mathrm{B}})$ for inelastic electron tunneling from an energy level ε of the substrate to an energy level ε' of the STM tip may be written with the use of general results for inelastic electron tunneling obtained in ref. 8. For the excitation process from the ground state we have

$$T^{(12)}(\varepsilon,\varepsilon';\varepsilon_{\mathrm{B}}) = 4\Delta_{\mathrm{B}}^{(1)}\Delta_{\mathrm{B}}^{(2)}\exp(-2\gamma_{\mathrm{B}\nu}^2/(\hbar\Omega)^2)\sum_{m=0}^{\infty}(\gamma_{\mathrm{B}\nu}/\hbar\Omega)^{2m}|D_m(\varepsilon;\varepsilon_{\mathrm{B}})|^2\delta(\varepsilon-\varepsilon'-\hbar\Omega) \tag{B.1}$$

where

$$D(\varepsilon;\varepsilon_{\mathrm{B}}) = \int_0^{\infty}\frac{\mathrm{d}t}{\hbar}(1-\mathrm{e}^{-\mathrm{i}\Omega t})^m\exp\{-\mathrm{i}(\varepsilon_{\mathrm{B}}-\varepsilon-\gamma_{\mathrm{B}\nu}^2/\hbar\Omega)t/\hbar - \Delta_{\mathrm{B}}t/\hbar + (\gamma_{\mathrm{B}\nu}/\hbar\Omega)^2\mathrm{e}^{-\mathrm{i}\Omega t}\} \tag{B.2}$$

Restricting ourselves to the one-quantum excitation process ($m = 1$) we obtain in the limit of small $\gamma_{B\nu}$

$$T_{\mathrm{exc}}^{(12)}(\varepsilon,\varepsilon';\varepsilon_{\mathrm{B}}) \approx \frac{4\pi\Delta_{\mathrm{B}}^{(1)}\Delta_{\mathrm{B}}^{(2)}}{\Delta_{\mathrm{B}}^2}\gamma_{\mathrm{B}\nu}^2\rho_{\mathrm{B}}(\varepsilon;\varepsilon_{\mathrm{B}})\rho_{\mathrm{B}}(\varepsilon-\hbar\Omega;\varepsilon_{\mathrm{B}})\delta(\varepsilon-\varepsilon'-\hbar\Omega) \tag{B.3}$$

where

$$\rho_{\mathrm{B}}(\varepsilon;\varepsilon_{\mathrm{B}}) = \frac{1}{\pi}\frac{\Delta_{\mathrm{B}}}{(\varepsilon-\varepsilon_{\mathrm{B}})^2+\Delta_{\mathrm{B}}^2} \tag{B.4}$$

The general expression for the transition probabilities w has the form

$$w(\varepsilon_{\mathrm{B}}) = \frac{1}{\hbar}\int \mathrm{d}\varepsilon\mathrm{d}\varepsilon' T(\varepsilon,\varepsilon';\varepsilon_{\mathrm{B}})f_1(\varepsilon)[1-f_2(\varepsilon-\hbar\Omega)] \tag{B.5}$$

For w_{exc} we have then

$$\begin{aligned} w_{\mathrm{exc}}(\varepsilon_{\mathrm{B}}) &= \frac{1}{\hbar}\int \mathrm{d}\varepsilon\mathrm{d}\varepsilon' T(\varepsilon,\varepsilon';\varepsilon_{\mathrm{B}})f_1(\varepsilon)[1-f_2(\varepsilon-\hbar\Omega)] \\ &= \frac{4\pi\Delta_{B}^{(1)}\Delta_{B}^{(2)}}{\hbar\Delta_{B}^2}\gamma_{\mathrm{B}\nu}^2\int \mathrm{d}\varepsilon\,\rho_{\mathrm{B}}(\varepsilon;\varepsilon_{\mathrm{B}})\rho_{\mathrm{B}}(\varepsilon-\hbar\Omega;\varepsilon_{\mathrm{B}})f_1(\varepsilon)[1-f_2(\varepsilon-\hbar\Omega)] \end{aligned} \tag{B.6}$$

In a similar way we obtain for the probability of absorption of a vibrational quantum by the tunneling electron

$$T_{\mathrm{abs}}^{(13)}(\varepsilon,\varepsilon';\varepsilon_{\mathrm{B}}) = 4\Delta_{\mathrm{B}}^{(1)}\Delta_{\mathrm{C}}\left(\frac{\gamma_{\mathrm{B}\nu}}{\hbar\Omega}\right)^2\frac{V_{\mathrm{BC}}^2}{|\Delta E(\varepsilon)|^2}\delta(\varepsilon-\varepsilon'+\hbar\Omega)|D_1^{\mathrm{abs}}(\varepsilon;\varepsilon_{\mathrm{B}})|^2 \tag{B.7}$$

where

$$|D_1^{\mathrm{abs}}(\varepsilon;\varepsilon_{\mathrm{B}})|^2 = \frac{\pi^2(\hbar\Omega)^2}{\Delta_{\mathrm{B}}\Delta_{\mathrm{C}}}\rho_{\mathrm{B}}(\varepsilon;\varepsilon_{\mathrm{B}})\rho_{\mathrm{C}}(\varepsilon+\hbar\Omega;\varepsilon_{\mathrm{B}}) \tag{B.8}$$

and

$$|\Delta E(\varepsilon)|^{-2} = \left|\int \mathrm{d}\varepsilon''\frac{\rho_{\mathrm{B}}(\varepsilon'')}{\varepsilon-\varepsilon''+\mathrm{i}\Delta_{\mathrm{B}}}\right|^2 \tag{B.9}$$

Substitution of eqns. (B.7)–(B.9) into eqn. (B.5) in the limit of narrow width of the energy level ε_{B} results in eqn. (14) for w_{abs} with

$$T_q = 4\frac{\pi\Delta_{\mathrm{B}}^{(1)}}{\hbar\Delta_{\mathrm{B}}}\gamma_{\mathrm{B}\nu}^2\frac{V_{\mathrm{BC}}^2}{|\Delta E(\varepsilon_{\mathrm{B}})|^2} \tag{B.10}$$

Acknowledgements

This work was started and partially performed when the author stayed at Imperial College London as Leverhulme Visiting Professor (grant No. F07058T). A financial support from Russian Foundation for Basic Research is also gratefully acknowledged (grant No. 03-03-32935).

References

1 J. W. Gadzuk, *Phys. Rev. B*, 1991, **44**, 13466.
2 R. C. Jacklevic, R. Wilkins, M. Amman and E. Ben-Jacob, *Phys. Rev. B*, 1991, **44**, 1407.
3 M. Jonson and A. Grincwajg, *Appl. Phys. Lett.*, 1987, **51**, 1729.
4 W. Schmickler, *Surf. Sci.*, 1993, **95**, 43.
5 A. D. Stone and P. A. Lee, *Phys. Rev. Lett.*, 1985, **54**, 1186.
6 R. E. Walkup, D. M. Newns and Ph. Avouris, *Phys. Rev. B*, 1993, **48**, 1858.
7 N. S. Wingreen, K. W. Jacobsen and J. W. Wilkins, *Phys. Rev. Lett.*, 1988, **61**, 1396.
8 N. S. Wingreen, K. W. Jacobsen and J. W. Wilkins, *Phys. Rev. B*, 1989, **40**, 11834.
9 H. Yuming, *J. Phys. C*, 1988, **21**, L23.
10 N. Tao, *Phys. Rev. Lett.*, 1996, **76**, 4066.
11 S.-W. Hla and K.-H. Rieder, *Annu. Rev. Phys. Chem.*, 2003, **54**, 307.
12 B. N. J. Persson and Ph. Avouris, *Surf. Sci.*, 1997, **390**, 45.
13 W. Ho, *J. Chem. Phys.*, 2002, **117**, 11033.

14 T. Mii, S. Tikhodeev and H. Ueba, *Surf. Sci.*, 2002, **502–503**, 26.
15 B. N. J. Persson and A. Baratoff, *Phys. Rev. Lett.*, 1987, **59**, 339.
16 B. N. J. Persson and H. Ueba, *Surf. Sci.*, 2002, **502–503**, 18.
17 S. Tikhodeev, M. Natario, K. Makoshi, T. Mii and H. Ueba, *Surf. Sci.*, 2001, **493**, 63.
18 H. Ueba and B. N. J. Persson, *Surf. Sci.*, 2004, **566–568**, 1.
19 M. Sumetskii, *J. Phys.: Condens. Matter*, 1991, **3**, 2651.
20 M. Sumetskii, *Phys. Rev. B*, 1993, **48**, 14288.
21 M. Sumetskii, *Phys. Rev. B*, 1993, **48**, 4586.
22 M. Sumetskii, *Appl. Phys. Lett.*, 1993, **63**, 3185.
23 A. M. Kuznetsov and J. Ulstrup, *J. Phys. Chem. A*, 2000, **104**, 11531.
24 A. M. Kuznetsov and J. Ulstrup, *Probe Microsc.*, 2001, **2**, 187.

General Discussion

Professor Lindsay opened the discussion of Professor Heath's paper: Please comment on the Hewlett-Packard report that switching is seen in systems that do not contain rotaxane molecules.

Professor Heath replied: This is an issue that has gotten a significant amount of attention. The HP group does see molecule-independent switching. When we utilized all-metal electrodes—which is how we began several years ago—we also saw molecule-independent switching. That is why we abandoned metal bottom electrodes and pursued silicon (or carbon nanotube) bottom electrodes.

The HP devices exhibit large amplitude switching that is temperature independent and only weakly (if at all) depends upon the actual molecule within the junction.[1] That group has postulated that a physical mechanism is responsible for the switching—namely the reversible, electrochemical growth of a single conducting filament that bridges the top and bottom electrodes. Indeed, when the HP devices switch from the low-conductance to the high-conductance state, a single 'hot-spot' of high conductance can be imaged, using scanning probe microscopy, on the top electrode surface.[2] This hot spot apparently correlates to the physical region in which the conducting filament is formed, and it disappears once the junction is electrically switched back to the low-conductance state. The CIT/UCLA devices, on the other hand, are observed to switch based upon a thermally activated process that depends strongly upon the nature of the molecule within the junction.[3–6] We have proposed a chemical mechanism for this switching, which I have described in my Faraday Discussion paper.

We have investigated this issue in some detail. In one set of experiments, we studied single molecule break junction field-effect transistors.[7] In those experiments, a single [2]rotaxane molecule was either chemisorbed to either end of the break junction, or physisorbed on one end and chemisorbed to the other end. We found that our device response was highly dependent upon the nature of the molecule/electrode contacts, but almost completely independent of the nature of the [2]rotaxane.

We also found that semiconducting, single-walled carbon nanotubes could be substituted for the silicon bottom electrode, and in those cases we were able to 'see' the molecular properties in the switching response.[8]

The result of all of this is that the electrodes make all of the difference in the world. One possibility is the image charge on the metal surface (which will be significantly greater than that on a semiconductor surface) and how that charge interacts with the electrochemical processes one is trying to realize on the molecular switches. The second possibility is that the metal/molecule contact is simply too ionic—*i.e.* it serves as a sort of Schottky barrier that constitutes the rate-limiting step to charge flow.

In Table 1 I compare the HP and Caltech/UCLA device types.

1. The switching amplitude can be both molecule and temperature dependent. For some rotaxanes investigated, the ground state co-conformation/ metastable state co-conformation (GSCC/MSCC) ratio is temperature dependent in acetonitrile solution. For those same molecules, the switching amplitude within an MSTJ is also temperature dependent.

2. The lifetime of the high-conductance state is both molecule and temperature dependent.

3. The sharp voltages at which the switch opens or closes are observed to shift to high magnitude bias with increased cycling of the device.

4. A current–voltage response similar to what is reported for the HP devices can be observed for the CIT/UCLA devices if those devices are operated at high voltages ($>|3\ V|$). SWNT-based MSTJs were demonstrated for [2]catenanes only.

5. Controls have included **Rblock**$^{4+}$, the dumbbell components of bistable [2]rotaxanes, degenerate [2]catenanes, and linear amphiphiles, such as eicosanoic acid.

6. Controls have included the dumbbell components of bistable [2]rotaxanes and linear amphiphiles, such as eicosanoic acid.

1 D. R. Stewart, D. A. A. Ohlberg, P. A. Beck, Y. Chen, R. S. Williams, J. O. Jeppesen, K. A. Nielsen and J. F. Stoddart, *Nano Lett.*, 2004, **4**, 133–136.

Table 1 A comparison of the differences of [2]rotaxane (and [2]catenane) containing devices reported by the HP groups and the CIT/UCLA groups

Property/ Device	Switching amplitude	High cond. state lifetime	Therm. activated?	Stable write voltages	# cycles	Bottom electrodes	Controls switch?
CIT/UCLA	3–10[1]	10 min–1 h[2]	YES	Yes	10^1–10^3	Si with native SiO_2 or SWNTs[4]	NO[5]
HP	10^2–10^5	> 125 days	NO	NO[4]	> 10^6	Metal (Pt)	YES[6]

2 C. N. Lau, D. R. Stewart, R. S. Williams and M. Bockrath, *Nano Lett.*, 2004, **4**, 569–572.
3 C. P. Collier, G. Mattersteig, Y. Li, E. W. Wong, K. Beverly, J. Sampaio, F. Raymo, J. F. Stoddart and J. R. Heath, *Science*, 2000, **289**, 1172–75.
4 D. W. Steuerman, H.-R. Tseng, A. J. Peters, A. H. Flood, J. O. Jeppesen, K. A. Nielsen, J. F. Stoddart and J. R. Heath, *Angew. Chem., Int. Ed.*, 2004, **43**, 2–7.
5 A. H. Flood, J. F. Stoddart, D. W. Steuerman and J. R. Heath, *Science*, 2004, **306**, 2055–2056.
6 J. W. Choi, A. Flood, D. W. Steuerman, S. Nygaard, A. Braunschweig, N. N. P. Moonen, B. W. Laursen, Y. Luo, E. DeIonno, A. J. Peters, Jan O. Jeppesen, J. F. Stoddart and J. R. Heath, *Chem. – Eur. J.*, DOI: 10.1002/chem.200500934.
7 H. Yu, K. Beverly, H.-R. Tseng, J. F. Stoddart and J. R. Heath, *Angew. Chem., Int. Ed.*, 2003, **42**, 5706–5711.
8 M. R. Diehl, D. S. Steuerman, H.-R. Tseng, S. A. Vignon, A. Star, P. C. Celestre, J. F. Stoddart and J. R. Heath, *ChemPhysChem*, 2003, **4**, 1335–1339.

Professor Lindsay: asked: The molecule contains polar groups—how do you know that no counterions are carried with it when the films are prepared?

Professor Heath replied: We don't know that and, in fact, it is very likely that the counterions are carried with the molecules—we see some evidence of that with XPS, *etc*. The molecules aren't just polar, they have a tetracationic group (the +4 charged cyclophane rings) and that group is accompanied by four PF_6^- anions. Whether there is some exchange of these counteranions with OH^- or not (from the aqueous subphase), I don't know—that is a hard question to answer. However, the electrochemical process itself doesn't really need counterions—the electrode materials themselves provide an image charge that is, in effect, a counterion. We reported evidence of that in our very first paper on reconfigurable switching—*i.e.* we showed that as the silicon electrode was cooled (and hence became less polarizable), the voltage required to cycle the device between the high and low conductance states shifted by nearly 1 V.[1] This would not be observed with metal electrodes, since the polarizability of a metal electrode is not really temperature dependent.

1 C. P. Collier, G. Mattersteig, Y. Li, E. W. Wong, K. Beverly, J. Sampaio, F. Raymo, J. F. Stoddart and J. R. Heath, *Science*, 2000, **289**, 1172–1175.

Professor McCreery commented: I agree that an image charge can act as a "counterion" in a device, as we have direct spectroscopic evidence for reduction of a molecule to an anion which is stabilised by a positive image charge in a gold counter electrode. Stated differently, a high imposed electric field can cause a redox reaction, analogous to that occurring in an electrochemical double layer.

Professor Heath said: Yes, I of course agree. It think that we made a pretty good case for this in our first paper on a reconfigurable switch.[1]

1. C. P. Collier, G. Mattersteig, Y. Li, E. W. Wong, K. Beverly, J. Sampaio, F. Raymo, J. F. Stoddart and J. R. Heath, *Science*, 2000, **289**, 1172–1175.

Professor Calvo asked: How many cycles have you tested your molecular devices before it fails?

Professor Heath replied: A good device will cycle for about 1000 times. Most devices will cycle for far fewer times. This is not something that we have worked hard on improving, although we will likely be thinking hard about this issue within the near future.

Dr Schuster asked: What are the differences between the switch and hysteresis of switch?

Professor Heath answered: A switch can be bistable—moving between a high and low conductance state upon the application of a stimulus. However, once the stimulus is removed, it may revert back to its original state (or ground state). A hysteretic response implies that a negative stimulus must be applied to switch the molecular back to the original state—similar to how a ferromagnetic bit responds the application of a magnetic field. Typically that implies that there is some sort of activation barrier involved in cycling between the two states. For our devices, the high-amplitude molecular mechanical motion provides that activation barrier—typically about 22 kcal mol^{-1} or so within a device.[1]

1. J. W. Choi, A. Flood, D. W. Steuerman, S. Nygaard, A. Braunschweig, N. N. P. Moonen, B. W. Laursen, Y. Luo, E. DeIonno, A. J. Peters, Jan O. Jeppesen, J. F. Stoddart and J. R. Heath, *Chem. – Eur. J.*, DOI: 10.1002/chem.200500934.

Professor Larsson asked: What makes you decide to use a particular molecule in your circuits? For example I recognise the molecule TTF on one of your pictures. This is a well-known molecule in conducting molecular crystals, but since you use a single TTF molecule, you are probably interested in other than conduction properties in this case?

Professor Heath replied: The TTF (tetrathiafulvalene) group is not the key part—the structure of the molecules that we use as switches is what matters most. TTF molecular crystals are good conductors, but that is also not the point. What we mostly care about are good switches—which means that we care about molecules that can change their conductance when they are electrochemically switched and the electrochemical switching process has a significant hysteresis.

The class of molecules that we use (and which contain TTF groups) are bistable, donor–acceptor, electrochemically switchable [2]catenanes and [2]rotaxanes. Our molecules are designed to do a number of things. First, they have to be amphiphilic so that they will orient at an air/water interface. It is from this interface (a Langmuir–Blodgett film) that we fabricate our devices. Second, the molecules should be bistable and electrochemically switchable between the two isomers (co-conformers). The TTF group, which is readily oxidized, does play an important role here. The third issue is that we need to control the potential energy surface that describes the bistability of the molecules, and we do that through the molecular structure of the switches.[1] I believe that all of this is pretty well-described in the Faraday Discussion paper that I have written.

1. J. W. Choi, A. Flood, D. W. Steuerman, S. Nygaard, A. Braunschweig, N. N. P. Moonen, B. W. Laursen, Y. Luo, E. DeIonno, A. J. Peters, Jan O. Jeppesen, J. F. Stoddart, J. R. Heath, *Chem. – Eur. J.*, DOI: 10.1002/chem.200500934.

Professor Weber asked: Is the switching in a solid state device happening under constant volume conditions or constant pressure? Is this important for the thermodynamic picture you presented?

Professor Heath answered: Typically constant volume and constant pressure conditions are most important for gas phase processes. Solution phase and other condensed phase process are almost always assumed to be at constant volume or constant pressure (which is the case for our systems). Thus, I don't think it is important. Constant temperature is important, and yes, we only varied temperature where it I specifically noted that we did.

Professor Weber asked: When the molecular film changes its structure during switching, would you consider the top layer–substrate distance as unchanged?

Professor Heath replied: Yes, I think that is probably the case. Thomas Bjørnholm's group has done some measurements (work in press to *Angewandte Chemie*, I think) that can speak to just this issue, although a truly direct quantitation has not been done. In addition, Goddard's group has calculated the structure of the ground state and metastable state co-conformers when they are sandwiched as a Langmuir monolayer between two electrodes.[1] Those results are, I believe, consistent with what Bjørnholm's group has observed.

1. S. S. Jang, Y. H. Jang, Y.-H. Kim, W. A. Goddard, III, A. H. Flood, B. W. Laursen, H.-R. Tseng, J. F. Stoddart, J. W. Choi, D. W. Steuerman, E. DeIonno, J. R. Heath, *J. Am. Chem. Soc.*, 2005, **127**, 1563–1575.

Professor Weaver said: The architectures proposed seem to be driven by the inadequacy of conventional lithography. As modern lithographic pattern definition can define structures to 23 nm period, does this have any impact on the methodology proposed? (using a periodically defined electrode array).

Professor Heath replied: Sure, but I have yet to see any sort of general lithographic method (*i.e.* can fabricate any arbitrary pattern) that can come close to a 20 nm pitch. Our SNAP process can exceed 20 nm pitch—we have demonstrated 16 nm,[1] and, as you have seen, it produces very high quality nanowires, including Si nanowires. I haven't seen any patterning method that can come close to this sort of perfection in fabrication. In fact, our colleagues at Hewlett Packard have repeatedly challenged several excellent groups who specialize in doing state-of-the-art electron-beam lithography to prepare patterns for them that could be utilized to fabricate crossbar memory circuits (out of metals, not silicon (silicon is tougher)) at pitches narrower than 60 nm. No one has yet been successful in meeting their challenge.

1. N. Melosh, A. Boukai, F. Diana, B. Geradot, A. Badolato, P. Petroff and J. R. Heath, *Science*, 2003, **300**, 112.

Dr Higgins communicated: Please could you outline briefly your evidence for the "innocence" of Ti deposition in these systems? Given DFT and experimental evidence for Al reactivity with conjugated organic materials in organic LED top contacts *etc.*, I find it very surprising that Ti does not affect these complex (and quite oxygen-rich) molecules.

Professor Heath communicated in reply: There is a growing body of literature indicating that Ti deposition can form a film that serves as a good electrical interface that is also largely non-destructive for certain molecules, although much depends on the nature of the molecular film and on the nature of the Ti deposition process. For low-density films such as certain SAMs assembled onto Au surfaces, Ti deposition is apparently destructive.[1] This is likely due to the much greater area per molecule for those SAMs. For example, a SAM formed from one of our [2]rotaxanes occupies just under 400 Å^2 molecule^{-1}, while a typical amphiphilic [2]rotaxane, compressed to π = 30 mN m^{-1} on a Langmuir trough (a typical transfer pressure for incorporating into a device) occupies 90–130 Å^2 molecule^{-1}.[2,3] Most of the molecular structure within a low-density [2]rotaxane SAM would be exposed to an incident flux of Ti, and thus destroyed by reactions with those atoms.

Highly compressed Langmuir monolayers, however, do apparently survive the deposition of Ti. Allara's group published some early work in which they carried out a series of X-ray photoelectron (XPS) studies on organic monolayers before and after Ti had been evaporated.[4] More recently Chang and co-workers published a reflectance-absorption infrared (RAIR) spectroscopic study of a molecular monolayer both before and after evaporation of 2 nm Ti.[5] Both groups found evidence that much of the molecular monolayer survived the deposition process. The most thorough study to date has been carried out by McCreery's group.[6] They investigated the nature of 1–3 nm thick Ti films evaporated on top of a thin nitroazobenzene (NAB) film (thickness controlled from a submonolayer to 4.5 nm) through the use of Raman spectroscopy. While much of the Raman spectrum of the 4.5 nm NAB film remained intact upon deposition of the Ti, they did find evidence for reaction between the Ti and the terminal NO_2 group, although those Raman signatures significantly recovered over a period of several days. For submonolayer films, the changes observed in the Raman spectrum were more significant, as would be expected from the above discussion of SAMs. Equally interesting is that no new spectral features were observed after Ti deposition on azobenzene. McCreery's group concluded that the reaction between Ti and the nitro group appears to prevent short circuits that can form from the inclusion of Ti into the molecular film, and that the Ti forms a covalent bond between the electrode and the molecular film. We have begun to carry out similar measurements on our [2]rotaxane and [2]catenane monolayers, and our results are so far consistent with what McCreery's group observed. We definitely observe that most of the RAIR spectral signatures of the molecular monolayer survives the deposition process, but we are still working out the details of what we believe happens.

We have reported that the titanium adhesion layer may be replaced with chromium,[7] although the resultant performance characteristics of MSTJs are inferior. Cr, like Ti, can react rapidly to form a Cr–C interface.[8] McCreery has investigated Cu, and many groups have investigated Au and other materials for different class of molecular electronic devices.

1 B. de Boer, M. M. Frank, Y. J. Chabal, W. Jiang, E. Garfunkel and Z. Bao, *Langmuir*, 2004, **20**, 1539–1542.
2 S. S. Jang, Y. H. Jang, Y.-H. Kim, W. A. Goddard, III, A. H. Flood, B. W. Laursen, H.-R. Tseng, J. F. Stoddart, J. W. Choi, D. W. Steuerman, E. DeIonno, J. R. Heath, *J. Am. Chem. Soc.*, 2005, **127**, 1563–1575.
3 S. S. Jang, Y. H. Jang, Y.-H. Kim, W. A. Goddard III, J. W. Choi, J. R. Heath, B. W. Laursen, A. H. Flood, J. F. Stoddart, K. Norgaard and T. Bjørnholm, "Molecular Dynamics Simulation of Amphiphilic Bistable [2]Rotaxane Langmuir Monolayer at the Air/Water Interface," *J. Am. Chem. Soc.*, in press, 9/2005.
4 K. Konstadinidis. P. Zhang, R. L. Opila and D. L. Allara, *Surf. Sci.*, 1995, **338**, 300.
5 S.-C. Chang, Z. Li, C. N. Lau, B. Larade and R. S. Williams, *App. Phys. Lett.*, 2003, **83,** 3198.
6 A. M. Nowak and R. L. McCreery, *Anal. Chem.*, 2004, **76**, 1089–1097.
7 C. P. Collier, G. Mattersteig, Y. Li, E. W. Wong, K. Beverly, J. Sampaio, F. Raymo, J. F. Stoddart and J. R. Heath, *Science*, 2000, **289**, 1172–75.
8 D. R. Jung and A. W. Czandema, *Crit. Rev. Solid State Mater. Sci.*, 1994, **19**, 1–54.

Professor Naaman opened the discussion of the papers by Professor Ashwell and Professor McCreery: There is an organization induced charge transfer when SAM is formed. Hence the dipole moment of the adsorbed molecule is not the same as that of the isolated one.

Professor: McCreery commented: Indeed, we have to keep in mind that these are not isolated molecules, and that their dipoles and/or molecular orbitals will be affected by both orientation and interactions with the contacts and each other. Our approach to this question is to use *in situ* spectroscopy when possible to probe electronic structure in completed junctions.

Professor Ashwell added: I agree but there are many possible structural forms and, when connected to the substrate *via* a flexible S–C_nH_{2n} bridge, overlap of the donor and acceptor groups of adjacent molecules is feasible if there is sufficient tilt. However, the direction of electron flow at forward bias from all reported rectifying SAMs is consistent with that predicted by Aviram and Ratner with rectification in opposite quadrants of the *I–V* plot when molecules are linked *via* the donor or acceptor end: Au–S–C_nH_{2n}–D–π–A and Au–S–C_nH_{2n}–A–π–D. It suggests the molecular origin of the rectification and, where protonation disrupts the donor–acceptor combination, it is also verified by the rectifying behaviour being respectively suppressed and restored upon exposure to acid and base.

Professor Lindsay said: Professor Naaman's comment about charge-transfer in response to accumulations of a surface dipole does not seem to be consistent with electric force microscopy results that show that in alkanethiol tethered systems, the change in dipole per unit area measured correlates precisely with the change in dipole of the head group.

Professor Naaman responded: It is important to realize that the cooperative effect I described is a result of not having contacts to the molecules on both sides. Namely, it is a result of the dipole layer having to hold a very large electrostatic energy. When a second contact is made, the polarization of the metal contact reduces the effect of the dipole. The cooperative effect is manifested clearly in the fact that the change of the work function as a result of the dipole layer is by far lower than expected based on the dipole layer of the isolated molecule. In addition it is important to realize that in the electric force microscopy studies a very large field is applied locally on the molecular system. This field perturbed the system so that one does not expect a relative weak cooperative effect to be expressed in these studies.

Dr Rampi addressed Professor Ashwell: Polar molecules organise in SAM with a tilt angle that depends on the polarity and tend to neutralise the dipole.

Professor Ashwell replied: It applies to all types of organised molecular films and in LB monolayers of the extensively studied zwitterion, $C_{16}H_{33}$-Q3CNQ, the molecules adopt three different structural forms dependent upon the conditions of formation. They exhibit different spectra with absorption maxima at 565 nm for an SHG-active purple form and at 610 and 670 nm for SHG-inactive turquoise forms. Only the non-centrosymmetric purple form exhibits rectifying behaviour.

Professor Heath asked Professor McCreery: Your Raman result showed that your molecule junctions are compromised by +3 V. Is this true for all top electrode materials? Is this a field or a voltage effect?

Professor McCreery replied: I would not use the word "compromised" as that implies degradation. We have observed redox activity with TiO_x and AlO_x top contact materials, some of which is reversible. With Cu or Au, the electron transport through the molecule is much faster, and apparently too fast for oxidation or reduction to occur. The oxides prevent rapid electron transport, and presumably a large electric field is developed across the molecule. We speculate that the redox reactions result from the electric field rather than the applied potential, since there is no reference electrode to establish the applied voltage relative to some vacuum reference level. The electric field across the molecule is in the region of 10^6 V cm^{-1}, which is certainly comparable to the electrochemical double layer that drives electrochemical oxidations and reductions. However, without having a reference potential, it is difficult to determine where the PPF potential lies relative to the redox level or LUMO of the adjacent molecule.

Professor Weber addressed Professor Ashwell: You showed I–V plots up to ±1 V, which were flat at positive bias and had a steep slope at ~−0.8 V. What happens at higher voltage? Is the rectification still there or is only the threshold asymmetric?

Professor Ashwell replied: The current–voltage plot characteristics were investigated from −1 V to +1 V. We have investigated other films to higher voltages and the rectification persists.

Dr Nichols asked: One may expect that the rectification ratio would depend strongly on tunnelling parameters (tip–surface distance in particular). Did you measure I–V curves at different tip–sample distances and did you see a dependence? Coupling at large tip–surface distance to the molecular state would be expected to be very weak.

Professor Ashwell replied: We have routinely investigated the tip–surface distance of a number of samples, both inherently rectifying and inherently non-rectifying. There are small variations in the behaviour and it is possible to induce slight electrical asymmetry but such variations are insignificant compared with the high rectification ratios now obtained.

Professor Hush said: The molecule displaying rectification has an aliphatic tail. Although this is shorter than in previous similarly described structures, it nevertheless introduces marked asymmetry. Theoretical considerations suggest that this feature will contribute to rectifying behaviour. What is known about the relative importance of this and a charge-transfer mechanism?

Professor Ashwell replied: The length of the aliphatic tail appears to have little or no effect on the rectification ratio for typical lengths used in organised molecular films. We have investigated the dependence for various Au–S–C_nH_{2n}–D–π–A and Au–S–C_nH_{2n}–A–π–D structures where $3 \leq n \leq 12$ and have found no significant change. The rectifying properties of these films may be suppressed by protonation, which disrupts the donor–acceptor combination, and then restored upon exposure to base. The I–V curves of the protonated form are symmetrical despite the aliphatic tail and thus, it is difficult to accept the recent theoretical considerations.

Professor Schiffrin commented: Asymmetry can easily result from the presence of an ionic junction (see for example, Fig. 7 of the paper). This could give rise to a large surface potential drop across the ionic junction. The current rectification ratio in Fig. 8 of the paper could simply result from a shift in surface potential of ~200 mV. What is the evidence that the properties measured in this case are due to the molecule and not to the ionic distribution within the junction?

Professor Ashwell replied: The cationic dye is aligned by an anionic surface: $Au–S–C_3H_6–SO_3^-$ $A^+–\pi–D$. Extended studies, as yet unreported, show the rectification ratio to be dependent upon the $A^+–\pi–D$ moiety being sterically hindered and therefore non-planar to maintain the integrity of the acceptor and donor. Ratios as high as 500 at ±1 V have been achieved for a sterically hindered cationic dye and significantly, non-polar cations exhibit symmetrical *I–V* curves when coupled in this way. The unreported data remove the ambiguity.

Dr Nichols asked: Have you considered whether reversible bias voltage induced changes in orientation could (in part) give rise to rectification of your molecular junctions. Could these play a role in the systems you present here or is this rather unlikely? I would imagine that the *in situ* Raman spectroscopic technique of Professor McCreery would be suitable for studying such molecular changes with bias voltage in your system.

Professor Ashwell replied: Bias induced changes of the molecular orientation are feasible for LB structures where molecules are untethered but, except for minor changes in tilt, unlikely for SAMs where they are chemisorbed. They could play a significant role but only for isolated or loosely packed molecules that are physisorbed, as close packing is likely to suppress molecular flipping as the bias is reversed.

Professor Kornyshev asked: What happens with the structure of the LB-film when you move along the rectification axis? Is there tilting or even corresponding reorientations of the molecules? It seems that you don't have any independent experimental characterisation of the structural changes on the film with potential bias. For theoretical interpretations of your rectification curve it would be very important to have such data.

Professor Ashwell replied: There is no independent experimental characterisation of the structural changes with potential bias but significantly, similar *I–V* characteristics have been obtained for multiple scans on the same spot. It suggests that the alignment is stable. Molecular diodes are still in their infancy and, at this stage, I believe that it is far more important to identify the criteria for rectification and experimentally enhance the rectification ratio. Our recent work has resulted in a ratio of 3000 at ±1 V, which is higher than many reported values for oxide-induced Schottky barriers. Theoretical interpretations will follow.

Dr Nichols asked Professors Ashwell and McCreery: Could you use a Raman spectroscopic probe, as applied by McCreery, to analyze V_{bias}-induced orientational switching?

Professor Ashwell replied: The majority of self-assembled structures investigated at Cranfield exhibit stable *I–V* characteristics for multiple scans on the same spot and there appears to be no significant bias-induced orientational switching.

Professor McCreery added: In principle, Raman can determine orientation on surfaces, as we have reported in a past publication,[1] and as many authors have described in the SERS literature. The requirements are that the molecule in question have a sufficiently large Raman cross section and number density to detect adequate signal. Without electromagnetic field enhancement (*i.e.* SERS), only resonance enhanced Raman scatterers are likely to meet these requirements. So I would say it is possible that Raman could be used in Langmuir–Blodgett and SAM structures, but it would be challenging. An example of applications to "switching" in carbon/nitroazobenzene/TiO_x/Au junctions in our lab has appeared recently.[2]

1 J. Zhao and R. L. McCreery, *Langmuir*, 1995, **11**, 4036.
2 A. M. Nowak and R. L. McCreery *J. Am. Chem. Soc.*, 2004, **126**, 16621–16631.

Professor Rovira asked: Can the change in the metal of the phthalocyanine affect the rectification behaviour due to the change in the redox behaviour? Could the huge rectification observed be due to a redox process in the phthalocyanine?

Professor Ashwell replied: It is too early to tell. The rectification ratio of 3000 at ±1 V was achieved about two months ago and the preliminary study has focused upon modifications to the cationic D–π–A dye rather than its copper phthalocyanine-3,4′,4″,4‴-tetrasulfonate anion.

Dr Nichols asked: Can you comment on the influence of other dipole barriers (*e.g.* surface dipole, water films on tip) on the recorded *I–V* response?

Professor Ashwell replied: These appear to have minimal effect on the shape of the *I–V* curve and rectification ratio. For example, isomeric D–π–A dyes with the self-assembling substituent located at different ends exhibit similar properties but with rectification in opposite quadrants of the *I–V* plot. Furthermore, the electrical asymmetry may be switched off and on by protonating and deprotonating the film, this reversible chemical switching confirming that the D–π–A chromophore dictates the properties.

Professor Fitzmaurice addressed Professor McCreery: How are the NAB molecules anchored on the carbon substrate?

Professor McCreery replied: By electrochemical reduction of a diazonium derivative of NAB, using the procedure originally described by Delamar *et al.*[1] We have studied this reaction extensively, as have several other groups, and characterized the resulting surface with Raman, XPS, AFM, *etc.* The bonding between the NAB and the carbon is most likely a phenyl–phenyl bond, and the coverage is very high due to the reactive nature of the phenyl radical generated by reduction of the diazonium reagent.

1 M. Delamar, R. Hitmi, J. Pinson and J. M. Saveant, *J. Am. Chem. Soc.*, 1992, **114**, 5883.

Professor Fitzmaurice asked: Should the model of the systems discussed explicitly include the two interfacial layers?

Professor McCreery answered: We strongly agree that the electronic response of the junction is dependent on the properties of both the contacts and the molecule. Strictly speaking, any model should consider the junction to be a single electronic system, with possibly strong interactions between the molecules and the conductors across the respective interfaces. Both the bonding and the electronic coupling across the various interfaces may have significant effects on junction behaviour, in addition to the structure of the molecule itself.

Professor Calvo said: My question refers to the interfaces in your experiments. The PPF surface seems to be pretty flat, though not as flat as HOPG since you have evidence of C–C bonds at the ring edges. The upper interface is produced by evaporation on e-beam definition. I wonder if you have evidence of the metal or oxide crystal size for the film making contact with the molecular layer. Can you compare the contact to the nanoparticle–molecule (in Professor Lindsay's paper) or to the Hg drop–molecule (in Professor Rampi's paper) interfaces?

Is there any evidence of Au evaporation penetrating the porous contact layer and making direct contact to the molecules?

Professor McCreery replied:

1. PPF resembles glassy carbon: it consists of "ribbons" of sp^2 carbon which form a disordered tangle of graphitic regions with crystallite sizes of ~50 Å. AFM shows that the PPF surface has an rms roughness of <5 Å, presumably due to the slow pyrolysis which forms the surface. This is not atomically flat like HOPG basal plane, but the disorder of PPF provides both higher conductivity normal to the carbon surface and many sites for the molecule to attach. Modification of HOPG basal plane by the same route produces "mushrooms" due to nucleation at defects and low basal plane reactivity.

2. We have no direct evidence for the structure of the electron-beam deposited layers. We do have evidence (XPS) for bonding of Ti or Cu to nitro-containing monolayers such as NAB. Lindsay's case is a Au–S bond, while Hg presumably forms a physisorbed contact. We presume that the

electron-beam metals form physisorbed contacts unless there is the possibility for a covalent interaction.

3. We have made PPF/NAB/Cu and PPF/NAB/Au junctions, so we know that direct metal contact with NAB produces much higher conductance and qualitatively different behaviour from the TiO_x or AlO_x junctions. So if Au were penetrating the oxide, we would expect electronic behaviour like these junctions, and that is not the case. Furthermore, we have also made PPF/AlO_x/Au and PPF/TiO_2/Au junctions, and these show high resistance. So there is no evidence for Au penetration through the oxide.

Professor Heath commented: We have made few-nanometre-wide trenches as model pinholes. We have found that it is very difficult to deposit metals like Ti into those pinholes—it typically forms a bridge over the pinhole, perhaps extending down 1 nm or less, even for 10 nm deep pinholes. We physically looked at the penetration of the metal using electron microscopy imaging of the side of the pinhole.

Professor Barbero addressed Professor Ashwell: It would make sense from the point of view of SHG, *i.e.* how the field will develop in the multilayer. A related question is: if metal nanoparticles could be assembled, would be possible to create nearly macroscopic contacts by self-assembly?

Professor Ashwell replied: It is possible to self-assemble nanoparticles on the surface as long as the molecule has an appropriate functional group to chemisorb the metal.

Professor Bjørnholm addressed Professors Ashwell and McCreery: What is the independent structural evidence you have that allows you to relate the molecular structures known from, say, NMR, to the *I–V* characteristics you measure in your devices?

Professor Ashwell responded: The orientation of D–π–A molecules is controlled by locating a thiol or thioacetate substituent on the donor or acceptor side for chemisorption on gold and similarly, a hydrophobic substituent for alignment and orientation in LB films with deposition on the upstroke resulting in the chromophore being adjacent to the substrate as long as the latter is sufficiently hydrophilic. The orientation is then simply indicated by the wetting contact angle and the forward bias direction of electron flow, in each case, is from the electrode to LUMO of the acceptor on one side of the device and from the HOMO of the donor to electrode on the other. Significantly, LB monolayers of differently oriented isomers in which a hexadecyl tail is attached on the acceptor side in one (D–π–A–$C_{16}H_{33}$) and on the donor side in the other ($C_{16}H_{33}$–D–π–A) exhibit rectification in opposite quadrants of the *I–V* plot. The direction of electron flow is as predicted by the Aviram and Ratner model.

Professor McCreery responded: NMR is exceedingly difficult on ~20 Å thick molecular layers on carbon, although we routinely use NMR to verify the structures of our precursors and diazonium reagents. For our modified PPF, we have used XPS, Raman, AFM, and FTIR to characterize molecular structure after surface bonding, and other groups have used these techniques plus STM and SIMS. Recent experiments with FTIR permit the orientation of the NAB to be determined on the PPF surface, and it has an average tilt angle of 31° relative to the surface normal. All these techniques confirm covalent bonding to primarily edge sites, with nearly perpendicular orientation, and with a thickness verified with AFM.

Dr Dryfe asked: Given that the nitroazobenzene reduction is chemically irreversible, does this irreversibility manifest itself as a change in the current–voltage response, *e.g.* if the sample bias were repeatedly cycled? Would there be a difference between this case and the non-nitro-substituted case?

Professor McCreery replied: As is apparent in Fig. 3B, the azobenzene junction does not differ significantly from that of the NAB junction. We did this experiment to determine if the nitro group was a primary determinant of junction conductance, and at least at first consideration it is not. However, we have not examined the *I*/*V* behaviour of NAB junctions following the irreversible reduction of the nitro group. The *I*/*V* curves do not differ dramatically from those observed with a fresh NAB junction, although there is a ~30% decrease in the current observed above +1 V.[1]

1 A. Pullen, PhD thesis, Ohio State University, Columbus, OH, 2004, p. 116.

Dr Dryfe asked: Could *in situ* EPR spectroscopy be used to detect radical anions formed in the outside junction cases?

Professor McCreery replied: EPR may have sufficient sensitivity, but I am unaware of its application to flat surfaces. Remember that disordered carbon materials often have a significant population of unpaired electrons, which show up as a broad EPR signal. So I think it would be hard to see a monolayer on top of a several micrometre thick PPF layer.

Professor Lindsay opened the discussion of Professor Remacle's paper: Just to clarify—you mean that the simulations do not meet the McConnell model, and not that there is real charge localization on the molecule?

Professor Remacle replied: The short answer is that whether the system is or is not in the McConnell super-exchange limit is, in our approach, not only a property of the system but is also determined by the particular experimental conditions. The primary variable of interest is the source–drain voltage. The higher it is the more likely is that the transferred charge is actually transmitted *via* the molecular backbone.

Let me elaborate. To determine the temporal history of the charge migration we start our time-dependent simulations with a non-stationary initial electronic state that is localized on one side of the contact. That is, the initial state is localized on those gold atoms that are directly coupled to the bulk electrodes and to the molecule. A non-stationary initial state must have a finite width in energy. The point is how large is that width in energy of the initial state.

When the non-stationary initial state is so broad in energy that it does span several MO's, our simulations show real charge localization on the molecule, as the electron (or hole) migrates between the two electrodes.

However, the superexchange limit can be realized if the initial state can be both localized and yet be not too broad in energy. Such is the case of saturated alkanes. A narrow in energy yet localized state spans only the two highest occupied MO's that are quasi-degenerate at zero field and localized at the two Au–S interface without significant charge density on the bridge (see Fig. 9 of our paper). In this case and when the source–drain voltage is low, so that only these two orbitals fall within the Fermi window, the population transfer will occur without any significant charge localized on the bridge. BUT, even for alkanes, if the initial state is broader in energy then, as we show in Fig. 4 of our paper, there will be charge localized on the bridge.

In conclusion. If a given junction is or is not in the superexchange limit depends also on how many and on the nature of the states that can be brought into the Fermi window.

Professor Lindsay then asked: Small molecules are in disagreement with Landauer-type calculations—Ghosh has calculated effects of electron–electron interactions, and concludes that they suppress transmission in small molecules—have you looked at many-body effects?

Professor Remacle responded: We have not looked yet at many-body effects on molecular transport. We however compute the electronic structure at DFT level which includes some of the electron–electron correlation in the exchange functional but amounts to treating electron–electron interactions as a self-consistent field.

Note however that we always compute the current in the same way, using the DFT functional B3LYP to determine the one electron molecular electronic structure. Yet when compared with experimental results, for some molecules, our computed current can be too low by one to three orders of magnitude while it can be too high by the same factor for others. It is therefore not clear that many-body effects can account for the discrepancies between computed and measured currents.

On the other hand, we have looked at many-body effects on the insulator to metal transition induced by mechanical compression in small arrays of self-assembled colloidal Ag quantum dots (QD)[1–3] as studied by Professor Heath and his group.[4–6] For these systems, we compute the many-electron electronic structure by a full-CI diagonalization of a semi-empirical Pariser–Parr–Pople electronic Hamiltonian which includes both on site and intersite Coulombic repulsion. We also include in the electronic Hamiltonian the fluctuations in the dot energies induced by the size dispersion of the dots. At short interdot separation, the exchange coupling is strong enough to overcome both the many-body Coulombic effects and the effects of the fluctuations in the site energies induced by the size dispersion of the dots. As a result, the array is metallic. At large interdot separation, the exchange coupling is not strong enough and the wave function is localized. The array becomes an insulator. Depending on which effect is the strongest, the fluctuation in the dot energies or the Coulomb blockade, the metal to insulator transition is of an Anderson type or of a Mott type. When the metal to insulator transition is of a Mott type, the current is suppressed by many-body effects.

1 F. Remacle and R. D. Levine, *J. Am. Chem. Soc.*, 2000, **122**, 4084.
2 F. Remacle and R. D. Levine, *ChemPhysChem*, 2001, **2**, 20.
3 K. C. Beverly, J. L. Sample, J. F. Sampaio, F. Remacle, J. R. Heath and R. D. Levine, *Proc. Natl. Acad. Sci., USA*, 2002, **99**, 6456.
4 C. P. Collier, R. J. Saykally, J. J. Shiang, S. E. Henrichs and J. R. Heath, *Science*, 1997, **277**, 1978.
5 S. Henrichs, C. P. Collier, R. J. Saykally, Y. R. Shen and J. R. Heath, *J. Am. Chem. Soc.*, 2000, **122**, 4077.
6 G. Markovich, C. P. Collier, S. E. Henrichs, F. Remacle, R. D. Levine and J. R. Heath, *Acc. Chem. Res.*, 1999, **32**, 415.

Professor Larsson asked a general question: What do you mean by superexchange?

Professor Remacle replied: Superexchange is a longer-range charge transfer mediated by a non-resonant bridge between the donating and accepting states. This limit is realized when the effective donor/molecular bridge and acceptor/molecular bridge couplings are small, but the donor and acceptor sites are energetically quasi-degenerate so that they are effectively coupled by higher order terms in perturbation theory.[1,2] When such is the case, the charge is transferred from the donor site to the acceptor site without significant population of the bridge. We have shown[3] that this limit was realized in self-assembled arrays of Ag colloidal QD's in the intermediate mechanical compression regime, which was later confirmed experimentally.[4] As we discussed in our paper and in our answer to Professor Lindsay's opening question, the superexchange mechanism is not only a property of the molecular system but also depends on the width in energy of the initial non-stationary state created by the transfer of an electron from the bulk electrode to the extended system.

For dithio molecular wires tethered between two small gold clusters, the effective coupling at the Au–S–C interface is typically too strong for the superexchange limit to be realized, except for saturated bridges when the Fermi window is narrow enough so that only the two highest occupied MO's localized at the interface at each end of the bridge participate in conduction.

1 H. M. McConnell, *J. Chem. Phys.*, 1961, **35**, 508.
2 G. C. Schatz and M. A. Ratner, *Quantum Mechanics in Chemistry*, Prentice-Hall, New York, 1993.
3 F. Remacle and R. D. Levine, *J. Phys. Chem. B*, 2001, **105**, 2153.
4 J. L. Sample, K. C. Beverly, P. R. Chaudhari, F. Remacle, J. R. Heath and R. D. Levine, *Adv. Mater.*, 2002, **14**, 124.

Professor Kornyshev asked: How is the difference between hopping and superexchange manifested in the nonlinear current–voltage plots?

Prfoessor Remacle responded: The nonlinearities in the current–voltage plots arise when molecular levels with different conducting character are brought in the Fermi window by varying the source–drain or the gate voltage. As such, they will arise in the I–V_{sd} curves both for conduction by hopping or superexchange. The main difference in the I–V_{sd} curves will be in the magnitude of the current. The superexchange limit is realized for conducting levels localized at the junction when the coupling at the junction is very weak, so that the magnitude of the current is very low, picoamperes in our simulations for saturated alkane dithiols (see Fig. 15 of our paper). On the other hand, the computed current is in the nanoampere to microampere regime when the hopping mechanism operates (see Fig. 12 of our paper for the I–V_{sd} curves computed for dithiolbutadiene). Note that the I–V_{sd} curves will be linear (Ohmic) in strong coupling limit.

The difference between the hopping and the superexchange mechanism should clearly appear if time-resolved current measurements were available. We show in our paper that the matrix element of the transition operator can be related by a Fourier transform relation to the crosscorrelation of the initial state created at the interface by the electron transfer from the electrode, see eqn. (16) of our paper. This was our motivation to compute the time-dependent charge transfer along the chain.

Dr Geskin asked: In superexchange there is exponential decay of conductance with distance. Is this dependence conserved in the sequential mechanism you find?

On the one hand, weak coupling leads to superexchange. On the other hand, weak coupling can lead to incoherent hopping or Coulomb blockade. Is it possible to predict which of these possibilities will be realized?

Professor Remacle replied:

1. As I discuss in my answer to Professor Lindsay's opening question, for a given molecule, superexchange or sequential mechanism depends on the width in energy of the localized non-stationary initial state that is created by the transfer of an electron from the bulk electrode to the gold atoms at the one side of the extended system.

The sequential mechanism operates when the width of the non-stationary initial state is broad enough to span several conducting MO's, and is accompanied by charge localization on the molecular backbone.

The superexchange limit is realized for a non-stationary initial state narrow in energy, and for conducting MO's localized on the gold atoms that are coupled to the electrodes. This is the case for saturated alkanes when the Fermi window is narrow enough so that the two states that participate in charge transport are the two highest occupied MO's, localized each at a given end of the extended system.

We have analyzed the length dependence of the current computed for dithiapropane, -hexane and -nonane in a previous paper,[1] and as can be seen from Fig. 1, in the low voltage limit where the superexchange mechanism operates, we recover an exponential decrease of the current with the length of the wire. Note however, that, as also found experimentally,[2,3] the value of the decay constant, β, decreases when the voltage increases. We get a value of β = 1.05 Å^{-1} at V_{sd} = −0.4 Volt. When V_{sd} increases further, more states participate to conduction and the I–V_{sd} curves exhibit nonlinearities. The mechanism for transport becomes sequential and because of the nonlinearities (see Fig. 1), it is no longer possible to define a value for β.

2. The temperature dependence of the current should be different in the two limits. We have shown in our work on transport properties of self-assembled arrays of QD's[4–6] that in the superexchange coupling limit the conductivity rises with temperature as $\exp(-(T_0/T)^{-1/2})$. In the Ohmic, activated regime, the dependence is $\exp(-T_{act}/T)$. In the limit of incoherent hopping, the current should decrease when increasing temperature.

1 F. Remacle, I. Willner and R. D. Levine, *J. Phys. Chem. B*, 2004, **108**, 18129.
2 W. Y. Wang, T. Lee and M. A. Reed, *Phys. Rev. B*, 2003, **68**, 035416.
3 X. D. Cui, A. Primak, X. Zarate, J. Tomfohr, O. F. Sankey, A. L. Moore, T. A. Moore, D. Gust, L. A. Nagahara and S. M. Lindsay, *J. Phys. Chem. B*, 2002, **106**, 8609.
4 F. Remacle, K. C. Beverly, J. R. Heath and R. D. Levine, *J. Phys. Chem. B*, 2002, **106**, 4116.
5 F. Remacle, *J. Chem. Phys.*, 2003, **119**, 4550.
6 F. Remacle, K. C. Beverly, J. R. Heath and R. D. Levine, *J. Phys. Chem. B*, 2003, **107**, 13892.

Professor Naaman asked:

1. How is the conduction affected by the fact that only a small gold cluster is included in the calculation?

2. Band structure has been observed in two photon photoemission from monolayers of alkylthiols. How does this observation coincide with the calculations?

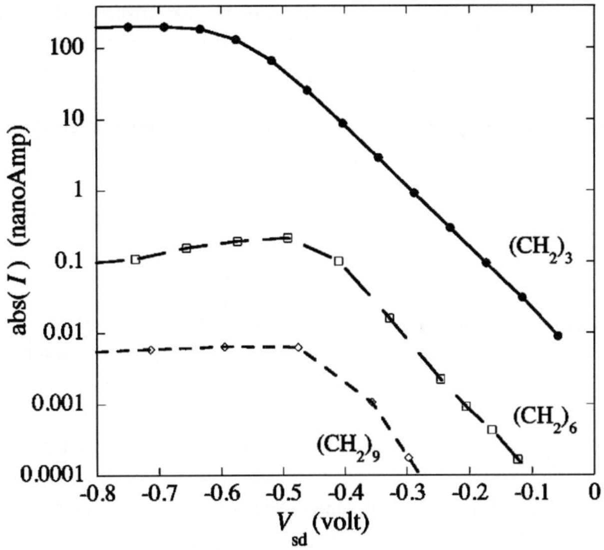

Fig. 1 Absolute value of the current (in nanoamperes on a log scale) *vs.* applied source–drain negative bias computed for dithiapropane, -hexane and -nonane tethered between two Au_3 clusters. Note how for $V_{sd} > -0.6$, the I–V_{sd} curves become nonlinear. For positive bias (not shown), the nonlinearities appear at lower values of V_{sd}.

Professor Remacle replied:

1. Including a small gold cluster on the left and on the right of the dithio molecule allows us to take into account the effect of the Au–S interface on the electronic structure of the molecule. For conjugated dithio molecules, we find that the highest occupied and lowest unoccupied MO's are well delocalized over the entire Au_3–S–molecule–S–Au_3 system while for saturated ones, the two highest MO's are localized at the Au–S–C interface. In that case, current flow is possible *via* a super-exchange mechanism because these two highest occupied MO's are coupled by the field induced by the source–drain bias.

When the cluster size increases sufficiently, the density of electronic states of the extended system increases into a quasi continuum of states but remains discrete: In addition to MO's localized at the interface or delocalized over the entire system, there are also MO's localized only on gold atoms, as was shown by Weber *et al.*[1] who investigated the electronic structure of conjugated aromatic dithio molecules bound to the left and to the right to a Au_{29} cluster. They found that conduction occurs through the HOMO, that is close in energy to the gold Fermi level and delocalized over the gold atoms and the molecule, which is similar to what we find with a smaller cluster size. In addition, their results show that the lowest unoccupied MO's above the HOMO are localized entirely on the gold cluster.

For the current to flow, whether the size of the gold cluster is big or small, one needs to couple the extended system to the continuum of the electrodes on the left and on the right. In our computation, we included this coupling in the weak coupling limit. When the coupling to the electrodes is strong, one needs to add a self-energy term to the electronic Hamiltonian. This term will both shift the position of the conducting levels and broaden them into Lorentzians. We expect that more than the size of the cluster, this term can induce significant changes in the conduction properties of the molecular wire. These effect are discussed in a Comment below.

2. This question is partially answered above. This observation suggests that the coupling to the electrodes is not weak. There are indeed others who say the same. In that case, one needs to include this coupling into the electronic Hamiltonian as a self-energy term. We discuss how to do so and the effect of a strong coupling to the electrodes in the following communication.

1. H. B. Weber, J. Reichert, F. Weigend, R. Ochs, D. Beckmann, M. Mayor, R. Ahlrichs and H. v. Lhoneysen, *Chem. Phys.*, 2002, **281**, 113.

Professor Remacle communicated: In this comment, we discuss the possible effects of a strong coupling between the extended system and continuous states of the electrodes and we derive the expression of the matrix elements of the transition operator for this case. In so far that in the contact between the extended system and the bulk electrode, the gold cluster is in physical contact with the gold metal without any protective ligands, that the coupling is strong is to be expected.

As shown in Fig. 2 of our paper, the entire junction can be divided into three subsystems, the left electrode, the right electrode and the extended system, which comprises the molecule itself tethered between two small gold clusters. The total Hamiltonian is:

$$H = H_0 + V \tag{1}$$

where the zero order Hamiltonian is $H_0 = H_{\text{left elec}} + H_{\text{ext}} + H_{\text{right elec}}$. V in eqn. (1) is the sum of the two coupling terms between the extended molecule and the right and the left electrode.

The expression for the current is given by (see eqn. (6) of our paper):

$$I = 2e\frac{\mathrm{d}\langle P\rangle}{\mathrm{d}t} = \frac{2e}{h}\int \mathrm{d}\varepsilon\, g(\varepsilon, V_{\text{sd}}, T)\langle \mathrm{F}|T(E^-)|\mathrm{I}\rangle\delta(E_{\mathrm{I}} - E_{\mathrm{F}})\langle \mathrm{I}|T(E^+)|\mathrm{F}\rangle \tag{2}$$

where $|\mathrm{I}\rangle$ and $|\mathrm{F}\rangle$ are the initial and final many electron states (see eqn. (5) of our paper). The transition operator $T(E^+)$ is given by

$$T(E^+) = V + VG(E^+)V \tag{3}$$

V is a one-electron operator that transfers an electron from the electrode to extended system or *vice-versa*. In our paper, we make the approximation of a weak coupling to the electrodes, so that $G(E^+)$ can be replaced by $G_0(E^+)$ in eqn. (3). We discuss here the situation when this weak coupling limit is not realized.

To do so, as in our paper, we use the partitioning technique[1] in a one electron description of the electronic structure to evaluate the matrix elements of the transition operator, $T(E^+)$. The two orthogonal subspaces are the subspace $\boldsymbol{P}$ of the one electron continuous states of the electrodes and the subspace $\boldsymbol{Q}$ of the one electron discrete states of the Hamiltonian of the extended system, H_{ext}. Applying the partitioning technique leads for the matrix elements of $T(E^+)$ between the two electrodes to:

$$\boldsymbol{P}T(E^+)\boldsymbol{P} = \boldsymbol{P}V\boldsymbol{Q}G(E^+)\boldsymbol{Q}V\boldsymbol{P},$$

$$\boldsymbol{Q}G(E^+)\boldsymbol{Q} = (E^+ - \boldsymbol{Q}H\boldsymbol{Q})^{-1} = [E^+ - (\boldsymbol{Q}H_{\text{ext}}\boldsymbol{Q} + \boldsymbol{Q}V\boldsymbol{P}(E^+ - \boldsymbol{P}H_{\text{elec}}\boldsymbol{P})^{-1}\boldsymbol{P}V\boldsymbol{Q})]^{-1} \tag{4}$$

where we have taken into account that the subspaces $\boldsymbol{Q}$ and $\boldsymbol{P}$ are othogonal and that V cannot induce transitions within these two subspaces.

The term, $\boldsymbol{Q}V\boldsymbol{P}(E^+ - \boldsymbol{P}H_{\text{elec}}\boldsymbol{P})^{-1}\boldsymbol{P}V\boldsymbol{Q}$, in eqn. (4) is the self-energy term that describes exactly the effect of the coupling to the continua of the electrodes on the discrete levels of the extended system, $\boldsymbol{Q}H_{\text{ext}}\boldsymbol{Q}$. It has a real part, given by the principal value of $\boldsymbol{Q}V\boldsymbol{P}(E^+ - \boldsymbol{P}H_{\text{elec}}\boldsymbol{P})^{-1}\boldsymbol{P}V\boldsymbol{Q}$, that shifts the energies of the levels of $\boldsymbol{Q}H_{\text{ext}}\boldsymbol{Q}$ and an imaginary part, $-\mathrm{i}\pi\boldsymbol{Q}V\boldsymbol{P}\delta(E^+ - \boldsymbol{P}H_{\text{elec}}\boldsymbol{P})\boldsymbol{P}V\boldsymbol{Q}$, which broadens the levels of $\boldsymbol{Q}H_{\text{ext}}\boldsymbol{Q}$ into a Lorentzian shape. $\boldsymbol{Q}H\boldsymbol{Q}$ in eqn. (4) is the effective Hamiltonian. It is a complex symmetric matrix that can be diagonalized self-consistently,[2,3] see also in another context refs. 4 and 5 or by an orthogonal transformation.[6] Its eigenvalues are complex and correspond to the energies of the extended system coupled to the continuum states of the electrodes.

As discussed in our paper, it is possible to include in the self-consistent diagonalization of the one electron Hamiltonian of the extended system, $\boldsymbol{Q}H_{\text{ext}}\boldsymbol{Q} = \boldsymbol{Q}(H_{\mathrm{Au}_n} + H_{\text{mol}} + H_{\mathrm{Au}_n} + V_{\text{Au-S}} - \boldsymbol{E}\cdot\boldsymbol{r})\boldsymbol{Q}$, the effects of two major perturbations, the coupling at the interface, $V_{\text{Au-S}}$, and the fields' effects (included as a dipolar term, see ref. 7) induced by the gating and the source–drain bias. Both lead to energy shifts of the MO's of the extended system and distortion of their charge densities.

When the coupling to the electrodes is included in the expression of the transition operator, each conducting state in the transmission function, eqn. (14) of our paper, is endowed with a finite width. The transmission function becomes a weighted sum of Lorentzian peaks and is no longer a weighted density of discrete conducting states. In other words, the conducting states contributing to the current are intrinsically broadened by the coupling to the continua of the electrodes. This intrinsic broadening comes in addition to the broadening by the Fermi window induced by the thermal averaging, see eqn. (13) of our paper. Its main effect will be to lead to smoother I–V curves with less pronounced non-linearities.

At 0 K, the Fermi window is a square function of the energy with a width equal to eV_{sd} and its wings are broadened as the temperature increases. Therefore, to significantly affect the current, the strength of the coupling to the continua of the electrodes must be of the order of eV_{sd}. Recent experimental and theoretical results suggest that it could indeed be so.

In the same way that we used to analyze the effect of the coupling V_{Au-S} at the interface in FD 131/4, we can define an effective coupling parameter, $V_{ext-elec}/\Delta E$, where $V_{ext-elec}$ is the strength of the coupling between the extended system and the electrodes, given by the matrix elements of $\boldsymbol{QVP}$ (eqn. (4)) and ΔE is the spacing between the MO's of the extended system.

If the matrix elements $\boldsymbol{QVP}$ are smaller that the energy spacings, ΔE, between the MO's of the extended system, one can use the Fermi Golden rule approximation[8] and just add an imaginary part, $-i\pi|\langle\lambda|V|\varepsilon\rangle|^2\delta(E - E_\lambda)$ to the discrete energy E_λ of the MO's. This is the limit known as that of isolated resonances. It could be realized for conjugated alkenes, since as shown in Fig. 10 of our paper, the molecular orbitals are about or more than one electronvolt apart and almost unshifted in energy by field effects.

On the other hand, when the matrix elements of $\boldsymbol{QVP}$ are not small with respect to the spacings, ΔE, transitions between the MO's of the extended system can be induced by their interaction with the continua of the electrodes. This will be the case when the MO's of the extended molecule are not well separated in energy. This situation is known as that of overlapping resonances. It may for example be realized when the external electric field effects induce avoided crossings between the MO's (see Figs. 13 and 16 of our paper). When such is the case, the coupling to the electrodes could modify significantly the transmission function and therefore the I–V curves. However, as discussed above, unless the strength of the source–drain bias is very small, these effects will be further thermally averaged by the Fermi window.

Note that when the subspace $\boldsymbol{Q}$ is written in the one electron eigenstates of the extended system, including the effects of the coupling at the interface and of external fields effects, one can in first approximation neglect the effects of the level shift operator in eqn. (4) because they are expected to be small compared to the shifts induced by the coupling at the interface and the external fields. The self-energy is then purely imaginary. In the absence of threshold effects for the continua, this approximation can be insured if the matrix elements of the coupling operator $\boldsymbol{QVP}$ do not vary with energy in which case the principal value part vanishes.

1 R. D. Levine, *Quantum Mechanics of Molecular Rate Processes*, New York, Dover, 1999.
2 Y. Q. Xue, S. Datta and M. A. Ratner, *Chem. Phys.*, 2002, **281**, 151.
3 M. Brandbyge, J.-L.Mozos, P. Ordejon, J. Taylor and K. Stokbro, *Phys. Rev. B*, 2002, **65**, 165401.
4 E. Brandas and P. Froelich, *Phys. Rev. A*, 1977, **16**, 2207.
5 Y. K. Ho, *Phys. Rep.*, 1983, **99**, 1.
6 F. Remacle and R. D. Levine, *Mol. Phys.*, 1996, **87**, 899.
7 F. Remacle, I. Willner and R. D. Levine, *J. Phys. Chem. B*, 2004, **108**, 18129.
8 G. C. Schatz and M. A. Ratner, *Quantum Mechanics in Chemistry*, Prentice-Hall, New York, 1993.

Professor Weber asked: Would the voltage dependence change if you consider only two pieces of gold without the molecule? In my opinion, the "HOMO" in such a finite system is representing electrons in the gold cluster and is unrelated to the HOMO of the molecule.

Professor Remacle agreed. The voltage dependence would certainly change for the case of conjugated wires, because as you also showed in your 2002 paper,[1] the HOMO involved in the conduction is delocalized over the entire system, that is, both on the gold atoms and on the atoms of the molecular wire.

For saturated dithioalkanes, we find that the two highest occupied MO's involved in conduction at low voltages are localized at the interface, with almost no electron density on the molecular backbone. In that case, conduction occurs *via* a superexchange mechanism and the molecular bridge is not accessed during the charge transfer. However, the energy of the two quasi degenerate MO's localized at the interface is lower by ≈0.4 eV than the Fermi level of bulk gold. So there is a threshold value of the V_{sd} below which conduction is exponentially small while there should not be any in the case of conduction between two pieces of bulk gold. Another difference that will arise in the voltage dependence is that even if the molecular bridge is not directly involved, it provides a dielectric medium for the charge transport.

1. H. B. Weber, J. Reichert, F. Weigend, R. Ochs, D. Beckmann, M. Mayor, R. Ahlrichs and H. v. Lhoneysen, *Chem. Phys.*, 2002, **281**, 113.

Mr Adisa asked:

1. How does varying the metal relate to the coupling and thus the superexchange properties *i.e.* if a different metal is used rather than Au, how does this affect the superexchange?

2. What specific properties of Au have interested you in using this metal for your studies?

Professor Remacle replied:

1. The superexchange mechanism is realized when the interface is very weakly coupled to the molecular backbone, so that the highest MO's involved in conduction are localized at the junction, with almost no electron density on the molecular backbone. In other words, this mechanism operates when the junction has a local ionization potential lower than the molecular backbone itself. It is therefore very much dependent on the value of the Fermi energy of the metal and on the characteristics of the bonding at the junction.

We did not investigate interfaces with another metal than gold but we looked at the effect of replacing the S atom by a O atom. We found that the interface Au–O has very similar properties to the Au–S one.

Ag and Au have similar Fermi energies (≈5.5 eV) and one would expect that the Ag–S interface behaves similarly to the Au–S one. Note however that the 4d orbitals of Ag are spatially less extended than the 5d orbitals of Au which leads to weaker bonds.[1]

2. We investigate the properties of the Au–S interface because it is widely used in experimental and in theoretical studies and therefore allows us to compared our results with those of others.

1 E. S. Kryachko and F. Remacle, *Chem. Phys. Lett.*, 2005, **404**, 142.

Dr Nichols asked: What effect does changing from single coordination geometry to multi-coordination geometry have?

Professor Remacle replied: We did not investigate the effect on the transport properties of the coordination geometry of gold at the interface in detail but experimental and theoretical studies show that it can be significant, with the intensity varying by more than an order of magnitude depending upon the coordination of the gold atom at the interface.

Professor McCreery opened the discussion of Professor Kuznetsov's paper: If tunnelling is not coherent, and involves a vibrational transition as a redox process, will it still occur at 0 K?

Professor Kuznetsov responded: In principle electron tunneling can occur at 0 K due to tunneling along the nuclear coordinates (the probability of which is small). If tunneling of nuclei is neglected, the electron tunneling can take place only if the electron energy level is shifted (by the application of the overpotential or by the variation of the bias voltage) to the activationless position, *i.e.* to the energy window region between the Fermi levels. Then many electrons can tunnel through this energy level while the latter relaxes towards the equilibrium position for the reduced state in the adiabatic limit. However, this will produce a pulse of the tunnel current. The final equilibrium position of the energy level must be located also within the energy window for a steady state tunnel current to flow.

Professor Lindsay asked: The timescale for inelastic tunnelling is presumably h/Δ where Δ is the electron–phonon coupling and not the phonon period, correct?

Professor Kuznetsov replied: There are several characteristic times for inelastic tunneling: $\hbar/\Delta$ (where Δ is the coupling of the electron with the electrodes), $\hbar/\gamma_{B\nu}$ (where $\gamma_{B\nu}$ is the electron–phonon coupling constant) and $1/W_{exc}$ (where W_{exc} is the probability per unit time for the tunneling with the excitation of the phonon). The latter is the most characteristic time for the whole process of inelastic tunneling.

Dr Haiss asked: Are the tunnelling time and the coupling constant equivalent parameters in relation to the probability of the excitation of phonons?

Professor Kuznetsov replied: No. The probability of tunneling with the excitation of phonons depends on both the tunneling time $\tau_{tunn} \sim \hbar/\Delta$ and the coupling constant $\gamma_{B\nu}$ which characterizes the interaction of the electron with vibrational modes.

Professor Kornyshev asked: What could enhance the visibility of the effect of phonon-assisted tunnelling in the three-terminal geometry?

Professor Kuznetsov answered: The higher is the frequency of the local mode, the better. Some conditions are conflicting. The resonances are sharper, the smaller is coupling of the electron states in the groups B and C with the electrodes, *i.e.* the smaller are Δ_B and Δ_C. However this results in smaller values of the currents.

The other conditions are as follows:

$$U_{12} > \Delta_B\ ,\ U_{13} \text{ is small, } \varepsilon_{F3} - \varepsilon_B \gg \Delta_{(1)}\ ,\ \varepsilon_C - \varepsilon_{F1} > V_{BC}.$$

The last two inequalities are the conditions for the elastic current to the third electrode to be small. The best possibility would be a variation of the position of ε_C at fixed potentials. However this is possible only using different systems which would make it impossible to compare the results.

Professor Kornyshev said: What are the temperatures that would make it impossible to see this effect?

Professor Kuznetsov responded: The effect of re-absorption of vibrational quanta is most pronounced at not very high temperatures when $k_BT < \hbar\Omega$ where Ω is the frequency of the local mode. If opposite inequality is fulfilled, the thermal population of the vibrational quanta will interfere with the effect although the current to the third electrode due to the absorptive tunneling will be even greater.

Professor Kornyshev asked: I guess your experiment cannot be realized in electrochemical set-up where the phonon spectrum would be quasi-continuous. Am I right?

Professor Kuznetsov replied: Yes, you are right. It is difficult to observe this effect in the presence of a strong interaction with classical phonons (polarization of the solvent) since usually the lifetime of the excited state of local vibrational mode is rather short. In this case the interaction with the classical vibrational modes will destroy the effect.

Professor Kornyshev commented: We are not speaking about temperature-excited phonons. In the Kuznetsov "setup" one phonon is emitted by electron tunnelling in one channel, and that very phonon helps to excite another electron to higher energies to tunnel in another channel. The effect will be strong if the latter cannot happen without phonon absorption.

Professor Kuznetsov agreed. The electron energy levels are arranged in such a way that the resonance elastic tunneling through the energy level of the group C to the third electrode is impossible. A current due to off-resonant tunneling is small and its dependence on the overpotential and bias voltage is rather weak and quite different from a resonance-like dependence of the current due to the effect of re-absorption of vibrational quanta excited by the electron tunneling in the different channel.

Dr Nichols asked: These effects should be most apparent for high frequency vibrational modes which are not strongly coupled to other modes. Is it likely that low frequency modes are too strongly coupled (with other modes and the environment) to produce a sharp resonance?

Professor Kuznetsov replied: It depends on the system. In a polar medium the interaction of the electron with the medium polarization (low frequency modes) itself is sufficiently strong to destroy the effect even without the interaction of the polarization with the high-frequency local mode. In vacuum or low polar medium the interaction of the medium with the electron and with the local mode may be rather weak and the effect could be observed.

Professor Barbero opened the discussion of Professor Larsson's paper: You compare polyacetylene, a $-(CH)_n-$ chain, with polysulfurnitride, a $-(SN)_n-$ chain, showing that the latter could have high conductivity, of metallic nature. Could you comment on the fact that metallic conductivity has been measured on polyaniline, a $-(N{=}C{-}C{=}C{-}C{=}N)-$ chain?

Professor Larsson replied: Polyacetylene, polyaniline, and polyphenylene are all insulators with a large band gap without doping. If electrons or holes are introduced into the chain, polaronic states are formed with some activation energy, no matter if we have oligomers or polymers. $(SN)_x$ is 3/4 filled and very different, since there is a near degeneracy that cannot be lifted by Jahn–Teller (or Peierls) effects.

Professor Kornyshev asked: Is there any way to detect experimentally the crossover from the spin-density (SDW) to charge-density waves (CDW)?

Professor Larsson answered: One example is ordinary crystallographic methods. In AgF_2 there are two phases with a complicated SDW structures at a low temperature and a very different CDW phase at a high temperature, with a well-defined transition temperature. Jérome discussed extensively the TTF-TCNQ system in his review (ref. 2 of our paper). Conductivity drops, X-ray diffuse scattering, elastic neutron scattering, and STM suggest CDW structures at low temperature. SDW is not seen, but SDW becomes a paramagnet above the Néel temperature anyway. Normally crossovers are not expected.

Professor Kornyshev said: This is just to summarize a few questions general to the area, in which all the three speakers, as well as many of us work, which have not been touched in depth in the previous discussion.

1. What is the effect of V_{sol} and V_g on transmission coefficients and coupling constants?
2. What is the role of classical bridge fluctuation?
3. What are the generic temperature effects on conductivity and rectification?

Professor Remacle replied:

1. We show in our paper that the V_{sd} and V_g can modify the electronic structure of the extended system, by shifting the position in energy of the conducting levels and also by distorting the spatial extension of their electronic density. These effects on the electronic structure lead in turn to changes in the transmission coefficients and affect the I–V_{sd} curves.

Conjugated π systems respond differently to external fields than saturated ones. Because of their symmetry, the π orbitals of conjugated system are very resilient to field effects, while the σ orbitals are strongly affected: small to medium applied voltage strengths shift their position in energy and distort their electronic density, leading to significant changes in the I–V curves.

2. We show that coherent ballistic transport occurs on a femtosecond time scale and therefore it is not very likely that it is strongly affected by bridge fluctuation. However, in cases where the molecular bridge is redox active, bridge fluctuations are expected to play a crucial role.

3. For coherent ballistic transport, the effect of temperature is to broaden the Fermi window and therefore to allow more conducting states to contribute to the current. Typically, the current increases with increasing temperature. When the states brought in the Fermi by increasing the

temperature conduct through a different mechanism, increasing the temperature can also lead to a switch in the temperature dependence of the *I–V* curves. We have shown in our work on transport properties of self-assembled arrays of Ag QD's[1–3] that at low temperatures, the mechanism for conduction is *via* superexchange (also know as variable range hopping, VRH), and characterized by a $\exp(-(T_{VRH}/T)^{-1/2})$ temperature dependence while at higher temperatures, conduction is through an activated mechanism and follow a $\exp(-T_{act}/T)$ law.

1 F. Remacle, K. C. Beverly, J. R. Heath and R. D. Levine, *J. Phys. Chem. B*, 2002, **106**, 4116.
2 F. Remacle and R. D. Levine, *Appl. Phys. Lett.*, 2003, **82**, 4543.
3 F. Remacle, K. C. Beverly, J. R. Heath and R. D. Levine, *J. Phys. Chem. B*, 2003, **107**, 13892.

Professor Kuznetsov replied:

1. I do not think that the bias voltage and overpotential (gate voltage) can produce a significant effect on electron–phonon and electron-electrode coupling constants. However they affect significantly the transmission function, first of all due to the shift of the positions of the electron energy levels.

2. Classical fluctuations in the bridge are of great importance and should be investigated in more details. They may change the geometry of the bridge and the properties of its contacts with the electrodes. They may produce large variations of the positions of the electron energy levels. The latter is crucial for the transmission function. In the case of the bridge with a redox group they result, in particular, in red-ox transformations.

3. There is no unique answer to this question. It depends on the system and the type of conductivity. For example, the temperature results in a decrease of conductivity if the band mechanism operates. Conversely, the conductivity increases with the temperature in the case of hopping mechanism.

Professor Lindsay addressed Professor Remacle: You use an initial state based on a small gold cluster, but had you used infinite slabs of metal, then the initial and final states are just plane waves on the left and right electrodes, is that not so?

Professor Remacle replied: There are plane waves on the left and on the right electrodes as long as the electrodes remain uncoupled to the molecule. On the other hand, the initial state created by the transfer of an electron from the bulk electrode to the extended system is non stationary. We describe this non stationary electronic initial state as the projection of the superposition of plane waves that describes the electron in the bulk electrode on the gold atoms of the extended system that are directly coupled to it. In other words, the initial state is the non stationary state localized at the interface of the extended system, that is created by the transfer of an electron from the bulk electrode. In our paper, see eqns. (16) and (17), we show how the matrix elements of the transition operator are related by a Fourier transform relation to the cross correlation function of that initial state.

Dr Macdonald asked a general question: Is there a difference in electron density along the bridge in the case of long-range electron transfer in DNA when compared with electron transfer in alkane or conjugated molecules?

Professor Larsson replied: Either you have tunnelling or you have conductivity with electrons located on the bridge. Probably in DNA and certainly in alkane bridges you have just tunnelling. Simulation of wave packet motion shows that the wave packet disappears on the donor and appears on the acceptor without visiting the bridge. In a conjugated bridge it happens frequently that the electron first transfers to the bridge and later to the acceptor.

Controlling charge transport in single molecules using electrochemical gate

Xiulan Li,[a] Bingqian Xu,[a] Xiaoyin Xiao,[a] Xiaomei Yang,[b] Ling Zang[b] and Nongjian Tao[a]

Received 22nd April 2005, Accepted 12th May 2005
First published as an Advance Article on the web 8th September 2005
DOI: 10.1039/b505666g

We have studied charge transport through single molecules covalently bound to two gold electrodes in electrolytes by applying a voltage between the two electrodes and a reference electrode (gate). This electrochemical gating can effectively control the current through the molecules, depending on the electronic properties of the molecules. For electrochemically inactive molecules, such as 4,4′-bipyridine and 1,4′-benzenedithiol, the gate voltage influences the transport current only slightly (less than 30%). This lack of significant gate effect is attributed to the large LUMO–HOMO gaps of the molecules and the screening of the gate field by the two electrodes. For nitro-oligo(phenylene ethynylene) (OPE-NO_2), which undergoes multiple irreversible reductions at negative gate voltages, the current through the molecules can be modulated several folds by the gate. This gate effect is irreversible and associated with the reduction of the NO_2 group to different products that have different electron withdrawing capabilities from the conjugate backbone of the molecule. The most interesting molecules are perylene tetracarboxylic diimide compounds (PTCDI), which exhibit fully reversible redox reactions. The current through PTCDI can be reversibly varied and controlled over three orders of magnitude with the gate. Such a large gate effect is related to a redox state-mediated electron transport process.

1. Introduction

The building blocks of silicon-based microelectronics are field effect transistors (FET) whose basic function is to control electrical current between two electrodes (source and drain) with a third electrode (gate). The ability to control electron transport through a single molecule is thus considered to be a crucial task in the development of molecular electronic devices.[1–3] This ability also provides one with a rather unique opportunity to study charge transport, a phenomenon that plays vital roles in many chemical, electrochemical and biological processes, on a single molecule basis. Most works in molecular electronics to date have focused on two-terminal devices.[4–7] A three-terminal FET-like device is highly desired, because it provides power gain, an essential requirement for large-scale integrated circuits. FET-like behavior has been demonstrated in carbon nanotubes[8,9] and semiconductor nanowires.[10] Interesting Coulomb blockade[11–14] and Kondo effects[12,13] have also been observed in molecular systems using a back gate at low temperatures. Theoretical models have predicted that the conductance of a single molecule can be modulated with a gate electrode in a fashion of conventional FET.[15,16] However, experimental demonstration of this FET behavior at single molecule level has been a challenge due to the following reasons:

First, the gate electrode has to be placed within a few angstroms distance from the molecule in order to achieve a large enough gate field, which is not an easy task using conventional solid-state device configurations. One method to overcome this difficulty is to use electrochemical gate in which the molecular junction is immersed in an electrolyte and the source and drain potentials are biased

[a] *Department of Electrical Engineering & The Center for Solid State Electronics Research, Arizona State University, Tempe AZ85287, USA*
[b] *Department of Chemistry and Biochemistry, Southern Illinois University, Carbondale IL 62901, USA*

with respect to a reference electrode inserted in the electrolyte. This reference electrode is called electrochemical gate. An applied gate voltage falls across the double layer at the reference electrode–electrolyte interface, the electrolyte between the gate electrode and the source/drain electrodes, as well as the double layer at the electrolyte-source (drain) electrode interface. Since the current between the gate and the source/drain electrode is negligible, the voltage drop in the electrolyte is practically zero. The effective gate–molecule distance is thus determined by the double layer thickness at the electrode–electrolyte interfaces, which is of the order of the size of a few solvated ions. This allows the electrochemical gate to create a rather large field in the molecule. Such a electrochemical gating method has been applied to control charge transport in conducting polymer films[17,18] and nanojunctions,[19,20] metal oxide semiconductors,[21] carbon nanotubes[22,23] and molecular junctions.[24–29]

Second, the screening of gate field by the source and drain electrodes must be minimized. This screening effect becomes serious when the distance between source and drain electrodes is small in comparison to the distance between the molecule and the gate electrode. For a molecular FET, the source–drain distance is determined by the molecule size, and thus it is a challenge to avoid the screening problem for small molecules using the solid state device configurations. Even for the electrochemical gate approach, the molecule used has to be longer than the thickness of the double layer. One obvious solution to this problem is to choose relatively long molecules. However, since the conductance of a molecule often decreases exponentially with its length, highly conductive conjugated molecules are clearly among the best choice.

Finally, a reliable method is needed to measure electron transport in single molecules connected to the source and drain electrodes. To date, many experimental methods have been developed, but most methods measure either a large or unknown number of molecules. A scanning tunneling microscope (STM) can perform electrical measurements on single molecules, but the measured conductance is complicated by the uncertain contact between the molecule and the STM tip. Nevertheless, these measurements, together with theoretical efforts, have provided the bulk of our current knowledge about electron transport in molecules. In order to reliably measure the conductance of a single molecule, one must: (1) provide reproducible contacts between the molecule and two probing electrodes,[30–33] (2) identify a "signature" that the measured conductance is due to not only the sample molecules but also a *single* sample molecule, and (3) perform statistical analysis, which is required because of the sensitive dependence of the molecular conductance on the microscopic details of the molecule–electrode contacts.

In this paper, we discuss electrochemical gating effects on the charge transport through single molecules covalently bound to two metal electrodes. We focus on three types of molecules: electrochemically inactive molecules, electroactive molecules that undergo irreversible redox reactions, and electroactive molecules that undergo reversible redox reactions. These molecular systems exhibit rather different electrochemical gating behaviors.

2. Experimental method

The experimental method used in this work is based on statistical analysis of many individual molecular junctions.[34] The individual molecular junctions are created using a modified scanning tunneling microscope (STM) which repeatedly moves a Au tip (source electrode) into and out of contact with a Au substrate (drain electrode) in a solution containing sample molecules (Fig. 1). The sample molecule chosen has two terminal groups that can bind strongly to Au electrodes, and thus has a chance to bridge the two electrodes and form a well defined molecular junction during the separation of the tip and the substrate electrode.

The process can be divided into three steps.

(1) A piezoelectric transducer drives the STM tip into contact with the substrate electrode.

(2) The tip is then pulled away from the substrate, during which a small Au neck forms between the tip and the substrate due to large metallic cohesion. The neck becomes thinner and thinner, and eventually approaches atomic scale. The conductance monitored during the pulling process decreases in discrete steps near integer multiples of a fundamental constant, $G_0 = 2e^2/h = 77.4$ μS, where e is the electron charge and h is the Planck constant (Fig. 1a and 1c). This phenomenon, called conductance quantization, occurs only when the neck is decreased to atomic scale.[35–41] As

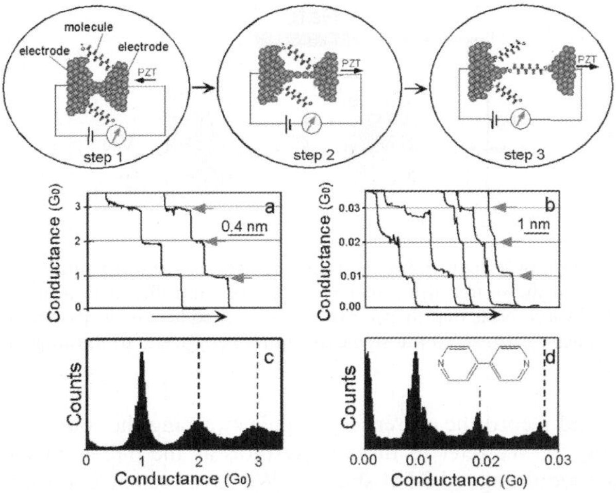

Fig. 1 Measurement of single molecule conductance. Step 1: A piezoelectric transducer (PZT) drives an electrode into contact with another electrode in a solution containing sample molecules. Step 2: The first electrode is then pulled back to form an atomic-scale neck between the two electrodes, which is signaled by conductance quantization (a,c). Step 3: After breaking the atomic-scale neck, the sample molecules start to bridge the two electrodes *via* chemical bonds. Further pulling breaks the individual molecules, which leads to a new series of steps in the conductance (b). Conductance histogram (d) constructed from more than a thousand measurements reveals well-defined peaks near integer multiples of a fundamental value ($0.01G_0$ for 4,4′-bipyridine shown here), which is identified as the conductance of a single molecule.

such, it is a useful indicator of when the tip and the substrate are connected by only a few Au atoms.[42]

(3) The tip is pulled further away to break the atomically thin neck, after which a new sequence of steps with much lower conductance values appears (Fig. 1b). In the case of 4,4′-bipyridine, a molecule with two nitrogen terminals that can bind to Au electrodes, the steps are near $0.01G_0$, $0.02G_0$, $0.03G_0$, . . . , two orders of magnitude lower than those due to the conductance quantization. We ascribe the new conductance steps that appear after breaking STM tip–substrate contact to the formation of molecular junctions, and the corresponding peaks in the histogram (Fig. 1d) to one, two and three molecules in the junctions.

This interpretation is supported by simultaneous measurement of the conductance and the force using a modified conducting atomic force microscope (AFM).[43] The measurement shows that each stepwise decrease in the conductance is accompanied by an abrupt drop in the force, corresponding to the breaking of contact between a molecule and the electrodes. Like the conductance histogram described above, the force histogram also reveals well-defined peaks at integer multiples of a fundamental force quantum, which corresponds to the force required to break a single molecule at the junction. For dithiol molecules that are bound to Au electrodes *via* S–Au bonds, the force is 1.5 nN. This force is the same as that required to break a Au–Au bond under the same conditions,[44] which indicates that the breakdown takes place likely at the Au–Au bond. The force to break a 4,4′-bipyridine obtained with the same pulling rate is only 0.8 nN, corresponding to the breakdown of a N–Au bond. In addition to AFM, several other control experiments also support the above conclusion. First, in the absence of molecules, no peaks below G_0 appear in the histogram. Second, in the presence of 2,2-bipyridine or alkanemonothiol molecules which cannot simultaneously bind to two electrodes, there are no well defined peaks in the conductance histogram either. Third, the peaks are located at different conductance values for different molecules. For instance, the conductance of alkanedithiol decreases exponentially with the molecule length.[34]

The histogram analysis shown in Fig. 1d provides an unambiguous determination of single molecule conductance averaged over a large number of measurements. Once the conductance of a single molecule is determined, pull the source and drain electrodes apart until the conductance drops to the lowest step, corresponding to the value of a single molecule conductance. We then

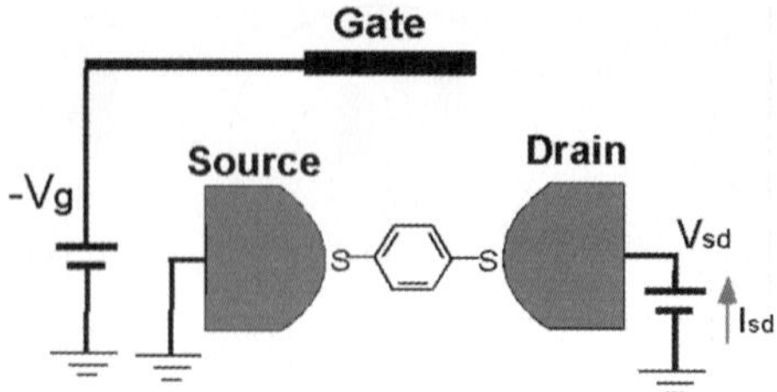

Fig. 2 Schematic illustration of a single molecule transistor with an electrochemical gate (a reference electrode in electrolyte). The gate and the source–drain bias voltages are controlled with a bipotentiostat (a Pt counter electrode not shown for clarity). Note that the gate voltage, V_g, cited here is the potential of the source electrode with respect to the reference electrode, so the sign convention is opposite to the solid state FET.

freeze the electrodes and record the current (I_{sd}) while sweeping the source–drain voltage (V_{sd}) or gate voltage (V_g) (Fig. 2). We refer to these procedures as the pull–hold–measure method. This approach is useful but requires good STM stability. We have been able to routinely hold the tip for many seconds to a few minutes using our setup.

The electrochemical gate was controlled by either a homemade or commercial bipotentiostat (PicoStat, Molecular Imaging Co.) using a Pt wire as counter electrode and a Ag wire as quasi-reference electrode (Fig. 2). The quasi reference electrode was calibrated against the more commonly used Ag/AgCl (in 3.5 M KCl) reference electrode. The substrate was prepared by thermally evaporating 100 nm gold on mica in a UHV chamber. Prior to each experiment, the substrate was briefly annealed in a hydrogen flame. The STM tip was prepared by cutting a 0.25 mm gold wire (99.999%), which was then coated with Apiezon wax in order to reduce ionic conduction and polarization currents. The leakage current, due to ionic conduction and polarization, was on the order of picoamperes. The STM cell was cleaned with piranha solution (98% H_2SO_4: 30% H_2O_2 = 3 : 1, v/v) and then sonicated in 18 MΩ water three times (**Caution**: piranha solution reacts violently with most organic materials and must be handled with extreme care). The electrochemical experiment was conducted in an argon atmosphere.

3. Results and discussion

3.1. Electrochemically inactive molecules

We have studied 4,4′-bipyridine and benzenedithiol molecules, both are electrochemically inactive within the potential windows studied here. These molecules can bind to two Au electrodes and form molecular junctions *via* N–Au and S–Au bonds, respectively. We have measured cyclic voltammograms of the two molecules adsorbed on Au electrodes in aqueous electrolytes, and found no electrochemical reactions within the potential windows studied here. The upper limit of the windows is fixed by the oxidation potential of gold, which is about 0.5 V, and the lower limit is determined by the desorption of the molecules, which is −0.5 V for 4,4′-bipyridine and −0.9 V for the thiol compounds.

To measure the gate effect of 4,4′-bipyridine, we first obtained a 4,4′-bipyridine molecular junction in 50 mM $NaClO_4$ supporting electrolyte and then we swept the electrochemical gate voltage while maintaining a constant bias voltage between the source and drain electrodes at 0.1 V. Fig. 3a is a typical source–drain current (I_{sd}) *vs.* gate voltage (V_g) curve. It shows a ~25% increase in the source–drain current as the gate voltage was swept negatively. To examine the reproducibility of the observation, we performed the same measurement on different molecular junctions prepared under the same conditions. Fig. 3b plots the histogram of the gate-induced current change, showing that the small gate effect is reproducibly observed despite of the run-to-run variations. We have performed the measurement by sweeping the gate voltage at different rates, and found no obvious dependence of the source–drain current on the sweep rate, which indicates that the gate effect is not due to a capacitive charging process. We can also rule out ionic leakage current as a possible cause for the gate effect because it is much smaller than the conduction current through the molecular junction.

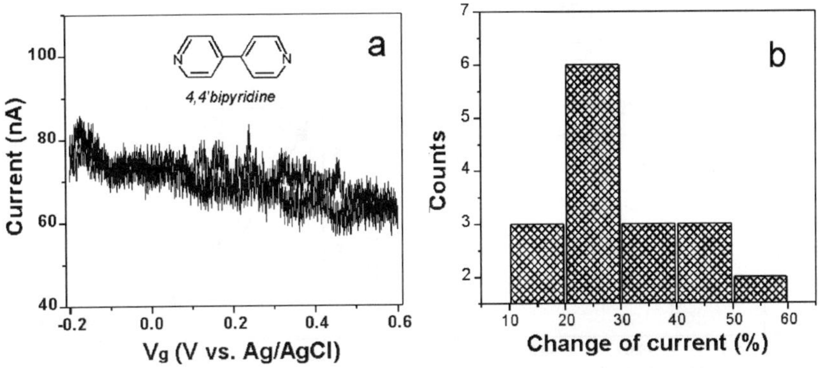

Fig. 3 (a) A typical source–drain current (I_{sd}) *vs.* gate voltage (V_g) curve of a 4,4′-bipyridine molecular junction in 0.05 M $NaClO_4$. The gate voltage is swept from −0.2 V to +0.6 V. (b) Statistic of the current change of 17 individual curves shown in (a). The source–drain bias is fixed at 0.1 V.

We have also studied the electrochemical gate effect on the conductance of benzenedithiol in a similar way. This molecule was used initially in the theoretical modeling of molecular FET.[15,16] The conductance of a single benzenedithiol bound to Au electrodes was previously determined in our lab using the STM break junction approach.[45] A typical source–drain current *vs.* gate voltage curve is plotted in Fig. 4. Within the electrochemical gate voltage window, the current through the benzenedithiol molecule changes only slightly (∼5%) with the gate voltage.

The lack of significant gate effect on the conductance of 4,4′-bipyridine and benzenedithiol are not fully understood yet. A possible reason is that both HOMOs and LUMOs of the two molecules are far away from the Fermi levels of the Au electrodes, and the applied gate voltage is not enough to shift the HOMOs or LUMOs close to the Fermi levels. This argument is consistent with the lack of electrochemical activity of these molecules. Another possible reason is the screening effect. In order to achieve effective gate control with a single gate electrode, the gate thickness has to be much smaller than the molecular length.[16] In the case of electrochemical gating, the gate thickness is given by the double layer thickness, which is comparable with the lengths of 4,4′-bipyridine and benzenedithiol (∼0.7 nm). This may result in substantial screening of the gate field.

3.2. Molecules with irreversible redox reactions

We have extended the above electrochemical gating measurement to oligo(phenylene ethynylene) (OPE) dithiolated molecules. In contrast to benzenedithiol and 4,4′-bipyridine, the OPE molecules are much longer and electroactive in their nitro-substituted form, which should lead to a large gate

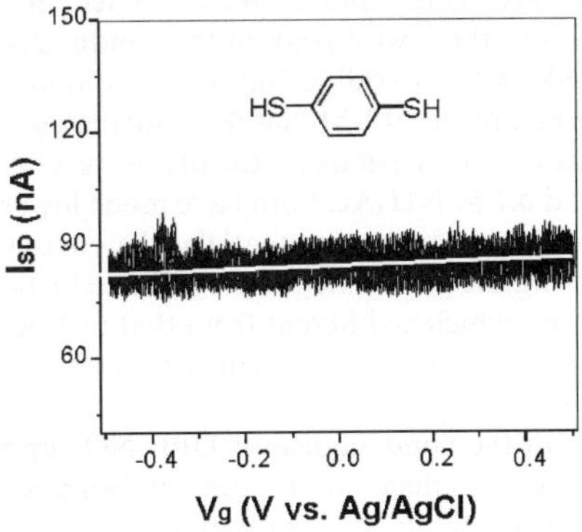

Fig. 4 A typical source–drain current (I_{sd}) *vs.* gate voltage (V_g) curve of a benzenedithiol molecular junction in 0.05 M $NaClO_4$.

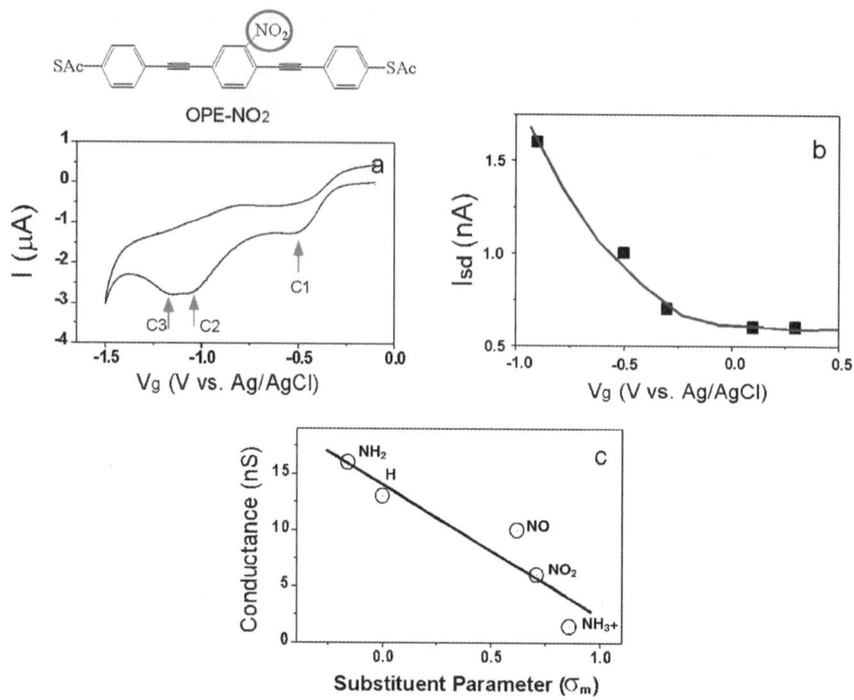

Fig. 5 (a) Cyclic voltammogram of thiol-deprotected OPE-NO_2 in $MeCN/H_2O$ containing 0.1 M TBAP and 0.001 M NaOH solution. (b) Source–drain current (I_{sd}) *vs.* gate voltage (V_g) for single OPE-NO_2. (c) Conductance *vs.* Hammett parameter (σ) plot. The location of the substituent is indicated with red circle in the structure of OPE-NO_2.

effect according to the above argument. Fig. 5a shows the cyclic voltammogram of nitro-substituted OPE (OPE-NO_2) on a gold electrode in anhydrous acetonitrile containing TBAP as supporting electrolyte (with a small amount of aqueous NaOH solution to remove the protection groups). It resolves three reduction peaks located near −0.5 (C1), −1.1(C2)[46] and −1.16 V (C3), respectively. The peaks, C_1 and C_2 are assigned to the electrochemical reduction of the nitro group, while third peak C_3 is, however, due to the reductive desorption of the adsorbed molecules because it diminishes with potential cycles.

Because the reduction of the OPE-NO_2 molecule is irreversible, we could not record the source–drain current by repeatedly sweeping the gate voltage. In order to study the gate effect, we determined the conductance histograms by holding the gate voltage at different values between −0.9 V and 0.3 V, starting at a positive gate voltage. We extracted the source–drain current *vs.* gate voltage by plotting the position of the lowest peak in the conductance histograms as a function of the gate voltage (Fig. 5b). Unlike 4,4′-bipyridine and benzenedithiol, the conductance of OPE-NO_2 is sensitive to the applied gate voltage. In 0.1 M NaOH, it increases from ~0.6 nA at 0.1 V to ~1.6 nA at −0.9 V (Fig. 5b). This gate effect depends on the pH of the electrolyte. We have measured the gate effect in 0.1 M $NaClO_4$ and 0.1 M NH_4Ac, both have much lower pH than 0.1 M NaOH. When holding the gate voltage above the reduction potentials, the conductance is the same in different electrolytes. However, if lowering the gate voltage below the reduction potentials, the conductance first increases and then decreases, which is different from that in NaOH solutions. This difference is related to the protonation of the amine group (reduced from the nitro group) in $NaClO_4$ and NH_4Ac.

We believe that the change in the conductance of OPE-NO_2 upon irreversible electrochemical reduction is due to the substituent effect on the central benzene ring based on the following considerations. The reduction processes in aqueous solution can be described as the following steps:

$$-NO_2 \rightarrow -NO \rightarrow -NH(OH) \rightarrow -NH_2.$$

The substituent group changes with the successive reductions. The chemical nature of a substituent can shift the frontier molecular orbital, and thus alter the electron transport efficiency through the molecule.[1,47] Fig. 5c shows that the measured conductance decreases linearly with the Hammett substituent parameter (σ), which is in good agreement with theoretical estimation by Vedova-Brook *et al.*[48] The decrease of the conductance with σ is expected because a larger σ corresponds to a higher electron withdrawing capability of the substituent.[49]

3.3. Molecules with reversible redox reactions

The gate effect observed in OPE-NO_2 is related to the electrochemical reductions of the NO_2 group, but the reactions are not reversible. Perylene tetracarboxylic diimide (PTCDI) is known to exhibit reversible redox reactions. To wire the molecule to two gold electrodes, we have synthesized two PTCDI derivatives, one is terminated with two thiol groups (PTCDI-dithiol) and the other with two pyridine groups (PTCDI-bipyridine). Both terminals can spontaneously bind to gold electrodes (Fig. 6a). The energy gap between LUMO and HOMO is only 2.5 eV. Such an energy gap implies a semiconductor-like molecule.

The cyclic voltammogram (Fig. 6b) of PTCDI-dithiol bound on gold electrodes shows two pairs of peaks at −0.55 V and −0.8 V, respectively, corresponding to two reversible reduction processes, so the levels of the empty molecular states are close to the Fermi level. The redox behavior of the surface bound molecule is similar to that in bulk solution, which indicates that the strong molecule-electrode binding does not significantly alter the molecular electronic states. The nominal lengths of the rigid molecules are about 2.3 nm, much greater than the gate thickness (~0.7 nm), so the screening of gate field due to the proximity of source and drain electrodes is significantly reduced, compared to the cases of 4,4′-bipyridine and benzenedithiol. These attributes make PTCDI an excellent candidate for molecular FET.

The conductance of the thiol-terminated PTCDI determined by the STM-break junction method is $\sim 1.2 \times 10^{-5}\ G_0$, while the conductance of the pyridine-terminated PTCDI is $\sim 5 \times 10^{-5}\ G_0$. The

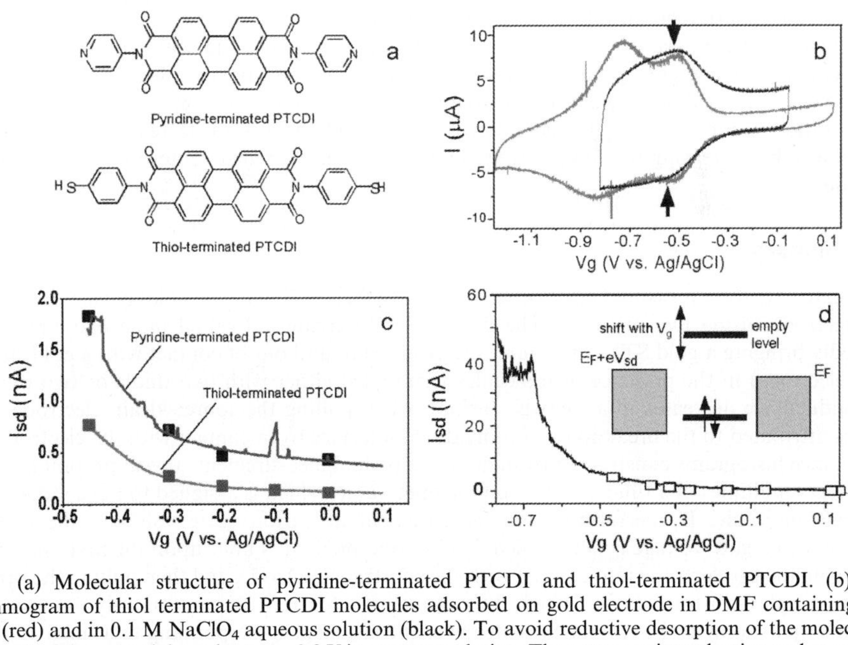

Fig. 6 (a) Molecular structure of pyridine-terminated PTCDI and thiol-terminated PTCDI. (b) Cyclic voltammogram of thiol terminated PTCDI molecules adsorbed on gold electrode in DMF containing 0.1 M TBAP (red) and in 0.1 M $NaClO_4$ aqueous solution (black). To avoid reductive desorption of the molecule, the lower limit of the potential was kept at −0.8 V in aqueous solution. The arrows point reduction and re-oxidation peaks of two redox processes. (c) Source–drain current (I_{sd}) *vs.* gate voltage (V_g) for single molecule transistors consisted of pyridine-terminated PTCDI and thiol-terminated PTCDI. The squares were determined from the peak position of the conductance histograms and the solid lines were obtained by directly recording the source–drain current while sweeping the gate voltage. (d) Sweeping the gate voltage to more negative values, the current through the thiol-terminated PTCDI increases by several orders of magnitude.

difference between the two derivatives may be attributed to the shorter length of the pyridine-terminated PTCDI. Although pyridine-terminated PTCDI molecule has same linkers as 4,4′-bipyridine does, the magnitude of the conductance is three orders lower than that of the latter. This indicates the molecular conductance does depend on the length. The ring structure in the middle of PTCDI molecule dominates the molecular resistivity. On the other hand, the charge transport through PTCDI molecule shows strong dependence on gate with the ring structure in PTCDI molecule. Fig. 6c plots the source–drain current *vs.* the gate voltage for both the thiol- and pyridine-terminated PTCDI in 0.1 M $NaClO_4$. Despite the run-to-run variations, all the curves reveal a rapid increase in I_{sd} when V_g is decreased below −0.1 V. Unfortunately, the pyridine-terminated PTCDI desorbs from gold electrodes below ∼−0.5 V, which prevents us from studying the gate effect below −0.5 V. For thiol-PTCDI, however, we were able to continuously record I_{sd}–V_g curves down to ∼−0.8 V (Fig. 6c). It reveals a large gate effect on the conductance of the molecule. For example, the conductance increases to 5×10^{-3} G_0 at $V_g = -0.8$ V, which is ∼500 times greater than that at 0 V. It also reveals a peak at $V_g \sim -0.65$ V.

The large gate effect obtained with these redox molecules resembles the *n*-type solid state FET (the gate sign convention here is opposite to that in solid state electronics), but the mechanisms are different. The current peak observed here is located at $V_g \sim -0.65$ V, close to the redox potential of the molecule. This leads us to believe that the current enhancement is due to an empty molecular state-mediated electron transport process.[50] One mechanism is resonant tunneling which predicts that the current reaches a peak when the empty state is shifted to the Fermi level by the gate (inset of Fig. 6d).[15,24,51] Since the V_{sd} is 0.1 V, the potential of the redox center is ∼−0.6 V at the resonance if considering the symmetry of the molecular junction. This potential is close to the first reduction potential (∼−0.55 V) of PTCDI adsorbed on gold electrodes (Fig. 6b). By considering the stochastic dynamics of solvent polarization and internal vibration modes, Kuznetsov and Schmickler[52] have shown recently that the resonant enhancement in the current through a redox molecule reaches the maximum near the redox potential, which supports the interpretation above. An alternative mechanism that predicts a maximum current near the reduction potential is a two-step electron transport in which an electron from one electrode tunnels into the empty state to reduce the molecule and then tunnel out into the second electrode.[53]

The on–off current ratio observed in the present system is $\sim 10^3$. The peak current is determined by the coupling of the empty states to the vibration modes and polarization of the surrounding environment.[54] So by designing a molecule to minimize the coupling, one may achieve higher on–off current ratio. Another way to further increase the on-off ratio is to decrease the off-resonance conductance by increasing the linker group length between the redox center and the source/drain electrodes.

4. Summary

In summary, we have measured electrochemical gate effect on single molecules covalently bound to gold electrodes using two methods. The first one is to create individual molecular junctions by repeatedly bringing a gold STM tip (source electrode) into and out of contact with a gold substrate (drain electrode) in the presence of molecules terminated either with two thiols or two pyridines. The conductance decreases in a stepwise fashion when pulling the source–drain electrodes apart, which is attributed to the breakdown of individual molecules from contact with the electrodes. The conductance histograms constructed from many individual measurements reveal pronounced peaks near integer multiples of a fundamental conductance value, which is assigned to the conductance of the single molecule. By performing the measurement at various gate voltages, the molecular conductance *vs.* gate voltage is determined. The second method is built upon the first one. It starts by bringing the tip electrode into contact with the substrate electrode, and then pulling the tip out of contact with the substrate during which the conductance (current) is continuously monitored. We freeze the tip as soon as the conductance drops to the last step, corresponding to a single molecular junction, and record the conductance as we sweep the gate voltage.

Using the two methods, we have studied the gate effect on three types of molecules: electrochemically inactive (4,4′-bipyridine and benzenedithiol), electrochemically active but with irreversible redox reactions (OPE-NO_2), and electrochemically active with fully reversible redox reactions (PTCDI). The gate effects on the first type of molecules are rather small. We attribute this weak gate effect to two possible reasons. The first one is that the LUMO and HOMO of these molecules are far away from Fermi energy levels of the drain and source electrodes. The second reason is the screening of the gate field from the proximity of the source and drain electrodes since both 4,4′ bipyridine and benzenedithiol are small in comparison to the double layer thickness. In the case of the second type of molecules, the conductance changes irreversibly by several folds, which is correlated with the irreversible reduction of the NO_2 moiety. The conductance decreases linearly with the Hammett parameter which describes the electron withdrawing capability from the conjugated backbone of the molecule. The third type of molecules is most interesting since the conductance can be reversibly controlled over nearly three orders of magnitude. The conductance also reaches a peak that resembles resonant tunneling *via* the empty state of the molecule. While the resonant tunneling model seems to explain qualitatively the experimental observation, more sophisticated theories are needed for a complete understanding of the large gate effect.

Acknowledgements

We thank Mark Ratner, Wolfgang Schmickler, Stuart Lindsay, Larry Nagahara and Salah Boussaad for helpful discussions, Nguyen Ly and Jin He for sample preparations, and DOE (DE-FG03-01ER45943) (Xu) and NSF(CHE-0243423) (Li and Xiao).

References

1 A. Aviram and M. Ratner, *Chem. Phys. Lett.*, 1974, **29**, 277–283.
2 R. l. Carroll and C. B. Gorman, *Angew. Chem., Int. Ed.*, 2002, **41**, 4379–4399.
3 C. Joachim, J. K. Gimzewski and A. Aviram, *Nature*, 2000, **408**, 541–548.
4 J. Chen, M. A. Reed, A. M. Rawlett and J. M. Tour, *Science*, 1999, **286**, 1550–1552.
5 C. P. Collier, E. W. Wong, M. Belohradsky, F. M. Raymo, J. F. Stoddart, P. J. Kuekes, R. S. Williams and J. R. Heath, *Science*, 1999, **285**, 391–394.
6 S. Datta, W. D. Tian, S. H. Hong, R. Reifenberger, J. I. Henderson and C. P. Kubiak, *Phys. Rev. Lett.*, 1997, **79**, 2530–2533.
7 R. M. Metzger, *Chem. Rev.*, 2003, **103**, 3803–3834.
8 S. J. Tans, A. R. M. Verschueren and C. Dekker, *Nature*, 1998, **393**, 49–52.
9 P. C. Collins, M. S. Arnold and P. Avouris, *Science*, 2001, **292**, 706.
10 Y. Cui and C. Lieber, *Science*, 2001, **291**, 851–853.
11 H. Park, J. Park, A. K. L. Lim, E. H. Anderson, A. P. Alivisatos and P. L. McEuen, *Nature*, 2000, **407**, 57–60.
12 J. Park, A. N. Pasupathy, J. L. Goldsmith, C. Chang, Y. Yaish, J. R. Petta, M. Rinkoski, J. P. Sethna, H. D. Abruna, P. L. McEuen and D. C. Ralph, *Nature*, 2002, **417**, 722–725.
13 W. J. Liang, M. Shores, M. Bockrath, J. R. Long and H. Park, *Nature*, 2002, **417**, 725–729.
14 S. Kubatkin, A. Danilov, M. Hjort, J. Cornil, J.-L. Bredas, N. Stuhr-Hansen, P. Hedegard and T. Bjornholm, *Nature*, 2003, **425**, 698–701.
15 M. Di Ventra, S. T. Pantelides and N. D. Lang, *Appl. Phys. Lett.*, 2000, **76**, 3448–3450.
16 P. Damle, T. Rakshit, M. Paulsson and S. Datta, *IEEE Trans. Nanotechnol.*, 2002, **1**, 145–1153.
17 E. W. Paul, A. J. Ricco and M. S. Wrighton, *J. Phys. Chem.*, 1985, **89**, 1441–1447.
18 H. S. White, G. P. Kittlesen and M. S. Wrighton, *J. Am. Chem. Soc.*, 1984, **106**, 5375.
19 H. X. He, C. Z. Li and N. J. Tao, *Appl. Phys. Lett.*, 2001, **78**, 811–813.
20 H. X. He, X. L. Li, N. J. Tao, L. A. Nagahara, I. Amlani and R. Tsui, *Phys. Rev. B*, 2003, **68**, 45302-1.
21 E. A. Meulenkamp, *J. Phys. Chem. B*, 1999, **103**, 7831–7838.
22 M. Krüger, M. R. Buitelaar, T. Nussbaumer, C. Schönenberger and L. Forró, *Appl. Phys. Lett.*, 2001, **78**, 1291-3.
23 S. Rosenblatt, Y. P. Yaish, Jiwoong, Gore, Jeff, V. Sazonova and P.L. McEuen, *Nano Lett.*, 2002, **2**, 869–872.
24 N. J. Tao, *Phys. Rev. Lett.*, 1996, **76**, 4066–4069.
25 D. I. Gittins, D. Bethell, D. J. Schiffrin and R. J. Nichols, *Nature*, 2000, **408**, 67–69.
26 E. Tran, M. A. Rampi and G. M. Whitesides, *Angew. Chem., Int. Ed.*, 2004, **43**, 3835–3839.
27 W. Haiss, H. Van Zalinge, S. J. Higgins, D. Bethell, H. Hoebenreich, D. J. Schiffrin and R. J. Nichols, *J. Am. Chem. Soc.*, 2003, **125**, 15294–15295.
28 F. Chen, J. He, C. Nuckolls, T. Roberts, J. E. Klare and S. Lindsay, *Nano Lett.*, 2005, **5**, 503–506.
29 B. Xu, X. Xiao, X. Yang, L. Zang and N. Tao, *J. Am. Chem. Soc.*, 2005, **127**, 2386–2387.
30 X. D. Cui, A. Primak, X. Zarate, J. Tomfohr, O. F. Sankey, A. L. Moore, T. A. Moore, D. Gust, G. Harris and S. M. Lindsay, *Science*, 2001, **294**, 571–574.
31 M. Magoga and C. Joachim, *Phys. Rev. B*, 1997, **56**, 4722–4729.
32 S. N. Yaliraki, M. Kemp and M. A. J. Ratner, *J. Am. Chem. Soc.*, 1999, **121**, 3428–3434.

33 F. Moresco, L. Gross, M. Alemani, K. H. Rieder, H. Tang, A. Gourdon and C. Joachim, *Phys. Rev. Lett.*, 2003, **91**, 36601.
34 B. Q. Xu and N. J. Tao, *Science*, 2003, **301**, 1221–1223.
35 J. K. Gimzewski and R. Moller, *Phys. Rev. B*, 1987, **36**, 1284–1287.
36 J. I. Pascual, J. Mendez, J. Gomez-Herrero, A. M. Baro, N. Garcia and V. T. Binh, *Phys. Rev. Lett.*, 1993, **71**, 1852–1855.
37 J. M. Krans, J. M. v. Ruitenbeek, V. V. Fisun, I. K. Yanson and L. J. d. Jongh, *Nature*, 1995, **375**, 767–769.
38 U. Landman, W. D. Luedtke, B. E. Salisbury and R. L. Whetten, *Phys. Rev. Lett.*, 1996, **77**, 1362–1365.
39 H. Hakkinen, R. N. Barnett and U. Landman, *J. Phys. Chem. B*, 1999, **103**, 8814–8817.
40 N. Agrait, J. G. Rodrigo and S. Vieira, *Phys. Rev. B*, 1993, **47**, 12345–12348.
41 A. I. Yanson, G. Rubio Bollinger, H. E. van den Brom, N. Agrait and J. M. van Ruitenbeek, *Nature*, 1998, **395**, 783–785.
42 H. Ohnishi, Y. Kondo and K. Takayanagi, *Nature*, 1998, **395**, 780–783.
43 B. Q. Xu, X. Y. Xiao and N. J. Tao, *J. Am. Chem. Soc.*, 2003, **125**, 16164–16165.
44 G. Rubio-Bolliger, S. R. Bahn, N. Agrait, K. W. Jacobsen and S. Vieira, *Phys. Rev. Lett.*, 2001, **87**, 26101.
45 X. Y. Xiao, B. Q. Xu and N. J. Tao, *Nano Lett.*, 2004, **4**, 267–271.
46 J. Chen, W. Wang, M. A. Reed, A. M. Rawlett, D. W. Price and J. M. Tour, *Appl. Phys. Lett.*, 2000, **77**, 1224–1226.
47 Y. Luo, C. P. Collier, J. O. Jeppesen, K. A. Nielsen, E. Delonno, G. Ho, J. Perkins, H. R. Tseng, T. Yamamoto, J. F. Stoddart and J. R. Heath, *ChemPhysChem*, 2002, **3**, 519.
48 N. Vedova-Brook, N. Matsunaga and K. Sohlberg, *Chem. Phys.*, 2004, **299**, 89–95.
49 C. Hansch, A. Leo and R. W. Taft, *Chem. Rev.*, 1991, **91**, 165–195.
50 L. Scudiero, D. E. Barlow, U. Mazur and K. W. Hipps, *J. Am. Chem. Soc.*, 2001, **123**, 4073–4080.
51 W. Schmickler and C. Widrig, *J. Electroanal. Chem.*, 1992, **336**, 213–221.
52 A. N. Kuznetsov and W. Schmickler, *Chem. Phys.*, 2002, **282**, 371–377.
53 A. M. Kuznetsov and J. Ulstrup, *J. Phys. Chem. A*, 2000, **104**, 11531–11540.
54 X. H. Qiu, G. V. Nazin and W. Ho, *Phys. Rev. Lett.*, 2004, **92**, 206102.

Two-dimensional assembly and local redox-activity of molecular hybrid structures in an electrochemical environment

Z. Li,[a] B. Han,[a] G. Meszaros,†[a] I. Pobelov,[a] Th. Wandlowski,[a] A. Błaszczyk‡[b] and M. Mayor§[b]

Received 10th May 2005, Accepted 30th June 2005
First published as an Advance Article on the web 15th September 2005
DOI: 10.1039/b506623a

The self-assembly and redox-properties of two viologen derivatives, *N*-hexyl-*N′*-(6-thiohexyl)-4,4′-bipyridinium bromide (HS-6V6-H) and *N,N′*-bis(6-thiohexyl)-4,4′-bipyridinium bromide (HS-6V6-SH), immobilized on Au(111)-(1 × 1) macro-electrodes were investigated by cyclic voltammetry, surface enhanced infrared spectroscopy (SEIRAS) and *in situ* scanning tunneling microscopy (STM). Depending on the assembly conditions one could distinguish three different types of adlayers for both viologens: a low coverage disordered and an ordered "striped" phase of flat oriented molecules as well as a high coverage monolayer composed of tilted viologen moieties. Both molecules, HS-6V6-H and HS-6V6-SH, were successfully immobilized on Au(poly) nano-electrodes, which gave a well-defined redox-response in the lower pA–current range. An *in situ* STM configuration was employed to explore electron transport properties of single molecule junctions Au(T)|HS-6V6-SH(HS-6V6-H)|Au(S). The observed sigmoidal potential dependence, measured at variable substrate potential E_S and at constant bias voltage ($E_T - E_S$), was attributed to electronic structure changes of the viologen moiety during the one-electron reduction/re-oxidation process $V^{2+} \leftrightarrow V^{+\bullet}$. Tunneling experiments in asymmetric, STM-based junctions Au(T)-S-6V6-H|Au(S) revealed current (i_T)–voltage (E_T) curves with a maximum located at the equilibrium potential of the redox-process $V^{2+} \leftrightarrow V^{+\bullet}$. The experimental $i_T - E_T$ characteristics of the HS-6V6-H–modified tunneling junction were tentatively attributed to a sequential two-step electron transfer mechanism.

1. Introduction

As the miniaturization of electronic components approaches the nanometer scale, new concepts to tailor structure, functionality and fabrication strategies are essential to overcome the fundamental physical and economic limitations of conventional inorganic silicon technology.[1] Bottom up (self-) assembly of well-defined nanoscale building blocks, such as molecules,[2] quantum dots,[3] and nanowires,[4] having key properties controlled by size, morphology and chemical composition represent an attractive alternative. The idea of building an electronic device using individual molecules was first proposed by Aviram and Ratner.[5] The molecular approach bears several unique opportunities: (1) custom-design of nanoscale molecular units, (2) high reproducibility due to

[a] *Institute of Surfaces and Interfaces ISG 3 and cni – Center of Nanoelectronic Systems for Information Technology, Research Center Jülich GmbH, D-52425, Jülich, Germany. E-mail: th.wandlowski@fz-juelich.de*
[b] *Institute for Nanotechnology, Research Center Karlsruhe GmbH, P. O. Box 3640, D-76021, Karlsruhe, Germany. E-mail: marcel.mayor@int.fzk.de*

† Home address: Institute of Materials and Environmental Chemistry, Chemical Research Center of the Hungarian Academy of Sciences, Pusztaszeri ut 59–67, H-1025 Budapest, Hungary.
‡ Home address: Faculty of Commodity Science, Al. Niepodległości 10, 60–967 Poznań, Poland.
§ Also: Department of Chemistry, University of Basel, St. Johanns-Ring 19, CH-4056 Basel, Switzerland.

self-organization and recognition properties of the molecular building blocks, (3) implementation of localized functions such as charge transfer, π–π stack conductivity or molecular rectification, *etc.*[6,7]

The development of several experimental techniques, including mechanically controlled break junctions,[8–10] nanopores,[11,12] nanowires,[4] crossed wires,[13] solid metal film[14] and mercury drop junctions,[15] and various scanning probes (SPM)[2,16–26] have enabled the ability to explore electrical properties of tailored individual molecules and/or molecular assemblies attached to conducting substrates. In SPM and surface nano-electrochemistry, the free or cluster tagged molecules[20,23] are typically in direct contact with one metal electrode. Bridging of individual molecules between a gold STM tip and a gold substrate was recently achieved in solution.[25–28] The statistical analysis of repeatedly created junctions provided access to single molecule conductivities.

Several groups explored solid state assemblies (metal|molecule|metal junctions) with redox-centers, exhibiting low-lying discrete HOMO respective LUMO levels, under *ex situ* or low temperature conditions (ambient, UHV) and discovered phenomena such as negative differential resistance (NDR),[29] rectification,[6] amplification,[30–33] stable switching and memory effects.[34,35] These observations were attributed to molecular (redox) energy states brought to resonance with the Fermi levels of the enclosing electrodes by the applied bias voltage or by nuclear configuration fluctuations (dominant at RT), respective vibrational relaxations (dominant at cryogenic temperatures). Examples of functional molecules used in these studies are 2′-amino-4,4′-di(ethynylphenyl)-5′-nitro-1-benzene thiolate,[11] metalorganic Co^{2+} and Fe^{2+}–terpyridine complexes,[36] oligothiophenes,[32] *p*-phenylenvinylen oligomers[31] and azurine.[33]

First examples for switching and rectification under electrochemical (and closely related) conditions were reported for porphyrins,[19,37] phthalocyanines,[38,39] metal proteins[21,40,41] and Ru-modified alkanethiols[15] deposited on HOPG or gold electrodes, gold nanoparticles tethered to a gold substrate with an alkylviologen redox group,[20] as well as for nanometer-size junctions bridged with heptaaniline oligomers[28] or polyaniline.[42] The molecular energetic states are coupled to the electrolyte environment, and to the metallic interfaces, characterized by electronic transmission ("coupling") coefficients. The following scenarios of interfacial electron transfer in a nanoscale electrochemical metal|redox molecule|metal configuration can be envisaged: [21] (1) *superexchange via off-resonance levels*,[43] (2) *resonance mode*,[44] (3) *coherent two-step electron transfer*,[21,43] (4) sequential two-step electron transfer.[21,43,45]

Working in an electrochemical environment has the advantage that two potential differences can be controlled individually: the bias voltage between two working electrodes, and the potential drop between one working electrode and a reference electrode. The latter may be considered as a "gate electrode". The current flow to the external circuit is assisted by a fourth electrode, the counter electrode (*cf.* 21). The idea of an electrochemical gate for applications in molecular electronics was introduced by Schönenberger *et al.*[46] and McEuen *et al.*[47] The field in the electrochemical double layer (EDL) is close to the gate field required to significantly change the current through a molecule, according to first principle calculations by di Ventra *et al.*[48]

Motivated by the pioneering work of the Liverpool group on single molecule conductivity[20,26,27] we report in this contribution a comparative study on the redox-activity of *N*-hexyl-*N*′-(6-thiohexyl)-4,4′-bipyridinium bromide (HS-6V6-H) and *N*,*N*′-bis(6-thiohexyl)-4,4′-bipyridinium bromide (HS-6V6-SH) immobilized and addressed on gold electrodes in an electrochemical environment. The tailored molecules are composed of a redox-active center, the viologen group, flexible alkyl-spacer units and one or two terminal SH anchor groups. Our approach combines classical and low current voltammetry (sub-picoampere resolution) with surface enhanced infrared spectroscopy (SEIRAS),[49] *in situ* scanning tunneling microscopy (STM) and spectroscopy (STS).[50] The main goal of the present paper is focused on measurements of electron transfer with redox-active, thiol-functionalized viologens as they are addressed by the electrochemical potential to explore relations between molecular and electronic structure in macroscale and nanoscale molecular assemblies approaching the single molecule level.

The redox-active viologens occur in three different oxidation states $V^{2+} \leftrightarrow V^{+\bullet}$ and $V^{+\bullet} \leftrightarrow V$.[51,52] They have been incorporated as "backbone"-components in self-assembled monolayers,[53–60] or in a nanometer-scale electronic switch[20,26] and various functional materials.[61] The self-assembly and redox-functionality of thiol-functionalized viologens on gold macro-electrodes was investigated by cyclic voltammetry,[53,54,56–60] quartz crystal microbalance,[53] IR[62,63] and Raman spectroscopy,[62] electro-reflectance,[57,59] *ex situ* STM[20,63] and STS.[26,63]

The present paper is organized as follows: Experimental details are summarized in Section 2. We report electrochemical studies on Au (111) single crystal macro-electrodes and gold nano-electrodes on the self-assembly and redox-activity of HS-6V6-H and HS-6V6-SH employing cyclic voltammetry, *in situ* SEIRAS and STM. The final paragraph addresses an attempt to explore single molecule electron transport and electron tunneling at a potential-controlled electrode/electrolyte interface.

2. Experimental

2.1 Synthesis

The synthetic strategy to the viologen derivatives HS-6V6-H (**1**) and HS-6V6-SH (**2**) is based on the nucleophilic substitution reaction between pyridines and terminal alkyl bromides (Scheme 1).[64]

HS-6V6-H (1). The viologen core was first mono-functionalized with a hexyl chain. 4,4′-Bipyridine and the corresponding 1-bromo-*n*-hexane **3** have been heated to 90 °C in chloroform in a pressure tube. The crude products were dissolved in dichloromethane and separated from the insoluble doubly substituted viologen derivatives by filtration to afford the desired mono-substituted viologen **4** in reasonable yields of 78%. The nucleophilic terminal sulfur groups are masked as thioacetates.[65] The required bifunctional hexyl chain with a bromine atom at one end and an acetyl-protected sulfur at the opposite end **7** has been synthesized from the bromohexene **6** and thioacetic acid in the presence of a radical source (yield 81%).[66] Bromohexene **6** has been converted to **7** in good yields (81%) by refluxing in toluene with thioacetic acid and 1,1′-azobis(cyclohexanecarbonitrile). The acetyl protected hexyl bromide **7** and the mono-hexylviologen derivative **4** have been heated to 110 °C in dimethylformamide (DMF) to form the acetyl-protected target structure **5** in very good yields (97%). In variation of the literature procedure[65] we replaced acetylchloride by acetylbromide to form *in situ* HBr in methanol under an argon atmosphere to deprotect the acetyl groups of **5**. HS-6V6-H (**1**) has been isolated as the pure bromide salt (yield 92%) as confirmed by elemental analysis.

HS-6V6-SH (2). 2.3 Equivalents of the terminally acetyl protected thiol-functionalized hexylbromide **7** have been used to substitute both nitrogens of 4,4′-bipyridine. The acetyl protected target structure **8** has been isolated in good yields (81%) after crystallization from ethanol. The acetyl-protected viologen dithiol **8** was dissolved in methanol and an excess of acetylbromide was added at −78 °C to afford the desired viologen dithiol **2** after stirring at room temperature (yield 95%).

***N*-Hexyl-*N*′-(6-thiohexyl)-4,4′-bipyridinium bromide, HS-6V6-H (1).** Yellow solid. Mp 274–275 °C; δ_H (300 MHz; MeOD; Me_4Si) 0.93 (t, 3H, CH_3, $J = 6.9$ Hz), 1.30–1.70 (m, 12H), 2.04–2.20 (m, 4H), 2.52 (t, 2H, SCH_2, $J = 6.4$ Hz), 4.77 (td, 4H, $J = 2.1$, $J = 6.8$ Hz), 8.70 (d, 4H, $J = 6.7$ Hz), 9.30 (d, 4H, $J = 6.1$ Hz); δ_C (75 MHz; MeOD; Me_4Si) 14.28, 23.49, 24.76, 26.65, 26.91, 28.67,

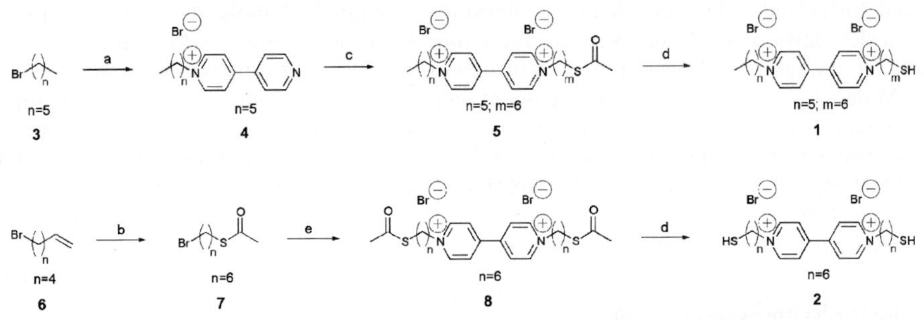

Scheme 1 Synthesis of the viologen monothiol **1** and dithiol **2**. (a) 4,4′-Bipyridine, $CHCl_3$, 90 °C, 16 h; (b) CH_3COSH, 1,1′-azobis(cyclohexanecarbonitrile), toluene, reflux, 1 h; (c) $Br(CH_2)_6SCOCH_3$, DMF, 110 °C, 24 h; (d) AcBr, MeOH, −78 °C to rt, argon atmosphere, 3 h; (e) 4,4′-bipyridine, DMF, 110 °C, 24 h.

32.32, 32.44, 32.54, 34.78, 63.20, 63.30, 128.33, 147.10, 151.31; Elemental analysis calcd (%) for $C_{22}H_{34}Br_2N_2S$: C 50.97, H 6.61, N 5.40, found: C 50.69, H 6.52, N 5.68; *m/z* (MALDI TOF) 357.5839 $[M^+ - 2Br]$.

***N,N′*-Bis(6-thiohexyl)-4,4′-bipyridinium bromide, HS-6V6-SH (2).** Yellow solid (95%). Mp 263–265 °C; δ_H (300 MHz; MeOD; Me_4Si) 1.38–1.58 (m, 8H), 1.64 (q, 4H, $J = 7.0$ Hz), 2.11 (q, 4H, $J = 7.6$ Hz), 2.51 (t, 4H, $J = 6.8$ Hz, SCH_2), 4.75 (t, 4H, $J = 7.5$ Hz, NCH_2), 8.69 (d, 4H, $J = 6.6$ Hz), 9.29 (d, 4H, $J = 6.6$ Hz); δ_C (75 MHz; MeOD; Me_4Si) 25.65, 27.45, 29.60, 33.21, 35.60, 64.10, 129.31, 148.10, 152.32; Elemental analysis calcd. (%) for $C_{22}H_{34}Br_2N_2S_2$: C 48.00, H 6.23, N 5.09, found: C 48.27, H 6.49, N 5.05; *m/z* (MALDI TOF) 390.0631 $[M^+ - 2Br]$.

2.2 Electrolyte solutions, electrodes and electrochemical measurements

The electrolyte solutions were prepared with Milli-Q water (18 MΩ, 2 ppb TOC), $KClO_4$ (Fluka puriss p.a., twice recrystallized from water), HCl (suprapure Merck), KOH (suprapure Merck), $Na_2HPO_4 \cdot 2H_2O$ (Fluka, puriss p.a.), $NaH_2PO_4 \cdot H_2O$ (Fluka, BioChem Ultra) and ethanol (p.a., KMF). All electrolytes were deaerated with argon before and during the experiments. The measurements were carried out at 20 ± 0.5 °C. The glassware was cleaned either in caroic acid or in a 1 : 1 mixture of hot H_2SO_4 (95–97%, pro-analysis, Merck) and HNO_3 (65% purissimum, Riedel-de-Haen), followed by extended rinsing with Milli-Q water.

Single crystal Au(111) electrodes were cylinders (EC, 4 mm height and 4 mm diameter) or discs (STM, 2 mm height and 10 mm diameter). They were flame-annealed with a hydrogen torch at red heat, and then cooled in high purity argon. Contact with the electrolyte was always established under potential control. Before modification with viologens island-free Au(111)-(1 × 1) surfaces were prepared by immersing a freshly flame-annealed electrode under potential control at 0.50 V (*vs.* SCE) into deaerated 0.1 M HCl. After a waiting time of 60 s, the electrode was removed and thoroughly rinsed with Milli-Q water.[67]

Gold nano-electrodes for electrochemical and STM experiments were prepared by etching a 0.25 mm diameter gold wire in a 1 : 1 mixture of ethanol and 36% HCl[68] and subsequent coating with polyethylene. The uncovered areas of these electrodes were typically 1 to 15 × 10^{-12} m^2.[69]

The preparation and characterization of quasi-single crystalline Au(111-25 nm) film electrodes used as substrate material in all SEIRAS experiments has been described previously in ref. 49.

Electrochemical experiments with macro-electrodes were carried out with a custom-made set-up described in ref. 70. A home-made low current potentiostat (resolution $<$ 10 fA) has been developed for the studies with nano-electrodes.[69] All potentials in this paper are quoted with respect to a saturated calomel electrode (SCE). Contact to the electrolyte was always established under strict potential control.

2.3 Electrode modification

Three different protocols have been developed to prepare gold electrodes modified with HS-6V6-H or HS-6V6-SH: (1) Low coverage *disordered* adlayers were obtained after immersion of a freshly prepared Au(111)-(1 × 1) electrode into a 50 μM deoxygenated ethanolic solution of the respective viologen derivative at room temperature for 1 min. (2) Low coverage *ordered* adlayers ("striped phases", *cf.* paragraph 3.4) were prepared by exposure of the gold substrates at room temperature to 50 μM deoxygenated ethanolic viologen-containing solution for 2 min, and subsequent annealing in pure ethanol at 70 °C for 12 h. (3) High coverage monolayers were obtained by immersion of the gold substrates into 1 mM ethanolic solution followed by thermal annealing in ethanol at 70 °C for 12 h. The temperature treatment of the samples was carried out in closed containers, which were deoxygenated. After incubation, the samples were removed from the solutions, rinsed with ethanol and carefully dried in a stream of argon.

2.4 Spectroelectrochemical set-up

The *in situ* SEIRAS experiments were carried out with a Bruker IF 66v/s Fourier Transform Spectrometer synchronized with a Heka PG 310 potentiostat employing a vertical spectroelectrochemical cell in Kretschmann ATR-configuration and *p*-polarized infrared radiation from a globar source.[49] The working electrode was a quasi-single crystalline Au(111-25 nm) film electrode prepared by electron beam evaporation of gold onto the (111) plane of a silicon hemisphere, and subsequent electrochemical annealing.[49] The spectroelectrochemical measurements were carried out under potential sweep or potential step conditions. In potential sweep experiments (10 or 50 mV s^{-1}), the spectrometer was operated in the slow scan mode. Typically, 80 sequentially acquired interferograms were co-added into each single beam spectrum. The spectra were plotted in absorbance units defined as $A = -\log(I/I_o)$, where I and I_o represent the intensities of the reflected radiation at the sample and the reference potentials, respectively. The complete description of the spectroelectrochemical ATR-SEIRAS experiment and of the sample preparation is given in ref. 49.

2.5 STM and STS measurements

The STM and STS experiments were carried out with a Molecular Imaging Pico-SPM which was positioned in a hermetically sealed container to prevent oxygen exposure of the samples. The STM tips were electrochemically etched tungsten or gold wires (0.25 mm diameter) coated with polyethylene. Platinum wires served as quasi-reference and counter electrodes. Experiments in neutral electrolytes were carried out with silver wires as quasi-reference electrodes. All STM experiments were performed at room temperature in constant current mode with tunneling currents ranging between 3 and 200 pA.

Local electron transfer characteristics of Au(T)|HS-6V6-SH(HS-6V6-H)|Au(S) junctions were explored by combining current distance measurements with *in situ* STM imaging.[50] A gold tip (T) coated with polyethylene (free area of 1 to 15 × 10^{-12} m^2) was brought to a predefined *xyz*-position and held at fixed bias ($E_T - E_S$). The distance z_o to the adsorbate-covered gold substrate (S) was fixed choosing a tunneling setpoint current i_o of 50 or 100 pA. After stabilization, the *z*-piezo feedback was disabled temporarily. The tip approached the surface up to $z_o - \Delta z_a$ with maximum rate, was held at this position for 100 ms, and subsequently retracted to $z_o + \Delta z_r$. The individual pulling traces were recorded (2000 data points, current range between 1 pA and 10 nA or, occasionally, between 10 pA and 100 nA). Typical values of Δz_a and Δz_r were 1.1 nm and 2 nm. i_o and Δz_a were carefully selected to ensure strong tip–adsorbate interactions, but not to drive the tip into the substrate surface.[25,28,50] The strategy allowed the spontaneous formation of molecular bridges between the gold STM tip and the substrate surface,[27] which were subsequently broken upon retraction. After enabling the *z*-feedback again, with the preset values of i_o, the surface was inspected by STM-imaging in constant current mode.

Current–voltage spectroscopy was carried out with viologen-modified gold tips. The redox-activity of these tips was tested by cyclic voltammetry. After mounting in the STM set-up the system was stabilized for several hours in a controlled argon atmosphere. The tunneling regime was established with set-point currents i_o ranging between 30 pA and 200 pA to preserve the integrity of the viologen derivative in the gap. After equilibration, the *z*-piezo feedback was switched off, and several current voltage traces were recorded at fixed bias ($E_T - E_S$) by sweeping simultaneously the potentials of tip (E_T) and substrate (E_S) with 2 V s^{-1}. We recorded individual single scans with 1000 data points. The cycle was repeated after stabilization at the preset values of i_o.

3. Results and discussion

3.1 Cyclic voltammetry with single crystal macro-electrodes

Fig. 1A shows a typical first scan voltammogram of an Au(111)-(1 × 1) electrode modified with a monolayer of HS-6V6-H, bromide salt, in 0.05 M $KClO_4$ adjusted to pH = 10 by addition of KOH. The solution resistance was carefully compensated by applying a positive feedback. The adlayer was prepared by immersion of the gold electrode in 1 mM ethanolic solution of HS-6V6-H followed by thermal annealing at 70 °C for 12 h (protocol 3). Contact with the electrolyte was established at $E = -0.700$ V. The electrode potential was first scanned in the positive direction. The adlayer capacitance was estimated to be 8.8 μF cm^{-2} in -0.200 V < E < -0.300 V. This value is rather large compared to alkanethiols[71] and indicates the incorporation of water molecules or ions into the film. Subsequent potential excursion reveals three characteristic pairs of current peaks labeled P1/P1′, P2/P2′ and P3/P3′. Restricting the potential range to $E \geq -1.050$ V results in two pairs

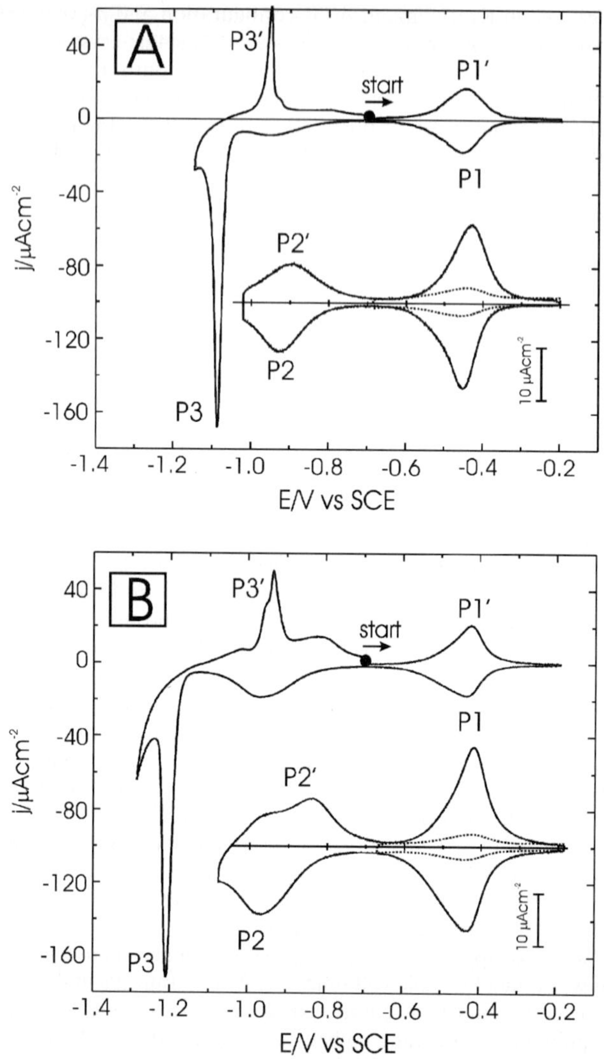

Fig. 1 First scan cyclic voltammograms (solid lines) of Au(111)-(1 × 1) electrodes in 0.05 M $KClO_4$, pH adjusted to 10 with KOH, modified with HS-6V6-H (A) and HS-6V6-SH (B) according to protocol 3, sweep rate 50 mV s^{-1}. The characteristic current peaks are labeled P1/P1′, P2/P2′and P3/P3′. The insets show first-scan voltammograms recorded in a limited potential range for high coverage (solid lines, protocol 3) and low coverage (dotted lines, protocol 2) viologen adlayers.

of well-defined current peaks (Fig. 1A). P1 ($E = -0.440$ V) and P1′ ($E = -0.428$ V) are assigned to the one-electron reaction $V^{2+} \leftrightarrow V^{+\bullet}$.[51,52] The peak-to-peak separation is rather constant up to 2 V s^{-1}, and the peak height scales linearly with the scan rate. The peak FWHM (P1, P1′) is 0.091 V. These observations indicate a fairly reversible process.[72] These observations reveal that the redox process is due to surface confined viologen moieties. The coverage was estimated to be $(3.5 \pm 0.3) \times 10^{-10}$ mol cm^{-2}, which corresponds to an area of 0.47 nm^2 per molecule. This value is in good agreement with literature data reported for related systems.[53,56–58,60] It is significantly smaller than the projected area of a planar oriented HS-6V6-H molecule (*cf.* crystal structure data in refs. 73 and 74) but larger than the estimated cross section of an all-*trans* oriented alkanethiol in a densely packed self-assembled monolayer.[75] This qualitative discussion points to a tilted orientation of the viologen moiety with respect to the surface normal as suggested by Sagara.[59] Attempts to estimate a rate constant from the variation of the peak separation on the scan rate and based on chronoamperometric experiments[72] with microelectrodes indicated that k_o (at $E_{P1/P1'}$) > 150 s^{-1}.[76] The second pair of peaks P2 ($E = -0.920$ V) and P2′ ($E = -0.900$ V) is assigned to $V^{+\bullet} \leftrightarrow V$.[51,52] The peak-to-peak separation is constant up to 1 V s^{-1} and the peak height scales linearly with the scan rate. The peak FWHM of P2 and P2′ are estimated to be 0.110 V and 0.132 V, respectively. The broadening indicates quasi-reversibility and higher disorder within the adlayer. The latter might be attributed partially to the superposition with the process at P3/P3′. The narrow peak P3 at $E = -1.080$ V, FWHM = 0.023 V, is assigned to the reductive desorption of the redox-active adlayer. Current integration, after correction of the double layer contribution, yields a desorption charge of approximately (160 ± 10) μC cm^{-2}. This value is considerably larger than 70 μC cm^{-2} typically reported for alkanethiols,[71] and indicates contributions of parallel processes such as bromide (perchlorate) desorption and cation (potassium) re-adsorption. The occurrence of the corresponding anodic peak P3′ at $E = -0.950$ V points to a partial oxidative re-adsorption of HS-6V6-H on Au(111)-(1 × 1). Multiple cycling in $-1.150\ V < E < -0.200$ V results in rather distorted current–voltage characteristics and shall not be considered further.

Systematic studies on the modification of Au(111)-(1 × 1) with HS-6V6-H revealed that a complete monolayer was obtained upon deposition from 1 mM ethanolic solution and annealing at 70 °C at incubation times longer than 1 h. No evidence for multilayer formation was obtained. Using lower concentrations of HS-6V6-H and shorter immersion times resulted, in a wide range of exposure times and after a subsequent thermal treatment in viologen-free ethanol, to rather stable coverages of the fully oxidized form of HS-6V6-H clustering around (0.9 ± 0.2) 10^{-10} mol cm^{-2}. The redox-response of a film prepared according to protocol 2 (*cf.* Section 2.3) is presented as a dotted line in Fig. 1A, for clarity only in a limited potential range.

Au(111)-(1 × 1) electrodes modified with the dithiol HS-6V6-SH, bromide salt, according to protocol 3 (Fig. 1B) gave results similar to those just described for HS-6V6-H: The double layer capacitance of a freshly prepared film amounts to 10 μF cm^{-2}. The first pair of redox peaks is rather reversible with P1 ($E = -0.438$ V, FWHM = 0.125 V) and P1′ ($E = -0.419$ V, FWHM = 0.095 V), and scales linearly with the scan rate. Careful exclusion of oxygen during the assembly and characterization steps resulted in a rather symmetric peak shape. The peak current scales linearly with scan rate. The monolayer coverage is estimated to be (4.5 ± 0.5) × 10^{-10} mol cm^{-2}, which corresponds to an area of 0.37 nm^2 per molecule. The second pair of peaks is less reversible, P2 ($E = -0.970$ V, FWHM = 0.136 V), P2′($E = -0.840$ V), and one often observes a splitting of P2′. The reductive desorption occurs at P3 ($E = -1.210$ V, FWHM = 0.028 V), and consumes a charge of 140 μC cm^{-2}. Partial oxidative re-adsorption is observed at P3′($E = -0.935$ V). The larger values of the FWHM of both oxidation/reduction peaks attributed to the viologen moiety in HS-6V6-SH indicate repulsive interaction and a higher disorder within the dithiol viologen adlayer as compared to the monothiol film. The higher value of the interfacial capacitance and the more negative reductive desorption peak P3 may represent a more hydrophilic electrolyte|sulfur end group interface than in the case of the CH_3-terminated monothiol. The latter trend was also observed in a comparative study of alkane mono- and dithiols on Au(111).[77] However, one important difference between HS-6V6-SH and HS-6V6-H needs to be mentioned: The presence of oxygen during the assembly process resulted in a distorted redox-response with indications of multi-layer formation. The strict absence of oxygen allowed us to prepare, for a wide range of conditions, adlayers with stable HS-6V6-SH coverages of $\Gamma = (1.1 \pm 0.2) \times 10^{-10}$ mol cm^{-2} according to protocol 2 (Section 2.3).

Comparing the potentials of the characteristic redox-processes $V^{2+} \leftrightarrow V^{+\bullet}$ and $V^{+\bullet} \leftrightarrow V$ for HS-6V6-SH and HS-6V6-H under the present experimental conditions with literature data[51–54,56–60,63] indicates that co-adsorbed bromide ions are partially replaced by perchlorate ions, which are in excess in the supporting electrolyte chosen in this study.

3.2 Cyclic voltammetry with gold nano-electrodes

The knowledge gained in the preparation and characterization of HS-6V6-H and HS-6V6-SH modified Au(111)-(1 × 1) macro-electrodes was applied to immobilize the viologen derivatives on electrochemically etched gold nano-electrodes (Fig. 2), which were coated with an insulating layer of polyethylene except at the apex. The electrode area exposed to the electrolyte after coating was

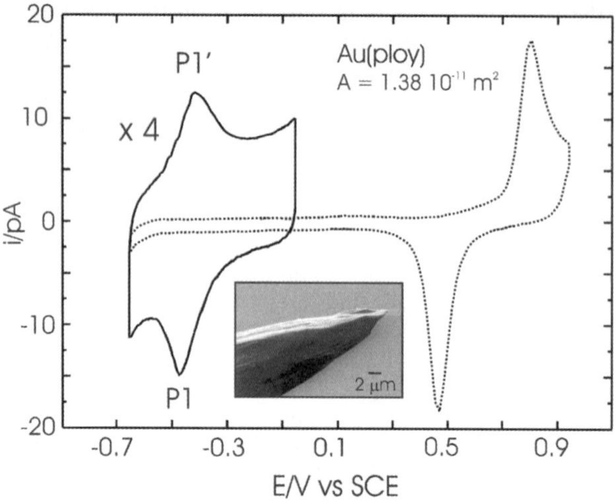

Fig. 2 Cyclic voltammograms of a bare (dotted line) and a HS-6V6-H-modified (solid line) gold nano-electrode in 0.05 M $KClO_4$, adjusted with phosphate buffer to pH 7, sweep rate 50 mV s^{-1}. The inset shows a SEM micrograph of a freshly etched, uncoated electrode.

estimated from calibration experiments with polycrystalline macro-electrodes of known geometry based on the charges consumed during the gold surface oxidation and reduction. The dotted trace in Fig. 2 shows the voltammograms of a typical gold nano-electrode in 0.05 M $KClO_4$, adjusted with phosphate buffer to pH 7. The effective free electrode area was estimated to 1.4×10^{-11} m^2. The SEM image of the uncoated tip is displayed as an inset in Fig. 2. Immobilization of the redox-active viologen was carried out by exposing the very end of the coated gold STM tip to a 1 mM ethanolic solution of HS-6V6-H or HS-6V6-SH in a controlled argon atmosphere for 1 h. Subsequently, the tip was carefully rinsed with pure ethanol, dried in a stream of argon, and immersed in a custom-made electrochemical cell to access its activity. The solid trace in Fig. 2 reveals, as an example, a well-behaved redox-response $V^{2+} \leftrightarrow V^{+\bullet}$ of surface immobilized HS-6V6-H. The oxidation and reduction peaks were found at $E = -0.428$ V and $E = -0.450$ V. The peak height scales rather linearly with scan rate. The amount of immobilized viologen was derived from the charges of the oxidation and reduction peaks, after background subtraction and assuming a one-electron redox-process, to 5×10^{-12} C. Referring to the free electrode area one obtains as an estimation of the surface coverage 3.7×10^{-10} mol cm^{-2}. Similar results were also obtained with HS-6V6-SH. These observations are in very good agreement with the results reported for Au(111)-(1 × 1) macro-electrodes indicating that the viologens retain their redox-functionality. Such modified nano-electrodes are suitable candidates to explore tunneling characteristics of redox-molecule modified nanogaps by current–(bias) voltage spectroscopy employing an electrochemical STM configuration.

3.3 SEIRAS experiments on self-assembly and redox-functionality

The assembly of HS-6V6-H and HS-6V6-SH on quasi-single crystalline Au(111–25 nm) film electrodes was first monitored in ethanolic solution without potential control. We ensured that the formation and characterization of the adlayers was strictly carried out in the absence of oxygen. Exposure to 50 μM solution of the respective viologen derivatives for 2 min, subsequent rinsing with pure ethanol to remove second-layer adsorbates, and drying of the adlayer in a stream of argon, all within the same spectroelectrochemical cell,[49] resulted only in rather weak bands in the C–H stretching region. No bands related to in-plane vibrations of the bipyridinium ring were observed. Application of the surface selection rules of SEIRAS indicates that the above preparation yields viologen adlayers, where the aromatic ring appears to be oriented close to parallel to the gold surface. Longer assembly times in 1 mM ethanolic viologen containing solution gave high coverage monolayers with prominent vibration modes in the 3200–2800 cm^{-1} and 1750–1100 cm^{-1} regions

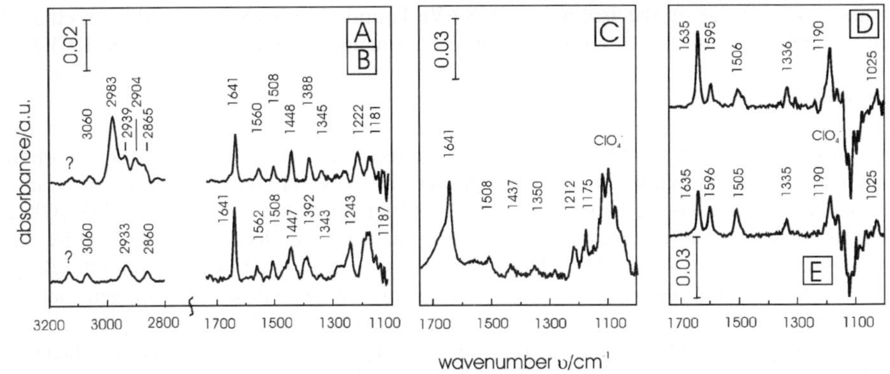

Fig. 3 SEIRAS spectra for both viologen derivatives on Au(111-25 nm) electrodes. Fig. 3A and Fig. 3B represent the spectral regions 3200 to 2800 cm^{-1} and 1750 to 1100 cm^{-1} for the assembly of high coverage monolayers from 1 mM ethanolic solutions of HS-6V6-H (A) and HS-6V6-SH (B) without potential control. The assembly time was 2 h. The plotted absorption spectra refer to the "dry" organic monolayers obtained after solution exchange, careful rinsing with bare ethanol and subsequent evaporation of the solvent in a gentle stream of argon. The reference spectrum was measured for the bare gold electrode in contact with dried argon. Fig. 3C represents the SEIRAS spectrum of a monolayer of the dication form of HS-6V6-H (V^{2+}) recorded at $E = -0.200$ V in 0.05 M $KClO_4$, pH adjusted to 7. The reference spectrum was obtained in the absence of viologen in 0.05 M $KClO_4$, pH = 7, at $E = -0.200$ V. The formation of the self-assembled adlayer was carried out from 1 mM ethanolic HS-6V6-H for 2 h. Fig. 3D and 3E represent SEIRAS spectra of the radical cations $V^{+\bullet}$ of HS-6V6-H (D) and HS-6V6-SH (E) obtained after a potential scan from $E_i = -0.200$ V to $E_f = -0.600$ V. The reference spectra were measured at $E_i = -0.200$ V.

(Fig. 3A and 3B). The reference spectra were acquired for the bare gold surface. The data for HS-6V6-H shall be described in detail. Differences of the HS-6V6-SH adlayer will be addressed comparatively. The bands in 3200–2800 cm^{-1} could be assigned to C–H stretching vibrations of the aromatic ring (υ_{CH} at 3060 cm^{-1}), the CH_3-asymmetric ($\upsilon_a(CH_3)$ at 2983 cm^{-1}) and symmetric ($\upsilon_s(CH_3)$ at 2904 cm^{-1}), respectively, the CH_2-asymmetric ($\upsilon_a(CH_2)$ at 2939 cm^{-1}) and the symmetric ($\upsilon_s(CH_2)$ at 2865 cm^{-1}) stretching modes.[78–80] The vibration modes $\upsilon_a(CH_3)$ and $\upsilon_s(CH_3)$ are absent for the HS-6V6-SH adlayer (Fig. 3B). The observed spectroscopic bands below 1750 cm^{-1} represent vibration modes of the bipyridinium ring having dipole moment components either parallel to the longer molecular axis, as defined by the 4,4′-*N*,*N*′- direction (B_{2u} in D_{2h} symmetry; 1641, 1508, 1222 and 1181 cm^{-1} with the latter dominated by the N-$(CH_2)_n$-R stretching vibration), or perpendicular to this axis (B_{3u} modes in D_{2h} symmetry; 1560, 1448, 1345 cm^{-1}).[78–83]

The positions of the methylene C–H-stretching modes provide a measure of the intermolecular environment of the alkyl chains in the molecular assembly. The peak frequencies of $\upsilon_s(CH_2)$ (2865 *vs.* 2850 cm^{-1}) and $\upsilon_a(CH_2)$ (2939 *vs.* 2920 cm^{-1}) in the liquid state are typically higher than those for a highly ordered crystalline polymethylene chain.[84] The applications of Porter's results to the present system leads to the conclusion that the methylene chain environment of the high coverage adlayers of HS-6V6-H and HS-6V6-SH are liquid-like, *i.e.*, there exists a considerable degree of conformational disorder in the two-dimensional arrangement of the alkyl chains. The observation of the aromatic C–H stretching mode at 3060 cm^{-1}, together with the simultaneous occurrence of the B_{2u} and B_{3u} in-plane bipyridinium ring vibrations, indicate a tilted orientation of both, the long and the short axis of the coplanar viologen moiety[73,78–80] with respect to the surface normal. Support for this conclusion is provided by recent results with related systems. Sagara *et al.*[57] concluded, using electro-reflectance spectroscopy that the long axis of the viologen group in *N*-butyl-*N*′-(4-thiobutyl)-4,4′-bipyridinium dihexaflourophosphate on Au(poly) exhibits an orientation angle of 70° with respect to the surface normal. Based on a combined infrared and near edge X-ray absorption fine structure spectroscopy study Rong *et al.*[85] reported tilting and leaning angles of 45° and 70°, respectively, for the aromatic moiety in monolayers of ω-(4′-methyl-biphenyl-4-yl)-alkanethiols with an even number of methylene spacer groups on Au(111).

The above knowledge on the assembly of HS-6V6-H and HS-6V6-SH on gold from ethanol was applied to explore the adlayer properties under electrochemical conditions in 0.05 M $KClO_4$ with a pH adjusted to either 7 or (in some cases) to 10. Because of stability reasons of the adlayer we restricted the spectroelectrochemical studies to the potential region $-0.700\ \text{V} \leq E \leq -0.200\ \text{V}$, *e.g.*, to $V^{2+} \leftrightarrow V^{+\bullet}$. High coverage monolayers on Au(111, 25 nm) were prepared according to protocol 3 in paragraph 2.3. The *in situ* SEIRAS spectra were recorded with reference to the adsorbate free electrolyte at −0.200 V. While the spectral signature in the C–H stretching region appears to be similar to the results in bare ethanol (Fig. 3A), we observed distinct changes in 1750–1000 cm^{-1}. The vibrations with net dipole moment changes parallel to the molecular axis defined by the 4,4′-*N*,*N*′-direction (B_{2u}, 1641, 1508, 1212, 1175 cm^{-1}) appear with a higher intensity than the B_{3u} modes (1437, 1350 cm^{-1}). This observation points to a stronger alignment of the viologen moiety with respect to the surface normal. Furthermore, we notice that the bipyridinium ring vibration at 1641 cm^{-1} is superimposed with a rather broad HOH bending mode of co-adsorbed interfacial water, δ_{HOH}. This finding supports our previous hypothesis based on values of the interfacial capacitance. The prominent positive peak at 1100 cm^{-1} is attributed to the symmetric Cl–O stretching mode of perchlorate ions incorporated into the HS-6V6-H respective HS-6V6-SH adlayer lattice. The presence of perchlorate ions increases the 2D order within the organic film as derived from peak shapes and intensities of the B_{2u} modes. The replacement of bromide or chloride ions in self-assembled monolayers of thiol-functionalized viologens by ions with a smaller dynamic hydration sphere, such as ClO_4^- and PF_6^-, was already suggested in voltammetry studies of Buttry[53,54] and Sagara.[58] The lower electron donor ability of ClO_4^- reduces the charge transfer between the viologen moiety and the co-adsorbed anion.[74]

Increasing the electrode potential from $E = -0.200$ V to $E = -0.600$ V triggers the one electron reduction of $V^{2+} \leftrightarrow V^{+\bullet}$. The simultaneously acquired SEIRAS difference spectra show the evolution of strong positive bands around 1635, 1595, 1506 (5), 1336 (5), 1190 and 1025 cm^{-1} for both viologen adlayers. The spectroscopic response is completely reversible, and agrees well with our voltammetry studies (paragraph 3.1.). The data acquired at $E = -0.600$ V are plotted in Fig. 3D and Fig. 3E. The prominent bands at 1635 and 1190 cm^{-1} can be assigned to the asymmetric in-plane vibrations of the monomer radical cation $V^{+\bullet}$. The others are attributed to "activated" totally symmetric (A_g) ring vibrations of the radical cation dimer $(V^{+\bullet})_2$, in which plane-to-plane arranged molecules are linked by π-bonding.[78–83] The IR spectrum of the electrochemically generated dimer species is interpreted by the vibronic coupling in a one-dimensional charge transfer complex.[82] The negative band around 1100 cm^{-1} indicates the release of perchlorate ions upon the reduction $V^{2+} \leftrightarrow V^{+\bullet}$. The radical cation monomer is stabilized in a hydrophobic environment[80] and by electrostatic attraction to co-adsorbed ClO_4^- anions. The spectra of HS-6V6-H (Fig. 3D) and HS-6V6-SH (Fig. 3E) demonstrate that the stability of the radical cation monomer appears to be higher in the monothiol adlayer. The formation of the radical cation dimer for the dithiol adlayer is facilitated by stronger π-interactions between the radicals compared to Coulomb interactions between radical monomers and anions,[82] as well as by the matching in the commensurability of the intra-assembly planes.[86]

A comprehensive analysis of the steady state and time-resolved ATR-SEIRAS experiments on the self-assembly and electroreduction of HS-6V6-H and HS-6V6-SH on gold Au(111, 25 nm) will be given elsewhere.[87]

3.4 Interfacial structure of HS-6V6-H and HS-6V6-SH

In situ STM was applied to monitor viologen adlayers assembled according to the preparation protocols (1) to (3). Exposure of an Au(111)-(1 × 1) electrode to a 50 μM ethanolic solution of HS-6V6-H or HS-6V6-SH for 1 min generated a low coverage, liquid-like disordered adlayer (protocol 1). Modification according to protocol 2 (2 min contact with 50 μM ethanolic viologen solution, 12 h annealing in pure ethanol at 70 °C) resulted in highly ordered, but extremely fragile adlayers. Both HS-6V6-H (Fig. 4A and 4B) and HS-6V6-SH (Fig. 4D and 4E) assemble on Au(111)-(1 × 1) in $-0.70\ \text{V} \leq E \leq -0.20$ V, *i.e.*, the stability ranges of V^{2+} and $V^{+\bullet}$ $(V^{+\bullet})_2$, in regular arrays of parallel stripes, which cover the entire substrate surface. Typical results for $E = -0.20$ V are summarized in Fig. 4. The stripes are not uniform. Bright and dark contrast patterns alternate. Individual domains extend 10 to 50 nm. Neighboring domains with sharp boundaries are mutually

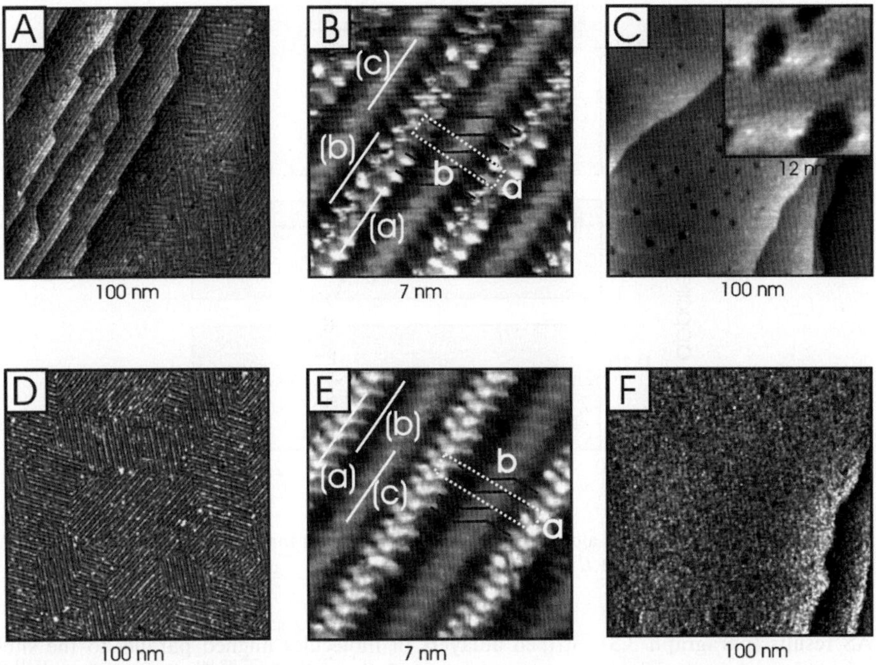

Fig. 4 *In situ* STM images of HS-6V6-H (Fig. A, B and C) and of HS-6V6-SH (Fig. D, E, F) on Au(111)-(1 × 1) in 0.05 M $KClO_4$, pH = 7, recorded at $E_S = -0.200$ V. The setpoint current was typically between 50 and 100 pA. The images represent the low-coverage "striped" adlayers (A, B, D and E) assembled according to protocol 2, and the high coverage monolayers formed according to protocol 3 (Fig. 4C and F) in paragraph 2.3 (see text for details). The high resolution images show distinct row patterns, which are indicated by the traces (a), (b) and (c). The typical zig–zag motif of the aligned segments is drawn as a sequence of black solid lines. The dotted rectangles illustrate an adlayer repeat pattern with $a = 2.6$ nm and $b = 0.5$ nm. Table 1 summarizes characteristic parameters of the viologen adlayers on Au(111)-(1 × 1).

rotated by multiples of 120° indicating registry of the adlayer with respect to the hexagonal substrate surface. This hypothesis is supported by the faceting of steps (Fig. 4A). Steps are aligned with characteristic angles of 120°. Step edges are decorated by parallel molecular rows. Occasionally one observes monatomically high gold and/or vacancy islands at domain boundaries. High resolution experiments such as shown in Fig. 4B and 4E reveal more details of the molecular adlayers. Rows composed of two (HS-6V6-H) or a more complex pattern (HS-6V6-SH) of bright dots (trace a) are separated by parallel rows of dark grey (trace b) respective light grey (trace c) ellipsoidal-like segments. Their axes exhibit characteristic tilt angles with respect to the main row direction, which were estimated to be 60 ± 5° (light grey rows) and 90 ± 10° (dark grey rows). The resulting, nearly equilateral zig–zag motif is indicated in both high resolution images. Parallel rows of the same type are separated by (2.6 ± 0.3) nm and (2.7 ± 0.3) nm for HS-6V6-H and HS-6V6-SH, respectively. Cross section profiles reveal a common periodicity of (0.50 ± 0.05) nm between individual features of each type of row, but distinct differences in the apparent corrugation height. Fig. 5 illustrates the corresponding results for HS-6V6-H. We estimated 0.14 to 0.19 nm (trace a), 0.03 to 0.06 nm (trace b) and 0.05 to 0.12 nm (trace c). Similar data were obtained for HS-6V6-SH. The above results suggest a commensurate repeat motif with $a = 0.50 \pm 0.05$ nm ($= \sqrt{3}\, a_{Au}$) parallel and $b = 2.6 \pm 0.3$ nm ($\sim 9\, a_{Au}$) perpendicular to the molecular rows.

In an attempt to derive structural models of the two adlayers we notice that the values of the apparent corrugation heights are similar to data reported for alkanethiols[88] aligned with their C–C–C backbone parallel, and for aromatic molecules[89] oriented with the π-system of the phenyl ring planar or slightly tilted to the Au(111) surface. This arrangement appears to be applicable for the two viologen adlayers prepared according to protocol 2. The conclusion is also supported by our

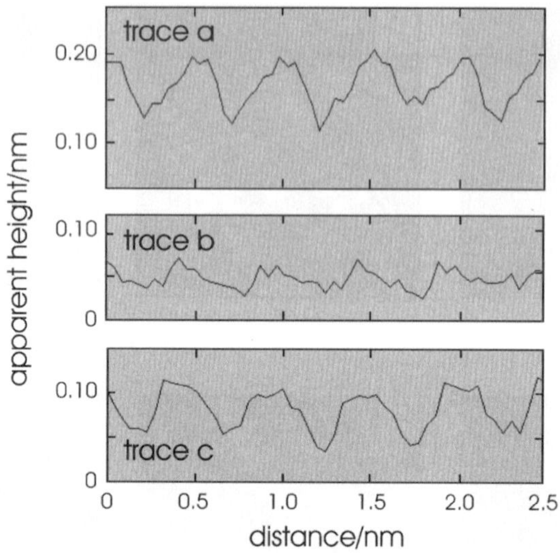

Fig. 5 Typical cross section profiles along the traces drawn in Fig. 4B for the striped adlayer of HS-6V6-H on Au(111)-(1 × 1) in 0.05 M $KClO_4$ at $E = -0.200$ V.

SEIRAS results (paragraph 3.3). Striped adlayers of molecules aligned parallel to the substrate surface have been described for low coverage phases of alkanethiols,[77,88,90] alkanedithiols[77,91,92] and several *n*-phenyl-alkanethiols on Au(111).[93–95] An enhancement of the tunneling current was observed in the vicinity of the sulfur group, often apparent as bright spots or slightly elongated features,[90–92] while the alkyl chains are observed as thin bands of rather dark contrast.[90,91] Aromatic groups, such as phenyl-, biphenyl- or bipyridine-units were also reported to enhance the tunneling contrast.[89,96–98] In applying this knowledge to the present case we notice that the width of the dark rows (trace b) increases with the length of the alkyl chains when replacing the hexyl groups by octyl (C_8) or decyl (C_{10}) in HS-6V6-H respective HS-6V6-SH, while all other features in the bright and light grey rows do not change.[87] The dimensions of the corresponding repeat motifs, which all indicate commensurability to the underlying Au(111) lattice, are compiled in Table 1. The values of b increase approximately 0.5 nm per $(CH_2)_2$-unit, which exactly scales with twice the distance between neighboring CH_2-groups.[99] In consequence, we assign the dark rows to the positions of the alkyl chains in a fully extended all-*trans* conformation with the molecular axis parallel to the surface plane. Although we could not resolve the gold lattice separately in these experiments, we assume, based on adsorption studies of alkanes on Au(111),[99] that the C–C–C axis aligns with the [1 $\bar{1}$ 0] direction of the hexagonal substrate. The light grey lines in Fig. 4B and 4E exhibit an internal structure, which resembles the arrangement of planar or extended stacking rows of bipyridine units in self-assembled monolayers.[97,98] In analogy to these reports we attribute the light grey rows in the high resolution STM images tentatively to the positions of the viologen moieties. The intermolecular distance between neighboring units, 0.43 nm ($= \sqrt{3}\, a_{Au} \sin 60°$) is too small to allow a plane–parallel arrangement to the substrate, but also too large for a perfect stacking structure. The latter is not expected to be stable due to electrostatic repulsion between 4,4′-bipyrdinium dications. Based on crystallographic studies with co-crystallized ClO_4^- ions[74] we propose a slightly tilted and twisted conformation of the two rings. The more open motif could also accommodate the co-adsorbed ClO_4^- ions. The exact arrangement could not be resolved in more detail. Despite the limited knowledge, we can assume a sufficiently open arrangement of the viologen units, which, most probably, is stabilized by co-adsorbed ClO_4^- (*cf.* paragraph 3.3). Finally, we assigned the bright rows to the positions of the terminal sulfur functionalities. The suggested structural models of HS-6V6-H and HS-6V6-SH on Au(111)/0.05 M $KClO_4$ are drawn in Fig. 6. We assume an antiparallel head-to-tail –S|CH_3– configuration for the monothiol and a head-to-head –S|S– configuration for the dithiol. The corresponding unit cells appear to be multiples of the characteristic repeat patterns (Fig. 4). They contain four (HS-6V6-H) respective two (HS-6V6-

Table 1 Repeat motif (a, b, α) and unit cell (a', b', α') parameters of the low coverage "striped" phases of viologen mono- and dithiols on Au(111)-(1 × 1)

Monothiol	Repeat a/nm	Motif b/nm	$\alpha/°$	Unit a'/nm	Cell b'/nm	$\alpha'/°$	$\Gamma/10^{-10}$ mol cm^{-2}
HS-6V6-H	2.6 ± 0.3	0.50 ± 0.05	90 ± 10	5.35	1	87	1.24 (1.10 ± 0.20)[b]
HS-8V8-H[a]	3.15 ± 0.20	0.48 ± 0.05	89 ± 8	6.50	1	87.5	1.02
HS-10V10H[a]	3.8 ± 0.3	0.50 ± 0.06	90 ± 7	7.65	1	88	0.87
Dithiol							
HS-6V6-SH	2.7 ± 0.3	0.50 ± 0.03	90 ± 8	5.35	0.5	87	1.24 (1.05 ± 0.30)[b]
HS-8V8-SH[a]	3.2 ± 0.3	0.50 ± 0.05	88 ± 7	6.50	0.5	87.5	1.02
HS-10V10-SH[a]	3.7 ± 0.5	0.50 ± 0.05	90 ± 10	7.65	0.5	88	0.87

[a] *cf.* Ref. 87; [b] Coverages estimated from electrochemical experiments such as plotted in Fig. 1.

SH) molecules. The estimated coverage corresponds to the values of the voltammetry experiments (Table 1). The STM experiments give slightly higher values because the molecular disorder existing at defects and domain boundaries is not accounted for in the calculations.

When changing the electrode potential to −0.600 V, *i.e.*, the radical cation $V^{+\bullet}$ $(V^{+\bullet})_2$ is present, we observed an increasing instability of the ordered molecular adlayers. While the main structural pattern is not changing, the light grey rows, previously attributed to the positions of the viologen moieties, could be easily distorted. This observation indicates a lower stability of the two-dimensional adlayer pattern.

We also report on the high coverage adlayers of HS-6V6-H and HS-6V6-SH formed according to protocol 3. We obtained monolayers for both systems. The monothiol forms an expanded hexagonal adlayer, which covers the entire substrate surface (Fig. 4C). The structure resembles a $(\sqrt{7} \times \sqrt{7})$ motif. The resulting coverage amounts to 3.3×10^{-10} mol cm^{-2}, which is in good agreement to the voltammetry data. The adlayer is composed of many, atomically deep vacancy islands with a typical size of 2 to 5 nm in diameter. The defects distort the molecular structure considerably. The open structure is expected due to the bulky size of the tilted viologen moiety (*cf.* paragraph 3.3), which prevents a $(\sqrt{3} \times \sqrt{3})$ arrangement. The latter is often observed for high coverage adlayers of thiols on Au(111).[88,90,95]

No ordered structure was found for the high coverage phase of upright HS-6V6-SH molecules on Au(111)-(1 × 1) (Fig. 4F). The difference to the monothiols nicely illustrates the structure-determining role of the alkyl chain above the aromatic group as already pointed out by Ulman *et al.*[100] We did not observe any evidence of multilayer adsorption. However, the latter was detected if the experiment was carried out in the presence of oxygen.

3.5 Addressing of local redox-activity

Employing an *in situ* STM configuration (four electrode arrangement) we recorded current–distance and current–voltage characteristics (*cf.* paragraph 2.5) to explore electron transport and tunneling properties of single and/or small assemblies of redox-active viologen derivatives immobilized on well-defined gold electrodes. Specifically, we aim to address the fundamental question how does the molecular conductivity change by tuning the redox state of the molecule (the "electrochemical gate effect").

Symmetric molecular junctions Au(T)|HS-6V6-SH|Au(S). The *first approach* involves the repeated formation and breaking of molecular junctions between a polyethylene coated gold tip and a Au(111)-(1 × 1) substrate electrode modified with HS-6V6-SH. The dithiol is composed of two SH-anchor groups, which could bridge the tip and the substrate electrodes and form molecular junctions. We have chosen a low-coverage, liquid-like phase to prevent two-dimensional long-

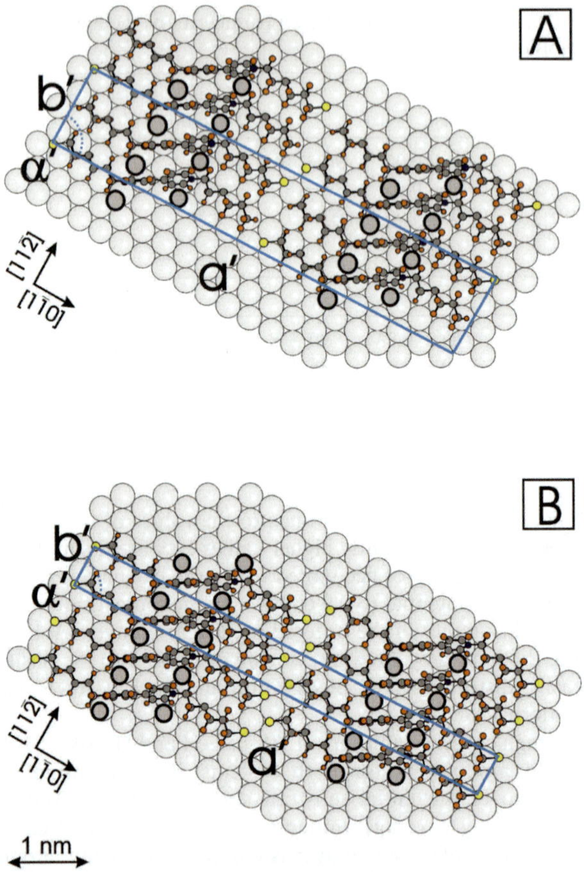

Fig. 6 Proposed packing models of the ordered striped phases of HS-6V6-H (A) and HS-6V6-SH (B) on Au(111)-(1 × 1). Co-adsorbed perchlorate ions are drawn as open circles, and they are tentatively located next to the ring nitrogen atoms. Sulfur atoms are represented as yellow circles. Their suggested positions are three-fold hollow sites of the substrate lattice. The suggested unit cells, given by $\begin{pmatrix} 18 & 1 \\ -2 & 4 \end{pmatrix}$(HS-6V6-H) and $\begin{pmatrix} 18 & 1 \\ -1 & 2 \end{pmatrix}$(HS-6V6-SH) in matrix notation with reference to the basic lattice vectors of the Au(111), are indicated by bold solid lines. The unit cell parameters a', b' and α' are summarized in Table 1, together with the dimensions of the experimentally estimated repeat motifs a, b, and α.

range order within the organic adlayer, and therefore to allow for different surface conformations of planar oriented HS-6V6-SH molecules. The study was carried out under potential control in 0.05 M $KClO_4$, pH adjusted to 7, in a chamber filled with argon to prevent exposure to oxygen. Attempts to create individual Au(T)|HS-6V6-SH|Au(S) junctions are based on slight, but distinct modifications of two techniques recently introduced by Tao[25] and Haiss.[26] The experiment started under feedback control with a setpoint current of 100 pA ensuring negligible interactions between the STM tip and the organic adlayer. The tip–sample distance was estimated, according to eqn. (1):[50]

$$|i_T| = G_o|E_T - E_S|\exp(-\kappa\sqrt{\Phi}z_o) \quad (1)$$

with $G_o = 2\,e_o^2/h$ as the conductance quantum, $\kappa = 10.12\ \text{eV}^{-1/2}\ \text{nm}^{-1}$ and assuming a rectangular potential barrier with a typical height $\Phi = 1.0$ eV,[50] one obtains for $|E_T - E_S| = 0.10$ V $z_o \sim 1.11$ nm. This value was considered as the upper limit of the approaching distance Δz_a. Typically, we have chosen $(z_o - \Delta z_a)$ with $\Delta z_a = 1.0$ nm to ensure sufficiently strong interaction between the uncoated very end of the gold tip and the disordered HS-6V6-SH adlayer, but also to prevent

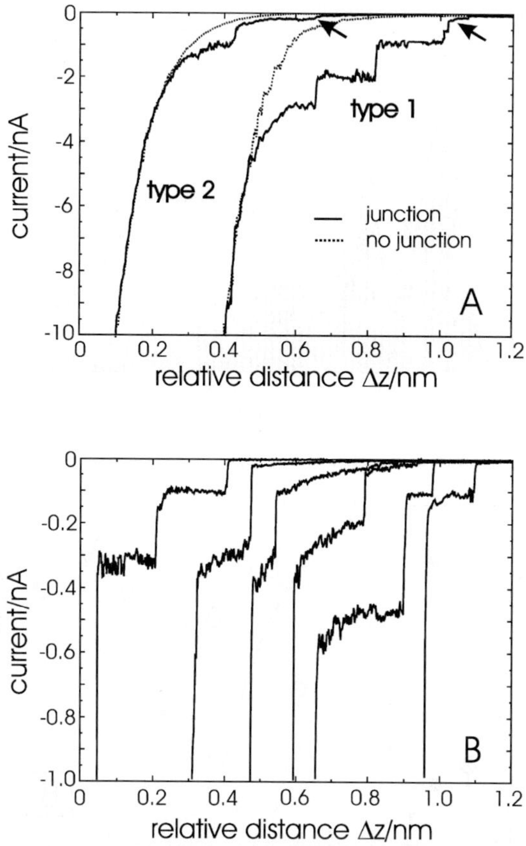

Fig. 7 Current–distance ($i - \Delta z$) "pulling" curves recorded with a gold STM tip, coated with polyethylene except for the very end of the apex, in the presence of a disordered, low coverage HS-6V6-SH adlayer (protocol 1) on Au(111)-(1 × 1) in 0.05 M $KClO_4$, pH = 7, at $E_S = -0.250$ V and $E_T = -0.150$ V. Individual traces were acquired according to a pre-defined loop described in detail in paragraph 3.5. The horizontal axes correspond to a relative distance scale. For clarity, individual traces are plotted with an offset. Fig. 7A illustrates representatives of two groups of $i - \Delta z$ curves characteristic for the breaking of molecular junctions Au(T)|HS-6V6-SH|Au(S). The solid traces are non-exponential with a series of plateaux, which are separated either by abrupt steps (type 1) or segments of gradually decreasing currents (type 2). The arrows indicate the positions of the lowest current plateaux. The dotted lines indicate i–Δz curves in the absence of molecular junctions between gold tip and substrate. Fig. 7B shows the low current segments of a sequence of i–Δz traces for the stretching of HS-6V6-SH junctions. The complete curves were recorded in $z_o - 1.0$ nm $< z < z_o + 2.0$ nm and $0 < i \leq 10$ nA; z_o was defined by the initial conditions $i_T = 100$ pA and $(E_T - E_S) = 0.100$ V.

mechanical contact to the gold surface. Individual traces were acquired according to the following loop: (1) feedback disabled at z_o (defined by $i_o = 100$ pA), (2) approach with a maximum rate to ($z_o - 1.0$ nm), (3) dwelling time of 100 ms at ($z_o - 1.0$ nm) to form molecular junctions, (4) recording of the current *vs.* distance characteristics upon retraction from ($z_o - 1.0$ nm) to ($z_o + 2.0$ nm) with a pulling rate of 6 nm s^{-1}, (5) stabilizing of the tip at z_o and switching on the feedback with $i_o = 100$ pA, (6) repetition of the cycle, and after every 10th trace inspection of the surface by *in situ* STM imaging. Fig. 7A shows the low current range of two characteristic stretching curves recorded at $E_S = -0.250$ V and $E_T = -0.150$ V in $0 < i \leq 100$ nA. The traces are non-exponential exhibiting a series of plateaux with a typical length of 0.2 to 0.3 nm which are separated by abrupt steps ("type 1") or by segments of a gradually decreasing current ("type 2"). The corresponding currents are rather low, and cannot be attributed to the well-known conductance quantization of a metal nanowire.[25] Instead, we ascribe these conductance steps to the breaking of individual, respective multi-molecular (HS-6V6-SH) junctions previously formed between the gold STM tip and the

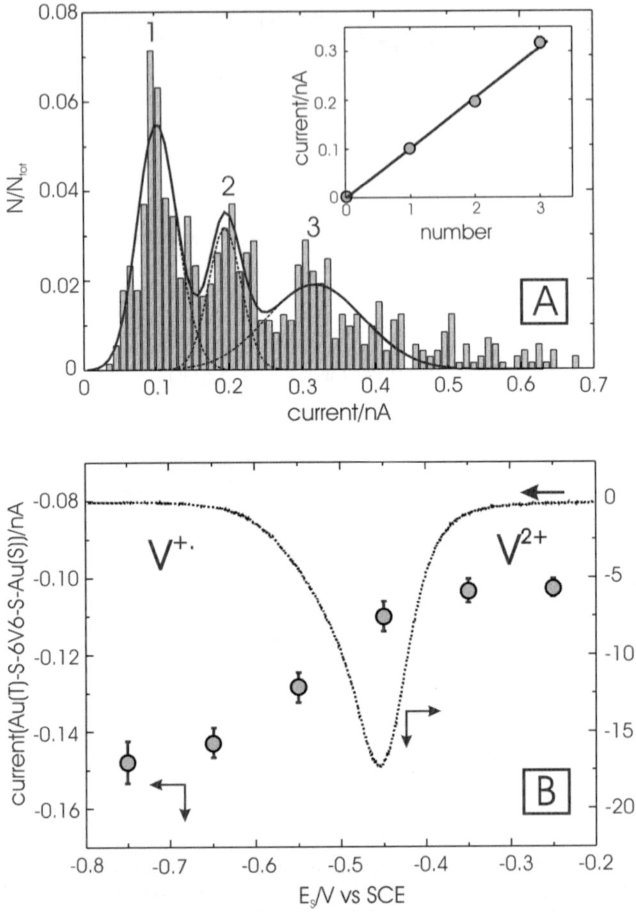

Fig. 8 (A) Histogram constructed from values of the plateau currents of a series of more than 500 individual i–Δz pulling curves for molecular junctions Au(T)|HS-6V6-SH|Au(S) recorded at $E_S = -0.250$ V and $E_T = -0.150$ V. Examples of individual traces were plotted in Fig. 7B. The peaks correspond to 1, 2 or 3 molecules in the gap between two gold electrodes. The peak positions were determined by Gaussian fits (dashed lines) to the histogram peaks (black solid line as the sum of the Gaussians). The inset shows the linear correlation between current and peak number. Fig. 9B represents a plot of the single molecule conductance currents of Au(T)|HS-6V6-SH|Au(S) junctions *vs.* the substrate potential E_S at fixed bias $(E_T - E_S) = 0.100$ V. The sign convention of the bias voltage gives the negative conduction currents. Each data point was obtained from histograms similar to the one in (A). The dotted line corresponds to the macroscopic current–voltage curve for the reduction of $V^{2+} \rightarrow V^{+\bullet}$ (*cf.* cyclic voltammograms in Fig. 1B).

substrate surface[25,26] This hypothesis is supported by the following observations: (1) Only exponentially decaying current–distance traces were found for the bare supporting electrolyte in $0 < i \leq 100$ nA (*cf.* dotted curves in Fig. 7A). (2) No current steps or extended plateaux were detected for the monothiol HS-6V6-H. This molecule contains only one HS–group, and is therefore not capable to form stable molecular bridges between the two gold electrodes. (3) There exist only discrete plateau currents in the i–Δz curves (Fig. 7B), which show a distinct dependence on the applied electrode potential (see Fig. 8).

We notice that non-exponential i–Δz characteristics were found in 30 to 40% of all the stretching transients. We attribute this observation to the small electrochemically active area of our coated gold STM tips and to the explicitly chosen low coverage of adsorbed HS-6V6-SH. Inspection of the electrode surface by *in situ* STM imaging provided a clue to the different nature of the two types of i–Δz traces. Individual transients with extended plateaux and rather large current steps ($\Delta i \ll 0.5$

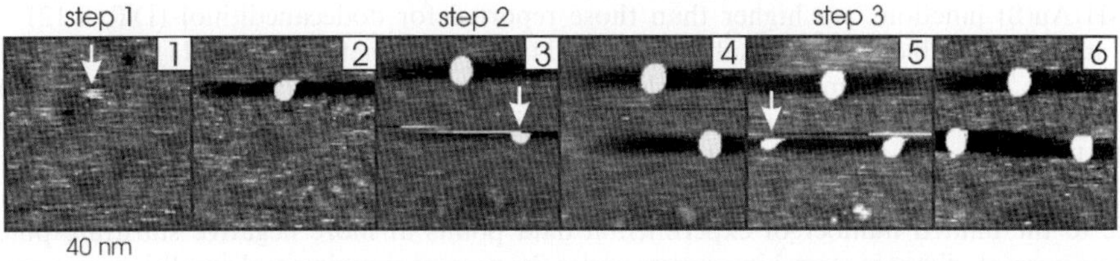

Fig. 9 Series of *in situ* STM images acquired simultaneously with three successive i–Δz traces of type 1 for a disordered adlayer of HS-6V6-SH on Au(111) in 0.05 M $KClO_4$. The white arrows in the images with odd numbers represent the formation of small gold islands immediately after the completion of an individual pulling event. The experiment was carried out at $E_S = -0.250$ V and $E_T = -0.150$ V, $i_T = 100$ pA. We notice, that the negative potential (and charge density) caused the partial reconstruction of the substrate surface (indicated by corrugations with a double row pattern), which could not be prevented by the low concentration of viologen.

nA) generated typically small gold islands on the Au(111) surface with a diameter of 3 to 5 nm. Fig. 9 illustrates a characteristic STM sequence acquired simultaneously with the recording of successive i–Δz traces of type 1. The x–y positions of the individual transients are indicated by the arrows in the frames 1, 3 and 5 while the even-numbered frames show the same surface area with the newly created monatomically high gold islands. We observed no indications of a tip crash. We favor the idea that these islands are the result of a substantial material transfer from the gold STM-tip accompanied with the rupture of a large number of local molecular junctions. No evidence of gold islands or of any other morphological changes of the substrate surface were detected for i–Δz traces with extended plateaux and gradually decreasing current segments ("type 2") as well as for "type 1" transients with small current steps ($\Delta i \ll 0.5$ nA).

In an attempt to explore the conductance properties of single molecule junctions Au(T)|HS-6V6-SH|Au(S) we carried out systematic stretching experiments in the low current regime ($i < 10$ nA) at various substrate potential E_S while keeping the bias voltage ($E_T - E_S$) fixed. In order to minimize contributions from direct tunneling between tip and substrate we restricted the quantitative analysis to $i < 1.0$ nA. Fig. 7B illustrates, as an example, the low current segments of i–Δz stretching transients recorded at $E_S = -0.250$ V and $(E_T - E_S) = 0.100$ V. Following a proposal by Haiss *et al.*,[26] we carried out a statistical analysis of the plateau currents and constructed histograms for substrate potentials in $-0.750 \text{ V} \leq E_S \leq -0.250$ V, *i.e.* ,within the range of the reduction process $V^{2+} \leftrightarrow V^{+\bullet}$. The histograms, which contain data of more than 500 individual transients, show a characteristic series of peaks corresponding to one, two or three molecules in the gap. A typical histogram, which was obtained from data acquired at $E_S = -0.250$ V, is plotted in Fig. 8A. Fitting these histograms with Gaussians determines the peak centers and the full width at half maximum. These peak values are then plotted *vs.* the peak number, and the linear slope yields the current of a single Au(T)|HS-6V6-SH|Au(S) junction at fixed bias. We observed linear slopes of the corresponding current–voltage curves for small bias values $|E_T - E_S| \leq 0.100$ V. The resulting single molecule conductance currents are plotted in Fig. 8B as a function of E_S at $(E_T - E_S) = 0.100$ V. The conductance is rather constant in the stability region of the oxidized viologen dication V^{2+}, increases at $E_S < -0.400$ V and appears to reach a plateau at $E_S \leq -0.700$ V, *i.e.*, past the formal potential of the reversible redox process $V^{2+} \leftrightarrow V^{+\bullet}$. Attempts to extend the experiment to potentials more negative than E_S gave rather scattered results due to the onset of hydrogen evolution at the chosen pH of 7 and/or due to the increasing reduction current of trace amounts of oxygen in the system. The change of the single-molecule conductance was also reported by Haiss,[26] however, no plateau was observed by these authors. The differences between the results of the Liverpool group and those reported in this paper might be attributed to the different electrolytes used and/or to the presence of oxygen which is known to significantly alter the association and redox-properties of dithiol-functionalized viologens.[52] We tentatively attribute the sigmoidal-type increase of the molecular conductance current to the higher electron density and the higher conjugation of $V^{+\bullet}$ compared to V^{2+}. This view is supported by force field calculations of Hester *et al.*[78] They reported that the inter-ring bond order increases when the viologen dication transforms into the radical upon reduction, which implies that the aromatic character extends across both rings for the latter compound. The values of the single-molecule conductivities of Au(T)|HS-

6V6-SH|Au(S) junctions are higher than those reported for dodecanedithiol (DT, 0.121 nS),[101] and are comparable with results for 11,11′-bis(4-thiomethylphenyl)-11,11′-diapocarotene (C, 2.06 nS).[102] All three molecules exhibit approximately the same lengths, and consist of two anchor groups separating an insulating aliphatic chain (DT), a conducting polyene backbone (C) or a dialkyl-viologen unit (this work). The comparison of these three systems clearly demonstrates the sensitivity of the chosen approach to the molecular structure.

Due to the limited number of experimental data points at more negative substrate potentials based on current–distance stretching curves under the present experimental conditions, we postpone the more detailed discussion of possible ET scenarios after reporting results of a second, rather complementary approach, which is also based on an *in situ* STM configuration.

Asymmetric molecular tunnel junction Au(T)-S-6V6-H|Au(S). The second approach to explore local electron transfer properties of alkylthiol-terminated viologens is based on a HS-6V6-H-modified gold nano-electrode, which was inserted as a tip in an *in situ* STM configuration. We have chosen the monothiol as the redox-active reactant because it is less sensitive to oxygen and chemically more robust than the dithiol. The adsorption on gold electrodes is well defined, and has been carefully investigated by *in situ* STM and SEIRAS employing macro-electrodes. HS-6V6-SH and HS-6V6-H exhibit similar assembly and redox-properties because both contain the functional viologen moiety, two symmetrically arranged alkyl spacer groups of the same length and identical bromide counter ions. The fabrication and characterization of the HS-6V6-SH-modified gold tips have been described before in paragraphs 2.3, 2.5 and 3.2. Pre-selected tips were mounted in the STM setup and stabilized in a carefully controlled argon atmosphere. Contact to the electrochemical STM cell was established under strict potential control of tip and substrate and in the absence of oxygen. The cell was filled with 0.05 M $KClO_4$. The pH of the electrolyte was adjusted to 7 with KOH to ensure the chemical integrity of the viologen and to provide a sufficiently wide potential range of ideal polarizability of tip and substrate. The latter was a viologen-free Au(111)-(1 × 1) electrode (*cf.* paragraph 3.1). The tip voltammogram (*i vs.* E_T) was recorded as a background curve, in the absence of tunneling, with 2 V s^{-1} in $-0.650 \text{ V} \le E \le -0.200 \text{ V}$ (dotted

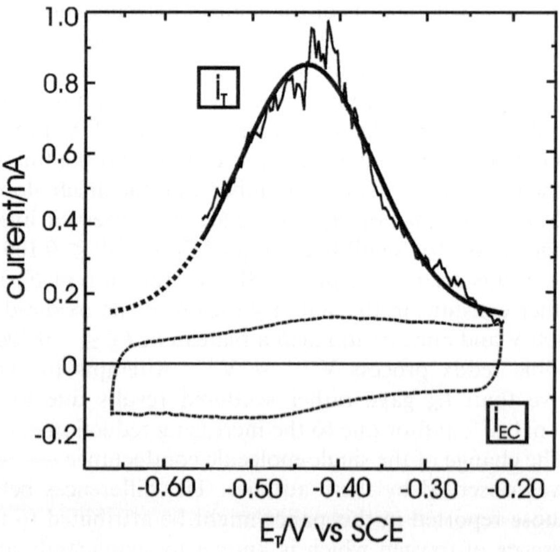

Fig. 10 Tunneling current *versus* tip potential recorded in 0.05 M $KClO_4$ at pH 7 for an asymmetric tunnel junction Au(T)-S-6V6-H|Au(S) at fixed bias $(E_T - E_S) = 0.100$ V. The solid black trace represents the average of 20 independent half cycles recorded when scanning the tip potential between −0.650 V and −0.200 V either in positive or in negative direction at 2 V s^{-1}, with the feedback being switched off during each individual cycle. The plotted curves are corrected for the electrochemical response of the HS-6V6-H-modified gold tip in the absence of tunneling. This trace is shown as a dashed line. The thick solid and dotted lines represent the result of a Gaussian fit to the redox-molecule-gated tunneling junction.

line in Fig. 10A). The observed response is determined by (i) the charge to build up the electrochemical double layer and (ii) the Faraday charge due to reduction and re-oxidation of immobilized HS-6V6-SH. The tunnel junction between the viologen-modified tip and the bare Au(111)-(1 × 1) electrode was formed by approaching the tip towards the surface under feedback control until its position was stabilized at a tunneling setpoint current $i_T = 100$ pA. Tip (E_T) and substrate (E_S) potentials were adjusted in the stability range of the oxidized viologen V^{2+}. These conditions minimized the interactions between substrate and modified tip. i_T *vs.* E_T curves were recorded with the feedback switched off during the negative and positive going scans in $-0.650 \text{ V} \leq E \leq -0.150$ V at 2.0 V s^{-1} and at constant bias $(E_T - E_S) = 0.10$ V. The latter entails parallel variation of substrate and tip potential relative to a common reference electrode. Both half-cycles exhibit characteristic maxima. After each complete cycle, the feedback was activated again, and the system was given sufficient time to stabilize. An average trace obtained from 20 independent negative and positive half-cycles, which were corrected for the corresponding electrochemical response of the gold tip in the absence of tunneling, is plotted in Fig. 10. The curve shows a well-defined maximum $i_M = (0.89 \pm 0.10)$ nA at $E_M = -(0.44 \pm 0.02)$ V, which is close to the equilibrium potential of the one-electron redox process $V^{2+} \leftrightarrow V^{+\bullet}$ (paragraphs 3.1. and 3.2.). i_T decreases significantly at $|E_S - E_M| > 0.10$ V and approaches the initially preset tunneling current of 100 pA. This trend illustrates the sufficient stability of the HS-6V6-H-modified tunnel junction during the sweep experiment. The $i_T - E_T$ characteristics can be modeled by a Gaussian peak function, which gives 0.14 ± 0.02 V as the FWHM. The current of the maximum at EM is significantly larger than the electrochemical background signal and can be attributed convincingly to an enhanced tunneling response. Control experiments with bare gold tips and tips modified with inactive dodecanethiol did not produce these features. We assume that the main contribution to the enhanced tunnel signal originates from one or a small number of redox-active molecules since only few of them will protrude deeper into the gap than others. Since the tunneling current decreases exponentially with distance, the contributions of molecules adsorbed further away from the tip apex are expected to be negligible small.

The appearance of the maximum in the tunnel current can be rationalized as follows:[21] The formation of a junction Au(T)-S-6V6-H|Au(S) implies that a molecule with discrete electronic levels is located in the gap between substrate and tip, both representing continuous distributions of electronic levels. The localized electronic states are broadened by interactions of the redox-center with the surrounding electrolyte[103,104] and with the substrate.[105] The interaction with the substrate shall be rather small due to the alkyl-spacer unit (C_6), which spatially separates the SH-anchor group and the redox-active viologen moiety. The broadening introduced by interactions with the solvent arises from the fluctuations of the polar molecules that constantly shift the energy level of the redox center *via* changing the potential experienced by the electrons in the redox-center. However, the single-electron transfer reorganization Gibbs free energy λ_R in the tunneling gap is thought to be significantly smaller than in the semi-infinite space in "classical electron transfer".[21] As the redox level is exposed to part of the potential drop $(E_T - E_S)$ a maximum is expected in the current–voltage characteristics. Variation of the overpotential at fixed bias $(E_T - E_S)$ is equivalent to a parallel shift of the Fermi levels of tip and substrate. The current first rises as the redox level approaches one of the Fermi levels (here the substrate Fermi level). Further increase leads to the current decrease as vacant levels of the positively biased electrode become increasingly thermally inaccessible. These expectations apply if $e \cdot (E_T - E_S) \leq \lambda_R$. The position of the tunneling current maximum provides criteria to distinguish between different mechanisms of electron transfer (ET).[19,21,43–45] Resonance tunneling[19,44] and a mechanism based on coherent two-step ET[21,43] predict a maximum in the tunneling current, but in both cases it is expected to be shifted from the equilibrium potential by the reorganization Gibbs free energy λ_R. A maximum located exactly at the equilibrium potential is predicted for a sequential two-step ET if the bias potential drop at the site of the redox-center is close to half the total potential drop $(E_T - E_S)$.[45] This theoretical treatment considers the ET in a redox-molecule functionalized tunneling junction as two consecutive interfacial single ET steps with vibration relaxation between the steps. The process comprises a cycle of consecutive molecular reduction and re-oxidation. A particularly interesting feature in the fully adiabatic limit is that each oxidation|reduction cycle is composed of manifolds of interfacial ET events at both electrodes enhancing the electron tunneling significantly, compared to the single ET.

Tentatively we favor the sequential two-step ET mechanism to explain the experimental i_T *vs.* E_T characteristics of the Au(T)-S-6V6-H|Au(S) tunnel junction. Experiments are in progress to systematically modify composition and structure of the redox-molecule functionalized tunneling junction.

Comparing the results of the current–distance characteristics recorded in stretching experiments with preformed symmetric Au(T)|HS-6V6-SH|Au(S) junctions with the current–voltage characteristics of asymmetric Au(T)-S-6V6-H|Au(S) junctions we may conclude that the former experiment involves dominant contributions from electron transport through the "molecular skeleton". The result of the second experimental approach yielded an enhanced tunneling response due to the activation of a molecular redox state, which appears to be strongly coupled to environmental (solvent) influences. Considerable experimental and theoretical efforts are anticipated to develop a more detailed understanding on the complementary nature of these two experimental approaches for addressing the local molecular redox-activity in an electrochemical environment.

4. Summary and conclusions

(1) Two redox-active viologen derivatives, *N*-hexyl-*N*′-(6-thiohexyl)-4,4′-bipyridinium bromide (HS-6V6-H) and *N,N*′-bis(6-thiohexyl)-4,4′-bipyridinium bromide (HS-6V6-SH), were synthesized based on a modification of literature procedures. The compounds were obtained in very high purity at good yields.

(2) The self-assembly and redox-properties of HS-6V6-H and HS-6V6-SH, immobilized on Au(111)-(1 × 1) macro-electrodes and Au(poly) nano-electrodes were investigated by cyclic voltammetry, surface enhanced infrared spectroscopy (SEIRAS) and *in situ* scanning tunneling microscopy (STM). The strict elimination of the oxygen exposure during the assembly and characterization steps enabled us to identify three distinct types of adlayers for both viologens: a low coverage *disordered* and a low coverage *ordered* "striped" phase of flat oriented molecules, and a high coverage monolayer composed of tilted viologen moieties. Using the primitive gold lattice as a basis one could represent the commensurate striped phases in matrix notation as $\begin{pmatrix} 18 & 1 \\ -2 & 4 \end{pmatrix}$ (HS-6V6-H) and $\begin{pmatrix} 18 & 1 \\ -1 & 2 \end{pmatrix}$ (HS-6V6-SH), respectively.

(3) The ATR-SEIRAS experiments revealed that the one-electron reduction $V^{2+} \rightarrow V^{+\bullet}$ yields, under our experimental conditions, preferentially the radical cation monomer ($V^{+\bullet}$) for HS-6V6-H, while a substantial amount of the radical cation dimer $(V^{+\bullet})_2$ was observed with (HS-6V6-SH).

The viologen adlayers appear to be stabilized by co-adsorption of ClO_4^-, which replace Br^- ions upon exposure of the modified gold electrodes to an aqueous electrolyte containing 0.05 M $KClO_4$.

(4) An *in situ* STM configuration was employed to explore electron transport properties of single molecule junctions Au(T)|HS-6V6-SH(HS-6V6-H)|Au(S) based on analyzing current–distance stretching characteristics. The observed sigmoidal potential dependence of the single molecule conductance, measured at variable substrate potential E_S and at constant bias voltage ($E_T - E_S$), was attributed to electronic structure changes of the viologen moiety during the one-electron reduction/re-oxidation process $V^{2+} \leftrightarrow V^{+\bullet}$. The simultaneous recording of *in situ* STM images and successive i–Δz traces revealed structural insight into the nature of the "pulling" experiment. Comparison with literature data shows that single molecular conductivities, as determined in Au(T)|HS-bridge-SH|Au(S) junctions, demonstrate convincingly a molecular structure dependence of the measured signals.

(5) Tunneling experiments in asymmetric, STM-based junctions Au(T)-S-6V6-H|Au(S) resulted in current (i_T)–voltage (E_T) curves with a well-defined maximum located at the equilibrium potential of the redox-process $V^{2+} \leftrightarrow V^{+\bullet}$, where the surface concentrations of the oxidized and of the reduced form of the reactants are equal. The experimental $i_T - E_T$ characteristics of the HS-6V6-H-modified tunneling junction were tentatively attributed to a sequential two-step electron transfer mechanism. This experimental result opens up a fascinating approach to explore local electron transfer, and will hopefully inspire further experimental and theoretical activities.

Acknowledgements

B. H., Z. L. and Th. W. acknowledge support of IFMIT, of the Volkswagen Foundation under contract I 77-116, and of the Research Center Jülich. G. M. thanks the German Academic

Exchange Service for a Fellowship and the Hungarian Scientific Research Fund OTKA (T42452). The work of A. B. and M. M. was supported by the German Ministry of Education and Research within the network project MOLMEM (BMBF-FZK 13 N 8360).

References

1 J. D. Meindel, Q. Chen and J. A. Davis, *Science*, 2001, **293**, 2044.
2 C. Joachim, J. K. Gimzewski and A. Aviram, *Nature*, 2000, **408**, 541.
3 D. L. Klein, R. Roth, A. K. L. Lim, A. P. Alivisatos and P. L. McEuen, *Nature*, 1997, **389**, 699.
4 J. Hu, T. W. Odom and C. M. Lieber, *Acc. Chem. Res.*, 1999, **32**, 435.
5 A. Aviram and M. Ratner, *Chem. Phys. Lett.*, 1974, **29**, 277.
6 (*a*) R. Metzger, *J. Solid. State Chem.*, 2002, **168**, 696; (*b*) R. Metzger, *Acc. Chem. Res.*, 1999, **32**, 950.
7 *Molecular Electronics II*, ed. A. Aviram, M. Ratner and V. Mujica, Ann. N.Y. Acad. Sci., New York, 2002, p. 960.
8 M. A. Reed, C. Zhou, C. Muller, T. P. Burgin and J. M. Tour, *Science*, 1997, **278**, 252.
9 C. Kergueris, J. P. Bourgoin, S. Palacin, D. Esteve, C. Urbina, M. Magoga and C. Joachim, *Phys. Rev. B*, 1999, **59**, 12505.
10 (*a*) J. Reichert, R. Ochs, D. Beckmann, H. B. Weber, M. Mayor and H. v. Löhneysen, *Phys. Rev. Lett.*, 2002, **88**, 176804; (*b*) J. Reichert, H. B. Weber, M. Mayor and H. v. Löhneysen, *Appl. Phys. Lett.*, 2003, **82**, 4137.
11 J. Chen, M. A. Reed, A. M. Rawlett and J. M. Tour, *Science*, 1999, **286**, 1550.
12 A. J. Storm, D. van den Broek and S. Lemay, *Biophys. J.*, 2002, **82**, 256 P 2.
13 J. G. Kushmerick, D. B. Holt, J. C. Yang, J. Narici, M. H. Moore and R. Shashiidhar, *Phys. Rev. Lett.*, 2002, **89**, 86801.
14 J. O. Lee, G. Lientsching, F. Wiertz, M. Strujik, R. A. J. Janssen, R. Egberink, D. A. Reinhoudt, P. Hadley and C. Dekker, *Nano Lett.*, 2003, **2**, 113.
15 E. Tran, M. A. Rampi and G. M. Whitesides, *Angew. Chem., Int. Ed.*, 2004, **43**, 3835.
16 Z. Donnhauser, B. A. Mantooth, K. F. Kelley, L. A. Bumm, J. D. Monnell, J. J. Stapleton, D. W. Price, D. L. Allara, J. M. Tour and P. S. Weis, *Science*, 2001, **292**, 2303.
17 S. G. Lemay, J. W. Jansen, M. van der Hout, M. Mooij, M. Bronikowski, P. A. Willis, R. E. Smalley, L. P. Kouwenhoven and C. Dekker, *Nature*, 2001, **412**, 6847.
18 (*a*) X. Lu, K. W. Hipps, X. D. Wang and U. Mazur, *J. Am. Chem. Soc.*, 1996, **118**, 7197; (*b*) Lscudiero, D. E. Barlow and K. W. Hipps, *J. Phys. Chem. B.*, 2000, **104**, 11899; (*c*) K. Lscudiero, D. E. Barlow and K. W. Hipps, *J. Am. Chem. Soc.*, 2001, **123**, 4073; (*d*) K. Lscudiero, D. E. Barlow and K. W. Hipps, *J. Phys. Chem. B*, 2002, **106**, 996.
19 N. J. Tao, *Phys. Rev. Lett.*, 1996, **76**, 4066.
20 D. I. Gittins, D. Bethell, D. J. Schiffrin and R. J. Nichols, *Nature*, 2000, **408**, 67.
21 J. Zhang, Q. Chi, A. M. Kuznetsov, A. G. Hansen, H. Wackerbarth, H. E. M. Christensen, J. E. T. Andersen and J. Ulstrup, *J. Phys. Chem. B*, 2002, **106**, 1131.
22 (*a*) D. J. Wold and C. D. Frisbie, *J. Am. Chem. Soc.*, 2001, **123**, 5549; (*b*) D. J. Wold, R. Haag, M. A. Rampi and C. D. Frisbie, *J. Phys. Chem. B*, 2002, **106**, 2813.
23 (*a*) X. D. Cui, A. Primak, X. Zarate, J. Tomfohr, O. F. Sankey, A. L. Moore, T. A. Moore, D. Gust, G. Harris and S. M. Lindsay, *Science*, 2001, **294**, 571; (*b*) X. D. Cui, X. Zarate, J. Tomfohr, O. F. Sankey, A. Primak, A. L. Moore, T. A. Moore, D. Gust, G. Harris and S. M. Lindsay, *Nanotechnology*, 2002, **13**, 5.
24 F. R. Fan, J. Yang, L. Cai, D. W. Price, S. M. Dirk, D. V. Kosynkin, Y. Yao, A. M. Rawlett, J. M. Tour and A. J. Bard, *J. Am. Chem. Soc.*, 2002, **124**, 5550.
25 B. Xu and N. J. Tao, *Science*, 2003, **301**, 1221.
26 W. Haiss, H. van Zalinge, S. J. Higgins, D. Bethell, H. Hörbenreich, D. J. Schiffrin and R. J. Nichols, *J. Am. Chem. Soc.*, 2003, **125**, 15294.
27 W. Haiss, R. J. Nichols, H. v. Zalinge, S. J. Higgins, D. Bethell and D. J. Schiffrin, *Phys. Chem. Chem. Phys.*, 2004, **6**, 4330.
28 F. Chen, J. He, C. Nuckolis, T. Roberts, J. F. Klare and S. Lindsay, *Nano Lett.*, 2005, **5**, 563.
29 J. Chen, M. A. Reed, A. M. Rawlett and J. M. Tour, *Science*, 1999, **286**, 1550.
30 J. Park, A. N. Pasupathy, J. I. Goldsmith, C. Chang, Y. Yaish, J. R. Petta, M. Rinkowski, J. Sithna, H. D. Abruna, P. M. McEuen and D. C. Ralph, *Nature*, 2002, **407**, 722.
31 S. Kubatkin, A. Danilov, M. Hjort, J. Cornil, J. L. Bredas, N. Suhr-Hansen, P. Hedegard and Th. Bjornholm, *Nature*, 2003, **425**, 698.
32 N. B. Zhitenev, H. Meng and Z. Bao, *Phys. Rev. Lett.*, 2002, **88**, 226801.
33 R. Rinaldi and R. Cingolani, *Physica E*, 2004, **21**, 45.
34 Y. Luo, C. P. Collier, J. O. Jeppesen, K. A. Nielswen, E. Delonno, G. Ho, J. Perkins, H. R. Tseng, T. Yamamoto, J. F. Stoddart and J. R. Heath, *ChemPhysChem*, 2002, **3**, 519.
35 X. Duan, Y. Huang and C. M. Liber, *Nano Lett.*, 2002, **2**, 487.
36 C. Li, W. Fan, D. A. Strauss, B. Lei, S. Asano, D. Zhang, J. Han, M. Meyyappan and Ch. Zhou, *J. Am. Chem. Soc.*, 2004, **126**, 7750.

37 S. Yoshimoto, A. Tada, K. Suto, R. Narita and K. Itaya, *Langmuir*, 2003, **19**, 672.
38 S. Yoshimoto, A. Tada, K. Suto and K. Itaya, *J. Phys. Chem. B*, 2003, **107**, 5836.
39 U. Mazur and K. W. Hipps, *J. Phys. Chem. A*, 1999, **103**, 9721.
40 A. Alessandrini, M. Gerunda, G. W. Canters, M. P. Verbeet and P. Facci, *Chem. Phys. Lett.*, 2003, **376**, 625.
41 J. J. Davis, *Philos. Trans. R. Soc. London, Ser. A*, 2003, **361**, 2807.
42 H. He, J. Zhu, N. J. Tao, L. A. Nagahara, L. Amlani and R. Tsui, *J. Am. Chem. Soc.*, 2001, **123**, 7730.
43 A. M. Kuznetsov and J. Ulstrup, *Electrochim. Acta*, 2000, **45**, 2339.
44 W. Schmickler and N. J. Tao, *Electrochim. Acta*, 1997, **42**, 2809.
45 (*a*) A. M. Kuznetsov and J. Ulstrup, *J. Phys. Chem. A*, 2000, **104**, 11531; (*b*) A. M. Kuznetsov and J. Ulstrup, *Probe Microsc.*, 2001, **2**, 187.
46 M. Krüger, M. R. Buitelaar, T. Nussbaumer, C. Schönenberger and L. Forro, *Appl. Phys. Lett.*, 2001, **78**, 1291.
47 S. Rosenblatt, Y. Yaishi, J. Park, J. Gore, V. Sazonova and P. L. McEuen, *Nano Lett.*, 2002, **2**, 869.
48 M. Di Ventra, N. D. Lang and S. T. Pantelides, *Chem. Phys.*, 2002, **281**, 189.
49 Th. Wandlowski, K. Ataka, S. Pronkin and D. Diesing, *Electrochim. Acta*, 2004, **49**, 1233.
50 G. Nagy and Th. Wandlowski, *Langmuir*, 2003, **19**, 10271.
51 C. L. Bird and A. T. Kuhn, *Chem. Soc. Rev.*, 1981, **10**, 49.
52 P. M. S. Monk, *The Viologens*, John Wiley & Sons, Chichester, 1998.
53 (*a*) H. C. DeLong and D. A. Buttry, *Langmuir*, 1990, **6**, 1319; (*b*) H. C. DeLong and D. A. Buttry, *Langmuir*, 1992, **8**, 2491.
54 S. L. Hiley and D. A. Buttry, *Colloids Surf., A*, 1994, **84**, 129.
55 J. Li, G. Chen and S. Dong, *Thin Solid Films*, 1997, **293**, 200.
56 S. A. John and T. Ohsaka, *J. Electroanal. Chem.*, 1999, **477**, 52.
57 T. Sagara, N. Kaba, M. Komatsu, M. Uchida and N. Nakashima, *Electrochim. Acta*, 1998, **43**, 2183.
58 T. Sagara, H. Tsuruta and N. Nakashima, *J. Electroanal. Chem.*, 2001, **500**, 255.
59 T. Sagara, H. Maeda, Y. Yuan and N. Nakashima, *Langmuir*, 1999, **15**, 3823.
60 R. J. Alvarado, J. Mukkerjee, E. J. Pacsial, D. Alexander and F. M. Raymo, *J. Phys. Chem. B*, 2005, **109**(13), 6164.
61 F. M. Raymo, R. J. Alvarado and E. J. Pacsial, *J. Supramol. Chem.*, 2002, **2**, 63.
62 X. Tang, Th. Chneider and D. Buttry, *Langmuir*, 1994, **10**, 2235.
63 W. Haiss, H. v. Zalinge, H. Hörbenreich, D. Bethell, D. J. S. Chiffrin, S. J. Higgins and R. J. Nichols, *Langmuir*, 2004, **20**, 7694.
64 J. L. Hartwell and M. A. Pogorelskin, *J. Am. Chem. Soc.*, 1950, **72**, 2040–2044.
65 D. I. Gittins, D. Bethell, R. J. Nichols and D. J. Schiffrin, *J. Mater. Chem.*, 2000, **10**, 79–83.
66 A. J. Pearson and J. J. Hwang, *J. Org. Chem.*, 2000, **65**, 3466–3472.
67 M. H. Hölzle, Th. Wandlowski and D. M. Kolb, *J. Electroanal. Chem.*, 1995, **394**, 271.
68 B. Ren, G. Piccardi and B. Pettinger, *Rev. Sci. Instrum.*, 2004, **75**, 837.
69 G. Meszaros, I. Pobelov, Z. Li and Th. Wandlowski, 2005, in preparation.
70 T. Pajkossy, Th. Wandlowski and D. M. Kolb, *J. Electroanal. Chem.*, 1996, **414**, 209.
71 C. A. Widrig, C. Chung and M. D. Porter, *J. Electroanal. Chem.*, 1991, **335**, 310.
72 A. J. Bard and L. Faulkner, *Electrochemical Methods*, Wiley, New York, 1980.
73 J. H. Russell and S. C. Wallwork, *Acta. Crystallogr., Sect. B: Struct. Crystallogr. Cryst. Chem.*, 1972, **B28**, 1527.
74 I. Yu. Polishchuk, L. G. Grineva, A. P. Polishchuk and A. N. Chernega, *Zh. Obshch. Khim.*, 1996, **66**, 1530.
75 A. Ulman, *Chem. Rev.*, 1996, **96**, 1533.
76 G. Meszaros, I. Pobelov, B. Han, Z. Li and Th. Wandlowski, 2005, unpublished.
77 M. J. Esplandiu, H. Hagenstrom and D. M. Kolb, *Langmuir*, 2001, **17**, 828.
78 R. E. Hester and S. Suzuki, *J. Phys. Chem.*, 1982, **86**, 4626.
79 S. Ghostal, T. Lu, Q. Feng and T. M. Cotton, *Spectroelectrochim. Acta*, 1988, **44A**, 651.
80 S. H. Brienne, R. P. Cooney and G. A. Bowmaker, *J. Chem. Soc., Faraday Trans.*, 1991, **87**, 1355.
81 M. Ito, H. Sasaki and M. Takahashi, *J. Phys. Chem.*, 1987, **91**, 3932.
82 M. Osawa and K. Yoshi, *Appl. Spectrosc.*, 1997, **51**, 512.
83 K. Arihara and F. Kitamura, *J. Electroanal. Chem.*, 2003, **550–551**, 149.
84 M. D. Porter, Th. B. Bright, D. L. Allara and C. E. D. Chidsey, *J. Am. Chem. Soc.*, 1987, **109**, 3559.
85 H. T. Rong, S. Frey, Y. J. Yang, M. Zharnikov, M. Buck, M. Wühn, Ch. Wöll and G. Helmchen, *Langmuir*, 2001, **12**, 1582.
86 A. Ulman and R. P. Scaringe, *Langmuir*, 1992, **8**, 894.
87 B. Han, Z. Li and Th. Wandlowski, 2005, in preparation.
88 G. E. Poirier, *Langmuir*, 1997, **13**, 2019.
89 M. Böhringer, K. Morgenstern, W. D. Schneider, R. Berndt, F. Mauri, A. de Vita and R. Car, *Phys. Rev. Lett.*, 1999, **83**, 324.
90 G. E. Poirier, *Langmuir*, 1999, **15**, 1167.
91 R. Staub, M. Toerker, T. Fritz, T. Schmitz-Hübsch, F. Sellam and K. Leo, *Langmuir*, 1998, **14**, 6693.

92 T. Y. B. Leung, M. C. Gerstenberg, D. J. Lavrich, G. Scoles, F. Schreiber and G. E. Poirier, *Langmuir*, 2000, **16**, 549.
93 T. Y. B. Leung, P. Schwarz, G. Scoles, F. Schreiber and A. Ulman, *Surf. Sci.*, 2000, **458**, 34.
94 L. Duan and S. J. Garnett, *Langmuir*, 2001, **17**, 2986.
95 G. Yang and G. Liu, *J. Phys. Chem. B*, 2003, **107**, 8746.
96 D. P. E. Smith, J. K. H. Hörber, G. Binning and H. Nejoh, *Nature*, 1990, **344**, 641.
97 Th. Dretschkow and Th. Wandlowski, *Top. Appl. Phys.*, 2003, **85**, 259.
98 S. de Feyter, J. H. van Esch, B. L. Ferringa and F. C. de Schryver, *Chem. Eur. J.*, 2004, **10**, 1124.
99 A. Marchenko, S. Lukyanets and J. Cousty, *Phys. Rev. B*, 2002, **65**, 45414-1.
100 S. D. Evan, E. Urankar, A. Ulman and N. Ferris, *J. Am. Chem. Soc.*, 1991, **113**, 4121.
101 X. D. Cui, A. Primak, X. Zarate, J. Tomfohr, O. F. Sankey, A. L. Moore, T. A. Moore, D. Gust, L. A. Nagahara and S. M. Lindsay, *J. Phys. Chem. B*, 2002, **106**, 8609.
102 J. He, F. Chen, J. Li, O. F. Sankey, Y. Terazono, Ch. Herrero, D. Gust, Th. A. Moore, A. L. Gust and S. M. Lindsay, *J. Am. Chem. Soc.*, 2005, **127**, 1384.
103 R. A. Marcus, *J. Chem. Phys.*, 1956, **24**, 979.
104 H. Gerischer, *Z. Phys. Chem., Neue Folge*, 1960, **26**, 223.
105 D. M. Newns, *Phys. Rev.*, 1969, **178**, 1123.

Measuring single molecule conductance with break junctions

Jin He,[a] Otto Sankey,[b] Myeong Lee,[b] Nongjian Tao,[c] Xiulan Li[c] and Stuart Lindsay*[abd]

Received 14th June 2005, Accepted 11th July 2005
First published as an Advance Article on the web 15th September 2005
DOI: 10.1039/b508434m

Single-molecule conductance measurements made under potential control provide a critical link between chemical and molecular electronic data. These measurements are made possible by the STM break-junction method introduced recently, but questions remain about its reliability. Here we report the use of a logarithmic current-to-voltage converter to examine a wide range of currents in an STM break junction study of octanedithiol, clearly showing both the gold-quantum wire regime and the single molecule conductance regime. We find two sets of molecular currents that we tentatively ascribe to different bonding geometries of the molecules in the break junction.

Introduction

A key problem in molecular electronics lies in the uncertain nature of the link between electrochemical and optical measurements of charge transfer in solvated molecules (the techniques used by chemists to test their molecules) and charge transfer in a 'dry' molecular electronic device where the surface potential is uncontrolled, and the solvent polarization that stabilizes charging is absent. The STM break-junction is a relatively new method for making measurements of single-molecule conductivity[1] and it lends itself to measurements in conducting solutions[2] because insulated probes, developed for electrochemical STM, are readily available.[3] The ability to measure single molecule conductance in conducting solution now enables direct comparison of electrochemical and molecular electronic properties. Using the technique, we have measured the single molecule conductance of oligoaniline molecules as a function of their oxidation state.[4] We have also measured changes in the local surface potential caused by the probe to substrate bias as manifested in a negative differential resistance when the applied bias changed the oxidation state of a previously conductive molecule. The effect vanished when the molecule was immersed in toluene, showing how this change of oxidation state was completely inhibited when ions are not present.[4] In addition, it has recently been demonstrated that non-reversible electrochemistry underlies the negative differential resistance reported at high bias in certain molecular films.[5,6] The STM break-junction method is a critical tool in such studies, so its reliability, and interpretation of results obtained using it, deserve careful examination. We have already shown elsewhere[7] that the STM break-junction method appears to be better than the self-assembled single molecule junctions we introduced some years ago[8] because it is free of artifacts associated with the gold nanoparticles used as contacts for the self-assembled junctions. However, the microscopic nature of the operation of the STM breakjunction is not obvious, because the atomic structure of the junction is not controlled. A recently-introduced technique in which molecules are trapped between fixed electrodes[9] gives values for the conductance and electronic decay constant (β) of alkanethiols that is much closer to what we

[a] *Biodesign Institute, Arizona State University, Tempe AZ 85287, USA. E-mail: Stuart.Lindsay@asu.edu*
[b] *Department of Physics and Astronomy, Arizona State University, Tempe AZ 85287, USA*
[c] *Department of Electrical Engineering and Center for Solid State Electronic Research, Arizona State University, Tempe AZ 85287, USA*
[d] *Department of Chemistry and Biochemistry, Arizona State University, Tempe AZ 85287, USA*

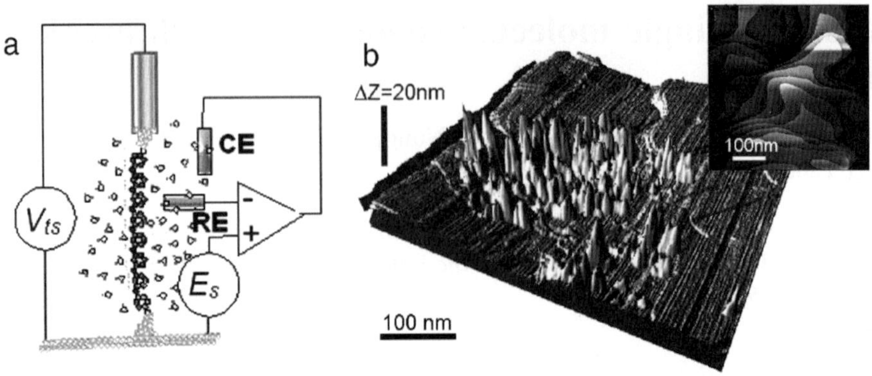

Fig. 1 (a) Showing the apparatus for STM break junction measurements schematically. The STM probe is insulated using techniques developed for electrochemical STM[3] enabling potential control of the probe and substrate. (b) The effects of repeated breakjunction formation. The inset (upper right) shows monatomic terraces on a clean gold substrate prior to measurements being made and the 3D image shows how filaments of *ca.* 20 nm height are pulled out of the surface after a series of measurements. This illustrates how the assumption of planar geometry currently used in the simulations may be invalid.

obtained with the nanoparticle method[8,10] than to the values obtained with the break-junction method.[1] Thus, it appears that the interpretation of STM break junction data is not as clear-cut as first thought, and it is our goal here to describe these newly-discovered complications. We also describe how we select data for analysis in the hope that this will be useful to other laboratories seeking to implement the method. We conclude that the method is reliable to within an order of magnitude or so, but that care is required when comparing different molecules with similar conductances because multiple values can be found for the single molecule conductance. Though simulations do not yet provide a good quantitative explanation of the multiple series of current steps, one explanation is that they arise from different attachments of the molecule to the electrodes.

Experimental

The STM break junction

The apparatus is illustrated schematically in Fig. 1a. A gold probe is pushed into a monolayer of molecules that are bis-functionalized with sticky ends (*e.g.*, dithiolates). As the probe is pulled out, the gold necks into a quantum wire showing Landauer steps in conductance.[11] Eventually the gold filament breaks, but the gap is often bridged by a molecule. The gold deforms more easily than the molecule[12] so a plot of current *vs.* time shows steps where the conductance remains relatively constant as the gold on either side of the molecule continues to deform as the gap is opened. As first demonstrated for the self-assembled nanojunctions,[8] a histogram of the gap conductances shows peaks corresponding to 1, 2, 3 *etc.* molecules trapped in the gap. Fig. 1a also illustrates how the measurement can be made with probe and substrate under potential control by insulating the probe almost all the way to its end.[3] This arrangement was used for measurements of single molecule conductance under potential control.[4,6] Fig. 1b shows STM images of the gold surface pre (inset) and post (3D image) a set of measurements, illustrating how gold filaments are drawn out of the surface.

The technique is remarkably simple and a number of molecules have been studied with it in the short time since it was introduced by Xu and Tao.[1] First-principles simulations are available for some of these molecules, and Table 1 shows a comparison of measured and calculated single molecule conductances. With the exception of benzenedithiol (where the disagreement is a factor 50), theory and experiment agree within a factor of about three or better over a range of measured conductances of 2500 : 1. This indicates that, with the possible exception of benzenedithiol, the technique gives reasonably reliable values for single molecule conductance. However, we recently found unexpected extra sets of peaks in the current distributions measured for short oligoanilines (Chen, Nuckolls and Lindsay, unpublished) and this raised questions about the uniqueness of the

Table 1 Comparison of measured and calculated conductances for three alkanedithiols (measurement[1] theory[29]), benzenedithiol (measurement,[30] theory[21]), four carotenoid polyenes[17] and an optically-switched photochromic molecule in two forms[18]

Molecule	G(meas.)/nS	G(theor.)/nS	Ratio
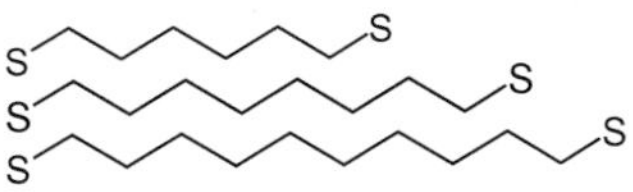	95 ± 6	185	0.51
	19.6 ± 2	25	0.78
	1.6 ± 0.1	3.4	0.47
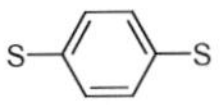	833 ± 90	47 000	0.02
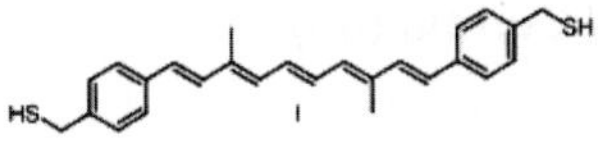 I	2.6 ± 0.05	7.9	0.33
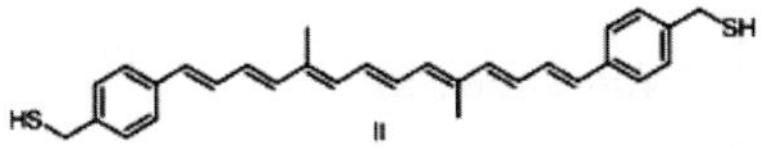 II	0.96 ± 0.07	2.6	0.36
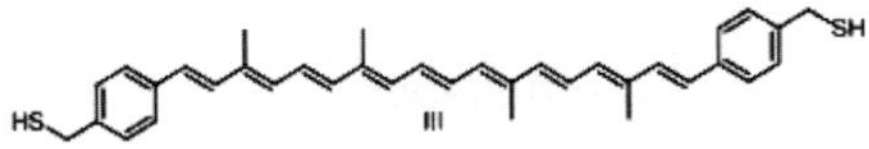 III	0.28 ± 0.02	0.88	0.31
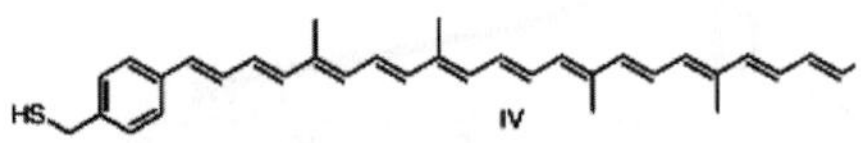 IV	0.11 ± 07	0.3	0.36
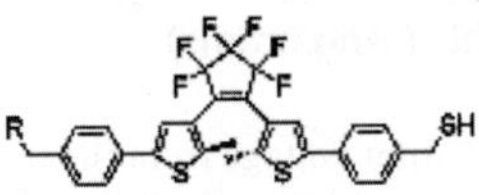	1.9 ± 3	0.8	2.4
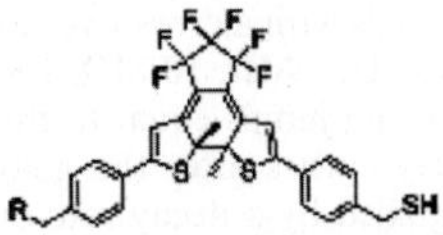	250 ± 50	143	1.74

data obtained from the STM break junction. The observation suggested that perhaps many other sets of peaks would be observed if a greater range of currents were recorded. We therefore decided to return to the simple case of octanedithiol, and to make measurements with a logarithmic current-to-voltage converter so that all possible features in the current-time curve could be revealed in a single experiment.

Data collection with a logarithmic amplifier

We used a logarithmic current to voltage amplifier that was built for our STM by Molecular Imaging (Tempe, AZ). It is based on the diode-feedback circuit described by Durig *et al.*[13] but uses the BAV199 low-leakage diode (Philips) and a contemporary operational amplifier. The amplifier feedback uses eight diodes (four for each direction of current) across a 1 GΩ resistor, giving rise to a linear response at low currents (<1 nA) a logarithmic response for currents >2 nA. Point by point calibration is required for currents between 1 and 2 nA. This series arrangement of four diodes is important for reducing the contribution of junction capacitance in the feedback loop—the circuit will not work at reasonable speeds with a single diode. The current to voltage conversion was calibrated with a series of resistors (see the figure caption) to yield the calibration curve shown in Fig. 2b. Measurements were made first on clean gold substrates[14] that were hydrogen-flame annealed and submerged in freshly-distilled toluene. The STM tips were formed by cutting a 0.25 mm-diameter Au (Alfa Aesar, 99.999%) wire. They were then immersed in Piranha solution (3 : 1 $H_2SO_4 : H_2O_2$ (30%) by volume) for 30 s, rinsed with DI water and dried in Ar stream prior to use. **Caution!** *Piranha solution is a very strong oxidant and is extremely dangerous to work with; gloves, goggles, and a face shield should be worn.* A typical plot of the output of the log amplifier *vs.* time

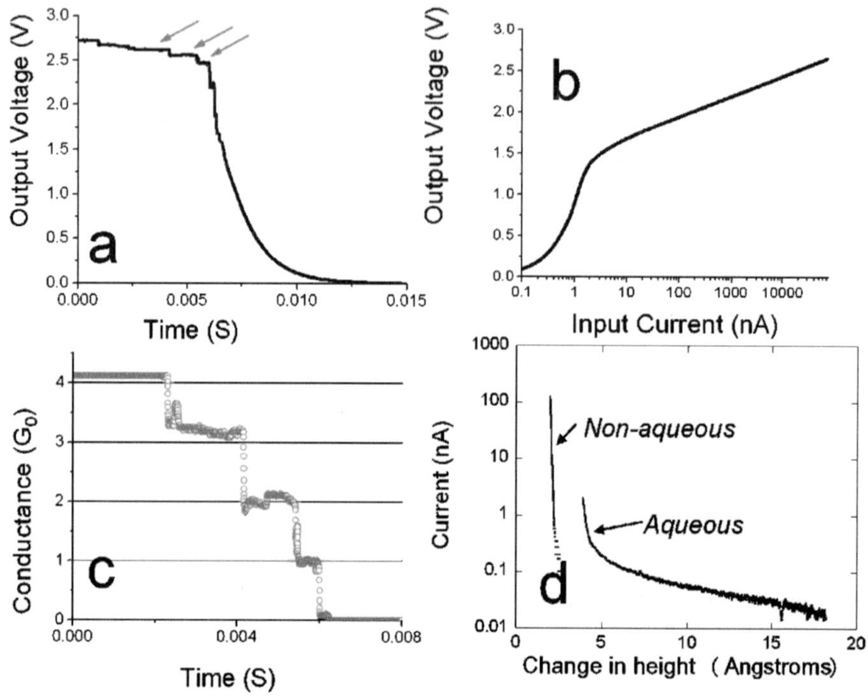

Fig. 2 Current *vs.* time using a logarithmic amplifier. (a) shows an example of raw output voltage from the log amplifier as a function of time. The structure pointed to by arrows is owing to quantized conductance of the gold wire. (b) Calibration curve for these data. The amplifier has a linear output for small signals with a cross-over to logarithmic behavior near 2 nA, and was calibrated using five different 1% resistors (15 kΩ, 1 MΩ, 10 MΩ, 100 MΩ and 1 GΩ). (c) Linearized data from (a) in the vicinity of the quantum steps showing jumps equal to the Landauer conductance ($e^2/2h$). (d) Examples of control (no molecules) current decay curves after the gold filament is broken for non-aqueous and aqueous environments. A clean surface is signaled by a decay rate of *ca.* a factor 10 Å in the region immediately following the break.

during a pull is shown in Fig. 2a. Features owing to the quantized conductance of the gold filament are clearly visible, and, when converted to current data using the calibration in Fig. 2b, they reveal the Landauer steps in the conductance, as shown in Fig. 2c.

Cleanliness is of critical importance, and never completely achieved. A measure of the state of the gap is given by the rapidity of the fall-off of current with distance once the wire breaks. This is illustrated in Fig. 2d which shows examples of the tails of the current-time plot for control experiments (no molecules) in toluene and also in aqueous electrolyte (50 mM H_2SO_4). Data obtained in aqueous electrolyte (or in the presence of dissolved molecules) usually decays more slowly at large distances corresponding to the longer electronic decay lengths observed in water[15] but close to point of breakage (where molecules are presumably excluded from the gap) the slope is on the order of a factor ten in current $Å^{-1}$, corresponding to the 5 eV workfunction of gold. Such sharp steps are a key indicator of clean data, and we use them as a criterion in selecting data.

Selection of data

The histograms compiled in Fig. 4 were compiled using a selected subset of all the curves. The fraction used varied from 47% to 80% (see the figure caption). In our study of oligoaniline, the fraction used was as small as 20% of the total data set. To retain curves in the final data set, we required that breaks between steps have a slope on the order of a decade $Å^{-1}$ (arrows on Fig. 3c). The rejected curves either have slopes much larger than this, or are so noisy that slopes cannot be determined. Sharp decay curves with no steps are also rejected because they contribute to the background that peaks at zero current. Noise owing to periodic stray signals can also generate what

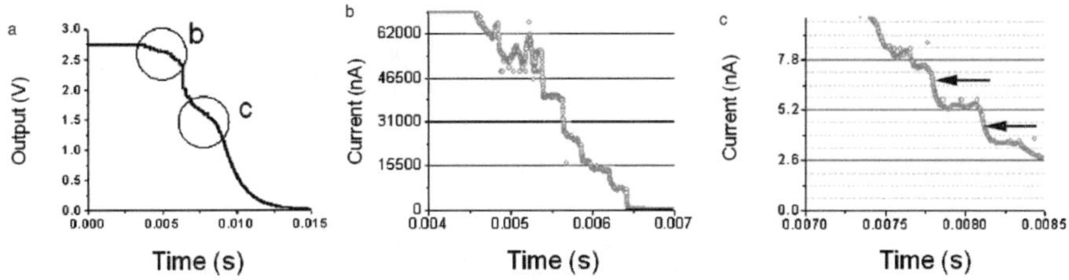

Fig. 3 Data showing both gold-wire quantized conductance and molecular conductance steps at $V = 0.2$ V. (a) Raw output over the entire current range showing distinct gold-wire (labeled b) and molecular (labeled c) steps. Converted using the calibration shown in Fig. 2b, these data yield current-distance data for the gold-wire steps (b) and the molecular steps (c).

appear to be steps, but these features can be rejected because (a) they occur at equal intervals of time and (b) there are no bias dependent features with this current separation.

Results

1,8-Octanedithiol was purchased from Aldrich and used without further purification. It was dissolved in freshly-distilled toluene to a concentration of 8 mM and introduced into the liquid cell of the microscope immediately after the 'clean' data shown in Fig. 2a were obtained. An example of raw data obtained after the molecules were introduced is shown in Fig. 3a. A new set of features (labeled 'c') has appeared below the Landauer quantum steps (labeled 'b'). Converted to current data, these regions are shown in Fig. 3b (Landauer steps) and 3c (molecular steps). The high

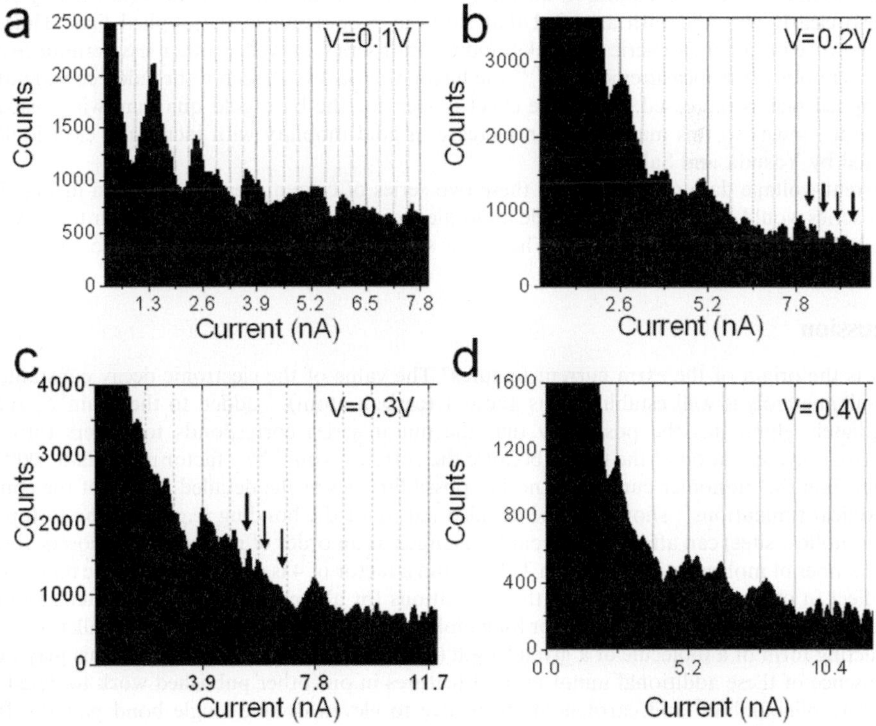

Fig. 4 Current histograms based on data selected as described in the text. Curves are for a tip substrate bias of 0.1 V (a: 132 out of 234 curves used), 0.2 V (b: 340 out of 720 curves used), 0.3 V (c: 357 out of 449 curves used) and 0.4 V (d: 179 out of 249 curves used). In addition to the main series of peaks falling on the gray vertical lines, a secondary series is evident (pointed to by arrows in (c) and (d)).

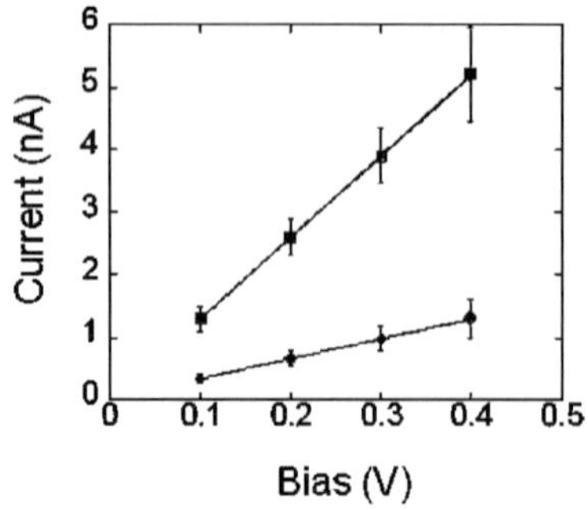

Fig. 5 Current–voltage data derived from histograms like those shown in Fig. 4. The major peak series yields the data shown by the squares, for which $G = 13 \pm 4$ nS. The minor series of peaks yield the data shown by the dots for which $G = 3.25 \pm 1$ nS. Uncertainties are based on an estimate of the relative the width of the current peaks in the histogram in order to give an indication of the spread of data. The slope of the current–voltage plots were fitted to a much higher precision.

current region shows steps corresponding to G_0 (with steps at multiples of G_0 washed out as previously reported for gold surfaces with thiol adsorbates[16]). Additional steps are now clearly visible in a much lower region of current (see Fig. 3b). Histograms of the current values obtained in the region labeled 'c' are shown in Fig. 4. At 0.1 V bias, a clear series of dominant peaks are observed as indicated by the vertical gray bars (which are coincident with the first three well-defined peaks in Fig. 4a). As the bias is increased, new peaks are found at proportionately higher currents as indicated by the vertical gray bars on Fig. 4b, c and d. However, there are also many features in the histogram that do not correspond to the major series in Fig. 4a. Some of these (labeled by arrows) are also clearly periodic with a period that shifts in proportion to the applied bias. Thus there appears to be both a major series of peaks, and a minor series, as well as other less periodic features. Here, as in our other measurements,[4,17,18] the histogram peaks are seen to broaden considerably as the applied bias is increased. The same effect is seen for the bare gold quantum-wire histograms (data not shown) so this may reflect an increase in gold mobility with increasing electric field as reported by Yasuda and Sakai.[19]

Current–voltage data obtained from these two series of current peaks are shown in Fig. 5. The major series would correspond to a single molecule conductance of 13 ± 4 nS, similar to the value of 19 ± 2 nS reported by Xu and Tao.[1] The minor series yields a conductance of 3.25 nS.

Discussion

What is the origin of the extra current features? The value of the electronic decay constant, β, in these alkanethiols is well established as about 1 (carbon atom)$^{-1}$ added to the chain[1,20] and this immediately eliminates the possibility that the minor series corresponds to dimers formed by oxidative polymerization of the thiols, because the currents would be a factor more than 2000 times smaller than the monomer currents. Another possibility lies in the detailed nature of the contacts. Theoretical simulations[21] show that the detailed nature of the bonding (*e.g.*, bonding to top atom sites *vs.* hollow sites) can affect the current by as much as an order of magnitude. Theoretical results for a number of molecules are shown in Table 2 and a factor of 4 is consistent with the magnitude of this effect in other molecules (but see the simulations for alkanedithiols below). Interestingly, the Table shows how the effect is smaller for long molecules (*cf.* the carotenoids) and smaller for the less conducting form of a molecule of a given length (*cf.* the photochromic molecule). This may explain the absence of these additional minor current features in our other published work to date (on an eight-ring oligoaniline[4] and carotenoids from five to eleven double–single bond pairs[17]). It also highlights the difficulty of using this technique on small molecules where this complication appears to be more significant (and, as a further complication, the conductance of dimers may not be negligible for small molecules). These difficulties may account for the significant discrepancy between theory and experiment in the case of benzenedithiol (Table 1).

Table 2 Single molecule conductances for on-top and hollow site bonding calculated for a series of carotenoids[17] (number of double-single bond pairs in parenthesis), benzenedithiol[21] and a photochromic molecule[18] in both open and closed forms

	G_{top} (nS)	G_{hollow} (nS)	Ratio
Carotene (5db)[17]	7.9	9.7	0.81
Carotene (7db)[17]	2.6	3.5	0.74
Carotene (9db)[17]	0.88	1	0.88
Carotene (11db)[17]	0.3	0.29	1.03
Benzenedithiol[21]	47 000	5000	9.4
Photochromic (open)[18]	0.84	0.64	1.3
Photochromic (closed)[18]	143	58	2.5

In addition to a series for top–top and hollow–hollow connected molecules, we might expect to observe features that correspond to top–hollow connections. Yet another type of step might consist of parallel combinations of molecules with different connections, for example a top–top connected molecule in parallel with a hollow–hollow connected molecule. Combinations like these might account for some of the extra features observed in the current histograms but better statistics and analysis will be required to clarify this.

Simulations

We have theoretically investigated the effect of binding site on the conductance of alkanedithiol molecules. Two sites were chosen; the hollow site (the thiol is above three Au atoms on a Au (111) surface), and the on-top site where the thiol directly bonds to a single Au atom in a Au surface (111) plane. The distance between the hollow site and S was 1.90 Å, and the distance between S and the on-top-Au was 2.42 Å.[22,23] The bond angle of S with the Au site and C on the chain was 119°. The theoretical method used was Green's function scattering theory for the Landauer transmission analysis.[21] The electronic structure method used was a local-orbital density functional method with pseudopotentials.[24–26]

We find the somewhat surprising result that the relative ordering of the conductance at the two sites depends on the length of the alkane chain. Table 3 lists the conductances for the two sites as a function of number of carbons N in the chain. The hollow site gives a larger conductance at short length, while the on-top site gives the larger conductance at large length. There is a crossover length where the two sites give an equal conductance, although in this specific case it happens for the relatively short $N = 4$ chain. For the specific chain of length eight carbon atoms (octanedithiol), there is only a 12% difference between the two conductances ($G_{top}/G_{hollow} = 1.12$). The calculated value of G_{top} is within a factor of two of the higher current data, and the calculated value of G_{hollow} is within a factor of eight. Thus, while the effect is present, these simulations fail to capture its full extent. One important factor may be the use of a planar Au(111) model for the contact; a glance at Fig. 1b shows how inappropriate this is. More realistic models of the contact geometry will require simulations of much greater complexity.

Nonetheless, these simulations point to some unexpected features. The crossover occurs because of a competition involving two effects. The first is the coupling of the molecule with the two Au contacts. We'll call these contributions G_c. The factor G_c includes the density of states of the metal and interaction couplings between the molecule and metal and at first glance one might expect that the ratio of the conductances would be the ratio of two values of G_c for the two geometries. The second effect is the difference in propagation through the molecule, which we will write simply as $e^{-\beta N}$. The decay parameter β is a property of the molecule. However, it depends on the relative alignment of molecular states with the Au Fermi level, $\beta(E_F)$. The Fermi level lineup with the molecular states depends on the bonding site of the molecule with the Au. The level shift is quite substantial; the Fermi level moves from away from near mid-gap toward the HOMO by about 0.5 eV in the on-top bonding geometry. In this specific molecule, such a large shift does not produce huge effects as might normally be expected since the band-gap of the molecule (in DFT) is about

Table 3 Calculated values of conductance for a series on alkanedithiols attached to on-top and hollow sites on Au(111). The present simulation produced a result about 5% different from the earlier simulation cited in Table 1

N	G_{top}/nS	G_{hollow}/nS	Ratio
4	1260	1290	0.98
6	208	174	1.20
8	26.4	23.6	1.12
10	4.29	3.15	1.36
12	0.67	0.43	1.56

10 eV.[27] The data shown in Table 2 yield $\beta_{top} = 0.95$ (carbon atom)$^{-1}$ and $\beta_{hollow} = 1$ (carbon atom)$^{-1}$ and $G_c = 71$ μS for the hollow site and 56 μS for the on-top site (from $G = G_c e^{-\beta N}$).

Comparison of results obtained with the various methods

Haiss *et al.*[9] use a technique in which the probe is brought very close to the surface and the current recorded as a function of time in the presence of dithiolated molecules. Jumps are observed in the current-time trace and when the frequency of these jumps is recorded as a function of their magnitude, peaks are found, presumably corresponding to 1, 2, 3 *etc.* molecules in the gap. The magnitudes of the currents are much closer to what was reported using the gold nanoparticle contacts[8,10] than to the results first reported by Xu and Tao using the break-junction method[1] (although Haiss *et al.* did report the observation of much higher current steps as well[9]). Furthermore, Haiss *et al.* found a value for β of 0.52 methylene^{-1}, close to the value reported using nanoparticle contacts[10] (0.57) and in disagreement with the result of *ca.* 1.0 reported by Xu and Tao,[1] found in simulations[27] and deduced from many other experimental methods (see references in Cui *et al.*[10]).

The agreement between the results of the break-junction method and theory is striking. Xu and Tao[1] reported that the low-bias single molecule conductance agreed well with simulations but this remarkable degree of agreement is better illustrated by plotting the theoretical results[27] together with the data[1] on the same plot (Fig. 6a). Of course, the smaller current series reported here would not be in such excellent agreement, but this agreement suggests that the major current series corresponds closely to what was modeled in the simulations.

The results obtained with a nanoparticle contact[8] are plotted in Fig. 6b as the thicker solid line. The data obtained by Haiss *et al.*[9] are shown by the squares. The two sets of data agree well at low bias, and the observation of a small value for β for both sets suggests a common origin. We

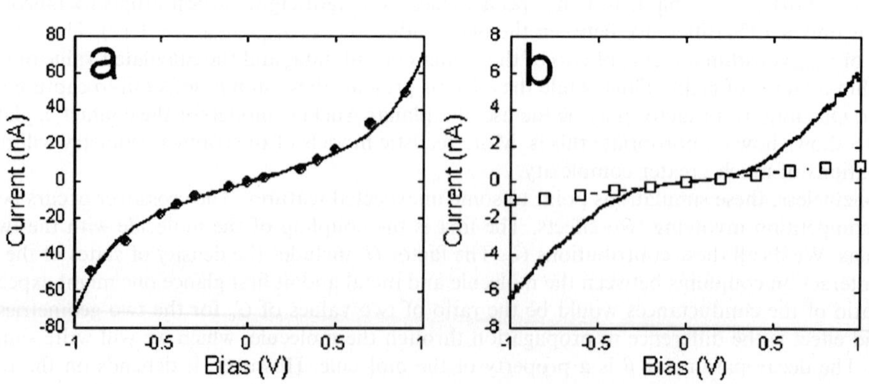

Fig. 6 (a) Dots show data for octanedithiol as reported by Xu and Tao.[1] The solid line shows the results of a first-principles simulation with *no* adjustable parameters.[27] Data and simulation are shown on the same scale with no adjustment or scaling of any sort. (b) The line shows average current–voltage data (*ca.* 1000 molecules) for octanedithiol on Au with a 1.5 nm diameter Au nanoparticle contact.[7] The squares are data for octanedithiol obtained by trapping the molecule in a fixed gap.[9]

explained our data qualitatively as the result of the electronic structure of the nanoparticle combined with charging of the nanoparticle.[7] We also showed that smaller nanoparticles extended the voltage range over which the current–voltage characteristic was linear.[7] It is therefore possible that the data of Haiss *et al.* correspond to molecules terminated in a gold atom (or small cluster) at one end. If this where the case, then the larger current jumps (*ca.* ×10) reported by them may correspond to direct connections to the electrodes without isolated atoms or clusters. Jumps between states where the molecule is either attached directly to the electrode, or parted from it, but still terminated by one or more gold atoms have been proposed as an explanation of so-called stochastic switching.[28] We note further that such charging is thermally-activated and thus might give rise to significant temperature dependence of the current.

Conclusions

We have shown here that the STM break junction provides data for single molecule conductance that appear to be reliable to within about an order of magnitude. By using a logarithmic current to voltage converter, we have shown that, in the case of octanedithiol, molecular conductance steps and gold-wire quantum conductance steps each occupy a distinct, well separated region of currents. There appear to be no spurious signals outside these regions, though the nature of the gold quantum wire steps is modified by the presence of a molecular adlayer. In addition to a major series of conductance peaks separated by 13 ± 4 nS, a second series was observed corresponding to 3.25 ± 1 nS and tentatively attributed to different atomic arrangements of the contacts, though simulations do not account for the full extent of the effect. Other features are observed in the histograms, but their assignment requires further study. The correct assignment of peaks is essential when comparing data from molecules of similar composition but slightly different length, as done when determining the electronic decay length. Multiple sets of current peaks were not a problem in our study of carotenoids[17] but they have complicated our analysis of data obtained from oligoanilines of various lengths (unpublished data).

Acknowledgements

We thank Bingqian Xu for useful discussions. Rich Nichols and David Schiffrin provided useful comments on our first draft of this ms. This work was supported by a NIRT grant of the NSF (ECS 01101175).

References

1 B. Xu and N. J. Tao, *Science*, 2003, **301**, 1221.
2 B. Xu, P. M. Zhang, X. L. Li and N. J. Tao, *Nano Lett.*, 2004, **4**, 1105.
3 L. A. Nagahara, T. Thundat and S. M. Lindsay, *Rev. Sci. Instrum.*, 1989, **60**, 3128.
4 F. Chen, J. He, C. Nuckolls, T. Roberts, J. Klare and S. M. Lindsay, *Nano Lett.*, 2005, **5**, 503.
5 J. He and S. Lindsay, *J. Am. Chem. Soc.*, 2005, submitted.
6 X. Xiao, L. Nagahara, A. Rawlett and N. J. Tao, *J. Am. Chem. Soc.*, 2005, in press.
7 J. Tomfohr, G. Ramachandran, O. F. Sankey and S. M. Lindsay, in *Introducing Molecular Electronics*, ed. G. Fagas and K. Richter, Springer, Berlin, 2005.
8 X. D. Cui, A. Primak, X. Zarate, J. Tomfohr, O. F. Sankey, A. L. Moore, T. A. Moore, D. Gust and S. M. Lindsay, *Science*, 2001, **294**, 571.
9 W. Haiss, R. J. Nichols, H. van Zalinge, S. J. Higgins, D. Bethell and D. J. Schiffrin, *Phys. Chem. Chem. Phys.*, 2004, **6**, 4330.
10 X. D. Cui, A. Primak, X. Zarate, J. Tomfohr, O. F. Sankey, A. L. Moore, T. A. Moore, D. Gust, L. A. Nagahara and S. M. Lindsay, *J. Phys. Chem. B*, 2002, **106**, 8609.
11 C. Z. Li, X. E. He and N. J. Tao, *Appl. Phys. Lett.*, 2000, **77**, 3995.
12 B. Xu, X. Xiao and N. J. Tao, *J. Am. Chem. Soc.*, 2003, **125**, 16164.
13 U. Durig, L. Novotny and B. Michel, *Rev. Sci. Instrum.*, 1997, **68**, 3814.
14 J. A. DeRose, T. Thundat, L. A. Nagahara and S. M. Lindsay, *Surf. Sci.*, 1991, **256**, 102.
15 A. Vaught, T. W. Jing and S. M. Lindsay, *Chem. Phys. Lett.*, 1995, **236**, 306.
16 C. Z. Li, H. Sha and N. J. Tao, *Phys. Rev. B*, 1998, **58**, 6775–6778.
17 J. He, F. Chen, J. Li, O. F. Sankey, Y. Terazono, C. Herrero, D. Gust, T. A. Moore, A. L. Moore and S. M. Lindsay, *J. Am. Chem. Soc.*, 2005, **127**, 1384.
18 J. He, F. Chen, P. A. Liddell, J. Andréasson, S. D. Straight, D. Gust, T. A. Moore, A. L. Moore, J. Li, O. F. Sankey and S. M. Lindsay, *Nanotechnology*, 2005, **16**, 695.

19 H. Yasuda and A. Sakai, 1997, **56**, 1069–1072.
20 X. D. Cui, X. Zarate, J. Tomfohr, A. Primak, A. L. Moore, T. A. Moore, D. Gust, G. Harris, O. F. Sankey and S. M. Lindsay, *Nanotechnology*, 2002, **13**, 5.
21 J. K. Tomfohr and O. F. Sankey, *J. Chem. Phys.*, 2004, **120**, 1542.
22 J. Tomfohr, *Electrontunneling transport theory for molecules*, PhD Thesis, Arizona State University, Tempe, AZ, 2002.
23 H. Kondoh, M. Iwasaki, T. Shimada, K. Amemiya, T. Yokoyama, T. Ohta, T. Shimomura and S. Kono, *Phys. Rev. Lett.*, 2003, **90**, 066102.
24 J. P. Lewis, K. R. Glaesemann, P. Voth, J. Fritsch, A. A. Demkov, J. Ortega and O. F. Sankey, *Phys. Rev. B*, 2001, **64**, 195103.
25 A. A. Demkov, J. Ortega, O. F. Sankey and M. P. Grumbach, *Phys. Rev. B*, 1995, **52**, 1618.
26 O. F. Sankey and D. J. Niklewski, *Phys. Rev. B*, 1989, **40**, 3979.
27 J. Tomfohr and O. F. Sankey, *Phys. Rev. B*, 2002, **65**, 245105.
28 G. K. Ramachandran, T. J. Hopson, A. M. Rawlett, L. A. Nagahara, A. Primak and S. M. Lindsay, *Science*, 2003, **300**, 1413.
29 J. Tomfohr and O. F. Sankey, *Phys. Status Solidi B*, 2002, **233**, 59.
30 X. Xiao, B. Q. Xu and N. J. Tao, *Nano Lett.*, 2004, **4**, 267.

DNA-templated assembly of nanoscale architectures for next-generation electronic devices

Sarmiza Elena Stanca,[a] Ramon Eritja[b] and Donald Fitzmaurice*[a]

Received 15th June 2005, Accepted 14th July 2005
First published as an Advance Article on the web 7th September 2005
DOI: 10.1039/b508471g

We report the assembly and structural characterization of a Y-shaped DNA template incorporating a central biotin moiety. We also report that this template may be used to assemble nanoscale architectures, which demonstrate the potential of this and related approaches to the fabrication of next-generation electronic devices. Of particular significance is the finding that it is possible to selectively metallize the above DNA template to obtain a three-electrode configuration. Also of particular significance is the finding that a biotin modified nanoparticle will recognize and bind selectively the central biotin moiety of the same template, once functionalized by the protein streptavidin.

1. Introduction

The demand for integrated circuits that will allow information be processed at even faster speeds remains undiminished. This is despite the fact that, as a result of miniaturization, the density of the wires and switches that comprise such circuits has doubled every eighteen months giving rise to Moore's Law.[1] While it appears certain that Moore's Law will hold true until 2016, it is not certain that it will hold true thereafter for two reasons.[2]

The first reason is that to build smaller wires and switches using established fabrication and materials technologies will require major scientific and technological advances. Specifically, it will require the development at great cost of new light sources and process tools; new mask and resist materials; and new high and low dielectric constant materials. The second reason is that as wires and switches become smaller the materials of which they are composed no longer exhibit bulk properties, but exhibit properties dominated by confinement and surface effects. As a consequence these wires and switches may exhibit novel characteristics. In other contexts this will represent an opportunity, in this context it will represent a major challenge.

There have been two principal responses of the related scientific and engineering communities. The first response has been to develop alternative fabrication and materials technologies. The second response has been to propose new integrated circuit architectures that can accommodate or even exploit the novel characteristics exhibited by these smaller wires and switches.

When contemplating alternative fabrication technologies, one is attracted to the self-assembly in solution and self-organization at a conventionally patterned silicon wafer substrate of nanoscale wires and switches. When contemplating alternative materials technologies, one is attracted to the use of biological molecules as templates and nanoparticles as building blocks.[3,4] It is noted, that there have been a number of recent reports that have demonstrated the potential of these and related approaches.[5–9]

It is in this context that we have recently reported the DNA templated assembly of a protein-functionalized 10 nm gap electrode from suitably modified gold nanoparticles on a silicon wafer

[a] *Department of Chemistry, University College Dublin, Belfield, Dublin 4, Ireland. E-mail: donald.fitzmaurice@ucd.ie; Fax: +353 1 716 2127*
[b] *Instituto de Biologia Molecular de Barcelona, Consejo Superior de Investigaciones Científicas, Jordi Girona 18-26, E-08034, Barcelona, Spain*

substrate.[10] We also reported that the above protein-functionalized electrode is recognized and bound selectively by a suitably modified gold nanoparticle, which is localized in the 10 nm gap.

Outlined here is the previously reported assembly and structural characterization of a Y-shaped DNA template incorporating a central biotin moiety. Also outlined here, is the previously reported use of this template to assemble model nanoparticle architectures. Of significance was the finding that a biotin modified nanoparticle will recognize and bind selectively the central located biotin moiety previously functionalized by the protein streptavidin.[11]

Detailed here is the previously unreported use of the above Y-shaped DNA template to assemble nanoscale architectures, which demonstrate the potential of this and related approaches to the fabrication of next-generation electronic devices. Of particular significance is the finding that it is possible to selectively metallize the above DNA template to obtain a three-electrode configuration. Also of particular significance is the finding that a biotin modified nanoparticle will recognize and bind selectively the central biotin moiety of the same template, once functionalized by the protein streptavidin.

2. Experimental

2.1 Preparation of gold nanoparticles

An aqueous dispersion of citrate-stabilized 5 nm diameter gold nanoparticles was prepared as described in detail elsewhere using reagents supplied by Aldrich.[12] The optical density of the above dispersion (0.53 au at 520 nm) was used to estimate the nanoparticle concentration (3.4×10^{14} particles dm^{-3}).

An aqueous dispersion of citrate-stabilized 16 nm diameter gold nanoparticles was prepared as described in detail elsewhere using reagents supplied by Aldrich.[13] A volume of this dispersion was subsequently modified by adsorption of a sub-monolayer of a biotin derivative at the surface of the gold nanoparticles also as described in detail elsewhere. The optical density of the above dispersion (0.49 au at 520 nm) was used to estimate the nanoparticle concentration (3.6×10^{12} particles dm^{-3}). Immediately prior to use an aliquot (20 μL) of the above dispersion was diluted in distilled deionized water (2 mL) yielding the working nanoparticle concentration (3.6×10^{10} particles dm^{-3}).

Biotin-labelled and albumin-passivated 5 nm diameter gold nanoparticles were purchased from Aldrich. The as-received nanoparticles were dispersed in a buffered water–glycerol mixture (1 : 1 by vol., pH 7.4, sodium phosphate). The optical density of the above dispersion (5.3 au at 520 nm) was used to estimate the nanoparticle concentration (4.0×10^{15} particles dm^{-3}). Immediately prior to use an aliquot (2 μL) of the above dispersion was diluted in distilled deionized water (2 mL) yielding the working nanoparticle concentration (4.0×10^{12} particles dm^{-3}).

A stable aqueous dispersion of 4-(dimethylamino)pyridine (DMAP) stabilized gold nanoparticles was prepared as described in detail elsewhere.[14] Briefly, DMAP was transferred from a chloroformic phase to an aqueous phase containing hydrochloroauric acid. Subsequent reduction, by the addition of sodium borohydride, led to nanoparticle formation. The nanoparticles were characterised by TEM and were found to have an average diameter of 3.6 nm and polydispersity of 1.10. The nanoparticle concentration of the as prepared dispersion was 17.6 μM. Prior to use, 10 μL of the above dispersion was diluted in 1 mL of water yielding a nanoparticle concentration of 176 nM at pH 8.0.

2.2 Synthesis of oligonucleotides

The oligonucleotides comprising the DNA template shown in Scheme 1 were synthesised using an Applied Biosystems 392 DNA synthesizer.

The oligonucleotide, which acted as the *anchoring element* (5′-thiolhexyl-CGA GTC ATT GAG TCA TCG AG-3′) was prepared on 1 μmol scale using controlled-pore glass (CPG) supports (Applied Biosystems) and standard phosphoramidites (Applied Biosystems) as described in detail elsewhere.[15] The above oligonucleotide was deprotected with concentrated ammonia (0.05 M DTT, 16 h, 55 °C) to obtain the thiol and desalted on a NAP-10 column.

The oligonucleotides, which acted as the two complementary *extension elements* (5′-CTA CGT CGC TGA CTA CCT GCG TAG GTC CCT AGA TGG CTA ACT CGG TGC ATC GCT CAC TGG ATA CAT CAG TCC ATG AAT GAC TCG ATG ACT CAA TGA CTC G-3′ and 5′-TCA

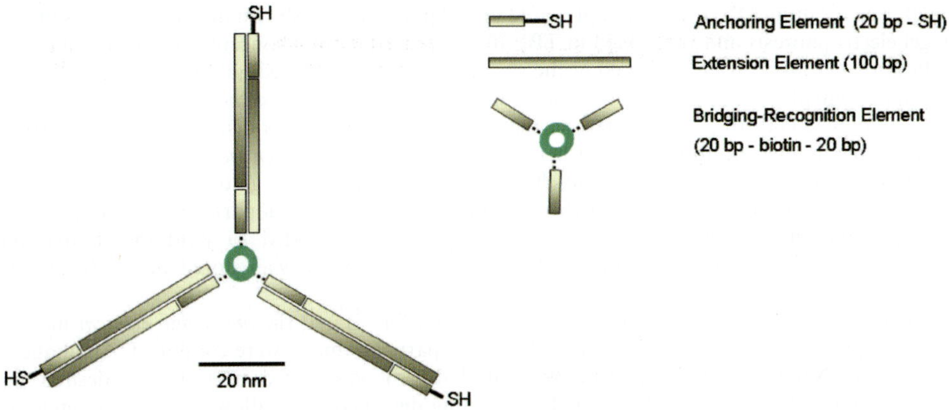

Scheme 1 DNA template incorporating *anchoring*, *extension* and *bridging–recognition* elements.

TTC ATG GAC TGA TGT ATC CAG TGA GCG ATG CAC CGA GTT AGC CAT CTA GGG ACC TAC GCA GGT AGT CAG CGA CGT AGG CAT TGA GCG ATG CAG GCA G-3′), were prepared on 0.2 μmol scale using polystyrene (LV200) supports (Applied Biosystems) and standard and reversed phosphoramidites (Applied Biosystems and Cruachem) as described in detail elsewhere.[15]

The *bridging–recognition element* (3′-CGT AAC TCG CTA CGT CCG TC-5′-heg-bio-bpp-[heg-5′-CTG CCT GCA TCG CTC AAT GC-3′]$_2$ was prepared on 0.2 μmol scale using polystyrene (LV200) supports (Applied Biosystems) and standard and reversed phosphoramidites (Applied Biosystems and Cruachem) as described in detail elsewhere.[16] Here heg denotes hexaethyleneglycol, bpp denotes $-[PO_3O(CH_2)_4CONHCH_2]CH_2OPO_3$ and bio denotes biotin–tetraethyleneglycol.

Following preparation of the above oligonucleotides, they were released form their supports by treatment with aqueous ammonia (32% by vol., 55 °C, 16 h). The resulting ammonia solutions were concentrated to dryness and the product purified by reverse-phase HPLC. The above oligonucleotides were prepared with the last DMT group at the 5′ end (DMT-on protocol) to facilitate reverse-phase purification.

The conditions for HPLC purification are outlined below. All purified products presented a major peak, which was collected. Solvent A: 5% acetonitrile in 100 mM triethylammonium acetate (pH 6.5). Solvent B: 70% acetonitrile in 100 mM triethylammonium acetate (pH 6.5). Columns: PRP-1 (Hamilton), 250 × 10 mm. Flow rate: 3 mL min^{-1}. A 30 min linear gradient from 10–80% B (DMT on), or a 30 min linear gradient from 0 to 50% B (DMT off). Yield: 10–20 OD at 260 nm after HPLC purification (200 nmol).

It should be noted that all the above oligonucleotides were analysed using a PCR primer design programme and the probabilities for formation of unwanted assemblies (hairpins *etc.*) were found to be small.

2.3 Streptavidin preparation and characterization

Streptavidin (Aldrich, lyophilized powder, essentially salt free, approx.14 unit mg^{-1}, protein molecular weight approximately 60 kDa) was dissolved as received in distilled deionized water (1.0 mg mL^{-1}). The activity of the resulting solution was determined by a spectrophotometric titration (1.15×10^{-6} mol dm^{-3} streptavidin equivalents).[17]

2.4 DNA templated assembly of model nanoscale architectures

A solution of *bridging–recognition* elements and a three-fold excess of pairs of complementary *extension elements* was hybridized by heating to 70 °C for 5 min and cooling to room temperature during 15 h in buffered aqueous solution (pH 7.4, phosphate, 0.3 M NaCl, 0.01% by wt sodium azide). The resulting DNA template, prepared on 1.0 μm scale and shown in Scheme 1, was purified by gel electrophoresis and redisolved in TBE (0.5 M) at a nominal concentration of nmol dm^{-3}.

Anchoring elements were adsorbed at the surface of citrate stabilized gold nanoparticles. Briefly, an aliquot (840 μL, 5×10^{-6} mol dm^{-3}) of an aqueous solution of the *anchoring elements* was added to an aliquot (2.000 mL, 3.42×10^{14} particles dm^{-3}) of an aqueous dispersion of nanoparticles (5 nm). This implies an approximately ten-fold excess of *anchoring elements*, assuming that each *anchoring element* occupies an area of 20 $Å^2$ on the surface of a gold nanoparticle and that approximately 400 oligomers are adsorbed at the surface of each nanoparticle.[18] An aliquot of an aqueous solution of NaCl (160 μL, 937 mmol dm^{-3}) was added to yield the desired final concentration of salt (50 mmol dm^{-3}). The resulting dispersion was stored in the fridge until required for use.

The above DNA template solution (10 μL, 30 nmol dm^{-3}) and the *anchoring element* modified gold nanoparticle dispersion (2.372 mL, 2.28×10^{14} particles dm^{-3}) were combined. An aliquot of an aqueous NaCl solution (343 μL, 397 mmol dm^{-3}) was added to yield the desired final concentration (4.97×10^{-9} mol dm^{-3}). The resulting dispersion was allowed stand overnight and stored in the fridge until required for use.

The templated assembly was immobilized on carbon-coated copper TEM grid or on a silicon wafer and exposed to an aliquot (10 μL) of the diluted stock dispersion (10 fmol dm^{-3}) of streptavidin (4 h), which is known to have a high affinity for biotin.[19] After washing, the templated assembly, now incorporating a streptavidin at the biotin site, it was exposed to an aliquot (10 μL) of the stock dispersions of 5 nm or 16 nm diameter biotin-modified gold nanoparticles (4 h).

Alternatively, the templated assembly was immobilized on carbon-coated copper TEM grid or on a silicon wafer and the gold nanoparticles enlarged by electroless deposition using GoldEnhance (Nanoprobes) for 1 min. The same assembly was then exposed to an aliquot (10 μL) of the diluted stock dispersion (10 fmol dm^{-3}) of streptavidin (4 h). The templated assembly, now incorporating a streptavidin at the biotin site, was subsequently exposed to an aliquot (10 μL) of the stock dispersion 5 nm diameter biotin-modified gold nanoparticles (4 h).

2.5 DNA templated assembly of single-electron transistor architecture

A solution of *bridging–recognition* elements and a three-fold excess of pairs of complementary *extension elements* and *anchoring elements* was hybridized by heating to 70 °C during 5 min and cooling to room temperature during 15 h in buffered aqueous solution (pH 7.4, phosphate, 0.3 M NaCl, 0.01% by wt sodium azide). The resulting DNA template, prepared on 1.0 μm scale, was purified by gel electrophoresis and redisolved in TBE (0.5 M) at a nominal concentration of 30 nmol dm^{-3}. An aliquot of an aqueous NaCl solution (343 μL, 397 mmol dm^{-3}) was added to the above DNA template solution (10 μL, 30 nmol dm^{-3}) to yield the desired final concentration (4.97×10^{-9} mol dm^{-3}). The resulting dispersion was allowed stand overnight and stored in the fridge until required for use.

Samples were prepared by placing a drop of the assembled and purified DNA template (10 μL of the 5 nM stock solution) on a polylysine modified carbon coated copper grid. Any excess solution was removed from the grid by wicking after 2 min. The above grid was placed on a filter paper, washed with a drop (20 μL) of distilled-deionised water (18.2 MΩ cm) and dried in air.

The DNA template immobilized on the carbon-coated copper grid was partially metallized with DMAP-modified gold nanoparticles. A drop of the DMAP-modified gold nanoparticle dispersion (10 μL of the 176 nM stock solution at pH 8.0) was deposited on the carbon-coated grid bearing the template. The excess gold nanoparticle dispersion was removed by wicking after 1 min. The above grid was placed on a filter paper, washed with a drop (20 μL) of distilled-deionised water (18.2 MΩ cm) and dried in air.

The DMAP-modified gold nanoparticles were displaced from the biotin moiety located in the middle of the thiol-DNA-biotin template by adsorption of a streptavidin. The carbon coated copper grid bearing the partially metallized template was placed on a clean silicon wafer. A drop (20 μl) of the streptavidin dispersion (0.01 mM in protein stock solution) was deposited on the grid. The silicon wafer, bearing the TEM grid, was then placed in a closed container for 4 h in order to slow solvent evaporation of the streptavidin dispersion. The TEM grid was placed on a filter paper, washed with a drop (20 μL) of distilled deionised water (18.2 MΩ cm) and dried in air. The streptavidin bound to the centrally located biotin moiety was visualized using a uranyl aciacetate.

To complete metallization of the DNA, Goldenhance™ solution was used in order to enlarge and enjoin the DMAP-modified gold nanoparticles. The enhancement procedure followed was that given by Nanoprobes™ (Cat. No. 2113). An incubation time of 1 min was employed. TEM grid was placed on a filter paper, washed with a drop (20 μL) of distilled-deionised water (18.2 MΩ cm) and dried in air.

Finally, the metallized thiol-DNA-biotin template was exposed to a dispersion of biotin-modified gold nanoparticles. As before the TEM grid was placed on a clean silicon wafer and a drop (20 μL) of the biotin modified gold nanoparticle dispersion (16 nm diameter, 0.36 nM in particle) was placed on the grid, completely wetting it. The silicon wafer with the carbon grid was placed in a closed container in order to slow solvent evaporation. After 4 h the TEM grid was placed on a filter paper, washed with a drop (20 μL) of distilled-deionised water (18.2 MΩ cm) and dried in air.

2.6 Transmission electron microscopy

Samples were deposited on treated carbon-coated 400 mesh copper grids. These grids had previously been treated by depositing a drop of polylysine solution (5 μg mL^{-1}) on the carbon film to increase the hydrophilicty of the surface and to improve DNA binding. All transmission electron micrographs (TEMs) were obtained using a JEOL 2000 FX TEMSCAN operating at an acceleration voltage of 80 kV.

In order to image the DNA components of the assembled nanoscale architectures, a drop (aprox. 50 μL) of an aqueous uranyl acetate solution (1%) was placed on a Teflon surface.[20] A carbon-coated copper grid was immersed sample-side down in the above drop (20 s) and any excess solution removed by wicking using filter paper. Finally, the grid was washed thoroughly with distilled-deionised water and dried in air.

2.7 Atomic force microscopy

Samples were deposited on *n*-type silicon wafers (2–4 Ω cm). These wafers had previously been heated to 350 °C in air for 4 h, cleaned using a piranha solution (70% H_2SO_4, 30% H_2O_2) and stored under ethanol until required for use. Immediately prior to use they were cleaned again using oxygen plasma (30 W, 50 Hz, 0.2–0.8 mbar O_2) for 5 min. All atomic force micrographs (AFMs) were obtained using a Digital Instrument's Multimode™ Nanoscope III instrument equipped with tapping mode etched silicon tips (Olympus Model OTESPA). Images were processed using Nanoscope Image Software (Ver. 4.23r6).

2.8 UV-visible spectroscopy

UV-visible spectra were recorded using a HP8452A diode array spectrophotometer and quartz cuvettes (1 cm).

3. Results and discussion

The three-branched DNA template shown in Scheme 1 is assembled from three oligonucleotides referred to as *anchoring elements* (20 bp), which have been modified to incorporate a thiol group at the terminal 5′-position. It is also assembled from a complex Y-shaped oligonucleotide, referred to as a *bridging–recognition element* (3 × 20 bp), which consists of three identical oligomers connected by a hexaethyleneglycol linker. Between each *anchoring element* and each branch of the *bridging–recognition element* there are two complementary oligonucleotides collectively referred to as *extension elements* (2 × 100 bp), which determine the length of the DNA backbone comprising each arm of the Y-shaped template. In this case the length of each arm is expected to be approximately 45 nm.

The template was first assembled and then deposited on a carbon-coated copper grid or on a silicon wafer for imaging by TEM and AFM, respectively. The template deposited on the copper grid was stained using an uranyl acetate solution. The resulting images, shown in Fig. 1, clearly show a Y-shaped DNA template with each branch the expected 45 nm in length. It should be noted

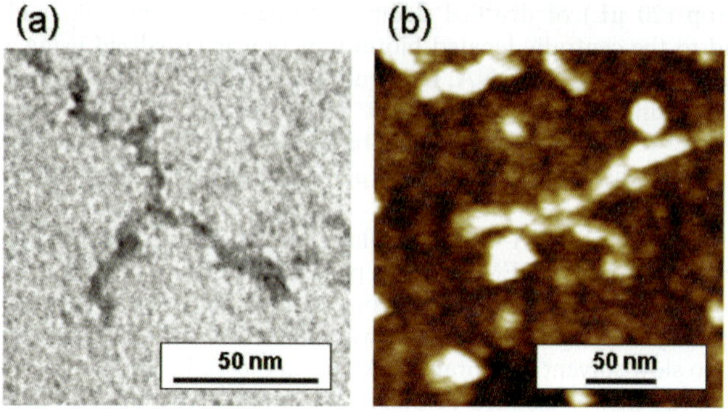

Fig. 1 (a) TEM image of the DNA template shown in Scheme 1, immobilized on a polylysine modified carbon-coated copper grid and stained with uranyl acetate 1%. (b) AFM image of the DNA Template in (a) immobilized (unstained) on a plasma-treated silicon wafer.

that because the persistence length of double-stranded greater than 45 nm the arm of the Y-shaped template are generally straight.[21] It should also be noted, however, that because the bridging element is flexible,[16] the relative orientations of the branches comprising the DNA template are not fixed and do vary from template to template. Parenthetically, some flexibility is not necessarily a disadvantage (for entropic reasons) when considering the subsequent organization of these templates at patterned substrates.

Shown in Fig. 2a is a TEM image of the nanoscale architecture depicted in Scheme 2, consisting of three 5 nm diameter gold nanoparticles attached to the thiol groups incorporated in the *anchoring elements*. It also consists of a central 5 nm diameter biotin-modified gold nanoparticle, which bound to the streptavidin protein incorporated in to the *bridging–recognition* element of the template. Of particular significance is the finding that a biotin modified nanoparticle will recognize and bind selectively the central biotin moiety, once functionalized by the protein streptavidin. The corresponding AFM image, shown in Fig. 2b, clearly shows four nanoparticles arranged as expected. It should be noted that the apparent diameters of these nanoparticles are larger than 5 nm due to the finite dimensions of the probe tip.

Clearly, the nanoscale architecture shown in Scheme 2 has been assembled using the template shown in Scheme 1. It is estimated that the overall efficiency of the assembly process, for a 1 μM scale preparation, is again 25% based on the assumptions that the assembly of the three terminal particles is as efficient as above and that the recognition directed assembly of the fourth particle is an efficient process. Building on this capability, a number of variations on this basic architecture

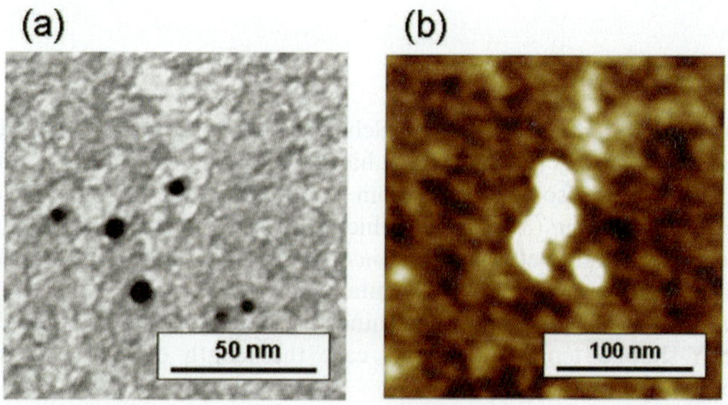

Fig. 2 (a) TEM image of the nanoscale architecture in Scheme 2 incorporating three citrate-stabilized gold nanoparticles (5 nm) and a central biotin-modified and albumin-passivated gold nanoparticle (5 nm), immobilized on a polylysine-modified carbon-coated copper grid. (b) AFM image of the nanoscale architecture in (a), immobilized on a plasma-treated silicon wafer.

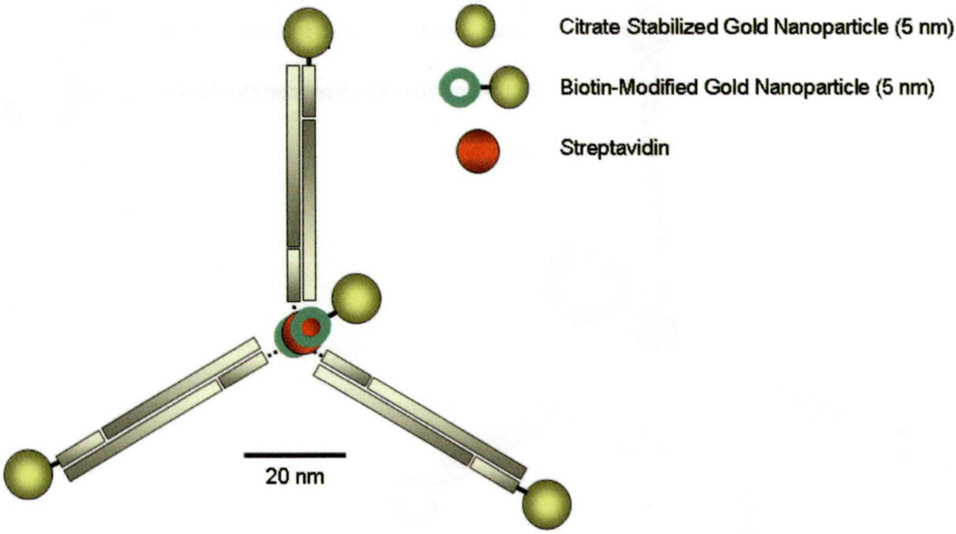

Scheme 2 Nanoscale architecture assembled by coupling three citrate-stabilized gold nanoparticles (5 nm) to the *anchoring elements* and by binding one biotin-modified and albumin-passivated gold nanoparticle (5 nm) to the biotin–streptavidin-modified *bridging–recognition Element* of the DNA Template in Scheme 1.

have been assembled using the same template, by using different nanoparticles and by enlarging some of the nanoparticles by electroless deposition.

Shown in Fig. 3 are a number of related nanoscale architectures, which have been assembled using the DNA template shown in Scheme 1. The principal difference in the case of the nanoscale architecture shown in Scheme 3 and Fig. 3a, is that the biotin labeled gold nanoparticle that recognizes and binds selectively the streptavidin is 16 nm in diameter. In the case of the nanoparticle architecture shown in Scheme 4 and Fig. 3b, the principal difference is that the 5 nm diameter gold nanoparticles linked to the *anchoring elements* are enlarged by electroless deposition. Of particular significance is the similarity of the nanoparticle architecture in Fig. 3b to a three electrode single particle device, a key milestone in the bottom-up assembly and organization of nanoscale electronic devices.

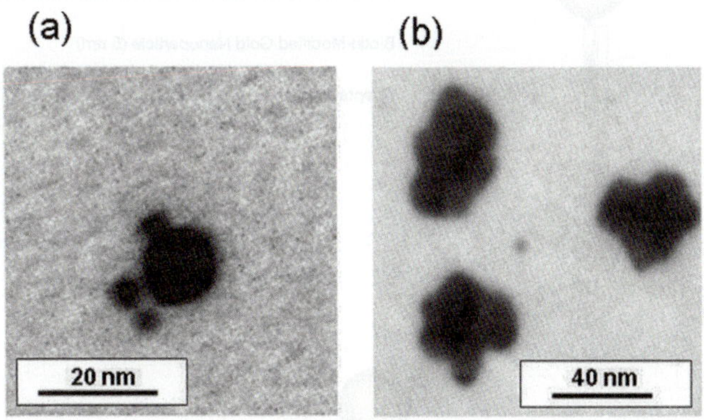

Fig. 3 (a) TEM image of the nanoscale architecture in Scheme 3 incorporating three citrate-stabilized gold nanoparticles (5 nm) and a central biotin-modified gold nanoparticle (16 nm), immobilized on a polylysine-modified carbon-coated copper grid. (b) TEM image of the nanoscale architecture in Scheme 4 incorporating three gold-enhanced citrate-stabilized gold nanoparticles (initially 5 nm) and a central biotin-modified and albumin-passivated gold nanoparticle (5 nm), immobilized on a polylysine-modified carbon-coated copper grid.

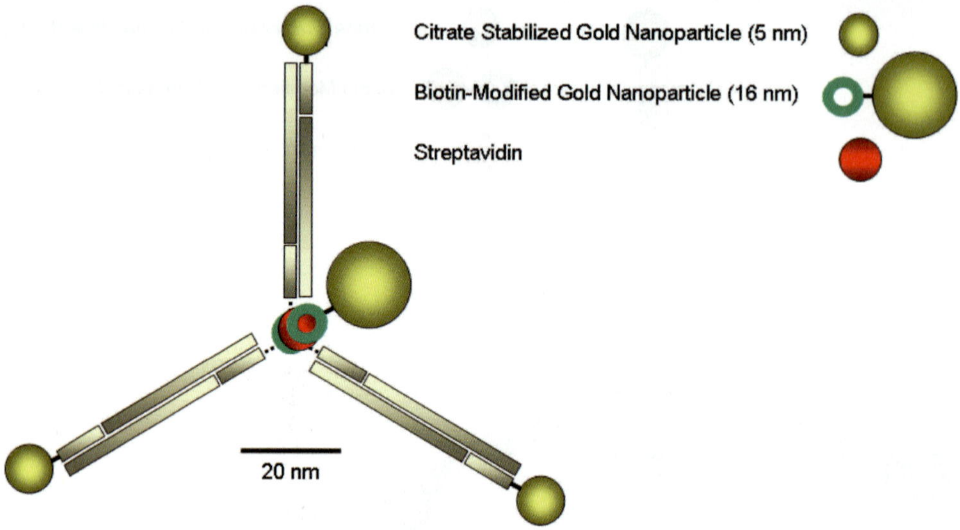

Scheme 3 Nanoscale architecture formed by coupling three citrate-stabilized gold nanoparticles (5 nm) to the *anchoring elements* and by binding one citrate-stabilized and biotin-modified gold nanoparticle (16 nm) to the biotin–streptavidin-modified *bridging–recognition element* of the DNA template in Scheme 1.

Accordingly, we have used the DNA template shown in Scheme 1 to template the assembly of DMAP stabilized gold nanoparticles. These nanoparticles, which are positively charged, are selectively adsorbed at the negatively charged backbone of the DNA template.[22] As a consequence, they organize in the Y-shaped pattern shown in Fig. 4a. Electroless deposition of gold enlarges and enjoins these nanoparticles as shown in Fig. 4b to yield a Y-shaped configuration of gold nanowires.[22]

If the partially metallized DNA templates shown in Fig. 4a is exposed to a dispersion of streptavidin, this tetrameric protein recognizes and binds selectively the biotin moiety incorporated in the bridging recognition element. As a consequence the gold nanoparticles adsorbed at the

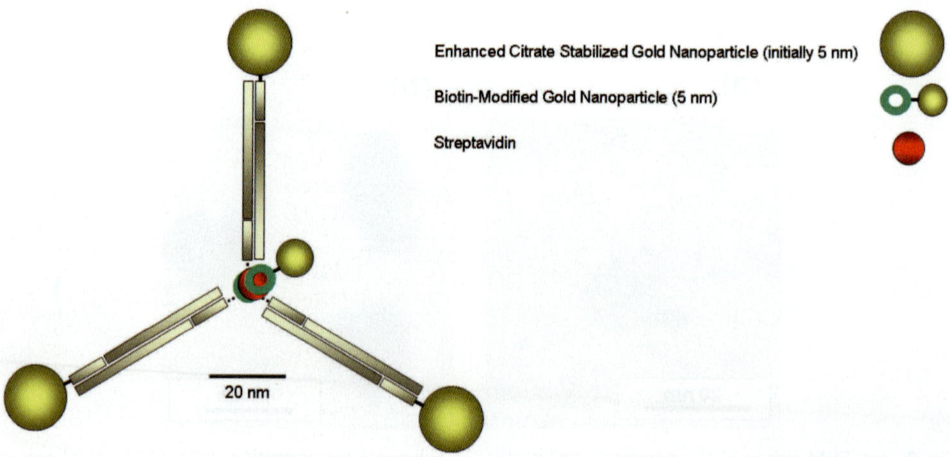

Scheme 4 Nanoscale architecture formed by coupling three citrate-stabilized gold nanoparticles (5 nm) to the *anchoring elements*, by enhancing these nanoparticles using electroless deposition of gold and by binding one biotin-modified and albumin-passivated gold nanoparticle (5 nm) to the biotin–streptavidin-modified *bridging–recognition element* of the DNA template in Scheme 1.

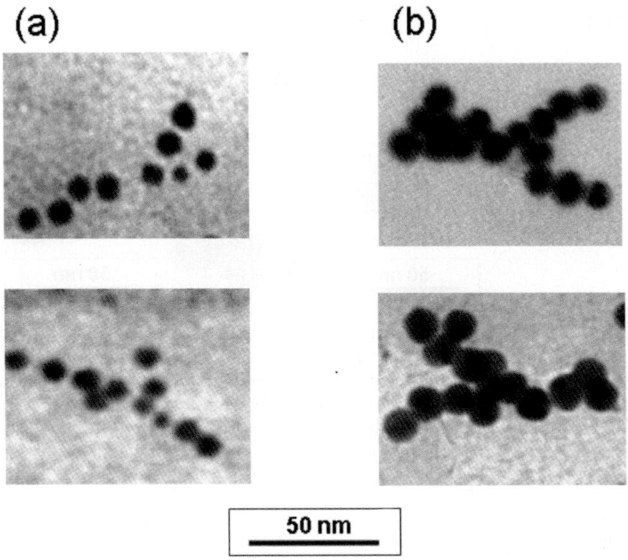

Fig. 4 (a) TEM image of the DNA template in Scheme 1 that has been recognized and bound selectively by DMAP-stabilized gold nanoparticles (4 nm). (b) As in (a) following enhancement of the gold nanoparticles by electroless deposition of gold.

centrally located bridging–recognition element are displaced opening a gap as shown in Fig. 5a. That there is a streptavidin located in this gap is confirmed by staining with uranyl acetate, which clearly visualizes this protein in Fig. 5b.

Taking the nanoscale architecture in Fig. 5b and enhancing and enjoining the gold nanoparticles by electroless deposition we obtain the nanoscale architecture shown in Fig. 6a. Subsequent exposure to a dispersion of biotin-modified 16 nm diameter gold nanoparticles, results in these particles recognizing and binding selectively the streptavidin in the gap. As a consequence there nanoparticles are localized in the gap as shown in Fig. 6b.

In short, the DNA template in Scheme 1 has been used to assemble the nanoscale architecture shown in Scheme 5. This nanoscale architecture is that of a single electron transistor. Specifically, it consists of three proximal nanowires, corresponding to a source–gate–drain architecture, and of a centrally located nanoparticle.

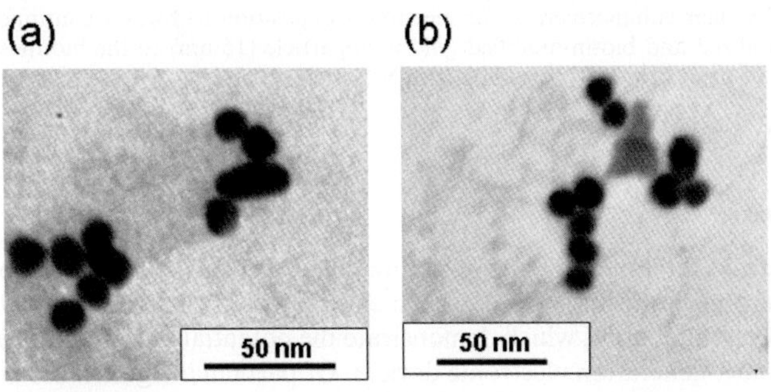

Fig. 5 (a) TEM image of the DNA template in Scheme 1 that has been recognized and bound selectively by DMAP-stabilized gold nanoparticles (4 nm) and subsequently treated with a dispersion of streptavidin to displace the nanoparticles adsorbed at the centrally-located biotin moiety of the *bridging–recognition element*. (b) As in (a) following staining with uranyl acetate 1% to visualize the bound streptavidin.

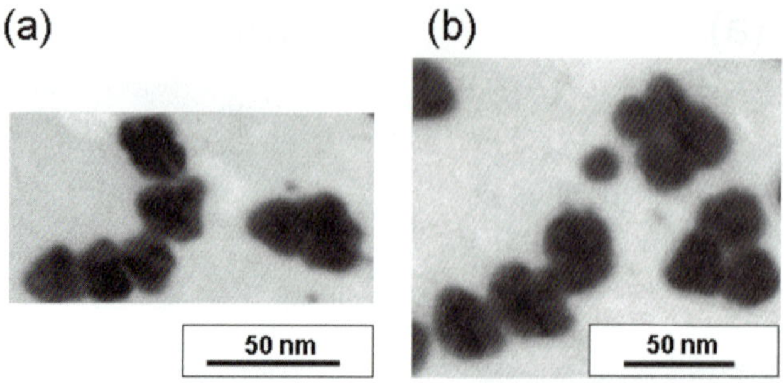

Fig. 6 (a) TEM image of the DNA template in Scheme 1 that has been recognized and bound selectively by DMAP-stabilized gold nanoparticles (4 nm), subsequently treated with a dispersion of streptavidin to displace the nanoparticles adsorbed at the centrally-located biotin, and enhanced by electroless deposition of gold to enlarge and enjoin the adsorbed nanoparticles. (b) As in (a) following exposure to a dispersion of biotin-modified gold nanoparticles (16 nm) which recognise and bind selectively the centrally-located biotin–streptavidin moiety of the *bridging–recognition element* to yield the nanoscale architecture in Scheme 5.

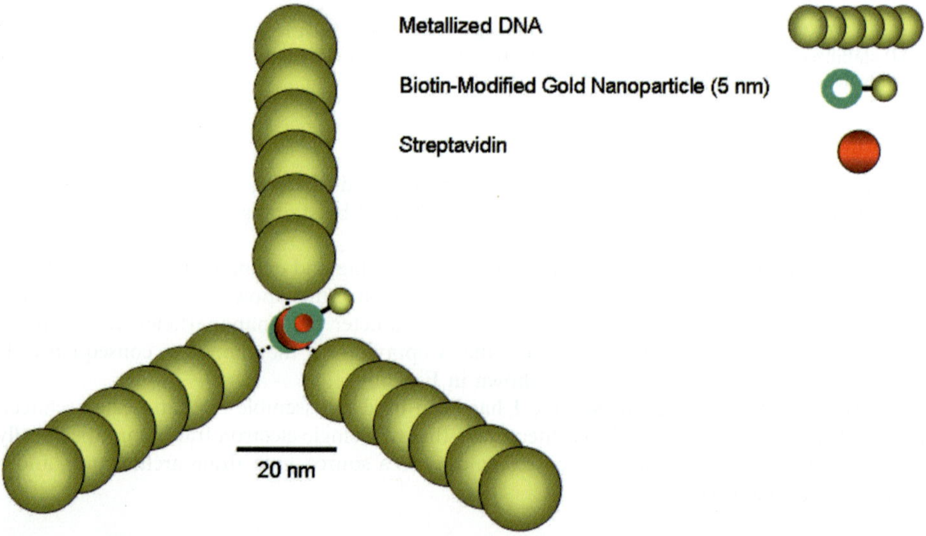

Scheme 5 Nanoscale architecture formed by binding of DMAP-stabilized gold nanoparticles (4 nm) to the *extension elements*, by their enhancement using electroless deposition to form a continuous nanowire and by binding a citrate-stabilized and biotin-modified gold nanoparticle (16 nm) to the biotin–streptavidin-modified *bridging–recognition element* of the DNA template in Scheme 1.

4. Conclusions

Here we have reported the assembly and structural characterization of a Y-shaped DNA template incorporating a central biotin moiety. We have also reported that this template may be used to assemble nanoscale architectures, which demonstrate the potential of this and related approaches to the fabrication of next-generation electronic devices. Of particular significance is the finding that it has been possible to selectively metallize the above DNA template to obtain a three-electrode configuration. Also of particular significance is the finding that a biotin modified nanoparticle recognizes and binds selectively the central biotin moiety of the same template, once functionalized by the protein streptavidin.

Acknowledgements

This study was supported by Science Foundation Ireland through the Centre for Research on Adaptive Nanostructures and Nanodevices.

References

1 G. Moore, *Electronics*, 1965, **38**.
2 *International Technology Roadmap for Semiconductors*, 2002, http://public.itrs.net/.
3 C. Niemeyer, *Angew. Chem. Int. Ed.*, 2001, **40**, 4128.
4 W. Parak, D. Gerion, T. Pellegrino, D. Zanchet, C. Micheel, S. Williams, R. Boudreau, M. Le Gros, C. Larabell C and P. Alivisatos, *Nanotechnology*, 2003, **14**, R15.
5 E. Braun, Y. Eichem, U. Sivan and G. Ben-Yoseph, *Nature*, 1998, **391**, 775.
6 C. Collier, E. Wong, M. Belohradsky, F. Raymo, J. F. Stoddart, P. Kuekes, R. Williams and J. Heath, *Science*, 1999, **285**, 391.
7 K. Keren, M. Krueger, R. Gilad, G. Ben-Yoseph, U. Sivan and U. Braun, *Science*, 2002, **297**, 72.
8 H. Yan, S. Park, G. Finkelstein, J. Reif and T. La Bean, *Science*, 2003, **301**, 1882.
9 H. Xin and A. Woolley, *J. Am. Chem. Soc.*, 2003, **125**, 8710.
10 A. Ongaro, F. Griffin, L. Nagle, D. Iacopino, R. Eritja and D. Fitzmaurice, *Adv. Mater.*, 2004, **16**, 1799.
11 S. Stanca, A. Ongaro, R. Eritja and D. Fitzmaurice, *Nanotechnology*, 2005, **16**, in press.
12 J. Slot and H. Geuze, *Eur. J. Cell Biol.*, 1985, **38**, 87.
13 S. Connolly, S. Cobb and D. Fitzmaurice, *J. Phys. Chem. B*, 2001, **105**, 2222.
14 F. Griffin and D. Fitzmaurice, *Langmuir*, 2005, manuscript submitted.
15 A. Ongaro, D. Iacopino, L. Nagle, R. Eritja and D. Fitzmaurice, *Nanotechnology*, 2003, **14**, 447.
16 M. Grimau, D. Iacopino, A. Avino, B. de la Torre, A. Ongaro, D. Fitzmaurice, J. Wessels and R. Erijta, *Helv. Chim. Acta*, 2003, **86**, 2814.
17 S. Connolly, R. Rao and D. Fitzmaurice, *J. Phys. Chem. B*, 2000, **104**, 4765.
18 T. Liu, T. Tang and L. Jiang L, *Biochem. Biophys. Res. Commun.*, 2004, **313**, 3.
19 N. Green, in *Spectrophotometric Determination of Avidin and Biotin in Methods in Enzymology*, ed. D. McCommick and D. Wright, Academic Press, New York, 1974, **ch. 18A**, pp. 418–425.
20 C. Gray, in *Electron Microscopy of Protein–Nucleic Acid Complexes in DNA-Protein Interactions*, ed. T. Moss, Humana Press, Totowa, NJ, 2001, pp. 1–638.
21 J. Bednar, P. Furrer, V. Katritch, A. Stasiak, J. Dubochet and A. Stasiak A, *J. Mol. Biol.*, 1995, **254**, 579.
22 A. Ongaro, F. Griffin, P. Beecher, L. Nagle, D. Iacopino, A. Quinn, G. Redmond and D. Fitzmaurice, *Chem. Mater.*, 2005, **17**, 1959.

Metalloprotein tunnel junctions: compressional modulation of barrier height and transport mechanism

Jason J. Davis,*[a] Nan Wang,[a] Ashley Morgan,[a] Tiantian Zhang[b] and Jianwei Zhao[ab]

Received 3rd June 2005, Accepted 13th June 2005
First published as an Advance Article on the web 14th October 2005
DOI: 10.1039/b507854g

Though the incorporation of sensory or potentially-switchable biological entities into electronic devices brings with it a number of complicating issues associated with hydration, structural complexity/delicacy, and low conductance, the possibility of resolving properties of fundamental importance (such as the influence of protein fold on conductance) at a molecularly-resolved level, are exciting. Our ability to analyse charge transport through a biological macromolecule remains, though, a significant practical and theoretical challenge. Though much information can be gained by carrying out such examinations at a molecular level, there exist few methods where such controlled analyses are, in fact, feasible. Here we report on the electron transport characteristics of a blue copper metalloprotein as characterized by conductive-probe atomic force microscopy. At very low imposed force, contact resistance is high, electrical contact unstable, and the junction undergoes dielectric breakdown at 1.1–1.5 GV m^{-1}. At increased applied force, the current–voltage characteristics are entirely reproducible and well-described by a Simmons (non-resonant) tunnelling model. Though highly resistive, observations demonstrate the ability of the protein matrix to mediate appreciable tunnelling current. Non-resonant behaviour is consistent with observations of bias-independent tunnelling imaging. In fitting observed transport characteristics to this model, it is possible to deconvolute barrier height and length at specific experimental conditions and, specifically, to monitor the modulation of these parameters by imposed compressional force. At higher field spectroscopic features assignable to metal based density of states are reproducibly observed. These vanish in a force regime where the tunnel barrier to direct tip–sample communication decreases.

Introduction

An understanding of charge transfer on the molecular scale lies central to a number of developments at the forefront of electronics, photonics, catalysis and sensory technology. Though electroanalytical investigations have provided much information on the process of electron transport across molecules confined to a planar electrode surface, the (inherently) stochastic characteristics are only resolved on scaling the molecule–surface contacts down to the nanometre scale. Though associated with considerable practical challenges, such "spatially resolved" experiments are also more applicable to the subsequent generation of truly single molecule devices. In view of their recognition and potentially switchable characteristics we, and others, have been interested in the interfacing of metalloproteins with nanometre-scale tunnelling junctions.[1–5]

In situ and *ex situ* scanning tunnelling microscopy (STM) experiments have alluded not only to the conductance characteristics of protein molecules but also the potential role of prosthetic groups in mediating electron flow.[5–9] Though conductance data can be obtained "more directly" from

[a] *Central Chemistry Laboratory, University of Oxford, South Parks Road, Oxford OX1 3TA. Fax: 01865 275914; Tel: 01865 275914*
b
[b] *Department of Chemistry, Nanjing University, Nanjing 210093, P.R. China*

scanning tunnelling spectroscopy (STS),[3,4] the location of the tip with respect to the molecule of interest is, essentially, unknown in such experiments and this makes quantitative determinations of conductance (as well as the relationship between conductance and potential molecular deformation) difficult. We have recently reported the use of conductive probe tunnel junctions in the spatially resolved analysis of metalloprotein tunnel conductance.[5,10] In this work, we have specifically sought to examine molecular conductance of a metalloprotein and the effects of calibrated compressive force on this. Though the mechanical unfolding of protein in an AFM configuration is now a highly developed area of research,[11–13] there have been, to our knowledge, no reports on molecular-scale protein compression. Currently there are no spatially resolved (structural) spectroscopic methods which would provide adequate information in this context. We have sought, however, to gain insight into this by modeling structure-related molecular conductance as a function of imposed vertical force.

The blue copper protein, azurin (Fig. 1), which plays an important role in respiratory and photosynthetic electron transport chains,[14,15] has been chosen as a model, redox-switchable biomolecule. The natural form of azurin contains a surface-exposed disulfide group that serves to anchor the protein to gold surfaces in the formation of robust arrays which may be reliably addressed electronically.[16,17]

Results and discussion

(i) Tunnel transport under conditions of low load

Azurin from *Pseudamonas aeuruginoas* has a molecular weight of some 13 kDa and, in an uncompressed state, would represent a tunnel length of some 35 Å when confined between two

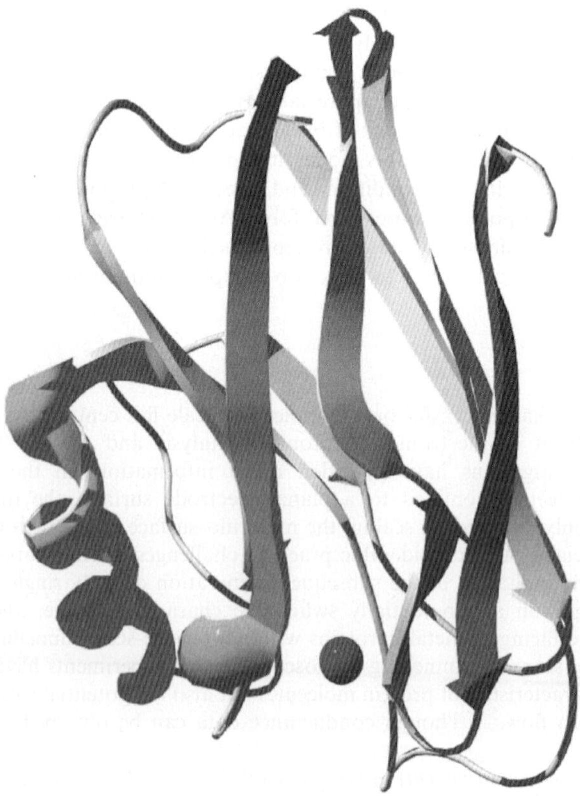

Fig. 1 Schematic presentation of the wild type azurin structure from *P. aeruginosa* showing its secondary structure elements. The copper ion is shown as a closed sphere.[15]

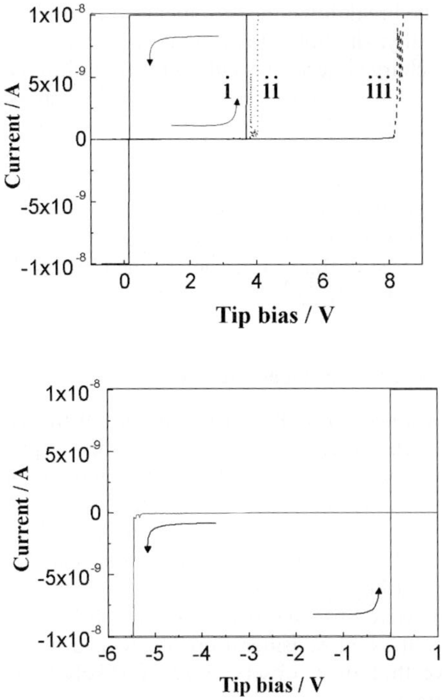

Fig. 2 Current–voltage characteristics at forces <2 nN. (a) Positive bias scan (i) 0.5 V s^{-1}; (ii) 10 V s^{-1}; (iii) 50 V s^{-1}. (b) Negative bias scan at the ramp speed of 10 V s^{-1}. The corresponding breakdown voltages are summarised in Table 1.

electrodes (Fig. 1). Typical current–voltage profiles recorded during positive bias scans under conditions of low (less than 2.5 nN) imposed compressional force are shown in Fig. 2 where it is apparent that negligible current flow is detected until a threshold applied voltage is reached. This threshold shows a sensitive dependency on voltage sweep rate (Table 1) and, once reached, the current is observed to jump to a value exceeding that measurable by the preamplifier (± 10 nA). These I-V relations are entirely reproducible but irreversible in the sense that the initial current–voltage curve is not retraced on reversing the applied field. Once the preamplifier current is saturated on the forward sweep, the current is maintained at values exceeding 10 nA on bias reversal before jumping to -10 nA at zero bias. This current hysteresis has been similarly observed in conductive AFM studies of self-assembled monolayers[18–23] where it has been assigned to dielectric breakdown of the adsorbate under high imposed field ($>10^9$ V m^{-1}).[24–26]

Previous investigations of dielectric breakdown in thin films have concluded that this only takes place when sufficient charge has accumulated at one, or both, sides of the film.[25–27] It seems reasonable to suggest that, under conditions of minimal compression (see below) and high field, charge accumulation on a ~ 3.5 nm diameter protein dielectric can occur. The time dependence of such accumulation is likely to be the origin of the threshold voltage scan rate dependency (a faster voltage scan, or shorter injection time, leading to a less efficient transfer of charge to the protein and, subsequently, an observation of breakdown only at comparatively higher applied field; Table 1).

Table 1 Ramp speed dependence of the metalloprotein junction breakdown voltage (force <2 nN)

Scan speed/Vs^{-1}	Breakdown voltage/V	Sample number
0.5 (positive bias scan)	3.32 ± 0.32	10
10 (positive bias scan)	3.87 ± 0.33	10
50 (positive bias scan)	8.12 ± 1.25	5
10 (negative bias scan)	-5.19 ± 0.34	10

It is also worth noting that the breakdown measurements at ramp speeds lower than 10 V s^{-1} have high reproducibility. Specifically, the breakdown limits were found to be largely invariant over a 0.5–10 V s^{-1} suggesting that charge injection is efficient (not limiting) across this range. Breakdown at negative tip bias has also been examined and, similarly, has an associated threshold (Fig. 2b), though this is somewhat larger than that observed at opposite field.

Assuming minimal protein compression at these forces, the breakdown voltage observations can lead to an approximation of the dielectric strength of the protein at between 1.1 ~ 1.5 GV m^{-1}. This range compares well to that shown by silicon dioxide SiO_2 (0.8–1.3 GV m^{-1}),[28] being somewhat smaller than the 2.0 GV m^{-1} determination for alkyl monolayers on either metal[24] or semiconductor,[22] surfaces but greater than values determined for alkyl bilayers sandwiched between a Hg drop and Ag plate (~0.5 GV m^{-1}).[25]

(ii) Tunnel transport under conditions of moderate load

Stable transport can be maintained through azurin junctions at forces greater than ~3 nN.[1,5,10] In the low bias regime (±0.5 V) the current–voltage relationship is approximately linear and can be used in an estimation of molecular resistance at specific imposed force (Fig. 3). At the minimal forces required to establish tunnelling without breakdown (3–5 nN) zero bias resistances are observed to be 45–60 GΩ, a range bearing good similarity to tunnel resistance set points used in the nondestructive STM imaging of these (and comparable) molecular adsorbates.[3,4,29] The tunnelling spectroscopic analyses reported here were highly reproducible in the low-medium (0 ± 1 V) bias regime, (resistance variations were observed to be less than 10% over >100 *I–V* analyses at each specified force). In view of this, and consideration of the relative magnitudes of tip radius and protein diameter, we believe that these observations are likely to be the result of current flow through a single metalloprotein.

In order to rule out the possibility that current could be dominated by conduction across the surface of a protein-associated water film, the effects of humidity on conductance were analysed. Within a 20–60% humidity range junction resistance values varied by less than 10%.

The resistance across, hexanethiol, octanethiol, decanethiol and dodecanethiol SAMs obtained under equivalent conditions of vertical force and humidity are 140 ± 50 MΩ, 1.5 ± 0.7 GΩ, 10 ± 5 GΩ and 40 ± 8 GΩ, respectively, values bearing close similarity to those previously determined.[19,30] Fitting these resistance variations to an exponential tunnelling equation, $R = R_0\exp(\beta d)$ (where R_0 is an effective contact resistance and d is the junction length) allows the exponential distance decay

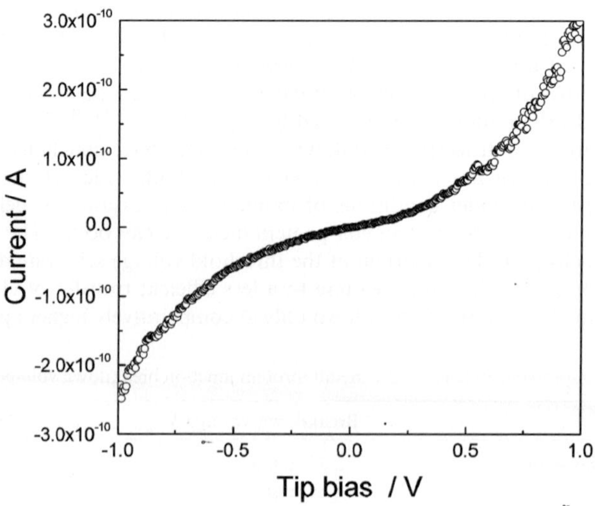

Fig. 3 A representative current–voltage curve obtained at a scan speed of 4 V s^{-1} and a junction force of 6 nN. Though the current shows some asymmetry about zero bias, this is minimal (less than 5%).

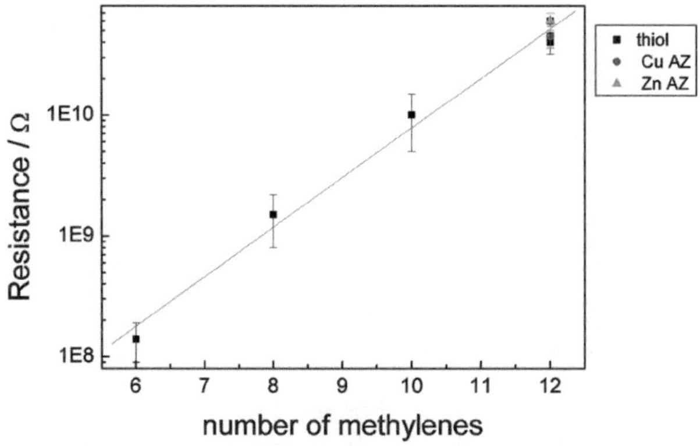

Fig. 4 Comparative force-resistance trends of azurin, ferritin (a 24-subunit iron storage protein) and a range of alkyl thiol SAMs. Data was obtained at junction forces in the 5–7 nN range at 40–50% humidity. Extrapolation of the data to zero chain length gives a contact resistance of some 100 kΩ in this force window.

factor β to be determined as 1.2 ± 0.1 per methylene group, a value in reasonable agreement with previous reports under comparable experimental conditions.[19]

Assuming that non-resonant superexchange tunnelling is the dominant mechanism of charge transfer across these junctions (see below), the greater resistance of the protein junctions over the saturated alkyl SAMs may be assignable to differences in both molecular height (tunnel distance) and tunnel barrier. If one assumes a tunnel barrier comparable to that through the saturated alkyl systems[31] (see below) extrapolation of the data in Fig. 4 leads to an estimated protein tunnelling distance of approximately 1.5 nm, suggesting that, even in this force regime, appreciable molecular compression takes place. This validity of this may be directly assessed through controlled force molecular imaging. Fig. 5 shows the force induced variation in azurin height as determined by fluid-phase (contact mode) AFM imaging under aqueous solution. Though the data shows some scatter, it is clear that the protein is compressed to something approaching half of its unperturbed height within force limits required is the establishment of reliable electrical contact. At the minimum force required for stable image acquisition (~300 pN), the measured protein height is ~2.8 nm.

A simple theoretical analysis of transport through these junctions has been carried out previously.[1,2] The electron transfer between two conducting electrodes separated by a thin dielectric layer can be well-described by the Simmons model[32] in which current density, i, is given by:

$$i = \frac{e^2}{2\pi h L^2}\left\{\left(\varphi_0 - \frac{V}{2}\right)\exp\left[-K\left(\varphi_0 - \frac{V}{2}\right)^{1/2}\right] - \left(\varphi_0 + \frac{V}{2}\right)\exp\left[-K\left(\varphi_0 + \frac{V}{2}\right)^{1/2}\right]\right\} \quad (1)$$

where

$$K = \frac{4\pi L}{h}(2me)^{1/2} \quad (2)$$

and L is the barrier length, φ_0 is the mean barrier height, e and m are the electron charge and mass, respectively. Due to the inherent asymmetry of conductive probe (CP-AFM) junctions, a more reasonable treatment is to divide the voltage drop into two non-equivalent components as follows:[2,33]

$$V = \alpha V + (1 - \alpha)V \quad (3)$$

where αV and $(1 - \alpha)V$ are the potential shifts at the tip and substrate, respectively. The net current density is then given by:

$$i = \frac{e^2}{2\pi h L^2}\{(\varphi_0 - \alpha V)\exp[-K(\varphi_0 - \alpha V)^{1/2}] - [\varphi_0 + (1 - \alpha)V]\exp[-K(\varphi_0 + (1 - \alpha)V)^{1/2}]\} \quad (4)$$

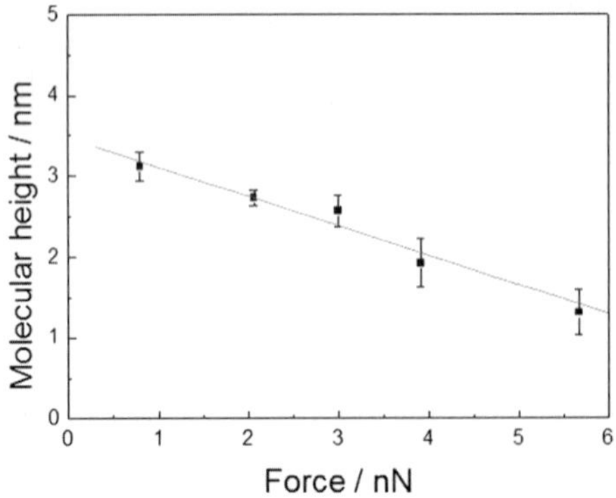

Fig. 5 The variation in molecular height with imaging force. Data taken from contact mode imaging under 18.2 MΩ cm water, with silicon nitride tip ($k = 0.06$). A ~ 5 μM solution of azurin in acetate buffer (pH 4.65) was spin coated onto a freshly annealed gold substrate, then rinsed with buffer and dried under nitrogen. Protein height was averaged over 90 individual molecules, with standard deviations below 10%. At forces in excess of 4 nN no images of sufficient clarity were obtainable.

Eqn. (4) provides a relationship between the variance of tunnelling current with bias and three structure-related variables, namely, φ_0, L, and the asymmetry factor α.

A comparison of experimental and simulated current–voltage data is shown in Fig. 6. At low force the data are highly symmetrical about the voltage origin (asymmetry typically $<4\%$) and well-described by eqn. (1) (the simulated data tracks the experimentally observed current variations to within 0.5% across the associated voltage range, demonstrating the applicability of a three-variable Simmons model to such configurations).[2] Best fits across 20 data sets at ~ 5 nN are obtained with $\alpha = 0.5$, $L = 12$ Å and $\varphi_0 = 1.1$ eV. This barrier height corresponds to a tunnelling coefficient (β) of 1.1 Å^{-1}, a value which compares reasonably well with those derived from bulk, solution-phase, studies.[34–36] As we have noted previously, current–voltage trends become noticeably more asymmetric on excursion to regions of higher compressional force where eqn. (4) becomes more appropriate.[2]

(iii) Modulation of conductance under moderate load

The inherent mechanical flexibility of AFM based molecular junctions can be utilized in an examination of compression-based changes in conductance. Stable and reproducible tunnelling

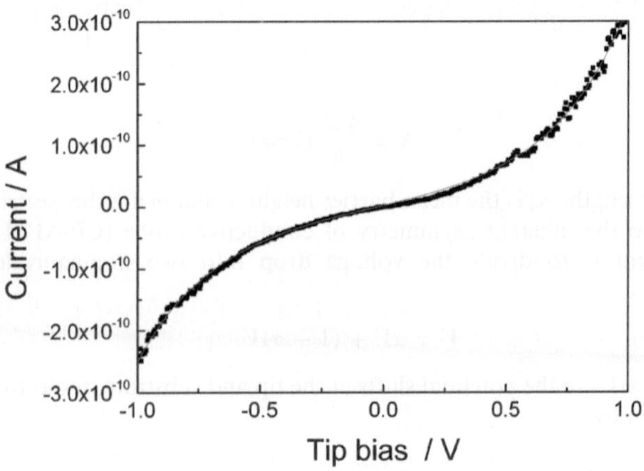

Fig. 6 A representative comparison of experimental (black) and Simmons simulated (grey) current–voltage characteristics at 6 nN.

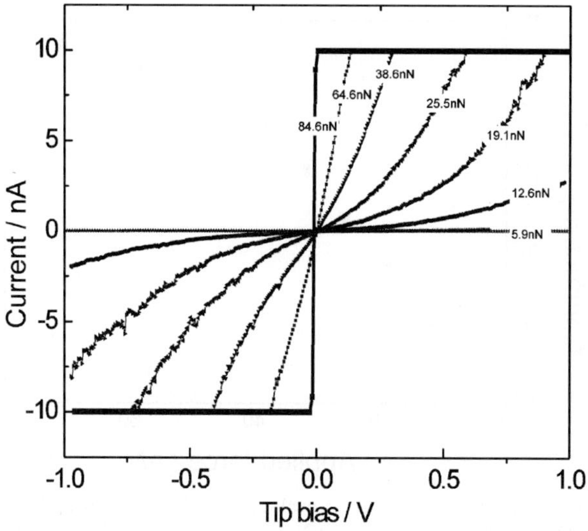

Fig. 7 Representative current–voltage behaviour obtained in the −1 to 1 V bias window (scan speed 4 V s^{-1}) at various applied force.

characteristics are attainable across a 3–80 nN range enabling the effects of variant applied load on the conductance of an azurin metalloprotein junction to be examined (Figs. 7 and 8). The highly reproducible increases in conductance with load can, in the first instance, be ascribed to either a load-dependent decrease in tunnelling barrier height, or tunnelling distance (or a combination of both).

Though it is impossible to deconvolute the force-dependent contribution of probe–probe contact area to the charge transfer characteristics of these junctions, variation in this parameter is expected to bring with it only minimal change in current.[2] Though the predictable variations in tunnel distance, L, with increasing load, can account for some of the conductance modulation observed, we have been unable to simulate these changes through variance of L or φ_0 alone. In considering barrier height changes under compression (Table 2) it is useful to make reference to previous work in which metalloprotein distance decay parameters have been determined in bulk, solution-phase, experiments.

Dutton *et al.* have considered the effects of atomic packing on the ability of a protein matrix to mediate electron transfer, and given an empirical expression to account for structure-dependent variations of the tunneling coefficient, β,[35]

$$\beta = 0.9\rho + 2.8(1 - \rho) \tag{5}$$

The association of electron transfer barrier height, φ_0 (in eV) with tunnel decay coefficient, β in Å^{-1}, is as through eqn. (6),[32,37]

$$\beta = 1.025{\varphi_0}^{1/2} \tag{6}$$

In principle, the atom packing density, ρ, can range from 1, corresponding to a fully packed medium ($\beta \sim 0.9$ Å^{-1}), to a value of 0, corresponding to the interstitial space in the protein structure ($\beta = 2.8$ Å^{-1}).[35,38–40] φ_0 is therefore correlated with the atomic packing density through eqn. (7):

$$\varphi_0 = (0.88\rho + 2.73(1 - \rho))^2 \tag{7}$$

i.e. the higher the packing density, the lower the barrier height. Under large constant force, the protein is compressed and this leads to an associated increase in atomic packing density and decrease in barrier height. In the regime of large molecular compression counterbalancing repulsive

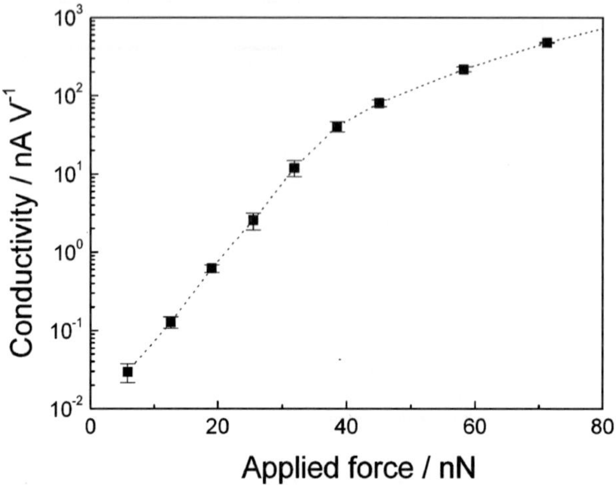

Fig. 8 Force dependence junction conductance as evaluated from the slopes of 20 *I–V* curves at each force (5.9 nN, 12.6 nN, 19.1 nN, 25.5 nN, 31.9 nN, 38.6 nN, 51.7 nN, 64.6 nN, 84.6 nN) in the low bias regime.

forces may lead to structural deformation in the horizontal plane where there is no geometric restriction. The associated increase in protein volume will be associated with a regime of decreasing atomic packing density/increasing barrier height. The trends here are qualitatively predictable, supported by molecular-dynamics simulations and have been previously outlined.[2] In summary, a Simmons analysis of the force-dependent current–voltage characteristics has suggested that the tunnel barrier and length vary from 1.1 to 0.7 eV and 12–10 Å, respectively across a 5–40 nN force range. In the regime of greater molecular compression, current–voltage trends are steep and cannot be adequately used in fitting.

(iv) Accessing the metallic states; negative differential resistance

A considerable number of tunnelling spectroscopy studies have now been published in which the ability of tunnel electrons to access molecular states in an STM configuration has, *apparently*, been evident. In tunnelling spectroscopic analyses, these molecular resonances are most usefully depicted within (dI/dV) *vs.* V plots in which the former parameter is directly related to the density of electronic states (DOS). By changing the energy range accessible to the tunnelling electrons (through the applied bias) it is, then, possible to modulate the ability of these to access orbital states within the junction and, therefore, the resistance. Of the spectroscopic features which may result, negative differential resistance (NDR), has, on the basis of its possible application within switchable devices, attracted much attention.[41–44] Despite the current understanding of the NDR phenomenon in conventional semiconductor resonant tunnelling diodes (RTDs), the phenomenon is less well-understood on the *molecular* scale.[45]

Table 2 Barrier height and length variations with imposed force as obtained from nonlinear least squares fitting of current–voltage trends by a Simmons model

Force/nN	φ_0/eV	L/Å
5.9366	1.11078	11.87979
10.38462	1.08265	11.5652
17.30769	1.00979	10.72098
24.53846	0.94241	9.78781
30.38462	0.77875	9.07742

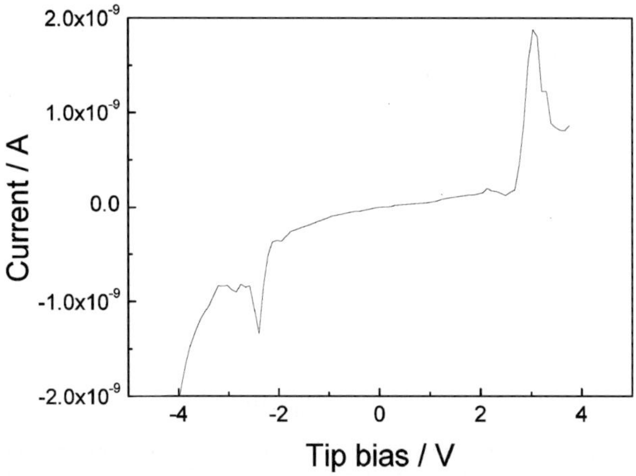

Fig. 9 Representative current–voltage behaviour at higher applied bias (scan speed 4 V s^{-1}; junction force 3–5 nN). The PVR (peak valley current ratios) lie in the range of 4–7.

Though NDR spectroscopic features have been assigned to numerous mechanisms,[42,46,47] those associated with Fermi level-frontier molecular orbital resonance are perhaps the most attractive from the perspective of relating tunnel transport to redox accessibility.[5,48] In imaging experiments, the effects of aligning the Fermi level of the underlying electrode to molecular orbital states on mediating increased current flow through the molecule has been strikingly evident in imaging experiments where the redox-active unit is observed to "light up" (become more conductive) at the appropriate potential.[49,50] The role, or otherwise, of metal-based orbitals in mediating current flow across these metalloprotein junctions is, then, worthy of some consideration. Though azurin contains a redox-accessible centre in these immobilized configurations[51] the current–voltage plots in the low (±1 V) bias regime discussed above are fully consistent with a non-resonant tunnelling process between the metallic contacts in which protein-based molecular states are not accessed at the bias voltages applied (no "molecular features" are evident in either raw data or derivative current–voltage plots in an analysis of many hundreds of scans). These spectroscopic characteristics are supportive of observations made in a tunnelling imaging mode; though increased current flow in the vicinity of the metal centre has occasionally been reported in metalloprotein tunnelling imaging, in our studies image contrast has been largely independent of overpotential across a broad range.[6,7,52] Though the possible reasons for this are difficult to deconvolute and may be associated with the experimental conditions relevant to controlled potential aqueous imaging, these observations are undeniably consistent with non-resonance. The same is true on expanding the energetic range (bias) available to the tunnelling electrons in a spectroscopic configuration to ±4 V *if* the compressional force lies in the 5–80 nN range. It is likely that, under such forces, molecular compression is significant (Table 2 and Fig. 5), robust electrical contact established at both interfaces, and both tunnel barrier and distance low. Direct tunnelling is then likely to be facile.[53] At reduced applied force (greater tunnel distance) it is reasonable to suggest that direct tunnelling will become less efficient. Fig. 9 shows a current–voltage plot of Cu-azurin obtained at a force of ~4 nN which displays NDR characteristics (an initial rise in current followed by a sharp decrease as bias is progressively augmented). Though the "peak heights" associated with the apparent resonances evident in Fig. 9 show some experimental variation they are invariably associated with a 300–700% increase in current above that extrapolated from the "non-resonant" regime. These "spectroscopic features" are often not symmetric about the voltage origin. Specifically, only ~30% of such spectra show apparent NDR at positive tip potential. A statistical analysis of "peak maxima" demonstrates considerable fluctuation in the voltage at which current maxima are observed; specifically, the peak centre at negative bias is observed at –2.2 ± 0.7 V and 2.75 ± 1.0 Vs at positive bias. We, and others, have noted these fluctuations previously.[48,54]

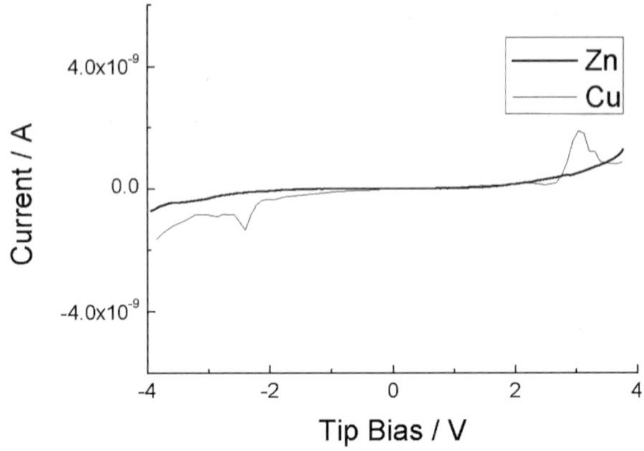

Fig. 10 Comparative current–voltage characteristics of the zinc and copper forms of azurin at junction forces in the 3–5 nN range. While zero bias resistances comparable, apparent NDR is observed only with the copper form.

The apparent NDR features were reproducible and only observable in the force window between breakdown (poor tip-molecule contact) and large molecular compression (the onset of direct tunnelling). With azurin metal substitution can be achieved with minimal structural perturbation and this, accordingly, facilitates studies in which the precise role of this site in mediating charge flow can be clarified. The assignment of these features to the copper states is, then, strengthened by observations made in comparative experiments carried out with the zinc protein (Fig. 10). In the low bias regime, an analysis of several hundred comparative current–voltage trends indicates, within experimental variation, no significant differences in Cu-azurin or Zn-azurin conductance. Significantly, the apparent resonances observed with the copper protein at high bias are not observed with the (redox-inactive) zinc protein.

The surface-confined (copper) azurin molecules display well-defined (if non-ideal) electrochemical responses (Fig. 11). Though the potential at which transmission through the confined molecule is observed to increase has, in some cases, been usefully related to the electrochemically-determined redox potential[50,50–57] we have, to date, been unable to find any sensible correlation between these potentials and those associated with the spectroscopic features in Figs. 9 and 10. Work is currently underway to enable these measurements to be obtained under electrolytic solution.

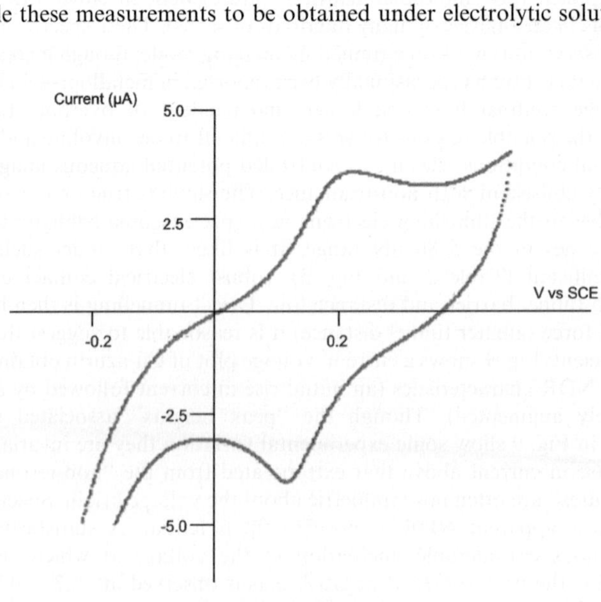

Fig. 11 Voltammetric response of an adsorbed adlayer of azurin on an evaporated gold electrode (100 mV s^{-1}).

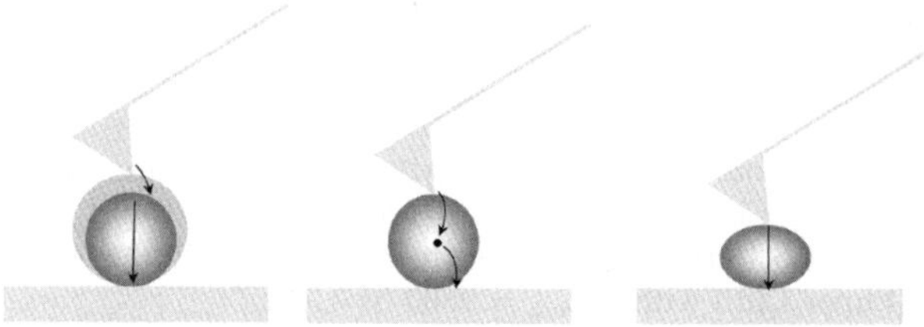

Fig. 12 A schematic representation of the compressional force-dependent tunnelling mechanism across a CP-AFM trapped redox protein. (a) At forces <2 nN the protein-electrode electrical contact is poor and charge build-up, leading to breakdown, is observed. (b) At forces in a 2 nN–5 nN range the protein redox site appears to be accessed by suitable energetic electrons. The tunnelling may then be describable as "two-step". (c) At forces exceeding 5 nN, direct tip–substrate tunneling is observed with associated pressure-dependent barrier height and length.

Conclusions

The difficulties associated with attaining robust (ideally chemisorbed) mechanical contact between nanometre-scale electrodes and molecule(s) of interest, and thereby acquiring reliable transport data, remain. Within this context, the scaling and mechanical flexibility provided by conductive AFM junctions is exceptionally useful. In chemisorbing a protein to a gold-coated probe, many of the effects of lateral drift on molecular analysis are removed. From an electronic perspective, it is clear that a protein fold, though classically highly-resistive, is able to act as a conduit for an appreciable current flow within such junctions. The measured gigaohm resistance of a $\sim$3 nm protein examined here lies in the range of a C_{12} saturated alkyl thiol SAM, the uncompressed height of which rests in reasonable agreement with simulation-determined (and imaging supported) protein compression data in this force regime.

At the minimum force required for stable feedback ($\sim$2 nN) the electrode-protein electrical contact is unstable and dielectric breakdown is observed at both bias extremes. Under conditions of very low force, it is reasonable to consider the junction as comprising two sequential tunnel barriers (one at the tip surface, the second at the substrate surface) across which most of the applied field will be dropped. If the associated interfacial resistances are high (in considerable excess of the quantum unit of resistance, 26 kΩ) charge accumulation on the molecule, followed by dielectric breakdown when the bias exceeds some critical value, occurs.[58] At higher force (5–80 nN), observations are consistent with a fall in contact resistance and the onset of direct, non-resonant, tip–substrate tunnelling with associated distance decay and tunnel lengths which are force dependent. In this regime, current–voltage trends are well-described by a Simmons model in which the protein is treated as a featureless (uniform) tunnel barrier. The simulated variance of tunnel distance with force correlates well with image height variations across an equivalent force range.

Transport across metal-molecule-metal junctions depends strongly on the position of the metallic Fermi energies with respect to the frontier molecular orbitals of the molecule. When the latter are not energetically accessible (the zinc protein), electron transport is expected to occur by non-resonant superexchange tunnelling. Though the copper-based orbital states are likely to be potentially accessible, it appears that this only occurs within a narrow force window where electrical contact is robust but (direct) superexchange tunnelling inefficient (the barrier to direct tunnelling is high). Specifically, in the regime of stable contact at low force, current–voltage trends show characteristics assignable to an ability of the electrons to access copper-based molecular states when the applied bias is high. Electronic movement then occurs *via* transient occupancy of copper-based DOS (Fig. 12). At higher compressional force, the barrier to direct tunneling falls and these spectroscopic features vanish. Under these conditions metal substitution in the protein leads to negligible change in current–voltage characteristics.

Though the subject of ongoing experimentation, the good correlation between barrier heights determined in this study and those resolved in bulk (solution phase) work, is interesting. The work described herein further sheds light on the important consideration of mechanical/environmental perturbation of subsequently reported "molecular" properties.

Experimental section

Materials

Azurin (*Pseudomonas aeruginosa* azurin wild-type) was kindly supplied by Professor Gerard Canters, Leiden Institute of Chemistry, Leiden. It was diluted in *ca.* 0.1 M acetate buffer (pH = 4.6, close to the protein isoelectric point) to form *ca.* 5 μM solution characterised by UV-Vis (Lambda 20, Perkin-Elmer Ltd.) with $\varepsilon_{279nm} = 9800\ M^{-1}\ cm^{-1}$.

Conducting AFM measurements

Gold-coated C-AFM probes (Si_3N_4 contact levels of 15 ± 5 nm tip radius, spring constant = 0.13 N m^{-1}, Mikromasch) were washed copiously with acetone, ethanol and deionised water (Millipore, 18.2 MΩ cm), and then modified with azurin by soaking in solution (~5 μM azurin, acetate buffer) for at least 3 h. After incubation, the probes were rinsed copiously with deionised water to remove any physically adsorbed material, blown dry with high purity argon, and placed in the microscope. The tip was then brought into low force (<1 nN) contact with freshly cleaved HOPG (Agar Scientific UK Ltd.). Current–voltage characterization was performed using a Molecular Imaging, Pico-SPM system in contact mode. Bias given throughout is the tip bias with respect to the substrate. All experiments were done under ambient conditions (22 ± 2 °C, 40–50% humidity).

Alkyl thiol SAMs were prepared on evaporated gold from 1 mM thiol solutions in ethanol over a period of at least 12 h. Before use, each sample was rinsed with ethanol and blown dry with high purity argon.

Numerical simulation

Nonlinear least squares fitting was performed with Origin 7.0/7.5 software. The fitting was terminated when a stable chi-squared value was achieved. The probe–protein contact area in these simulations was assumed to be fixed at a value defined by the approximate diameter of the undeformed protein (3 nm).

Acknowledgements

The authors wish to acknowledge the financial support of the EPSRC, and to thank Prof. Allen Hill FRS and Mr Danny Axford for helpful suggestions. Prof. Gerard Canters, Leiden Institute of Chemistry, The Netherlands, is thanked for supplying azurin.

References

1 J. Zhao and J. J. Davis, *Colloids Surf., B*, 2005, **40**, 189.
2 J. Zhao, J. Davis, M. Sansom and A. Hung, *J. Am. Chem. Soc.*, 2004, **126**, 5601.
3 J. J. Davis, *Philos. Trans. R. Soc. London, Ser. A*, 2003, **361**, 2807.
4 J. J. Davis, C. L. Wrathmell, J. Zhao and J. Fletcher, *J. Mol. Recognit.*, 2004, **17**, 167.
5 J. J. Davis, D. A. Morgan, C. L. Wrathmell, D. N. Axford, J. Zhao and N. Wang, *J. Mater. Chem.*, 2005, **15**, 2160.
6 J. Davis, H. Hill, A. Kurz, C. Jacob, W. Maret and B. Vallee, *J. Phys. Chem. Commun.*, 1998, 12.
7 E. P. Friis, J. E. T. Andersen, Y. I. Kharkats, A. M. Kuznetsov, R. J. Nichols, J. D. Zhang and J. Ulstrup, *Proc. Natl. Acad. Sci. USA*, 1999, **96**, 1379.
8 A. M. Kuznetsov and J. Ulstrup, *J. Electroanal. Chem.*, 2004, **564**, 209.
9 J. Zhang, Q. Chi, A. M. Kuznetsov, A. G. Hansen, H. Wackerbarth, H. E. M. Christensen, J. E. T. Andersen and J. Ulstrup, *J. Phys. Chem. B*, 2002, **106**, 1131.
10 J. W. Zhao and J. J. Davis, *Nanotechnology*, 2003, **14**, 1023.
11 H. Clausen-Schaumann, M. Seitz, R. Krautbauer and H. E. Gaub, *Curr. Opin. Chem. Biol.*, 2000, **4**, 524.
12 M. Seitz, *Nanobiotechnology*, 2004, 404.
13 M. Rief and H. Grubmuller, *ChemPhysChem*, 2002, **3**, 255.
14 E. T. Adman, *Adv. Protein Chem.*, 1991, **42**, 145.
15 H. Nar, A. Messerschmidt, R. Huber, M. v. d. Kamp and G. W. Canters, *J. Mol. Biol.*, 1991, **221**, 765.
16 J. J. Davis, C. M. Halliwell, H. A. O. Hill, G. W. Canters, M. C. van Amsterdam and M. P. Verbeet, *New J. Chem.*, 1998, **22**, 1119.
17 J. J. Davis, D. Bruce, G. W. Canters, J. Crozier and H. A. O. Hill, *Chem. Commun.*, 2003, 576.

18 S. B. Sachs, S. P. Dudek, R. P. Hsung, L. R. Sita, J. F. Smalley, M. D. Newton, S. W. Feldberg and C. E. D. Chidsey, *J. Am. Chem. Soc.*, 1997, **119**, 10563.
19 D. J. Wold and C. D. Frisbie, *J. Am. Chem. Soc.*, 2000, **122**, 2970.
20 X. D. Cui, A. Primak, X. Zarate, J. Tomfohr, O. F. Sankey, A. Moore, D. Gust, G. Harris and S. M. Lindsay, *Science*, 2001, **294**, 571.
21 X. Cui, X. Zarate, T. Tomfohr, O. SankeY, A. Primak, A. Moore, T. Moore, D. Gust, G. Harris and S. Lindsay, *Nanotechnology*, 2002, **13**, 5.
22 J. W. U. Zhao, *App. Phys. Lett.*, 2003, **83**, 2034.
23 J. M. Beebe, V. B. Engelkes, L. L. Miller and C. D. Frisbie, *J. Am. Chem. Soc.*, 2002, **124**, 11268.
24 D. Wold and C. Frisbie, *J. Am. Chem. Soc.*, 2001, **123**, 5549.
25 R. Holmlin, R. Haag, M. Chabinye, R. Ismagilov, R. Terfort, M. Rampi and G. Whitesides, *J. Am. Chem. Soc.*, 2001, **123**, 5075.
26 S. O. Mbarga, D. Malec, N. Zebouchi and H. The-Giam, *J. Electrost.*, 1977, **40–41**, 355.
27 Y. Selzer, A. Solomon and D. Cahen, *J. Am. Chem. Soc.*, 2002, **124**, 2886.
28 C. Raynaud, *J. Non-Cryst. Solids*, 2001, **280**, 1.
29 J. J. Davis and H. A. O. Hill, *Chem. Commun.*, 2002, 393.
30 J. Zhao and K. Uosaki, *Nano Lett.*, 2002, **2**, 137.
31 D. J. Wold, R. Haag, M. A. Rampi and C. D. Frisbie, *J. Phys. Chem. B*, 2002, **106**, 2813.
32 J. G. Simmons, *J. Appl. Phys.*, 1963, **34**, 1793.
33 W. D. Tian, S. Datta, S. H. Hong, R. Reifenberger, J. I. Henderson and C. P. Kubiak, *J. Chem. Phys.*, 1998, **109**, 2874.
34 H. B. Gray and J. R. Winkler, *Annu. Rev. Biochem.*, 1996, **65**, 537.
35 C. C. Page, C. C. Moser, X. Chen and P. L. Dutton, *Nature*, 1999, **402**, 47.
36 K. Fujita, N. Nakamura, H. Ohno, B. S. Leigh, K. Niki, H. B. Gray and J. H. Richards, *J. Am. Chem. Soc.*, 2004, **126**, 13954.
37 V. Mujica and M. Ratner, *Chem. Phys.*, 2001, **264**, 365.
38 C. C. Moser, J. M. Keske, K. Warncke, R. S. Farid and P. L. Dutton, *Nature*, 1992, **355**, 796.
39 F. J. S. Guo and G. Mclendon, *J. Phys. Chem.*, 1995, **99**, 8458.
40 D. N. Beratan, J. N. Onuchic, J. R. Winkler and H. B. Gray, *Science*, 1992, **258**, 1740.
41 J. Chen, W. Wang, M. A. Reed, A. M. Rawlett, D. W. Price and J. M. Tour, *App. Phys. Lett.*, 2000, **77**, 1224.
42 F. Fan, R. Lai, J. Cornil, Y. Karzazi, J. Brédas, L. Cai, L. Cheng, Y. Yao, D. P. Jr, S. Dirk, J. Tour and A. Bard, *J. Am. Chem. Soc.*, 2004, **126**, 2568.
43 A. M. Rawlett, T. J. Hopson, L. A. Nagahara, R. K. Tsui, G. K. Ramachandran and S. M. Lindsay, *Appl. Phys. Lett.*, 2002, **81**, 3043.
44 R. A. Wassel, G. M. Credo, R. R. Fuierer, D. L. Feldheim and C. B. Gorman, *J. Am. Chem. Soc.*, 2004, **126**, 295.
45 P. Zhao, H. L. Cui, D. Woolard, K. L. Jensen and F. A. Buot, *J. Appl. Phys.*, 2000, **87**, 1337.
46 J. Gaudioso, L. Lauhon and W. Ho, *Phys. Rev. Lett.*, 2000, **85**, 1918.
47 N. Sutin, B. S. Brunschwig and C. Creutz, *J. Phys. Chem. B*, 2003, **107**, 10687.
48 C. Gorman, R. Carroll and R. Fuierer, *Langmuir*, 2001, **17**, 6923.
49 N. J. Tao, *Phys. Rev. Lett.*, 1996, **76**, 4066.
50 W. Han, E. Durantini, T. Moore, A. Moore, D. Gust, P. Rez, G. Leatherman, G. Seely, N. Tao and S. Lindsay, *J. Phys. Chem. B*, 1997, **101**, 10719.
51 L. Andolfi, D. Bruce, S. Cannistraro, G. W. Canters, J. J. Davis, H. A. O. Hill, J. Crozier, M. P. Verbeet, C. L. Wrathmell and Y. Astier, *J. Electroanal. Chem.*, 2004, **565**, 21.
52 P. Facci, D. Alliata and S. Cannistraro, *Ultramicroscopy*, 2001, **89**, 291.
53 J. Zhao and K. Uosaki, *J. Phys. Chem. B*, 2004, **108**, 17129.
54 C. B. Gorman, Y. He and R. L. Carroll, *Langmuir*, 2001, **17**, 5324.
55 R. L. McCreery, *Chem. Mater.*, 2004, **16**, 4477.
56 W. Haiss, H. V. Zalinge, S. J. Higgins, D. Bethell, H. Hoebenreich, D. J. Schiffrin and R. J. Nichols, *J. Am. Chem. Soc.*, 2003, **125**, 15294.
57 U. Mazur and K. W. Hipps, *J. Phys. Chem.*, 1995, **99**, 6684.
58 K. K. Likharev, *Proc. IEEE*, 1999, **87**, 606.

Long-range interfacial electron transfer of metalloproteins based on molecular wiring assemblies

Qijin Chi,* Jingdong Zhang, Palle S. Jensen, Hans E. M. Christensen and Jens Ulstrup*

Received 4th May 2005, Accepted 16th May 2005
First published as an Advance Article on the web 28th September 2005
DOI: 10.1039/b506136a

We address some physical features associated with long-range interfacial electron transfer (ET) of metalloproteins in both electrochemical and electrochemical scanning tunneling microscopy (ECSTM) configurations, which offer a brief foundation for understanding of the ET mechanisms. These features are illustrated experimentally by new developments of two systems with the blue copper protein azurin and enzyme nitrite reductase as model metalloproteins. Azurin and nitrite reductase were assembled on Au(111) surfaces by molecular wiring to establish effective electronic coupling between the redox centers in the proteins and the electrode surface for ET and biological electrocatalysis. With such assemblies, interfacial ET proceeds through chemically defined and well oriented sites and parallels biological ET. In the case of azurin, the ET properties can be characterized comprehensively and even down to *the single-molecule level* with direct observation of redox-gated electron tunnelling resonance. Molecular wiring using a π-conjugated thiol is suitable for assembling monolayers of the enzyme with catalytic activity well-retained. The catalytic mechanism involves multiple-ET steps including both intramolecular and interfacial processes. Interestingly, ET appears to exhibit a substrate-gated pattern observed preliminarily in both voltammetry and ECSTM.

1. Introduction

Integration of biological molecules into specific matrices (*e.g.* electrodes, polymers, and nanoparticles) to assemble hybrid structures with specific functions (*e.g.* biosensing molecular devices represented by biosensors and biofuel cells), has been one of the most intriguing subjects in molecular bioelectronics.[1–3] Attractive target biomolecules are redox proteins and enzymes that play vital roles in biological ET processes of respiration and photosynthesis, because functional responses (*e.g.* electron transfer and catalysis) can be detected rapidly by electrical signals. Immobilization of proteins on solid surfaces or at liquid/solid interfaces is essential in achieving these objectives. Consequently, understanding and control of charge transfer through the molecule and across the interface become crucial for the rational design of efficient functional systems. Redox proteins should, in principle, be assembled in such a way that appropriate molecular orientation is achieved with full retention of their biological functions and effective electronic coupling between the protein redox centre and the addressable electrode. This, however, remains a daunting challenge for most proteins due to their structural complexity and conformational flexibility. To some extent, molecular wiring assembly has provided solutions to coupling the biological redox centre physically and electronically to the electrode surface. Along with advances in genetic protein engineering, modern surface science technologies (*e.g.* scanning probe microscopies), and theoretical insights, this has promoted molecular bioelectronics to a new phase in the past decade.[4]

Department of Chemistry and Nano · DTU, Building 207, Technical University of Denmark, 2800, Kgs. Lyngby, Denmark. E-mail: cq@kemi.dtu.dk; ju@kemi.dtu.dk.

Molecular wiring assembly refers to the use of small organic molecules to confine protein molecules on electrode surfaces for two-dimensional protein array formation. Small organic molecules, serving as both a linker and an ET-assisting bridge, should possess functional groups at both ends of the molecule. One group can link to the electrode surface, while the other one can interact with the protein. The most common way to organize small organic molecules into an ordered monolayer on metal surfaces is based on gold–thiol chemistry, *i.e.* thiolate self-assembled monolayers (SAMs).[5,6] SAMs have been proved to display functionally versatile and flexible features for a wide range of potential applications in nanoscale science and technology. Specific interactions between the protein molecule and the linker can in principle be created by tailoring terminal groups of the SAM, which could in turn control molecular orientations of the protein. This strategy has been extensively employed to different types of redox proteins and enzymes, such as *cytochrome c*,[7–14] *azurin*,[15–19] *nitrite reductase*,[20,21] *amine oxidase*,[22] and *ferredoxin*.[23] Interfacial ET was detected electrochemically in all cases, but electronic structures and ET mechanisms of protein monolayers have been resolved at quite different levels.

In the present work, we focus on azurin and nitrite reductase cases with the general objective toward observing long-range interfacial ET at *the nanoscale and single-molecule levels*. We have employed a combination of single-crystal electrochemistry and ECSTM to reveal both surface electronic structures and interfacial ET patterns of protein monolayers. Use of single-crystal metals (*e.g.* Au(111) and Pt(111)) as working electrodes can significantly improve the signal/noise ratio due to the well-defined surface. This is crucial for electrochemical detection of protein monolayer ET because the signal is usually weak and could be buried in background voltammetric responses. Furthermore, results from electrochemical measurements can be reasonably compared with those from STM as the electrodes used in both measurements have the same surface quality. When STM operates in an electrochemical mode, the energy states of both the substrate and tip are under potential control relative to a common reference electrode in aqueous buffer environment. This configuration is particularly suitable for *in situ* mapping of electronic structures and properties of redox proteins during their biological action such as ET or/and electrocatalysis, as aqueous solution is indispensable for almost all biological processes in nature.

Fig. 1 shows the three-dimensional structures of *Pseudomonas aeruginosa* azurin (Fig. 1A) and *Alcaligenes xylosoxidans* nitrite reductase (Fig. 1B). Azurin is a blue single-copper protein which functions as electron carrier physiologically associated with oxidative stress responses in many bacteria.[24] This protein is relatively simple in structure, but an ideal model for exploring single-electron tunneling because of its high stability and excellent redox properties. In contrast, nitrite reductase is structurally and functionally more composite. As a homotrimer with two copper centres in each subunit, this enzyme is involved in the bacterial denitrification in the global nitrogen cycle.[25] The functional pathway of enzyme monolayers rests on multiple-ET steps including both intramolecular and interfacial processes. This enzyme thus provides an opportunity, although challenging, to address more complex interfacial ET processes accompanied by biological electrocatalysis in similar well-defined microenvironments.

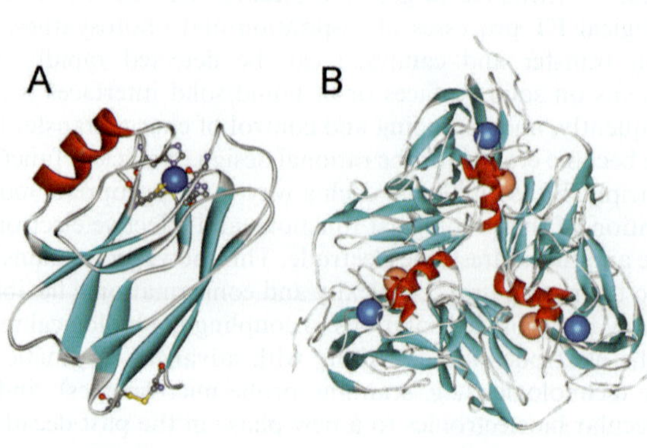

Fig. 1 Three-dimensional crystallographic structures of (A) *Pseudomonas aeruginosa* azurin and (B) *Alcaligenes xylosoxidans* blue copper nitrite reductase. Type I Cu ions are stressed in blue and type II Cu ions in orange colour.

2. Experimental section

2.1. Materials and chemicals

The Au(111) electrodes used in both electrochemical and STM experiments were prepared from polycrystalline gold wires ($\geq$ 99.99%, Goodfellow Inc., UK) by hydrogen flame according to the procedures detailed previously.[26–28] The quality of the electrodes were regularly checked by recording cyclic voltammograms in 0.5 M H_2SO_4 and/or 0.1 M $HClO_4$ or/and STM imaging. All electrodes were subjected to annealing and quenching with hydrogen flame before use.

A series of straight-chain alkanethiols $CH_3(CH_2)_{n-1}SH$ (n = 4,6,8,10,12,14,16,18) with the highest available purity (from Sigma-Aldrich), and 4-methylbenzenethiol (MBT) ($\geq$ 98%, Sigma-Aldrich) were used as received. Azurin was purified from *Pseudomonas aeruginosa* and its concentration determined by UV-Vis spectrometry.[16] Blue copper-containing nitrite reductase from *Alcaligenes xylosoxidans* (*Ax*CuNiR) was lab-produced by genetic recombinant methods.[29] Briefly, the encoding gene of this enzyme (DSM 2402) was cloned into a pET22b vector and overexpressed in the cytoplasm of *E. coli* BL21 (DE3). The expressed holoenzyme (*ca.* 110 kDa) was purified to apparent homogeneity using cation-exchange chromatography and specific catalytic activity assayed (typically *ca.* 220 U mg^{-1}). The recombinant enzyme is indistinguishable in identity from the native, as confirmed by various spectroscopic measurements.[29]

The NH_4Ac buffer (20 mM, pH 4.6) was prepared from 5 M stock solution (Fluka, ultrapure), and the solution pH adjusted with HAc (99.7%, Aldrich). The NaAc buffers (10 mM, various pHs) were prepared by mixing sodium acetate (superpur, >99%, Merk) with HAc, and solution pHs adjusted further with NaOH (ultrapure, Merck). Potassium nitrite (>98%, Fluka) was used as the substrate of *Ax*CuNiR without further purification. Milli-Q water (Millipore-Housing, 18.2 MΩ) was used throughout cleaning and solution preparations.

2.2. Preparations of protein monolayers

The SAMs of alkanethiols were obtained by immersing freshly-quenched Au(111) electrodes in ultrapure ethanol (Merck) containing 1 mM or less alkanethiol overnight (*ca.* 22–24 h adsorption). After thorough rinsing with ethanol and Milli-Q water, the electrodes were transferred to 50–100 μM azurin solutions incubated at 4 °C or room temperature for an appropriate time to form azurin submonolayes or monolayers. The resulting electrodes were carefully rinsed with Milli-Q H_2O and NH_4Ac buffer prior to measurements. These azurin-containing electrodes are denoted as Az-$CH_3(CH_2)_{n-1}S$-Au(111) (n = 4, 6, 8, 10, 12, 14, 16, 18, 20). Similar processes were used for preparations of *Ax*CuNiR monolayers, except that the enzyme concentration used is much lower (1–10 μM). The enzyme-containing electrode is denoted as *Ax*CuNiR-CH_3BT-Au(111).

2.3. Electrochemical measurements

All measurements were carried out using an Autolab system (Eco Chemie, Netherlands) controlled by general purpose electrochemical system software at room temperature (22 ± 2 °C). A three-electrode system, consisting of a platinum coiled wire as counter electrode (CE), a reversible hydrogen electrode (RHE) as reference electrode (RE) and a Au(111)-based working electrode (WE), was used with the WE in a hanging-meniscus configuration.[18] The RHE was checked regularly against a saturated calomel electrode (SCE) after each measurement. All electrode potentials are reported *vs.* SCE. Purified argon (Chrompack, 5 N) was applied to purge dioxygen from electrolyte solutions before the measurements, and the gas stream maintained over the solution during the measurements.

2.4. Electrochemical STM imaging

STM measurements were performed with a PicoSPM system (Molecular Imaging Co., USA) equipped with a bipotentiostat for potential control of both substrate and tip. Electrochemical control was conducted in a lab-designed cell with a three-electrode system as for normal electrochemical measurements. The tips were prepared from tungsten or Pt/Ir wires (∅ 0.25 mm) by electrochemical etching and insulated with apiezon wax to reduce Faradaic currents to 10 pA or less.

2.5. Data analysis and computation

Electrochemical data were analyzed largely by using accessory software of the Autolab system. Mathcad Professional 2001 was used in computations to simulate experimental data.

3. Results and discussion

3.1. Physical features of long-range interfacial ET of proteins

Interfacial ET is encountered in a wide range of systems represented by electrochemical ET (including ET reactions across the liquid/liquid and liquid/solid interfaces) and interfacial protein ET. While ET mechanisms in homogeneous conditions have been well understood after the past fifty-year pursuit of both theories and experiments, similar details of interfacial ET need more to be worked out.[30–33] This is due to the complexity of the nonuniform interfacial environment. In other words, the complications arise from the nature of the hybrid geometrical conditions (or boundary conditions) at electrochemical interfaces and of inhomogeneous hydrophobic/hydrophilic regions in proteins. The protein ET pathways can span both hydrophobic and hydrophilic regions, but dielectric constants in two regions are dramatically different (*e.g.* close to 80 in the hydrophilic region but only *ca.* 2 in the hydrophobic one). These general conditions show that understanding of interfacial ET mechanisms (*e.g.* quantifying reorganization energies) is more complex compared with homogeneous cases.

Fig. 2 shows a comparison of a simple form of the free energy surfaces in principle for any ET reaction (Fig. 2A) with a more detailed illustration of the electrochemical ET process represented by

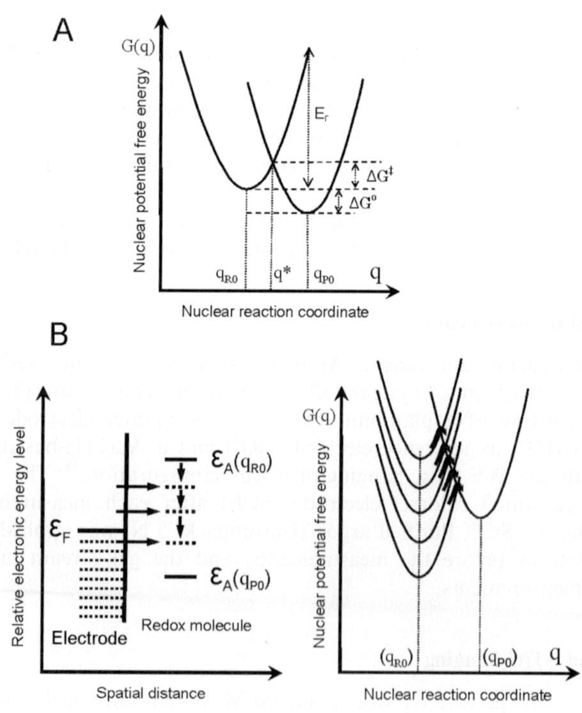

Fig. 2 One-dimensional schematic representations of (A) a simple form of free-energy surfaces of ET reactions, and (B) relative electronic energy levels (left) and free energy surfaces (right) for electrochemical ET.

cathodic process at a metal electrode (Fig. 2B). A general form of the rate constant corresponding to Fig. 2A is[33]

$$k_{\mathrm{ET}} = A\kappa_{\mathrm{el}} \exp\left(-\frac{(\Delta G^\circ + E_{\mathrm{r}})^2}{4E_{\mathrm{r}}k_{\mathrm{B}}T}\right) \tag{1}$$

where k_{ET} is the rate constant, A a pre-factor associated with an effective frequency for nuclear motion, κ_{el} the electronic transmission coefficient, ΔG° the reaction free energy, E_{r} the reorganization energy, k_{B} Boltzmann's constant, and T the temperature. This equation quantifies the relations of several major factors governing the ET kinetics.

In electrochemical ET, the reaction can be directly monitored by the current flow. The current is in principle composed of contributions from all electronic levels of the electrode (Fig. 2B), but predominantly from levels around the Fermi level.[4] The redox level of the molecule, with the initial level at equilibrium above the Fermi level (ε_{F}), is strongly coupled to the environment. By nuclear configurational fluctuations the initial level is lowered to match populated levels in the electrode where ET proceeds, and subsequently the electron is trapped on the molecule. When the redox molecules are immobilized on the electrode surface, the following form holds broad validity for the relation of the (cathodic) current density with effect factors.

$$j_{\mathrm{cath}}(\varepsilon, \eta) = e\Gamma_{\mathrm{ox}} \int \mathrm{d}\varepsilon\, \rho(\varepsilon) f(\varepsilon) k_{\mathrm{ET}}(\varepsilon, \eta) \tag{2}$$

$$k_{\mathrm{ET}}(\varepsilon, \eta) = \kappa_{\mathrm{el}}(\varepsilon, \eta)\frac{\omega_{\mathrm{eff}}}{2\pi} \exp\left(-\frac{[E_{\mathrm{r}} + e\eta - (\varepsilon - \varepsilon_{\mathrm{F}})]^2}{4E_{\mathrm{r}}k_{\mathrm{B}}T}\right) \tag{3}$$

$$\kappa_{\mathrm{el}}(\varepsilon, \eta) = [T_{\varepsilon\,\mathrm{A}}(\varepsilon, \eta)]^2 \sqrt{\frac{4\pi^3}{E_{\mathrm{r}}k_{\mathrm{B}}T\hbar^2\omega_{\mathrm{eff}}}} \tag{4}$$

where ε is the energy level, η the overpotential, Γ_{ox} the population of adsorbed molecules in their oxidized form, $\rho(\varepsilon)$ the electronic level density, $f(\varepsilon)$ the Fermi function, ω_{eff} the effective vibrational frequency of all the nuclear modes contributing to E_{r}, $T_{\varepsilon\mathrm{A}}(\varepsilon,\eta)$ the electron exchange factor describing electronic coupling between the energy level of the electrode and the molecular redox level. Other symbols have their usual meaning. There are *three conspicuous differences* in comparison with ET in homogeneous conditions.[30,31] *First*, the electrode incorporates a continuous manifold of electronic states (Fig. 2B), whereas only a single pair of states is involved in homogenous ET (Fig. 2A). As a consequence, some phenomena (*e.g.* an inverted free energy region) are absent from the electrochemical ET. *Second,* the nature of the electronic wave functions is different. These are spatially localized for ET in homogeneous solution, but two-dimensionally delocalized in electrochemical ET. The electrochemical tunneling factor can thus be conveniently approached by invoking electronic density rather than individual level wave functions. *Finally*, the driving force is tuned by the overpotential applied to the electrode, being zero at the equilibrium redox potential. This is a considerable advantage for investigations of ET properties, especially in the ECSTM configuration.

A combination of electrochemistry and STM has provided a novel means to study the nature of interfacial ET, by which both molecular electronic images with unprecedented resolution and scanning tunneling spectra can be recorded. This requires immobilization of redox molecules on the electrode surface in well-defined organization. The theoretical framework for understanding of the STM redox process has been pursued over the past decade.[4,34–39] The two-step mechanism, which can be viewed as an evolution of electrochemical ET notions to the STM configuration, has emerged as one model suitable to interpret experimental observations of various systems. Fig. 3 shows a schematic energy diagram to illustrate the relative energy levels of the STM substrate (*i.e.* working electrode), the STM tip and a redox molecule in the substrate–tip gap. The energy levels are all modified by changing either the substrate potential or the bias voltage. Here we focus on the case of constant bias voltage. Under this condition, the substrate and tip levels are shifted in parallel. As a result, the redox level is shifted by the overpotential relative to the substrate and tip Fermi levels.

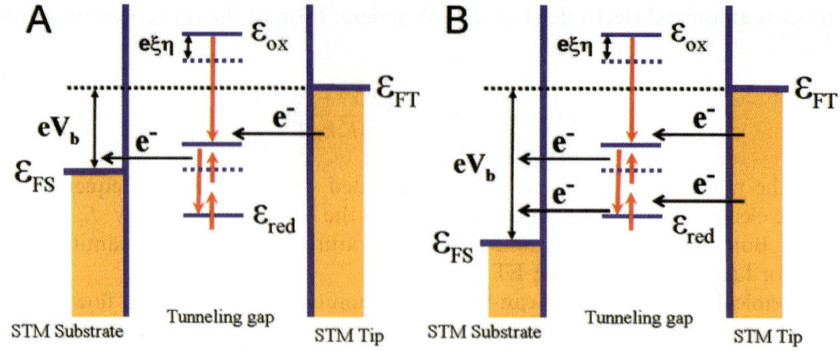

Fig. 3 Schematic energy diagrams of two-step ET mechanisms in electrochemical STM redox processes: (A) and (B) correspond, respectively, to the cases of *small* and *large negative* bias voltages between the STM substrate and tip.

The bias voltage can either be small (defined as $|e\gamma V_{\text{bias}}| < (E_r + e\xi\eta)/(1-\gamma)$) or large ($|e\gamma V_{\text{bias}}| > (E_r + e\xi\eta)/(1-\gamma)$). In the case of small bias voltage (negative), an initially oxidized (vacant) level is above the tip Fermi level and an initially reduced (occupied) level below both the substrate and tip Fermi levels (Fig. 3A). As for electrochemical ET, nuclear configurational fluctuations bring the redox level into the energy region close to the Fermi level of the electrodes. This induces a two-step ET process. Interfacial ET proceeds first from the tip to the vacant redox level and then from the temporarily occupied redox level to the substrate (Fig. 3A). A corresponding ET sequence in the opposite direction would occur at positive bias voltage (not shown). The tunneling current (I_t) is thus gated by the redox level, which can be directly displayed by changes either in the STM image contrast or in the current-overpotential relations. In the case of large bias voltage (negative), the reduced level of the molecule at equilibrium remains above the Fermi level of the substrate (Fig. 3B). The situation becomes more complicated because the large bias voltage notion is determined not only by the ratio of $e\gamma V_{\text{bias}}/E_r$ but also by the overpotential. The tunneling current is expected first to rise with increasing overpotential, but the reduced level is still trapped in the region between the Fermi levels and transmits electrons in an activationless mode. Upon further increasing overpotential, both the oxidized and reduced levels are in the energy window with coherent and activationless electron tunneling, independent of both the overpotential and bias voltage. If the overpotential continues to rise, the reduced level will eventually traverse the substrate Fermi level and the current drop. The overall current-overpotential relation thus consists of an ascending branch at small overpotential and a descending branch at large overpotential, bridged by a tunnelling current plateau.[39]

The nonadiabatic limit of weak electronic coupling between the redox center and the enclosing electrodes applies for long-range interfacial ET of proteins. At small bias voltages the dependence of tunneling (I_t) on the *effective* overpotential ($\xi\eta$) and the bias voltage (V_{bias}) is governed by two-step ET rate constants in the following combination.

$$I_t = e\frac{\vec{k}^{\text{o/r}}\vec{k}^{\text{r/o}}}{\vec{k}^{\text{o/r}} + \vec{k}^{\text{r/o}}} \tag{5}$$

where $\vec{k}^{\text{o/r}}$ and $\vec{k}^{\text{r/o}}$ are the rate constants for ET between the tip and the protein and between the protein and the substrate, respectively; an arrow indicates the electron flow direction following the pathway shown in Fig. 3A. $\vec{k}^{\text{o/r}}$ and $\vec{k}^{\text{r/o}}$ have the following forms:

$$\vec{k}^{\text{o/r}} = \kappa_t \rho_t \frac{\omega_{\text{eff}}}{2\pi}\frac{2k_B T}{\alpha_t}\exp\left(-\frac{(E_r - e\xi\eta - e\gamma V_{\text{bias}})^2}{4\pi k_B T}\right) \tag{6}$$

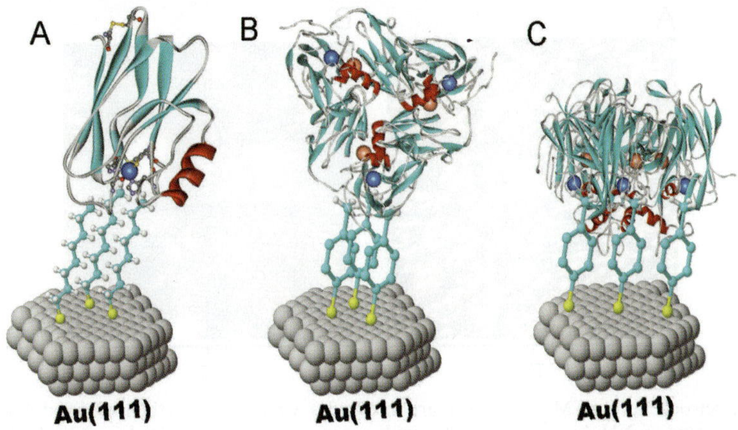

Fig. 4 Schematic representations of molecular wiring assemblies of (A) azurin, and (B) and (C) nitrite reductase in two different orientations on Au(111) surfaces. Not drawn to scale.

$$\vec{k}^{\mathrm{r/o}} = \kappa_{\mathrm{s}}\rho_{\mathrm{s}}\frac{\omega_{\mathrm{eff}}}{2\pi}\frac{2k_{\mathrm{B}}T}{\alpha_{\mathrm{s}}}\exp\left(-\frac{(E_{\mathrm{r}} - eV_{\mathrm{bias}} + e\xi\eta + e\gamma V_{\mathrm{bias}})^2}{4\pi k_{\mathrm{B}}T}\right) \quad (7)$$

where κ_t and κ_s are electronic transmission coefficients for ET between the tip and the protein, and between the protein and the substrate, respectively; ρ_t and ρ_s are the electronic level densities of the tip and the substrate; ω_{eff} is the effective nuclear vibrational frequency; α_t and α_s are ET transfer coefficients, respectively, between the tip and the protein and between the protein and the substrate; ξ is the fraction of the substrate-solution potential drop and γ the fraction of the bias voltage drop at the molecular redox center. Other symbols have their usual meaning. These equations are used in theoretical computations in comparisons with experimental observations.

In summary, physical features relevant to long-range interfacial ET of protein monolayers might include: (1) ET taking place in the nonadiabatic regime with weak electronic coupling and indirect interactions between the redox centre and the enclosing electrodes; (2) tunneling pathways playing a dominant role in the ET process; (3) the driving force being tuned by the overpotential, being zero at the equilibrium redox potential; (4) ET kinetics being free of diffusion effects, and with the device-like behaviour anticipated; and (5) molecular orientation of the protein relative to the electrode surface being significant in determining ET properties. Among these some are general of any electrochemical interfacial ET.

3.2. Molecular wiring assemblies and surface electronic images

In the present cases, molecular wiring assemblies of protein molecules are based on non-covalent interactions. In this context azurin has two appropriate structural features, *i.e.* a surface disulfide Cys3Cys26 group and a hydrophobic patch around the copper centre, located at opposite ends of the molecule; and they can be exploited to confine protein molecules on Au(111) surfaces with controlled orientations. The hydrophobic patch is spatially and chemically suitable for coupling to the terminal methyl group of alkanethiols. Hydrophobic interaction is proved to be sufficiently strong to wire azurin molecules with the copper centre facing the electrode surface (Fig. 4A). Topographic electronic structures of azurin monolayers are well resolved at the molecular level by STM images (*e.g.* Fig. 5) acquired in buffer solutions. Each protein molecule is revealed as a contrasted spot. The average size of *ca.* 4 nm in diameter, estimated from the cross sections of a large number of molecules, is close to the expectation from the X-ray crystallographic dimensions and consistent with those of azurin molecules assembled directly on bare Au(111) surfaces *via* its the surface disulfide group.[27] The distribution of molecules is quite uniform over the surface (Fig. 5A).

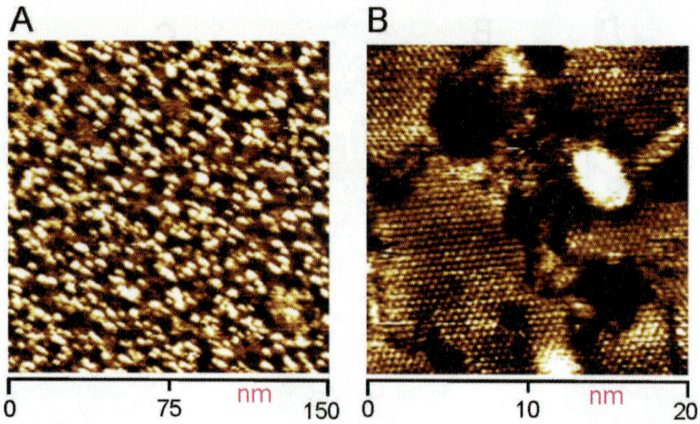

Fig. 5 Two electrochemical STM images of azurin molecules wired on octanethiol-modified Au(111) surfaces. The images recorded in NH_4Ac buffer (pH 4.6) with a constant current mode: $I_t = 0.1$ nA, $V_b = -0.25$ V, $E_w = +0.20$ V (*vs.* SCE). Scan area 150 nm × 150 nm (A), and 20 nm × 20 nm (B).

Notably high-resolution STM images, where the ordered octanethiol is clearly seen underneath the azurin molecules (Fig. 5B), demonstrate a well-organized assembly in both organic and protein adlayers. This is crucial for observations of single-molecule protein ET (see below). Another important issue is that the protein population can be estimated from the STM images, and the results compared with those obtained from electrochemical measurements. This can serve as a suitable way to quantify the percentage of biologically active molecules over the total amount of immobilized protein. Such an approach is of general importance in protein adsorption chemistry, particularly suitable for redox proteins.

Molecular orientation of *Ax*CuNiR on 4-methylbenzenethiol (MBT)-modified Au(111) surfaces is more complex due to the presence of multiple redox centres in the *Ax*CuNiR molecule. Each enzyme subunit has two copper ions, one of which is type I copper (*i.e.* blue copper). The local surface structure around the type I copper center resembles that of the azurin copper centre. Hydrophobic interaction is thus expected in molecular wiring assemblies. At least two possibilities can be envisaged: one would involve only one of the three type I copper ions in each enzyme molecule interacting with the methyl group of MBT (Fig. 4B); in the other mode all three type I copper ions interact with the linker molecules (Fig. 4C). We have not been able to distinguish between these two possible orientations from electrochemical and STM measurements. Fig. 6 shows a STM image of an *Ax*CuNiR submonolayer. Individual enzyme molecules are resolved, although the resolution is not as high as for the azurin case. This is most likely attributed to the size of this enzyme, resulting in a larger tunneling barrier. It is noted that the presence of nitrite (*i.e.* the substrate of this enzyme) is essential to obtain detectable STM contrast. STM observations thus suggest that nitrite serves as a gate for long-range interfacial ET (*i.e.* substrate-gated ET). This view is supported by voltammetric measurements detailed in section 3.4.

3.3. Distance-controlled ET and redox-gated tunneling resonance of azurin

In the assembly shown in Fig. 4A, hydrophobic interactions are chemically responsible for immobilization of azurin molecules. Mutagenesis experiments suggest that the tryptophan 48 residue, which is not likely to interact physically with the methyl group, could be important in establishing electronic coupling between the copper center and the electrode.[19] This molecular wiring assembly leads to fast and reversible interfacial ET by such an effective electronic coupling. Interfacial ET is detected by well-defined cyclic voltammograms (CVs) with low background charging. High stability of this system further allows fast speed CVs to be acquired, by which rate constants can be precisely estimated by Laviron's method.[40] These features make it possible that ET properties of this system are investigated in detail.

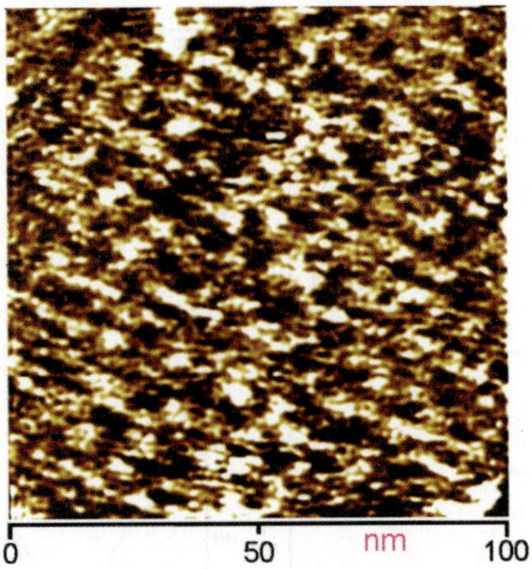

Fig. 6 An electrochemical STM image of *Ax*CuNiR–CH_3BT–Au(111) recorded in NaAc/HAc buffer (pH 6.0) in the presence of *ca.* 200 μM KNO_2. $I_t = 0.2$ nA, $V_b = -0.52$ V, $E_w = 0.04$ V (*vs.* SCE). Scan area 100 nm × 100 nm.

The alkanethiol in the present ET system serves as both a linker molecule and an ET-assisting bridge. The tunneling barrier is thus adjusted by varying the bridge length. The equilibrium redox potential of azurin is largely independent of the alkane chain length. The ET kinetics is governed by the reorganization free energy (E_r), the bridge length, and conformational flexibility of the protein-bridge interface. Fig. 7 shows a series of CVs of Az–$CH_3(CH_2)_{n-1}$S–Au(111) (n = 4,6,8,10,12,14,16,18). The increase of the peak separation with increasing bridge length indicates a direct observation of distance-controlled kinetics. The distance is calculated for the bridge length as described previously.[41] The reorganization energy can be estimated from chronoamperometric (CA) measurements. The method was described previously in detail.[16] Briefly, CA measurements were performed by applying symmetric overpotentials *vs.* the formal potentials of adsorbed azurin. The rate constants at given overpotential are obtained from the slope of current-time curves, and E_r is estimated by fitting experimental data. The value is in the range 0.55–0.30 eV corresponding to the distance range of 8–26 Å. Notably the reorganization energy thus appears to increase with decreasing bridge length, most likely due to more flexible conformation of the protein-bridge interface at shorter bridges. The rate constant becomes virtually independent of the distance in the range of 4–8 methylene units, where the ET kinetics is controlled mainly by configuration-gating. Exponential distance decay is observed with longer bridges with a distance-decay factor (β) of 0.83 $Å^{-1}$ (Fig. 8). The decay factor is consistent with that for charge transfer through pure saturated hydrocarbons,[41,42] indicative of efficient electronic coupling between the copper center and the gold electrode.

Azurin–$CH_3(CH_2)_7$S–Au(111) with sub-monolayer azurin population was used to observe single-molecule ET by tuning the overpotential. STM observations were first performed with a larger scan area (*e.g.* 200 × 200 nm^2) to obtain images of the azurin sub-monolayer with molecular resolution, and then focused on a few individual molecules by reducing the scan area. By keeping a constant bias voltage between the substrate and tip, STM imaging started with the substrate potential set at the equilibrium redox potential of azurin (*i.e.* zero overpotential). Imaging was continued towards either positive or negative overpotentials by adjusting the substrate and tip potentials in parallel (*i.e.* at constant bias voltage), and finally returned to the equilibrium potential. Both the octanethiolate monolayer and the azurin sub-monolayer are sufficiently robust to withstand ECSTM scanning. As a result a series of STM images were acquired at various overpotentials. The single-molecule contrast is clearly tuned by the redox state of azurin. The current contrast-overpotential relation is shown Fig. 9A, with a maximum around the equilibrium redox potential. The current contrast decreases upon applying either positive or negative overpotentials, but the effect is asymmetric, the decay being faster at negative overpotentials. The maximum current on/off ratio is *ca.* 9, which is comparable with the cases of small redox molecules.[43–46]

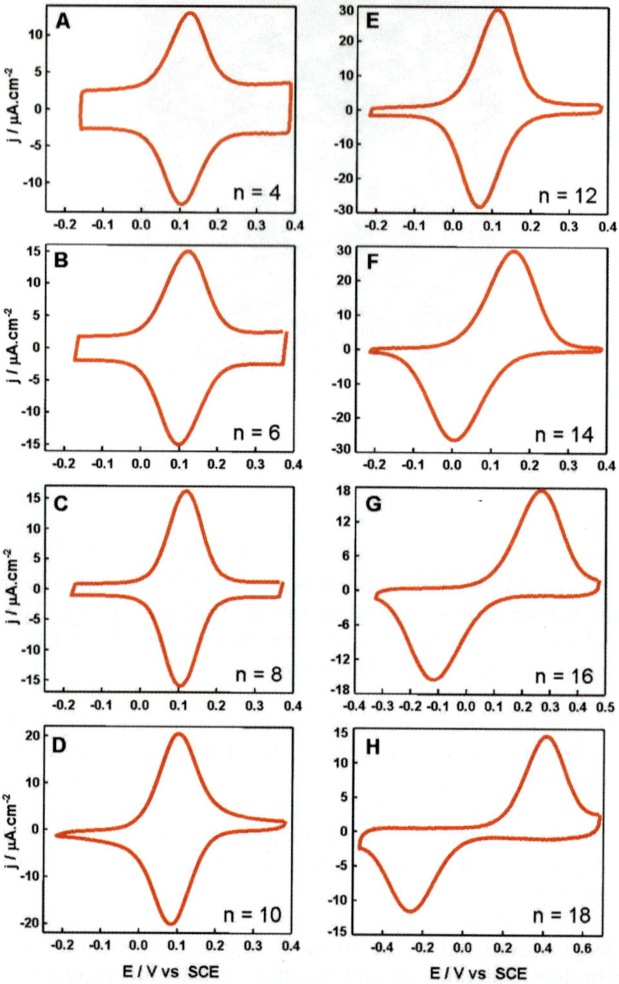

Fig. 7 Distance-dependent cyclic voltammograms of Az–$CH_3(CH_2)_{n-1}$S–Au(111) electrodes (n = 4,6,8,10,12,14,16,18) recorded in 20 mM NH_4Ac buffer (pH 4.6) at the same scan rate of 2 V s^{-1}. The bridge length is indicated by the number of methylene groups (n).

The asymmetric overpotential dependence can be represented formally by introducing an exponential overpotential-dependent pre-factor in the $\vec{k}^{r/o}$ expression form (eqn. (7)) to obtain:

$$\vec{k}^{r/o} = \kappa_s \rho_s \frac{\omega_{eff}}{2\pi} \frac{2k_B T}{\alpha_s} \exp(e\theta\eta) \exp\left(-\frac{(E_r - eV_{bias} + e\xi\eta + e\gamma V_{bias})^2}{4\pi k_B T}\right) \quad (8)$$

θ can represent either overpotential dependence of the tunneling barrier or electrochemical double-layer effects, in the range of 3–6 eV^{-1} for the present case. Fig. 9B shows a calculated tunneling current-overpotential relation using eqns. (5), (6) and (8), giving an asymmetric overpotential-dependence as in the experimental observations (Fig. 9A). We can thus obtain the main parameters that reflect the ET nature of this system in the ECSTM configuration. These parameters provide the following information: (1) the position of the resonance maximum is determined by all the factors; in the present case the maximum appears in a narrow positive overpotential region close to the equilibrium redox potential. (2) The rate-limiting step is ET between the tip and the azurin

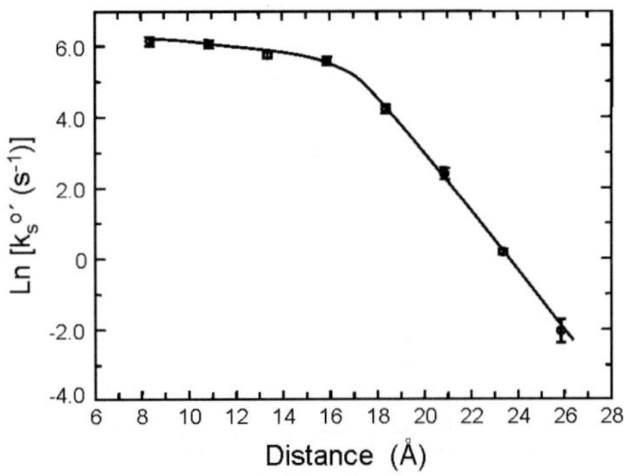

Fig. 8 Distance-dependent electron transfer rates of Az–$CH_3(CH_2)_{n-1}S$–Au(111) systems illustrated by a plot of the interfacial standard ET rate constants (in natural logarithm) *vs.* the distance.

molecule, because at least an order of magnitude difference between the two transmission coefficients is required. This accords with the fact that interfacial ET between the azurin molecule and the substrate is relatively fast (*ca.* 500 s^{-1} estimated from electrochemical measurements). (3) The reorganization free energy is estimated as 0.35 to 0.45 eV, implying that E_r is not significantly different in STM and electrochemical ET. (4) The distribution of the bias voltage in the tunneling gap is asymmetric with optimal values of $\gamma \approx 0.15$–0.20 at the site of the redox center (*i.e.* the copper ion in azurin), according with the asymmetric location of this center both physically and electronically. (5) The distribution of the overpotential between the tunneling gap also significantly affects the tunneling resonance. However, the fraction of the substrate-solution potential drop (ξ = 0.7–0.75) deviates less from the ideal expectation (ξ = 1.0). In short, the two-step ET model has offered a comprehensive explanation of detailed ET mechanisms of the STM redox process for the present case.

3.4. Substrate-gated ET and electrocatalysis of nitrite reductase monolayers

The overall long-range ET of *Ax*CuNiR involves the following steps completing an electrocatalytic cycle.

$$Cu_I^{2+} + e \rightarrow Cu_I^{+} \qquad (I)$$

$$Cu_{II}^{2+} + Cu_I^{+} \rightarrow Cu_{II}^{+}Cu_I^{2+} \qquad (II)$$

$$\left\{ \begin{array}{lcl} NO_2^- + H_2O + e & \rightarrow & NO + 2OH^- \\ Cu_{II}^+ - e & \rightarrow & Cu_{II}^{2+} \end{array} \right\} \qquad (III)$$

In the fully oxidized enzyme resting state, type I copper (T1Cu) is first reduced by an electron injected from the electrode. This is followed by intramolecular electron exchange between T1Cu and type II copper (T2Cu). T2Cu is reoxidized by nitrite which is reduced to nitric oxide. In the natural protein ET processes, T1Cu serves as an entry site of electrons from the enzyme redox partner (*e.g.* cytochrome c_{551} or cupredoxins), while the T2Cu site functions for binding and reduction of the substrate nitrite. The tunneling pathway of intramolecular ET consists of eleven peptide bonds with a distance about 12.6 Å. Intramolecular ET is thus crucial in the catalytic cycle and could be a rate-limiting step in nature. In the electrochemical ET system, the electrode acts as an artificial redox partner to inject electrons to the T1Cu site, and intramolecular ET is a relay step in the overall ET process. With further consideration of noncovalent interactions between the enzyme and the wiring molecule, this system holds biomimetic features and ET should parallel biological ET. If diffusion effects of nitrite ions in solution are ignored, the rate-determining step could be either interfacial ET between T1Cu and the electrode (step (I)) or intramolecular ET (step (II)), depending on electronic coupling efficiency between the enzyme and the electrode through the wiring molecule. Due to the large size and multiple redox centres of this enzyme (compared to azurin), interfacial ET kinetics is proved to be very sensitive to the bridge length of the wiring

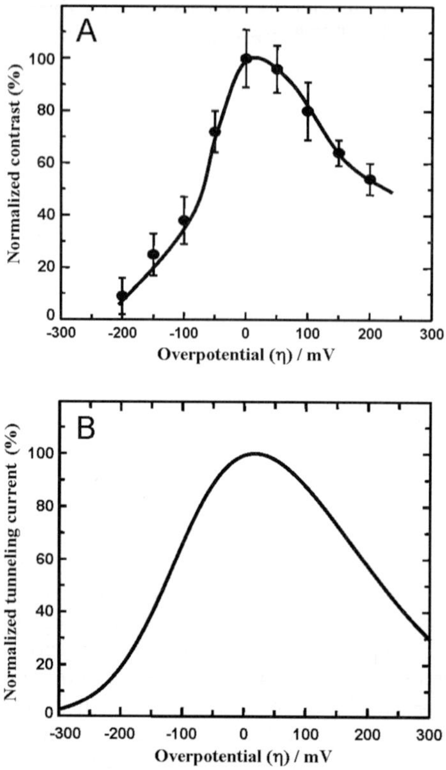

Fig. 9 Redox-gated electron tunneling resonance spectra of individual azurin molecules observed by ECSTM and calculated by eqns. (5)–(8): (A) experimental results of relative electronic contrast *vs.* overpotential, and (B) computed tunneling current *vs.* overpotential.

molecule. A short π-conjugated thiol is employed here to achieve the mostly possible effective electronic coupling.

No Faradaic response from the enzyme monolayer is detected in the absence of nitrite (Fig. 10A, dotted curve). A pair of redox peaks with a middlepoint potential around 0.1 V (*vs.* SCE) is invoked by addition of nitrite controlled at low concentrations (≤ 1 μM), (Fig. 10A, solid curve). Anodic and cathodic peaks are largely symmetric, suggesting noncatalytic voltammetric character. An apparent rate constant, estimated from the peak separation, is approximately 0.3 s^{-1}. The cathodic peak potential is determined as 0.03 V *vs.* SCE at pH 6.0 (*i.e. ca.* 0.27 V *vs.* NHE). This value is close to the expected reduction potential of T1Cu (0.29 V *vs.* NHE at the same pH), but different from that of T2Cu (0.34 V).[47,48] This might be viewed as an indication of voltammetric signals indeed arising from interfacial ET between T1Cu and the electrode. However, this does not exclude that the electrochemical potential represents a mixture of T1Cu and T2Cu. Long-range interfacial ET is only detected in the presence of nitrite for the linker molecule used here. That is, interfacial ET is substrate-gated. Further experiments were performed toward better understanding of the role of nitrite in the noncatalytic ET response. After the measurements shown in Fig. 10A (solid curve), the enzyme-containing electrode was rinsed to remove free nitrite and re-inserted in the nitrite-free buffer solution for voltammetric detections. The redox peaks remain largely unchanged (Fig. 10B). This could imply that only binding of nitrite to the enzyme is necessary to induce ET by opening the tunneling pathway, rather than involving catalytic reduction of the substrate. Studies of *Ax*CuNiR ET in homogeneous solutions suggest that the rate of intramolecular ET is virtually modified by the substrate binding. The binding of nitrite triggers changes in the T2Cu redox potential, which in turn results in a modification in the driving force of the ET reaction.[49–51] On the other hand, the binding of the substrate nitrite could also enhance *the conductivity of enzyme molecules*. The substrate-gated

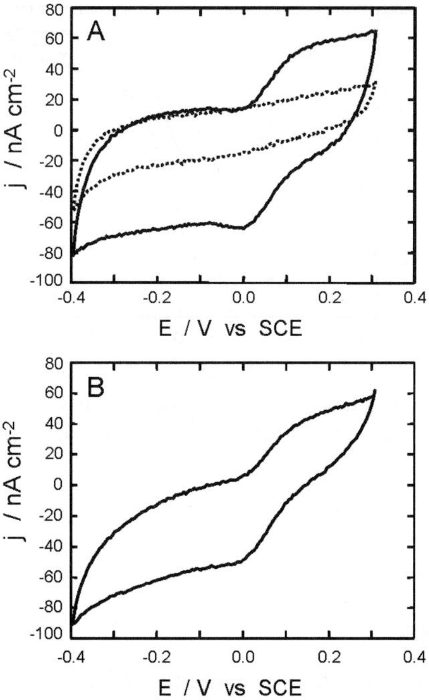

Fig. 10 Cyclic voltammograms of *Ax*CuNiR-CH_3BT-Au(111) in 10 mM NaAc/HAc buffer (pH 6.0): (A) without (dotted lines) and with 1 μM (solid line) KNO_2, and (B) obtained in KNO_2-free buffer after measurements in (A) and the electrode was rinsed with Milli-Q H_2O and NaAc/HAc buffer. Scan rate 10 mV s^{-1}.

interfacial ET could thus be the reason why the presence of nitrite is essential for STM electronic imaging of this enzyme (*cf.* Fig. 6 and text).

The voltammetric responses at higher concentrations of nitrite (≥ 5 μM) exhibit typical electrocatalytic features with dominant cathodic currents (Fig. 11). Electrocatalytic acitivity is affected dramatically by solution pH with an optimal pH around 6.0, similar to observations for pH dependent catalytic activity in homogeneous solution.[52] Suzuki and coworkers also showed that the fastest intramolecular ET occurs at pH 5.8–6.1 and in the presence of nitrite.[47] This suggests that the effects of solution pH on electrocatalytic activity arise mainly from intramolecular ET kinetics. The following approximate electrochemical version of the Michaelis–Menten equation applies for the relation between electrocatalytic current density (j) and the substrate concentration ([S]):

$$j = nF\Gamma \frac{k_{cat}[S]}{K_m + [S]} \tag{9}$$

where n is the number of electrons transferred, F the Faraday's constant, Γ the surface population of the enzyme, k_{cat} the apparent turnover rate constant, and K_m the apparent Michaelis–Menten constant.

The surface coverage is estimated either from noncatalytic voltammetry (*e.g.* Fig. 10A) or STM images as $7.6 \pm 0.8 \times 10^{-13}$ mol cm^{-2} (equivalent to *ca.* 16% of the closely-packed monolayer). Good agreement is observed from both methods, an indication that biological ET function of the enzyme is well retained in the assembly. From the fitting of experimental data using eqn. (9) K_m and k_{cat} were estimated as 44 μM and 30 s^{-1}. K_m is close to that for the free mobile enzyme (34 μM),[52] suggesting that there is no significant effect of immobilization on the enzyme–substrate affinity or structural conformations. k_{cat} is, however, clearly smaller than reported rate constants of intramolecular ET (185–1900 s^{-1}) observed by pulse radiolysis in homogeneous solution.[50,52] This could be attributed to the differences in ET rate-determining steps and microenvironment. While the catalytic rate in homogeneous solution is predominantly determined by intramolecular ET, the

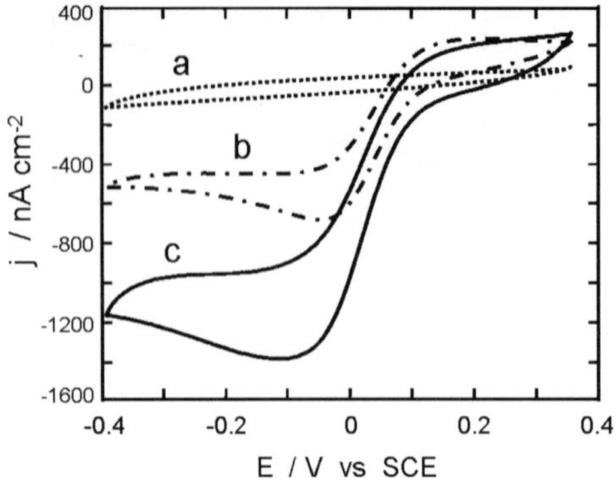

Fig. 11 Electrocatalytic cyclic voltammograms of *Ax*CuNiR–CH_3BT–Au(111) in 10 mM NaAc/HAc buffer (pH 6.0) with 20 (curve b) and 100 (curve c) μM KNO_2. For comparison, a CV (curve a) in the absence of KNO_2 is also shown. Scan rate 10 mV s^{-1}.

electrocatalytic efficiency is governed by concerted effects of intramolecular ET and interfacial ET between T1Cu and the electrode.

4. Summary and outlook

We have presented observations of long-range interfacial ET of metalloproteins from a concerted effort of theoretical notions and experimental developments. These include some general physical features and experimental results of azurin and nitrite reductase by mainly employing electrochemistry and STM. Molecular wiring provides a suitable bottom-up pathway to assemble the proteins on Au(111) surfaces where biological functions are well-protected. Molecular orientations of proteins can be controlled *to some extent* so that effective electronic couplings for long-range ET are established. Investigations of the azurin case have reached the nanoscale and single-molecule levels, gaining comprehensive understanding of the ET mechanisms. Particularly, single-molecule ET tunneling resonance is *in situ* observed. Experimental observations of the more composite metalloenzyme *Ax*CuNiR system have not reached the same level of detail. This is due to its large size, multiple-step ET nature, and less robust character. Improvements and outlooks could include: (1) to approach high-resolution STM observations with more efficient electronic coupling between the enzyme and the electrode; (2) to explore the substrate binding by acquiring high-resolution three dimensional structures of the enzyme-substrate complex, with possible implications for the relation to the molecular conductivity; and (3) to improve generally the electric contact between the large-size protein molecule and the enclosing electrodes in STM, so that single-molecule enzymatic electrocatalysis can be really approached by electrochemical STM.

Acknowledgements

This work was supported by the Danish Technical Research Council, NanoScience Center at the University of Copenhagen, and Brødrene Hartmann's Foundation.

References

1 I. Willner and E. Katz, *Angew. Chem., Int. Ed.*, 2000, **39**, 1180.
2 F. A. Armstrong and G. S. Wilson, *Electrochim. Acta*, 2000, **45**, 2623.
3 E. Katz and I. Willner, *Angew. Chem., Int. Ed.*, 2004, **43**, 6042.
4 J. Zhang, Q. Chi, A. M. Kuznetsov, A. G. Hansen, H. Wackerbarth, H. E. M. Christensen, J. E. T. Andersen and J. Ulstrup, *J. Phys. Chem. B*, 2002, **106**, 1131.
5 M. D. Porter, T. B. Bright, D. L. Allara and C. E. D. Chidsey, *J. Am. Chem. Soc.*, 1987, **109**, 3559.
6 H. O. Finklea, in *Electroanalytical Chemistry*, ed. A. J. Bard, Marcel Dekker, New York, 1996, 19, p. 109.
7 M. J. Tarlov and E. F. Bowden, *J. Am. Chem. Soc.*, 1991, **113**, 1847.
8 S. Song, R. A. Clark, E. F. Bowden and M. J. Tarlov, *J. Phys. Chem.*, 1993, **97**, 6564.
9 S. Terrettaz, J. Cheng and C. J. Miller, *J. Am. Chem. Soc.*, 1996, **118**, 7857.

10 Z. Q. Feng, S. Imabayashi, T. Kikiuchi and K. Niki, *J. Chem. Soc., Faraday Trans.*, 1997, **93**, 1367.
11 A. El. Kasmi, J. M. Wallace, E. F. Bowden, S. M. Binet and R. J. Linderman, *J. Am. Chem. Soc.*, 1998, **120**, 225.
12 A. Avila, B. W. Gregory, K. Niki and T. M. Cotton, *J. Phys. Chem. B*, 2000, **104**, 2759.
13 J. Wie, H. Liu, A. R. Dick, H. Yamamoto, Y. He and D. H. Waldeck, *J. Am. Chem. Soc.*, 2002, **124**, 7705.
14 K. Niki, W. R. Hardy, M. G. Hill, H. Li, J. R. Sprinkle, E. Margoliash, K. Fujita, R. Tanimura, N. Nakamura, H. Ohno, J. H. Richards and H. B. Gray, *J. Phys. Chem. B*, 2003, **107**, 9947.
15 A. K. Gaigalas and G. Niaura, *J. Colloid Interface Sci.*, 1997, **193**, 60.
16 Q. Chi, J. Zhang, J. E. T. Andersen and J. Ulstrup, *J. Phys. Chem. B*, 2001, **105**, 4669.
17 L. J. C. Jeuken, J. P. McEvoy and F. A. Armstrong, *J. Phys. Chem. B*, 2002, **106**, 2304.
18 F. A. Armstrong, N. L. Barlow, P. L. Burn, K. R. Hoke, L. J. C. Jeuken, C. Shenton and G. R. Webster, *Chem. Commun.*, 2004, 316.
19 K. Fujita, N. Nakamura, H. Ohno, B. S. Leigh, K. Niki, H. B. Gray and J. H. Richards, *J. Am. Chem. Soc.*, 2004, **126**, 13954.
20 J. Zhang, A. C. Welinder, A. G. Hansen, H. E. M. Christensen and J. Ulstrup, *J. Phys. Chem. B*, 2003, **107**, 12480.
21 A. G. Hansen, J. Zhang, H. E. M. Christensen, A. C. Welinder, H. Wackerbarth and J. Ulstrup, *Isr. J. Chem.*, 2004, **44**, 89–100.
22 C. R. Hess, G. A. Juda, D. M. Dooley, R. N. Amii, M. G. Hill, J. R. Winkler and H. B. Gray, *J. Am. Chem. Soc.*, 2003, **125**, 7156.
23 J. Zhang, H. E. M. Christensen, B. L. Ooi and J. Ulstrup, *Langmuir*, 2004, **20**, 10200.
24 E. Vijigenboom, J. E. Busch and G. W. Canters, *Microbiology*, 1997, **143**, 2853.
25 W. G. Zumft, *Microbiol. Mol. Biol. Rev.*, 1997, **61**, 533.
26 Q. Chi, J. Zhang, E. P. Friis, J. E. T. Andersen and J. Ulstrup, *Electrochem. Commun.*, 1999, **1**, 91.
27 Q. Chi, J. Zhang, J. U. Nielsen, E. P. Friis, Ib Chorkendorff, G. W. Canters, J. E. T. Andersen and J. Ulstrup, *J. Am. Chem. Soc.*, 2000, **122**, 4047.
28 J. Zhang, Q. Chi, J. U. Nielsen, E. P. Friis, J. E. T. Andersen and J. Ulstrup, *Langmuir*, 2000, **16**, 7229.
29 W. H. Ho, B. L. Ooi, A. M. Jørgensen, L. Borg, L. L. Jespersen and H. E. M. Christensen, *Protein Expression Purif.*, 2003, **32**, 288.
30 A. M. Kuznetsov, *Charge Transfer in Physics, Chemistry, and Biology*, Gordon & Breach, Amsterdam, 1995.
31 A. M. Kuznetsov and J. Ulstrup, *Electron Transfer in Chemistry and Biology: An Introduction to the Theory*, John Wiley & Sons Ltd, New York, 1999.
32 P. F. Barbara, T. J. Meyer and M. A. Ratner, *J. Phys. Chem.*, 1996, **100**, 13148.
33 R. A. Marcus and N. Sutin, *Biochim. Biophys. Acta*, 1985, **811**, 265.
34 A. M. Kuznetsov, P. Sommer-Larsen and J. Ulstrup, *Surf. Sci.*, 1992, **275**, 52.
35 W. Schmickler and C. Widrig, *J. Electroanal. Chem.*, 1992, **36**, 213.
36 E. P. Friis, Y. I. Kharkats, A. M. Kuznetsov and J. Ulstrup, *J. Phys. Chem. A*, 1998, **102**, 7851.
37 E. P. Friis, J. E. T. Andersen, Y. I. Kharkats, A. M. Kuznetsov, R. J. Nichols, J. D. Zhang and J. Ulstrup, *Proc. Natl. Acad. Sci. USA*, 1999, **96**, 1379.
38 A. M. Kuznetsov and J. Ulstrup, *J. Phys. Chem. A*, 2000, **104**, 11531.
39 J. Zhang, A. M. Kuznetsov and J. Ulstrup, *J. Electroanal. Chem.*, 2003, **541**, 133.
40 E. Laviron, *J. Electroanal. Chem.*, 1979, **101**, 19.
41 D. J. Wold, R. Haag, M. A. Rampi and C. D. Frisbie, *J. Phys. Chem. B*, 2002, **106**, 2813.
42 O. S. Wenger, B. S. Leigh, R. M. Villahermosa, H. B. Gray and J. R. Winkler, *Science*, 2005, **307**, 99.
43 N. J. Tao, *Phys. Rev. Lett.*, 1996, **76**, 4066.
44 D. I. Gittins, D. Bethell, D. J. Schiffrin and R. J. Nichols, *Nature*, 2000, **408**, 67.
45 W. Haiss, R. J. Nichols, S. J. Higgins, D. Bethell, H. Höbenreich and D. J. Schiffrin, *J. Am. Chem. Soc.*, 2003, **125**, 15294.
46 T. Albrecht, A. Guckian, J. Ulstrup and J. G. Vos, *Trans. IEEE*, 2005, **4**, 430.
47 K. Kobayashi, S. Tagawa, Deligeer and S. Suzuki, *J. Biochem. (Tokyo)*, 1999, **126**, 408.
48 F. K. Yousafzai and R. R. Eady, *J. Biol. Chem.*, 2002, **277**, 34067.
49 S. Suzuki, T. Kohzuma, Deligeer, K. Yamaguchi, N. Nakamura, S. Shidara, K. Kobayashi and S. Tagawa, *J. Am. Chem. Soc.*, 1994, **116**, 11145.
50 O. Farver, R. R. Eady, H. L. Abraham and I. Pecht, *FEBS Lett.*, 1998, **436**, 239.
51 O. Farver, R. R. Eady, G. Sawers, M. Prudencio and I. Pecht, *FEBS Lett.*, 2002, **561**, 173.
52 Z. H. L. Abraham, B. E. Smith, B. D. Howes, D. J. Lowe and R. R. Eady, *Biochem. J.*, 1997, **324**, 511.

Gating current flowing through molecules in metal–molecules–metal junctions

Elizabeth Tran,[a] Marco Duati,[b] George M. Whitesides*[a] and Maria A. Rampi*[b]

Received 27th April 2005, Accepted 28th June 2005
First published as an Advance Article on the web 11th October 2005
DOI: 10.1039/b505860k

We have assembled two junctions that incorporate redox sites between Hg electrodes by different interactions. In the first junction, Hg-SAM-R//R-SAM–Hg, the redox site (R) are covalently linked to each electrode in self assembled monolayers (SAM-R). In the second junction, Hg–SAM//R//SAM–Hg, the redox sites dissolved in solution are trapped by electrostatic interaction at the SAM formed at the electrodes. The current flowing through these junctions can be controlled by adjusting the potential applied at the electrodes with respect to the redox potential of the species by using an electrochemical system. The current flowing in these two junctions is mediated by the redox sites through different mechanisms. In particular, the current flowing through the Hg–SAM-R//R-SAM–Hg junction occurs through a self exchange mechanism between the redox sites organized at each electrode, while the current flowing through the Hg–SAM//R//SAM–Hg junction is dominated by a redox-cycling mechanism. The systems described here are easy to assemble, well-characterized, yield reproducible data and make it easy to modify the electrical properties of the junctions by changing the nature of the redox centres. For these characteristics they are well suited for collecting fundamental information relevant to the fabrication of molecular switches.

1. Introduction

In the last decade, the number of studies of electron transport through molecules by using metal–insulator–metal (MIM) junctions has been growing exponentially.[1] The most interesting results of these studies indicate that: (i) there is a correlation of electron transfer rate with the molecular structure,[2] similar to that widely investigated by using the molecular donor–bridge–acceptor (DBA system),[3] (ii) the features of the *I–V* curves measured for some molecules show abrupt increases and decreases of the current. This phenomenon, called "negative differential resistance" (NDR) is of great interest in testing theories of electron transport and perhaps for electronic devices.[4–14] Organic molecules are, in general, poor conductors of electrons, and "conduct" by tunnelling. However, when molecules have low-lying orbitals (usually the HOMO or LUMO), applying a potential across the electrodes of a metal–molecule–metal junction can cause the energy of the molecular orbital to fall between the Fermi levels of the electrodes (Fig. 1). Under these conditions, the current flowing through the molecule shows a large increase in current as a result of a change in the electron transport mechanism.[15,16] In principle, molecules with well-defined and readily accessible redox states are, thus, attractive candidates to demonstrate that it is possible to obtain an NDR signal.

Here we show that, in junctions incorporating redox sites (Fig. 2) is possible *to control* the current flowing through a metal–molecules–metal junctions and *to predict* at which potential the junction will switch from an "off" to an "on" state. By placing the junctions, together with a Ag/AgCl reference electrode and a platinum counter electrode, in electrolyte solution, and by using a bipotentiostat, it is possible to control the potentials of the two electrodes independently with respect to the reference electrode (Figs. 3 and 4) and therefore to control the potential of the

[a] *Department of Chemistry and Chemical Biology, Harvard University, Cambridge, MA 02138, USA*
[b] *Dipartmento di Chimica, Università di Ferrara, 44100 Ferrara, Italy*

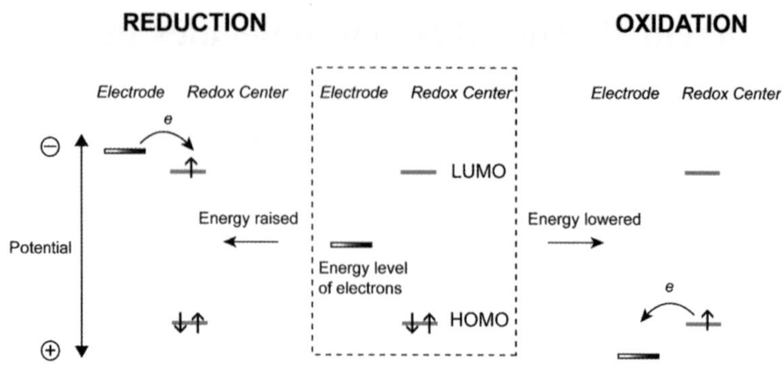

Fig. 1 Scheme of the electron transport mediated by molecules incorporating redox centers. The molecular orbitals HOMO (or LUMO) can donate (or accept) electrons as a function of the electrode Fermi levels under applied voltage.

electrodes with respect to the potential of the redox centre (E_0). We achieve this control over potential by carrying out two types of experiments. In one experiment (experiment A), we keep the potential of one electrode constant, and change that of the second electrode. In a second experiment (experiment B), the potentials of both electrodes are varied simultaneously, while maintaining a constant difference in potential between them.

We report a study of electron transfer by using two junctions that incorporate redox sites *via different kinds of interactions*. In the first junction Hg–SAM-R//R-SAM–Hg, the redox sites (R) are covalently linked to the electrodes being incorporated in self assembled monolayers (SAM-R) that cover the Hg-drop electrodes (Fig. 2a). The junction consists of two mercury drop electrodes, both of which are coated with a self-assembled monolayer (SAM) of the Ru-terminated thiol $HS(CH_2)_{10}CONHCH_2pyRu(NH_3)_5(PF_6)_2$ (henceforth abbreviated as HS-C_{10}-Ru).[17,18] The thiol HS-C_{10}-Ru readily absorbs from 1 mM acetonitrile solutions onto mercury electrodes and gives electroactive monolayers. After the formation of the SAMs on each electrode, the two Hg electrodes are brought in contact to form the junction Hg–SAM-R//R-SAM–Hg.

In the second junction Hg–SAM//R//SAM–Hg, the redox sites are incorporated between the electrodes by trapping them from a solution *via* electrostatic interaction with the SAMs formed on each electrode (Fig. 2b). The junction is formed by using two mercury drop electrodes, whose surfaces are covered by SAMs carrying charged terminating groups so that oppositely charged redox centres in solution to be trapped between the two electrodes by electrostatic interaction. We used self-assembled monolayers of alkanethiols terminated with carboxylic acid groups (11-mercaptodecanoic acid, henceforth abbreviated as $HSC_{10}COOH$) for two reasons. Carboxylic acid groups of this SAM on gold can be ionized at pH > 4, and the COO^- groups allow positively charged $Ru(NH_3)_6^{3+}$ ions in solution to be trapped between the two electrodes and to access the electrode surfaces *via* defect sites. This feature allows redox centres to be incorporated into metal–SAM//SAM–metal junctions without involvement of chemical synthesis and would be highly attractive for its simplicity and versatility.

By measuring the current flowing through the junctions by standard electrochemical techniques we demonstrate that these junctions become conductive when the electrode potentials are adjusted

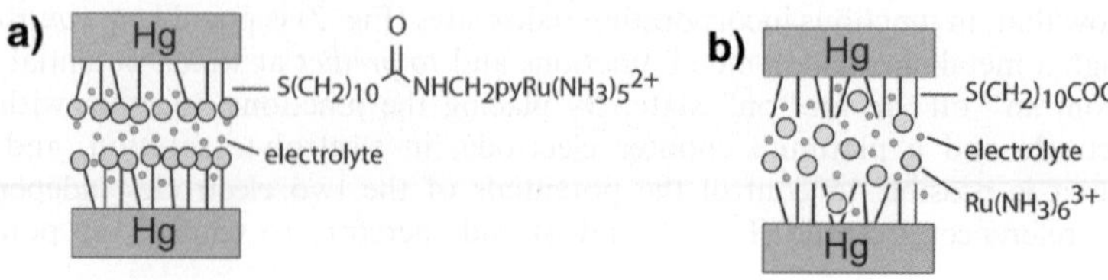

Fig. 2 Schematic representations of interfaces of (a) the junction Hg–SAM-R//R-SAM–Hg and (b) the junction Hg–SAM//R//SAM–Hg.

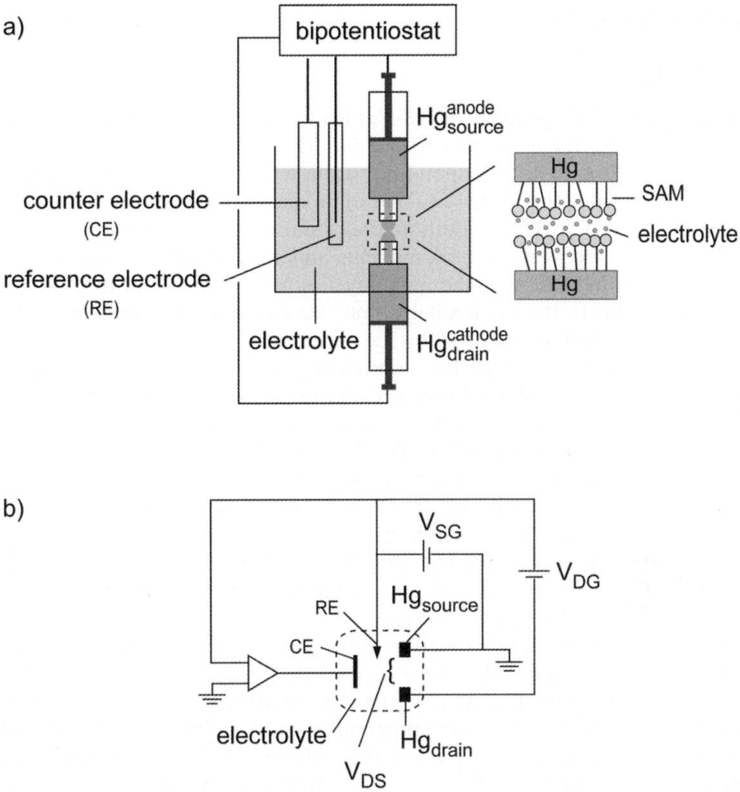

Fig. 3 Scheme of (a) the electrochemical system used to characterize the Hg–SAM-R//R-SAM–Hg junction and (b) the electrical circuit for the experimental setup. V_{SG} and V_{DG} are the potentials of the source and drain electrodes with respect to the reference electrode, respectively. The electrolyte is the gate whose potential is established by the voltage applied to the reference electrode relative to ground. I_D and I_S are the currents at the drain and source electrodes, respectively.

to the formal potential of the redox centres, and that they show diode like characteristics analogous to those of solid-state devices. We also show that the current flowing through the two junctions are dominated by two different electron transfer mechanisms.

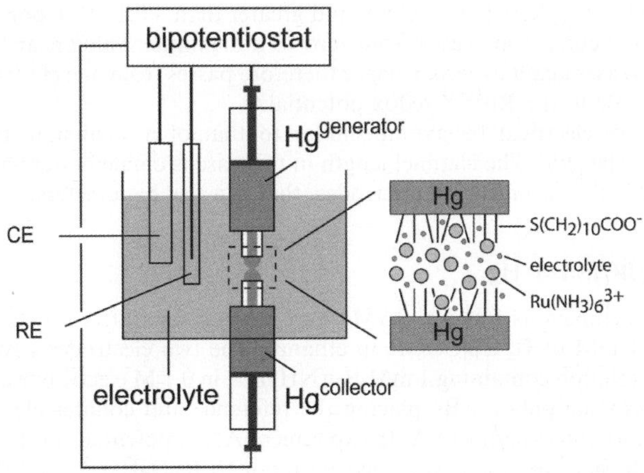

Fig. 4 Scheme of the electrochemical system used to characterize the junction Hg–SAM//R//SAM–Hg.

2. Results and discussion

Junction Hg–SAM-R//R-SAM–Hg

To test the organization of monolayers of HS-C_{10}-Ru on Hg drops, we run cyclic voltammograms of a monolayer of HS-C_{10}-Ru in contact with a 0.2 M aqueous Na_2SO_4 electrolyte (pH 4). The results show a stable redox wave corresponding to the Ru(II) $\leftrightarrows$ Ru(III) interconversion at $E^{o\prime}$ = −0.04 V *vs.* Ag/AgCl, a value that is close to that of an analogous dissolved compound in the same medium ($E^{o\prime}_{solu}$ = +0.04 V *vs.* SCE). In addition, the oxidation and reduction peaks are symmetrical, with a $<$10 mV difference between the peak maxima, and an ideal full width at half maximum of 90 mV. These observations indicate that the monolayer is well-ordered and that the $pyRu(NH_3)_5{}^{2+}$ moieties are homogeneous in their redox behaviour; we infer that they are located at the external surface of the monolayer and are solvated by the electrolyte.

To assemble the junction, we aligned two microsyringes on a vertical line, in opposite directions (Fig. 3a). Two Hg drops were extruded inside a solution of a thiol HS-C_{10}-Ru 1 mM acetonitrile solutions. After the formation of the SAMs on each electrode, the thiol solution was removed and replaced with an electrolyte solution of Na_2SO_4. The two Hg–SAM-R electrodes are brought into contact and placed together with the reference and the counter electrodes into the electrolyte solution, then we run experiments A and B. In the experiment A, the potentials of the mercury electrodes were controlled such that one (cathode) acted as the electron donor and the other (anode) as the electron acceptor. We designated the cathode and anode as the drain and source electrodes (Fig. 3b), respectively, by analogy to the convention used in semiconductor devices, where current is considered to flow from a more positive region to a more negative region.[19] We measured the conductance through the junction as a function of the potentials of the drain and source electrodes with respect to the reference electrode (*i.e.*, V_{DG} and V_{SG}, respectively) and as a function of the potential between the source and drain (V_{DS}), using the electrolyte solution as a gate. We controlled the potential of the gate by controlling the potential applied to the reference electrode relative to ground. Since the source electrode is at ground, this voltage is V_{SG}. Fig. 5a shows the drain/source currents when V_{DG} is fixed at −0.20 V, where the attached ruthenium is in its +2 oxidation state, and V_{SG} is varied. For $V_{SG} \leq V_{DG}$, the currents are negligible and the junction is non-conducting. Increasing V_{SG} to values more positive than −0.14 V results in an anodic current flow corresponding to the oxidation of Ru^{II} to Ru^{III} at the source and a cathodic current flow corresponding to the reduction of Ru^{III} to Ru^{II} at the drain. The anodic and cathodic currents are equal and increase to a plateau with a half-wave (half-maximum) potential, −0.04 V, that is near the formal potential, $E^{o\prime}$, of the $Ru^{II/III}$ couple. Based on these observations, we propose that charge transport through the junction occurs as a result of oxidation of Ru^{II} to Ru^{III} at the source, electron exchange between Ru^{III} at the source and Ru^{II} at the drain, and reduction of Ru^{III} at the drain back to Ru^{II} as key steps. In the experiment B, the conductance of the Hg–SAM//SAM–Hg junction was measured operating at fixed drain-source potentials. Fig. 5b shows that at fixed V_{DS} = +0.10 V, the drain current is negligible for V_{SG} less than −0.25 V and greater than +0.15 V. Upon scanning V_{SG} from −0.25 to +0.15 V, the current increases from zero to a maximum value near the $E^{o\prime}$ of the redox couple and then decreases again to zero. Charge therefore passes from one electrode to another only when V_{SG} is at or close to the $Ru^{II/III}$ redox potential.[20]

The junction shows electrical behaviour similar to that of a solid-state transistor when the electrolyte is used as the gate. The channel length in this case is defined essentially by the thickness of the SAMs between the two mercury electrodes, that is a few nanometers.

Junction Hg–SAM//R//SAM–Hg

The junction was assembled as follows. SAMs were formed separately on the two Hg electrodes from a solution of 1 mM of $HSC_{10}COOH$ in ethanol. The two electrodes covered by the SAMs, were immersed in a solution containing 1 mM $Ru(NH_3)_6{}^{3+}$ in 0.1 M electrolyte at pH 9 and brought in contact by a micromanipulator. By placing the reference and counter electrodes to create an electrochemical cell, we run experiment A. In experiment A, the potentials of the mercury electrodes were controlled such that one acts as an electron donor to $Ru(NH_3)_6{}^{3+}$ and the other acts as an electron acceptor from $Ru(NH_3)_6{}^{2+}$ (Fig. 4). It is well known that electroactive species confined

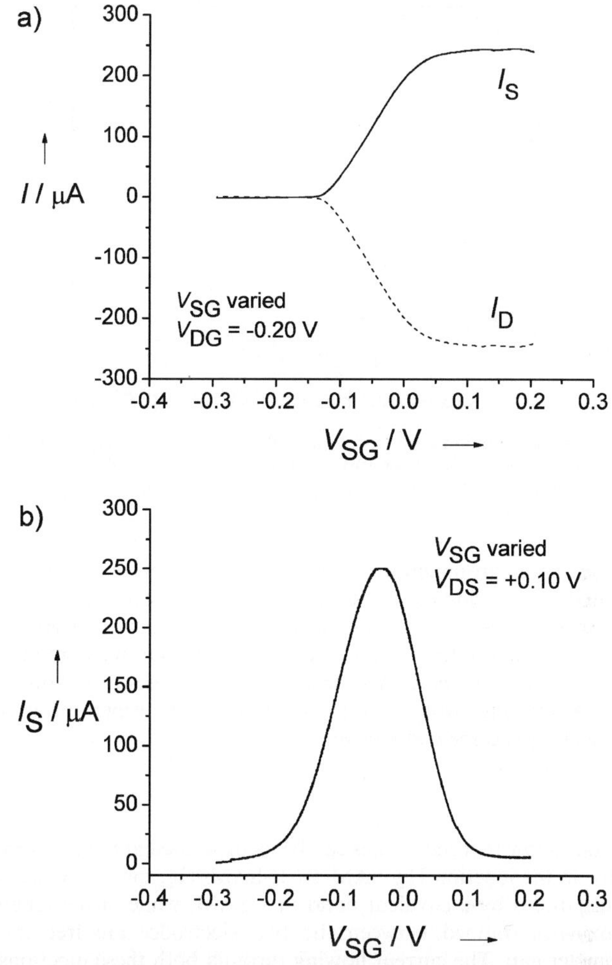

Fig. 5 Current–voltage characteristics of the Hg–SAM-R//R-SAM–Hg junction showing diode like behaviour. All experiments were carried out in 0.2 M aqueous Na_2SO_4 at pH 4. (a) I_D and I_S as a function of V_{SG}. V_{DG} was fixed at −0.20 V and the scan rate was 50 mV s^{-1}. (b) I_S as a function of V_{SG} for the same junction as in (a). V_{DS} was fixed at +0.10 V.

between two electrodes closely spaced, generate a current when the potentials of the electrodes are controlled such that one donates electrons and the other accepts them.[21–24] The species generated at one electrode (the "generator" electrode) can diffuse to the second electrode (the "collector")[21,22] and generate current *via* a mechanism called redox cycling.[23] The potential of the collector electrode was held at +0.10 V *vs.* Ag/AgCl, while the generator electrode was swept towards negative potentials past $E^{o\prime}$ (−0.21 V) for the $Ru(NH_3)_6{}^{3+} \rightarrow Ru(NH_3)_6{}^{2+}$ reaction and back again (Fig. 6). Fig. 6 show a typical *I–V* curve measured after cycling the potential across the junction. The current–voltage curves show that at the generator and at the collector the current initially increases slowly, and then typically becomes sigmoidal and increases rapidly until a steady-state current–voltage curve (*i.e.* the currents of the forward and reverse scans are about the same). This result indicates that (i) ruthenium hexaamine is trapped inside the junction, and (ii) that the two mercury electrodes are close enough to one another for redox cycling to occur.

We also observe a negative shift of the half-wave (half-maximum) potential $E_{1/2}$ relative to $E^{o\prime}$ for the $Ru(NH_3)_6{}^{3+} \leftrightarrows Ru(NH_3)_6{}^{2+}$ reaction. The value of $E_{1/2}$ corresponds to the formal potential for reduction of $Ru(NH_3)_6{}^{3+}$ trapped inside the junction. The negative shift in $E_{1/2}$ is indicative of the

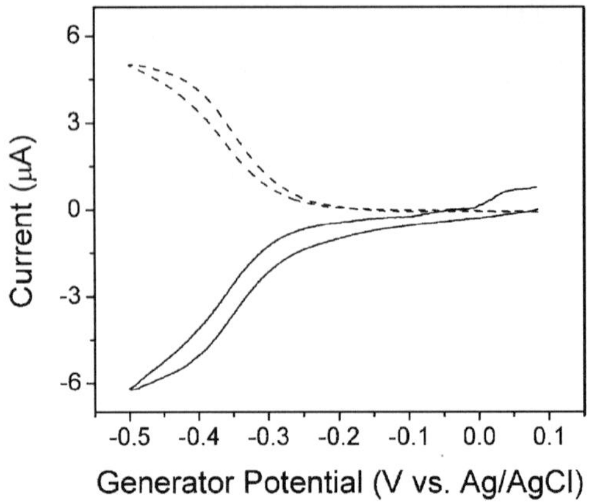

Fig. 6 Generation–collection voltammogram of 1 mM $Ru(NH_3)_6^{3+}$ in a Hg–SAM//R//SAM–Hg junction at pH 9. The collector potential was held at +0.10 V, and the generator potential was cycled between +0.1 and −0.5 V at a scan rate of 50 mV s^{-1}. The areas of the junctions were *ca.* 0.20 mm^2.

generation of a more polar environment inside the junction. We propose that under applied voltage scans across the junction, a tilting/compression of the SAMs can occur[25,26] with a consequent expulsion of the electrolyte. Such a process would cause the two SAM-covered electrodes to approach one another and allow the interaction between the surface-immobilized COO^- groups and the $Ru(NH_3)_6^{3+}$ couple to increase. Also this junction switches from "off" to "on" by varying the potential applied at the generator: when the potential of the generator approaches the formal potential $E_{1/2}$, the junction became conductive.

3. Conclusions

The results of this study show that redox sites can be incorporated at the interface of the junction by different interactions: in the junction Hg–SAM-R//R-SAM–Hg, the redox sites are immobilized at the surface of the electrodes by a covalent bond in a SAM, while in the junction Hg–SAM//R//SAM–Hg the redox sites trapped between the two electrodes are free to diffuse inside the interelectrode nanometer gap. The current flowing through both these junctions can be controlled by controlling the potential of the electrodes by conventional electrochemical measurements: the current flows only when the potential of the electrodes are controlled so that one can oxidize the redox species and the other can reduce it. Significantly, the current flow is generated in these two junctions by a different mechanism. The junction Hg–SAM-R//R-SAM–Hg switches from *off* to *on* when the electronic states of the redox centres on the two monolayers falls between the Fermi levels of the electrodes. The intensity of the current depends on the rates of (i) electron tunnelling between the mercury electrodes and the surface-bound Ru^{II}/Ru^{III} centres and (ii) electron hopping between adjacent Ru^{II} and Ru^{III} centres.

In the junction Hg–SAM//R//SAM–Hg the current flows through the junction only when the electrode potentials are adjusted so that one donates electrons and the other accepts them. Under these conditions, the current is dominated by a mechanism of redox cycling. The intensity of the current depends on the distance between the electrodes. It is important to stress that, in such a confined space, electron hopping between adjacent species in different oxidation states, could become operative. In both systems, the current flowing through the junction depends on the number of redox centres trapped in the area in which the two compliant mercury electrodes are in contact. The systems described here are easy to assemble and make it easy to modify both the area and electrical properties of the metal–metal junctions. The area can be modified by changing the pressure applied to the mercury electrodes, while tuning of the electrical properties may be achieved by changing the nature of the redox centre. Because these junctions are well-characterized and yield reproducible data, they are well suited for collecting fundamental information relevant to the fabrication of molecular switches.

Acknowledgements

This work is supported by NSEC (award no. NSF-PHY-0117795), NSF (award no. CHE-0101432), and NSERC (in the form of a postdoctoral fellowship to E.T.), by the EU: Project Contract GRD1-2001-40772 (MWFM) and Project Contract IST-2001-35503 (LIMM).

References

1 A. Nitzan and M. A. Ratner, *Science*, 2003, **300**, 1384.
2 R. E. Holmlin, R. F. Ismagilov, R. Haag, V. Mujica, M. A. Ratner, M. A. Rampi and G. M. Whitesides, *Angew. Chem.*, 2001, **40**, 2316V.
3 *Electron Transfer in Chemistry*, ed. Balzani, vols. 1–5, Wiley-VCH, Weinheim, 2001.
4 W. Han, E. N. Durantini, A. L. Moore, D. Gust, P. Rez, G. Leatherman, G. R. Sealey, N. J. Tao and S. M. Lindsay, *J. Phys. Chem. B*, 1997, **101**, 10719.
5 D. I. Gittins, D. Bethell, D. J. Schiffrin and R. J. Nichols, *Nature*, 2000, **408**, 67.
6 J. Park, A. N. Pasupathy, J. I. Goldsmith, C. Chang, Y. Yaish, J. R. Petta, M. Rinkoski, J. P. Sethna, H. D. Abruna, P. L. McEuen and D. C. Ralph, *Nature*, 2002, **417**, 722.
7 W. Liang, M. P. Shores, M. Bockrath, J. R. Long and H. Park, *Nature*, 2002, **417**, 725.
8 J. Chen, M. A. Reed, A. M. Rawlett and J. M. Tour, *Science*, 1999, **286**, 1550.
9 N. J. Tao, *Phys. Rev. Lett.*, 1996, **76**, 4066.
10 D. L. Feldheim and C. D. Keating, *Chem. Soc. Rev.*, 1998, **27**, 1.
11 U. Mazur and K. W. Hipps, *J. Phys. Chem. B*, 1999, **103**, 9721.
12 R. A. Wassel, G. M. Credo, R. R. Fuierer, D. L. Feldheim and C. B. Gorman, *J. Am. Chem. Soc.*, 2004, **126**, 295.
13 W. Haiss, H. van Zalinge, S. J. Higgins, D. Bethell, H. Höbenreich, D. J. Schiffrin and R. J. Nichols, *J. Am. Chem. Soc.*, 2003, **125**, 15294.
14 R. Rinaldi, A. Biasco, G. Maruccio, R. Cingolani, D. Alliata, L. Andolfi, P. Facci, F. De Rienzo, R. Di Felice and E. Molinari, *Adv. Mater.*, 2002, **14**, 1449.
15 A. Nitzan and M. A. Ratner, *Science*, 2003, **300**, 1384.
16 D. Cahen and G. Hodes, *Adv. Mater.*, 2002, **14**, 789.
17 H. O. Finklea and D. D. Hanshew, *J. Am. Chem. Soc.*, 1992, **114**, 3173.
18 H. O. Finklea and D. D. Hanshew, *J. Electroanal. Chem.*, 1993, **347**, 327.
19 P. Horowitz and H. Winfield, *The Art of Electronics*, Cambridge University Press, New York, 1995.
20 E. Tran, M. A. Rampi and G. M. Whitesides, *Angew. Chem.*, 2004, **43**, 3835.
21 B. Fosset, C. A. Amatore, J. E. Bartelt, A. C. Michael and R. M. Wightman, *Anal. Chem.*, 1991, **63**, 306.
22 A. J. Bard, J. A. Crayston, G. P. Kittlesen, T. V. Shea and M. S. Wrighton, *Anal. Chem.*, 1986, **58**, 2321.
23 P. G. Pickup and R. W. Murray, *J. Am. Chem. Soc.*, 1983, **105**, 4510–4514.
24 O. Niwa, M. Morita and H. Tabei, *Anal. Chem.*, 1990, **62**, 447.
25 X. D. Cui, A. Primak, X. Zarate, J. Tomfohr, O. F. Sankey, A. L. Moore, T. A. Moore, D. Gust, L. A. Nagahara and S. M. Lindsay, *J. Phys. Chem. B*, 2002, **106**, 8609.
26 Y. Selzer, A. Salomon and D. Cahen, *J. Phys. Chem. B*, 2002, **106**, 10432.

General Discussion

Professor Naaman opened the discussion of Dr Tao's paper: How did you define the position of the HOMO and LUMO?

Professor Tao replied: We do not really know the relative alignment of the HOMO or LUMO with respect to the Fermi energy levels of the electrodes. However, some rough idea may be obtained for redox active molecules. For example, if a molecule can be easily reduced, it indicates that its LUMO is too far away from the Fermi energy levels.

Professor Schiffrin said: You mentioned noise in your conductivity measurements. We did observe "noise" in the current under some conditions for conduction through substituted bipyridyl compounds. The probable origin of this is inelastic tunnelling that may result in excitation of vibrational modes that change within the Au–S contact or the conductivity through conformational changes. What is the origin of the "noise" in your experiments? Do you observe a structure in this "noise"?

Professor Tao replied: We observed "noise" in some redox molecules. For example, in the case of oligothiophenes, the conductance fluctuates with rather large amplitude when we shifted the potential towards the redox potential (equilibrium potential). The correlation of the "noise" with the redox reaction suggests that in addition to the mechanism you pointed out, *i.e.*, instability of the Au–S contact due to inelastic tunnelling or "heating" effect, intrinsic redox processes, such as random switching of the molecule between oxidized and reduced states, may also cause large conductance fluctuations.

Professor Rampi asked: From your experimental data, which mechanism of electron transfer seems to be operating? The single step or the two step one?

Professor Tao said: In the case of PTCDI, a peak-like feature is observed near the equilibrium potential, which suggests a two-step process. However, since the equilibrium potential depends on the local environment, the exact equilibrium potential of the molecule is unknown, and the question whether it is a "single" step or "two" steps is still open.

Professor Kuznetsov commented: In your paper you mentioned two mechanisms of tunnelling with reference to two of my papers (refs. 52 and 53). In fact there is the same mechanism in both cases, *viz*., so-called redox-catalysed electron tunnelling. The electron tunnelling takes place in the process of oxidation/reduction transformation of the bridge group. The difference is in the strength of coupling with the electrodes and in some details of theoretical description. The case of weak coupling corresponds to non-adiabatic limit in terms of ordinary electrochemical reactions. This is really a two-step process. In the case of strong coupling (adiabatic limit) the process also involves two steps: reduction and oxidation. However, here multiple electron tunnelling takes place in each reduction/oxidation step.

Professor Tao replied: After re-reading your papers, I have a better understanding of your work. We are carrying out further experiments in order to determine whether the redox systems belong to strong or weak coupling situations.

Dr Haiss remarked: If molecules are tethered between tip and sample having a chemical contact at both ends, small tip deformations may not significantly influence the potential dependence of the current since it is mainly determined by through bond tunnelling but in conventional tunnelling spectroscopy where dI/dz is large, small deformations of the tip are expected to influence the experimental results significantly.

I have performed calculations of the stress-induced deformation of an STM tip during potential scans, which will be subject of a forthcoming publication. An estimation of the stress induced length change for a set of parameters, which are typically encountered in an electrochemical experiment yields values of $\Delta z > 0.1$ nm for a realistic tip geometry, where only the outermost 50 nm are uncoated. It is obvious that such a change in tip sample separation will induce artefacts in electrochemical gating spectroscopy and hence the tunnelling current will increase during positive going potential sweeps due to the compressive surface stress which is induced by a positive charge on the metal surface.

Surface stress effects at the apex of an STM tip may influence electrochemical gating spectroscopy since potential sweeps with concomitant changes of the surface charge at the tip may result in an elastic deformation of the tip. For this reason it is important to coat the tip perfectly. To what degree can you control the coating of the STM tip?

Professor Tao replied: We coated our tips with an insulating layer, which reduces the leakage current below 1 pA, so the exposed surface area should be rather small. Even though, there should still be some potential-induced stress, causing changes in the tip–substrate separations. Since the individual conductance traces reveal steps with lengths well over 0.1 nm, a small change in the tip–substrate separation does not significantly change the molecular conductance. The reason, we believe, is that most of the stress effect is at the contacts involving Au–Au bonds (softer than C–C bonds in the molecules), which tend to minimize the stress on the measured conductance.

Dr Nichols asked: Have you examined the influence of electrolyte concentration? You should be able to tune the gating effect out (particularly for short molecules) by reducing electrolyte concentration.

Professor Tao replied: Not yet, but I agree that this will be an important experiment to carry out.

Professor Lindsay addressed Dr Haiss: There appear to be three distinct processes for conduction in octanedithiol, depending on contact geometry. Presumably, all these measurements suffer from these complications. The mechanisms need to be sorted out for more complex experiments to be understood.

Dr Haiss replied: Using the *I*(s)[1] and *I*(t)[2] techniques, we observe only one fundamental current peak in the conductivity histograms and integer multiples of this single molecule conductivity in case of the *I*(s) technique. Both, the *I*(s) and the *I*(t) method result in the same value of single molecule conductivity. The break junction technique[3] on the other hand results in larger conductivity values, a finding which we have attributed to a different attachment geometry of the alkanedithiols, alongside the gold 'chains', as discussed in detail in ref. 2.

1 W. Haiss, H. van Zalinge, S. J. Higgins, D. Bethell, H. Höbenreich, D. J. Schiffrin and R. J. Nichols, *J. Am. Chem. Soc.*, 2003, **125**(50), 15294–15295.
2 W. Haiss, R. J. Nichols, H. van Zalinge, S. J. Higgins, D. Bethell and D. J. Schiffrin, *Phys. Chem. Chem. Phys.*, 2004, **6**(17), 4330–4337.
3 B. Q. Xu and N. J. J. Tao, *Science*, 2003, **301**(5637), 1221–1223.

Dr Wandlowski commented: The typical free area of a well coated STN tip is $< 1\ \mu m^2$, which gives leakage current much below 1 pA in the double layer region, and a few picoamperes in the gold oxidation/reduction region. Such types of STN tip are very suitable for high quality spectroscopic studies.

Dr Haiss replied: I agree that it is possible to reduce the exposed area of an STM tip so far, that electrochemical currents, induced at the tip as a result of a potential scan, are so small, that they do not significantly influence electrochemical STS measurements. However, there are other considerations, which have to be taken into account in this type of experiments. Surface stress induced deformations may lead to artefacts in STS performed in an electrochemical environment as detailed in my first comment.

Professor Porath addressed Professor Tao: In the paper (section 3.3) it says: "... the strong molecule electrode binding does not significantly alter the molecular electronic states."

I would expect that attaching the molecule to the electrode would lead to E_F equalisation, charge transfer and level rearrangement. Can you please comment?

Professor Tao replied: "Molecular electronic states" here are referred to "redox states", which do not change much since the cyclic voltammograms of molecules in bulk solution and immobilized on the electrodes are similar. The binding does in general change other molecular electronic states, such as S groups, and also affects the alignment of the redox states to the Fermi energy levels.

Dr Rovira asked about the PTCDI compounds: Do you think that the different conductance that SH and pyridine substituted compounds present can be due to the different N–Au and S–Au contacts?

Which contact do you think is better as an electrode–molecule interface?

Professor Tao replied: Yes, but it is difficult to say which provides a better electrode–molecule interface because (1) we do not have systematic data for other molecules and (2) the pyridine-terminated and SH-terminated PTCDI compounds have somewhat different lengths, which may also contribute to the difference in the measured conductance values.

Dr Nichols commented: You have argued for the OPE based systems that changes in electron density in the aromatic system for different substituents drive the changes in conductance. Is this an important factor or is it the energy shift in frontier molecular orbitals (within the Fermi window) that is important? Or are both models self consistent?

Professor Tao replied: I think that "changes in electron density" and "energy shift in frontier orbitals" are related, and I can imagine that one will change with the other. In other words, they may not be considered as two separate mechanisms.

Professor Bryce mentioned the possibility of other terminal groups, especially selenols (SeH) which may be stable when assembled on Au over a wider potential window than the analogous thiol or pyridyl system.[1]

1 Y. Sato and F. Mizutani, *Phys. Chem. Chem. Phys.*, 2004, **6**, 1328.

Professor Tao replied: It would be indeed nice if someone could synthesize selenol-terminated molecules and perform the conductance measurements.

Professor Naaman remarked: The local field at the molecule is very different from the field you applied. This must be taken into account. The local field will depend on the exact structure of the contacts and the size of the molecule.

Professor Tao answered: Yes, I agree that the local field will depend on the exact structure of the contacts and the size of the molecule. It should also depend on the electrolytes. However, it is interesting to point out that several redox systems (studied by the Liverpool group, Dr Wandlowski's, Professor Lindsay's and Dr Rampi's groups, and by our own experiments) have conductance changes that are correlated with the redox potentials measured in electrochemical experiments. This seems to indicate that the dependence of the local field on the exact structure of the contact is not as large as one might think.

Professor Lindsay commented: Electrochemical control of the surface potential removes this problem by referring the potential at the molecule to a surface of standard polarization.

Professor Bethell communicated: I am impressed by the remarkably good linear correlation between conductance and the Hammett substituent parameter σ_m in substituted OPE dithiols (Fig. 5c of the paper), especially bearing in mind that these molecules are not symmetrical. Such a good correlation would not be found using other substituent parameters. I should like to ask whether any deeper significance can be attached to the observed correlation with the meta-substituent constant than a simple electrostatic influence. For example, could the correlation imply preferential alignment of the molecules in one direction relative to the field between the contacts that are generated in the break?

Professor Tao communicated in response: It is possible that one might find a better interpretation of the observed substituent effect. However, it is not obvious to me that the orientation of the molecule can lead to the correlation. We determined the conductance with a relative small bias voltage at which the current–voltage curves are rather symmetric.

Dr Higgins communicated: In the experiment shown in Fig. 5 of your paper, what happens to I_{sd} if the potential sweep is reversed? Does it stay at the high value seen at ~ -0.9 V (as one would expect from the irreversible reduction wave)?

Professor Tao communicated in reply: It is highly reversible—the reverse curves overlap with the forward sweep curves. Yes, it does stay there (until the molecular junction breaks).

Dr Davis communicated: Can the authors comment on any apparent correlation between the quality of the voltammetry and the sensitivity of source–drain current to gate voltage?

Professor Tao communicated in response: We have not observed any apparent correlation between the two. For example, the cyclic voltammogram of ferrocene compounds is very nice and reversible, but the source–drain current changes with the gate voltage by only a factor of <10. In contrast, PTCDI's cyclic voltammogram is not as nice but the current increases by nearly three orders of magnitude.

Professor McCreery opened the discussion of Dr Wandlowski's paper: The reduction from V^{+2} to V^{+1} is accompanied by formation of a quinoid species with a significantly smaller HOMO–LUMO gap. For example, biphenyl has a 5.3 eV gap while its anionic, quinoid radical is $\sim$2.5 eV. Such a change should result in a large conductance increase, yet you see only about a factor of two. Why?

Dr Wandlowski replied: Our experimental study documented convincingly the formation of single molecule junctions Au(T) |HS–6V6–SH| Au(S) under strict environmentally controlled conditions. The observed conductance increase upon reduction $V^{2+} \rightarrow V^{+\bullet}$ is attributed to the higher electron density and the higher conjugation of the $V^{+\bullet}$ moiety. However, the overall conductance of the junction contains additional contributions of the electrically rather insulating alkyl chains. Lindsay *et al.* estimated the conductance of a single molecule junction Au(T) |dodecanthiol| Au(S) to 0.121 nS.[1] The alkyl chains appear to reduce the contributions of the structurally modified central viologen unit to the overall junction conductance. A current effort based on electronic structure and transport calculations aims to address this hypothesis in more detail.[2]

1 X. D. Cui, A. Primak, X. Zarate, J. Tomfohr, O. F. Sankey, A. L. Moore, T. A. Moore, D. Gust, L. A. Nagahara and S. M. Lindsay, *J. Phys. Chem. B*, 2002, **106**, 8609.
2 F. Evers and Th. Wandlowski, in preparation.

Professor Schiffrin said: Fig. 8 clearly shows that the potential and not the interfacial field is responsible for the changes in conductivity. The sigmoidal shape of the conductivity does not match the electrochemical transformation $V^{2+}/V^{\cdot+}$. Electrochemical equilibrium should prevail considering the way in which these experiments are carried out.

Dr Wandlowski replied: The single molecule conductance experiment reported in Fig. 8 represents the potential-induced transition between two equilibrium states. In the two limiting cases the junction contains either the "dication species V^{2+}" or the "monomer radical cation $V^{+\bullet}$". The conductance of the entire junction changes upon negative polarization from a low conductance (oxidized) state to a high conductance (reduced) state. The inflection point of the plot conductance current *vs.* substrate potential for the symmetric junction Au(T)|HS–6V6–SH|Au(S) (strong coupling, Fig. 8) is close to the position of the maximum tunnelling current observed in the single molecule tunnelling experiment of the asymmetric junction Au(T)-S–6V6–H|Au(S) (weak coupling, Fig. 10). The different shapes of the current voltage characteristics of the two experimental approaches reflect the different coupling properties of both junctions. Ongoing electronic structure and transport calculations will address this aspect in detail.[1]

1 F. Evers and Th. Wandlowski, in preparation.

Dr Nichols asked: In Fig. 10 of the paper, how does the position of resonance depend on parameters such as tip bias potential or proximity of viologen coated tip to surface?

Dr Wandlowski said: The position of the resonance is rather independent on the absolute position of the viologen-modified tip with respect to the bare Au(111)-(1 × 1) surface provided that strong physical interference is avoided. The peak position is linearly dependent on bias potential. Investigations were carried out in $-0.20\ \mathrm{V} < (E_T - E_S) < 0.20\ \mathrm{V}$.

Dr Higgins communicated: In section 3.3 of your paper, you refer to vibrational modes characteristic of radical cation–radical cation dimers. Would you not expect to see splitting in the redox wave for the first reduction, as I think has been seen by other authors where this phenomenon has occurred?

Dr Wandlowski communicated in reply: The SEIRAS experiments show that, under our experimental conditions, the monomer radical cation $V^{+\bullet}$ appears to be the dominant product upon reduction of HS–6V6–H ($V^{2+} \rightarrow V^{+\bullet}$). The intensities of the bands attributed to the radical cation dimer ($(V^{+\bullet})_2$) are significantly smaller. The slow scan voltammetric curve (Fig. 1A) is rather symmetrically for $V^{2+} \rightarrow V^{+\bullet}$ ($(V^{+\bullet})_2$) with an ideal value of 0.091 V FWHM. These qualitative criteria indicate the preference of one single component in the monolayer of HS–6V6–H on Au(111)-(1 × 1). In case of HS–6V6–SH both, spectroscopic and electrochemical experiments (Figs. 1B and 3) point to the formation of a larger amount of the radical cation dimer. The shape of the *I–E* curves upon reduction $V^{2+} \rightarrow V^{+\bullet}$ ($(V^{+\bullet})_2$) is not symmetric. The tail towards more negative potentials reflects the formation of a second species. On the other hand the formation potentials of $V^{+\bullet}$ and $(V^{+\bullet})_2$ appear to be not very different. In consequence, we do not see a clear separation of the two waves employing 50 mV s^{-1} as scan rate. Extended kinetic studies with nanoelectrodes and high scan rates might allow studying both processes separately. This topic was not the aim of the present study.

Finally we would like to comment on the role of oxygen. We discovered in our experiments that the presence of oxygen strongly favours the formation of the radical cation dimer. Oxygen contamination was strictly controlled (eliminated) in all of our experiments reported in this paper. Careful inspection of voltammetric data reported in the literature clearly indicated the presence of trace amounts of oxygen in several earlier studies. This fact might explain some differences between our experimental data and some observations reported before in the literature.

Dr Davis communicated: The authors extrapolate a "two step electron transfer mechanism" from the shown data; can they explain the origin of this conclusion?

Dr Wandlowski communicated in response: The conclusion on the mechanism is based on criteria derived in theoretical papers of Kuznetsov and Ulstrup.[1–3] As already discussed in the last chapter of our paper, resonance tunnelling and a mechanism based on coherent two-step electron transfer predict a maximum of the tunnelling current shifted from the equilibrium potential by the reorganization energy λ_R. A maximum located near the equilibrium potential is predicted for a sequential two-step electron transfer if the bias potential drop at the site of the redox center is close to half the potential drop ($E_T - E_S$). The latter situation is assumed to be applicable for the present case. A more comprehensive answer will be given in a forthcoming contribution, which combines our experimental work with theoretical model studies on electron transport in the present metal–molecule–metal junction.[4]

1 J. Zhang, Q. Chi, A. M. Kuznetsov, A. G. Hansen, H. Wackerbarth, H. E. M. Christensen, J. E. T. Andersen and J. Ulstrup, *J. Phys. Chem. B*, 2002, **106**, 1131.
2 A. M. Kuznetsov and J. Ulstrup, *Electrochim. Acta*, 2000, **45**, 2339.
3 (*a*) A. M. Kuznetsov and J. Ulstrup, *J. Phys. Chem.*, 2000, **104**, 11531; (*b*) A. M. Kuznetsov and J. Ulstrup, *Probe Microsc.*, 2001, **2**, 187.
4. F. Evers, A. Kuznetsov and Th. Wandlowski, in preparation.

Dr Davis communicated: Can the authors comment on the magnitude of the barrier height Φ used in eqn. (1) of the paper?

Dr Wandlowski communicated in reply: The barrier height $\Phi \sim 1.0$ eV represents a typical average value observed for free and adsorbate-covered electrified solid–liquid interfaces. For a detailed study we refer to our recent work in *Langmuir*[1] and to a forthcoming publication.[2]

1 G. Nagy and Th. Wandlowski, *Langmuir*, 2003, **19**, 10271.
2 I. Pobelov, G. Nagy and Th. Wandlowski, in preparation.

Dr Davis communicated: Can the authors clarify the aims and conclusions associated with the "nanoelectrode voltammetry"?

Dr Wandlowski communicated in response: We refer in our paper to the expression "voltammetry with nanoelectrodes". The so-called nanoelectrodes were prepared by electrochemical etching of gold wires and subsequent coating with polyethylene, except the very end. The effective free area of the electrode was estimated from the charges consumed during surface oxidation and reduction, and their comparison with macroscopic gold electrodes of known geometry. Assuming a conical shape of the uncoated very end of the electrode, high resolution SEM allowed estimating an apex radius of less than 20 nm.

The electrochemical experiments were carried out with a home-build low current potentiostat (current resolution <10 fA). The performance of the instrumental set-up is illustrated in Fig. 1, which shows the CV of an etched and coated gold STM tip in phosphate buffer recorded with 10 mV s^{-1}. A detailed description of our instrumental set-up and several applications will be given in ref. 1.

The high current resolution of our electrochemical experiment allowed monitoring the extend of surface modification with HS–6V6–H quantitatively, and to obtain exact background data for single molecule tunneling experiments like the one shown in Fig. 8 of our paper. An additional application of our nanoelectrodes and custom-made instrumental set-up includes the estimation of kinetic data of fast electron transfer reactions, such as $V^{2+} \leftrightarrow V^{+}$. Here large perturbation techniques and macroelectrodes give rather distorted and sometimes completely wrong results.

1 G. Meszaros, I. Pobelov, Z. Li and Th. Wandlowski, 2005, in preparation.

Professor Bryce commented: Alkylation of pyridine is reversible, under certain conditions. It is, therefore, important to ensure that for studies on viologen dications that the molecule is still intact, especially under electrochemical recycling conditions.

Dr Wandlowski answered: The stability of the viologen derivatives was carefully checked by *in situ* IR-spectroscopy and cyclic voltammetry employing exactly the same conditions as in the STM-based tunnelling and transport experiments (electrolyte, absence of oxygen). No degradation was observed in acidic and neutral electrolyte upon potential excursions in -0.20 V $< E < -1.00$ V. However, extended cycling at potentials $E < -1.00$ V caused irreversible modifications of the electrochemical response. We attribute these changes to the quasi-reversibility of the second redox process $V^{+\bullet} \rightarrow V^{\bullet}$ in aqueous electrolyte under our experimental conditions, and to the onset of the

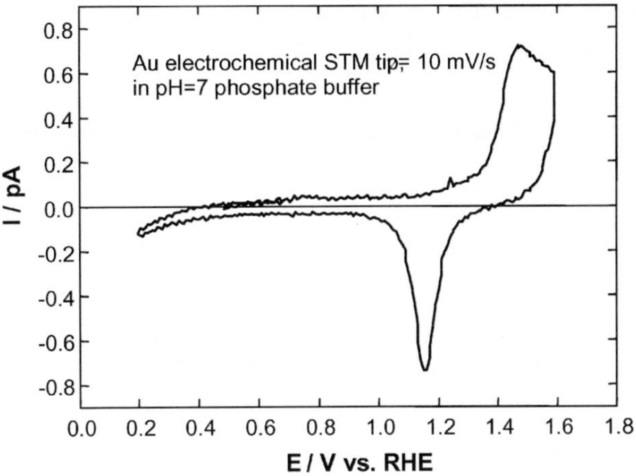

Fig. 1 Unfiltered low current voltammogram of a PE-coated Au-nanoelectrode showing typical features of gold oxidation and reduction, and a double layer current below 100 fA.

reductive desorption of the surface bound viologen molecules. Experiments in aqueous alkaline electrolytes, pH $\gg 7$, showed slow (>1 hour) degradation of the electrochemical response, which is attributed to the cleavage of the N–C bond of the viologen moiety.[1]

1 P. M. S. Monk, *The Viologens*, John Wiley & Sons, Chichester, 1998.

Professor Kornyshev opened the discussion of Professor Lindsay's paper: A way to distinguish the conducting or semiconducting behaviour in the bulk is to study the temperature dependence of conductance. Have you tried to do this to determine the mechanism of single molecule conductance?

Professor Lindsay replied: Temperature dependence has been measured for some of the features observed in the break junction measurements of single octanedithiol molecules. The ones well-fitted by tunnelling theory are not temperature dependent. The low conductance (1 nS) feature is.

Professor Kornyshev asked: When you compare the type of conductance, are the statements made for the same environment?

Professor Lindsay answered: Yes. And environment does change the response as we showed for oligoaniline.

Professor Porath commented on the first slide of Professor Lindsay's talk, which stated that DNA had been found to be a conductor, an insulator, a semiconductor and a superconductor in different experiments: Some of the electrical transport measurements were never reproduced. The other measurements results depend strongly on the measurement configuration: surface contact along the molecule, covalent attachment, single molecules *vs.* bundles and networks. Therefore, most results reflect the specific experiment rather than the general behaviour of DNA. What we do know by now is that in an optimal configuration (no non-specific interaction and covalent attachment) the DNA can transport relatively high currents.

The use of terms like conductor, semiconductor, insulator *etc.* should be done with care regarding 1-D wires since one defect is enough to govern the measured behaviour. The issue is whether or not the wire is able to transport charges. How? And how fast?

Professor Lindsay replied: Your comment embodies what I am trying to convey—without reliable contacts and a reliable count of the number of molecules measured, no reliable data are obtained.

Professor Porath commented: The mechanisms for charge transport in solid state experiments (electrodes) are different than those for charge transfer in the solution chemistry experiments. In the latter case (donor–bridge–acceptor systems) one charge carrier is injected into the chain and the electronic configuration is altered as the carrier is moving along the chain. In the former case the molecule is attached to a metal, such that the Fermi levels should equalize, the molecular levels get broadened such that the injection barrier is lowered, charge is transferred and the electronic levels are reorganised. Then, on voltage application, the Fermi levels are fixed, the barriers along the molecule become charged and also a very large number of change carriers is available for the transport. All the above means that probably different and faster mechanisms take over in the solid state experiments than in the solution chemistry experiments.

Professor Lindsay replied: The environment plays a key role in charge transport. But so do the nature of the contacts and the length of the molecule. Dry DNA is structurally different from wet DNA, and that will make a difference too.

Dr Nichols said: On the timescale of the experiment I would expect many sites (top and hollow) to be sampled due to rapid surface diffusion so the distinction between top and hollow sites of binding should then not be so apparent.

Professor Lindsay replied: The gold surface is certainly mobile, but I would expect large changes in connection geometry to result in breaking of the molecular junction as the molecule is under tension. For the special case of octanedithiol though, this distinction does not seem to be so important, at least as simulated.

Dr Haiss asked: How would you explain the pronounced differences in single molecule conductivity measurements using different experimental methods (*i.e.* the break junction method as introduced by Professor Tao and the $I(s)$ method as introduced by Haiss).

Professor Lindsay responded: The forming of clean metal surfaces seems to be very important in getting good bonds to the molecule. It therefore appears that forming, and then breaking a metallic junction is required. This is not done in your method. On the other hand, we also find data similar to yours (*e.g.* the 1 nS feature for octanedithiol)) in our results, just as you see higher current steps in yours. We accounted for the 1 nS feature we saw using attached Au nanoparticles by invoking a poor contact in that case, so it seems possible that the poorer contacts give the 1 nS features and the better contacts the 4 and 20 nS features (in octanedithiol).

Professor Kuznetsov made a general comment: I would like to attract your attention to the system which has been investigated experimentally by the Liverpool group. We have analysed it theoretically. The system involves a nanoparticle linked to the electrode by a molecule which can be reduced or oxidized. The tunnel current in this system is probed by an STM tip. In fact we have double tunnel contact with an ordinary tunnel contact and bridged contact. The total current is determined by the interplay of both contacts. If the first contact dominates we have the behaviour typical for bridged contact with redox transformations. Tunnelling in this case occurs due to redox transformations of the bridge group. In the case of weak coupling with the electrodes the transfer of each electron is a two-step process (reduction/re-oxidation). In strong-coupling limit each step (both reduction and re-oxidiation) results in the transfer of a large number of electrons. The current is affected by the over potential. For symmetric system the dependence of the current on the over potential exhibits a maximum at equilibrium potential. This dependence is weaker, the larger bias voltage. This affect is rather general and independent of the reorganisation energy. The current/bias voltage has the typical sigmoidal form. Other details may be found in ref. 1.

1 A. A. Kornyshev, A. M. Kuznetsov and J. Ulstrup, *ChemPhysChem.*, 2005, **6**, 583–586.

Dr Schuster opened the discussion of Professor Fitzmaurice's paper with a question about gold wires: The electroless growth of the wires seems to produce discrete particles. How does this affect the conductivity of the resulting wire?

Professor Fitzmaurice replied: Following electroless deposition the nanowires are no longer discrete but have conjoined—that this is the case is clear from the low resistivities we have measured for these nanowires. Nevertheless, it is clear that such wires are more 'granular' and the extent to which this contributes to the measured resistivities has not been investigated in detail. However, it must be assumed that there will be some additional electron scattering at the surface of such wires and at the defects that exist in the regions between particles as they grow and are joined together, and that will result in an increased resistivity. This will be investigated by examining in detail the structure and the temperature dependence of the electrical properties of these wires before and after annealing.

Professor Porath asked:

1. How would you achieve wires with fewer defects?
2. Assuming that perfect wires will be obtained, how would you control the junction details between the arm wires and the nanoparticle that acts as the device, to have reproducible and controlled behaviour?

Professor Fitzmaurice replied:

1. It will be possible to reduce the defects present in these wires by optimizing the electroless enhancement of the nanoparticles attached to the DNA template and by annealing these wires using a range of techniques (thermal, electrical, laser *etc.*).
2. It is clear that the reproducible positioning of the nanoparticle in the nanogap will require very precise control over the metallization of the template, specifically over the dimensions of the gap and over the position of the anchoring point for the nanoparticle. The use of a large protein, which at once defines the gap and provides an anchoring point is apparently an effective approach. We will, however, not be in a position to be more definitive until the electrical characterization is complete, which in itself will provide information about the dimensions of the gap and the position of the nanoparticle.

Professor Fermin asked: Previous works have shown that the average interparticle distance and the distribution of nearest neighbour distances vary as a function of time (coverage) during the electrostatic adsorption of nanoparticles on polypeptide modified surfaces.[1] In relation to Fig. 4a of your work, I have two questions: What is the average interparticle distance between the DMAP stabilized gold nanoparticles? Can the number of particles in a DNA segment be controlled by the adsorption time and/or particle size?

1 (*a*) J. J. Kakkassery, J. P. Abid, M. Carrara and D. J. Fermín, *Faraday Discuss.*, 2004, **125**, 157–149; (*b*) D. J. Fermín, *Faraday Discuss.*, 2004, **125**, 206–207.

Professor Fitzmaurice responded: The average interparticle distance is approximately 3 nm. The number of adsorbed nanoparticles can be controlled by adjusting the particle size and the exposure time.

Professor Kornyshev asked: Concerning your mentioned size effects on resistivity—related to electron scattering from the surfaces, you can verify their existence and even characterise them *via* varying the size of the gold particles that cover the DNA. Can you vary the size of the particles?

Secondly, if you can vary the particle size, why don't you make a dispersion of different sizes? Percolation theory[1] tells us that in a mixture of particles of different sizes the percolation threshold is lower than in a mono-disperse system. The reason is simple, small particles fill the gaps between larger particles and you get as a result more pathways for the conductivity.

1 A. S. Ioselevich and A. A. Kornyshev, *Phys. Rev. E*, 2002, **65**, 021301.

Professor Fitzmaurice responded: The suggestion that one might, based on insights provided by percolation theory, be better served by using mixtures of nanoparticles of different sizes to partially metallize the DNA template is an excellent suggestion and one which we will pursue. In this context we can systematically vary the size of these nanoparticles by thermally promoted ripening. These findings will be reported soon. The study of electron scattering in these wires will be involved and rely on a number of complimentary approaches including the following: temperature dependent transport studies in wires of different diameters prior to and following annealing. Such studies are currently underway.

Professor Schiffrin addressed Professors Fitzmaurice and Heath: During this meeting we have hard of two approaches for the construction of devices at the nanoscale—"hard" wired or deterministic (Professor Heath) or "floppy" or poorly defined as you have described. Could you comment on some likely futures, for example, for computational devices for these two approaches?

Professor Fitzmaurice replied: First, would wish to be clear that we are not proposing our approach in opposition to the approach proposed by Professor Heath—but as one of the possible approach, which at this stage very early in the development of bottom-up assembly-manufacturing technologies, should also be actively considered. Second, we would not equate floppy with poorly defined. The templates may indeed be flexible, but once they have been pre-organized at a patterned substrate, they will no longer be floppy. Furthermore, the very fact that they are floppy may well assist in their pre-organization in a cooperative process.

In respect of the devices that might be realized, we are non-committal at this time. We are trying to understand how any nanoscale device, including wires and switches, might be assembled from the bottom-up using biological templates and would hope that our findings would be sufficiently general that they would be relevant for whatever architectures will need to be realized in the future and for application inside and outside nanoscale electronics.

Professor Heath replied: I took this question to mean "Where are we going to find useful technologies between these various fabrication concepts." I think that the earliest applications will be ones that are unique to molecular electronics—*i.e.* in the areas of sensing, photonics, actuation, *etc.* I don't think it is necessary for us to look to compete with the Intels of the world to fabricate better logic, memory, *etc.*, circuits. On the other hand, if we can learn how to fabricate such circuits at the types of densities I referred to in my talk, then I think we have a terrific handle on a manufacturing method that can enable to us develop many different technologies.

Second, I think it is OK for the stuff that we are working on to be simply interesting, and not useful.

Professor Fitzmaurice addressed Professor Heath: The recent paper by Heath and Stoddart—demonstrating a switch based on the mechanical motion of a single molecule (or a few molecules)—and the very recent paper by Leigh (demonstrating a motor based on the biased molecular motion of many molecules) reveal a rich opportunity space based on one or many rigid or flexible molecules. This would, in turn, suggest that it is unlikely that we have identified all that is important at this time.

Professor Heath agreed.

Dr Brust asked: There are many examples in the literature that indicate that nanoparticles assembled using DNA linkers get very close to each other when samples for TEM are prepared, *i.e.* the dimensions of the DNA template in solution are not conserved. What do you observe in your system (Schemes 2–4)?

Professor Fitzmaurice replied: We observe that in respect of the linear and non-linear (trigonal) templates there is a distribution of configurations. These distributions have been characterized in each case in the related papers.[1,2] In respect of the linear template this should give rise to nanoparticle pairs that are 90 nm apart. In fact, we observe that 10% are greater than 90 nm and 10% are less than 40 nm. On average the nanoparticles are 69 nm apart. This is due, less to flexibility of the template (the persistence length of duplex DNA is over 50 nm), and more to the fact that the particles can lie on top of the template and the template can lie on top of the particles. A similar analysis has been undertaken for the trigonal template.

1 D. Iacopino, A. Ongaro, L. Nagle, R. Eritja and D. Fitzmaurice, *Nanotechnology*, 2003, **14**, 447.
2 S. E. Stanca, A. Ongaro, R. Eritja and D. Fitzmaurice, *Nanotechnology*, 2005, **16**, 1905.

Professor Lindsay opened the discussion of Dr Davis's paper: Were your measurements made in ambient conditions?

Dr Davis replied: Yes, as I think is detailed in the paper, the CPAFM measurements were carried out under (controlled) ambient environment. We have a body of correlative ECSTM data in electrolyte and are aware of the issues and limitations of ambient "electronics". We are making progress with electrolytic CPAFM.

Dr Nandhakumar asked: Please can you clarify the experimental conditions used for molecular height measurements? According to the paper all measurements were done at ambient conditions in contact mode.

Dr Davis responded: As is detailed in the paper, the molecular height measurements to which I think you refer are under fluid (fluid contact), where forces can be tightly controlled. We have exhaustively characterised these adlayers by a variety of methods, including fluid phase AFM and STM.[1–4]

1 L. Andolfi, D. Bruce, S. Cannistraro, G. W. Canters, J. J. Davis, H. A. O. Hill, J. Crozier, M. P. Verbeet, C. L. Wrathmell and Y. Astier, *J. Electroanal. Chem.*, 2004, **565**(1), 21–28.
2 J. J. Davis, C. L. Wrathmell, J. Zhao and J. Fletcher, *J. Mol. Recognition* 2004, **17**(3), 167–173.
3 J. J. Davis, D. Bruce, G. W. Canters, J. Crozier and H. A. O. Hill, *Chem. Commun.*, 2003, **5**, 576–577.
4 J. J. Davis and H. A. O. Hill, *Chem. Commun.*, 2002, **5**, 393–401.

Professor McCreery said: This is an old question, but you have provided a new context. When a Simmons model yields a low barrier height ($\sim$1 eV), various mechanisms such as diffusive tunnelling are invoked to explain the low apparent barrier. Might such mechanisms apply in your case?

Dr Davis answered: I'm not sure I have any new answers to this "puzzle". The frontier orbital gap of saturated alkyls, for example, as resolved by PES, is some 8 eV. An initial first approximation of the "gap" between a Fermi energy and molecular states will be at least 5 eV. Despite this, a typical tunnel barrier resolved in a Simmons analysis of CPAFM data lies, from memory, close to 1.1 eV. The simple schematics usually invoked in such work are highly unlikely to be accurate. It is probable that molecules analysed in these configurations are subject to orbital mixing with metallic density of states (DOS), field induced perturbation and energetic broadening. The effective tunnel barrier resolved is likely to be an integrated sum over all the barriers that realistically exist to tunnelling and may be far from that expected based on PES data and a simple square tunnelling barrier.

Calculations[1] put the HOMO–LUMO gap of azurin at $\sim$1.4 eV, close to the tunnel barrier we compute but I suspect this is fortuitous.

1 S. Corni, *Chem. Phys. Lett.*, 2004, **393**, 118–123.

Professor McCreery asked: What energy levels does the observed tunnelling barrier correspond to?

Dr Davis replied: I think this is the same question I attempted to answer previously. In the largest force window we appear to be in the non-resonant regime so there are no specific levels being addressed.

Professor McCreery addressed Professor Naaman: Has anyone measured the electron affinity of a molecule on a surface, in order to determine its LUMO?

Professor Naaman replied: Yes. The LUMO energies can be measured either by inverse photoemission[1] or by two-photon-photoemission spectroscopy.[2]

1 See for example: H. Vazques, W. Gao, F. Flores and A. Kahn, *Phys. Rev. B*, 2005, **71**, 041306 and references cited therein.
2 See for example: T. Vondrak, C. J. Cramer and X.-Y. Zhu, *J. Phys. Chem. B*, 1999, **103**, 8915–8919.

Professor Kornyshev addressed Professor Naaman: Chemisorption theory allows us to estimate the broadening of the electronic energy levels of the adsorbed molecule, as well as their shifts due to the interaction with the substrate. These effects are, naturally, the stronger the stronger the chemisorption bond, and they can easily be stronger than half of an electronvolt. In the 80s, these effects were "massively" studied for many chemisorbed molecules both theoretically and experimentally. How much does the attachment of a long chain to electrodes on its opposite poles affect its electronic structure? Is this known for the molecules under study?

Professor Naaman responded: There are four effects that must be discussed when one deals with adsorption of molecules on substrates.

1. The formation of a chemical bond.
2. The equalization of the electrochemical potential between the molecule and the substrate. This effect may lead to either charge redistribution on the molecule or charge transfer between the molecule and the substrate.
3. Charge redistribution on the molecule due to dipole–dipole interaction and due to the polarizability of neighbouring adsorbed molecules.
4. In the case of charged molecule systems one has to include the polarization of the surface in the energetics.

So from the above it is clear that energy levels on the molecules can be shifted by more than an electronvolt (for example due to the image charge). In addition, when discussing molecules it is important to realize that, unlike atoms, some of the molecular orbitals may interact strongly with the substrate while others interact weakly.

The question is about what we call strong effects: if we discuss the nature of the LUMO and HOMO and their position relative to the vacuum level, clearly even weak electronic coupling with the substrate can induce large changes in these properties.

Dr Bronowska asked Dr Davis: I have a question regarding the way the force was applied on the protein during the MD simulation: how was the direction of the force been selected?

Dr Davis replied: The simulations were carried out by a collaborator of ours (Professor Mark Samson, Oxford) in a way detailed in ref. 1. It is not possible to quantify force in such calculations only the separation between the two electrodes the protein bridges. These computations, then, give a qualitative prediction of protein structural change as the electrodes approach.

1 J. Zhao, J. Davis, M. Sansom and A. Hung, *J. Am. Chem. Soc.*, 2004, **126**, 5601–5609.

Dr Nichols asked: Can you be more specific about the negative differential resistance (NDR) mechanism for the azurin junction?

Dr Davis responded: It is currently very difficult to be specific about NDR mechanisms as there is no general consensus or practically-accessible way of confirming a model. These spectroscopic features are inherently stochastic and the potential at which they appear is also somewhat variable (though these facts are, of course, understood). We, and others, have observed "NDR" in adlayers which have redox-accessible states and accordingly assign the associated spectroscopic features to resonance between the Fermi states and molecular DOS. In the case of azurin, specifically, 'NDR' is observed most frequently when the tip is negatively biased; we assign this to resonance between the probe Fermi level and the Cu(II) DOS. When the current–voltage trends are symmetric it appears that both electrodes are able to access this DOS.

Professor Lindsay commented: The existence of a peak in the DOS on a metal surface is, in general, implausible (it might exist for some very special geometries). However, NDR can also arise from reactions with *e.g.* ambient oxygen. Were your data reversible?

Dr Davis said: I am not sure I entirely agree that DOS peaks are implausible or necessarily geometry-specific. Specific examples of bonding at surfaces can generate "DOS spikes". If one takes the view that NDR, under these conditions at least, is redox-related then an "EC mechanism" may result in irreversible electrochemical characteristics and a subsequent drop in (resonant) tunnel current. I am not sure this adequately covers all published observations of redox-ascribed NDR in the literature. We are currently looking at the reversibility of these spectroscopic features but I can say, in our hands, irreversibility does not appear to be the default.

Dr Nichols asked: Can you comment on the symmetry of NDR effect at positive or negative tip bias in for instance Fig. 9 of the paper? Does the NDR effect remain with repeated cycling?

Dr Davis replied: I think we mention the symmetry of "NDR" about the voltage origin in the text. In essence, these features are observed predominantly at negative tip bias though, approximately 10–15% of the time they are observed at both bias potentials. In the latter case the symmetry about the voltage origin is by no means perfect. If we assign these features to resonance with molecular states then the electrode potential at which this occurs will be a sensitive function of field and this, of course, need not necessarily be the same for both tip and planar substrate.

We are currently looking at the stability of these spectroscopic features with respect to repeated cycling. It is clear that, with some redox systems, the response appears to be stable.

Dr Davis communicated a question on the subject of Dr Chi's paper: In Fig. 9, what is the SAM and associated rate of electron transfer? This is not clear from the text.

Dr Chi communicated in response: The SAM is a $CH_3(CH_2)_7S$ monolayer. With this anchor molecule, the ET rate between the azurin redox center and the Au(111) is *ca.* 500 s^{-1}. However, the ET rate between the azurin redox center and the tip is expected to be smaller but unknown (there is no straightforward way to measure this step directly).

Professor Tao opened the discussion of Dr Chi's paper: The electron transfer rate you extracted from the cyclic voltammograms is ~ 500 s^{-1}, which would give rise to a very small transport current through the molecule if one uses a naïve model to calculate it. In your paper, you have a more sophisticated model that relates the electron transport current to electron transfer rate. Have you estimated the magnitude of the current in addition to the shape of the current *vs.* potential curves?

Dr Chi replied: We have not done this in detail, but the current generated by a single-electron-transfer process at this rate (500 s^{-1}) is in principle very small. However, the single-molecule tunneling current could be significantly raised by the redox reaction, due to a possible multiple-electron transfer in the ECSTM configuration.

Professor Ulstrup addressed Dr Chi and Professor Tao: The relatively high tunnelling currents for azurin immobilized on alkanethiol modified Au(111)-electrode surfaces could appear puzzling in view of the modest values of the interfacial electrochemical rate constants. This is different from the behaviour of the Os-complexes reported in our paper, for which these rate constants are much higher. The fact that a "spectroscopic" tunnelling feature still appears for azurin could be a result of the confined environment in the *in situ* STM configuration. This difference from the semi-infinite electrochemical environment is more important for the composite monolayer/protein system than for the Os-complexes giving, perhaps, a smaller environmental reorganization free energy or longer environmental relaxation time. For such effects, there is, however, presently no direct evidence. They could increase the interfacial rate constants and the tunnelling currents.

Professor Tao replied: The point regarding local environment in the STM measurement for the molecules is indeed interesting and important one, which needs to be studied. Some theoretical modelling from you or other groups will be a great help for those wish to investigate it experimentally.

Dr Chi added: Professor Ulstrup's comments are also interesting for further studies in the future.

Dr Larsson asked: When the enzyme operates in nature (photosynthesis) it docks carefully to the substrate and transfers electrons over quite a short distance (15–20 Å). You have tunnelling transfer over considerably larger distances.

1. Is it possible to measure such small tunnelling currents?

2. Would it be possible to make a "real wire" in protein? By that I mean a molecular wire that is capable of accepting or donating electrons, so that the electrons move in the wire?

Dr Chi replied: These two questions are very interesting.

1. It is definitely possible, and we are currently working on a similar issue to your question. However, to our best knowledge no successful example is available yet. This is due to more complicated processes involved in the case of enzymes, not only pure electron transfer but also other chemical reactions (*e.g.* substrate binding and catalysis). In addition, experimentally it is not easy to make a good electric contact between the enzyme molecule and the probe.

2. I am not sure what is the exact meaning of "a real wire". I believe the polypeptide chain is a real wire for electron-transfer (ET) in proteins. Many studies of long-range protein ET have shown that the ET pathway (particularly intramolecular ET) is optimized and proceeds along the specific amino acid sequence. Therefore, biomimetic synthesis can in principle prepare a real wire for ET.

Professor Lindsay asked: Does the use of a specific molecular anchor increase electron transfer enough to account for the fact that you see redox-mediated conductance changes while Dr Davis does not?

Dr Chi replied: I believe so, although additional causes cannot be ruled out completely. Our results show that the interfacial ET rate between the substrate and the protein molecule is crucial for direct observations of redox-gated tunneling resonance. This is in fact not only for the protein but also for other redox couples (*e.g.* the transition metal complexes, *cf.* Professor Ulstrup's paper).

Professor Schiffrin commented: It is difficult to see why the reorganisation energy should change with distance. The only possible reason is the influence of image charges on the metal, which are often ignored in electrochemical rate constants calculations. The changes observed must be unrelated to the reorganisation energy but to electron transfer through the bridge as indeed indicated in section 3.2 of the paper.

Dr Chi answered: This is a good point. The intrinsic reorganisation energy (λ) of protein ET should not change with the length of anchor thiols, but the environmental energy could change. What we observed is the total reorganisation energy including both. Interfacial ET rate of azurin anchored by short-chain thiols ($n < 6$) is in fact limited by the configuration-rearranged process. The phenomena were observed in the Cytochrome *c* case as well.

Dr Nichols commented: The modelling of Fig. 9A in Fig. 9B is an impressive fit. There are a lot of adjustable parameters in eqn. (8) of the paper (*e.g.* electrochemical potential profile or tip bias distribution, γ and ξ)—how do you choose realistic values for all the adjustable parameters?

Dr Chi replied: Many thanks for your generous appreciation. It is absolutely true that there are quite a lot of adjustable parameters in eqn. (8). We chose appropriate parameters in the modelling by both information obtained from the electrochemical measurements (*e.g.* reorganization energy and possible potential profile) and a trial-and-error way (*e.g.* to identify which parameter is most sensitive to the asymmetric observation).

Dr Davis asked: Can you elaborate on the apparent sensitivity of image contrast with NiR on substrate binding? Though the voltammetry is clearly sensitive to nitrite, the coupling of the binding site to the electrode is likely to be weak whether or not nitrite is present. Is there any evidence for an electrostatic, as opposed to resonance, effect?

Dr Chi replied: That the presence of substrate nitrite is essential for interfacial ET detection and STM imaging seemingly implies that the substrate-binding enhances the electronic coupling between NiR and the working electrode (Au(111)), which in turn raises the ET rate. If we compare this to the Azurin case, apparently it is understandable. However, the detail of studies of the NiR case is not comparable with the Azurin case. Before we can make an unambiguous conclusion, therefore, several respects need to be improved in the near future as we described in section 4 of the paper.

Professor Lindsay asked a general question: A large range of parameters have been used in the various experiments reported at this meeting, so one might think that this included conditions where conversion between oxidized and reduced forms was slow, yet none of the data presented show evidence of populations of different species. Could the transition on the electrode surface be continuous?

Dr Haiss responded: For redox transitions of molecules at metal surfaces, the exchange of electrons between the molecule and the metal surface is fast compared to the sampling time of the current in an STM. For that reason I would expect that only the time averaged ratio between oxidised and reduced forms of the molecule can be seen in $I(V)$ or $I(E)$ spectroscopy performed with an STM and hence the redox transition appears continuous in this type of experiment. In addition, broadening of the redox transition may occur due to the presence of a second metal surface in close proximity to the molecule and due to the presence of chemical contact on both sides of the molecule.

Professor McCreery commented: We see a significant red shift for chemisorbed molecules bonded to graphitic carbon. This red shift and broadening implies significant modification of HOMO and LUMO by the surface. Also, we have observed a broadening of the nitroazobenzene reduction on a glassy carbon surface, compared to that in solution.[1] The molecular electronics field needs to be concerned about strong coupling between the conductor and the molecule, as that is very desirable for real applications.

1 T. Itoh and R. L. McCreery, *J. Am. Chem. Soc.*, 2002, **124**, 10894–10902.

Professor Kuznetsov commented: The reasons for the impossibility of the inverted region in ordinary electrochemical systems and in STM systems under discussion are clear. The maximum of tunnelling current near equilibrium potential is independent of the value of reorganisation energy (if not extremely small) and of the coupling with the electrodes provided the molecule has two stable states, oxidized and reduced (at a given overpotential). This is a rather general result emerging from the theory.

Professor Naaman added: There is a difference between the optical HOMO–LUMO gap and the electric one, due to the effect of polarization. In the case of charged molecule near the surface, due to image charge, the energetic of the LUMO will be significantly modified.

Dr Davis communicated: The authors have acquired some beautiful images and contrast variance and I congratulate them on the quality of this. Can they clarify the basis of the comment near the end of section 3.3 that "at least an order of magnitude difference between the two transmission coefficients is required".

Dr Chi communicated in reply: We would like to thank you for your generous appreciation. In the ECSTM configuration of the present system, the two-step ET model (provided a negative bias is applied) that the first ET step proceeds between the tip and the azurin redox center, and the second step ET between the azurin redox center and the working Au(111) electrode. The modelling, consistent with experimental observations, suggests that the first step is rate-limiting process (*i.e.* a much smaller transmission coefficient which is defined as usual).

Dr Davis communicated: The authors state that tip electronic coupling is weak and (importantly) rate limiting. If this is the case can they explain the clear dependence of conductance on coupling to the substrate (with its associated variance on surface potential)?

Dr Chi communicated in response: What we can say right now is that the electronic coupling between the redox center and the working electrode seems to play more crucial roles in determining redox-gated tunneling resonance. However, there are some details remaining open at present.

Nanoscale aryleneethynylene oligomers incorporating fluorenone units as electron-dopable molecular wires

Changsheng Wang, Andrei S. Batsanov and Martin R. Bryce*

Received 12th May 2005, Accepted 13th June 2005
First published as an Advance Article on the web 29th September 2005
DOI: 10.1039/b506712j

The first general synthetic procedures to obtain fluorenone-containing aryleneethynylene oligomers have been developed. The strategy involves stepwise Sonogashira cross-coupling methodology, with 2,7-diethynylfluorenone and 1,4-diiodo-2,6-di(hexyloxy)benzene as key building blocks, with terminal benzenethiol functionality protected as cyanoethyl derivatives. A second family of compounds contains a central 9-[(4-pyridyl)methylene]fluorene or 9-[di(4-pyridyl)methylene]fluorene unit in the backbone. UV-Vis absorption studies in solution establish that the increase of molecular lengths from *ca.* 4 nm (compounds **15** and **16**) to *ca.* 7 nm (compounds **17**, **27** and **28**) results in only a very small red shift, with the effective conjugation length slightly longer than that of the 1,4-di(phenylethynyl)benzene (PEPEP) subunits. The pyridyl groups in **27** and **28** are only weakly conjugated to the π-electron backbone (UV-Vis data) which is consistent with the twisted conformation observed in the X-ray crystal structures of model compounds **20** and **24**. Cyclic voltammetric studies reveal that the reduction waves of the fluorenone, 9-[(4-pyridyl)methylene]fluorene and 9-[di(4-pyridyl)methylene]fluorene units endow these oligomers with *n*-doping characteristics, with reversible reduction waves being observed for some compounds. The new aryleneethynylene oligomers have potential applications as molecular wires in practical devices.

Introduction

Molecular-scale electronics comes to the rescue

The microelectronics industry is approaching an intrinsic limit for the miniaturisation of conventional, solid-state, semiconductor components.[1] In the mid-1960s, Gordon Moore predicted that the minimum device feature size on a semiconductor chip would halve approximately every 18–24 months and this has proved to be the case (and is now known as Moore's Law). Molecular-scale electronics adds an exciting new dimension as it involves entirely different concepts from traditional inorganic electronics.[2] Chemical synthesis provides precisely-structured nanometre-sized organic molecules with tailored optoelectronic properties. These molecules can be functionalised for interconnecting into an electronic circuit using techniques such as self-assembly at metal or semiconductor surfaces with the goal of providing ultra-dense, low-power, low-cost circuits. The idea that a few organic molecules, or even a single molecule, when aligned between electrodes would behave as a diode with the organic bridge acting like a wire, was first proposed by Aviram and Ratner in 1974.[3]

Single molecules have the advantage of being *ca.* 10^6 times smaller in area than modern silicon devices, thereby offering great advantages as key components in future computational devices. Synthetic chemistry gives unprecedented functional group diversity, and it is possible to prepare *ca.* 1 mol of many electroactive molecules under laboratory conditions, all with identical structures, corresponding to *ca.* 10^{23} devices!

Department of Chemistry and Centre for Molecular and Nanoscale Electronics, University of Durham, Durham, UK, DH1 3LE . E-mail: m.r.bryce@durham.ac.uk

Within contemporary molecular electronics,[4] molecules which display wire-like behaviour are at the forefront of attention. The term "molecular wire" describes an organic or organometallic system which facilitates intramolecular electron or charge transport from one site to another, usually under the control of an external electrical, electrochemical or photonic stimulus.[5] Functional, monodisperse, π-conjugated monomers and oligomers with rod-like backbones, which are available in incremental lengths from *ca.* 1 to >10 nm, are of particular interest from experimental and theoretical viewpoints. Crucial for the advent of working devices are the new fabrication methods and probes which enable the molecules to be connected in a controlled manner into hybrid organic/semiconductor device architectures.[6] While synthetic chemistry can provide organic wires with reliable control over HOMO–LUMO levels suitable for conduction over distances of several nanometres,[7] there is considerable debate over the extent and mode of mixing of the orbitals at the termini of the wire with those at the surface of the metal contacts.[8] To enable the covalent anchoring of wire molecules to metal surfaces, thiols (or protected thiols) are the most widely-used terminal groups,[9] although there are viable alternatives which have received less attention, *e.g.* nitriles,[10] isocyanides[11] and pyridyl units.[12]

The versatility of ethynylated aromatic bridges

Ethynylated aromatic/heteroaromatic systems, with the generic structure $(\text{–aryl–C}{\equiv}\text{C–})_n$, are versatile components for advanced materials applications.[13] Predictably, the systems with lower aromatic stabilisation energy, *e.g.* thiophene, favour more extensive delocalisation through the bridge, and display stronger intramolecular coupling effects.[14] Specific examples include: (a) 2,5-thienylethynylene oligomers of up to 10 nm length, with protected thiol termini, for conduction studies;[15] (b) *p*-phenyleneethynylene oligomers with 5,5′-diethynyl-2,2′-bipyridine moieties in the backbone for studies of photoluminescence derived from MLCT states;[16] (c) combined 5,5′-diethynyl-2,2′-bipyridine and thiophene units for electrochemical and spectroscopic studies;[17] (d) oligo(9,9-dihexyl-2,7-fluorenylethynylene)s as blue emitters for light emitting devices;[18] (e) co-oligomers possessing phenylene- and heterocyclic-enediyne units in the backbone which are strongly fluorescent;[19] (f) a poly(*p*-phenyleneethynylene) derivative bearing macromolecular polyester substituents which shows unusual thermochromic behaviour;[20] (g) co-crystals of selectively fluorinated and non-fluorinated 1,4-bis(phenylethynyl)arenes which display liquid crystalline behaviour;[21] (h) ethynylene derivatives of porphyrins,[22] tetrathiafulvalenes[23] and organometallic complexes[24] which are redox-active and possess unusual optoelectronic and structural properties.

An attraction of aryleneethynylene motifs as molecular wires in metal–molecule–metal configurations is that they are more rigid π-systems and lack *Z*/*E* isomerism, compared to their arylenevinylene counterparts. However, the barrier to rotation about the aryl–ethynyl bond in aryleneethynylene units is very low, estimated at <1 kcal mol^{-1}, so although co-planar conformations of the arylene rings (sometimes stabilised by strong intermolecular π-stacking interactions) are generally observed in the crystalline state, their structures in solution or in molecular junctions may be less well-ordered.[25]

Harnessing redox-activity in molecular wires

Reversible redox chemistry should be a desirable feature for a molecular wire, as *irreversibility* implies that the molecule either decomposes or undergoes a chemical reaction under bias: in either case a molecular junction could be more easily broken when a charge is injected, thereby reducing the stability and operating lifetime of the device. Tour *et al.*,[26] and more recently Tao *et al.*[27] noted that a nitro substituent facilitates the electrochemical reduction of oligo(phenyleneethynylene)s, a factor which may contribute to the negative differential resistance (NDR) observed in self-assembled monolayers (SAMs) of compounds **1a** and **1b** (Chart 1) (although more recent data for **1a** have been ascribed to nanofilamentary metal switching mechanisms).[28] Functioning single-molecule (SM) transistors have been fabricated using electrochemically-reversible molecules, *e.g.* C_{60}[29] and an organocobalt complex.[30] Schiffrin *et al.* elegantly exploited the two-terminal 4,4′-bipyridinium derivative **2** tethered between gold electrodes. The conductance of single molecules of **2**, monitored as a function of an applied electrode potential, increased upon one-electron reduction to the radical cation. These data were explained by the increase in electron density at the

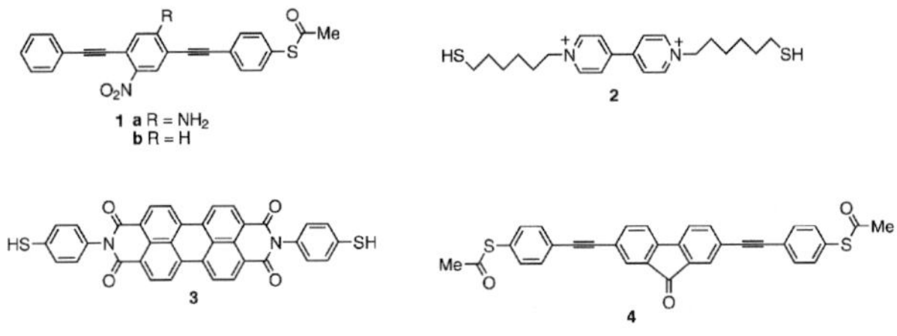

Chart 1

bipyridinium moiety upon reduction, concomitant with the alignment of the Fermi level of the gold electrode relative to the LUMO of the bipyridinium group, *i.e.* resonant tunelling.[31] More recently, Tao *et al.* have shown that the current through the n-channel perylene tetracarboxylic diimide molecule **3** attached to gold electrodes can be reversibly controlled over nearly three orders of magnitude by utilising a gate electrode.[32] The transistor properties of an oligo(phenylenevinylene) oligomer in various redox states (n- and p-doped) have been reported, using chemical vapour deposition to place the molecules in a gap between the source and drain electrodes.[33] Even though redox-switchable integrated nanoelectronic circuits may not be commercially-viable in the foreseeable future, these studies are important for understanding the fundamental relationship between structure and electronic function in device architectures.

From this viewpoint we have explored the incorporation of 9-fluorenone units into the backbone of aryleneethynylene oligomers. Fluorenone possesses the desirable features of good chemical stability to a range of synthetic transformations, combined with the reversible formation of radical anions at accessible potentials under electrochemical conditions. Two other groups have reported 2,7-diethynylfluoren-9-one derivatives. Bunz *et al.* studied the luminescent properties of poly(*p*-phenyleneethynylene) and poly[9,9-bis(dodecyl)-2,7-fluorenylethynylene] derivatives doped with a low level (2–10%) of fluorenone units.[34] Tour *et al.* incorporated a single 2,7-diethynyl-fluoren-9-one unit into short wire-type molecules, *e.g.* **4**.[35]

We now report the first general procedures for the synthesis of fluorenone-containing aryleneethynylene oligomers of *ca.* 4 nm and 7 nm lengths (compounds **15**, **16** and **17**) with protected thiol functionality for assembly onto metal electrodes.[36] A second series of 7 nm molecules is reported which possess additional pendant pyridyl substituents (compounds **27** and **28**). The pyridyl units could serve as a prototype intramolecular gating feature (by virtue of protonation or metal coordination to the nitrogen atoms) or as an additional anchor point in a three-terminal device. Solution spectroscopic and redox properties of the new wire molecules are discussed, along with three X-ray crystal structures (compounds **15**, **20** and **24**).

Results and discussion

Synthesis

Our general approach uses an iterative palladium-mediated Sonogashira strategy,[37] with compound **7** serving as the key fluorenone building block for the compounds shown in Scheme 1. As reported previously,[35] compound **7** was readily-obtained from 2,7-diiodofluorenone **5** *via* the bis(trimethylsilylethynyl) intermediate **6**. The *n*-hexyloxy chains of compound **8** ensured good solubility in organic solvents at all stages of the sequences shown in Schemes 1 and 2. Two-fold cross-coupling reaction of **7** with **8** (optimised molar ratio of 1 : 5) gave the product **9** in 43% yield. For access to terminal thiol functionality, we chose the cyanoethyl-protected reagent **13** which was readily obtained in two steps from 4-iodothiophenol.[38] This protecting group has been widely used for tetrathiafulvalene thiolate derivatives,[39] but is essentially unexplored in other contexts. A disadvantage of using an acetyl protecting group for thiophenol derivatives is that it can be too easily removed, giving rise to unwanted side reactions during multi-step syntheses. We note that Tour *et al.* observed that hydrolysis of the thioacetate group of **1b** under basic conditions prior to SAM formation lead to oxidation of the S atoms and reduction of the nitro group to amine, such that *ca.* 30% of the derived SAM was an amino-substituted derivative.[40] The cyanoethyl group can be removed from protected thiophenol derivatives under mild conditions (sodium methoxide or

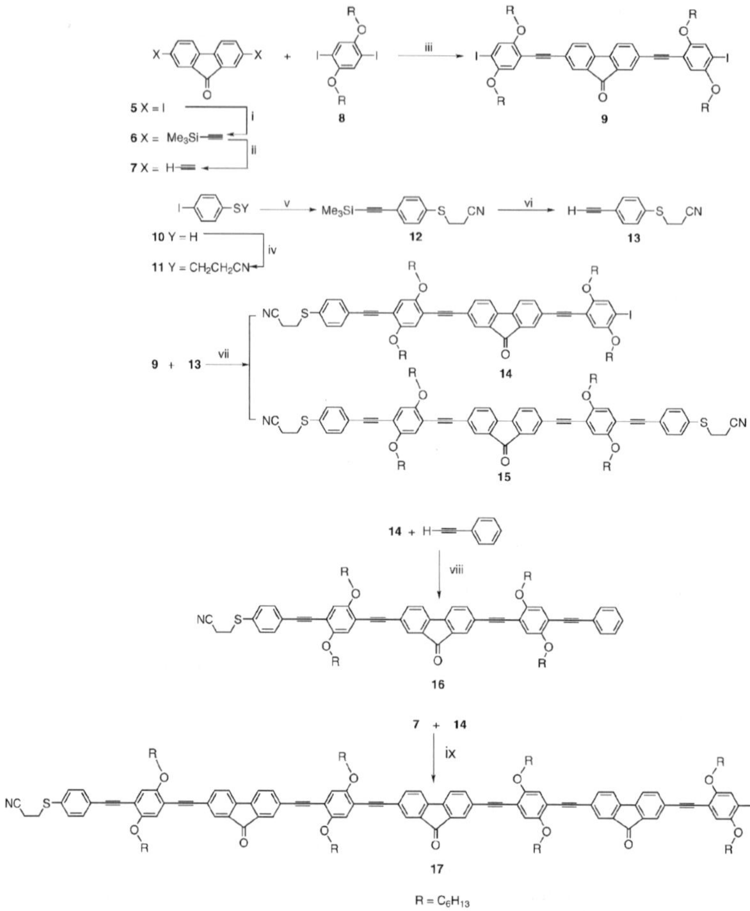

Scheme 1 *Reagents, conditions and yields*: *i*, Trimethylsilylacetylene (TMSA) (4 eq.), THF, $[PdCl_2(PPh_3)_2]$ (3%), CuI, TEA, rt 2 h, then 50 °C, 2 h, 98% yield; *ii*, 1 : 1 (v/v) DCM-methanol, K_2CO_3, rt 12 h, 76% yield; *iii*, **7** and **8** (5 eq.), THF, $[PdCl_2(PPh_3)_2]$ (3%), CuI, TEA, rt 12 h, then 50 °C, 2 h, 43% yield; *iv*, **10**, 3-bromopropionitrile, dry DMF, K_2CO_3, 100 °C, 24 h, 93% yield; *v*, **11**, TMSA (1.5 eq.), THF, $[Pd(PPh_3)_4]$ (3%), CuI, TEA, rt 3 h, 88% yield; *vi*, HOAc, Ac_2O, TBAF, rt, 0.5 h, 76% yield; *vii*, **9**, **13**, THF, $[PdCl_2(PPh_3)_2]$, CuI, TEA, rt 12 h, 50 °C, 2 h, 49% (for **14**) and 26% (for **15**); *viii*, **14**, phenylacetylene, THF, $[PdCl_2(PPh_3)_2]$, CuI, TEA, rt 12 h, 50 °C, 2 h, 98%; *ix*, **7**, **14**, THF, $[Pd(PPh_3)_4]$, CuI, TEA, rt 0.5 h, then 50 °C, 3 h, 54%.

tetrabutylammonium fluoride in THF at room temperature).[36] The increased stability of the 2-cyanoethylthio group (compared to thioacetate) is advantageous in the Sonogashira reactions in Schemes 1 and 2, enabling the use of higher temperatures, longer reaction times and increased amounts of triethylamine.

The next step was to apply the cross-coupling protocol to a mixture of **9** and **13**. Two products, *viz*. **14** and **15**, resulting from mono- and di-coupling processes, respectively, were obtained in 49% and 26% yields. We did not attempt to optimise the formation of **15** as separation of the two products was straightforward and compound **14** was used as a key reagent in Scheme 2; see below.

The second series of compounds is depicted in Scheme 2. The key feature here is that one or two pyridyl substituents are incorporated as pendant groups in the central unit of the wires **27** and **28**. The pyridyl units were attached to the fluorene core at the initial step of the synthesis, using either 4-pyridinecarboxaldehyde or di(4-pyridyl)ketone in a base-initiated condensation with 2,7-dibromo- or 2,7-diiodofluorene. For subsequent steps the strategy is conceptually similar to that in Scheme 1, with mono-iodo compound **14** serving in two-fold reactions with the 2,7-diethynylfluorene reagents **23** and **26** to give the target molecules **27** and **28**, respectively.

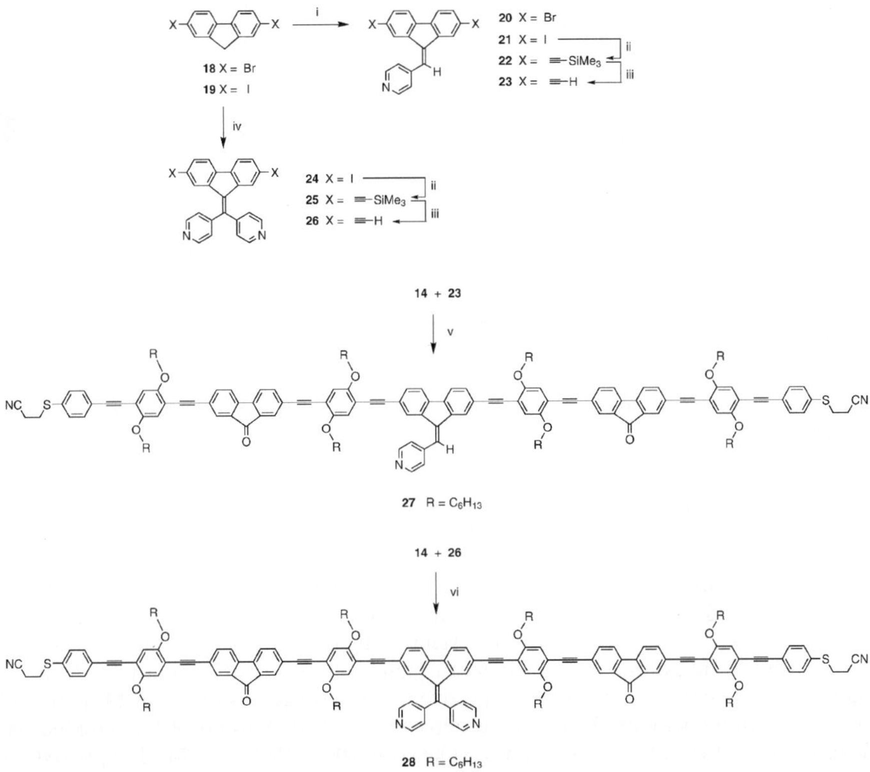

Scheme 2 *Reagents, conditions and yields*: *i*, **18**, 4-pyridinecarboxaldehyde, ethanol, NaOH (aq.), rt 4 h, then 0.5 h reflux, 86% yield of **20**; **19**, KOBu^t, 4-pyridinecarboxaldehyde, THF, rt, 80% yield of **21**; *ii*, **21**, trimethylsilylacetylene (TMSA) (4 eq.), THF, [$PdCl_2(PPh_3)_2$] (3%), CuI, TEA, rt 12 h, then 50 °C, 0.5 h, 96% yield; *iii*, 1 : 1 (v/v) DCM-methanol, K_2CO_3, rt 2 h, 95% yield; *iv*, **19**, KOBu^t, di(4-pyridyl)ketone, THF, rt 0.5 h, then 50 °C, 1 h, 80% yield; *v*, **14**, **23**, THF, [$PdCl_2(PPh_3)_2$], CuI, TEA, rt 5 min, then 50 °C, 5 h, 41% yield (≡61% yield, excluding recovered **14**); *vi*, **14**, **26**, THF, [$Pd(PPh_3)_4$], CuI, TEA, rt 1 h, then 50 °C, 4 h, 40% yield (≡92% yield, excluding recovered **14**).

X-ray crystal structures of 15, 20 and 24

The X-ray crystal structure of compound **15** (Fig. 1)[36] revealed an almost planar conformation for the molecule, with all the rings parallel to within 13°. As expected, the 2,7-diethynylfluorenone unit is not linear, the two –C≡C-aryl-C≡C-aryl-S rods forming an angle of *ca.* 160°. The intramolecular S···S′ distance in the crystal structure of **15** is 37.4 Å. This value is close to the MM$^+$ calculated value of 35.8 Å, obtained using HyperChem 6.03. Considering the additional Au···S

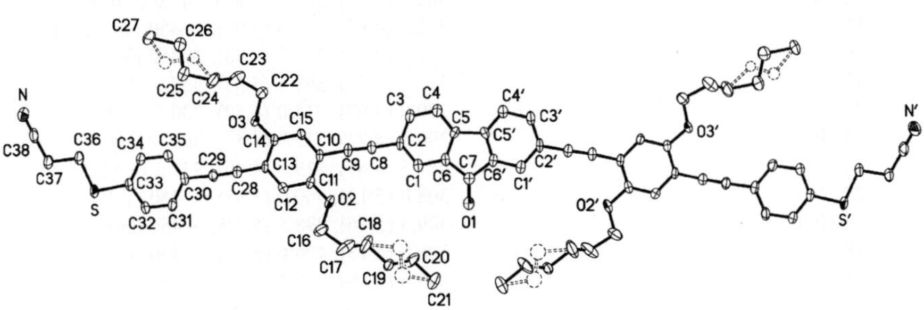

Fig. 1 Molecular structure of **15** determined by X-ray diffraction.

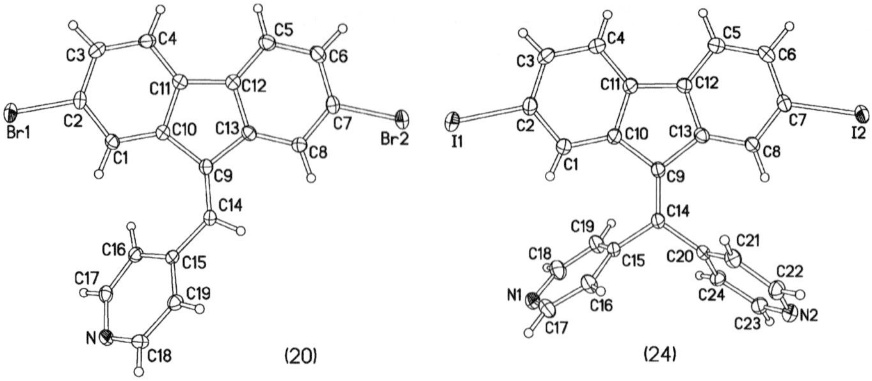

Fig. 2 Molecular structures of **20** and **24** determined by X-ray diffraction. Atomic displacement ellipsoids are drawn at the 50% probability level.

bonds in a Au···**15**···Au junction, a gap of *ca.* 4 nm would be required for two-terminal device fabrication. The S···S′ distances in **17**, **27** and **28** were calculated to be 68.6 Å, which allows a corresponding electrode gap of *ca.* 7 nm.

The crystal structures of **20** and **24** (Fig. 2) were obtained to determine the conformations adopted by these building blocks which provide the central core of the *ca.* 7 nm molecules **27** and **28**. For both **20** and **24** the fluorenylmethylene system is slightly warped. The carbon atoms C(1) to C(14) show average deviations from their mean planes of 0.02 Å in **20** and 0.05 Å in **24**, with maximum deviations of 0.05 and 0.09 Å, respectively. The stronger distortion in **24** is probably due to steric overcrowding between the pyridyl groups and the *peri*-H atoms of the fluorene, as is the slight lengthening of the C(9)=C(14) bond [1.345(3) Å in **20**; 1.358(3) Å in **24**]. The pyridyl rings are inclined to the coordination plane of the C(14) atom: in **20** by 46.0°; in **24** by 71.5° (ring i) and 62.3° (ring ii), the dihedral angle between the two pyridyl rings being 78.6°.

UV-Vis absorption spectra

The UV-Vis absorption spectra of chloroform solutions of representative compounds are collated in Table 1. The spectra of compounds **16**, **17**, **26** and **28** are shown in Fig. 3. There are three points worth noting.

(1) The significant increase of molecular lengths from *ca.* 4 nm (compounds **15** and **16**) to *ca.* 7 nm (compounds **17**, **27** and **28**) results in only a very small red shift. This saturation effect at relatively short conjugation lengths is similar to that observed in the oligo(9,9-dihexyl-2,7-

Table 1 Solution UV-Vis absorption data for compounds **14–28** in chloroform

Compound	λ_{max}/nm (log ε)
14	302.5 (4.76), 383.0 (4.88), 447.0 (4.07)
15	313.0 (4.85), 392.5 (4.95), ~450 (sh, 4.17)
16	311.5 (4.85), 392.5 (4.98), ~452 (sh, 4.18)
17	312.0 (5.15), 399.5 (5.33), 457.0 (4.75)
23	279.0 (4.67), 315.0 (4.52), 330.0 (4.59)
23-HCl	290.5 (4.72), 332.0 (4.41), 369 (4.20)
26[a]	281.5 (4.67), 319.0 (4.46), 331.0 (4.52)
27	308.0 (5.13), 400.0 (5.32), ~450 (sh)
27-HCl	309.5 (5.16), 398.5 (5.33), ~450 (sh)
28	308.0 (5.13), 400.0 (5.33), ~450 (sh)
28-HCl	309.5 (5.16), 398.5 (5.33), ~450 (sh)

[a] The salts of **26** were only very sparingly soluble in chloroform.

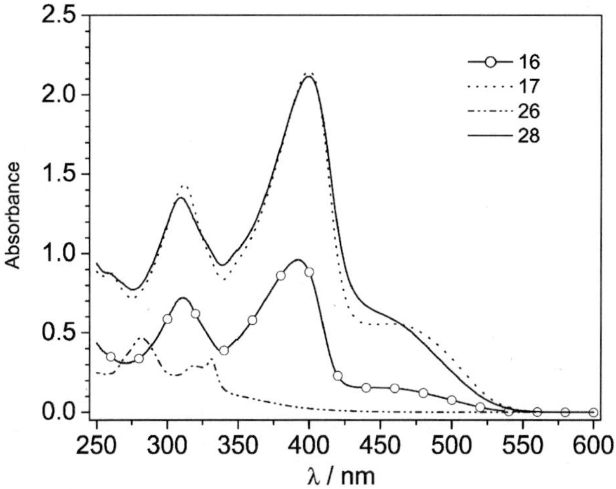

Fig. 3 UV-Vis absorption spectra of 1×10^{-4} M solutions of compounds **16**, **17**, **26** and **28** in chloroform (1 mm cell).

fluorenyleneethynylene) series.[18] Increasing the length of the molecules shown in Table 3 beyond *ca.* 4 nm will not iteratively red shift their absorption spectra: the *de facto* conjugation length in this class of oligomers is approximately the length of the 1,4-di(phenylethynyl)benzene (PEPEP) unit.

(2) The molar extinction coefficients of the molecules in Table 3 are proportional to the number of PEPEP repeating units contained in them. For instance, molecules **17**, **27** and **28** each possess four PEPEP moieties and solutions of these three molecules at the same concentration have almost superimposable absorption profiles. However, solutions of **15** and **16**, which possess two PEPEP units, absorb at half the intensities of **17**, **27** or **28** at the same concentration (Fig. 3).

(3) Replacing the central fluorenone oxygen atom of **17** with a (4-pyridyl)methylene group (**27**) or a di(4-pyridyl)methylene group (**28**) has only a very small effect on the two primary absorption bands at *ca.* λ_{max} 310 and 400 nm. Similarly, the addition of gaseous HCl to the solutions of **27** and **28**, which protonates the pyridine nitrogen(s) (see CV data below) caused no significant change in the absorption spectra. This can be explained by the weak conjugation of the pyridyl group to the π–electron backbone due to the twisted conformation, as revealed in the crystal structures of model compounds **20** and **24** (see the discussion above). Nevertheless, these structural changes and protonation did affect the ketone n–π* transition at *ca.* 450 nm (Fig. 3).

Table 2 Reduction peak potentials of 9-fluorenone, compounds **7**, **15–17**, **23**, and **26–28**[a]

Compound	E_{red} (V)
9–Fluorenone	−1.74 (rev)
7	−1.51 (rev), −2.25 (rev)
15	−1.47 (irr), −2.10 (irr), −2.23 (irr)
16	−1.48 (rev), −2.13 (irr), −2.36 (irr)
17	−1.52 (rev), −2.30 (irr)
23	−1.75 (irr), −2.12 (irr)
23–HCl	−1.12 (irr), −1.69 (irr), −2.46 (irr)
26	−1.53 (rev), 1.70 (rev)
26–HCl	−0.89 (irr), −1.48 (rev), −1.60 (rev), −2.46 (irr)
27	−1.49 (irr), −2.17 (irr), −2.42 (irr)
27–HCl	−0.88 (irr), −1.50 (rev), −2.19 (irr)
28	−1.51 (rev), −2.12 (irr), −2.41 (irr)
28–HBF_4	−0.91 (irr), −1.39 (irr), −1.75 (irr) −2.13 (irr)

[a] For experimental conditions see the Experimental section. Solvent: DMF, except compound **17** (DMF-THF mixture, 2 : 1 v/v).

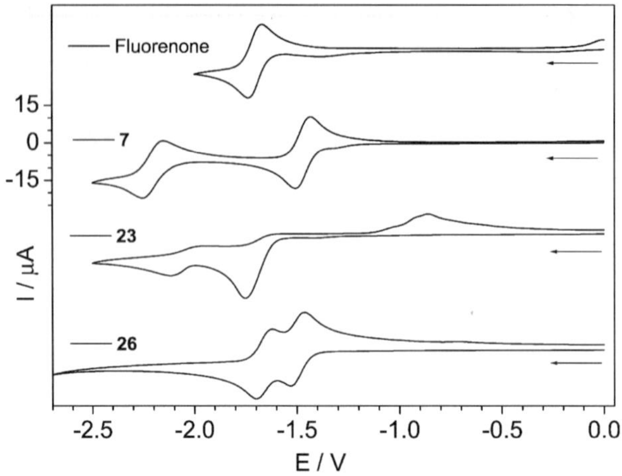

Fig. 4 Cyclic voltammograms of diethynyl building blocks **7**, **23** and **26**. The CV of 9-fluorenone, under the same experimental conditions, is shown for comparison.

Solution electrochemistry

The redox chemistry in solution was studied by cyclic voltammetry (CV) and the data are presented in Table 2 and Figs. 4–6. The fluorenone units endow these compounds with *n*-doping characteristics. A clear and predictable trend is that extending the conjugation with ethynyl and arylenethynyl substituents leads to a positive shift in the reduction potential compared to 9-fluorenone, which reversibly forms the radical anion at E_{red} −1.74 V (*vs.* Ag/Ag^+). This is evident in Fig. 4 by a comparison of fluorenone with compound **7**: the latter compound showed two one-electron reductions. The radical anion peak is shifted by 230 mV; the second wave at more negative potentials is assumed to be associated with reduction of the terminal ethynyls. Similar data were obtained for the π-extended wires **15–17**, although for all these systems the second and third waves are irreversible, probably involving reduction processes at the alkyne bonds.

The introduction of central 4-pyridyl units resulted in significant differences in the CVs of **23** and **26**. For **23** both the reduction waves of the fluorenone analogue and the terminal ethynyls became irreversible due to the presence of the (4-pyridyl)methylene moiety. Obviously, this C=C bond is harder to reduce than a carbonyl, and the derived methylene radical anion is less stable, possibly undergoing side-reactions at the exocyclic methylene carbon. In contrast, the dipyridyl analogue **26**

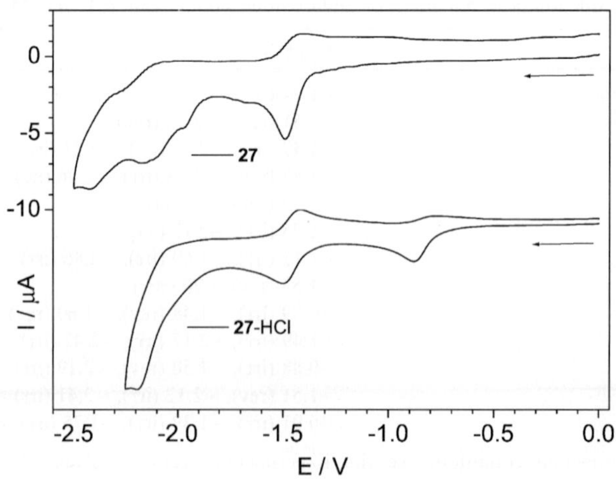

Fig. 5 Cyclic voltammograms of compound **27** and its protonated derivative.

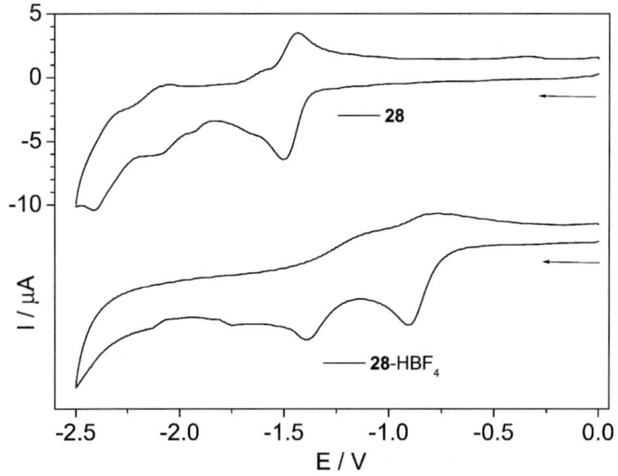

Fig. 6 Cyclic voltammograms of compound **28** and its protonated derivative.

displays two, fully-reversible, single-electron reductions (Fig. 4). The stabilized dianion (presumably comprising a fluorenyl anion and a dipyridylmethylene anion species) predictably makes the ethynyl units harder to reduce and hence no further reduction waves were observed on scanning cathodically to −2.75 V (*vs.* Ag/Ag$^+$). For the pyridyl-functionalised wires **27** (Fig. 5) and **28** (Fig. 6) the redox behaviour is less clean at more negative potentials than *ca.* −1.6 V, where irreversible processes were observed. For compounds **15**,[36] **17**, **27** and **28** if the reductive scan was limited to −1.8 V the first wave was cleanly reversible in all cases. Upon protonation of the pyridyl moieties of **23**, **27** and **28** with HCl or HBF_4 a new irreversible wave appeared at considerably more positive potentials (by *ca.* 500–600 mV) compared to that of the neutral precursor. This is consistent with the one-electron reduction of the pyridinium moieties.

Table 3 Crystal data

Compound	**20**	**24**
Formula	$C_{19}H_{11}Br_2N$	$C_{24}H_{14}I_2N_2$
Formula weight	413.11	584.17
T/K	120	120
Symmetry	Orthorhombic	Monoclinic
Space group	$Pca2_1$ (#33)	$P2_1/n$ (#14)
a/Å	33.409(3)	12.903(1)
b/Å	3.9299(3)	8.684(1)
c/Å	11.400(1)	19.495(2)
β/°	90	107.40(1)
V/Å^3	1496.8(2)	2084.4(4)
Z	4	4
μ/mm^{-1}	5.41	3.03
Refls collected	21504	43989
Unique refls	4367	7539
R_{int} (%)	8.9, 3.7[a]	11.4, 3.2[a]
Refls $I > 2\sigma(I)$	4204	6099
$R(F)$, $I > 2\sigma(I)$	2.2%	2.7%
w$R(F^2)$, all data	5.2%	7.2%

[a] After absorption correction.

The absolute values of redox potentials will differ between solution phase, traditional SAM and single molecule device configurations due to the very different molecular environments. For the latter two it is necessary to consider that desorption processes may occur. Recent studies on reductive desorption of thiolates on gold surfaces show that the potential for the onset of desorption differs substantially depending on the structure of the gold surface.[41] The reduction potentials of the fluorenone systems reported herein lie close to the edge of the potential window for gold-thiolate SAMs, although there are precedents for clean reduction processes of SAMs of other redox-active moieties in this cathodic region. For example, Mirkin *et al.* observed reversible reduction waves for SAMs of a thiolated-fullerene derivative on gold at $E_{1/2}$ −1.03 V and −1.56 V (in THF, *vs.* Fc/Fc^+). Both waves were ascribed to surface-confined species.[42] Similarly, Echegoyen *et al.* have recently reported reversible reduction processes for *non*covalently immobilized C_{60} ($E_{1/2}$ −0.93 V and −1.34 V in 1,2-dichlorobenzene, *vs.* Ag/Ag^+; ≡−0.97, −1.38 V *vs.* Fc/Fc^+) in SAMs of thiolated-cyclotriveratrylene derivatives on gold surfaces.[43] The waves were assigned to surface-confined species. The authors noted that an unusual feature of these data was the very close similarity between the potentials of the surface-confined C_{60} and the solution values, and considered a possible contribution from intermolecular donor-acceptor interactions.

Clearly there is scope for further detailed studies of the electrochemically-controlled desorption processes of organics with varying anchor groups on a range of surfaces. A recent experimental study has concluded that a benzeneselenolate monolayer was more stable than a benzenethiolate monolayer on gold: the reductive desorption peak of benzeneselenolate was 0.21 V more negative than that of benzenethiolate.[44]

Conclusions

This work comprises the first systematic study involving the incorporation of fluorenone, 9-[(4-pyridyl)methylene]fluorene or 9-[di(4-pyridyl)methylene]fluorene as electron-accepting units in aryleneethynylene oligomers. New highly-functionalised synthons, *e.g.* **9**, **14** and **26**, have been utilised in the construction of a range of nanoscale wires of *ca.* 4 nm (compounds **15** and **16**) and *ca.* 7 nm lengths (compounds **17**, **27** and **28**). Long alkoxy chains attached to the phenyl units circumvent the solubility problems often encountered with extended rigid rod aryleneethynylene systems. A special feature of our methodology is the use of the cyanoethyl protecting group for the benzenethiol termini. This end-group served well in the synthetic steps in Schemes 1 and 2: notably it is entirely compatible with Sonogashira cross-coupling conditions, and may become established as a versatile alternative to the traditional thioester protecting group. CV data established that fluorenone and 9-[di(4-pyridyl)methylene]fluorene are suitable electron acceptor components for *n*-dopable molecular wire structures, whereas 9-[(4-pyridyl)methylene]fluorene appears to be less suitable due to its irreversible electrochemical reduction properties. The thiol termini will provide the anchor points in two-terminal devices. Moreover, there is a fascinating possibility that the pyridyl substituents of **27** or **28** could serve as additional attachment points in three-terminal device structures.

Current work involves the characterisation of the electrical properties of these wires and their use as active components in two-terminal junction devices. Preliminary studies have been performed by our collaborators at Cranfield University. Compounds **15** and **17** were deprotected and assembled on gold substrates. Scanning tunneling microscopy of the monolayers contacted by a gold tip showed symmetrical current-voltage characteristics.[45] This work demonstrates the potential for these systems to act as molecular wires, and the results will be reported in detail in due course.

Experimental

General

Elemental analyses were obtained on a Carlo-Erba Strumentazione instrument. Melting points were determined in open-end capillaries using a Stuart Scientific melting point apparatus SMP3 at a ramp rate of 5 °C min^{-1} near the melting temperatures, without calibration. Solution ^{1}H NMR and ^{13}C NMR spectra were recorded on Varian Unity 300, Bruker Avance 400 and Varian Inova 500 spectrometers operating at (^{1}H) 299.91, 400.13, 499.99 and (^{13}C) 75.42, 100.62, 124.99 MHz respectively. Chemical shifts are reported in ppm downfield of TMS. EI mass spectra were obtained on a VG7070E instrument operating at 70 eV and ES+ spectra were obtained on a Micromass LCT spectrometer. MALDI-TOF spectra were obtained on an Applied Biosystems Voyager-DE STR using dithranol matrix operating in reflector mode. UV-Vis absorption spectra were recorded on a Cary 5E spectrometer. Cyclic voltammograms were recorded using a BAS CV50W electrochemical analyser and the potentials quoted were referenced to Ag/Ag^+ non-aqueous reference electrode, which contains a solution of 0.01 M $AgNO_3$ and 0.1 M $TBAPF_6$ in acetonitrile, unless otherwise stated in Table 2. Pt disk ($\varnothing$ = 1.6 mm) was used as the working electrode and a Pt wire was used as counter electrode. In all cases, the potentials were scanned at 100 mV sec^{-1}. The hydrochloric acid salts of **23**, **26**, and **27** used for CV measurements were prepared by bubbling HCl gas into the DCM solutions of the corresponding compounds, followed by evaporation of the solvent to afford a product which was then dried under high vacuum (*ca.* 2×10^{-2} mbar). The tetrafluoroborate salt of **28** was prepared in a similar way by adding tetrafluoroboric acid etherate into the DCM solution of **28** followed by evaporation and vacuum drying. DMF was purified with a PureSolv solvent purification system and argon was bubbled through the solutions for 20 min prior to CV measurements. Under these conditions, the background from 0 to −2.75 V was transparent to reductions and ferrocene was oxidised at a half-wave potential of −0.06 V.

THF for cross-coupling reactions was purified using a PureSolv solvent purification system. Triethylamine (TEA) was distilled over sodium in an argon atmosphere. $[PdCl_2(PPh_3)_2]$ was purchased from OMG AG & Co KG; $[Pd(PPh_3)_4]$ was prepared in our laboratory using a reported method.[46] Other chemicals and routine solvents were used as purchased without further purification. All reactions were performed under an argon atmosphere.

Synthetic procedures and characterisation data for selected compounds

Compound 9. Compound **7**[35] (0.57 g, 2.5 mmol) and 1,4-dihexyloxy-2,6-diiodobenzene **8**[47] (6.63 g, 12.5 mmol) were dissolved in THF (120 cm^3) and the solution was degassed by bubbling with argon for 0.5 h. $[Pd(PPh_3)_4]$ (175 mg), CuI (90 mg) and triethylamine (10 cm^3) were added sequentially and the mixture was stirred under Ar for 12 h at rt then 2 h at 50 °C. The resultant orange suspension was concentrated by vacuum evaporation and the solid residue was chromatographed on silica (eluent 1 : 1 v/v DCM/petroleum ether bp 40–60 °C) to afford **9** as an amorphous orange solid (1.10 g, 43%), which crystallised as orange needles after slow evaporation of its solution in DCM-hexane mixture; mp 106–107 °C. Anal. calcd for $C_{53}H_{62}I_2O_5$: C, 61.63; H, 6.05; found: C, 61.76; H, 6.11. ^1H NMR ($CDCl_3$, 400 MHz): δ 0.91 (m, 6 H), 1.36 (m, 8H), 1.53 (m, 4H), 1.83 (m, 4H), 3.97 (q, J = 7.2 Hz, 4H), 6.89 (s, 1 H), 7.30 (s, 1H), 7.46 (dd, J_{12} = 8.0 Hz, J_{14} = 0.8 Hz, 1H), 7.62 (dd, J_{12} = 7.8 Hz, J_{13} = 1.4 Hz, 1H), 7.77 (dd, J_{13} = 1.6 Hz, J_{14} = 0.8 Hz, 1H). ^{13}C NMR ($CDCl_3$, 100 MHz): δ 22.56, 22.61, 25.66, 25.71, 29.1, 29.2, 31.46, 31.52, 69.8, 70.0, 87.8, 88.2, 93.3, 112.9, 115.8, 120.5, 123.7, 124.6, 127.2, 134.4, 137.6, 143.1, 151.8, 154.3, 192.0. MALDI-TOF MS (*m*/*z*): 1032.21 (M^+, 100%).

Compounds 14 and 15. A mixture of compound **9** (0.86 g, 0.83 mmol), compound **13** (0.188 g, 1.00 mmol), $[PdCl_2(PPh_3)_2]$ (54 mg) and CuI (30 mg) in THF (5 cm^3) was stirred under argon to give a red-orange solution. Triethylamine (15 cm^3) was added and the mixture was stirred at rt for 12 h and then at 50 °C for an additional 2 h. The solvents were removed *in vacuo* and the red residue was column chromatographed on silica (eluent 3 : 1 v/v DCM/petroleum ether bp 40–60 °C) to obtain **14** (R_f 0.5) as an oil which on storage for a few days became a red amorphous solid (0.45 g, 49%) and **15** (R_f 0.3) as an orange solid (0.25 g, 26%).

14. mp 85–110 °C. Anal. calcd for $C_{64}H_{70}INO_5S$: C, 70.38; H, 6.46; N, 1.28; found: C, 70.16, H, 6.19; N, 2.00. ^1H NMR ($CDCl_3$, 400 MHz): δ 0.90 (m, 12H), 1.37 (m, 16H), 1.54 (m, 8H), 1.84 (m, 8H), 2.63 (t, J = 7.4 Hz, 2H), 3.17 (t, J = 7.2 Hz, 2H), 3.96–4.05 (m, J = 7.4 Hz, 8H), 6.90 (s, 1H), 7.007 (s, 1H), 7.009 (s, 1H), 7.31 (s, 1H), 7.36 (d, J = 8.4 Hz, 2H), 7.49 (d, J = 8.4 Hz, 2H), 7.47–7.52 (m, 2H), 7.64 (dd, J_{12} = 8.0 Hz, J_{13} = 1.6 Hz, 1H), 7.65 (dd, J_{12} = 7.8 Hz, J_{13} = 1.6 Hz, 1H), 7.80 (m, 2H). ^{13}C NMR ($CDCl_3$, 75 MHz): δ 14.0, 18.1, 22.56, 22.60, 22.62, 25.70, 25.72, 29.09, 29.19, 29.23, 29.6, 31.45, 31.51, 31.55, 69.5, 69.7, 70.0, 87.2, 87.8, 88.1, 93.3, 94.1, 94.2, 112.9, 113.5, 113.9, 115.8, 116.6, 116.7, 117.7, 120.5, 122.5, 123.6, 124.5, 127.13, 127.16, 130.2, 132.2, 133.8, 134.4,

137.58, 137.62, 143.1, 151.7, 153.5, 153.6, 154.3, 192.0. MALDI-TOF MS (*m*/*z*): 1091.4 (M^+, 100%). (Compound **14** was pure by ^{1}H NMR and ^{13}C NMR spectroscopy).

15. mp 116.2–116.8 °C; Anal. calcd for $C_{75}H_{78}N_2O_5S_2$: C, 78.22; H, 6.83; N, 2.43; found: C, 78.16; H, 6.75; N, 2.48. ^{1}H NMR ($CDCl_3$, 500 MHz): δ 0.91 (m, 6H), 1.38 (m, 8H), 1.56 (m, 4H), 1.86 (m, 4H), 2.63 (t, $J = 7.3$ Hz, 2H), 3.18 (t, $J = 7.3$ Hz, 2H), 4.04 (m, 4H), 7.01 (s, 1H), 7.02 (s, 1H), 7.37 (d, $J = 8.5$ Hz, 2H), 7.49 (d, $J = 8.5$ Hz, 2H), 7.53 (d, $J = 8.0$ Hz, 1H), 7.66 (dd, $J_{12} = 7.5$ Hz, $J_{13} = 1.5$ Hz, 1H), 7.82 (s, 1H). ^{13}C NMR ($CDCl_3$, 125 MHz): δ 14.1, 18.2, 22.64, 22.65, 25.72, 25.75, 29.3, 29.7, 31.56, 31.57, 69.55, 69.57, 87.2, 88.1, 94.1, 94.2, 113.5, 114.0, 116.7, 116.8, 117.8, 120.6, 122.6, 124.6, 127.3, 130.3, 132.3, 133.8, 134.5, 137.8, 143.2, 153.6, 153.7, 192.2. MALDI-TOF MS (*m*/*z*): 1150.6 (M^+, 100%).

Compound 17. A mixture of compound **14** (0.199 g, 0.182 mmol), compound **7** (20.5 mg, 0.09 mmol), [Pd(PPh_3)$_4$] (10 mg) and CuI (5 mg) and TEA (40 cm^3) was stirred at rt for 0.5 h and at 50 °C for an additional 3 h to afford a red–orange suspension. The solvent was removed *in vacuo* and the red solid residue was column chromatographed on silica (DCM eluent) to yield a red solid. The solid was dissolved in chloroform (5 cm^3) and the solution was added dropwise to ethanol (50 cm^3) with stirring. The suspension was boiled for 2 min then cooled to rt and suction filtered to afford compound **17** as a red solid (104 mg, 54%); mp 258–260 °C. Anal. calcd for $C_{145}H_{146}N_2O_{11}S_2$: C, 80.75; H, 6.82; N, 1.30; found: C, 80.37; H, 6.61; N, 1.16; ^{1}H NMR ($CDCl_3$, 500 MHz): δ 0.94 (m, 12H), 1.40 (m, 16H), 1.56 (m, 8H), 1.87 (m, 8H), 2.63 (t, $J = 7.5$ Hz, 2H), 3.17 (t, $J = 7.5$ Hz, 2H), 4.03 (m, 8H), 6.98 (s, 2H), 7.00 (s, 2H), 7.35 (d, $J = 8.5$ Hz, 2H), 7.48 (m, 5H), 7.62 (dd, $J_{12} = 7.5$ Hz, $J_{13} = 1.5$ Hz, 3H), 7.76 (s, 1H), 7.77 (s, 1H), 7.78 (s, 1H); ^{13}C NMR ($CDCl_3$, 125 MHz): δ 14.06, 14.08, 18.2, 22.65, 22.67, 22.69, 25.73, 25.76, 25.78, 29.3, 29.7, 31.59, 31.60, 69.5, 87.2, 88.2, 94.2, 94.3, 113.6, 113.7, 114.0, 116.65, 117.78, 120.5, 122.6, 124.7, 127.3, 130.3, 132.3, 133.8, 134.5, 137.7, 143.2, 153.63, 153.67, 192.2; MALDI-TOF MS (*m*/*z*): 2155.879 (M^+, 100%).

Compound 27. To the stirred mixture of **14** (0.594 g, 0.544 mmol), **23** (87 mg, 0.287 mmol), [$PdCl_2(PPh_3)_2$] (20 mg) and CuI (10 mg) in THF (10 cm^3) was added TEA (20 cm^3). The mixture was stirred at rt for 5 min then heated to 50 °C and stirred for 5 h to afford an orange suspension. The solvents were removed by vacuum evaporation and the orange solid residue was chromatographed on silica (eluent: 3% diethyl ether in DCM) to afford **27** as a red amorphous solid (0.25 g, 41% yield; ≡61% yield, excluding recovered **14**, 0.19 g). Compound **27**: mp 202–222 °C. Anal. calcd for $C_{151}H_{151}N_3O_{10}S_2$: C, 81.26; H, 6.82; N, 1.88; found: C, 81.09; H, 6.80; N, 1.94. ^{1}H NMR ($CDCl_3$, 400 MHz): δ 0.91 (m, 24H), 1.38 (m, 32H), 1.56 (m 16H), 1.85 (m, 16H), 2.63 (t, $J = 7.4$ Hz, 4H), 3.18 (t, $J = 7.4$ Hz, 4H), 4.04 (m, 16H), 6.96 (s, 1H), 6.99 (s, 1H), 7.01 (s, 4H), 7.02 (s, 1H), 7.05 (s, 1H) 7.36 (d, $J_{12} = 8.0$ Hz, 4H), 7.49 (d, $J = 8.4$ Hz, 4H), 7.50 (s, 1H), 7.53 (s, 4H), 7.56 (s, 2H), 7.60 (d, $J = 8.0$ Hz, 2H), 7.65–7.70 (m, 7H), 7.81 (s, 4H), 7.93 (s, 1H), 7.78 (d, $J = 4.0$ Hz, 2H). MALDI-TOF MS (*m*/*z*): $C_{151}H_{151}N_3O_{10}S_2$–$H^+$ requires: 2232.0953; found: 2232.1525.

Compound 28. By analogy with the synthesis of compound **27**, the mixture of **14** (0.251 g, 0.23 mmol), **26** (43 mg, 0.113 mmol), [Pd(PPh_3)$_4$] (15 mg) and CuI (5 mg) in THF (10 mL) and TEA (20 cm^3) was stirred at rt for 1 h then at 50 °C for 4 h. The crude product mixture was column chromatographed on silica (eluent: 1 : 4 v/v diethyl ether-chloroform) to afford **28** as a red amorphous solid (105 mg, 40% yield; ≡92% yield, excluding recovered **14**, 130 mg). Compound **28**: mp 143–163 °C. Anal. calcd for $C_{156}H_{154}N_4O_{10}S_2$: C, 81.14; H, 6.72; N, 2.43; found: C, 80.90; H, 6.74; N, 2.33. ^{1}H NMR ($CDCl_3$, 500 MHz): δ 0.92 (m, 12H), 1.37 (m, 16H), 1.56 (m, 8H), 1.85 (m, 8H), 2.62 (t, $J = 7.3$ Hz, 2H), 3.16 (t, $J = 7.3$ Hz, 2H), 4.02 (m, 8H), 6.82 (s, 1H), 6.89 (s, 1H), 6.95 (s, 1H), 6.99 (s, 2H), 7.34 (m, 4H), 7.48 (m, 5H), 7.63 (d, $J = 7.5$ Hz, 3H), 7.78 (s, 2H), 8.76 (d, $J = 5.5$ Hz, 2H). MALDI-TOF MS (*m*/*z*): 2309.0 ($M^+ + 1$).

Crystallographic studies. X-Ray diffraction experiments were carried out on a three-circle Bruker diffractometer with a SMART 6000 CCD area detector, using graphite-monochromated Mo Kα radiation ($\bar{\lambda} = 0.710\ 73$ Å) and a Cryostream open-flow N_2 cryostat (Oxford Cryosystems). Full sphere of reciprocal space was covered by three runs of narrow (0.3°) ω scans, each run at a different φ angle. Absorption corrections: for **20** by numerical integration based on crystal face-indexing (transmission $T_{max} = 0.713$, $T_{min} = 0.187$), for **24** by a semi-empirical method based on Laue equivalents ($T_{min}/T_{max} = 0.799$).[48] The structures were solved by direct methods and refined by full-matrix least squares against F^2 of all data, using SHELXTL software.[49] All non-H atoms were refined in anisotropic, H atoms in isotropic approximation. The absolute structure (polarity) of **20** was determined from anomalous scattering by Br atoms: Flack parameter −0.002(6).[50] Crystal data and experimental details are listed in Table 3. Full crystallographic data, excluding structure factors have been deposited at the Cambridge Crystallographic Data Centre, CCDC-276241 and 276242. See http://dx.doi.org/10.1039/b506712j for crystallographic data in CIF or other electronic format. (The data for **15** have been reported previously, CCDC-231673).[36]

Acknowledgements

We thank EPSRC and QinetiQ for funding this work.

References

1 M. Lundstrom, *Science*, 2003, **299**, 210.
2 *An Introduction to Molecular Electronics*, ed. M. C. Petty, M. R. Bryce and D. Bloor, Oxford University Press, Oxford, 1995. The terminology "molecular electronics" is often used very loosely in the literature. In this book we suggested two sub-categories, namely "molecular materials for electronics" which is relevant for bulk applications, and "molecular-scale electronics" for single molecule applications.
3 (*a*) A. Aviram and M. Ratner, *Chem. Phys. Lett.*, 1974, **29**, 277; (*b*) For molecules which exhibit molecular rectification under experimental conditions inspired by the Aviram and Ratner proposal, see: R. M. Metzger, B. Chen, U. Höpfner, M. V. Lakshmikantham, D. Vuillaume, T. Kawai, X. Wu, H. Tachibana, T. V. Hughes, H. Sakurai, J. W. Baldwin, C. Hosch, M. P. Cava, L. Brehmer and G. J. Ashwell, *J. Am. Chem. Soc.*, 1997, **119**, 10455; (*c*) G. Ho, J. R. Heath, M. Kondratenko, D. F. Perepichka, K. Arsenault, M. Pézolet and M. R. Bryce, *Chem. Eur. J.*, 2005, **11**, 2914.
4 Reviews on molecular electronics: (*a*) A. Nitzan and M. A. Ratner, *Science*, 2003, **300**, 1384; (*b*) J. R. Heath and M. A. Ratner, *Phys. Today*, 2003, **56**, 43; (*c*) G. Maruccio, R. Cingolani and R. Rinaldi, *J. Mater. Chem.*, 2004, **14**, 542; (*d*) R. A. Wassel and C. B. Gorman, *Angew. Chem. Int. Ed.*, 2004, **43**, 5120.
5 (*a*) N. Robertson and C. A. McGowan, *Chem. Soc. Rev.*, 2003, **32**, 96; (*b*) A. C. Benniston, *Chem. Soc. Rev.*, 2004, **33**, 573.
6 Reviews: (*a*) C. Joachim, J. K. Gimewski and A. Aviram, *Nature*, 2000, **408**, 541; (*b*) D. K. James and J. M. Tour, *Chem. Mater.*, 2004, **16**, 4423; (*c*) R. L. McCreery, *Chem. Mater.*, 2004, **16**, 4477.
7 D. M. Adams, L. Brus, C. E. D. Chidsey, S. Creager, C. Creutz, C. R. Kagan, P. V. Kamat, M. Lieberman, S. Lindsay, R. A. Marcus, R. M. Metzger, M. E. Michel-Beyerle, J. R. Miller, M. D. Newton, D. R. Rolison, O. Sankey, K. S. Schanze, J. Yardley and X. Zhu, *J. Phys. Chem. B*, 2003, **107**, 6668.
8 (*a*) M. Di Ventra, S. T. Pantelides and N. D. Lang, *Phys. Rev. Lett.*, 2000, **84**, 979; (*b*) J. Chen and M. A. Reed, *Chem. Phys.*, 2002, **281**, 127; (*c*) N. Agraït, A. L. Yeyati and J. M. van Ruitenbeek, *Phys. Rep.*, 2003, **377**, 81.
9 (*a*) J. M. Tour, A. M. Rawlett, M. Kozaki, Y. Yao, R. C. Jagessar, S. M. Dirk, D. W. Price, M. A. Reed, C. -W. Zhou, J. Chen, W. Wang and I. Campbell, *Chem. Eur. J.*, 2001, **7**, 5118; (*b*) L. Wei, K. Padmaja, W. J. Youngblood, A. B. Lysenko, J. S. Lindsey and D. Bocian, *J. Org. Chem.*, 2004, **69**, 1461.
10 S. M. Dirk and J. M. Tour, *Tetrahedron*, 2003, **59**, 287.
11 D. Bong, I. Tam and R. Breslow, *J. Am. Chem. Soc.*, 2004, **126**, 11796.
12 B. Xu and N. J. Tao, *Science*, 2003, **301**, 1221.
13 Reviews: (*a*) U. H. F. Bunz, *Chem. Rev.*, 2000, **100**, 1065; (*b*) J. M. Tour, *Acc. Chem. Res.*, 2000, **33**, 791.
14 S. Frayesse, C. Coudret and J. -P. Launay, *J. Am. Chem. Soc.*, 2003, **125**, 5880.
15 D. L. Pearson and J. M. Tour, *J. Org. Chem.*, 1997, **62**, 1376.
16 K. D. Ley, Y. Li, J. V. Johnson, D. H. Powell and K. S. Schanze, *Chem. Commun.*, 1999, 1749.
17 S. Goeb, A. De Nicola and R. Ziessel, *J. Org. Chem.*, 2005, **70**, 1518.
18 S. H. Lee, T. Nakamura and T. Tsutsui, *Org. Lett.*, 2001, **3**, 2005.
19 (*a*) N. F. Utesch and F. Diederich, *Org. Biomol. Chem.*, 2003, **1**, 237; (*b*) Y. Nakano, K. Ishizuka, K. Muraoka, H. Ohtani, Y. Takayama and F. Sato, *Org. Lett.*, 2004, **6**, 2373.
20 Y. Wang, B. Erdogan, J. N. Wilson and U. H. F. Bunz, *Chem. Commun.*, 2003, 1624.
21 (*a*) C. Dai, P. Nguyen, T. B. Marder, A. J. Scott, W. Clegg and C. Viney, *Chem. Commun.*, 1999, 2493; (*b*) T. M. Fasina, J. C. Collings, D. P. Lydon, D. Albesca-Jove, A. S. Batsanov, J. A. K. Howard, P. Nguyen, M. Bruce, A. J. Scott, W. Clegg, S. W. Watt, C. Viney and T. B. Marder, *J. Mater. Chem.*, 2004, **14**, 2395.
22 (*a*) N. P. Redmore, I. V. Rubstov and M. J. Therien, *J. Am. Chem. Soc.*, 2003, **125**, 8769; (*b*) T. O. Screen, J. R. G. Thorn, R. G. Denning, D. G. Bucknall and H. L. Anderson, *J. Mater. Chem.*, 2003, **13**, 2796.

23 M. B. Nielsen, N. F. Utesch, N. N. P. Moonen, C. Boudon, J. -P. Gisselbrecht, S. Concilio, S. P. Piotto, P. Seiler, P. Günter, M. Gross and F. Diederich, *Chem. Eur. J.*, 2002, **8**, 3601.
24 R. Dembinski, T. Bartik, B. Bartik, M. Jaegerr and J. A. Gladysz, *J. Am. Chem. Soc.*, 2000, **122**, 810.
25 A. Beeby, K. Findlay, P. J. Low and T. B. Marder, *J. Am. Chem. Soc.*, 2002, **124**, 8280.
26 (*a*) J. M. Seminario, A. G. Zacarias and J. M. Tour, *J. Am. Chem. Soc.*, 2000, **122**, 1964; (*b*) J. Chen, W. Wang, M. A. Reed, A. M. Rawlett, D. W. Price and J. M. Tour, *Appl. Phys. Lett.*, 2000, **77**, 1224.
27 X. Xiao, L. A. Nagahara, A. M. Rawlett and N. Tao, *J. Am. Chem. Soc.*, 2005, DOI: 10.1021/ja050381m.
28 J. M. Tour, L. Cheng, D. P. Nackashi, Y. Yao, A. K. Flatt, S. K. St. Angelo, T. E. Mallouk and P.D. Franzon, *J. Am. Chem. Soc.*, 2003, **125**, 13279.
29 H. Park, J. Park, A. K. L. Lim, E. H. Anderson, A. P. Alivisatos and P. L. McEuen, *Nature*, 2000, **417**, 57.
30 J. Park, A. N. Pusupathy, J. I. Goldsmith, C. Chang, Y. Yaish, J. R. Petta, M. Rinkoski, J. P. Sethna, H. D. Abruna, P. L. McEuen and D. C. Ralph, *Nature*, 2002, **417**, 722.
31 (*a*) D. I. Gittins, D. Bethell, D. J. Schiffrin and R. J. Nichols, *Nature*, 2000, **408**, 67; (*b*) W. Haiss, H. van Zalinge, S. J. Higgins, D. Bethell, H. Höbenreich, D. J. Schiffrin and R. J. Nichols, *J. Am. Chem. Soc.*, 2003, **125**, 15294.
32 B. Xu, X. Xiao, X. Yang, L. Zang and N. Tao, *J. Am. Chem. Soc.*, 2005, **127**, 2386.
33 S. Kubatkin, A. Danilov, M. Hjort, J. Cornil, J.-L. Brédas, N. Stuhr-Hansen, P. Hedegard and T. Bjørnholm, *Nature*, 2003, **425**, 698.
34 N. G. Pschirer, K. Byrd and U. H. F. Bunz, *Macromolecules*, 2001, **34**, 8590.
35 D. W. Price, Jr. and J. M. Tour, *Tetrahedron*, 2003, **59**, 3131.
36 For a preliminary report on the synthesis of some of the compounds in Scheme 1 see: C. Wang, A. S. Batsanov, M. R. Bryce and I. Sage, *Org. Lett.*, 2004, **6**, 2181.
37 K. Sonogashira, *J. Organomet. Chem.*, 2002, **653**, 46, and references therein.
38 C. Wang, A. S. Batsanov, M. R. Bryce and I. Sage, *Synthesis*, 2003, 2089.
39 N. Svenstrup, K. M. Rasmussen, T. K. Hansen and J. Becher, *Synthesis*, 1994, 809.
40 J. J. Stapleton, P. Harder, T. A. Daniel, M. D. Reinard, Y. Yao, D. W. Price, J. M. Tour and D. L. Allara, *Langmuir*, 2003, **19**, 8245.
41 K. Arihara, T. Ariga, N. Takashima, K. Arihara, T. Okajima, F. Kitamura, K. Tokuda and T. Ohsaka, *Phys. Chem. Chem. Phys.*, 2003, **5**, 3758.
42 X. Shi, W. B. Caldwell, K. Chen and C. A. Mirkin, *J. Am. Chem. Soc.*, 1994, **116**, 11598.
43 S. Zhang, A. Palkar, A. Fragoso, P. Prados, J. de Mendoza and L. Echegoyen, *Chem. Mater.*, 2005, **6**, 2063.
44 Y. Sato and F. Mizutani, *Phys. Chem. Chem. Phys.*, 2004, **6**, 1328.
45 G. J. Ashwell and W. D. Tyrrell, unpublished results.
46 L. S. Hegedus, in *Organometallics in Synthesis*, ed. M. Schlosser, Wiley, Chichester, 1994, p. 448.
47 Z. Peng and L. Yu, *J. Am. Chem. Soc.*, 1996, **118**, 3777.
48 *SADABS*, version 2.03, Bruker, AXS, Madison, WI, 2001.
49 *SHELXTL*, version 5.10, Bruker, AXS, Madison, WI, 1997.
50 H. D. Flack, *Acta Crystallogr., Sect. A*, 1983, **39**, 876.

Functionalised conjugated materials as building blocks of electronic nanostructures

Diego F. Acevedo, Juan Balach, Claudia R. Rivarola, María C. Miras and César A. Barbero*

Received 20th April 2005, Accepted 30th June 2005
First published as an Advance Article on the web 21st September 2005
DOI: 10.1039/b505585g

Two different approaches towards conjugated material (carbon nanotubes, conjugated polymers) functionalisation are presented: covalent bonding of functional groups and non covalent interaction with soluble polymers. Covalent functionalisation of carbon nanotubes is made by reaction of the aromatic ring with aryl radicals, produced by reduction of diazonium ions. In the case of conducting polymers, covalent functionalisation is brought about by reaction of polyanilines with diazotized aromatic amines (including amino terminated azo dyes). The non covalent functionalisation of carbon nanotubes is made by wrapping the nanotubes with soluble conducting polyanilines. The functionalised materials are characterised by FTIR spectroscopy, X-ray diffraction, dynamic light scattering, ultraviolet-visible absorption and emission spectroscopy, transmission electron microscopy, cyclic voltammetry, differential electrochemical mass spectroscopy and conductivity measurements. The materials are used to build ionic self assembled multilayers using a layer-by-layer deposition process. The charge transport and electrocatalytic behaviour of the assemblies, relevant to the application of the assemblies in nanostructured electrochemical biosensors, are evaluated using different redox molecules and/or its intrinsic electroactivity as probes.

1. Introduction

Conjugated carbon based materials, such as carbon nanotubes and conducting polymers could be used to build electronic devices. While it is possible that some carbon nanotube devices could compete with silicon based devices (*e.g.* in power transistors), it is more likely that wet processing of carbon nanotubes will be used to produce simple devices of technological interest. In that sense, they are closer to conjugated polymers which have been applied in low cost electronic devices ("plastic electronics").[1] An impediment to do that is the low solubility of pristine carbon nanotubes. The low solubility is due to weak interaction with solvent molecules and strong π–π and dispersive interaction between individual nanotubes. A quite similar effect exists in conjugated polymers due to similar interactions. Material processability could be achieved by attachment of functional groups which interact efficiently with solvents and, due to steric effects, decrease the interaction between neighbouring chains or nanotubes. Even more, carbon nanotubes and conducting polymers could be combined to create materials with new properties.

Self assembly of multilayers is a way to create nanostructures where at least one dimension, perpendicular to the multilayer assembly, is in the nanometric range. The self assembly from solution requires overcoming the entropy decrease, due to complex formation, involved in the adsorption process. This could be performed allowing strong interaction (electrostatic, hydrogen bonding, *etc.*) between layers and by depositing species with low translational (macromolecules, nanostructures) and/or configurational (rigid structures) entropy. A less understood factor is the so

Departamento de Química, Universidad Nacional de Rio Cuarto, Agencia Postal No 3, 5800 Rio Cuarto, Argentina. E-mail: cbarbero@exa.unrc.edu.ar

called "hydrophobic effect" which is related with the change of entropy due to water and/or small ion displacement upon bilayer formation.[2]

The formation of ionic self-assembled multilayers by a layer-by-layer procedure was introduced by Decher in 1991.[3] The technique is based on the alternating adsorption of multiply charged cationic and anionic species from solutions (usually aqueous) of each species. The popularity of the method is due to the ability to create highly tailored polymer thin films with a nearly unlimited range of functional groups incorporated within the structure of the film. This flexibility is achieved through the use of a simple adsorption process that is readily accessible. The area has progressed to include the incorporation of electroactive polymers, organic dyes, semiconductor quantum dots, electrochemically active species, and a wide variety of proteins and other biosystems. Ionic self assembled multilayers could be built using conjugated materials, like carbon nanotubes and conducting polymers. Both are rigid and have high molar mass but they are insoluble in common solvents and carbon nanotubes bear no charge. Carbon nanotubes[4] are relatively new nanomaterials that display attractive structural, mechanical, and electronic properties, including improved electrochemical activity.[5] The inherent insolubility of pristine carbon nanotubes in solvents, in particular in water, has complicated their handling as chemical reagents and has limited their integration into nanostructured systems. The challenge of solubilizing carbon nanotubes in water has been addressed through non-covalent modification with surfactants[6] or polymers[7] or their covalent modification with hydrophilic functionalities.[8] Since the goal of the work is to use conjugated materials as building blocks of electrostatic polyelectrolyte multilayers, it is desirable to solubilize the carbon nanotubes by adding charged functionalities. To do that, both covalent modification and polyelectrolyte based solubilization could be used. The number of polymers that has been used to render carbon nanotubes soluble by non covalent interaction, has been rather limited.[6,9] On the other hand, carbon nanotube modified electrodes have been used in electrochemical sensors.[10] Different groups have built ionic self-assembled multilayers using carbon nanotubes which are shortened and oxidized by treatment with chemical oxidants (*e.g.* HNO_3).[11] Such treatments could alter the electronic properties of the carbon nanotubes because the chemical structure is changed. Conducting polymers are well known conducting materials[12] which have been used for various applications.[13] Among others, those materials have been used to build sensors[14] and biosensors[15] based on the effect of analytes on the electronic properties (*e.g.* conductivity). Conducting polymers have been previously used to build ionic self-assembled multilayers.[16] Rubner and coworkers[17] prepared polyaniline (emeraldine base) solutions in water/dimethylacetamide at pH 2.5 (protonated form = polycation) and self assembled it with poly(styrenesulfonate) (polyanion). Since polyaniline is insoluble in water, a cosolvent is required which interacts strongly with the polymer chains. Mattoso and coworkers used poly(alkoxyanilines), which are soluble in water in doped state to build multilayers.[18] An alternative is to covalently link charged functional groups to promote interaction with other polyelectrolytes in the multilayer. The presence of charges could also render the polymer soluble in aqueous media due to ion–dipole interaction with the solvent. Sulfonated polyanilines,[19] suitable to build ionic self-assembled multilayers, were produced by electrophilic sulfonation,[20] copolymerization[21] or nucleophilic addition.[22] The resulting polyanion was self-assembled with polycations.[23] Altogether, the variety of groups which have been linked to polyaniline to build ionic self-assembled multilayers has been somewhat limited.

In the present communication we describe the synthesis of soluble charged conjugated materials based on conducting polymers and carbon nanotubes. The materials are used to build electrostatic self assembled multilayers. The electronic properties of the ionic self-assembled multilayers, relevant to their applications in electrochemical biosensors, are tested using electrochemical probes.

2. Experimental

Aniline (Merck) was purified by distillation at reduced pressure. Multiwall carbon nanotubes (Sun Natotech, China) were produced by CVD, using Ni as catalyst. The nanotubes are 10–30 nm in diameter and 200 nm long (TEM). Poly(diallyldimethylammonium chloride) (PDAMAC, Aldrich) has a MW = 400.000 g mol^{-1}. All aqueous solutions were prepared with ultrapure water. All other reagents were of analytical quality.

2.1 Polyaniline synthesis

Polyaniline (emeraldine form) was prepared, as described in the literature,[24] by oxidation of aniline (0.1 mol L^{-1}) in 1 mol L^{-1} HCl with ammonium persulfate (equimolar) at temperatures below 5 °C. The temperature was monitored during the polymerization. When the reaction was completed, after the maximum of the temperature has occurred, the polymer was filtered out under suction and washed with 1 mol L^{-1} HCl solution (0.5 L) and water (1 L). The polymer was then converted into its base form by stirring for 24 h in a 0.1 mol L^{-1} ammonia solution. The emeraldine base form (50% oxidation) used in all reactions had an intrinsic viscosity $[\eta]$ of 1.12.

2.2 Diazonium ion synthesis and coupling to polyaniline (PANI)

The amines were diazotized with sodium nitrite and concentrated HCl in an ice bath.[25] PANI was suspended in TRIS® buffer (pH = 8) and mixed with the diazonium salt solution in an ice bath. The solid was filtered under vacuum and washed first with 1 L of 1 mol L^{-1} HCl solution and then with 1 L distilled water. The products were filtered out of the mixture under vacuum and dried (dynamic vacuum for 48 h).

2.3 Covalent modification of carbon nanotubes

Diazotized amines were mixed with carbon nanotubes in an ultrasonic bath. Then, the temperature was raised to 60 °C and heated for 1 h. The reaction mixture was acidified with concentrated HCl and centrifuged (10,000 rpm) to separate the modified nanotubes. The precipitate was washed several times with water and dried under vacuum. The powder was dissolved in 1 M NaOH for multilayer assembly.

2.5 Non covalent solubilization of carbon nanotubes

Carbon nanotubes were mixed with 0.5% (w/v) solution of the soluble polyaniline in 1 M NaOH. The solution was stirred overnight and any precipitate present filtered out.

2.6 Procedure for self-assembly of the multilayers

The electrodes were immersed (15 min each) successively in solutions of the conjugated material (0.5% w/v in 1 M NaOH) and PDAMAC (2% w/v in water). Between immersions, the plates were washed by immersing for 15 min in ultra pure water, without stirring. Since we used different electrode materials, we did not want to use different covalent modification procedures of the electrode surface for each material. It is possible to adsorb spontaneously PDAMAC on the electrode surface to produce a first charged layer on GC, ITO and Au. Therefore, no covalent modification of the electrode surface was necessary. All operations were performed at room temperature.

2.7 Ultraviolet-visible spectroscopy

To obtain the ultraviolet-visible spectra, clear solutions were prepared by dissolving the polymers in aqueous solutions. The spectra were taken in 1 cm path length quartz cells using a diode array UV-visible spectrophotometer (HP 8452A). The fluorescence measurements of solutions, in NaOH 1 M, were made in a Spex Fluoromax fluorometer.

2.8 FTIR spectroscopy

Fourier Transform Infrared Spectroscopy (FTIR) measurements of chemically prepared polymers were performed on the powder dried under vacuum at ambient temperature for 48 h. The polymers were mixed with KBr and pellets for transmission measurements were prepared. The spectra were taken in an Impact 400 FTIR spectrometer (Nicolet) with a resolution of 4 cm^{-1}, using 200 scans.

2.9 Conductivity

Conductivity measurements were carried out using the Van der Pauw method on polymer films deposited onto LDPE. The samples were first doped by a 1 h immersion in aqueous 1 M HCl and then dried *in vacuo* over 24 h. Four contacts were painted on the films with silver paint. The current was applied with a galvanostat (Amel 2049) while the voltage was measured with a 6 1/2 digit multimeter (HP). The thickness of the films, necessary to calculate the conductivity was estimated using the relationship proposed previously.[26]

2.10 Electrochemistry

Electrochemical measurements were carried out in a conventional three electrode cell, controlled by a computerised potentiostat (GAMRY PC4). Glassy carbon electrodes are 3 mm thick plates of V10 (Carbone Lorraine). The plates are polished with a diamond paste of diminishing particle size (10–0.3 μm). The GC plates are tested running GC of the couples of interest before self-assembling. ITO electrodes (Blazers) are thin films of indium doped tin oxide on glass with a surface resistivity of 10 Ω. All potentials are quoted against the saturated calomel electrode (SCE).

2.11 X-Ray diffraction

X-Ray diffractograms were recorded on an automatic Rigaku DII X-ray diffractometer using CuKα radiation of wavelength, $\lambda = 1.540\ 60$ and $\lambda = 1.544\ 34$ nm. The diffractometer was operated at 40 kV and 20 mA.

2.12 Dynamic light scattering (DLS)

Dynamic light scattering measurements were performed in a Malvern 4700 DLS. The measurements were made with an angle of 90°, at 25 °C.

2.13 Transmission electron microscopy

TEM samples were obtained by evaporation of a nanotube dispersion onto a gold grid. The images were taken with a Philips EM 301 microscope.

3. Results and discussion

3.1 Synthesis of functionalized polyanilines

Since the goal is to be able to build ionic self-assembled multilayers using conjugated polymers, it is necessary to synthesize charged soluble conducting polymers. It has been shown[27–29] that coupling of diazonium ions (with polyaniline) could render a variety of functionalised polyanilines using the same modification reaction (Scheme 1).

In that way, it is also possible to covalently link charged groups such as sulfonate ($-SO_3^-$) and carboxylate ($-COO^-$), to the polyaniline backbone. The FTIR spectrum of a typical modified polyaniline (Fig. 1), is compared with that of PANI, and reveals the presence of new bands which can be seen at 1040 and 997 cm^{-1}. Those are assigned to the anti-symmetric and symmetric stretching of the S=O group.[30] The results confirm that the reaction allows the synthesis of polyanilines with linked charged groups.

The conductivity of polyaniline decreases upon modification due to a reduction in π-conjugation of the chain by steric effects from the attached groups.[31] However, the conductivity decreases upon modification (Table 1) is of less than an order of magnitude. This is quite acceptable for electrochemical applications, where thin films are used. The small decrease is true only if conductivity of materials in the fully protonated state, which were equilibrated with 1 M HCl, are compared. In Table 1 are also described the conductivity values obtained with films less protonated, equilibrated with pH 4 solutions. It is apparent that deprotonation has a clear effect on conductivity.

The conducting polyelectrolytes are all soluble ion aqueous media ($pH > 9$), giving clear solutions, because ion–dipole interactions are strong enough to overcome the chain to chain

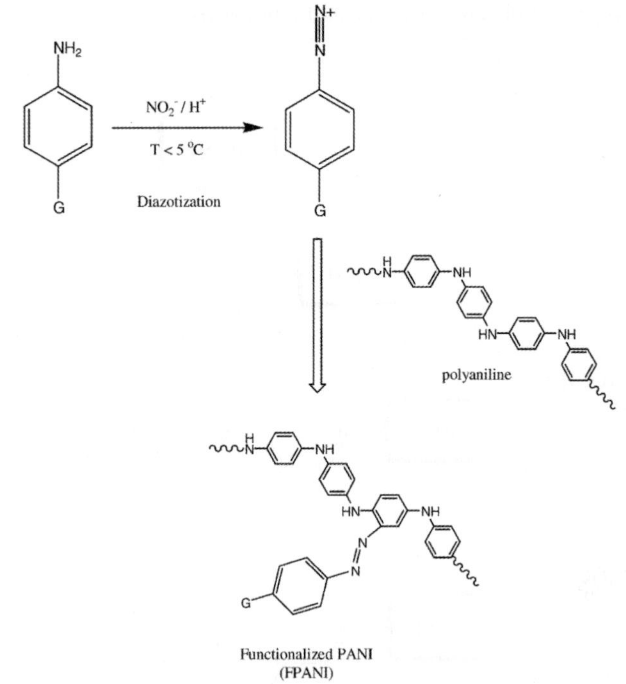

Scheme 1 Reaction scheme of diazonium ion synthesis and reaction with polyaniline.

interactions. Indeed, dynamic light scattering measurements of solutions do not show particle presence (above 1 nm). In the case of conducting polyelectrolytes bearing $-COO^-$ groups, at lower pHs, the $-COO^-$ groups become protonated (to form $-COOH$) and the polyelectrolyte precipitates

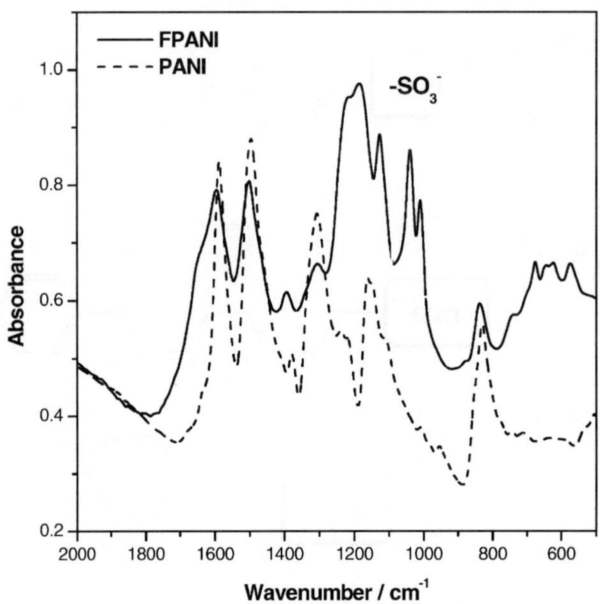

Fig. 1 FTIR spectrum of polyaniline (dashed line) and a polyaniline modified by reaction with *p*-sulfobenzenediazonium (full line).

Table 1 Conductivity values of polyanilines modified with charged groups

Polymer	Structure	ρ^a/S cm^{-1}
PANI	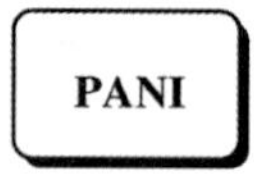	3.78
PANI–N=N–C_6H_5COOH	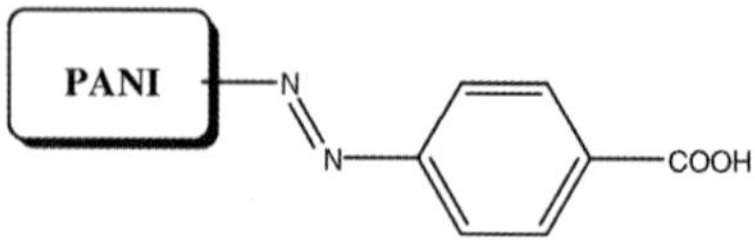	0.96 (0.15)
PANI–N=N–$C_6H_3(OCH_3)$–N=N–C_6H_5–COOH	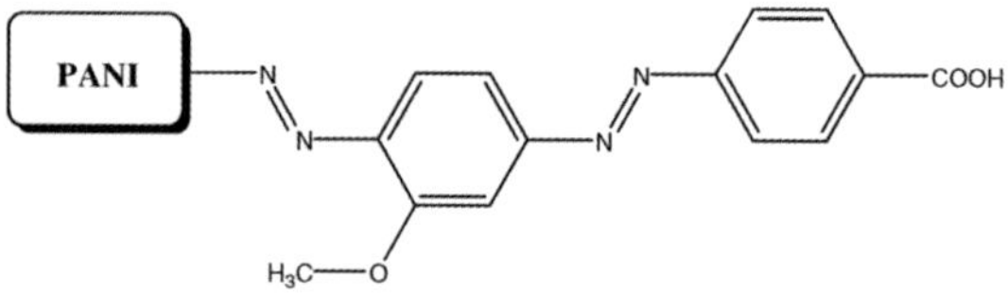	0.14 (0.05)
PANI–N=N–$C_6H_3(OCH_3)$–N=N–C_6H_5–SO_3H	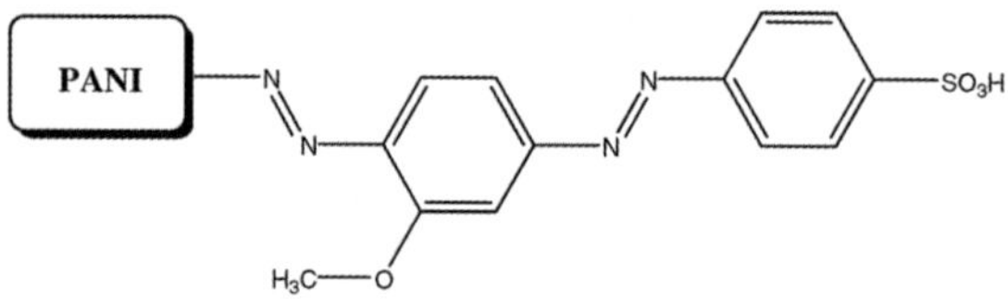	0.15 (0.028)
PANI–N=N–C_6H_4–N=N–C_6H_5–SO_3H	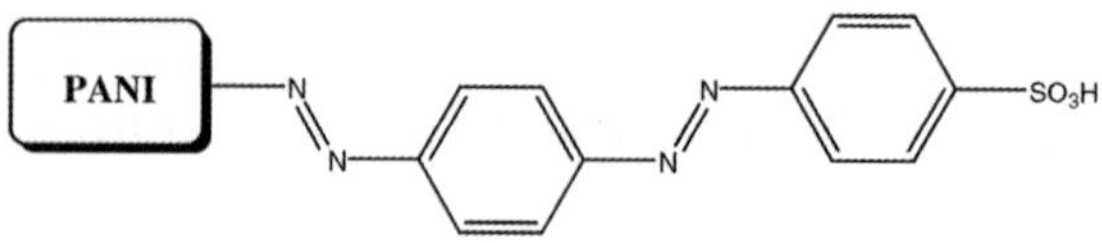	0.20 (0.07)
PANI–N=N–$C_{12}H_6$–N=N–C_6H_5–SO_3H	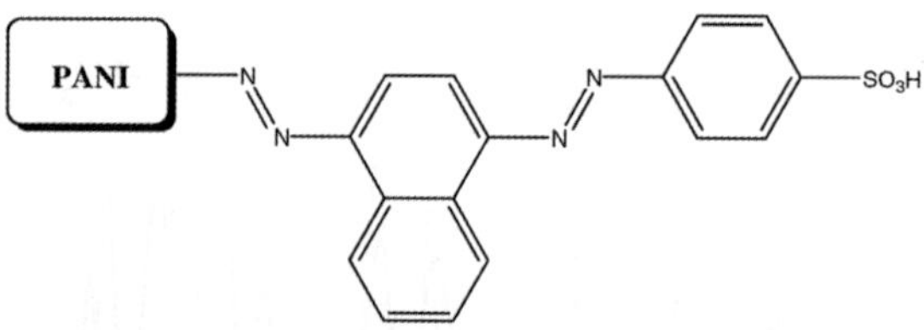	0.57
PANI–N=N–C_6H_4–N=N–C_6H_5–COOH	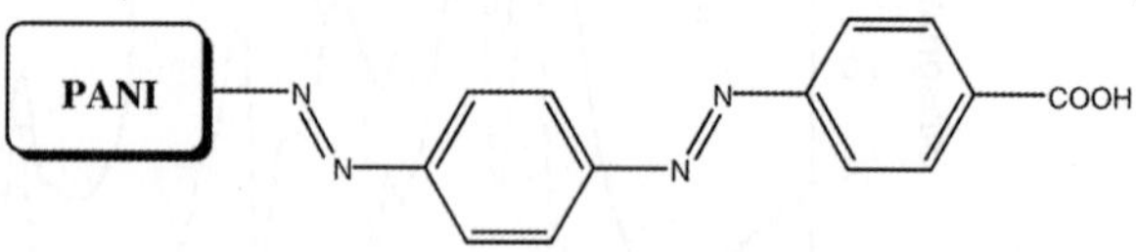	0.59 (0.10)
PANI–N=N–C_6H_5–SO_3H	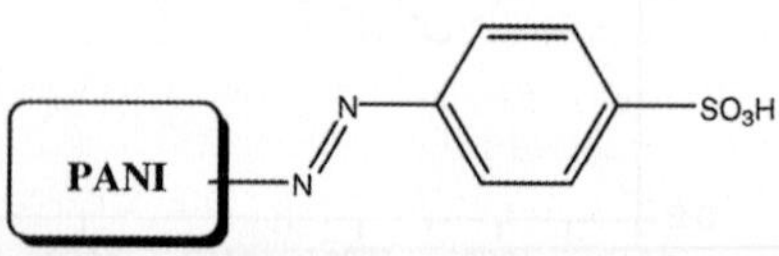	2.58 (0.43)

[a] Conductivity measured on films equilibrated with 1 M HCl. The conductivity of films equilibrated with a buffer solution (pH = 4) is stated between brackets.

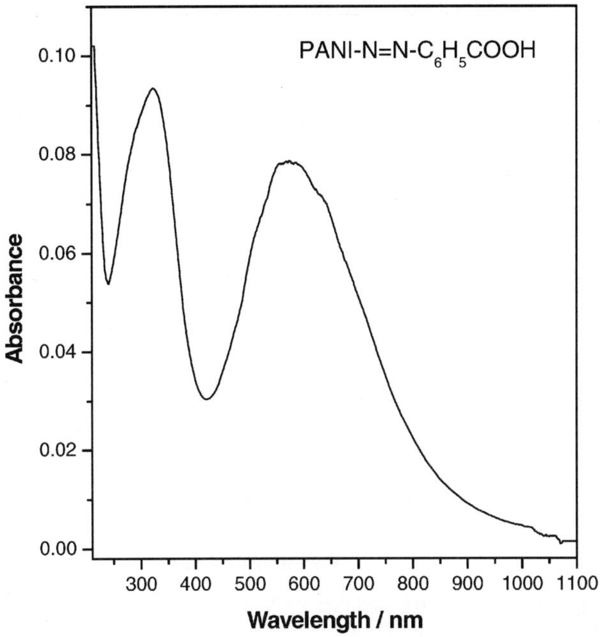

Fig. 2 UV-visible spectra of a polyaniline modified by reaction with *p*-carboxybenzenediazonium salt. The spectra are obtained with a 0.5% solution in 1 M NaOH.

because the ion–dipole interactions are strongly decreased. On the other hand, the polyelectrolytes containing $-SO_3^-$ groups are more difficult to precipitate by decreasing the pH because the group is not protonated up to very high proton activities.

In Fig. 2 is shown the UV-visible absorption spectrum of a soluble polyaniline (Fig. 2). While it is not possible to obtain a similar spectrum of PANI, due to its insolubility in aqueous solution, it is noteworthy that a similar one could be obtained for the base form of polyaniline (emeraldine base) in solvents like *N*-methylpyrrolidone.

While semiconducting polymers present clear fluorescence, conductive polymers do not usually present luminescence. This is due to an internal redox quenching of the excited units by the oxidised units. Accordingly, polyaniline presents luminescence only in the reduced state (leucoemeraldine).[32] However, it has been reported that the emeraldine form of polyaniline could present luminescence when polymer films are segregated into fully reduced and fully oxidized domains.[33] In solution, it has been shown that polyaniline block copolymers also present luminescence.[34] It was suggested that different comonomer blocks isolate the fluorophores (aromatic amine units) from the quenchers (quinoneimine units). The most soluble polyanilines present fluorescence (Fig. 3), suggesting that quinonimine and amine units are isolated by the steric effect of the moieties attached to the backbone. The fluorescence could be used to study non covalent interactions of conducting polyelectrolytes with other species (*e.g.* carbon nanotubes, see below).

The modification reaction is quite versatile because it could also link groups possessing other properties (fluorescence, optical absorption, complex formation, magnetism, *etc.*) to the polymer backbone, producing conducting polyelectrolytes with special properties. Those could then be used to build multilayers with more complex properties. It is noteworthy to mention that the modification reaction could be used to attach more than one group in sequence to the polymer backbone. In that way, conducting polyanilines with charged groups and other functionalities in the same chain could be synthesized.

Thin films could also be deposited by drop or spin coating of soluble polyaniline solutions in aqueous ammonia. The solvent is evaporated and the ammonium salt of the polyelectrolyte decomposed by heating with an infrared lamp. The films are stable in acid media (up to pH 5–6)

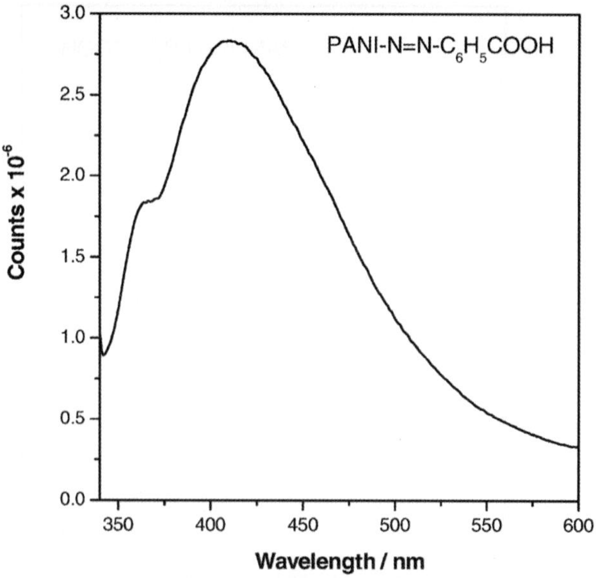

Fig. 3 Fluorescence spectrum of –COOH modified PANI (solution in 1 M NaOH/H_2O). Excitation wavelength = 315 nm. Emission range = 340 to 600 nm.

and show two redox processes, in a similar way to the parent polyaniline polymer, albeit with the peak separation between both processes reduced due to steric and inductive effects. The fact that modified polyanilines are electroactive furnishes a way to detect the incorporation of the conducting polymer in the multilayers.

3.2 Synthesis of functionalised carbon nanotubes

Two methods of adding functionalities to carbon nanotubes are described: covalent reaction with diazonium ions and non covalent interactions with soluble polyanilines.

3.2.1 Covalent reaction with diazonium ions. The covalent reaction chosen is based on the method of Tour and coworkers.[35] It involves the formation of highly reactive aryl radicals by charge transfer from diazonium ion to the nanotube, with loss of N_2. The highly reactive species attach to carbon nanotubes (Scheme 2).

The method has its drawbacks. The most important is that it produces a highly reactive species which could react not only with the nanotubes but with other molecules, such other diazonium ion

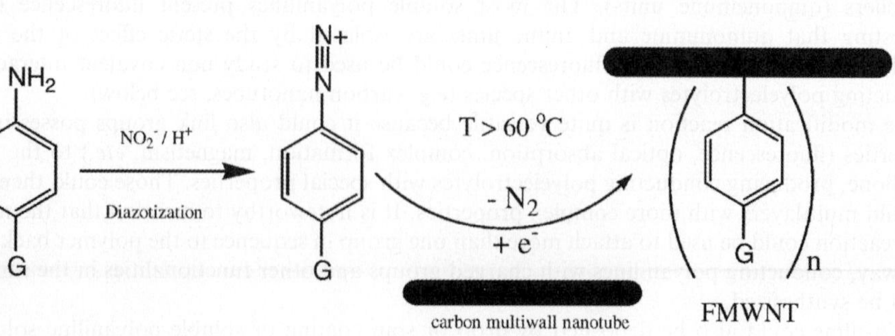

Scheme 2 Reaction scheme of diazonium ion synthesis and carbon nanotube modification.

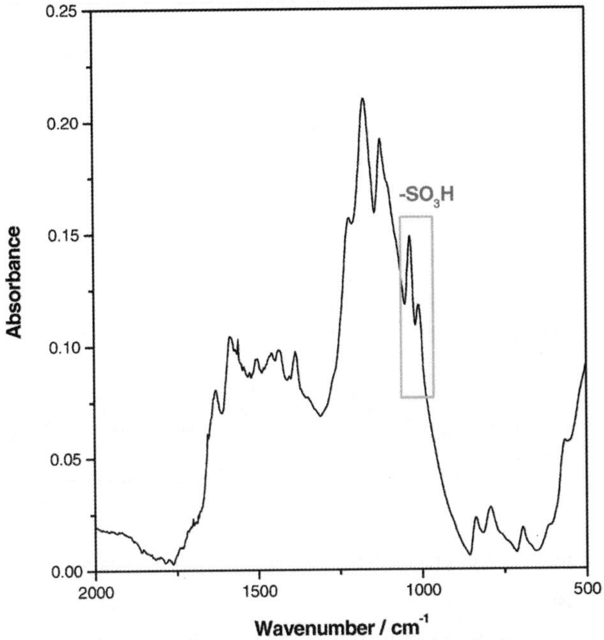

Fig. 4 FTIR spectroscopy of a modified multiwall carbon nanotube.

molecules, giving undesirable side products. Indeed, in our hands it works well only when the aromatic ring is strongly deactivated towards electrophilic substitution, due to the presence of an electron withdrawing group ($-NO_2$, $-COOH$, $-SO_3H$). This limitation makes it difficult to link more complex azo dyes to the carbon nanotubes. Therefore, the diversity of the functional groups is limited. As we will see, such a limitation is not present when noncovalent interactions are used. While the original method uses organic solvents (or no solvent)[36] and organic nitrites, diazonium ions, produced in aqueous solution by the well known method of aromatic amine diazotization with nitrite ion, could also be used to modify carbon nanotubes.[37] Evidence of the attachment of the groups to the carbon nanotubes is obtained by FTIR spectroscopy (Fig. 4). The presence of bands assigned to characteristic vibrations of the attached functionalities (*e.g.* $-SO_3^-$), confirm the linking of the functional groups to the nanotube wall.

Additional information is obtained from X-ray diffraction measurements of the functionalized carbon nanotubes (Fig. 5). The diffraction pattern of carbon nanotubes shows peaks characteristic of the graphitic planes. Those peaks are still present in the modified carbon nanotubes suggesting that the modification reactions do not significantly disrupt the walls of the nanotubes. However, new peaks appear indicating that the reaction produce new covalent bonds in the structure.

The covalent modification of carbon nanotubes could increase their solubility not only by incorporation of groups (*e.g.* $-COO^-$) possessing strong interactions (*e.g.* ion–dipole) with the solvent but also by decreasing the interaction between neighbouring nanotubes. Transmission electron microscopy results confirm that pristine carbon nanotubes, which are bundled together (Fig. 6a), could be separated into isolated nanotubes by modification (Fig. 6b).

Dynamic light scattering measurements of carbon nanotube dispersions reveal a clear decrease in the size of nanotube aggregates after modification (Fig. 7). The mean size of the smaller particles decreases from 866 to 181 nm. The size of bigger aggregates also decreases upon modification. It is noteworthy to mention that modified nanotube dispersions are stable for months while the dispersion of pristine nanotubes show precipitates in less than 1 h. Higher aggregates, however, remain in the dispersions. It has been recently suggested that the reaction in organic solvents is unable to dislodge the bundles and only by using oleum (H_2SO_4/20% SO_3) as solvent is it possible to produce free modified nanotubes.[38]

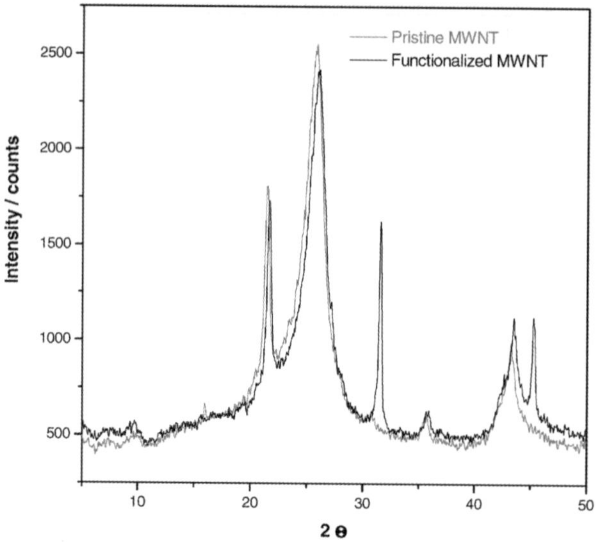

Fig. 5 X-Ray diffraction of pristine (grey line) and modified carbon nanotubes (black line).

Since electrochemical methods are going to be used to study the electronic properties of ionic self-assembled multilayers, it is necessary to know the electrochemical properties of functionalised carbon nanotubes. To do that, glassy carbon (GC) electrodes were modified with functionalised carbon nanotubes films by drop coating. The $\Delta E\text{p}$ obtained (at 50 mV s^{-1}) is 130 mV, revealing a quasi-reversible behaviour of the redox couple at the modified electrode. The results contrast with the peak separation ($\Delta E\text{p} = 549$ mV at 50 mV s^{-1}), obtained at the polished GC electrode, which has a more sluggish charge transfer. The results show less reversibility than that observed when nanotube bundles grown on a surface were used as the electrode,[39] suggesting that chemical functionalisation and/or contact between nanotubes and the base electrode affects somewhat the charge transfer. However, the improvement in the charge transfer is clear. As it was pointed out by Compton and coworkers,[40] such effect does not mean that carbon nanotubes have unique catalytic properties other than those of graphite edge planes. For our purposes, however, it means that we could evaluate the presence of functionalised carbon nanotubes on less active electrodes measuring

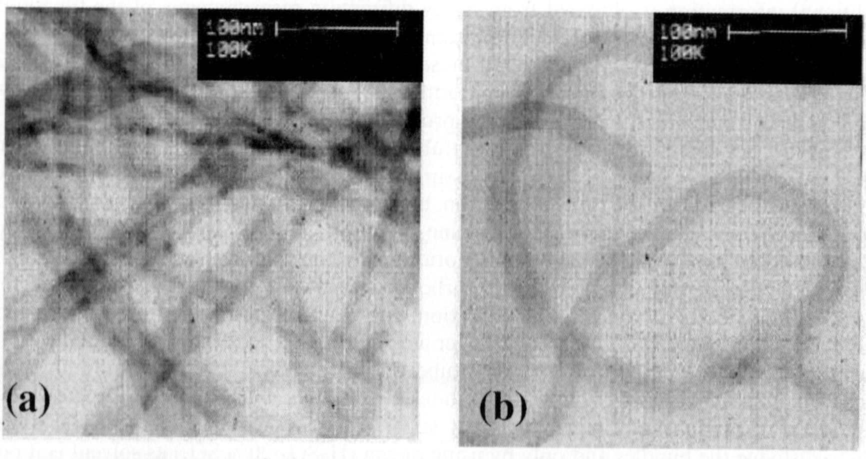

Fig. 6 Transmission electron micrographs of: (a) pristine multiwall carbon nanotubes and (b) modified carbon nanotubes (covalent bonding of $-C_6H_5COO^-$).

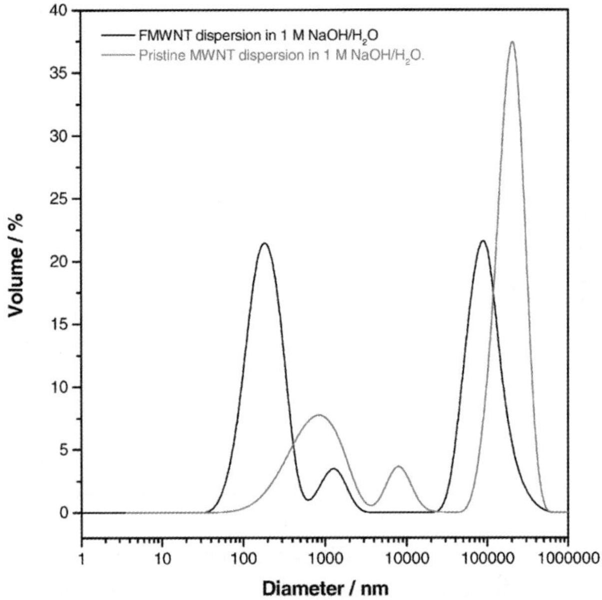

Fig. 7 Dynamic light scattering measurement of dispersions in water: (grey line) pristine multiwall carbon nanotubes; (black line) multiwall carbon nanotubes modified with $-C_6H_5COO^-$ functionalities.

the electrochemical current of ferrocyanide oxidation. A redox couple more relevant to the application of those nanostructures in biosensors is hydrogen peroxide. It is known that multiwall carbon nanotubes could electrocatalyze the oxidation and reduction of H_2O_2.[41] Similar results are obtained using functionalised carbon nanotubes. However, while on oxidation only the current increases, a clear electrocatalytic effect is observed for H_2O_2 reduction. Therefore, functionalised carbon nanotubes could be used to detect hydrogen peroxide. As the electrode potentials are quite anodic, closer to those of water oxidation, it is important to check if significant degradation of the nanotubes occurs. Differential electrochemical mass spectroscopy measurements of the same electrochemical system reveals that O_2 is formed ($m/z = 32$) during hydrogen peroxide oxidation but no CO_2 ($m/z = 44$) is detected. This result suggests that no significant nanotube oxidation is produced during H_2O_2 oxidation.

3.2.2 Non-covalent interaction of soluble polyanilines with carbon nanotubes. As it was pointed out, covalent modification of carbon nanotubes seems to be only possible with some groups. An alternative procedure consists of dispersing the carbon nanotubes using non-covalent interactions with conducting polymers for dispersion stabilization.[42] While it is possible to use non conducting soluble polymers[43] for that goal, the fact that both conducting polymers and carbon nanotubes has extended electronic clouds allows the creation of π–π interactions between them. We have used soluble polyanilines to interact with multiwall carbon nanotubes. The X-ray diffraction pattern of pristine multiwall carbon nanotubes is not altered by solubilization with conducting polymers, showing only a superposition of the broad bands due to the amorphous polyaniline and the peaks due to the carbon nanotubes. This result is reasonable, since no covalent linkages are formed during solubilization. The DLS measurements of a typical aqueous dispersion of multiwall carbon nanotubes using functionalized polyanilines (Fig. 8) show that the polydisperse multiwall carbon nanotubes sample (peaks at 800 nm, 7800 nm and 20 μm) changes to particles with a mean size of 210 nm. Such dispersions of multiwall carbon nanotubes using functionalized polyanilines are stable for months.

While it is clear that the presence of a conducting polyelectrolyte improves the dispersion of the carbon nanotubes, the nature of the polymer–carbon nanotube interaction is less clear. To try to find out if there are specific π–π interactions, we use fluorescence quenching measurements. As can

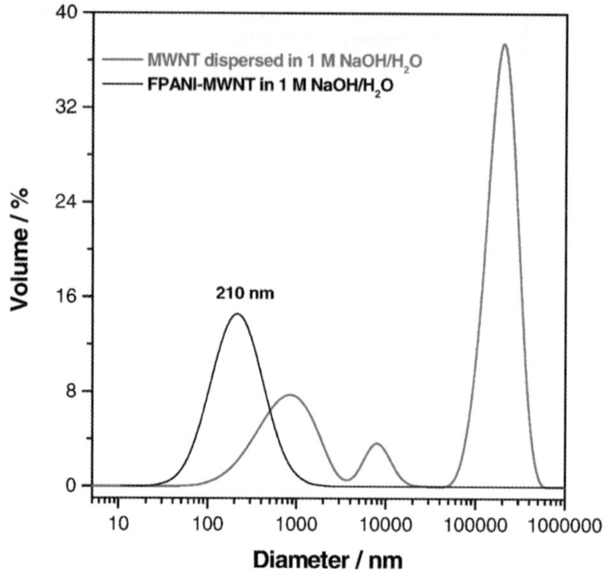

Fig. 8 DLS measurement of: (black line) pristine multiwall carbon nanotubes dispersed in water; (grey line) multiwall carbon nanotubes dispersed using 1% of a soluble conducting polyelectrolyte (PANI–N=N–$C_6H_5SO_3^-$).

be seen (Fig. 9), the conducting polymer fluorescence decreases by successive additions of carbon nanotubes to the conducting polymer solution.

The existence of fluorescence quenching strongly suggests that electron or charge transfer is possible between conducting polymer chains and carbon nanotubes. Therefore, it is likely that π–π interactions are present. Whether it is electron or charge transfer, is a question under current study.

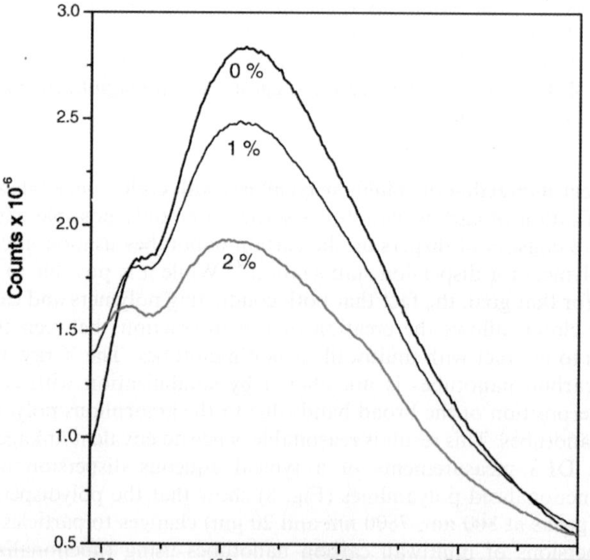

Fig. 9 Quenching of the fluorescent emission of a modified polyaniline (PANI–N=N–$C_6H_5COO^-$) by increasing amounts (see text) of multiwall carbon nanotubes.

Thin films could be deposited by drop or spin coating of multiwall carbon nanotubes dispersed using functionalized polyanilines dispersions in aqueous ammonia. The solvent evaporates and the ammonium salt of the polyelectrolyte decomposes, by heating using an infrared lamp. The films are stable in acid and base and show the same redox processes as the soluble polyaniline used to stabilize the dispersions. Therefore, electrochemical techniques could be used to detect the incorporation of the multiwall carbon nanotubes, dispersed using functionalized polyanilines, into the multilayers.

3.3 Ionic self-assembly of multilayer nanostructures

3.3.1 Ionic self-assembled nanostructures built using functionalised carbon nanotubes. There have been reports of using layer by layer self assembly to build ionic self-assembly of multilayer nanostructures with carbon nanotubes using heavily oxidized and shortened multiwall carbon nanotubes.[44] We have used the functionalised multiwall carbon nanotubes, described above, to build the multilayers. As it has been shown above, some electrochemical reactions show faster electron transfer on carbon nanotube modified electrodes than on glassy carbon electrodes. Therefore, the current due to a redox reaction, such as $Fe(CN)_6^{4-}$ oxidation, could be used to measure the amount of functionalised multiwall carbon nanotubes deposited at each deposition step during the layer by layer process. The cyclic voltammograms (Fig. 10) show increasing current maxima with each bilayer deposited. Additionally, the peak separation decreases slightly with the number of layers. This indicates that charge transfer seems to be at least as fast in the outer layers as in the inner layers. Otherwise, if a significant resistance would be added by each layer, the peak separation should increase.

The fact that the current increases evenly with the number of layers suggest that: (i) all functionalised multiwall carbon nanotubes layers are connected electrically between them; (ii) the redox probe ($Fe(CN)_6^{4-}$) is able to penetrate the whole multilayer assembly. The combination of both conditions implies that a three-dimensional electrode, with nanometric definition, has been assembled. The first condition seems difficult to achieve if ionic self-assembled multilayers are constituted by rigid layers. Since the polyelectrolyte used to build the multilayer (PDAMAC) is not

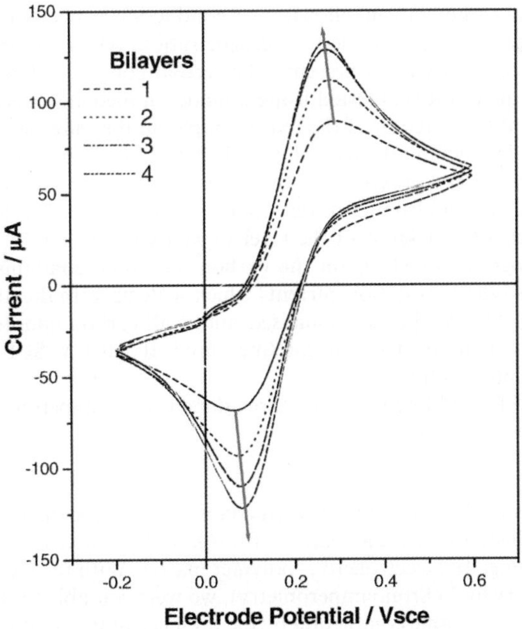

Fig. 10 Cyclic voltammograms of potassium ferrocyanide (2.13 mM) on self-assembled multilayer of functionalised multiwall carbon nanotubes and a polyelectrolyte (PDAMAC). Electrolyte = 1 M KCl. Scan rate = 50 mV s^{-1}. The multilayers are deposited on ITO/glass as base electrode.

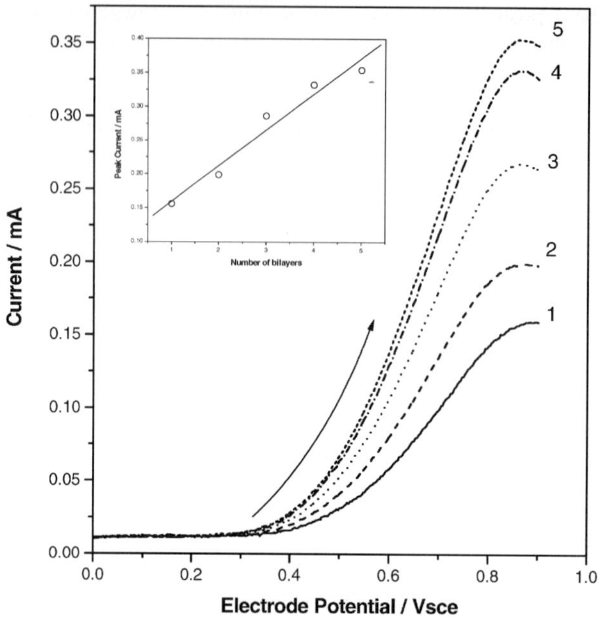

Fig. 11 Cyclic voltammograms of hydrogen peroxide (10 mM) on self-assembled multilayer of functionalised multiwall carbon nanotubes and a polyelectrolyte (PDAMAC). Electrolyte = Phosphate buffer (0.1 M, pH = 7). Scan rate = 1 mV s^{-1}. The multilayers are deposited on polished glassy carbon as base electrode.

conductive, it should inhibit electronic conduction between the conductive layers. However, it has been shown[45] that the conductivity of PANI/polyelectrolyte assemblies increase with the number of layers, until a constant value is reached. It is thought that the layers in the ionic self-assembled multilayers interpenetrate among them, allowing for conduction paths to be built. In the case of the carbon nanotubes, one could expect less interpenetration owing to the rigid nature of the nanotubes, but some seems to be still present. The second condition depends on the porosity of the layer and could be quantitatively tested using a model of mediated electrocatalysis. The model should take into account the reaction of the redox couple on the outer layer, the whole multilayer and/or the base electrode. Such mathematical models have already been proposed by Saveant and coworkers[46] and Albery and coworkers,[47] to explain redox catalysis mediated by redox polymers. It is apparent that ionic self-assembled multilayers could provide an excellent test for the theory due to the possibility of varying the thickness of the layer in well known steps.[48]

The cyclic voltammograms of H_2O_2 on the carbon nanotube containing ionic self-assembled multilayers (Fig. 11) also show oxidation currents which scale up with the number of bilayers in the assembly (insert in Fig. 11). At the functionalised multiwall carbon nanotubes based multilayers, peroxide could be also reduced. The current (measured at −0.4 V SCE) increases in absolute magnitude with the number of bilayers.

Therefore, the assembly could be used as a three-dimensional amperometric sensor of H_2O_2, to assemble peroxide based biosensors.

3.3.2 Ionic self-assembled multilayers nanostructures built using carbon nanotubes solubilized by mean of soluble polyanilines. One special feature of this kind of soluble nanotubes is the presence of an intrinsically conducting, and electroactive, polymer matrix. With the techniques used up to now by us (cyclic voltammetry and chronoamperometry), we were not able to detect any improvement on the overall multilayer electronic conduction by the presence of the conductive polymer. It is quite reasonable to think that intermixing of carbon nanotubes from neighbouring layers gives enough conductivity. On the other hand, the electrocatalytic properties of the multiwall carbon nanotubes should be present. Indeed, the ionic self-assembled multilayers built using carbon nanotubes

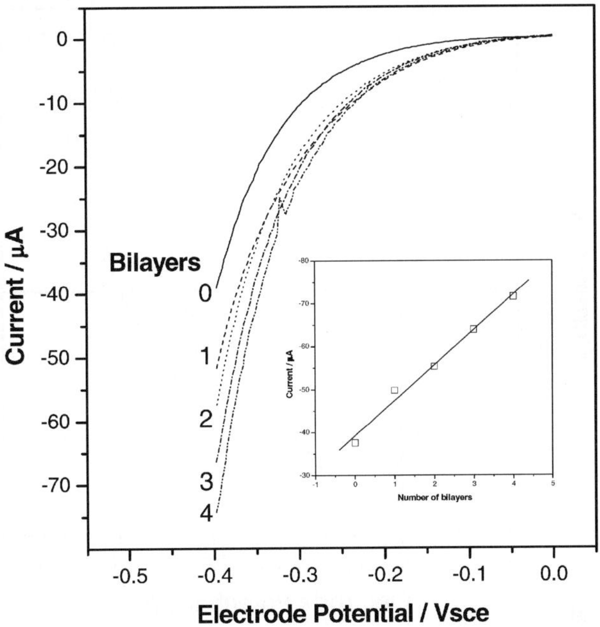

Fig. 12 Cyclic voltammograms of hydrogen peroxide (10 mM) on self-assembled multilayers of multiwall carbon nanotubes, dispersed with functionalised polyanilines, and a polyelectrolyte (PDAMAC). Electrolyte = phosphate buffer (0.1 M, pH = 7). Scan rate = 1 mV s^{-1}. The multilayers are deposited on polished glassy carbon electrodes.

solubilized by non-covalent interactions with soluble polyanilines show similar electrocatalytic behaviour towards ferrocyanide oxidation than functionalised carbon nanotube based ionic self-assembled multilayers. Using a less anodic extreme of the potential scan, to avoid degrading the polyaniline matrix, it is also observed that hydrogen peroxide could be reduced (Fig. 12). The current (measured at a given potential) increases with the number of bilayers (see insert in Fig. 12).

Recently it has been shown that the wiring of redox enzymes by redox hydrogels is improved by linking carbon nanotubes to the enzyme.[49] Therefore, multiwall carbon nanotubes dispersed using functionalised polyanilines could be useful as building blocks of biosensors based on enzyme/redox hydrogel multilayers[50]

3.3.3 Ionic self-assembled multilayers nanostructures built using soluble polyanilines. In the case of ionic self-assembled multilayers built only with soluble polyanilines as electroactive component, the probe used to test the multilayer is the intrinsic electroactivity of the soluble polyaniline. In Fig. 13 are shown the cyclic voltammetric responses of a typical functionalised PANI based ionic self-assembled multilayers, during the build up of the multilayer. As it can be seen the redox charge increases with the addition of bilayers. The peak current scales linearly with the number of bilayers. It is noteworthy that the same multilayer built on GC shows a different behaviour. The voltammetric current signal increases abruptly after the first immersion step. Afterwards, it changes only *ca.* 10% for each layer. It seems that strong adsorption of the functionalised PANI on the GC surface creates a relatively thick polymer layer during the first immersion. Afterwards, the layer by layer procedure built up the multilayer in a normal way.

The cyclic voltammograms are quite similar to those observed when polyaniline is produced by electrochemical polymerization.[51] The peak potential does not shift significantly when the number of layers is increased, suggesting that the resistance through the multilayer is small. It is noteworthy that the electrochemistry is sensitive enough to detect a few layers of electroactive material.

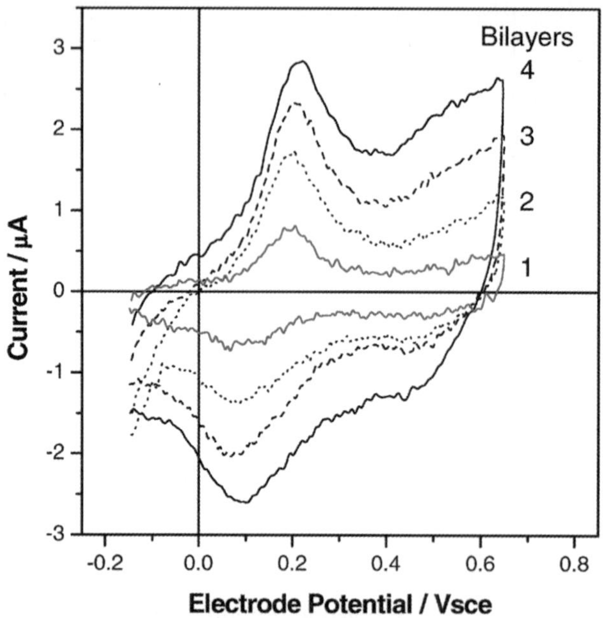

Fig. 13 Cyclic voltammograms of self-assembled multilayers made of functionalised polyanilines and a polyelectrolyte (PDAMAC). Electrolyte = 1 M HCl. Scan rate = 50 mV s^{-1}. The multilayers are deposited on ITO/glass plates as base electrodes.

4. Conclusions

Functionalised conjugated materials are produced using covalent modification and non covalent interactions. Soluble polyanilines, bearing charged groups, are produced by reaction of diazonium ions with polyaniline. The polymers have lower conductivity than the parent polymer and show clear fluorescent emission in solution. Covalent modification of carbon nanotubes is achieved by reaction of diazonium ions with pristine carbon nanotubes. In that way, charged functionalities (–COO–, SO_3^-) are linked to multiwall carbon nanotubes, rendering them soluble. The modification does not alter significantly the nanotube graphitic structure. Non covalent solubilization of carbon nanotubes is made by interaction of soluble polyanilines with pristine multiwalled carbon nanotubes. The electrochemical reaction of a redox couple: $Fe(CN)_6^{4-}/Fe(CN)_6^{3-}$ is more reversible at ionic self-assembled multilayers, built using functionalised carbon nanotubes or carbon nanotubes dispersed with functionalised PANI, than on unmodified GC electrodes. The ionic self-assembled multilayers, containing functionalised carbon nanotubes, also electrocatalyze hydrogen peroxide oxidation/reduction. The amperometric signal scales with the number of layers, in both cases, suggesting that all layers are connected and there is three-dimensional access to the multilayer by the redox probes. When the multilayers are built using functionalised polyanilines or carbon nanotubes dispersed using functionalised polyanilines, the assemblies show intrinsic electroactivity. The peak current of the cyclic voltammograms depends on the number of layers. The nanostructures described here only present nanometric scaling in one dimension. However, the conductive building blocks described here could be easily used to build three-dimensional ionic self-assembled multilayers by depositing the layers onto solid nanoparticles (*e.g.* polystyrene latex), as it has been carried out with other materials.[52] In any case, these kind of nanostructures make a bridge between subnanometric species (chemical compounds such as polymers) and nano-objects like nanoparticles, nanotubes, *etc.* Both conducting polymers chains and carbon nanotubes are effective nanowires. Self-assembly of those materials into multilayers are an easy way to produce practical devices. Moreover, the synthesis and functionalisation of the conjugated materials as well as the self assembly of the devices are bottom-up procedures which could be extended to practical applications. In that sense, conjugated materials could be used as building blocks of assembled electronic nanostructures. The materials furnish electronic conductivity and/or other properties to the assemblies, playing the role of a "molecular ink" to wire nanostructures. While this could be considered a simple role, there are few alternative materials which could be used. Moreover, an effective wiring of the assemblies at the molecular level is usually a critical factor on device performance.

Acknowledgements

M. Tagliazuchi, G. A. Planes and R. Echevarria are thanked for measuring TEM images, DEMS profiles and the X-ray diffractograms, respectively. Fruitful discussions with E. J. Calvo and P. N. Bartlett are gratefully acknowledged. D. F. Acevedo and C. Rivarola thank CONICET for graduate and postgraduate research fellowships, respectively. J. Balach thanks EDUMAT for an undergraduate research fellowship. This work was funded by FONCYT, CONICET, Fundación Antorchas, Agencia Cordoba Ciencia, TENARIS and SECYT-UNRC. C. Barbero is a permanent research fellow of CONICET.

References

1 G. B. Blanchet, Y.-L. Loo, J. A. Rogers, F. Gao and C. R. Fincher, *Appl. Phys. Lett.*, 2003, **82**, 463.
2 H. H. Rmaile and J. B. Schlenoff, *Polym. Mater. Sci. Eng.*, 2001, **84**, 678.
3 (*a*) G. Decher and J.-D. Hong, *Macromol. Chem. Macromol. Symp.*, 1991, **46**, 321; (*b*) G. Decher and J.-D. Hong, *Ber. Bunsen-Ges Phys. Chem.*, 1991, **95**, 1430.
4 S. Iijima, *Nature*, 1991, **354**, 56.
5 B. S. Sherigara, W. Kutner and F. D'Souza, *Electroanalysis*, 2003, **15**, 753.
6 S. Bandow, A. M. Rao, K. A. Williams, A. Thess, R. E. Smalley and P. C. Eklund, *J. Phys. Chem. B*, 1997, **101**, 8839.
7 M. J. O'Connell, P. Boul, L. M. Ericson, C. Huffman, Y. Wang, E. Haroz, C. Kuper, J. Tour, K. D. Ausman and R. E. Smalley, *Chem. Phys. Lett.*, 2001, **342**, 265–271.
8 (*a*) Y.-P. Sun, W. Huang, Y. Lin, K. Fu, A. Kitaygorodskiy, L. A. Riddle, Y. J. Yu and D. L. Carroll, *Chem. Mater.*, 2001, **13**, 2864; (*b*) H. Peng, L. B. Alemany, J. L. Margrave and V. N. Khabashesku, *J. Am. Chem. Soc.*, 2003, **125**, 15174.
9 (*a*) R. Bandyopadhaya, E. Nativ-Roth, O. Regev and R. Yerushalmi-Rozen, *Nano Lett.*, 2002, **2**, 25; (*b*) A. Star, D. W. Steuerman, J. R. Heath and J. F. Stoddart, *Angew. Chem., Int. Ed.*, 2002, **41**, 2508; (*c*) J. Wang, M. Musameh and Y. Lin, *J. Am. Chem. Soc.*, 2003, **125**, 2408.
10 B. S. Sherigara, W. Kutner and F. De Souza, *Electroanalysis*, 2003, **15**, 753.
11 (*a*) K. Bumsu, H. Park and W. M. Sigmund, *Langmuir*, 2003, **19**, 2525; (*b*) K. Bumsu and W. M. Sigmund, *Langmuir*, 2003, **19**, 4848; (*c*) A. A. Mamedov, N. A. Kotov, M. Prato, D. M. Guldi, J. P. Wicksted and A. A. Hirsch, *Nat. Mater.*, 2002, **1**, 190; (*d*) J. H. Rouse and P. T. Lillehei, *Nano Lett.*, 2003, **3**, 59.
12 (*a*) H. Shirakawa, *Angew. Chem., Int. Ed.*, 2001, **40**, 2575; (*b*) A. G. McDiarmid, *Angew. Chem., Int. Ed.*, 2001, **40**, 2581; (*c*) A. J. Heeger, *Angew. Chem., Int. Ed.*, 2001, **40**, 2591; (*d*) J. D. Stenger-Smith, *Prog. Polym. Sci.*, 1998, **23**, 57; (*e*) T. Yamamoto, *Macromol. Rapid Commun.*, 2002, **23**, 583.
13 (*a*) T. A. Skotheim, R. L. Elsenbaumer and J. R. Reynolds, *Handbook of Conducting Polymers*, 2nd edn., Marcel Dekker, New York, 1998; (*b*) J. W. Schultze and H. Karabulut, *Electrochim. Acta*, 2005, **50**, 1739.
14 (*a*) L. H. Dall Antonia, M. E. Vidotti, S. I. Cordoba de Torresi and R. M. Torresi, *Electroanalysis*, 2002, **14**, 1577; (*b*) L. L. Miller, J. S. Bankers, A. J. Schmidt and D. C. Boyd, *J. Phys. Org. Chem.*, 2000, **13**, 808; (*c*) T. A. Dickinson, J. White, J. S. Kauer and D. R. Walt, *Tibtech*, 1998, **16**, 250; (*d*) E. Shoji and M. S. Freund, *J. Am. Chem. Soc.*, 2002, **124**, 12486.
15 M. Gerard, A. Chaubey and B. D. Malhotra, *Biosens. Bioelectron.*, 2002, **17**, 345.
16 L. G. Paterno, L. H. C. Mattoso and O. N. Oliveira, Jr., *Quim. Nova*, 2001, **24**, 228.
17 (*a*) J. H. Cheung, W. B. Stockton and M. F. Rubner, *Macromolecules*, 1997, **30**, 2712; (*b*) J. H. Cheung, A. F. Fou and M. F. Rubner, *Thin Solid Films*, 1994, **244**, 985; (*c*) J. H. Cheung, A. F. Fou, M. Ferreira and M. F. Rubner, *Polym. Preprints*, 1993, **34**, 757; (*d*) A. C. Fou, D. Ellis, M. Ferreira and M. F. Rubner, *Polym. Preprints*, 1994, **35**, 221; (*e*) M. Ferreira, J. H. Cheung and M. F. Rubner, *Thin Solid Films*, 1994, **244**, 806.
18 L. H. C. Mattoso, L. G. Paterno, S. P. Campana Filho and O. N. Oliveira, Jr., *Synth. Met.*, 1997, **84**, 123.
19 M. Raposo, R. S. Pontes, L. H. C. Mattoso and O. N. Oliveira, Jr., *Macromolecules*, 1997, **30**, 6095.
20 (*a*) M. Raposo, L. H. C. Mattoso and O. N. Oliveira, Jr., *Thin Solid Films*, 1998, **327**, 739; (*b*) L. H. C. Mattoso, V. Zucolotto, L. G. Paterno, R. Van Griethuijsen, M. Ferreira, S. P. Campana Fo and O. N. Oliveira, Jr., *Synth. Met.*, 1995, **71**, 2037; (*c*) L. F. Malmonge and L. H. C. Mattoso, *Polymer*, 1995, **36**, 245; (*d*) L. H. C. Mattoso, C. E. Martinez, L. G. Borato, L. G. Patterno, P. S. P. Herrmann, O. N. Oliveira, Jr. and L. A. Colnago, in *International Conference on Science And Technology of Synthetic Metals*, Montpellier, France, Le Corum, 1998, p. 34.
21 A. Malinauskas, *J. Power Sources*, 2004, **126**, 214.
22 C. Barbero, M. C. Miras, B. Schnyder and R. Kötz, *J. Mater. Chem.*, 1994, **4**, 1775, and references therein.

23 (*a*) C. Barbero and R. Kötz, *Adv. Mater.*, 1994, **6**, 577; (*b*) R. M. Q. Mello, R. M. Torresi, S. I. Cordoba de Torresi and E. A. Ticianelli, *Langmuir*, 2000, **16**, 7835.
24 H. Salavagione, G. M. Morales, M. C. Miras and C. Barbero, *Acta Polym.*, 1999, **50**, 40.
25 (*a*) A. C. Fou, O. Onitsuka, M. Ferreira, M. F. Rubner and B. R. Hsieh, *J. Appl. Phys.*, 1996, **79**, 7501; (*b*) O. Onitsuka, A. C. Fou, M. Ferreira, B. R. Hsieh and M. F. Rubner, *J. Appl. Phys.*, 1996, **80**, 4067; (*c*) M. Onoda and K. Yoshino, *J. Appl. Phys.*, 1995, **78**, 4456.
26 A. G. MacDiarmid, J. C. Chiang, A. F. Richter, N. L. D. Somarsiri and A. J. Epstein, in *Conducting Polymers*, ed. L. Alcacer, Reidel Publishing Co., Dordrecht, Holland, 1987, p. 105.
27 B. D. Furniss, A. J. Hannaford, P. W. G. Smith and A. R. Tatchell, *Vogel's Textbook of Practical Organic Chemistry*, 5th edn., Prentice Hall, Harlow, 1989.
28 I. Sapurina, A. Riede and J. Stejskal, *Synth. Met.*, 2001, **123**, 503.
29 C. Barbero, G. M. Morales, D. Grumelli, G. Planes, H. Salavagione, C. R. Marengo and M. C. Miras, *Synth. Met.*, 1999, **101**, 694.
30 C. Barbero, H. J. Salavagione, D. F. Acevedo, D. E. Grumelli, F. Garay, G. A. Planes, G. M. Morales and M. C. Miras, *Electrochim. Acta*, 2004, **49**, 3671.
31 D. F. Acevedo, H. J. Salavagione, M. C. Miras and C. A. Barbero, *J. Braz. Chem. Soc.*, 2005, **16**, 259.
32 R. Conley, in *Infrared Spectroscopy*, Allyn and Bacon Inc, Boston, 1966, pp. 144–146.
33 I. Mav and M. Zigon, *Synth. Met.*, 2001, **119**, 145.
34 R. G. Thorne, J. G. Master, S. A. Williams, A. G. MacDiarmid and R. M. Hochstrassner, *Synth. Met.*, 1992, **49/5**, 423.
35 J. Y. Shimano and A. G. MacDiarmid, *Synth. Met.*, 2001, **123**, 251.
36 H. J. Salavagione, D. F. Acevedo, M. C. Miras, A. J. Motheo and C. A. Barbero, *J. Polym. Sci., Part A1*, 2004, **42**, 5587.
37 (*a*) C. A Dyke and J. M. Tour, *J. Am. Chem. Soc.*, 2003, **5**, 1156; (*b*) J. L. Bahr and J. M. Tour, *Chem. Mater.*, 2001, **11**, 3823.
38 (*a*) J. L. Bahr, J. Yang, D. V. Kosynkin, M. J. Bronikowski, R. E. Smalley and J. M. Tour, *J. Am. Chem. Soc.*, 2001, **27**, 6536; (*b*) C. A. Dyke and J. M. Tour, *Chem. Eur. J.*, 2004, **10**, 812.
39 C. Zhou, S. Kumar, C. D. Doyle and J. M. Tour, *Chem. Mater.*, 2005, **17**, 1997.
40 J. L. Hudson, M. J. Casavant and J. M. Tour, *J. Am. Chem. Soc.*, 2004, **126**, 1158.
41 J. M. Nugent, K. S. V. Santhanam, A. Rubio and P. M. Ajayan, *Nano Lett.*, 2001, **1**, 87.
42 (*a*) R. R. Moore, C. E. Banks and R. G. Compton, *Anal. Chem.*, 2004, **76**, 2677; (*b*) C. E. Banks, T. J. Davies, G. G. Wildgoose and R. G. Compton, *Chem. Commun.*, 2005, **7**, 829.
43 J. Wang, M. Musameh and Y. Lin, *J. Am. Chem. Soc.*, 2003, **125**, 2408.
44 A. B. Dalton, J. N. Coleman, M. in het Panhuis, B. McCarthy, A. Drury, W. J. Blau, B. Paci, J.-M. Nunzi and H. J. Byrne, *J. Photochem. Photobiol., A*, 2001, **144**, 31.
45 M. Zhang, A. Smith and W. Gorski, *Anal. Chem.*, 2004, **76**, 5045.
46 M. Zhang, Y. Yan, K. Gong, L. Mao, Z. Guo and Y. Chen, *Langmuir*, 2004, **20**, 8781.
47 (*a*) J. Paloheimo, K. Laakso, H. Isotaalo and H. Stubb, *Synth. Met.*, 1995, **68**, 249; (*b*) P. Granholm, J. Paloheimo and H. Stubb, *Phys. Status Solidi B*, 1998, **205**, 315.
48 C. P. Andrieaux, J. M. Dumas-Bouchiat and J. M. Saveant, *J. Electroanal. Chem., Interfacial Electrochem.*, 1984, **169**, 9.
49 W. J. Albery and A. R. Hillman, *J. Electroanal. Chem., Interfacial Electrochem.*, 1984, **170**, 27.
50 E. J. Calvo, *Electrodos Modificados con Sistemas Enzimaticos y Cables Moleculares Integrados*, XIV Congreso Argentino de Fisicoquimica y Quimica Inorganica 11–14 de Abril, Termas de Rio Hondo, Argentina, 2005.
51 P. P. Joshi, S. A. Merchant, Y. Wang and D. W. Schmidtke, *Anal. Chem.*, 2005, **77**, 3183.
52 E. J. Calvo, C. Danilowicz and A. Wolosiuk, *J. Am. Chem. Soc.*, 2002, **124**, 2452.

Thermal gating of the single molecule conductance of alkanedithiols

Wolfgang Haiss,[a] Harm van Zalinge,[a] Donald Bethell,[a] Jens Ulstrup,[b] David J. Schiffrin[a] and Richard J. Nichols[a]

Received 26th May 2005, Accepted 13th June 2005
First published as an Advance Article on the web 4th October 2005
DOI: 10.1039/b507520n

The temperature dependence of the single molecule conductance (SMC) of α,ω-alkanedithiols has been investigated using a scanning tunnelling microscopy (STM) method. This is based on trapping molecules between a gold STM tip and a gold substrate and measuring directly the current across the molecule under different applied potentials. A pronounced temperature dependence of the conductance, which scales logarithmically with T^{-1}, is observed in the temperature range between 293 and 353 K. It is proposed that the origin of this dependence is the change in distribution between molecular conformers rather than changes in either the conduction mechanism or the electronic structure of the molecule. For alkanedithiols the time averaged conformer distribution shifts to less elongated conformers at higher temperatures thus giving rise to higher conductance across the molecular bridges. This is analysed by first calculating energy differences between different conformers and then calculating their partition distribution. A simple tunnelling model is then used to calculate the temperature dependent conductance based on the conformer distribution. These findings demonstrate that charge transport through single organic molecules at ambient temperatures is a subtle and highly dynamic process that cannot be described by analysing only one molecular conformation corresponding to the lowest energy geometry of the molecule.

1. Introduction

The temperature dependence of the electrical properties of solids is understood in most instances in some detail. The same level of understanding has not been achieved yet for molecular devices and although there has been a substantial theoretical effort[1–3] it has only been recently possible to measure thermal effects in molecular junctions down to the single molecule level. In a recent publication, Selzer *et al.*[4] have shown that it is possible to measure the temperature dependence of the current-voltage (*I*–*V*) response of a single molecule immobilised in a break junction. It was shown that electron transport in a "molecular wire', characterised by a conjugated π system and having a relatively rigid-rod structure, could be switched from coherent superexchange tunnelling at low temperatures to incoherent temperature dependent hopping at higher temperatures.[4] However, other important contributions to the temperature dependence of single molecule conductance such as molecular conformation have been generally neglected in the past, and this forms the motivation for the current manuscript. We show that thermal effects are highly significant at ambient temperatures and as such they would need to be considered in the design of future nano-electronic devices based on molecular electronics.

It has been previously shown that electrochemically gated electron transport across redox active molecules can be investigated in nanoscale structures[5–8] and that these measurements can be carried out for single molecules.[9–12] In the present work, it is shown that temperature can have a gating

[a] *Centre for Nanoscale Science, Chemistry Department, University of Liverpool, Liverpool L69 7ZD, UK*
[b] *Department of Chemistry, Technical University of Denmark, DK-2800, Lyngby, Denmark*

action in charge transfer across molecules through control of the molecular conformation. This has been investigated by measuring the temperature dependence of the molecular conductance for conformationally non-rigid alkanedithiol molecules. The observed behaviour is analysed in terms of the distribution of molecular conformers. The understanding of the effect of molecular conformation on the electrical characteristics of single molecules is of great importance in the emerging field of unimolecular electronics. As pointed out by Troisi and Ratner,[13,14] control of molecular conformation at the metal-molecule interface offers possibilities to realise molecular rectifiers through, for instance, electrical field or current driven switching between conformers. For instance, Donhauser *et al.*[15] have observed conductance switching through thermal fluctuations at the single molecule level, while Gaudioso and Ho[16] have demonstrated that in an STM, tunnelling electrons can be used to excite vibrations.

In order to analyse ambient temperature effects on single molecule conductance we have recorded and analyzed data over the temperature range between 293 and 353 K. The present work demonstrates the importance of conformational effects on molecular conductance in this temperature range and attempts to quantify these effects through single molecule measurements. We show that thermal effects are very significant at ambient temperatures and can be described in terms of the temperature dependence of conformer distribution and a tunnelling barrier model.

2. Methods

Several approaches have been described in the literature for measuring transport through either single or small groups of molecules, including mechanically formed break junctions,[17–20] break junctions formed by electromigration methods[21] and a variety of scanning probe microscopy techniques based on either STM[22–24] or conducting AFM.[25,26] In this work an STM method recently described, based on trapping molecules between a gold STM tip and a gold substrate, has been used.[9] Electrical contact between the target molecule and the gold probes was achieved by the use of thiol groups present at either end of an α,ω-alkanedithiol molecule. The spontaneous attachment and detachment of molecular wires was easily monitored by recording the tunnelling current as a function of time with the STM feedback loop disabled, the so called "$I(t)$ method".[9] A Pico2000 system (Molecular Imaging) STM using the PicoScan 4.19 software was used for these measurements. The relevant tunnelling parameters are the tip potential (U_t) and the setpoint current (I_0). Following disengagement of the feedback loop the current jumps ($I - I_0$, I = tunnelling current) resulting from the spontaneous attachment and detachment of molecules between the tip and the substrate were recorded. The current jumps were then analysed as histograms in order to distinguish between single and multiple molecule attachment or detachment events. The conductance of the molecule in the gap was calculated from these characteristic histograms.

Gold films employed as substrates were prepared by the evaporation of 2 nm of chromium followed by 200 nm of gold onto Tempax® glass (Berliner Glas KG) under high vacuum. These were flame-annealed at approximately 800–1000 °C with a Bunsen burner immediately prior to use; a procedure that is known to result in atomically flat Au(111) terraces.[27]

The $I(t)$ method requires the attachment of the target molecule on the substrate at low coverage. This was achieved by immersion of the gold films in 0.05 mM methanolic solutions of the α,ω-alkanedithiols (Aldrich, reagent grade) for 15 s. After adsorption, the samples were thoroughly washed in ethanol and blown dry with nitrogen. It has been previously shown that such low exposures favour a flat-lying adsorbate orientation from which it is possible to form a molecular bridge between tip and substrate.[9,10] The Au tips were carefully flame-annealed prior to the experiments to remove any organic contamination. Measurements were carried out in air between 293 and 353 K.

The gold on glass samples were mounted on a metal plate that was heated with a variable temperature resistive heating source. The temperature was measured with two thermocouples simultaneously, one placed directly on the heating stage and the other pressed on top of the sample with a rubber seal ring. The thermocouples were calibrated with a precision thermometer.

Theoretical analyses in Section 4 were performed with SPARTAN PRO©. A MMFF basis set was used for the molecular modelling whereas a 6-31G* basis set was used for the Hartree–Fock calculations. The starting point for the H–F calculations for the alkanedithiols was the generation of the corresponding alkane conformers using molecular mechanics (*e.g.* HS-C_6H_{12}-SH was represented by C_8H_{18}). The outermost CH_3 groups were then replaced by S–H groups assuming a linear C–S–H geometry.

Structure optimisation at this level of theory for all possible conformers was outside the possibility of the computation facility available and for this reason single point energy calculations were then carried out after energy minimisation by freezing the symmetry of each conformer. The energies thus obtained were used to calculate the corresponding partition functions describing conformer distribution. Spatial and electronic boundary conditions imposed by the enclosing substrate and tip electrodes were not included in these computations. Focus is thus on the conductance properties of the molecules themselves.

3. Results

Fig. 1a shows a typical example of a current jump (current step = I_w) for 1,8-octanedithiol (ODT) on Au(111) at 65 °C obtained with the $I(t)$ technique.[9] The downwards current jump corresponds to a detachment event. These events were statistically analysed as histogram plots (Fig. 1b). The shape of the histograms obtained at 65 °C (Fig. 1b) was similar to those obtained at room temperature and allowed the determination of the molecular conductance (σ_M) with a standard deviation of 10 to 20%. The current jumps were not dependent on the tunnelling current for $I_0 > I_w$ (see Fig. 1c) and I_w (height of the current jump) scaled linearly with the applied voltage for measurements at 0.2, 0.3 and 0.4 V. The behaviour of the system was very similar to that previously observed for room temperature measurements.[9]

A pronounced increase of conductance at elevated temperatures was found for all the alkanedithiols studied. For example, the conductance of nonanedithiol (NDT) increased from (0.52 ±

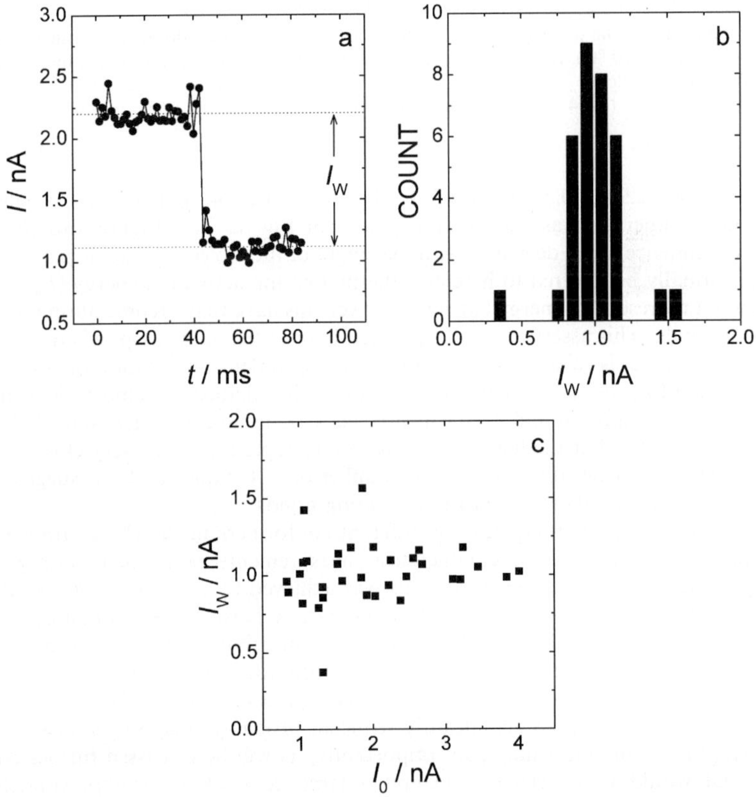

Fig. 1 $I(t)$ measurements at elevated temperatures; 1,8-octanedithiol, 65 °C, U_t = +0.2 V. (a) Typical current jump (height = I_w); (b) Histogram of $I(t)$ current jumps. (c) Plot of the jump height as a function of set-point current I_0.

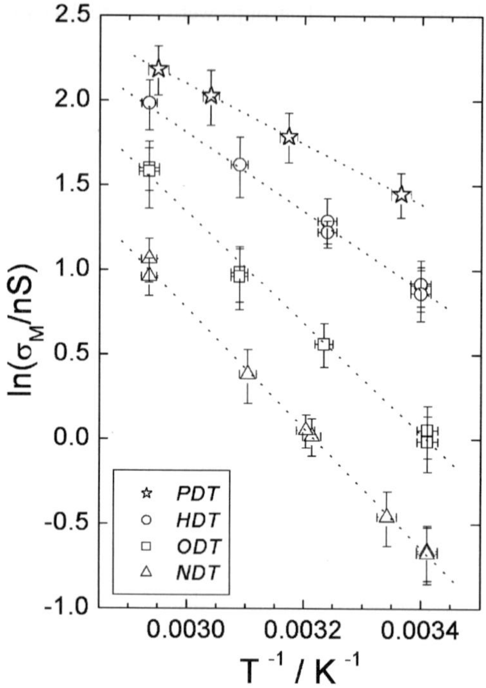

Fig. 2 Logarithmic plot of the single molecule conductance of several alkanedithiols as a function of the inverse temperature (T range: 20–70 °C); dashed lines are linear fits; NDT: 1,9-nonanedithiol, slope: $-(3533 \pm 71)$ K; ODT = 1,8-octanedithiol, slope: $-(3271 \pm 118)$ K; HDT = 1,6-hexanedithiol, slope: $-(2325 \pm 104)$ K; PDT = 1,5-pentanedithiol, slope: $-(1774 \pm 6)$ K; measurements were taken in air at $U_t = +0.2$ and $+0.4$ V; error bars = ±sd.

0.09) nS at 20 °C to (2.90 ± 0.37) nS at 67 °C. The logarithm of σ_M is linearly dependent on the inverse absolute temperature as can be clearly seen in Fig. 2. This Figure also shows that the temperature dependence of σ_M decreases with the length of the molecule. The slopes of the linear fits in Fig. 2 will formally be referred to here as a thermal gating activation energy (E_g) although this quantity does not represent an energy barrier that electrons have to overcome during their transport through the molecule. This assertion will be justified later in the manuscript where it is shown that the thermal gating effect is related to the temperature dependent distribution of conformers.

The dependence of E_g on the number of C–C bonds in the alkanedithiol molecule is shown in Fig. 3. An almost linear relation is found from which E_g per C–C bond can be determined, $E_{C-C} = (39 \pm 2)$ meV. It is remarkable that the linear fit in Fig. 3 intercepts the y-axis very close to zero (intercept = 0.04 ± 0.08). E_g scales directly with the number of C–C bonds and it is suggested that this is a consequence of the number of conformer-creating bonds.

Further evidence for the participation of different conformers in the charge transport through alkanedithiols was obtained from conductance measurements performed at large tip-sample separations (*i.e.* small tunnelling currents). This was achieved by altering the set point current I_0. Larger values of I_0 result in a smaller tip-sample separation. It was expected that only at relatively short tip-sample separations all the conformers would contribute to the observed current. Fig. 4 shows the dependence of the current jump values for pentanedithiol (PDT) on I_0. If I_0 is smaller than a critical current (I_c), that is to say the tip-sample separation (s) exceeds a critical distance (s_c), the value of the current jumps is significantly reduced, probably because for $s > s_c$, not all the conformers can bridge the tunnelling gap. Importantly, as will be discussed further on, the short conformers that would be expected to exhibit a large conductance cannot contribute to the measured current. In contrast, for $s < s_c$, the current jumps showed little dependence on I_0 for values up to 30 times that of I_w. It is proposed, therefore, that for $s < s_c$ the different conformers of a molecule present in the gap can establish chemical contact across the two contacts and therefore,

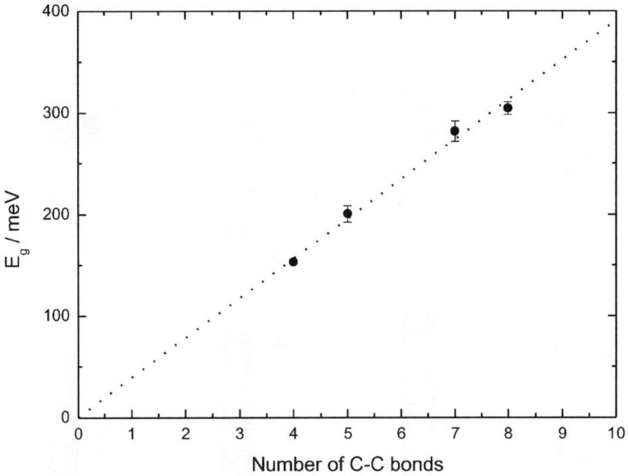

Fig. 3 Dependence of E_g (in meV) on the number of C–C bonds in the molecule; error bars = ±sd. Note that the linear fit (dotted line) crosses the *y*-axes close to 0.

all conformers can contribute to I_w for this condition. It has been previously shown that only molecules establishing chemical contact between the metal probes contribute significantly to the observed current.[25]

4. Theoretical aspects of the temperature dependence of single molecule conductance

The temperature dependence of the conformer distribution and its influence on electron transport through molecules will be discussed first for pentane as a simple example to describe the computational strategy. Fig. 5 shows ball and stick models of the four energetically different conformers of pentane. These structures correspond to energy minima obtained from a molecular modelling calculation. It is assumed that tunnelling occurs across a barrier the width of which is related to the distance between the terminal groups. In this approximation temperature affects tunnelling through the temperature dependent distribution of conformers. However, the tunnelling

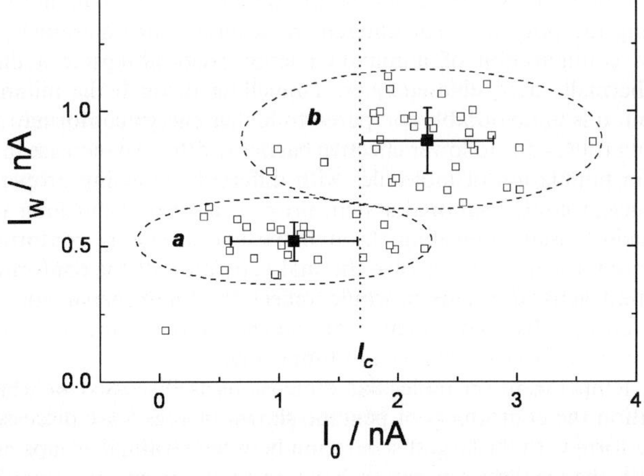

Fig. 4 Dependence of the current jumps in the $I(t)$ spectra of 1,5-pentanedithiol, on the set-point tunnelling current (I_0); $T = 24$ °C; $U_t = +0.2$ V; group a: $I_w = (518 \pm 71)$ pA at $I_0 = (1127 \pm 529)$; group b: $I_w = (890 \pm 120)$ pA at $I_0 = (2250 \pm 553)$ pA.

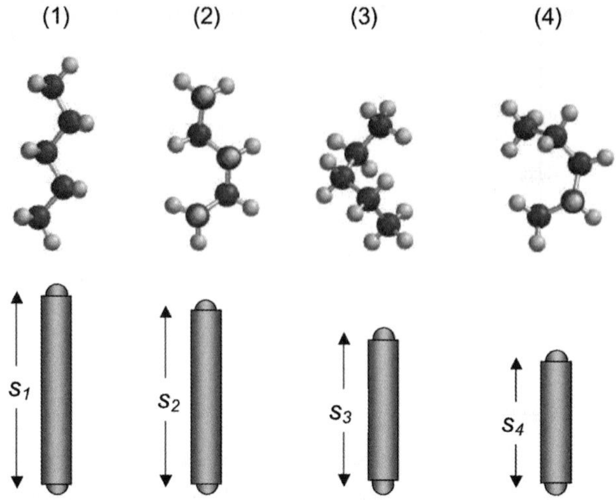

Fig. 5 Ball and stick models of four energetically different pentane conformers; numbering from 1 (all *trans*) to 4 (all *gauche*) as in Tables 1 and 2. The lower diagram shows the effective barrier model, where the conformers are replaced by a medium with the barrier width (s_x) being determined by the length of the conformer (the distance between C_1 and C_5 in this case).

probability will also depend on conformation through changes with temperature of the energy of the HOMO and LUMO. This question is discussed further on.

In what follows, a qualitative discussion of these contributions is presented. Alkanethiols do not have a low lying LUMO that can be temporarily populated or depopulated by electron transfer and therefore, the temperature dependence of the single-molecule conductance cannot originate from multiphonon electron transfer steps such as those observed for single-molecule electronic conductance where low-lying redox centres are involved as electronic relays. Sequential electron transfer through localized redox centres operates in the STM tunnelling mechanism of the viologen-based redox bridge molecule studied recently.[7–11] The redox transmitter group used in that work was linked to the substrate and tip electrodes by thiol-linker groups similar to those addressed here. Bridge group molecules based on transition metal complexes are other systems which also exploit redox transmitter groups.[28,29]

The temperature dependence of tunnelling through a polymethylene chain can be regarded as an example of "gated" electron tunnelling through the alkanedithiol molecules ("gated" in this context means regulating the flow of current across the molecular wire by changing the temperature and hence, by controlling the population of different molecular conformations). This implies that a nuclear equilibrium configuration of minimum energy exists alongside a distribution of other configurations at thermally accessible energies. Tunnelling through the minimum energy configuration is feasible but it is unfavourable compared to higher energy conformations, which can offer more facile tunnelling routes due to lower effective barrier widths. An increase in temperature would result in a change in population of molecules with different tunnelling properties. Thus, thermal activation of the nuclear conformational system prior to electron tunnelling would lead to more facile electron tunnelling than tunnelling through the lowest energy conformer. Gated electron tunnelling is therefore most likely to involve thermal activation of the conformer distribution and the observed apparent activation energy would reflect this pre-organisation. In this model, the apparent activation energy observed is then a consequence of the changes in the distribution of the structural identity of the molecules involved in tunnelling.

The influence of temperature on molecular dimensions is discussed in what follows. For the purposes of illustration the conformers of pentane shown in Fig. 5 are discussed. Conformer 1 in Fig. 5 is the all *trans* form with the largest separation between terminal groups, while conformer 4 is the most folded with the smallest separation between terminal groups. For the purposes of our calculations we make the simplest approximation that the barrier width for tunnelling between the terminal groups is related to the separation between the two terminal groups. Hence, our

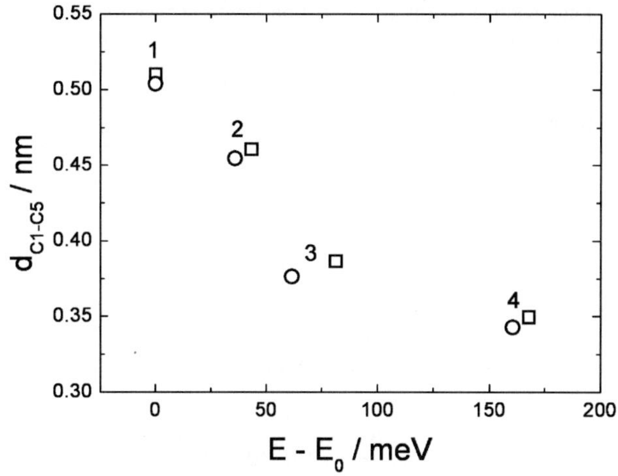

Fig. 6 Dependence of the distance between C_1 and C_5 on the relative energy for the four energetically different conformers of pentane. Molecular modelling calculations (circles) are compared with Hartree–Fock calculations (squares); the conformer number shown next to the symbols relates to the geometry of the molecule shown in Fig. 5 and Tables 1 and 2.

approximation is that conformer 1 will have the largest barrier width, while conformer 4 has the smallest. It is proposed to the first approximation that the conductance scales with the barrier width according to a simple 1-D tunnelling barrier model (see below). Using this assumption each conformer is replaced by a barrier of effective width s_x (where x refers to conformer x), while the other barrier properties such as barrier height remain unchanged. Hence the effective barrier medium for all 4 conformers is assumed to be the same, with the width progressively decreasing from conformer 1 to 4. This model is also illustrated in the lower half of Fig. 5 for each of the conformers.

The influence of temperature on conformer distribution and tunnelling probability was estimated by calculating first the energy of the conformers relative to the conformation of lowest energy. Fig. 6 shows the results of such calculations. The relative energy ($E - E_0$) increases when the distance between C_1 and C_5 decreases. Qualitatively, it can be seen that the average length of the molecule is reduced notably upon an increase of temperature and hence the tunnelling probability from C_1 to C_5 should increase significantly with increasing temperature. To quantify the influence of this effect, the partition function for the conformers was determined using a statistical mechanics formulation: A molecular state with relative energy ε_j and degeneracy g_j has a probability P_j to be occupied at a temperature T given by:[30]

$$P_j = \frac{g_j}{q}\exp(-\varepsilon_j/kT) \tag{1a}$$

with the molecular partition function q given by:

$$q = \sum_j g_j \exp(-\varepsilon_j/kT) \tag{1b}$$

ε_j is the energy of the molecule in the quantum state j. Table 1 shows the probability of the molecule to be present as a particular conformer (eqn. (1)) for 20 and 70 °C and the corresponding molecular dimensions. The probability for a molecule to be present as a conformer of higher energy, with a resulting decrease of average length, increases significantly in this temperature range. As schematically shown in Fig. 5 and in Table 1, the coiled conformations dominate as the temperature is raised. The observed temperature dependence would therefore most likely arise from increasing contributions from more contracted conformers.

The tunnelling process can be described by a simple model, assuming one-dimensional tunnelling through a rectangular barrier, where the exponential decay of the wave function in the barrier is

Table 1 Dependence of the relative energy ($E_x - E_1$) on length change ($d_x - d_1$) of the four energetically different pentane conformers determined from molecular mechanics calculations. The energy increases with decreasing length of the molecule. d_x denotes the distance between the two outermost C atoms in conformer x. P denotes the probability distribution of the conformers calculated from the partition function, taking into account the degeneracy (Deg.)

Conformer number	Deg.	$E_x - E_1$/meV	$d_x - d_1$/Å	d_x/Å	P/% 20 °C	P/% 70 °C
1 (*a*,*a*)	1	0	0	5.04	46.7	41.0
2 (*g*,*a*)	4	36	−0.50	4.54	45.0	48.5
3 (*g*,*g*)	2	62	−1.28	3.76	8.13	10.2
4 (*g*,*g*)	2	160	−1.61	3.43	0.16	0.36

characterised by the decay constant κ (in nm^{-1}) with

$$\kappa = \sqrt{\frac{2m_e^*\phi}{(h/2\pi)^2}} \tag{2}$$

and

$$-2\kappa = \frac{d(\ln(I))}{ds} \tag{2a}$$

where h is Planck's constant, ϕ is the effective barrier height (in J) and m_e* is the effective electron mass. If ϕ is expressed in meV, κ in nm^{-1} and m* is taken as the rest mass of the electron then eqn. (2) transforms into the simple expression:

$$\kappa/nm^{-1} = \sqrt{\frac{\phi/meV}{38}} \tag{2b}$$

A change of the separation between the ends of the molecule attached to the contacts, $\Delta s = s_x - s_1$ consequently results in a change of the tunnelling current given by:

$$\frac{I_x}{I_1} = \exp(-2\kappa(s_x - s_1)) \tag{3}$$

where the subscript "1" denotes the lowest energy conformer and "x" corresponds to conformer x. I_x then refers to the current through conformer x and I_1 is the current through the lowest energy conformer.

Assuming $\phi_x = \phi_1 = 5$ eV (valid for small applied potentials) and m_e* = m_e, (m_e = rest mass of the electron) eqn. (3) yields:

$$\frac{I_x}{I_1} = \exp(-22.94(s_x - s_1)) \tag{4}$$

with s in nm. From eqn. (4), the relative current (I_x/I_1) for the different conformers can be calculated. For the lowest energy conformation (conformer 1; all anti) $I_x/I_1 = 1$. The total tunnelling current (I_T) is given by

$$I_T \propto \sum (P_x I_x/I_1) \tag{5}$$

The sum over all conformers of the probability-weighted currents $\sum(P_x I_x/I_1)$ is a measure of the temperature dependent tunnelling probability. Table 2 shows numerical values of the quantities described above for tunnelling between C_1 and C_5 in a pentane molecule.

Table 2 also shows the percentage current fractions (I_{fra}) for the different conformers. It is interesting to note that the energetically higher lying conformers 2 and 3 (compare with the results in Table 1) give the largest contributions to the tunnelling current. It is important to notice that the transition between the different conformers of alkanes (at least in the gas phase) is very fast since these involve rotations of C–C bonds.[31] Therefore, in STM experiments, where the tunnelling current through a molecule is measured in a time scale of milliseconds, the probability-weighted

Table 2 Relative currents (I_x/I_1) for tunnelling between the outermost C atoms of pentane calculated from eqn. (4) for the four different pentane conformers. The sum of the probability-weighted currents ($\sum(P_xI_x/I_1)$) shown at the bottom of columns 6 (for 20 °C) and 7 (for 70 °C) is a measure of the tunnelling probability between C_1 and C_5. Its value increases by approximately 15% upon an increase of temperature from 20 to 70 °C

Conformer number	$d_x - d_1$ /Å	I_x/I_1	P_x/% 20 °C	P_x/% 70 °C	P_xI_x/I_1 20 °C	P_xI_x/I_1 70 °C	I_{fra}/% 20 °C	I_{fra}/% 70 °C
1 (*a,a*)	0	1	46.73	40.95	0.47	0.410	13.5	10.3
2 (*a,g*)	−0.50	3.12	44.98	48.51	1.40	1.514	40.7	38.2
3 (*g,g*)	−1.28	18.7	8.12	10.16	1.52	1.897	43.9	47.8
4 (*g,g*)	−1.61	40.4	0.16	0.36	0.07	0.146	1.9	3.7
Sum:			100	100	3.45	3.97	100	100

d_x is the length of conformer x. P_x is the probability (%) of the molecule to be present as conformer x. I_x is the current through conformer x, while I_1 is the current through conformer 1, the ground state. I_{fra} are the percentage current fractions for the different conformers.

average of the conductivities of the different conformers is measured. For these reasons σ_M should be seen as the time average of the conductance of an ensemble of conformers in thermal equilibrium.

To study the chain length dependence of $d(\ln(\sum(P_xI_x/I_1))/dT^{-1}$ calculations of $\sum(P_xI_x/I_1)$ as a function of temperature have been carried out for hexane and octane (Fig. 7). The results of these

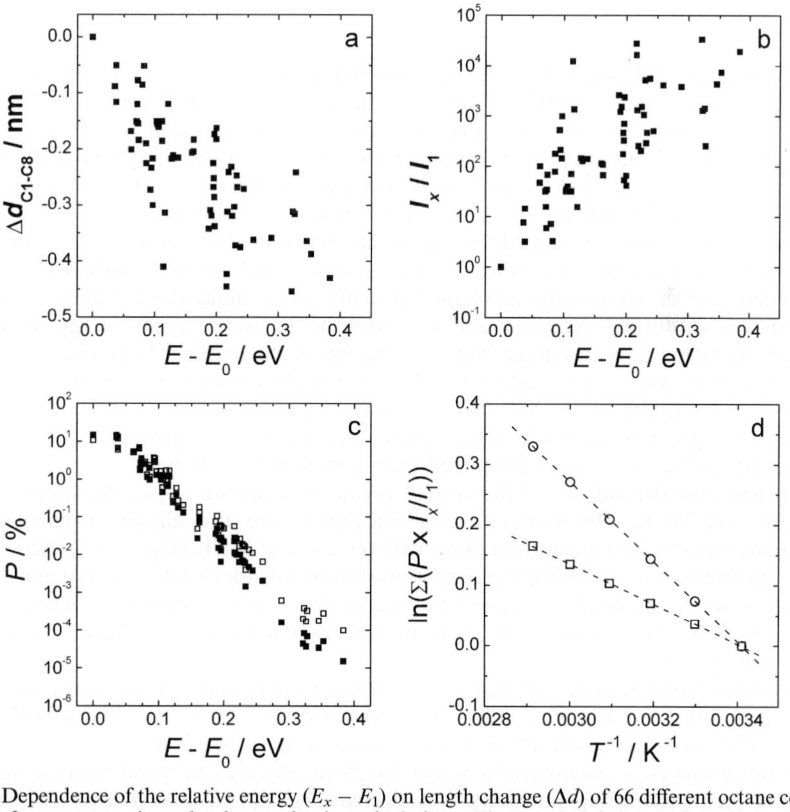

Fig. 7 (a) Dependence of the relative energy ($E_x - E_1$) on length change (Δd) of 66 different octane conformers determined from systematic molecular mechanics calculations. The energy of the conformers increases with decreasing length of the molecule. (b) Relative conductance (I_x/I_1) of the conformers of octane. (c) Probability of occupation for 20 °C (filled squares) and 70 °C (open squares) for octane. (d) T-dependence of $\ln(\sigma_M)$ for both hexane and octane; the value of $\sum(PI_x/I_1)$ has been normalised with respect to its value at 20 °C. The logarithm of the conductance scales linearly with $1/T$ and the slope increases with increasing chain length.

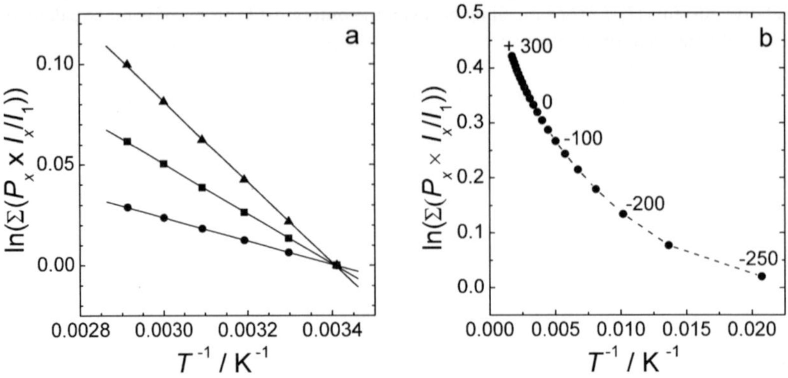

Fig. 8 (a) The dependence of the logarithm of the calculated tunnelling probability on T^{-1} for different alkanedithiols: 1,3-propanedithiol (circles; slope = −46 K), 1,4-butanedithiol (squares; slope = −124 K) and 1,5-pentanedithiol (triangles; slope = −201 K); the lines are linear fits. Hartree–Fock methods were used for the calculations of the tunnelling probability between the two terminal sulfur atoms. Note that the conductance scales linearly with $1/T$ in the temperature range between 290 and 350 K and the slope increases with increasing chain length as was found in the experimental study. For each molecule, the value of $\sum(P_x \times I_x/I_1)$ has been normalised with respect to its value at 20 °C. (b) Hartree–Fock calculations for 1,3-Propanedithiol covering a wider temperature range; the numbers next to the symbols correspond to the temperature in °C. The value of $\sum(P_xI_x/I_1)$ has been normalised with respect to its value at 0 K.

calculations are in broad agreement with the experimental results (Fig. 2), since $\ln(\sum(P_xI_x/I_1))$ and hence $\ln(\sigma_M)$ scales linearly with $1/T$ and the slope of $d(\ln(\sum(P_xI_x/I_1))/dT^{-1}$ increases with increasing chain length. To enable direct comparison between theory and experiment, calculations have also been carried out for propanedithiol, butanedithiol and pentanedithiol, where a realistic value of the barrier height (1.2 eV [32]) and an effective electron mass ($m_e^* = 0.4m_e$,[32]) have been used (Fig. 8). The numerical results for the slope in Fig. 8 were similar for different calculation methods for the case of the simple propanedithiol molecule, whereas the results for the more complex butanedithiol and pentanedithiol molecules on the other hand, were strongly dependent on the particular choice of the computational method (Hartree–Fock, molecular mechanics or DFT) used to calculate the conformer geometry and its corresponding single point energy. If all possible conformers, including those resulting from the rotation of the terminal H atoms, were taken into account, unrealistic values were calculated due to a wide variance of the different conformer energies of similar length. To avoid these problems the number of conformers was reduced by assuming a linear geometry for the C–S–H groups for each different conformation of the S–$(CH_2)_n$–S skeleton. Hence the calculations proceeded from a straight C–S–H bond.

Fig. 8a shows the dependence of the current predicted from eqn. (4) in the temperature range between 290 and 350 K. The results of these calculations are in qualitative agreement with the experimental results in the temperature range between 293 and 353 K (Fig. 2). Fig. 8(b) shows that, for 1,3-propanedithiol as an example, the relationship between $\ln(\sum(P_x \times I_x/I_1))$ and $1/T$ is non-linear over a wider temperature range (not accessible in our experiments). Therefore at low temperatures where the ground state conformer is the main state populated, the temperature dependent gating is much weaker.

The calculated slope of $d(\ln(\sum(P_xI_x/I_1))/dT^{-1}$ for tunnelling between the two sulfur atoms[33] in pentanedithiol (201 K; see Fig. 8(a)) differs from the experimental value of $d(\ln(\sigma_M))/dT^{-1}$ (1774 K; see Fig. 2). The difference can be attributed to a breakdown of the effective barrier height at small substrate–tip distances, a phenomenon which has been observed at metal–vacuum–metal[34] and metal–water–metal tunnelling junctions.[35] A decrease of the effective barrier height at small separations will increase the slope of $d(\ln(\sum(I_x/I_1))/dT^{-1}$ since shorter conformers are more prevalent at higher temperatures (*vide infra*).

The model and calculations above represent straightforwardly tunnelling through a temperature dependent barrier width and the strong temperature dependence clearly indicates the gated nature

Table 3 Influence of the effective barrier height on the temperature dependent tunnelling probability (P_xI_x/I_1) in propane; see text for details

Conformer number	$\Delta\phi$/meV	I_x/I_1	P_x(%) 20 °C	P_x(%) 70 °C	P_xI_x/I_1 20 °C	P_xI_x/I_1 70 °C
1 (*a*,*a*)	0	1	46.73	40.96	46.73	40.95
2 (*g*,*a*)	−23	1.03	44.98	48.52	46.42	50.07
3 (*g*,*g*)	−145	1.22	8.125	10.17	9.936	12.43
4 (*g*,*g*)	−129	1.20	0.163	0.361	0.195	0.432
Sum			100	100	1.033	1.039

of the conduction mechanism; *i.e.* shorter conformers present a narrower barrier width. However, through bond tunnelling would also include a gating element, considering the structural dependence of the energetics and electronic overlaps in the appropriate HOMO's or LUMO's. The temperature dependence of these effects might be significant but are not expected to be nearly as strong as the effective barrier width considerations, as the bonds contributing to the HOMO and LUMO remain the same in different conformations. The effects can be illuminated by the following simple calculation. The effective barrier height for through-bond tunnelling in a symmetric metal-molecule-metal junction is given by half the difference between the HOMO and the LUMO energy if the Fermi level of the metal is placed half way between the energy of these molecular orbitals.[36] The tunnelling probability for the different conformers is then influenced by a change in barrier height ($\Delta\phi$) relative to the barrier height of the lowest energy conformation (ϕ_1). Eqns. (2b) and (3) and $\phi_x = \phi_1 + \Delta\phi$ yield:

$$I_1 = I(\phi_1) = \exp(-2d(\phi_1/38)^{1/2}) \tag{6}$$

$$I_x = I(\phi_1 + \Delta\phi) = \exp(-2d((\phi_1 + \Delta\phi)/38)^{1/2}) \tag{7}$$

From eqns. (6) and (7), the current ratio is given by

$$I_x/I_1 = \exp(-2d(((\phi_1 + \Delta\phi)/38)^{1/2}) - (\phi_1/38)^{1/2})) \tag{8}$$

Table 3 shows numerical results calculated from eqn. (8) for pentane using the *P* values from Table 1 and assuming $\phi_1 = 5$ eV and $d = 0.6$ nm = constant. $\Delta\phi$ has been calculated for the various conformers from the HUMO and LUMO energies. The current increases only by 0.6% for an increase of temperature from 20 to 70 °C. The increase is small. This conclusion should, however, be substantiated by '*ab initio*' calculations for the situation encountered in the experiments, *i.e.* with the alkanedithiols incorporated between two suitably represented gold electrodes and including all the different conformers *and* different adsorption geometries. Such computations are feasible but beyond the scope of the present work.

The analysis presented in this section has demonstrated that conformational changes are the most likely reason for the pronounced temperature dependence observed for single molecule conductance of alkanedithiols. The molecule trapped between tip and support has freedom to adopt many conformations. In support of the proposed mechanism, it is noteworthy that the conductance of self assembled monolayers of alkanedithiols sandwiched between two metal electrodes show only little temperature dependence.[32] This is probably a result of the absence of conformational freedom for molecules present in a densely packed layer.

5. Conclusions

The pronounced temperature dependence of the conductance of single alkanedithiol molecules observed experimentally can be understood semi-quantitatively by taking into account the temperature dependent conformer distribution and its influence on the tunnelling probability in a simple barrier tunnelling model. This model also predicts a pronounced increase of the temperature dependence of σ_M with increasing chain length, which was observed experimentally. These findings demonstrate that charge transport through organic molecules at ambient temperatures is a subtle and highly dynamic process, which cannot be described by the analysis of only one conformation locked in the lowest energy geometry.

The model presented is a simplification of a complex problem in gated tunnelling since the interplay between through-bond and through-space tunnelling requires a more extensive analysis. The main point of this work, however, is to have demonstrated that single molecule conductance is strongly temperature dependent whereas the energies of the HOMO and LUMO are not. The implications of these observations are that we are dealing here mainly with conformation effects, which have been generally neglected in the past.

Acknowledgements

This work was carried with financial support from the EPSRC under grants GR/R07684/01 (Scanning Tunneling Spectroscopy of Nanoscale Structures) and EP/C00678X01 (Mechanisms of Single Molecule Conductance) and the European Union (SUSANA project (HvZ), Human Potential Program (HPRN-CT-2002-00185)).

References

1 D. Segal and A. Nitzan, *Chem. Phys.*, 2002, **281**, 235.
2 A. Nitzan, *Annu. Rev. Phys. Chem.*, 2001, **52**, 681.
3 D. Segal, A. Nitzan, M. Ratner and W. B. Davis, *J. Phys. Chem. B*, 2000, **104**, 2790.
4 Y. Selzer, M. A. Cabassi, T. S. Mayer and D. L. Allara, *J. Am. Chem. Soc.*, 2004, **126**, 4052.
5 D. I. Gittins, D. Bethell, R. J. Nichols and D. J. Schiffrin, *Adv. Mater.*, 1999, **11**, 737.
6 D. I. Gittins, D. Bethell, R. J. Nichols and D. J. Schiffrin, *J. Mater. Chem.*, 2000, **10**, 79.
7 D. I. Gittins, D. Bethell, D. J. Schiffrin and R. J. Nichols, *Nature*, 2000, **408**, 67.
8 W. Haiss, R. J. Nichols, S. J. Higgins, D. Bethell, H. Hobenreich and D. J. Schiffrin, *Faraday Discuss.*, 2004, **125**, 179.
9 W. Haiss, R. J. Nichols, H. van Zalinge, S. J. Higgins, D. Bethell and D. J. Schiffrin, *Phys. Chem. Chem. Phys.*, 2004, **6**, 4330.
10 W. Haiss, H. van Zalinge, S. J. Higgins, D. Bethell, H. Hobenreich, D. J. Schiffrin and R. J. Nichols, *J. Am. Chem. Soc.*, 2003, **125**, 15294.
11 W. Haiss, H. van Zalinge, H. Hobenreich, D. Bethell, D. J. Schiffrin, S. J. Higgins and R. J. Nichols, *Langmuir*, 2004, **20**, 7694.
12 B. Q. Xu, X. Y. Xiao, X. M. Yang, L. Zang and N. J. Tao, *J. Am. Chem. Soc.*, 2005, **127**, 2386.
13 A. Troisi and M. A. Ratner, *Nano Lett.*, 2004, **4**, 591.
14 A. Troisi and M. A. Ratner, *J. Am. Chem. Soc.*, 2002, **124**, 14528.
15 Z. J. Donhauser, B. A. Mantooth, K. F. Kelly, L. A. Bumm, J. D. Monnell, J. J. Stapleton, D. W. Price, A. M. Rawlett, D. L. Allara, J. M. Tour and P. S. Weiss, *Science*, 2001, **292**, 2303.
16 J. Gaudioso and W. Ho, *J. Am. Chem. Soc.*, 2001, **123**, 10095.
17 M. A. Reed, C. Zhou, C. J. Muller, T. P. Burgin and J. M. Tour, *Science*, 1997, **278**, 252.
18 C. Kergueris, J. P. Bourgoin, S. Palacin, D. Esteve, C. Urbina, M. Magoga and C. Joachim, *Phys. Rev. B*, 1999, **59**, 12505.
19 H. B. Weber, J. Reichert, R. Ochs, D. Beckmann, M. Mayor and H. von Lohneysen, *Physica E*, 2003, **18**, 231.
20 J. Reichert, R. Ochs, D. Beckmann, H. B. Weber, M. Mayor and H. von Lohneysen, *Phys. Rev. Lett.*, 2002, **88**, art no.
21 H. Park, A. K. L. Lim, A. P. Alivisatos, J. Park and P. L. McEuen, *Appl. Phys. Lett.*, 1999, **75**, 301.
22 A. Yazdani, D. M. Eigler and N. D. Lang, *Science*, 1996, **272**, 1921.
23 B. Q. Xu and N. J. J. Tao, *Science*, 2003, **301**, 1221.
24 M. Dorogi, J. Gomez, R. Osifchin, R. P. Andres and R. Reifenberger, *Phys. Rev. B*, 1995, **52**, 9071.
25 X. D. Cui, A. Primak, X. Zarate, J. Tomfohr, O. F. Sankey, A. L. Moore, T. A. Moore, D. Gust, G. Harris and S. M. Lindsay, *Science*, 2001, **294**, 571.
26 G. Leatherman, E. N. Durantini, D. Gust, T. A. Moore, A. L. Moore, S. Stone, Z. Zhou, P. Rez, Y. Z. Liu and S. M. Lindsay, *J. Phys. Chem. B*, 1999, **103**, 4006.
27 W. Haiss, D. Lackey, J. K. Sass and K. H. Besocke, *J. Chem. Phys.*, 1991, **95**, 2193.
28 T. Albrecht, A. Guckian, J. Ulstrup and J. G. Vos, *Trans. IEEE Nanotechnol.*, 2005, in press.
29 T. Albrecht, A. Guckian, J. Ulstrup and J. G. Vos, *Nano Lett.*, 2005, in press.
30 P. W. Atkins, *Physical Chemistry*, Oxford University Press, Oxford, 1st edn., 1978.
31 P. Y. Bruice, in *Organic Chemistry*, Prentice Hall, Englewood Cliffs, NJ, 2004.
32 W. Y. Wang, T. Lee and M. A. Reed, *J. Phys. Chem. B*, 2004, **108**, 18398.
33 For completeness it should be mentioned that $\mathrm{d}(\ln(\sum(P_xI_x/I_1))/\mathrm{d}T^{-1}$ is similar for other pairs of atoms.
34 S. Yagyu and M. Yoshitake, *Surf. Interface Anal.*, 2004, **36**, 1110.
35 J. R. Hahn, Y. A. Hong and H. Kang, *Appl. Phys. A*, 1998, **66**, S467.
36 X. D. Cui, A. Primak, X. Zarate, J. Tomfohr, O. F. Sankey, A. L. Moore, T. A. Moore, D. Gust, L. A. Nagahara and S. M. Lindsay, *J. Phys. Chem. B*, 2002, **106**, 8609.

In situ scanning tunnelling spectroscopy of inorganic transition metal complexes†

Tim Albrecht,[a] Kasper Moth-Poulsen,[b] Jørn B. Christensen,[b] Adrian Guckian,[c] Thomas Bjørnholm,[b] Johannes G. Vos[c] and Jens Ulstrup[a]

Received 18th April 2005, Accepted 16th May 2005
First published as an Advance Article on the web 12th October 2005
DOI: 10.1039/b505451f

Redox molecules with equilibrium potentials suitable for electrochemical control offer perspectives in nanoscale and single-molecule electronics. This applies to molecular wiring, but also towards higher sophistication such as transistor or diode function. Most recent nanoscale or single-molecule functional systems are, however, fraught with operational limitations such as cryogenic temperatures and ultra-high vacuum, or lack of electrochemical potential control. We report here cyclic voltammetry (CV) using single-crystal Au(111)- and Pt(111)-electrodes and electrochemical *in situ* scanning tunnelling microscopy (STM) of a class of Os(II)/(III)- and Co(II)/(III)-complexes, the former novel in molecular electronics. The complexes are robust, with ligand groups suitable for linking the complexes to the Au(111)- and Pt(111)-surfaces *via* N- and S-donor atoms. The CV data reflect monolayer behaviour. Interfacial ET of the Os-complexes is fast, $k^0_{ET} \geq 10^6$ s^{-1}, while the Co-complex reacts much more slowly, $k^0_{ET} \approx (1-3) \times 10^3$ s^{-1}. *In situ* STM of the Os-complexes shows a maximum in the tunnelling current/overpotential relation at constant bias voltage with up to 50-fold current rise. The peak position follows systematically the bias voltage and equilibrium potential, in keeping with theoretical frames for two-step electron transfer (ET) of *in situ* STM of redox molecules. The molecular conductivity behaves broadly similarly. The Co-complex also shows a tunnelling spectroscopic feature but much weaker than the Os-complexes. This can be ascribed to the much smaller interfacial ET rate constant, again caused by large intramolecular nuclear reorganization and weak electronic coupling to the substrate electrode. Overall the study has mapped the properties of target molecules needed for stable electronic switching, of possible importance in molecular electronics towards the single-molecule level, in room temperature condensed matter environment.

1. Introduction

Emerging interfacial chemistry at the nanoscale and towards the single-molecule level offers electronically functional molecular systems of increasingly high sophistication. Molecular conductivity and interfacial electron transfer (ET) between the functional target molecules and enclosing electrodes are central notions. Overarching objectives are, to design and prepare strategic molecules and supramolecular systems, to map and control the molecular and supramolecular organization and function at the monolayer and single-molecule levels, and to insert the organized or single-molecule systems into mesoscopic and nanoscale environments. These can be scanning

[a] *Technical University of Denmark (DTU), Dept. Chemistry, Nano-DTU, Kemitorvet, Bldg. 207, Dk-2800, Kgs Lyngby, Denmark*
[b] *University of Copenhagen, Nano-Science Center, Universitetsparken, Dk-2100, Copenhagen, Denmark*
[c] *Dublin City University (DCU), School of Chemical Sciences, Dublin 9, Ireland*

† Electronic supplementary information (ESI) available: Details of the synthesis of the Co complex. See http://www.rsc.org/suppdata/fd/b5/b505451f/

tunnelling microscopy (STM) or other configurations such as nanogap electrode assemblies.[1–6] The former offers clues to the molecular *mechanisms* of interfacial ET, molecular conductivity, and electron tunnelling through molecules. The latter, fabricated by lithographic methods,[7] break-junction,[8] and electromigration[9] may also represent steps towards "circuits" of nanoscale, and ultimately single-molecule electronic components.[10–12] Recently reported nanoscale and single-molecule electronic systems testify that these objectives are being approached. Single-molecule components include different types of molecular wiring extending to wiring based on biomolecules,[13–18] molecular rectifiers,[6,19,20] molecular "switches",[21–23] *i.e.* electronic systems with large on-off current or conductivity ratios in narrow voltage ranges, molecular "transistors",[24–26] and approaches to "circuits" of these.[10–12,27]

Molecular rectification and "switch" features are based on interfacial electron transfer between the redox target molecules and the enclosing electrodes, which also connect the molecules to external circuits. Limitations of a number of single-molecule devices—the molecular "transistors" in particular—are, however, that they can only be brought to operate at cryogenic temperatures and/or in ultra-high vacuum.[25–27] Room temperature and condensed matter conditions are, on the other hand, needed for practical devices. Nanoscale and single-molecule *electrochemically* based ET systems here offer constructive approaches to the operational principles and molecular mechanisms of single-molecule devices, and guidance to the construction of such devices. In particular, the three(four)-electrode electrochemical *in situ* STM configuration shows close analogies to three-electrode molecular transistors where the source and drain electrodes correspond to the substrate and tip electrodes, and the gate electrode to the electrochemical reference electrode.

State-of-the-art interfacial electrochemical ET systems can, first, be brought to operate in well-defined environments (ultra-pure conditions, atomically flat electrochemical surfaces). Spatial and functional resolution can, further, be at the nanoscale and single-molecule levels.[28,29] This level of resolution is also finding its way into bioelectrochemistry.[1,30] Several types of electrochemical single-molecule redox switching systems have, secondly been reported. One is Tao's study of Fe–protoporphyrin IX on highly oriented pyrolytic graphite (HOPG) by electrochemical *in situ* STM.[31] This molecule showed a tunnelling current peak around the equilibrium redox potential, *i.e.* a narrow potential region with high "switching" current or "negative differential resistance". Another system is the viologen/Au nanoparticle combination reported by Schiffrin and associates.[32] Theoretical frames for single-molecule electrochemical redox switching are, finally, available.[1,30,33]. These are rooted in interfacial electrochemical ET theory[34,35] but incorporate environments and new ET phenomena specific to *in situ* STM or electrochemical nanogap electrode configurations appropriate to the nanoscale dimensions.

We report here a comprehensive electrochemical and *in situ* STM study of a class of transition metal complexes novel in molecular *in situ* STM and molecular electronics. The specific complexes are the Os-complexes $[Os(bpy)_2(p2p)_2]^{2+/3+}$ and $[Os(bpy)_2(p0p)Cl]^{+/2+}$. These are compared with a similar Co-complex, $[Co(terpy)(terpy–O–(CH_2)_6–SAc]^{+/2+}$, Fig. 1 (bpy = 2,2′-bipyridine, p0p = 4,4′-bipyridine, p2p = 1,2-bis(-4-pyridyl)ethane, terpy = 2,2′,2″-terpyridine, Ac = acetyl). These molecular systems offer the following merits as molecular electronics targets:

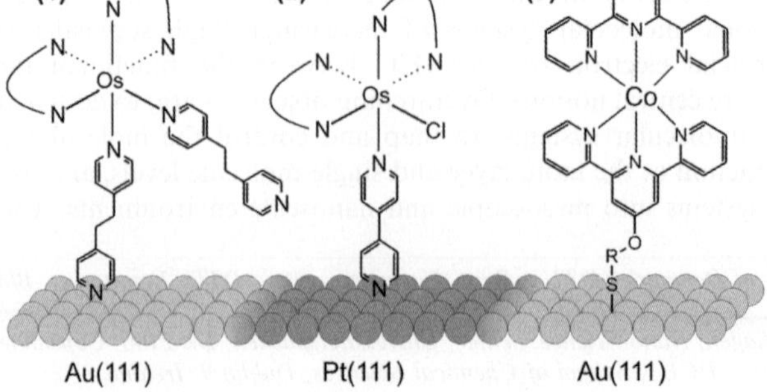

Fig. 1 Molecular structures of compounds (**1**), (**2**), and (**3**) bound to their respective substrates. The arcs in (**1**) and (**2**) stand for 2,2′-bipyridine.

• They exist in two accessible robust oxidation states.

• The equilibrium electrochemical redox potentials are in ranges suitable for electrochemical characterization using single-crystal Au(111)- and Pt(111)-electrodes, as well as for insertion into *in situ* STM and possibly nanogap electrode configurations.

• They offer broad versatility as to ligand types, linker groups, solubility in different solvents, and other "technologically" important properties.[36,37] This is illustrated presently by the different redox potentials of the two Os-complexes and the different Os- and Co-metal centres.

• The complexes and the whole system class are supported comprehensively by established coordination chemistry including chemical synthesis, spectroscopy, thermodynamic properties, ET and ligand substitution kinetics *etc.*,[36,37] as well as by theoretical frames based on ligand field, molecular orbital, and molecular charge transfer theory.[34,35,38]

The transition metal compounds have been addressed by single-crystal voltammetry and *in situ* electrochemical STM in the same microscopic environments. As novel features, the complexes show a tunnelling current spectroscopic dependence on the substrate electrode potential, bias voltage, and equilibrium redox potential all consistent with recent theoretical frames of the three-level system nature and the three-electrode nanoscale environment in which they are brought to operate.[30,33] The Os- and Co-complexes show, moreover, pronounced differences, which can be related to different local mode reorganization and electronic coupling features.

2. Experimental

2.1 Synthesis and characterization of the Os complexes (1) and (2)

$[Os(bpy)_2(p2p)_2](PF_6)_2$ and $[Os(bpy)_2(p0p)Cl](PF_6)$ were prepared by refluxing $[Os(bpy)_2Cl_2]\cdot 2H_2O$ and $AgNO_3$ in 1 : 1 ethyleneglycol : water solution until a grey precipitate formed. After removing the precipitate, p2p or p0p was added and refluxing continued for 96 h. Unreacted ligand was removed and a concentrated solution of NH_4PF_6 added. The precipated product was washed with water, dried with diethylether, and purified by column chromatography. The purity was checked by NMR spectroscopy.

2.2 Synthesis and characterization of the Co complex (3)

The monoterpy complex (**4**) was prepared by partial sublimation of terpyridine from the bisterpycobalt(II) complex as reported by Hogg and Wilkins, *cf.* Fig. 2.[39] Thermogravimetric analysis showed that the ideal sublimation temperature was 175 °C. The alkyl substituted terpy (**7**) was prepared by alkylation of 2,6-*bis*-(pyrid-2-yl)-4-pyridone (**5**) with 6-acetlylthio-1-bromohexane (**6**) in DMF using potassium carbonate as a base.[40] The heteroleptic cobalt(III) complex (**3**) was prepared by mixing equimolar amounts of (**4**) and (**7**) in DMF containing hydrogen peroxide, followed by precipitation of the complex using aqueous KPF_6 (see electronic supplementary information (ESI) for details†).

2.3 Surface and monolayer preparation

Au(111) and Pt(111) electrodes were prepared as described.[41,42] Monolayers of the respective compound were prepared by soaking the electrodes overnight in 10–50 μM water–dimethylformamide solutions (Os complexes (**1**) and (**2**): H_2O : DMF = 2 : 1 (m/m); Co complex (**3**): H_2O : DMF = 1 : 2 (m/m)), followed by rinsing with Millipore water and 0.1 M $HClO_4$ or 0.1 M $NaClO_4$.

2.4 Electrochemistry on Au(111)- and Pt(111)-electrodes

An electrochemical Autolab system with SCANGEN/ADC750 modules controlled by the GPES software package (Ecochemie, Netherlands) was used. Electrochemical (glass) cells were custom-made. All cyclic voltammograms were recorded in linear scan mode. A freshly prepared reversible hydrogen electrode (RHE) and a Pt wire served as reference and counter electrode, respectively. The reference potential was checked with a saturated calomel electrode (SCE) after each experiment. Electrodes were employed in a hanging-meniscus configuration (area: $\sim$0.1 cm^2 for Au(111) and $\sim$0.03 cm^2 for Pt(111)). Prior to each experiment, the electrochemical cell was deoxygenated by

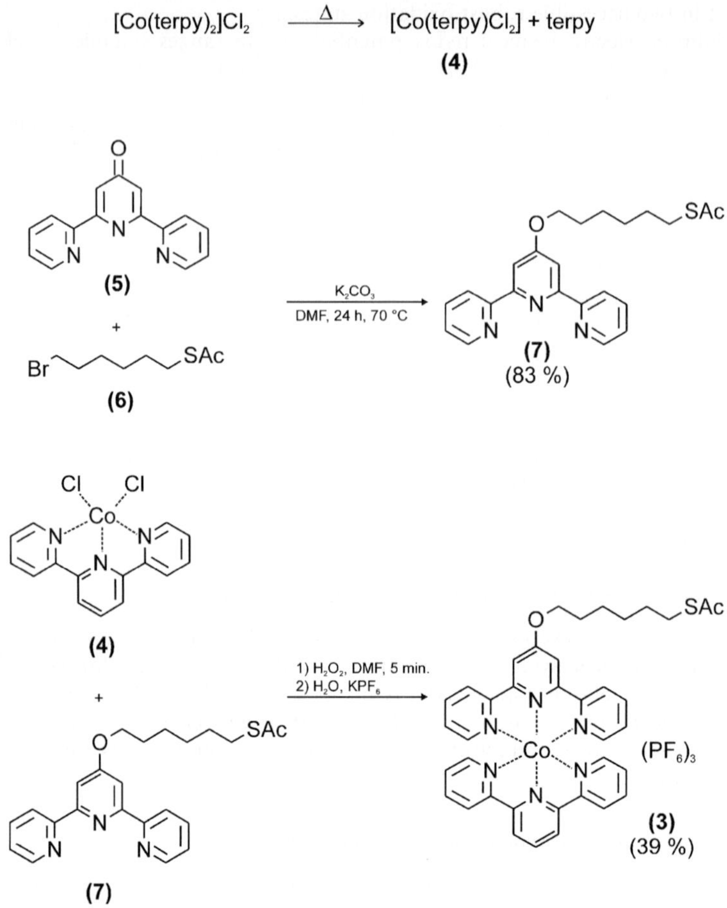

Fig. 2 Synthesis of the Co-complex (**3**) (schematic, see ESI† for details).

purified Ar (Chrompack O_2 filter) and a steady flow of Ar was maintained at all times during operation. Electrolytes were 0.1 M $HClO_4$ for **1** and **3**, and 0.1 M $NaClO_4$ for **2**. In the case of **1**, the solution contained 1–10 μM of free complex.

2.5 *In situ* STM and STS

A PicoSPM instrument (Molecular Imaging Co., USA) with a bipotentiostat for independent control of substrate and tip potential was used in constant-current mode for STM imaging. The custom-made three-electrode Teflon cell was equipped either with a Pt wire or a leak-free Ag/AgCl electrode (Harvard Apparatus Inc., USA) as reference electrode. Pt wire was also used as counter electrode. In either case the reference potential was checked with a SCE reference electrode after each experiment because also the Ag/AgCl reference electrode was found to vary within ±15 mV from one experiment to another. Electrolytes were as in electrochemical experiments, *cf.* above.

For experiments with the feedback loop switched off, *i.e.* for $I_t(E_s)$ spectroscopy at constant bias voltage, the tip was approached to a certain setpoint current inside the tunnelling range. The feedback loop was then switched off and the substrate potential E_s cycled once in a potential range sufficiently wide to pass the equilibrium redox potential of the respective complex with both E_s and the tip potential E_t. The bias voltage between tip and substrate, $V_{bias} = E_t - E_s$, was kept constant. The initial substrate potential of each scan, $E_{s,i}$ thus equals its final potential, $E_{s,f}$. The difference in tunnelling current before and after one scan was used as an indicator for lateral and/or vertical drift and, thus, to assess the quality of an $I_t(E_s)$ scan. From STM imaging it was concluded that lateral drift usually amounted to about one molecular diameter for typical sweep durations between 1–10 s.

Distance-dependent conductivity measurements were performed by approaching the tip to a setpoint current of 1.0 nA at a constant value of E_s. The tip–substrate bias voltage was then varied, by changing $E_s \pm 50$ mV at constant tip potential E_t and a sweep duration of typically 0.2 s. Ten subsequent scans were averaged. The tip was then retracted by $\Delta z = 0.1$ nm and the bias sweep repeated. At each substrate potential, 21 data sets were obtained in this way, so the total distance change amounted to 2.0 nm. The height of a gold terrace was used ($\sim$0.24 nm) for an approximate calibration of the z-direction.[43] The slope of the $I_t(V_{bias})$ curve at $V_{bias} = 0$ corresponds to the conductivity and was determined by linear regression.

For tunnelling current–distance spectroscopy, the tip was approached to a setpoint current of 1.0 nA ($V_{bias} = -0.05$ V), at a constant substrate potential E_s, and then retracted by 5.0 nm. The semi-logarithmic plot of $I_t(\Delta z)$ yielded a straight line for 0.05 nm $< \Delta z <$ 0.55 nm. The slope was used to calculate an approximate tunnelling barrier height within the WKB approximation according to eqn. (1).[44]

$$\frac{\mathrm{dln}(I_t)}{\mathrm{d}z} = \frac{2\sqrt{2m_e}}{\hbar}\sqrt{\phi_{t,av}} \tag{1}$$

$\phi_{t,av}$ is the average formal tunnelling barrier height, m_e the electronic mass, and $\hbar$ Planck's constant divided by 2π.[44]

3. Theoretical treatment

The tunnelling spectroscopic patterns of the *in situ* STM configuration of the Os- and Co-complexes follow expectations from recent theoretical concepts and formalism,[30,33] with some interesting differences between Os and Co. Fig. 3 summarizes some central notions. The redox molecule is characterized by a single electronic energy level separated from all other molecular electronic levels. The level is, moreover, strongly coupled to the nuclear environment, so that thermal nuclear activation and relaxation are crucial. The redox level can exist in an oxidized, ε_{ox}, and a reduced state, ε_{red}, the equilibrium positions of which are separated by twice the environmental reorganization free energy, λ, *i.e.* $\varepsilon_{ox} - \varepsilon_{red} = 2\lambda$. In multi-phonon transitions such as presently considered λ is

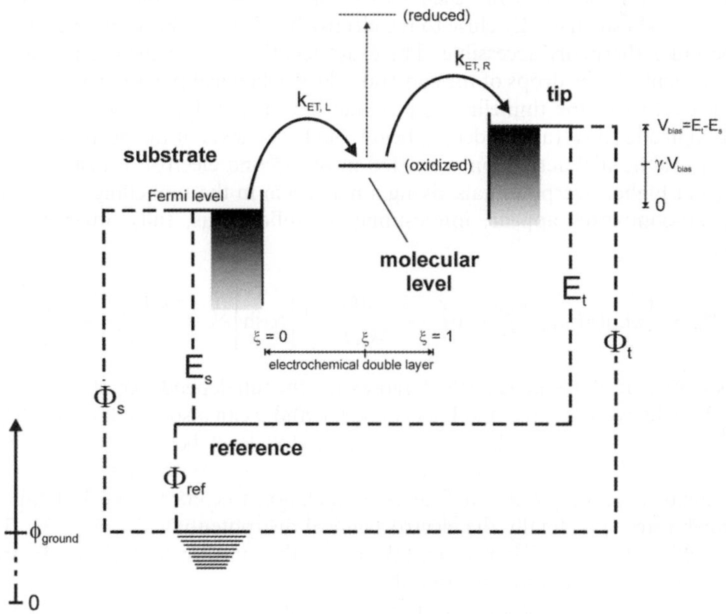

Fig. 3 Potential distribution in the STM tunnelling gap including parameters in eqns. (2) and (3).

a large quantity, *i.e.* a considerable fraction of an electronvolt. The electronic energies of the enclosing substrate and tip electrodes are represented by two metallic continua, populated up to the Fermi levels, ε_{Fs} and ε_{Ft}. These are separated by the energy associated with the bias voltage, V_{bias}, *i.e.* $\varepsilon_{Ft} - \varepsilon_{Fs} = eV_{bias}$, where V_{bias} is given by the substrate, E_s, and tip electrochemical potentials, E_t, *i.e.* $V_{bias} = E_t - E_s$. e is (numerically) the electronic charge. Fig. 3 thus corresponds to a positive bias voltage.

The overall ET from substrate to tip (or *vice-versa* if the bias voltage is negative) is mediated by the redox level in a sequential two-step mode with temporary redox level population and depopulation. The level is initially empty (oxidized) and located above both Fermi levels. Electron tunnelling in this configuration is feasible but unlikely due to low electron population in the tip electrode at these high energies, or inefficient superexchange tunnelling. When substrate and tip potentials are tuned close to resonance with the molecular level, and the latter brought further close to ε_{Fs} and ε_{Ft} by environmental configurational fluctuations, interfacial ET from substrate (negatively biased electrode) to the empty molecular level, however, occurs. Vibrationally assisted, multi-phonon electron transfer through the redox level in the "Fermi window" between ε_{Fs} and ε_{Ft}, *i.e.* relaxation of the, now occupied or reduced redox level towards a new position below both Fermi levels follow subsequent to the first electron transfer step. In a second, also thermally assisted electron transfer step the electron is transmitted from the occupied molecular redox level to the positively biased electrode, here the tip. Analogous steps in the opposite direction proceed at negative bias voltage.

Fig. 3 and the sequence of electron transfer steps noted are general features of vibrationally assisted electron tunnelling in the molecular *in situ* STM patterns when localized redox levels are close enough to the Fermi levels of the enclosing electrodes. The following more subtle differences, however, also arise and correspond to limits of "strong" and "weak" interactions between the molecular redox level and the enclosing electrodes, respectively.[30,33,45] These limits accord, interestingly, with the behaviour of the Os-complexes and the Co-complex, respectively. Two "relaxation times" can be introduced to illustrate the difference. One is the "relaxation time" for electron transfer, given by $\tau_{ET} \approx \hbar/\Delta$ where Δ is the energy "broadening" caused by the interaction with the electrodes (here taken to be the same for both electrodes). The other one is the environmental nuclear relaxation time, τ_{rel}. In the limit of strong electronic interactions, $\tau_{ET} \ll \tau_{rel}$. A multitude of electrons, in a number of about $n_{el} \approx 2e\kappa\rho|V_{bias}| \gg 1$, where κ is the electronic transmission coefficient and ρ the metallic electronic energy level density, is then transferred in a time window τ_{rel}. This number can reach two or three orders of magnitude. As a result, the tunnelling current rises significantly close to the electrochemical equilibrium potential E^0, where the redox level becomes thermally accessible. The exact location of the current rise also depends on other details, particularly the drops of the substrate electrochemical potential and bias voltage at the site of the redox centre in the tunnelling gap, ξ and γ, respectively. As the overpotential is raised further, the occupied level is trapped deeper below the Fermi level of the positively biased electrode (the tip) and larger thermal activation is needed in the second electron transfer step. The current therefore drops at higher overpotentials giving a maximum in the tunnelling current-overpotential relation. The Os-complexes appear, interestingly to follow this, fully adiabatic multi-electron transfer pattern.

$$i_{tunn}^{ad} = \frac{1}{2} e\kappa\rho(eV_{bias}) \frac{\omega_{eff}}{2\pi} \exp\left(-\frac{\lambda - eV_{bias}}{4k_BT}\right) \left[\cosh\left[\frac{(\frac{1}{2} - \gamma)eV_{bias} - e\xi\eta}{2k_BT}\right]\right]^{-1} \quad (2)$$

The pattern is summarized in eqn. (2), which represents the full dependence of the tunnelling current on both the substrate overpotential and bias voltage. κ and ρ can also be related to the energy range $\Delta\varepsilon \approx (\kappa\rho)^{-1}$, traversed by the relaxing molecular redox level between successive "individual" electron transfers. In the adiabatic limit of strong interactions $\Delta\varepsilon \ll k_BT$. ω_{eff} is a characteristic nuclear vibrational frequency, λ the nuclear reorganization free energy, k_B Boltzmann's constant and T the temperature. η is, finally the electrochemical overpotential, $\eta = E_s - E^0$. The electronic transmission coefficient and level density are taken to be the same for both electrodes in eqn. (2), but this constraint can be relaxed straightforwardly.

A qualitatively different behaviour emerges in the opposite limit of weak molecule-electrode interactions, the fully diabatic limit. In this limit $\tau_{ET} \gg \tau_{rel}$ and $\Delta\varepsilon \gg k_BT$. The occupied redox level relaxes fully subsequent to the first electron transfer step. This is again followed by the second step but only a single electron is now transferred in the overall process, *i.e.* within τ_{rel}, and the tunnelling enhancement close to energy resonance is much weaker than before. The Co-complex seems to follow this pattern. The tunnelling current is given by

$$i_{tunn}^{dia} = e\frac{k_{ET,L}k_{ET,R}}{k_{ET,L} + k_{ET,R}} \tag{3}$$

with

$$k_{ET,L} = \kappa_L \rho_L \frac{2k_B T}{\alpha_L} \exp\left[-\frac{(\lambda + e\xi\eta - e\gamma V_{bias})^2}{4\lambda k_B T}\right]$$

and

$$k_{ET,R} = \kappa_R \rho_R \frac{2k_B T}{\alpha_R} \exp\left[-\frac{(\lambda - eV_{bias} - e\xi\eta + e\gamma V_{bias})^2}{4\lambda k_B T}\right]$$

where α_L and α_R are the electrochemical transfer coefficients for interfacial electron transfer at the substrate and tip electrode, respectively.

4. Results

4.1 Electrochemical characterization of (1)–(3)

Electrochemical investigations of the self-assembled monolayers (SAM) of complexes (**1**), (**2**), and (**3**) on Au(111) and Pt(111), were performed to determine their equilibrium redox potentials and to optimize the preparation conditions for the respective SAMs prior to *in situ* STM/STS studies. In the case of (**3**), kinetic studies were also undertaken as ET kinetics could here be separated from charging currents with the macroscopic electrodes used (*vide infra*).

All complexes form densely packed, electroactive SAMs, Fig. 4A–C. The peak intensities I_p increase linearly with scan rate ν indicating that ET indeed originates from surface-immobilized species in all three cases.

The cyclic voltammograms of (**1**) display a notable asymmetry. The full-width-at-half-maximum (FWHM) is smaller than the ideal value of 90 mV ($z = 1$, $T = 298$ K) for both anodic and cathodic branches at low scan rates <0.3 V s^{-1}, but reaches constant values of ~90 mV (anodic peak) and ~70 mV (cathodic peak) at $\nu > 0.3$ V s^{-1}. FWHM < 90 mV have been assigned to attractive electrostatic interactions in the monolayer which may, in the present case, originate from perchlorate ions ClO_4^-.[46] The formal potential E^0 is +0.62 ± 0.01 V (on Au(111), 0.1 M $HClO_4$). Peak charges were determined to 4.7 ± 0.3 μC cm^{-2} (49 ± 3 pmol cm^{-2}, (18.5 ± 0.3)^2 Å^2 molecule^{-1}) and 3.0 ± 0.2 μC cm^{-2} (31 ± 3 pmol cm^{-2}, (23.1 ± 0.8)^2 Å^2 molecule^{-1}) for anodic and cathodic peaks, respectively. The calculated area per molecule fits reasonably well with the geometrical size of (**1**) if only one p2p linker is connected to the surface while the other one is oriented approximately parallel to the surface. This is in line with *in situ* STM imaging results[47] and previous electrochemical data for a comparable complex.[48] The difference in peak charge may, however, also be partly due to difficulties in defining a baseline for the cathodic part. The cyclic voltammograms remained unchanged over several hundreds of scans in the potential windows depicted in Fig. 4A, testifying to high monolayer stability.

The origin of the pair of small peaks at 0.2 V and 0.4 V is not completely clear, but could be related to a partly hydrolyzed complex, *e.g.* $[Os(bpy)_2(p2p)(H_2O)]^{2+/3+}$. Haddox and Finklea found an equilibrium redox potential of about +0.3 V for $[Os(bpy)_2(4\text{-AMP})(H_2O)]^{2+/3+}$, bound to Au (pH 1, 4-AMP: 4-aminomethylpyridine).[49]

The CVs of (**2**) are more symmetric, but the peaks are broader than expected for a surface-immobilized redox species with FWHM of 0.14 ± 0.1 V, Fig. 4B. The average peak charge is 6.0 ± 0.3 μC cm^{-2} (62 ± 3 pmol cm^{-2}) which corresponds to (16.3 ± 0.5)^2 Å^2 molecule^{-1}. The formal potential E^0 is +0.23 ± 0.01 V in 0.01 M $HClO_4$ (0.1 M $NaClO_4$) and is shifted slightly upwards to 0.24 ± 0.01 V in 0.1 M $NaClO_4$ (on Pt(111)).

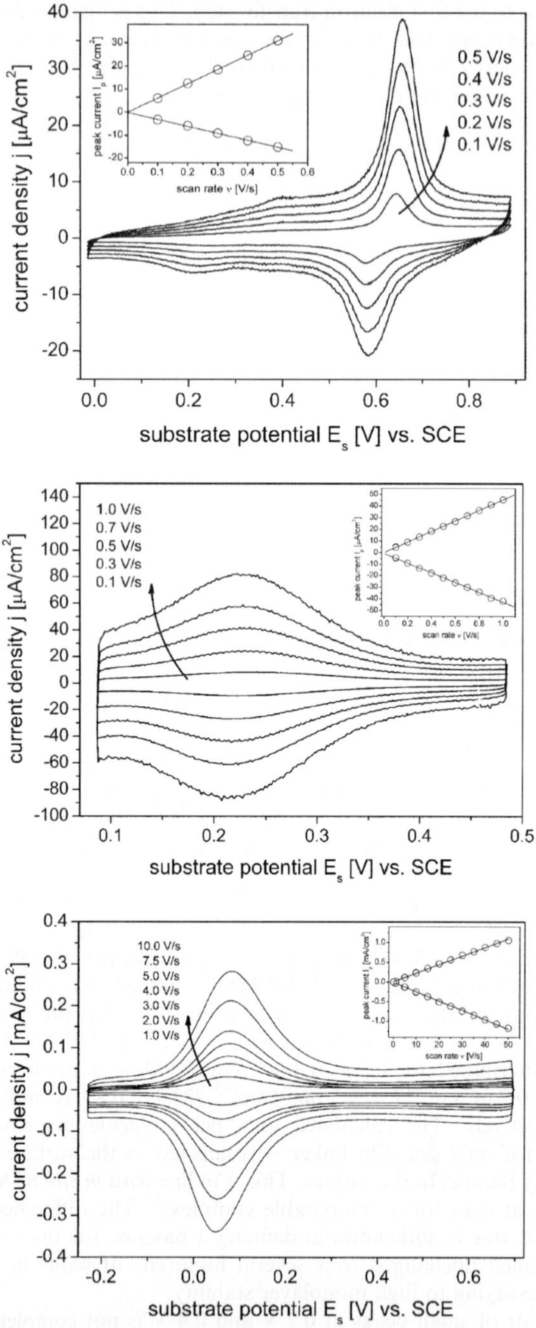

Fig. 4 A (top): CV data for (**1**) on Au(111) in 0.1 M $HClO_4$ at different scan rates. B (middle): CV data for (**2**) on Pt(111) in 0.01 M $HClO_4$ + 0.1 M $NaClO_4$ at different scan rates. C (bottom): CV data for (**3**) on Au(111) in 0.1 M $HClO_4$ at different scan rates; insets: peak current density *vs.* scan rate; all compounds (**1**)–(**3**) display a linear correlation between peak current I_p and scan rate ν.

Cyclic voltammetry data for (**3**) yielded a coverage of 3.9 ± 0.3 μC cm^{-2} (40 ± 3 pmol cm^{-2}) and an area of $(20.3 \pm 0.7)^2$ Å^2 molecule^{-1}, Fig. 4C. The peak width, FWHM was 0.14 ± 0.1 V, again larger than 90 mV, possibly indicating repulsive interactions on the surface. The formal potential E^0

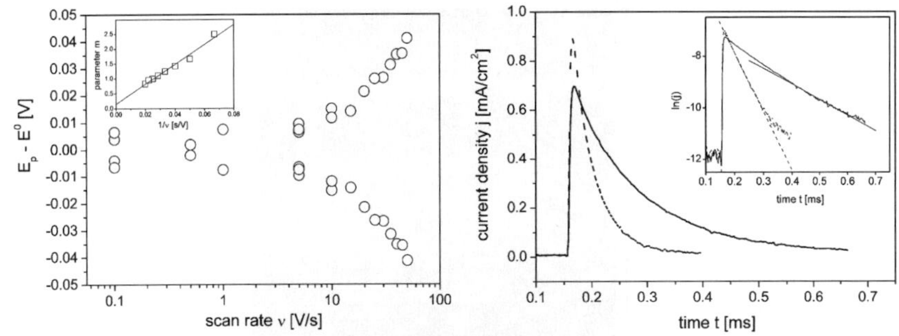

Fig. 5 A (left): Peak separation ΔE_p *vs.* scan rate ν for (**3**); ΔE_p clearly increases for $\nu > 10$ V s^{-1}; inset: Results of a Laviron-type analysis. B (right): Chronoamperometric data for (**3**); E_s-jump to E^0 (solid line) and at a redox-inactive E_s (dashed line); inset: logarithmic plot including linear fit curves.

is $+0.05 \pm 0.01$ V (on Au(111), 0.1 M $HClO_4$), and accords well with the electrochemical data from $[Co(terpy)_2]^{2+/3+}$ in solution.[50]

4.2 Kinetic analysis of ET for the Co complex

Interfacial ET kinetics of Os complexes such as (**1**) and (**2**) are known to be very fast. Forster *et al.* reported a rate constant of k^0_{ET} of 3×10^5 s^{-1} for a compound similar to (**1**), *i.e.* [Os(bpy)$_2$(p2p)Cl](PF$_6$)$_2$, in 0.1 M $NaClO_4$ on Pt microelectrodes.[51] For (**3**), however, capacitive and ET currents could be separated with the millimetre-sized electrodes used. The peak–peak separation increased notably for $\nu > 10$ V s^{-1} and up to $\nu \sim 50$ V s^{-1}, no significant distortion in the CVs occurred. The observed effect is thus due to ET kinetic control rather than time constant limitations of the cell. Fig. 5A plots the peak position E_p, relative to the formal potential E^0, *vs.* the scan rate ν. Peak separations for scan rates 15 V s^{-1} $\leq \nu \leq$ 50 V s^{-1} were analyzed according to Laviron's method and the resulting parameter $m = (RT/F)k^0_{ET}/\nu$ plotted *vs.* $1/\nu$ to obtain $k^0_{ET} = 1.3 \times 10^3$ s^{-1}, Fig. 5A inset.[52]

In a second approach, chronoamperometry was used. Fig. 5B shows typical current decays for small potential jumps of 10 mV in the capacitive potential region ($E_i = -0.11$ V, $E_f = -0.10$ V, dashed line) and to the formal potential E^0 ($E_i = +0.04$ V, solid line). The former yields a time constant RC $= 4 \times 10^{-5}$ s. The latter current decay was analyzed from 5RC to 6×10^{-4} s resulting in a decay constant $1/\tau_{ET}$ of 5.6×10^3 s^{-1}, Fig. 5B inset. As $E_f = E^0$, $k^0_{ET} = 1/(2\tau_{ET}) = 2.8 \times 10^3$ s^{-1}.

This value compares well with the rate constant determined from the Laviron analysis. As an average, we obtained $(2 \pm 1) \times 10^3$ s^{-1} which is two to three orders of magnitude slower than the Os complexes such as (**2**). The ET distances are comparable for all three complexes. The difference between the Os- and Co-complexes presumably originates from different electronic coupling and reorganization energy. Further investigations using microelectrodes are currently on the way to differentiate between these two factors and the role of ion pairing in ET.

4.3 *In situ* STM imaging of $[Os(bpy)_2(p0p)Cl]^{2+/3+}$

The complexes (**1**)–(**3**) form stable, disordered monolayers on the respective metal surface, *i.e.* Au(111) for (**1**) and (**3**), and Pt(111) for (**2**). Local domain formation was observed only for (**1**), but the domains disappeared after repetitive cycling of the substrate potential.[47] As an example, an *in situ* STM image of (**2**) is shown in Fig. 6. The disorder in the adlayer makes it difficult to perform a statistical size analysis, but the bright features in Fig. 6 roughly correspond to the expected molecular size of (**2**).

4.4 *In situ* scanning tunnelling spectroscopy

The tunnelling current is expected to increase when the molecular redox level is tuned into resonance with the Fermi levels of the electrodes, *i.e.* when the molecular energy level is brought

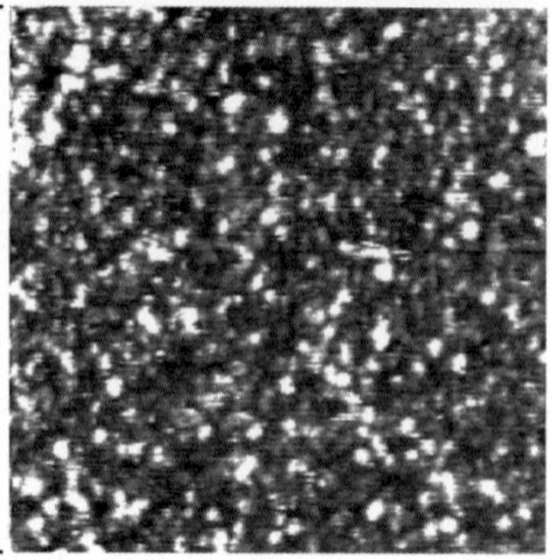

Fig. 6 *In situ* STM image of (**2**) on Pt(111) in 0.1 $NaClO_4$; $E_s = 0.33$ V, $E_t = 0.13$ V, $V_{bias} = -0.2$ V, $I_{set} = 0.08$ nA (50 nm × 50 nm).

within the energy window of substrate and tip Fermi levels. As described in the experimental section, the tunnelling current was measured during a sweep of the substrate potential E_s at constant tip–substrate bias voltage V_{bias} with the tip at tunnelling distance and the feedback loop switched off (constant tip–surface distance). The increase in tunnelling current is indeed observed for all three compounds (**1**), (**2**), and (**3**), but the enhancement is much stronger for the former two.[47] Fig. 7A depicts a typical cathodic $I_t(E_s)$ scan for (**2**), together with a cyclic voltammogram obtained under the same experimental conditions. The increase in tunnelling current around the equilibrium redox potential is clearly seen with an intensity in the nA range corresponding to a 5–10-fold increase of the tunnelling current away from the equilibrium redox potential (off resonance). The position depends, further linearly on tip/substrate bias voltage, eqn. (2), with a slope of –0.5. A slight dependence on E_s scan direction is also observed, Fig. 7C, possibly due to dielectric relaxation in the tunnelling gap.

$$\eta_{max} = -\frac{1}{\xi}\left(\frac{1}{2} - \gamma\right) V_{bias} \tag{4}$$

The sign of the tunnelling current peak does not depend on sweep direction, but is reversed when the sign of the applied tip/substrate bias voltage is changed. Furthermore, if the tip is retracted out of tunnelling range, the feature disappears abruptly. The observed current increase cannot therefore, be due to Faradaic electrochemical currents, but must originate from the tunnelling process. Contributions from lateral electron hopping through the monolayer cannot be ruled out, but are not expected to be of major importance.[53]

In the case of (**3**), a peak feature around the equilibrium redox potential is also observed, but the intensity of 5–10 pA remains close to the present instrumental noise level, and care must be taken in order discern the true signal from random background fluctuations. Still the feature depicted in Fig. 7B is reproducible and follows a two-step ET behaviour, *cf.* Section 5. We therefore conclude that the observed feature is due to the same effect as for (**2**). The reason for the approximately two orders of magnitude smaller intensity is not completely clear, but can be explained by different ET kinetics between the molecular level and the substrate, *cf.* Fig. 3. We return to this point in Section 5.

Other ways of characterizing the tunnelling process are based on tunnelling-distance spectroscopy, where either the tunnelling current or the conductivity at different distances z and substrate potentials E_s is recorded. According to the $I_t(E_s)$ data above, the conductivity is expected to peak around the equilibrium redox potential of the immobilized complex. This is indeed the case for distances $\Delta z \geq 0.6$ nm, whereas for smaller Δz, measured conductivities scatter substantially and no notable trend is observed. A representative data set for the former case is shown in Fig. 8 for (**2**) ($\Delta z = 2.0$ nm).

The conductivity peaks at a tip potential $E_t = 0.25$ V which is close to the electrochemical redox potential $E^0 = 0.24$ V (0.1 M $NaClO_4$, *cf.* above). The maximum conductivity decreases from approximately 8 nA V^{-1} for $\Delta z = 0.6$ nm to 1.5 nA V^{-1} for $\Delta z = 2.0$ nm, but the ratio of the

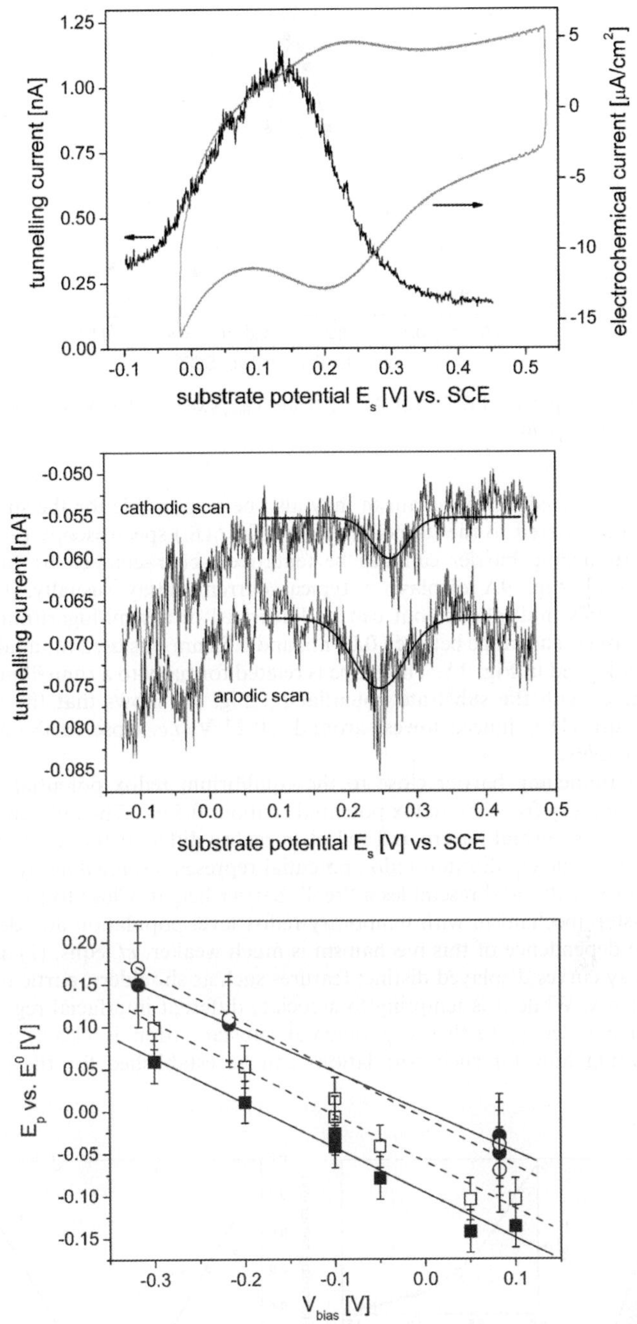

Fig. 7 A (top): Tunnelling current I_t *vs.* substrate potential E_s (black, left scale) and CV (grey, right scale) for (**2**) (Pt(111), 0.1 M $NaClO_4$); $V_{bias} = +0.05$ V. B (middle): $I_t(E_s)$ data for (**3**) (Au(111), 0.1 M $HClO_4$), anodic and cathodic scan; $V_{bias} = -0.3$ V. C (bottom): Peak position of $I_t(E_s)$ relative to E^0 *vs.* V_{bias} for (**2**) (squares) and (**3**) (circles); closed symbols cathodic, open symbols anodic scans.

conductivities for tip potentials at the peak and away from the peak remains approximately constant at a value of 3. As the conductivity changes with E_s, the tip/substrate distance is, however, different for each potential at given setpoint current, and measured conductivities cannot be

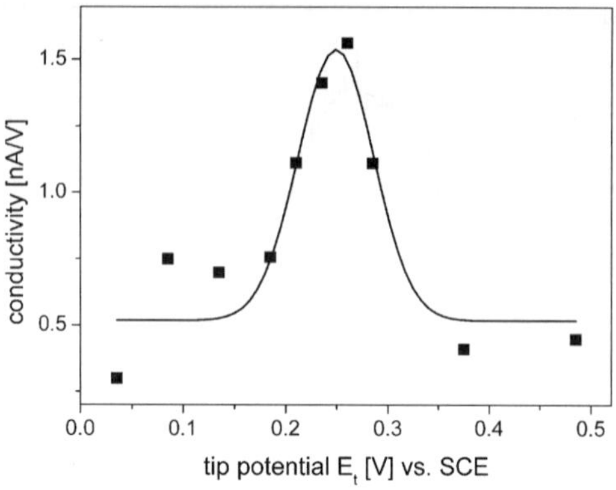

Fig. 8 Conductivity *vs.* tip potential E_t for (**2**), $\Delta z = 2.0$ nm. V_{bias} sweep $\pm$ 0.05 V (E_t = const.) (Pt(111), 0.1 M $NaClO_4$); solid line: Gaussian fit.

compared directly. The feedback mechanism may also be responsible for the smaller conductivity enhancement ($\sim$3) compared to the tunnelling current in $I_t(E_s)$ spectroscopy ($\sim$5–10).

An "effective" tunnelling barrier can still be tentatively characterized by tunnelling current–distance spectroscopy. Fig. 9A displays a typical current decay (usually, 10–20 scans were averaged). The decay is multiphasic, but can be linearized in a semi-logarithmic plot for $\Delta z <$ 0.55 nm, Fig. 9A (inset). The range between 0.05 nm and 0.55 nm was used for analyzing the slope m of $\ln(I_t)$ *vs.* Δz as indicated in Fig. 9A. This slope is related formally to a tunnelling barrier height in the gap, and changes with the substrate potential E_s. Fig. 9B shows that the tunnelling barrier determined from eqn. (1) is indeed lowest around +0.27 V, *i.e.* close to the equilibrium redox potential of the complex.

The decrease in tunnelling barrier close to the equilibrium redox potential compared to off-resonance tunnelling away from the redox potential is about 0.5 eV. The different apparent barrier heights are, however, of formal nature and reflect in reality different tunnelling mechanisms. The barrier height far from the equilibrium redox potential represents tunnelling by superexchange *via* coupling to the redox centre and resembles a "real" barrier height. Close to the equilibrium redox potential the two-step mechanism with temporary redox level population and depopulation takes over. The distance dependence of this mechanism is much weaker, *cf.* eqns. (2) and (3).

Some of the decay curves displayed distinct features such as shoulders, particularly at tunnelling currents below 0.1 nA. While it is tempting to associate different interfacial regions with different phases in the current decay, further experimental substantiation is clearly needed.[54] We are currently investigating how far such correlations can be established for the present molecular system.

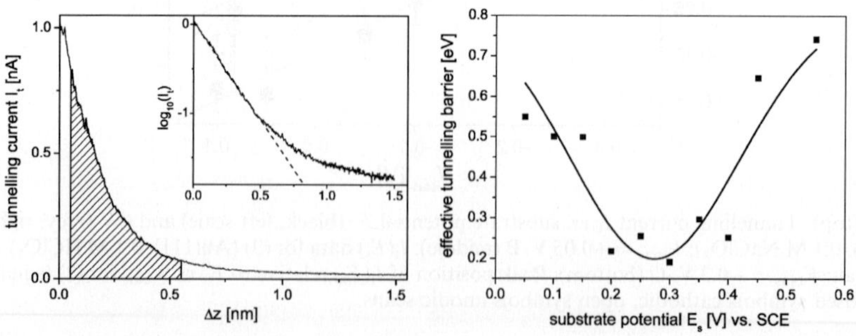

Fig. 9 A (left): Representative tunnelling current–distance curve for (**2**) (Pt(111), 0.1 M $NaClO_4$); $E_s = 0.1$ V, $V_{bias} = -0.05$ V; hatched section used for analysis. Inset: Logarithmic plot of $I_t(\Delta z)$ and linear fit curve. B (right): Effective tunnelling barrier for (**2**) according to eqn. (1), at different substrate potentials E_s (symbols); solid line: Gaussian fit.

5. Discussion

The most important result of the present study is that electron tunnelling through a class of robust transition metal complexes in electrochemical *in situ* STM has been clearly disclosed and shown to proceed by vibrationally assisted two-step ET. This is reflected, first and foremost in a strong tunnelling spectroscopic feature in the *in situ* STM tunnelling current/overpotential relation close to the equilibrium potential. This result is itself novel but is also supported comprehensively by systematic overpotential and bias voltage variation as well as by addressing several target complexes with different equilibrium redox potentials and metal centres. The latter extends the comparison to nuclear reorganization and electronic coupling features. The combined data, supported by electrochemical data show a consistent pattern which can be framed by recent theoretical formalism appropriate to single-molecule redox processes in the three-electrode electrochemical *in situ* STM configuration.

This accordance is emphasized by several specific properties of both the target molecules and the external parameter correlations. The *in situ* STM current-overpotential relations of the two Os-complexes are thus shifted systematically with the bias voltage and the equilibrium redox potentials as anticipated from the intrinsic physical mechanism based on sequential two-step multi-phonon ET and the accompanying theoretical frames. The observed tunnelling current also displays significant differences associated directly with the different chemical and physical properties of the complexes. In addition to the characteristic peak shifts with the bias voltage and equilibrium redox potentials, the tunnelling current enhancement at the peak potential is thus much larger for the Os-complexes than for the Co-complex. This difference seems to correlate, at least qualitatively with the interfacial electrochemical ET rate constants, which are again determined crucially by the nuclear reorganization (free) energy and the electron exchange factors for coupling between the molecular redox centre and the substrate. These differences can be illuminated further by the following observations.

To our knowledge, there are no kinetic data for (**1**), but for a closely related compound, termed "p2pCl" hereafter, in which one p2p linker is replaced by a chloride ligand. The rate constant of this complex has been determined to k_{ET}(p2pCl; 0.1 M $NaClO_4$) = 3×10^5 s^{-1} (Pt microelectrodes).[48]

There are no kinetic data for (**2**) in aqueous electrolyte, but k_{ET} can be estimated based on its value in DMF by comparing to data for p2pCl in 0.1 M $NaClO_4$ and DMF.[46,51] In this way, k_{ET}(p0pCl; aqueous solvent) $\approx 9.3 \times 10^6$ s^{-1} is obtained, *i.e.* approximately 30 times higher than k_{ET}(p2pCl; 0.1 M $NaClO_4$). These values are again two to three orders of magnitude larger than k_{ET} of the Co complex (**3**) (*cf.* Section 4.2).

The tunnelling current enhancement around the equilibrium redox potential is approximately 1–2 nA for (**1**) and (**2**), but only about 5–10 pA for (**3**), so the enhancement is about 100–400 times stronger for the Os- than for the Co-complex.[47] This is in accordance with electrochemical kinetics patterns and would imply that the ET step from the substrate to the molecule is in fact rate determining for the overall process. ET between the redox center and the tip should therefore be comparatively fast for (**1**) and (**2**).[47] The much faster interfacial ET of the Os-complexes than of the Co-complex follows patterns for electron exchange between analogous Os- and Co-complexes in homogeneous solution.[55] The much slower electron self-exchange in Co-complexes such as $[Co(phen)_3]^{2+/3+}$ compared to their Os-analogues is due both to a significant intramolecular contribution to the reorganization energy and most likely also a much smaller electron exchange factor for the Co-complexes. The Co-complexes may indeed accord with the weak-coupling, diabatic limit of ET, eqn. (3), whereas the Os-complexes would accord with the strong-coupling, adiabatic limit, eqn. (2). For this reason the tunnelling current enhancement close to the redox potential is also expected to be much smaller for the Co-complex.

There is, thus no significant difference in the tunnelling current enhancement between (**1**) and (**2**), even though their electrochemical rate constants may differ by about an order magnitude. This is understandable if the Os-complexes follows the pattern of fully adiabatic ET, but additional effects also need to be considered. Ion pairing is, for example known to occur for Os complexes similar to (**1**) and (**2**) and may contribute significantly to the ET kinetics.[46] At the same time, this would affect the molecular conductivity.[56] A direct experimental determination of k_{ET}(**2**) under the same solution conditions, possibly varying the perchlorate concentration, is therefore required. This would help to clarify how far electrochemical data carry over to the tunnelling gap. In principle, both reorganization free energy and electronic coupling will be affected, but the importance of such changes is not clear at present.

The study has illuminated and mapped working principles and theoretical frames of molecular redox switching and amplification ("transistor-like" behaviour) at a level close to that of the single molecule. In contrast to most previously reported molecular transistors, the present systems operate at room temperature and in condensed matter environment. This may hold promise for use as real forthcoming molecular electronics components, for example in nanogap electrode arrays.

6. Summary

A class of robust Os- and Co-complexes has been characterized by cyclic voltammetry and electrochemical *in situ* STM at single-crystal Au(111)- and Pt(111)- electrodes, in both cases in aqueous perchloric acid solution. The Os-complexes are linked to the surfaces by a 4,4′-bpy or a 1,2-bis(-4-pyridyl)ethane linker, the Co-complex *via* the S-atom of 6-acetylthiohexyl substituted terpyridine ligand (**7**). Cyclic voltammetry reflects monolayer behaviour. Electrochemical ET of the Os-complexes is too fast for direct measurement ($k^0_{\mathrm{ET}} \geq 10^6\ \mathrm{s}^{-1}$), while the Co-complex reacts much more slowly, with $k^0_{\mathrm{ET}} = (1-3) \times 10^3\ \mathrm{s}^{-1}$.

In situ STM of the Os-complexes shows a clear tunnelling current maximum close to their equilibrium redox potentials. The ratio between the current at the peak overpotential and suitably remote from the peak maximum reaches values up to 50 (for (**1**)).[47] The peak position follows systematically the bias voltage and the equilibrium redox potential, in keeping with expectations from theoretical frames for *in situ* STM processes of two-step sequential ET. Broadly similar behaviour is exhibited by the molecular conductivity. The Co-complex also shows a tunnelling spectroscopic feature but much weaker than the Os-complexes. The difference can be referred to larger intramolecular nuclear reorganization and weaker electronic coupling to the enclosing electrodes, and therefore much smaller interfacial ET rate constants. This class of robust transition metal complexes hold perspectives as components in forthcoming molecular electronics in nanogap electrode arrays.

Acknowledgements

Financial support from the Danish Technical Science Research Council, the EU Training and Mobility programme SUSANA, the Danish nanoscience initiative and the Nano-Science Center, University of Copenhagen is acknowledged. Theis Brock-Nannestad is acknowledged for valuable assistance.

References

1 J. Zhang, Q. Chi, T. Albrecht, A. M. Kuznetsov, M. Grubb, A. G. Hansen, H. Wackerbarth, A. C. Welinder and J. Ulstrup, *Electrochim. Acta*, 2005, **50**, 3143–3159.
2 A. M. Kuznetsov and J. Ulstrup, *J. Electroanal. Chem.*, 2004, **564**, 209–222.
3 *Molecular Devices and Machines: A Journey into the Nanoworld*, ed. V. Balzani, M. Venturi and A. Credi, Wiley-VCH, Weinheim, 2003.
4 *Nanoelectronics and Information Technology: Advanced Electronic Materials and Novel Devices*, ed. R. Waser, Wiley-VCH, Weinheim, 2003.
5 J. Zhang, Z.-L. Wang, J. Liu, S. Chen and G.-Y. Liu, *Self-assembled Nanostructures*, Kluwer, New York, 2003.
6 R. M. Metzger, *Chem. Rev.*, 2003, **103**, 3803–3834.
7 S. E. Kubatkin, A. V. Danilov, H. Olin and T. Claeson, *J. Low-Temp. Phys.*, 2000, **118**, 307–316.
8 J. Moreland and J. W. Ekin, *J. Appl. Phys.*, 1985, **58**, 3888–3895.
9 H. Park, A. K. L. Lim, A. P. Alivisatos, J. Park and P. L. McEuen, *Appl. Phys. Lett.*, 1999, **75**, 301–303.
10 J. C. Ellenbogen and J. C. Love, *Proc. IEEE*, 2000, **88**, 386–426.
11 C. P. Collier, E. W. Wang, M. Belohradsky, F. M. Raymo, J. F. Stoddart, P. J. Kuekes, R. S. Williams and J. R. Heath, *Science*, 1999, **275**, 391–394.
12 S. Ami, M. Hliva and C. Joachim, *Nanotechnology*, 2003, **14**, 283–289.
13 W. Haiss, R. J. Nichols, H. Van Zalinge, S. J. Higgins, D. Bethell and D. J. Schiffrin, *Phys. Chem. Chem. Phys.*, 2004, **6**, 4330–4336.
14 B. Xu and N. Tao, *Science*, 2003, **301**, 1221–1223.
15 W. Y. Wang, T. Lee and M. A. Reed, *J. Phys. Chem. B*, 2004, **108**, 18398–18407.
16 X. D. Cui, A. Primak, X. Zarate, J. Tomfohr, O. F. Sankey, A. L. Moore, T. A. Moore, D. Gust, G. Harris and S. M. Lindsay, *Science*, 2001, **294**, 571–574.
17 B. Q. Xu, P. M. Zhang, X. L. Li and N. Tao, *Nano Lett.*, 2004, **4**, 1105–1108.

18 H.-W. Fink and C. Schöneberger, *Nature*, 1999, **398**, 407–410.
19 M.-K. Ng, D.-C. Lee and L. Yu, *J. Am. Chem. Soc.*, 2002, **124**, 11862–11863.
20 M. Fujihira, *Ann. N. Y. Acad. Sci.*, 1998, **852**, 306–329.
21 J. Chen, W. Wang, M. A. Reed, A. M. Rawlett, D. W. Price and J. M. Tour, *Appl. Phys. Lett.*, 2000, **77**, 1224–1226.
22 C. B. Gorman, R. L. Caroll and R. R. Fuierer, *Langmuir*, 2001, **17**, 6923–6930.
23 A. R. Pease, J. O. Jeppesen, J. F. Stoddart, Y. Luo, C. P. Collier and J. R. Heath, *Acc. Chem. Res.*, 2001, **34**, 433–444.
24 N. B. Zhitenev, H. Meng and Z. Bao, *Phys. Rev. Lett.*, 2002, **88**, 226801/1.
25 J. Park, A. N. Pasupathy, J. I. Goldsmith, C. Chang, Y. Yaish, J. R. Petta, M. Rinkoshi, J. P. Sethna, H. D. Abruña, P. L. McEuen and D. C. Ralph, *Nature*, 2002, **417**, 722–725.
26 S. Kubatkin, A. Danilov, M. Hjort, J. Cornil, J.-L. Brédas, N. Stuhr-Hansen, P. Hedegaard and T. Bjørnholm, *Nature*, 2003, **425**, 698–701.
27 K. Nørgaard and T. Bjørnholm, *Chem. Commun.*, 2005, 1812–1823.
28 *Electrochemical Nanotechnology: In situ Local Probe Techniques at Electrochemical Interfaces*, ed. W. Lorenz and W. Plieth, Wiley-VCH, 1998.
29 D. M. Kolb, *Angew. Chem.*, 2001, **40**, 1162–1181.
30 J. Zhang, Q. Chi, A. M. Kuznetsov, A. G. Hansen, H. Wackerbarth, H. E. M. Christensen, J. E. T. Andersen and J. Ulstrup, *J. Phys. Chem. B*, 2002, **106**, 1131–1152.
31 N. Tao, *Phys. Rev. Lett.*, 1996, **76**, 4066–4069.
32 D. L. Gittins, D. Bethell, D. J. Schiffrin and R. J. Nichols, *Nature*, 2000, **208**, 67–69.
33 (*a*) A. M. Kuznetsov and J. Ulstrup, *J. Phys. Chem. A*, 2000, **104**, 11531–11540, and references there; (*b*) A. M. Kuznetsov and J. Ulstrup, *J. Phys. Chem. A*, 2001, **105**, 7494, erratum.
34 A. M. Kuznetsov, *Charge Transfer in Physics, Chemistry and Biology: Physical Mechanisms of Elementary Processes and an Introduction to the Theory*, Gordon and Breach, Reading, MA, 1995.
35 A. M. Kuznetsov and J. Ulstrup, *Electron Transfer in Chemistry and Biology: An Introduction to the Theory*, Wiley, Chichester, 1999.
36 R. J. Forster, E. Figgemeier, P. Loughman, A. Lees, J. Hjelm and J. G. Vos, *Langmuir*, 2000, **16**, 7871.
37 E. M. Kober, J. L. Marshall, W. L. Dressick, B. P. Sullivan, J. V. Caspar and T. J. Meyer, *Inorg. Chem.*, 1985, **24**, 2755.
38 E. M. Kober, J. V. Caspar, B. P. Sullivan and T. J. Meyer, *Inorg. Chem.*, 1988, **27**, 4587.
39 R. Hogg and R. G. Wilkins, *J. Chem. Soc.*, 1962, 341.
40 P. R. Andres, H. Hofmeier, G. G. B. Lohmeijer and U. S. Schubert, *Synthesis*, 2003, 2865–2871.
41 J. Clavilier, R. Faure, G. Guinet and R. Durand, *J. Electroanal. Chem.*, 1980, **107**, 205–209.
42 A. Hamelin, *J. Electroanal. Chem.*, 1996, **407**, 1–11.
43 S. Sugita, T. Abe and K. Itaya, *J. Phys. Chem.*, 1993, **97**, 8780–8785.
44 R. J. Hamers and D. F. Padowitz, *Methods of tunneling spectroscopy with the STM, in Scanning Probe Microscopy and Spectroscopy*, ed. D. Bonnel, Wiley-VCH, Weinheim, 2nd edn., 2001.
45 A. M. Kuznetsov and J. Ulstrup, *Electrochim. Acta*, 2000, **45**, 2339–2361.
46 R. J. Forster and L. R. Faulkner, *J. Am. Chem. Soc.*, 1994, **116**, 5444–5452.
47 T. Albrecht, A. Guckian, J. Ulstrup and J. G. Vos, *IEEE Trans. Nanotechnol.*, 2005, **4**, 430–434.
48 R. J. Forster and J. P. O'Kelly, *J. Phys. Chem.*, 1996, **100**, 3695–3704.
49 R. M. Haddox and H. O. Finklea, *J. Phys. Chem. B*, 2004, **108**, 1694–1700.
50 D. Cummins and H. B. Gray, *J. Am. Chem. Soc.*, 1977, **99**, 5158–5167.
51 R. J. Forster and L. R. Faulkner, *Anal. Chem.*, 1995, **67**, 1232.
52 E. Laviron, *J. Electroanal. Chem.*, 1979, **101**, 19–28.
53 F. R. F. Fan, J. P. Yang, L. T. Cai, D. W. Price, S. M. Dirk, D. V. Kosynkin, Y. X. Yao, A. M. Rawlett, J. M. Tour and A. J. Bard, *J. Am. Chem. Soc.*, 2002, **124**, 5550–5560.
54 G. Nagy and Th. Wandlowski, *Langmuir*, 2003, **19**, 10271–10280.
55 E. D. German, *Rev. Inorg. Chem.*, 1983, **5**, 123.
56 B. Xu, H. He and N. J. Tao, *J. Am. Chem. Soc.*, 2002, **124**, 13568–13575.

Fast temporal fluctuations in single-molecule junctions

Rolf Ochs,[a] Daniel Secker,[ac] Mark Elbing,[a] Marcel Mayor*[ab] and Heiko B. Weber*[ac]

Received 6th May 2005, Accepted 13th June 2005
First published as an Advance Article on the web 7th September 2005
DOI: 10.1039/b506390f

The noise within the electrical current through single-molecule junctions is studied at cryogenic temperature. The organic sample molecules were contacted with the mechanically controlled break-junction technique. The noise spectra refer to a system where only few Lorentzian fluctuators occur in the conductance. The frequency dependence shows qualitative variations from sample to sample.

1. Introduction

We have seen in the past decades a breathtaking development of continuous shrinking of the device feature size in the semiconductor technology. This process is still going on, but will soon be slowed down by some fundamental limitations when the size of the active devices reaches the molecular scale. Then, presumably, similar reproducibility will not be in reach and the investment for new fabrication facilities will be enormous. Therefore, it makes sense to try out new concepts for future nanoelectronics and see how far one can drive alternative technologies. Among the different concepts proposed, *molecular electronics* provides a unique richness of conceivable functionalities, which is given by the enormous number of possible molecules. As first demonstrators, diodes[1] and negative differential resistance devices[2] have been fabricated on the base of molecular monolayers. Further, devices that switch their resistance as a consequence of voltage pulses have been shown,[3–5] which may be employed for both data storage and logic devices. In a research-oriented effort, we are interested in the investigation of charge transport across single-molecule junctions, which is in many respects a simpler system than a molecular film device. We have developed a technique that was able to demonstrate that indeed single molecules could be contacted, and further showed how the electronic properties can purposefully be controlled by a proper choice and synthesis of organic molecules.[6–9] However, some limitations of single-molecule junctions became also evident: the electronic properties displayed sample-to-sample fluctuations, which are not surprising when the single-molecule level is reached. They reflect the uncontrolled local environment of the molecular structure and also structural disorder in the gold tips, which cannot be controlled reproducibly. Similar scattering of results have been reported for transistor geometries, based on a single molecule as the active channel.[10–12]

In addition to sample-to-sample fluctuations, changes of the electronic current–voltage characteristics as a function of time are quite probable, which correspond to the rearrangement of the local environment of the actively conducting molecular bridge, or to a change in the microscopics of the contact bond.[6,13,14] If they are sufficiently slow so they can be independently observed, such abrupt changes are often termed *switching* and the corresponding two-level conductance signal as a function of time is frequently called *telegraph noise*, both being ubiquitous phenomena in solid state physics.[15,16] Such effects were, for example, reported in STM experiments, where the tip was placed on top of a molecule (or few molecules) and the tunneling current was continuously switching

[a] *Forschungszentrum Karlsruhe, Institut für Nanotechnologie, Postfach 3640, D-76021, Karlsruhe, Germany*
[b] *University of Basel, Department of Chemistry, St. Johannsring 19, CH-4056, Switzerland. E-mail: marcel.mayor@unibas.ch*
[c] *Universität Erlangen-Nürnberg, Staudtstr. 7, D-91058, Erlangen, Germany. E-mail: heiko.weber@physik.uni-erlangen.de*

between two or three different conductance states.[17] They are usually attributed to single bonds, atoms or groups of atoms which tunnel or move (over a low activation barrier) between two nearly degenerate spatial configurations. Further sources of noise are (i) the thermal Johnson–Nyquist noise, which corresponds to fluctuations in the quantum mechanical occupation of electron states, (ii) shot noise, which has its origin in the discrete nature of charge carriers, and (iii) electronic noise of the detection circuit. Whereas (i) and (iii) play a role in our measurements and are appropriately subtracted, (ii) comes only into play at higher frequencies and is disregarded.

In this paper, we report on frequency-dependent measurements of the conductance in a frequency range of 100 Hz to 100 kHz. The data show a power-law decrease of noise towards high frequencies, which can be related to fluctuating degrees of freedom of the microscopic configuration in the junction. Interestingly, we observe again sample-to-sample fluctuations not only in the amplitude, but also in the exponent of the power law. Possible mechanisms are discussed.

2. Experimental

2.1. Molecules

Prototypical candidates for the investigation of temporal fluctuations in a single-molecule junction are the molecular rods **1** and **2**, which are displayed in Fig. 1. Their rigid rod-like structure together with their terminal acetyl protected sulfur groups make them well suited for single-molecule investigations in a mechanically controlled break junction. For a defined voltage drop within the molecular structure, the molecular rods are separated into two π-systems by considerable torsion angles in their central biphenyl units induced by steric repulsion of two methyl groups in *ortho* position of their biphenylic C–C bonds. Thereby the overlap of the π-orbitals is decreased resulting also in a reduced electronic coupling between both adjacent phenyl-rings in the biphenyl core. To investigate potential effects of the electronic properties of the rod on the temporal fluctuations, the molecular rod **2** was further functionalised with four fluorine atoms on each terminal phenyl subunit. It is noteworthy that these fluorine atoms alter the electronic properties considerably by their strong negative inductive effect and their moderate positive mesomeric effect, while the structural features of the rod **2** remain basically unchanged compared to **1**. Both molecular rods have been designed and synthesized as control samples of experiments investigating an asymmetric rod as a single molecule rectifier.[9]

2.2. Formation of a gold–molecule–gold contact

The electronic measurements are carried out with the established mechanically controlled break-junction technique.[18–20] A thin gold wire is lithographically structured on a flexible substrate such as phosphorus bronze with an insulating layer of polyimide. At its narrowest constriction, the wire features a width of approximately 50–80 nm which is underetched in an oxygen plasma. The result is

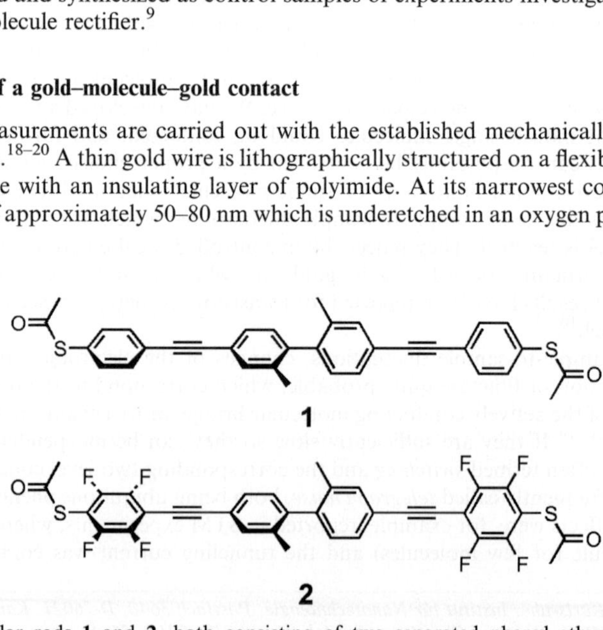

Fig. 1 The molecular rods **1** and **2**, both consisting of two separated phenyl–ethynyl–phenyl π-systems. Effective electronic barriers are created in both molecular rods by considerable torsion angles between both phenyl rings of their central biphenyl-subunits. While structural features remain equal, electronic properties of **2** are considerably varied compared to **1** by fluorine substituents.

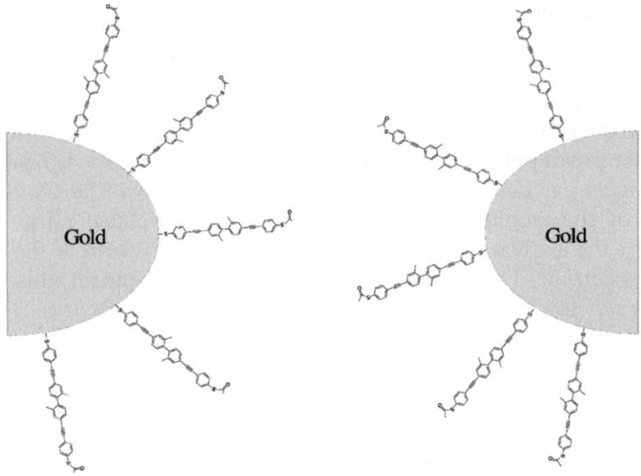

Fig. 2 Schematic view of the gold electrodes and some molecules on the surface. A gold–sulfur bond is established by cleaving of an acetyl protection group.

a freestanding, suspended gold bridge. The substrate is mounted in a three-point bending mechanism and the gold wire is subsequently stretched until it is broken, yielding a pair of opposing electrodes with a spacing that can be varied with an accuracy on the order of 10^{-10} m, when atomic rearrangements are disregarded.

The open electrode pair is immersed in a droplet of a tetrahydrofurane (THF) solution containing the organic molecules under investigation. The solvent evaporates rapidly and a covalent sulfur–gold bond is formed (Fig. 2).

The entire setup is mounted in a vacuum chamber which is equipped with a ^{4}He continuous flow cryostat that allows for cooling down the sample to around 20 K. In a simple two-wire configuration the conductance is monitored and several current–voltage characteristics are recorded. At a pressure of approximately 2×10^{-8} mbar the electrode distance is reduced until the first molecule reaches the opposite electrode and the second sulfur–gold bond is established bridging the gap. While closing the electrode gap further, additional molecules are contacted. Then, the contact is reopened, *i.e.* the electrode distance is increased again. Just before the conductance suddenly drops to zero, the *I–V*-curves are very stable and reproducible. At this point, stable single- or few-molecule contacts are established. This has already been demonstrated in a previous paper employing molecules **1** and **2** (Fig. 1) in comparison with a third molecule[9] and with other molecular rods consisting of comparable subunits in ref. 6. The stability of such a junction is attributed to a fixed molecule–gold contact realisation.

2.3. Noise detection

As in any measurement of electric conductance, the current is not constant in time but exhibits stochastic fluctuations around a mean value. When all external contributions, such as *e.g.* radiation, are eliminated by screening, the remaining fluctuations are referred to as noise. The corresponding physical quantity (at constant voltage) is the autocorrelation function of the current $K_{II}(t)$ and accordingly its Fourier transform, the spectral density $S_{II}(f)$:[15,21]

$$K_{II}(t) = \langle \Delta I(t') \Delta I(t' + t) \rangle$$

$$S_{II}(f) = 2 \int_{-\infty}^{\infty} K_{II}(t) \mathrm{e}^{-i2\pi f t} \, \mathrm{d}t$$

The following experimental setup is used to measure $S_{II}(f)$. A shunt resistance R_S is inserted into the circuit for the current measurement, converting the current into a voltage signal. It is chosen such that the voltage is high enough to be measured with sufficient accuracy while the current measurement is not influenced significantly. For the molecules studied here which have resistances in the 100 MΩ range, $R_S = 100$ kΩ proves to be a good trade-off.

Before the voltage signal is detected by a spectrum analyzer, electronic amplification is required. In order to avoid the the first amplification step adding more noise than the intrinsic signal, we have chosen two low-noise preamplifiers in parallel. The spectrum analyzer (Agilent 89410A), equipped with two input channels, calculates the cross-correlation of these two signals. By this procedure the uncorrelated noise of the amplifiers is eliminated and the correlated noise of the signal at R_S remains. Technically, the analyzer calculates first the Fourier transform of the correlation function and then the cross spectrum. The final result is the frequency-dependent noise density $S_{II}(f)$.

All spectra shown below are calibrated and corrected by the frequency dependence of the transmission line. For this purpose, the signal of a white noise source, *i.e.* a noise source with a constant frequency dependence, is sent through the setup. The measured spectrum is not frequency-independent and is used to calibrate the noise measurements.

3. Noise-generating mechanisms

The mechanisms of noise generation can be distinguished by two different phenomena: the spectral density $S_{II}(f)$ can either depend on the frequency, or not.

3.1. Lorentz oscillator

The frequency-dependent noise is the so-called flicker noise, often referred to as 1/*f*-noise. In contrast to the frequency-independent noise (described below) there is no physical theory that describes the flicker noise in a global way. It occurs though in all electronic devices, resistances, semiconductor devices, and so on. There are several processes that can be responsible for this noise: fluctuating particle numbers, *e.g.* generation–recombination processes in a p–n transition,[22] varying mobility of charge carriers,[23] or fluctuating impurities.[24,25]

Their common properties are fluctuations of the resistance and, as a consequence at constant voltage, current noise. In this case the following is valid:

$$K_{II}(t) = \langle \Delta I(t')\Delta I(t' + t)\rangle \propto I^2$$

and so

$$S_{II}(f) \propto I^2$$

In order to explain the flicker noise behaviour, one uses the model of a Lorentz oscillator. The fluctuations arise from elementary processes each having their own time-constant τ. A single process leads to an auto-correlation function proportional to $e^{-t/\tau}$.[26] Hence its Fourier transform represents a Lorentz curve:

$$S_{II}(f) \propto \frac{1/(\pi\tau)}{4\pi^2 f^2 + (1/\tau)^2}$$

This shows that the noise density of a single fluctuator is for $f \gg 1/\tau$ proportional to f^{-2}.

For a finite number of fluctuators, one may assume that their activation energies W_n have a certain distribution. This, however, will result in a broad distribution of time constants τ, following the relationship $1/\tau \propto \exp(-\beta W)$. Already for few fluctuators the summation leads approximately to $S_{II} \propto f^{-1}$ (Fig. 3). For a continuously and uniformly distributed W_n, the integration leads equivalently to $S_{II} \propto f^{-1}$, which is the textbook result for a large ensemble.[15]

3.2. Thermal noise and shot noise

There are two further sources of noise, where the spectral density is independent of the frequency.

First, there is so-called thermal noise, also known as Johnson–Nyquist noise. It already occurs at equilibrium, *i.e.* when no current flows. This noise results from fluctuations in the quantum mechanical occupation of electronic states and, as a consequence, from the temperature-dependent statistical distribution of the kinetic energy of the electrons. It leads to a variance of the current

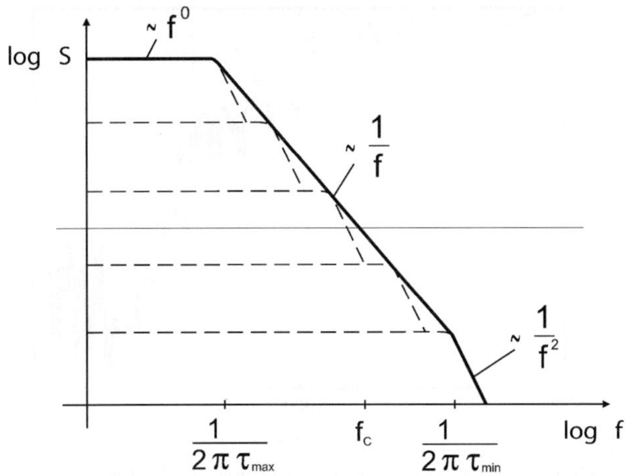

Fig. 3 Schematic view of the superposition of several Lorentz fluctuators. The $1/f^2$-dependence of the noise of a single process with time-constant τ leads effectively to a $1/f$-dependence of the superposition.[15]

despite the fact that the mean value of the current is zero. The resulting spectral density of the auto-correlation is constant for the frequencies relevant in our experiments. This is the well known Johnson–Nyquist theorem:

$$S_{II}(f) \simeq \frac{4k_{\mathrm{B}}T}{R}$$

In contrast to thermal noise, shot noise is a non-equilibrium effect, *i.e.* it occurs only when a net current flows. It has its origin in the discrete nature of charge and thus occurs in situations where the current is flowing across a defined barrier. This is the case for example in diodes or vacuum tubes.[27] When an electron hits the barrier, it is either transmitted or reflected, there is no third possibility. As a consequence, the probability distribution of the number of transmitted electrons is binomial and this fact leads to $S_{II}(f) = 2eI(1 - \alpha)$.[28] For the case of a very small transmission coefficient α, one reaches the Poisson regime and the spectral density is described by the well known Schottky relation:

$$S_{II}(f) = 2eI$$

4. Results

4.1. Noise spectra and current-dependence

The measurements on the immobilised molecules **1** and **2** in a MCB are performed in the following way: a voltage sweep is applied between the contacts and the current is measured. In parallel, at each measuring point a complete noise spectrum is recorded by the spectrum analyzer. So, altogether, we gain both the frequency dependence of the noise density $S_{II}(f)$ and its current dependence $S_{II}(I)$.

As we are not interested in the thermal noise which is neither frequency- nor current-dependent, the measured value at $I = 0$ is subtracted. Additionally, we get rid of undesired external perturbations by that scaling. As already mentioned in Section 2.3 the data are calibrated with the response to a white noise source signal.

Figs. 4 and 5 show the noise spectra of two measurements at a current of $I = 5$ nA in the frequency range [1 kHz,100 kHz], recorded at $T \approx 30$ K. One immediately sees that beyond ~5 kHz the noise level is decreasing, which looks very similar to flicker noise. The linearity in the double-logarithmic plot indicates a power-law behaviour $S_{II}(f) \propto f^{\xi}$. The lines correspond to exponents of $\xi = -1$ for Fig. 4, whereas for another junction (Fig. 5) $\xi = -2$ is better applicable.

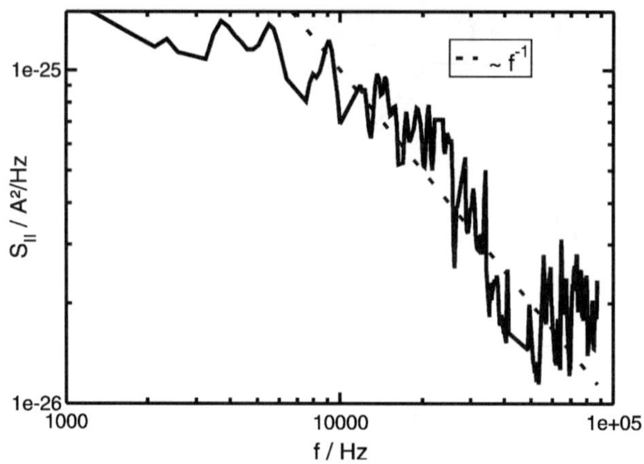

Fig. 4 Noise density $S_{II}(f)$ of a gold–molecule–gold contact (molecule **2**) measured at $T \approx 30$ K and $I = 5$ nA. The frequency dependence shows a power-law behaviour approximately $S_{II} \propto f^{-1}$ at higher frequencies. Similar results are obtained with molecule **1**.

According to the previous explanations, these values correspond to the expectations for a finite number of Lorentz oscillators and a single Lorentz oscillator for the different samples, respectively.

A closer look at the current dependence of the noise density shown in Fig. 6. At a fixed frequency of 5 kHz S_{II} shows a nearly quadratic dependence of I. Hence, the noise is generated by resistance fluctuations of the junction.

The shot noise at a current of $I = 5$ nA has its upper limit at the Poisson value $S_{II} = 2eI = 1.6 \times 10^{-27}$ A^2/Hz. In our frequency range the noise density is much larger and the shot noise can be neglected.†

4.2. Sample-to-sample variations

For the conductance measurements obtained with our technique, we observe sample-to-sample fluctuations in the stable I–V characteristics, which we attribute to be caused by an uncontrolled, disordered local environment of the sample molecule.[6,9] Here, the noise measurements show power-law behaviour of $S_{II}(f)$ which matches well to the limiting cases $\xi = -1$ and $\xi = -2$. Further data not presented here show effective exponents between these two values. Hence, we find again sample-to-sample variations, which occur here as different frequency dependencies. Again, we attribute this to local variations of the environment of the bridging molecule, in paricular different numbers of fluctuators which cause the noise as it is described by the Lorentz-oscillator model in section 3.1.

But what are the elementary processes which cause such resistance fluctuations?

First, the molecule itself could flicker. Both molecular rods consist of four phenyl subunits which are linked with each other by C–C bonds or by ethynyl linkers. The electronic transparency is assumed to depend sensitively on the overlap of adjacent systems.[29] While the rotation of the central C–C bond is limited by the steric demands of two methyl groups, the axial rotation along the ethynyl C≡C triple bond is not hindered. Axial rotations of the subunits of the molecular rod may be a source of resistance fluctuations, which may be excited by the flow of current in addition to thermal excitation. It should be stressed that molecular rotations typically lie in frequency ranges far beyond the upper frequency limit of our experiments. However, in a packed environment with other molecules which come close to the bridging molecule, all degrees of freedom could eventually be slowed down considerably. The fact that differences in the fluctuations between both molecules **1**

† In order to characterize shot noise, the measurements would have to be performed up to higher frequencies to make the shot noise contribution visible. This is indeed very challenging due to RC-depression of the transmission at higher frequencies.

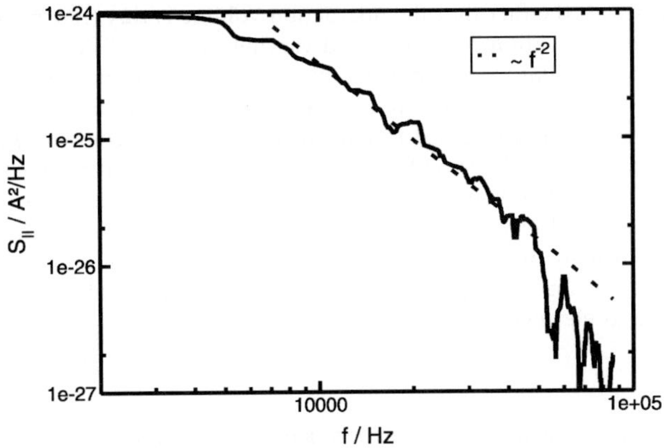

Fig. 5 Noise density $S_{II}(f)$ of a gold–molecule–gold contact (molecule **1**) measured at $T \approx 30$ K and $I = 5$ nA. The frequency dependence shows a power-law behaviour approximately $S_{II} \propto f^{-2}$ at higher frequencies. Similar results are obtained with molecule **2**.

and **2** have not been observed in these investigations rather points to processes independent from the electronic structure of the molecule.

Another possible fluctuator could be the gold–molecule contact. Whereas it is presumably not fully released (then the gold electrode tips would immediately be destabilised and retracted), it could be that there are fluctuations between two (or more) metastable configurations. A similar mechanism has been proposed to explain telegraph-like fluctuations in STM experiments.[30]

Further, the resistance is influenced by the atoms which form the gold tip. Due to the covalent bond between the molecule and the leads, the gold atoms are a part of the contact's wave function. Instabilities which happen in the electrode tips, or, more likely, at its surface could serve as Lorentz fluctuators. In addition, the molecule is sensitive to the local environment. Other molecules or ions which come close to the bridging molecule can electrostatically polarise the molecule's backbone and affect the conductance.

To summarise, not only fluctuations within the molecule, but also in its local environment can affect the conductance of the junction. Hence the noise is a probe for the fast dynamics of the contact molecule and/or its local environment. Due to the fact that many mechanisms could create

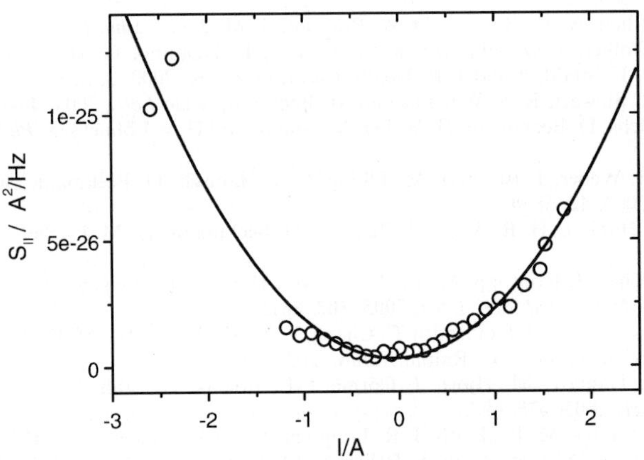

Fig. 6 Noise density $S_{II}(I)$ of a gold–molecule–gold contact (molecule **2**) measured at $T \approx 30$ K and $f = 5$ kHz. The fitted curve represents a quadratic regression of the data (circles). Similar results are obtained with molecule **1**.

such noise, combined with the fact that the interaction volume is tiny (few nanometers around the molecular bridge), the observation of a small and varying number of fluctuators is not surprising.

Because the Lorentz oscillators are presumably driven by thermal activation, a continuous temperature dependence would be desirable. This, however, is difficult to achieve, because a variation of the temperature is connected to rearrangements in the contact, driven by the thermal expansion of the electrodes. We have performed few measurements at higher temperatures, which clearly show that the noise level increases substantially (by roughly one order of magnitude at $T = 150$ K). This fits well to the observation that conductance peaks which are rather sharp at cryogenic temperatures appear strongly blurred at room temperature.[20,7] Such inhomogeneous broadening of the peaks corresponds to an averaging over many configurations, when many more degrees of freedom are activated at elevated temperatures.

5. Outlook

These phenomena are rather general in our experiments and are not specially related to molecules **1** or **2**. For example, we have seen qualitatively similar behaviour for a molecule without the central tilted unit, *I–V* characteristics of which have been reported in.[7] This underscores the notion that the environment could play an important role. For molecules with a switching functionality, which will play a key role for molecular electronic devices, this probe of the fast dynamics could be suitable to detect precursor instabilities before and after switching occurs. But also for stiff molecules, flicker noise is a probe which unambiguously shows that the local environment of a single molecule contact is of critical importance. If there was only one molecule spanned between two stiff and stable electrodes, we would not detect flicker noise. Hence, this observation confirms a picture that has been previously developed:[6,20] there is a disordered environment of the bridge molecule, which results in both static sample-to-sample fluctuations in the current–voltage characteristics[6] as well as dynamical flicker noise.

Acknowledgements

We thank for helpful discussions with Michelle Di Leo, Jan Würfel, Joachim Reichert, Matthias Hettler, Peter Hänggi, Detlef Beckmann, Daniel van der Veer, Ralph Krupke and Hilbert von Löhneysen. Financial support from the Volkswagen foundation and from the network project MOLMEM of the German Ministry of Education and Research (BMBF-FZK 13 N 8360) is gratefully acknowledged.

References

1 R. M. Metzger, *Acc. Chem. Res.*, 1999, **52**, 950–957.
2 J. Chen, M. A. Reed, A. M. Rawlett and J. M. Tour, *Science*, 1999, **286**, 1550–1552.
3 M. A. Reed, J. Chen, A. M. Rawlett, D. W. Price and J. M. Tour, *Appl. Phys. Lett.*, 2001, **78**, 3735.
4 Y. Luo, C. P. Collier, J. O. Jeppesen, K. A. Nielsen, E. Delonno, G. Ho, J. Perkins, H. -R. Tseng, T. Yamamoto, J. F. Stoddart and J. R. Heath, *ChemPhysChem*, 2002, **3**, 519.
5 C. N. Lau, D. R. Stewart, R. S. Williams and M. Bockrath, *Nano Lett.*, 2004, **4**(4), 569–572.
6 J. Reichert, R. Ochs, D. Beckmann, H. Weber, M. Mayor and H. v. Löhneysen, *Phys. Rev. Lett.*, 2002, **88**, 176804.
7 M. Mayor, H. B. Weber, J. Reichert, M. Elbing, C. v. Hänisch, D. Beckmann and M. Fischer, *Angew. Chem. Int. Ed.*, 2003, **42**, 5834.
8 M. Mayor, C. v. Hänisch, H. B. Weber, J. Reichert, D. Beckmann and M. Fischer, *Angew. Chem. Int. Ed.*, 2002, **41**, 1183.
9 M. Elbing, R. Ochs, M. Köntopp, M. Fischer, C. von Hänisch, F. Weigend, F. Evers, H. B. Weber and M. Mayor, *Proc. Natl. Acad. Sci., USA*, 2005, **102**, 8815.
10 J. Park, A. N. Pasupathy, J. I. Goldsmith, C. Chang, Y. Yaish, J. R. Petta, M. Rinkoski, J. P. Sethna, H. D. Abruña, P. L. McEuen and D. C. Ralph, *Nature*, 2002, **417**, 722.
11 S. Kubatkin, A. Danilov, M. Hjort, J. Cornil, J.-L. Brédas, N. Stuhr-Hansen, P. Hedegård and T. Bjørnholm, *Nature*, 2003, **425**, 698.
12 W. Liang, M. P. Shores, M. Bockrath, J. R. Long and H. Park, *Nature*, 2002, **417**, 725.
13 F. Zahid, A. W. Ghosh, M. Paulsson, E. Polizzi and S. Datta, *Phys. Rev. B*, 2004, **70**, 245317.
14 S. N. Yaliraki, M. Kemp and M. A. Ratner, *J. Am. Chem. Soc.*, 1999, **121**, 3428.
15 R. Müller, *Rauschen*, Springer-Verlag, Berlin, 1990.
16 S. Brouër, G. Weiss and H. B. Weber, *Europhys. Lett.*, 2001, **54**, 654–660.

17 Z. J. Donhauser, B. A. Mantooth, K. F. Kelly, L. A. Bumm, J. D. Monnell, J. J. Stapleton, D. W. Price, A. M. Rawlett, D. L. Allara, J. M. Tour and P. S. Weiss, *Science*, 2001, **292**(5525), 2303–2307.
18 J. M. van Ruitenbeek, A. Alvarez, I. Pineyro, C. Grahmann, P. Joyez, M. H. Devoret, D. Esteve and C. Urbina, *Rev. Sci. Instrum.*, 1996, **67**, 108–111.
19 N. Agraït, A.L. Yeyati and J. M. van Ruitenbeek, *Phys. Rep.*, 2003, **377**(2–3), 81–279.
20 J. Reichert, H. B. Weber, M. Mayor and H. von Löhneysen, *Appl. Phys. Lett.*, 2003, **82**, 4137–4139.
21 A. Blum, *Elektronisches Rauschen*, Teubner, Stuttgart, 1996.
22 A. L. McWorther, in *Semiconductor Surface Physics*, ed. R. H. Kingston, University Press, Philadelphia, 1957.
23 F. N. Hooge, *Physica B*, 1990, **162**, 344.
24 J. L. Black, *Low-Energy Excitations in Metallic Glasses*, Springer-Verlag, Berlin, 1981.
25 S. Kogan, *Electronic Noise and Fluctuations in Solids*, University Press, Cambridge, 1996.
26 Y. Imry, *Introduction to Mesoscopic Physics*, Oxford University Press, New York, 1997.
27 C. Schönenberger and S. Oberholzer, *Proc. SPIE*, 2004, **5469**, 233–243.
28 Y. M. Blanter and M. Büttiker, *Phys. Rep.*, 2000, **336**, 15.
29 J. Chen and M. Reed, *Chem. Phys.*, 2002, **281**, 127–145.
30 Z. J. Donhauser and P. S. Weiss, unpublished.

Influence of bridge topology and torsion on the intramolecular electron transfer

Vega Lloveras,[a] José Vidal-Gancedo,[a] Daniel Ruiz-Molina,[a] Teresa M. Figueira-Duarte,[b] Jean-François Nierengarten,[b] Jaume Veciana*[a] and Concepció Rovira*[a]

Received 11th May 2005, Accepted 8th June 2005
First published as an Advance Article on the web 12th October 2005
DOI: 10.1039/b506678f

To study molecules able to act as good "molecular wires", intramolecular electron transfer between two triphenylmethyl redox centers connected by bridges with different topologies and substituents have been studied in solution, both by UV-Vis-NIR and EPR spectroscopies. The synthetic methodology used allows a complete control of the geometry of polychlorotriphenylmethyl diradicals **1** and **2**, which have *para* and *meta* topologies, respectively, as well as of their *E*/*Z* isomerism. This fact is used to show the influence of the different topologies in the ease of electron transfer, which is larger for the *para* than for the *meta* isomer where a small or negligible electronic coupling is observed. The related diradical **3** that have the same topology as the *para* isomer **1** but bearing two substituents on the central phenyl ring shows similar ease of electron transfer, that the *para* isomer **1**.

1. Introduction

The enormous interest at present for molecular wires arises from the potential use of such systems on integrated molecular-sized devices, to scale down the size of the devices and integrate molecular switches with molecular-scale wires.[1] The interest and the foreseen technical advantages of such integrated molecular devices are significant.[2,3] First, the ever-increasing miniaturization of the architectural components of microchips reverts in the reduction size of computational systems. Second, the time it takes for an electron to travel through the circuit can be minimized using such molecular-scale electronic architectures.[3] On the other hand, the main drawbacks are: (i) the severe limitations existing for manipulating objects at the molecular level and for transferring the information from the macroscopic world to molecular electronic devices and (ii) the high conformational requirements that the molecule to be used must fulfil. For instance, an average interelectrode distance of 100 Å obtained using "top-down" techniques; *i.e.*, lithography, requires the synthesis of molecular systems that, in addition to a π-conjugated pathway, have similar dimensions to the interelectrode distance and restricted conformational geometries.

The first step to progress on this field is the understanding of rules for the prediction and control of the electron propagation in organic "molecular wires". Electron transfer in donor–bridge–acceptor (D–B–A) molecules has been actively investigated, in part to determine which aspects of the bridge structure control the intramolecular electron transfer (IET).[4] Mixed-valence compounds are excellent candidates for such studies since IET phenomena can be easily determined from the position, intensity and width of the so-called intervalence transition band that usually appears in the near-infrared region. In addition, in those systems that generate stable radical mixed valence species, the dynamics of the thermally activated IET process can also be studied from the thermal variation of the EPR spectrum of the mixed valence species.[5,6]

Even thought the conclusions from the solution studies do not necessarily correlates for the same molecules in other environment as is the solid state, in electrochemical experiments in which the donor or acceptor is replaced by a conducting solid also seek to reveal how the ET rate depends on

[a] *Institut de Ciència de Materials de Barcelona (CSIC), Campus Universitari de Bellaterra, 08193, Cerdanyola, Spain. E-mail: cun@icmab.es, vecianaj@icmab.es*
[b] *Laboratoire de Chimie de Coordination du CNRS, 205 route de Narbonne, 31077, Toulouse Cedex 4, France*

the "bridge" structures.[7] On the other hand, the values of the "decay factors" β obtained by measurements of current density through saturated and unsaturated molecules of different length, using junctions of the type Hg–SAM(1)/SAM(2)–Ag, are in good agreement with corresponding values obtained in solution.[8]

A prime objective of our research has been the synthesis and characterization of pure organic mixed valence systems using polychlorinated triphenylmethyl radical units as electroactive centers.[9] Such radicals are particularly interesting not only because they have a large persistency displaying high thermal and chemical stabilities but also because they are electroactive species giving rise, either chemically or electrochemically, to the corresponding anions and cations which are also quite stable species.[10] As a consequence, it seemed interesting to obtain and study symmetrical molecules consisting of two of these radicals linked with different types of bridges.

In order to maximize the electron transfer rates in mixed valence compounds, the bridge choice is crucial. Both its electronic structure and the effective distance between the redox centers are known to play a critical role in determining the ease of the electron transfer.[4] For these reasons, we have designed a series of structurally well defined diradicals that incorporate phenylene-vinylene units as a bridge since homo- and heteronuclear metallic complexes with phenylene and/or vinylene units promoting through-bond photoinduced energy or electron transfer over considerable distances (*i.e.* 20 Å) had already been described.[11]

We have studied both by UV-Vis-NIR and EPR spectroscopies the intramolecular electron transfer phenomena in the radical anions derived from the partial reduction of pure organic diradicals **1** and **2** that varies in the topological substitution of the bridge.

These diradicals, which exhibit a well defined topology and isomerism, consist of two polychlorinated triphenylmethyl radicals linked by divinylphenylene bridges. The difference with other previous work using these bridges relays in the bulk chlorine substituents in the two phenyl rings linked to the electroactive center (the radical central carbon) that can promote big changes in the angle of rotation due to steric hindrance. The synthetic methodology used allows a complete control of the *E*/*Z* isomerism of diradicals **1** and **2**, which have *para* and *meta* topologies, respectively. We will show that the topology clearly influences the ease of electron transfer. Indeed, small V_{ab} coupling are expected to appear when the triphenylmethyl radical units are connected in the *meta* position, in contrast to the case where the two units are in the *para* position.

By other hand, we also present the study of the intramolecular electron transfer phenomena in the radical anion derived from the partial reduction of diradical **3** with the same *para* topology as **1** but bearing substituents in two positions of the central phenyl ring of the bridge. We want to determine if the possible rotation around the σ bonds promoted by the steric hindrance between these substituents and the bulky chlorine atoms on the neighboring phenyl rings could originate a severe loss of coplanarity of the bridge and consequently, a decrease of the long-distance electronic coupling between the terminal sites.

2. Results and discussion

2.1. Synthesis

The synthetic route for preparing radicals **1–3** is based on two main steps. The first step is the synthesis of their triphenylmethane precursors by a Wittig reaction while the second step is the generation of the corresponding dicarbanions by an acid–base reaction followed by the subsequent oxidation of such carbanions to the corresponding diradicals (Scheme 1).

Reaction of the α-*H*-4-(bromomethyl)tetradecachlorotriphenylmethane derivative (**4**),[12] with triphenylphosphine yielded the starting substituted phosphonium bromide **5** which was coupled with the suitable bisaldehyde under Wittig conditions to give the corresponding polychlorinated bis-triphenylmethanes **6**, **7** and **8**. This reaction is strongly stereoselective since it yields exclusively the *E*/*E*-isomers, as ascertained by NMR spectroscopy.[9c,13] Such stereoselectivity is justified if we consider that the ylide derived from the phosphonium bromide **5** is stabilized by the presence of the polychlorinated aromatic ring. In fact, the natural preference of ylides with electron withdrawing groups that stabilize the betaine form has been experimentally shown to give mainly the *E*-isomer.[14] In addition, the *Z*/*E*-isomer distribution of the Wittig products is also strongly influenced by the

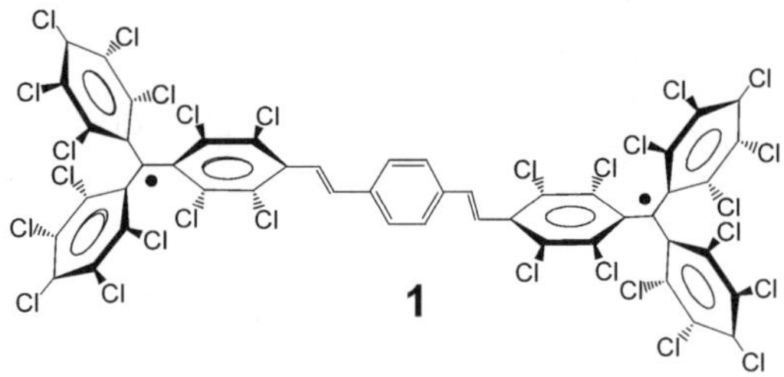

1

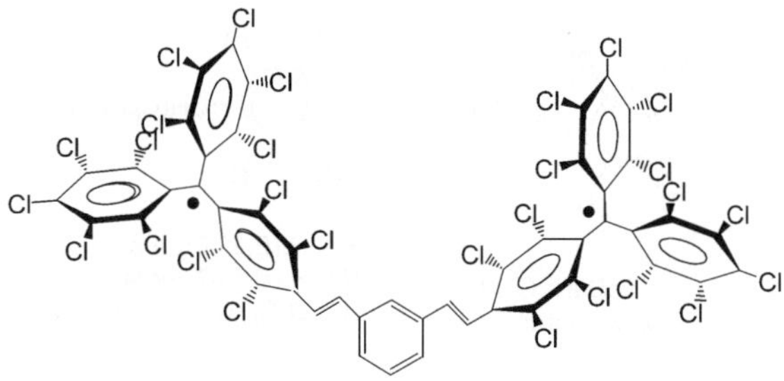

2

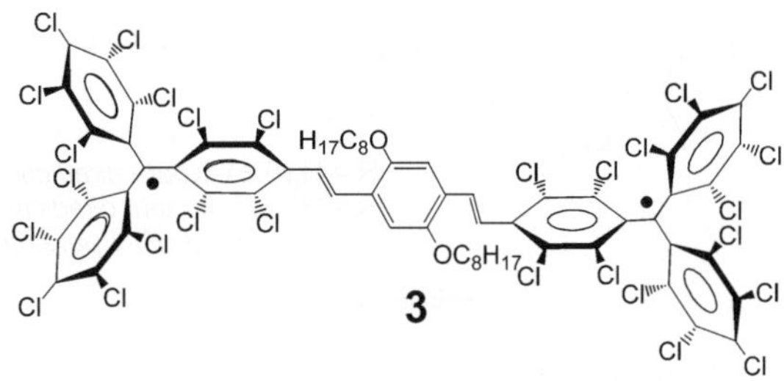

3

nature of the base used in the preparation of the ylide;[15] being the potassium *tert*-butoxide, used in this work, the base of choice for maximizing the yields of *E*-olefins.[16]

Afterwards, the synthesis of the diradical species **1**, **2** and **3** was undertaken by treatment of the corresponding triphenylmethane precursors **6**, **7** and **8** with an excess of *n*-$Bu_4N^+OH^-$ and the subsequent oxidation of the resulting dianions with *p*-chloranil in a "*one-pot reaction*". The complete synthesis can also be performed as a "*one-pot reaction*" without the isolation of the intermediate triphenylmethane derivatives **6–8** obtaining in this case better yields. Diradicals **1**, **2** and **3** were isolated as dark brown (75%), green (68%) and dark brown (64%) solids respectively. Characterization of all compounds was performed by elemental analysis and spectroscopic techniques.[17] As described for a related monoradical with *p*-bromovinylenephenylene substituent, bis(pentachlorophenyl)[4-(4-bromophenyl-β-styryl)-2,3,5,6-tetrachlorophenyl]-methyl radical (**9**),[9c] no isomerization of double bonds was noticed for diradicals **1–3** by spectroscopical techniques. Finally, it is worth noticing that the diradicals are highly persistent and thermally stable in the solid state and also in diluted solutions, even when exposed to air. No enhanced reactivities of the diradicals with respect to their monoradical counterparts were observed.

2.2 *Meta versus para* bridge topology

2.2.1 Study of diradicals 1 and 2. Electrochemical studies in CH_2Cl_2, with *n*-Bu_4NPF_6 (0.1 M) as supporting electrolyte (*vs.* SCE) and using a Pt wire as a working electrode, were done at room temperature. The CV of diradicals **1** and **2** show only one reversible reduction process at a constant potential of approximately −0.12 V (*vs.* Ag/AgCl), in spite of the presence of two electronically active triphenylmethyl units. This result suggests the presence of very weak or negligible electronic interactions between the triphenylmethyl units of diradicals **1** and **2**, since in the case of a strong or moderate electronic interaction between the two units, two electrochemical waves are anticipated.

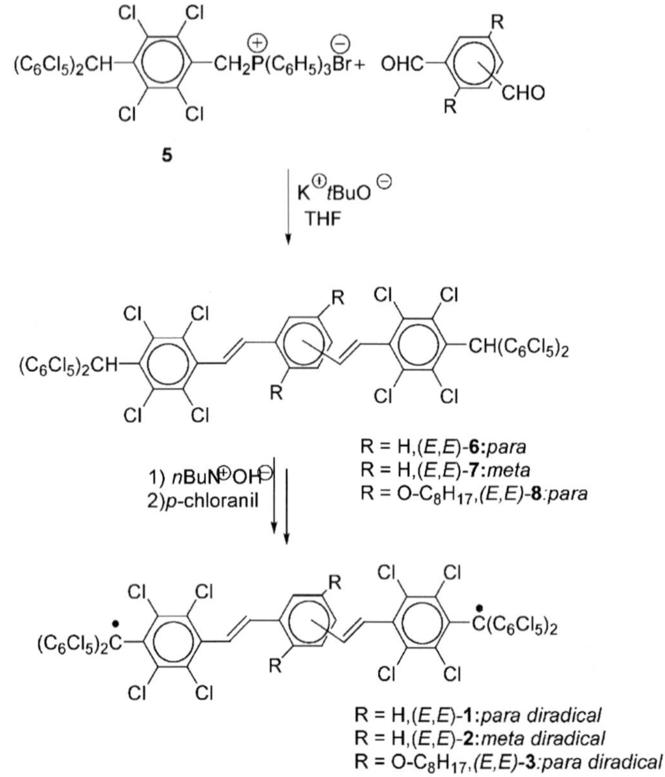

Scheme 1

Only if both units exhibit very weak or negligible electronic interactions, the two standard redox potentials must be very close and a single two-electron wave will be observed, with minor differences in shape with respect to a true two-electron process. Cyclic voltammetric peak separation of the reversible reduction waves for the diradicals **1** and **2** are 83 and 78 mV, respectively, which are similar to that observed for the related monoradical **9** (80 mV).[9c] Such separation serves as one criterion for electrochemical reversibility and those observed for radicals **1** and **2** are closed to the expected theoretical separation of 59 mV. The slightly higher peak separation may arise from the high resistance of the solutions used for the measurements.

The magnetic susceptibility of powder samples of biradicals **1** and **2** was measured in the temperature range of 4–300 K. Diradicals **1** and **2** follow the Curie–Weiss law, down to low temperatures, with $\theta = -4$ K and -0.4 K respectively, due to weak intra- and/or intermolecular antiferromagnetic interactions, being as expected larger for the *para* isomer **1**. The effective magnetic moments found at room temperature were 2.41 and 2.43 μ_B for diradicals **1** and **2**, respectively. These values are in excellent agreement with the theoretical effective magnetic moment expected for vanishing or non-interacting pair of doublets (2.45 μ_B).

UV-Visible spectra of diradicals **1** and **2** were recorded in dichloromethane and relevant data are shown in Table 1. Visible spectra of the polychlorotriphenylmethyl radicals usually show an intense absorption band around 386 nm and two weaker bands centered around 565–605 nm, all of which are assigned to the radical character of the triphenylmethyl units.[10] In the case of diradicals **1** and **2**, due to the presence of a certain degree of electronic delocalization, bathochromic shifts with enhanced absorptivities, compared to unsubstituted chlorinated triarylmethyl radicals, would be expected. This behaviour is indeed observed for diradical **2**. For diradical **1** substantially larger bathochromic shifts and enhanced absorptivities are observed as expected from the larger electronic delocalization present in this compound due to its *para* connectivity.

Table 1 Relevant UV-Vis and EPR data

Compound	EPR[ab] $\Delta H_{1/2}$	a_H	$a_{C,\alpha}$	$a_{C,arom}$	D/hc	E/hc	UV-Vis[e] λ/nm ($10^{-3}\varepsilon$)
1	0.95	0.95(2H)	14.0	5.9, 6.4	3.9×10^{-4}	[c]	600(3.2), 440(20.7), 386(57.4), 339(33.1)
$\mathbf{1}^{-\bullet}$	0.85	1.90(1H)	[d]	[d]	—	—	[f]
$\mathbf{1}^{-2}$	—	—	—	—	—	—	606(46.4), 522(49.3), 339(32.3)
2	0.98	0.90(2H)	13.9.0	5.5, 6.5	2.3×10^{-4}	[c]	573(3.1), 415(26.0), 385(33.7)
$\mathbf{2}^{-\bullet}$	0.58	1.80(2H) 0.60 (2H) 0.60 (2H)	[d]	[d]	—	—	[f]
$\mathbf{2}^{-2}$	—	—	—	—	—	—	570(49.4), 525.4(51.4), 293(39.2)
3	0.96	0.88(2H)	14.8	5.4, 6.5	3.9×10^{-4}	[c]	680(3.8), 503(11.6), 418(sh), 388(52.3) 321(15.7)
$\mathbf{3}^{-\bullet}$	0.88	1.80(1H)	[d]	[d]	—	—	1420 (0.26), 615 (36.5), 523 (38.9), 388 (49.6)
$\mathbf{3}^{2-}$	—	—	—	—	—	—	611(62.5), 525(60.3), 318(30.1)

[a] In CH_2Cl_2 solutions at rt; except for $\mathbf{1}^{-\bullet}$ and $\mathbf{2}^{-\bullet}$ and $\mathbf{3}^{-\bullet}$ that were taken at 200 K (see text). [b] Linewidths and hyperfine coupling constants (in Gauss) are computer simulated values; g-values found were similar, 2.0027(3), for all open-shell species. Zero-field splitting parameters (in cm^{-1}) observed in frozen solutions at 120 K. Principal components of g-tensors for $\mathbf{1}^{-\bullet}$ and $\mathbf{2}^{-\bullet}$ are, respectively, $g_x = 2.0033$, $g_y = 2.0033$, $g_z = 2.0010$ and $g_x = 2.0029$, $g_y = 2.0029$, $g_z = 2.0025$. These values were determined by simulation of frozen EPR spectra. [c] Negligible. [d] Not observed. [e] In CH_2Cl_2 solution at r.t. [f] See text.

X-band EPR spectra of radicals **1** and **2** were recorded in CH_2Cl_2 : toluene solution in the temperature range of 180–293 K. The isotropic EPR spectra at 300 K were fairly well simulated by using the parameters given in Table 1. The experimental and simulated spectra of diradical **1** is shown in Fig. 1 and is almost identical to the one of biradical **2**. At room temperature both spectra showed the lines corresponding to the coupling of the unpaired electrons with the different nuclei with nonzero magnetic moments; *i.e.*, with ^1H and the naturally abundant ^{13}C nuclei. Computer simulation gave the isotropic g-values (g_{iso}) and the isotropic hyperfine coupling constants (a_i). The obtained g_{iso}-values (2.0022 and 2.0024 for radicals **1**, **2** respectively) are very close to those observed for other polychlorotriphenylmethyl radicals.[10] More interesting is the analyses of the isotropic hyperfine coupling constant values with the hydrogen nuclei of the vinylene moieties and with some of the carbon nuclei of the triphenylmethyl unit. For instance, the EPR spectrum of diradicals **1** and **2** at 300 K display three main overlapping symmetrical lines originating from the hyperfine coupling with two hydrogen atoms of the ethylene moiety. The values of the coupling constants for diradicals **1** and **2** are approximately half than those found for the related monoradical **9**.[9c] It is then possible to conclude that the two electrons in diradicals **1** and **2** are magnetically interacting with a magnetic exchange coupling constant J, that fulfils the following condition, $J \gg a_i$.

Furthermore, the absolute values of zero-field splitting parameters, $|D/hc|$ and $|E/hc|$, for diradicals **1** and **2** were obtained from the simulated spectra of both diradicals in frozen solutions and are given in Table 1. These parameters arise from the dipolar magnetic interactions between the two unpaired electrons and can be used to calculate the average interspin separation in a given compound. Therefore, from the zero-field splitting parameter $|D/hc|$, given in cm^{-1}, and using eqn. (1),[18] an average interspin separation of 15.0 and 17.5 Å was found for diradicals **1** and **2**, respectively.

$$r = \left[\frac{3g^2\beta^2}{2hc}\frac{1}{|D/hc|}\right]^{1/3} \quad (1)$$

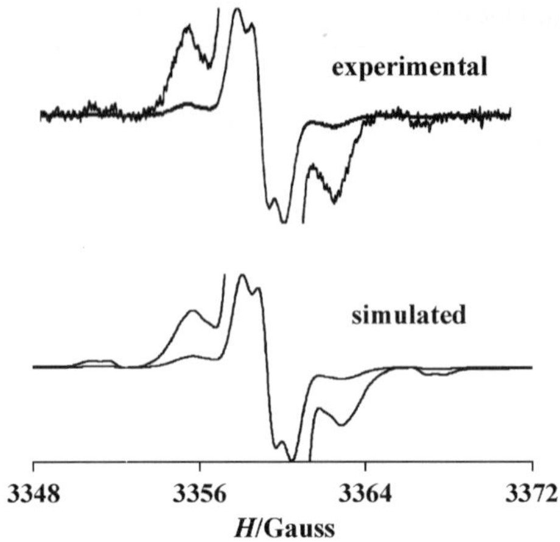

Fig. 1 Simulated and experimental isotropic solution EPR spectra of diradical **1** in CH_2Cl_2 : toluene (1 : 1) at 285 K.

The average interspin separation found for diradical **1** is smaller than the nominal separation between the two alpha carbon atoms, determined by semiempirical AM1 calculations,[19,20] where most of the spin density of polychlorinated triphenylmethyl radicals is localized.[10] Indeed, the through-bond distance between the two electron active sites; *i.e.*, the sum of bond lengths corresponding to the conjugated pathway is 24.1 Å and the through space distance is 19 Å. This result is in agreement with the existence of a certain degree of electronic delocalization due to the *para*-connectivity of the divinylphenylene bridge that reduces the effective separation of the two spins in **1**. As expected, the average interspin separation found for diradical **2** is closer to the nominal separation between the two α-carbon atoms (through bond, 22.7 Å and through space 17 Å) due to the lower degree of electronic conjugation in the *meta*-isomer in comparison with the isomer with *para*-connectivity. Finally, EPR measurements of frozen dilute solutions of both diradicals in the 4–100 K range revealed that there exists a very weak magnetic interaction between the two unpaired electrons since the signal intensity for both diradicals follows the Curie law. This result, along with the $J \gg a_i$ condition, permit to estimate the limits for the intramolecular exchange coupling constant for diradicals **1** and **2**; that are 15 $cm^{-1} \gg J/hc \gg 10^{-3}$ cm^{-1}.[21]

2.2.2 Intramolecular electron transfer process. Determination of the experimental electronic coupling parameter V_{ab} has been undertaken by spectroelectrochemistry using a previously described methodology.[22] Such a methodology is based on the reduction of diradicals **1** and **2** and the simultaneous observation of the corresponding electronic absorption spectra, regularly recorded during the reduction process for different values of the average number (n) of electrons added; where $0 \leq n \leq 2$. This situation allows us to control the evolution of the electrochemical reaction simply by monitoring the evolution of the characteristic radical and anion bands. The different reduced species that can be derived from both diradicals are depicted schematically in Scheme 2. As can be observed there, diradicals **1** and **2**, as well as the dianion $\mathbf{1}^{-2}$ and $\mathbf{2}^{-2}$, are homovalent species whereas the radical anions $\mathbf{1}^{-\bullet}$ and $\mathbf{2}^{-\bullet}$ should be regarded as a mixed-valence species. Accordingly, only the last species may show intervalence transition bands that usually appear at low energies; *i.e.*, at the near-infrared region.

During the reduction of diradical **1** the sharp band at 386 nm, characteristic of the radical chromophore, decreases until it completely disappears for $n = 2$, while the band at 339 nm does not change at all. At the same time, two new broad and intense bands at 522 and 606 nm grow in, which correspond to the continuous formation of the dianion $\mathbf{1}^{-2}$. Similar behavior was observed for diradical **2**. Fig. 2 shows the evolution of the UV-Vis spectra during the course of the reduction of diradical **1** until reaching the highest concentration of the mixed valence species $\mathbf{1}^{-\bullet}$. The observation of some well-defined isosbestic points during reduction of diradicals **1** and **2** indicate

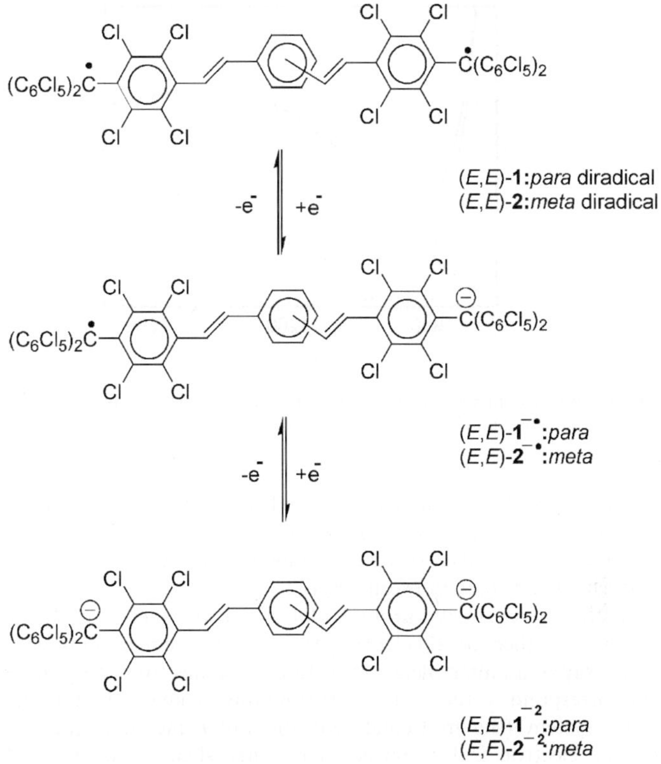

Scheme 2

that no by-products due to decomposition processes are originated during such electrochemical reductions.

The main difference in the stepwise coulometric titration of diradicals **1** and **2** is the observation of a new band centered around 1400 nm for the radical-anions $\mathbf{1}^{-\bullet}$, shown in Fig. 3. Indeed, during the reduction process of diradical **1**, a weak and broad band centered around 1400 nm appears and develops until the complete formation of $\mathbf{1}^{-\bullet}$. Afterwards, the intensity of this band decreases until

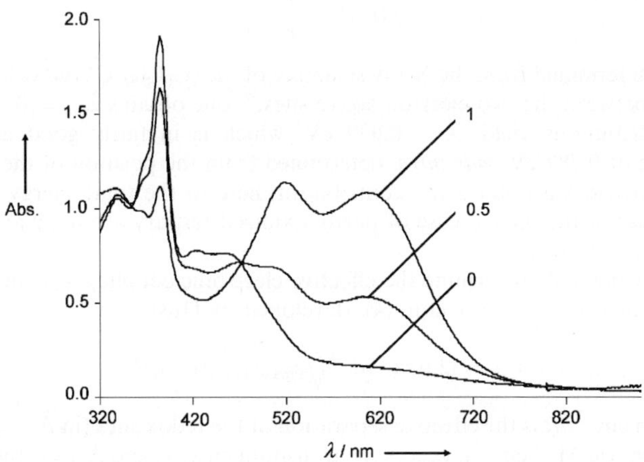

Fig. 2 Evolution of the UV-Vis spectrum during the course of the reduction of diradical **1**. The number of added electrons is given on each spectrum.

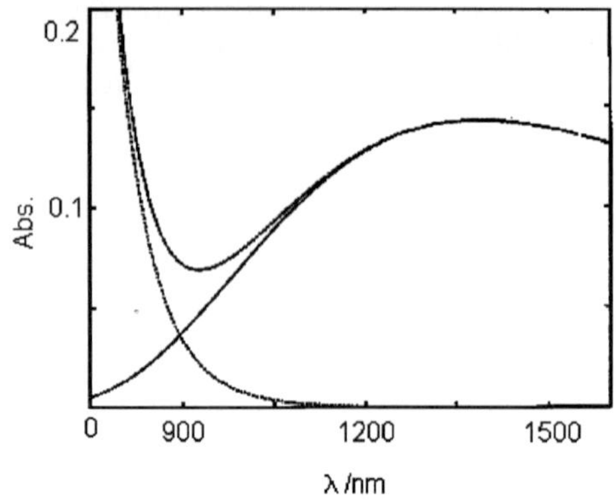

Fig. 3 Electronic spectrum of the mixed-valence species $\mathbf{1}^{-\bullet}$ (dashed line), sum of the intervalence transition band (dashed–dotted line) and the band corresponding to the radical and/or anion chromophore (dotted line).

it disappears when the dianion $\mathbf{1}^{-2}$ is completely formed. This is the typical behavior expected for an intervalence band originated by the presence of an intramolecular electron transfer phenomenon in the mixed-valence species $\mathbf{1}^{-\bullet}$. On the other side, no intervalence band was detected for the radical anion $\mathbf{2}^{-\bullet}$ formed during the reduction of diradical **2**.

The assignment of the band centered around 1400 nm in the spectra of $\mathbf{1}^{-\bullet}$ as an intervalence transition was confirmed by theoretical calculations following the method previously employed by Nelsen *et al.*[23] The energy of an intervalence transition, λ, comes from two different contributions, λ_{in} and λ_{out}, which correspond to the energy contributions produced by the arrangements of the internal part of the system (the molecule) and the outer medium (the solvating medium), respectively, when the electron is transferred. The inner shell, or internal Marcus, term, for polychlorinated triphenylmethyl radicals is $\lambda_{in} = 0.214$ eV and was obtained by quantum mechanical calculations at the AM1 level performed on the $(C_6Cl_5)_3C^\bullet$ fragment.[19] The outer shell term was estimated using the Marcus dielectric continuum model shown in eqn. (2),

$$\lambda_{out} = 14.41g\ (r,R)\gamma \tag{2}$$

where γ is the Marcus solvent parameter, which is 0.380 for CH_2Cl_2, $g(r,R)$ is the geometrical parameter of the electron active center and λ_{out} is given in eV. The $g(r,R)$ parameter is calculated using eqn. (3) from the effective radius, r, of the localized charge -represented as a sphere centered at the carbon atom- and R, the distance between the centers of the two electron active centers, R,[24]

$$g(r, R) = \frac{1}{r} - \frac{1}{R} \tag{3}$$

with $r = 5.4$ Å, as determined from the X-ray structure of the $(C_6Cl_5)_3C^\bullet$ radical,[10c] and $R = 19$ Å as the limit distance between the two electron active sites,[24] one obtains $\lambda_{out} = 0.726$ eV. Finally, the sum of both contributions yields $\lambda = 0.939$ eV; which is in fairly good agreement with the experimental value of 0.787 eV: *vide infra*, determined from the position of the intervalence band. Interestingly, λ_{in} provides a noteworthy contribution here to the total energy of the intervalence transition λ, whereas in the related case of partly oxidized tertiary aromatic amines this term was almost negligible relative to λ_{out}.[25]

In order to experimentally determine the effective electronic coupling V_{ab} (in cm^{-1}) between the redox sites of diradical **1** we can use eqn. (4), developed by Hush.

$$V_{ab} = [2.05 \times 10^{-2}\sqrt{\varepsilon_{max}\bar{\nu}_{max}\Delta\bar{\nu}_{1/2}}]R^{-1} \tag{4}$$

Here V_{ab} is given in cm^{-1}, R is the effective separation of the redox sites (in Å), ε_{max} is the maximum extinction coefficient (in M^{-1} cm^{-1}), $\bar{\nu}_{max}$ is the transition energy, and $\Delta\bar{\nu}_{1/2}$ is the full-width at half-height (both in cm^{-1}) of the intervalence absorption band. After correction for the comproportionation equilibrium between the different redox forms of **1**, $\mathbf{1}^{-\bullet}$ and $\mathbf{1}^{-2}$, and the deconvolution of the experimental spectrum,[26] the following parameters for the intervalence absorption band were obtained: $\bar{\nu}_{max} = 6349\ cm^{-1}$ (*i.e.* $\lambda = E_{opt} = 0.787$ eV), $\Delta\bar{\nu}_{1/2} = 2940\ cm^{-1}$, and $\varepsilon_{max} = 677\ M^{-1}\ cm^{-1}$. Application of these parameters to eqn. (4) provides an effective electronic coupling for $\mathbf{1}^{-\bullet}$ of $V_{ab} = 121\ cm^{-1}$ (0.015 eV), when $R = 19$ Å is used.

We also studied the thermally activated electron transfer process in $\mathbf{1}^{-\bullet}$ by EPR spectroscopy. For this purpose, the electrochemical reduction of **1** was performed until an almost complete reduction to $\mathbf{1}^{-2}$ is achieved. The resulting solution contains, besides the desired mixed-valence $\mathbf{1}^{-\bullet}$ species, mainly the EPR-silent $\mathbf{1}^{-2}$ species and, therefore, the resolution of the spectra is good. The EPR spectrum of $\mathbf{1}^{-\bullet}$ at 200 K displays two symmetrical lines arising from the hyperfine coupling with one hydrogen atom of the ethylene moiety (Fig. 4) and the ^{1}H hyperfine coupling constant is very close to that of the related monoradical **9**.[9c] This result clearly demonstrates that at this temperature the unpaired electron of radical anion $\mathbf{1}^{-\bullet}$ is localized on the EPR timescale on only one-half of the molecule; *i.e.* on one stilbene-like moiety.

As depicted in Fig. 4, when the temperature is increased, a new central line gradually emerges between the two initial ones. This evolution is consistent with the increase of the electron jump rate on going from 200 K (the slow-exchange limit) to 300 K (the fast-exchange limit) due to a thermally activated electron transfer between the two equivalent sites, that lead to the coupling of the unpaired electron of $\mathbf{1}^{-\bullet}$ with two equivalent ^{1}H nuclei. The EPR spectra of $\mathbf{1}^{-\bullet}$ at various temperatures were fairly well simulated by using the rates given in Fig. 4 and the program of Heinzer.[27] The best fit was made by varying the rate constant k_{th} for the thermally activated electron transfer between the two equivalent sites. The experimental and simulated EPR spectra are compared in Fig. 4. The resulting k_{th} values were plotted using a linear Eyring plot [$\ln(k_{th}/T)$ *vs.* $1/T$] from where a value for the energy barrier to the thermal electron transfer, ΔH^*, of 0.117 eV was obtained.

According to Hush theory, E_{opt} corresponds to the energy for the vertical electron transfer between the sites. In the limit of a vanishingly small electronic interaction (*i.e.*, $V_{ab} \approx 0$), the thermal barrier to electron transfer is then predicted to be one-quarter of E_{opt}, because of the usual parabolic shape of the energy wells, according to the following relationship,

$$\Delta H^* = (E_{opt}/4) - V_{ab} \qquad (5)$$

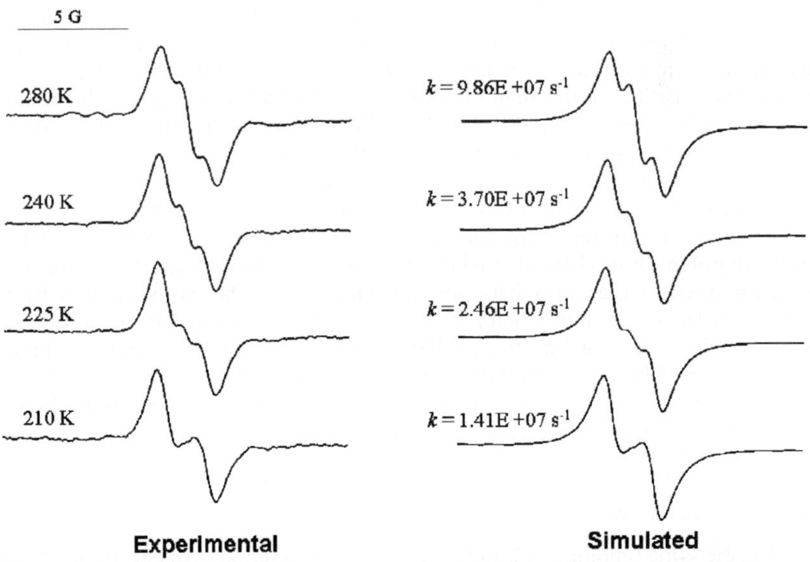

Fig. 4 Experimental (left) and simulated (right) EPR spectra of $\mathbf{1}^{-\bullet}$ at different temperatures in CH_2Cl_2 with 0.1 M of $n\mathrm{Bu_4NPF_6}$. Spectra recorded at the very end of the reduction process. At this stage, the major species present in the solution is the EPR-silent dianion $\mathbf{1}^{-2}$. Most of the observed EPR signal is due to $\mathbf{1}^{-\bullet}$, since for statistical reasons diradical **1** must have negligible concentration. See text for details on simulation.

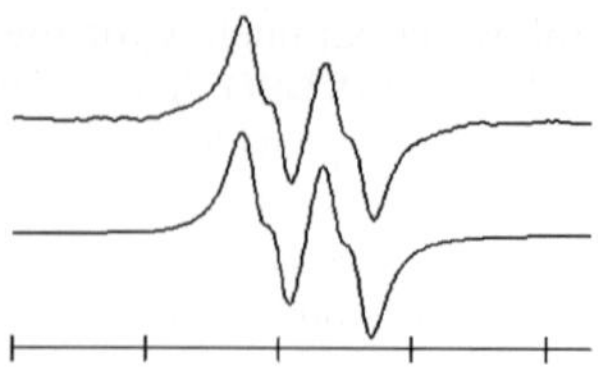

Fig. 5 Experimental (top) and simulated (bottom) EPR spectra of the mixed-valence species $\mathbf{2}^{-\bullet}$ in a frozen solution of toluene : CH_2Cl_2 (1 : 1) at 200 K.

In the present case, this relation is not fulfilled, because ΔH^* (0.117 eV) is appreciably lower than one-fourth of the optical energy (0.787/4 = 0.197 eV). A naïve argument would be to assign the difference to the presence of a significant electronic coupling. This would give $V_{ab} = 0.080$ eV, a value much higher than the one found experimentally from the intensity of the intervalence transition with eqn. (4); *i.e.*, 0.015 eV. Several hypotheses can be considered to explain the difference in V_{ab} values obtained by both methods: (i) The determination of V_{ab} by a difference between two terms of comparable magnitude from eqn. (5), is certainly highly inaccurate. (ii) Some processes such as nuclear tunneling could contribute to lower the effective thermal energy barrier. (iii) The use of eqn. (4) to calculate V_{ab}, may be not correct for organic compounds where there is a certain degree of electronic delocalization. Actually, the separation between the redox sites, R, which is employed in eqn. (4), is not perfectly defined because in the ground state the wavefunction of the electron is not fully localized on the alpha carbon atom, but has a non negligible extension on the adjacent phenyl rings. This fact has already been pointed out by Nelsen *et al.*[6c] and by Launay *et al.* for organic nitrogen compounds.[25]

The EPR spectrum of $\mathbf{2}^{-\bullet}$ at 200 K displays two symmetrical lines, each one with a shoulder, originated by the hyperfine coupling with the two hydrogen atoms of one of the ethylene moieties. The EPR spectrum at 200 K was fairly well simulated by using the parameters given in Table 1. The experimental and simulated EPR spectra are compared in Fig. 5. One of the resulting ^{1}H hyperfine coupling constants is very close to that observed for the mixed-valence radical-anion species $\mathbf{1}^{-\bullet}$, whereas the second one is considerably smaller. Similar behavior has been observed in other polychlorinated bis(triphenylmethyl) systems.[28] When the temperature is increased no evolution is observed confirming that, at least in the temperature range studied, intramolecular electron transfer is not observed because it is slower than the timescale of the EPR experiment. This result, together with the lack of an intervalence band transition in the absorption spectra, suggests that there is a localization effect in the radical-anion species $\mathbf{2}^{-\bullet}$. Previous experimental and theoretical studies performed on diferrocenylbenzenes showed that such localization occurs when the two electroactive centers are connected in the *meta* position, whereas if they are connected in the *para* position, intramolecular electron transfer across the phenylene ring is observed.[29] Similar results have also been detected in polynorbornyl and divinyl alkane systems. Indeed, recent studies have shown that V_{ab} values are markedly larger for polynorbornyl dienes than for the divinyl alkanes, by as much as 35% for the shorter members and 20% for the larger members. These surprising results indicate that simple *n*-alkyl bridges are more effective mediators of through-bond coupling over long distances than the polynorbornyl bridges, even though the chromophore units in the latter are connected by two alkyl relays, rather than one. According to the authors, these results imply the presence of interferences occurring between the various coupling pathways in the polynorbornyl bridge.[30]

2.3 Ring substituent effects

Molecule **3** has the same topology as **2** and they differ only in the substituents on the phenyl ring in the bridge having two octyloxy groups in molecule **3**. These groups were added to impart solubility to the higher members of a series with increasing number of vinylene phenylene groups in the bridge. Before developing the complete series it is important to asses if the rotation around the σ-bonds promoted by the steric hindrance between the substituents in the central rings and the chlorine substituents in the phenyl rings directly attached to the electroactive site, could originate a decrease of the long-distance electronic coupling between the terminal sites.

2.3.1. Study of diradical 3. The cyclic voltammetry of diradical **3** shows, as expected, only one reversible reduction process at a slightly lower potential ($\Delta E_{1/2} = -0.18$ V *vs.* Ag/AgCl) than biradicals **1** and **2**, which is accounted for by the electron donating character of the substituents in the central ring of **3**.

The UV-Vis spectrum of biradical **3**, recorded in the same conditions as biradicals **1** and **2** is similar to that of the topologically related radical **1**, showing bathochromic shifts of the lowest energetic bands corresponding to the triphenylmethyl radical. Some differences on those bands attributed to the central phenyl ring of the bridge are also observed (See Table 1).

Also similar are the EPR characteristics (Table 1). Indeed, in isotropic CH_2Cl_2 solution, birradical **3** shows three overlapped lines due to the hyperfine coupling with two hydrogen atoms. This fact and the values of the coupling constants are characteristic of a biradical with a magnetic interaction between the two electrons with a magnetic exchange coupling constant J, that fulfils the following condition, $J \gg a_i$.

The absolute values of zero-field splitting parameters, $|D/hc|$ and $|E/hc|$, for diradical **3** obtained from the spectra in frozen toluene : CH_2Cl_2 mixture, are the same of biradical **1** (Table 1) denoting identical extent of dipolar magnetic interactions between the two unpaired electrons in both biradicals. The same results were also obtained by variable temperature measurements in the range between 4 and 100 K of this frozen diluted solution.

2.3.2 Intramolecular electron transfer process. As in the previous cases, both the optically activated and the thermally activated electron transfer processes on the mixed valence species derived from biradical **3** were studied by Vis-NIR and EPR spectroscopies respectively.

As shown in Fig. 6 during the reduction of diradical **3** the sharp band centered at 388 nm, characteristic of the radical, decreases while two new broad and intense bands centred at 523 and 615 grown in, indicating the formation of the anion species. In addition, during the reduction process, a weak and broad band centred at 1420 nm increases until the complete formation of the mixed-valence radical anion species $\mathbf{3}^{-\bullet}$, from where the intensity decreases until it disappears completely when the dianion $\mathbf{3}^{2-}$ is formed. This band was, as previously done, assigned to an intervalence transition. Using the previously described methodology and equations, a V_{ab} coupling parameter of 100 cm^{-1} was determined when the effective separation of the redox sites is considered

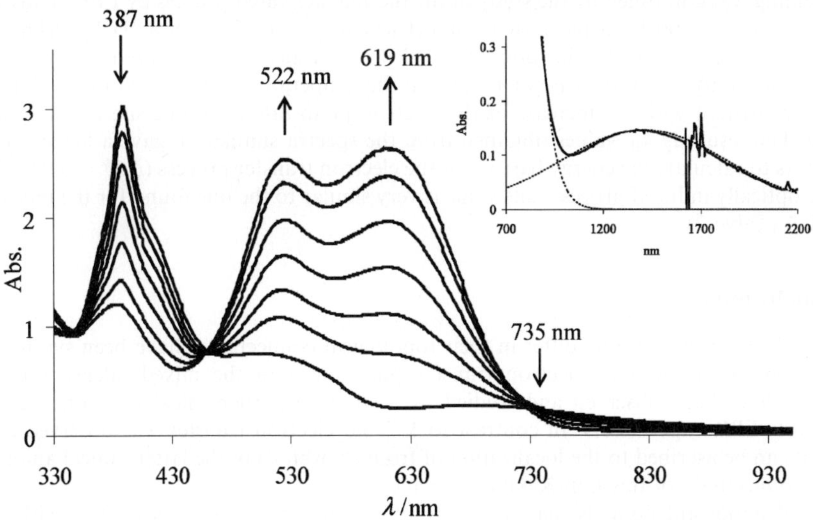

Fig. 6 Evolution of the UV-Vis during the course of the reduction of diradical **3**. Inset shows the intervalence band in the NIR.

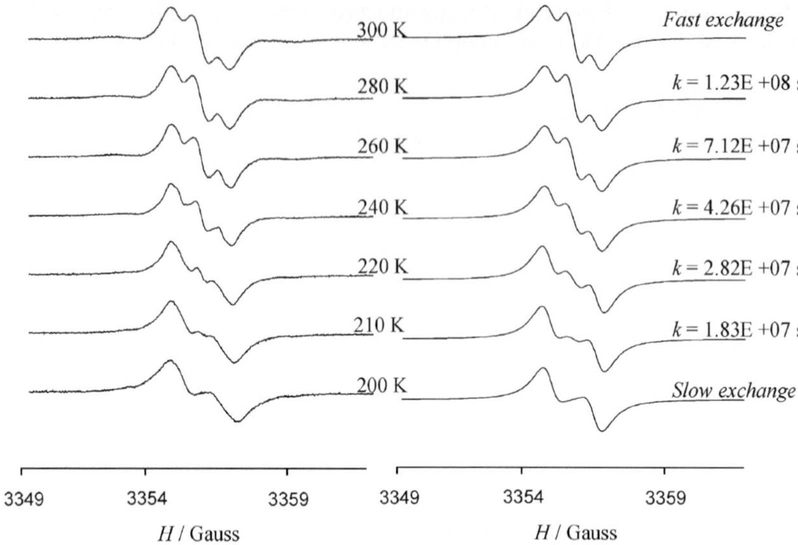

Fig. 7 Experimental (left) and simulated (right) EPR spectra of $\mathbf{3}^{-\bullet}$ at different temperatures in CH_2Cl_2 with 0.1 M of $n\mathrm{Bu_4NPF_6}$. Spectra recorded at the very end of the reduction process. See text for details.

Table 2 Experimental coupling parameters and activation energies corresponding to the electron transfer processes of mixed-valence species derived from biradicals **1** and **3**

Compound	V_{ab}/cm^{-1}	ΔG^*/kcal mol^{-1}
$\mathbf{1}^{-\bullet}$	121	6.0
$\mathbf{3}^{-\bullet}$	100	5.9

to be 19 Å.[31] The V_{ab} value is very close to that of the topologically related mixed valence compound $\mathbf{1}^{-\bullet}$ which suggest that ring substituents effects on electron transfer-rates in this class of "molecular wires" bridging molecules is very small.

This finding was confirmed by the study of the thermal activated process by EPR. Thus, the EPR spectrum of $\mathbf{3}^{-\bullet}$ at 200 K displays two symmetrical lines (Fig. 7) with a ^{1}H hyperfine coupling constant consistent with the localization, at the EPR time scale, of the unpaired electron on one site of the molecule only. As in the case of $\mathbf{1}^{-\bullet}$, when the temperature is increased the evolution of the spectra is consistent with the increase of the electron jump rate from one site to the other of the molecule. The resulting k_{th} values obtained from the spectra simulation gave a linear Eyring plot that permits to calculate the energy barrier for the electron transfer process ($\Delta G^* = 5.9$ kcal mol^{-1}). As in the optically induced process, this value is very similar to the one found for the mixed valence species $\mathbf{1}^{-\bullet}$ (Table 2).

3. Conclusions

Diradicals **1** and **2**, which only differ in their topological connectivity, have been synthesized and characterized. Intramolecular electron transfer phenomena in the mixed-valence radical-anion species $\mathbf{1}^{-\bullet}$ have been observed and studied by a spectroelectrochemical titration and temperature-variable EPR experiments. In contrast to $\mathbf{1}^{-\bullet}$, no electron transfer was observed for $\mathbf{2}^{-\bullet}$, a result that can be ascribed to the localization of frontier orbitals in the latter radical anion because the *meta* connectivity of this non-Kekulé molecule.

A second important point is that the study of the mixed valence species $\mathbf{3}^{-\bullet}$, with the same topology as $\mathbf{1}^{-\bullet}$, reveals that the ring substitution effects on electron transfer rates are very small. This finding is significant since it suggest that adding substituents to the bridges in order to impart different functionality or simply higher solubility could be readily made without dramatically affecting the electronic coupling along the bridges between the electroactive moieties.

Interestingly, $\mathbf{1}^{-\bullet}$ and $\mathbf{3}^{-\bullet}$ constitutes one of the rare examples of a mixed valence species for which the determination of both the optical and thermal energies for electron transfer is possible.

Acknowledgements

This work was supported by grants from the DGI (Spain), project MAT2003-04699, DGR (Catalunya), project 2001SGR00362, Catalan Network CeRMAE, and COST Chemistry D14 Action "Functional Materials".

References

1 (*a*) R. Metzger, *Acc. Chem. Res.*, 1999, **32**, 950; (*b*) J. M. Tour, *Acc. Chem. Res.*, 2000, **33**, 79; (*c*) R. Lloyd and C. B. Gorman, *Angew. Chem., Int. Ed.*, 2002, **41**, 4378; (*d*) D. M. Adams, L. Brus, C. E. D. Chidsey, S. Creager, C. Creutz, C. R. Kagan, P. V. Kamat, M. Lieberman, S. Lindsay, R. A. Marcus, R. M. Metzger, M. E. Michel-Beyerle, J. R. Miller, M. D. Newtown, D. R. Rolison, O. Sankey, K. S. Schanze, J. Yardley and X. Zhu, *J. Phys. Chem. B*, 2003, **107**, 6668.

2 J. M. Tour, M. Kozaki and J. M. Seminario, *J. Am. Chem. Soc.*, 1998, **120**, 8486.

3 J. M. Seminario and J. M. Tour, in *Molecular Electronics – Science and Technology*, ed. A. Aviram and M. A. Ratner, New York Academy of Science, New York, 1998, p. 69.

4 (*a*) *Electron Transfer in Chemistry*, ed. V. Balzani, Wiley, Weinheim, 2001, vol. I–V; (*b*) M. A. Fox, *Chem. Rev.*, 1992, **92**, 365–490, special issue on electron transfer; (*c*) M. N. Paddon-Row, *Acc. Chem. Res.*, 1994, **27**, 18; (*d*) P. F. H. Schab, M. D. Levin and J. Michl, *Chem. Rev.*, 1999, **99**, 1863; (*e*) J. P. Launay, *Chem. Soc. Rev.*, 2001, **30**, 386.

5 (*a*) J. C. Smart and B. L. Pinsky, *J. Am. Chem. Soc.*, 1980, **103**, 1009; (*b*) J. A. Kramer and D. N. Hendrickson, *Inorg. Chem.*, 1980, **19**, 3330.

6 (*a*) S. F. Rak and L. L. Miller, *J. Am. Chem. Soc.*, 1992, **114**, 1388; (*b*) J. Bonvoisin, J. P. Launay, C. Rovira and J. Veciana, *Angew. Chem., Int. Ed. Engl.*, 1994, **33**, 2106; (*c*) S. F. Nelsen, R. F. Ismagilov and D. A. Trieber, II, *Science*, 1997, **278**, 846; (*d*) S. V. Lindeman, S. V. Rosokha, D. Sun and J. K. Kochi, *J. Am. Chem. Soc.*, 2002, **124**, 843; (*e*) N. Gautier, F. Dumur, V. Lloveras, J. Vidal-Gancedo, J. Veciana, C. Rovira and P. Hudhomme, *Angew. Chem., Int. Ed.*, 2003, **42**, 2765.

7 (*a*) S. Creager, C. J. Yu, C. Bamdad, S. O'Connor, T. MacLean, E. Lam, Y. Chong, G. T. Olsen, J. Luo, M. Gozin and J. F. Kayyem, *J. Am. Chem. Soc.*, 1999, **121**, 1059; (*b*) J. J. Sumner, K. S. Weber, L. A: Hockett and S. E. Creager, *J. Phys. Chem. B*, 2000, **104**, 7449; (*c*) H. D. Sikes, J. f. Smalley, S. P. Dudek, A. R. Cook, M. D. Newton, C. E. D. Chidsey and S. W. Feldberg, *Science*, 2001, **291**, 1519.

8 (*a*) R. E. Homlin, R. F. Ismagilov, R. Haag, V. Mujica, M. A. Ratner, M. A. Rampi and G. M. Withesides, *Angew. Chem., Int. Ed.*, 2001, **40**, 2316; (*b*) R. E. Homlin, R. Haag, R. F. Ismagilov, M. A. Rampi and G. M. Withesides, *J. Am. Chem. Soc.*, 2001, **123**, 5075.

9 (*a*) J. Bonvoisin, J.-P. Launay, C. Rovira and J. Veciana, *Angew. Chem., Int. Ed. Engl.*, 1994, **33**, 2106; (*b*) J. Sedo, D. Ruiz, J. Vidal-Gancedo, C. Rovira, J. Bonvoisin, J -P. Launay and J. Veciana, *Adv. Mater.*, 1996, **8**, 748; (*c*) C. Rovira, D. Ruiz-Molina, O. Elsner, J. Vidal-Gancedo, J. Bonvoisin, J.-P. Launay and J. Veciana, *Chem. – Eur. J.*, 2001, **7**, 240; (*d*) O. Elsner, D. Ruiz-Molina, J. Vidal-Gancedo, C. Rovira and J. Veciana, *Nano Lett.*, 2001, **1**, 117.

10 (*a*) M. Ballester, *Acc. Chem. Res.*, 1985, **297**, 131; (*b*) J. Veciana, J. Riera, J. Castañer and N. Ferrer, *J. Organomet. Chem.*, 1985, **297**, 131; (*c*) O. Armet, J. Veciana, C. Rovira, J. Riera, J. Castañer, E. Molins, J. Rius, C. Miravitlles, S. Olivella and J. Brichfeus, *J. Phys. Chem.*, 1987, **91**, 5608; (*d*) I. Ratera, D. Ruiz-Molina, J. Ensling, K. Wurst, C. Rovira, P. Gutlich and J. Veciana, *J. Am. Chem. Soc.*, 2003, **125**, 1462; (*e*) M. Ballester, C. Miravitlles, E. Molins and C. Carreras, *J. Org. Chem.*, 2003, **68**, 2748; (*f*) D. Maspoch, N. Domingo, D. Ruiz-Molina, K. Wurst, G. Vaughan, J. Tejada, C. Rovira and J. Veciana, *Angew. Chem. Int. Ed.*, 2004, **43**, 1828.

11 (*a*) V. Grosshenny, A. Harriman and R. Ziessel, *Angew. Chem., Int. Ed. Engl.*, 1995, **34**, 1100, and references therein; (*b*) C. Patoux, J.-P. Launay, M. Beley, S. Chodorowski-Kimmes, J.-P. Collin, S. James and J.-P. Sauvage, *J. Am. Chem. Soc.*, 1998, **120**, 3717; (*c*) F. Barigelletti, L. Flamigni, V. Balzani, J.-P. Collin, J.-P. Sauvage, E. C. Constable and A. M. W. Cargill Thompson, *J. Chem. Soc., Chem. Commun.*, 1993, 942; (*d*) F. Barigelletti, L. Flamigni, V. Balzani, J.-P. Collin, J.-P. Sauvage, E. C. Constable and A. M. W. Cargill Thompson, *J. Am. Chem. Soc.*, 1994, **116**, 7692; (*e*) A. C. Benniston, V. Goulle, A. Harriman, J. M. Lehn and B. Marczinke, *J. Phys. Chem.*, 1994, **98**, 7798; (*f*) R. Ziessel, M. Hissler, A. El-Ghayoury and A. Harriman, *Coord. Chem. Rev.*, 1998, **178–180**, 1251.

12 M. Ballester, J. Veciana, J. Riera, J. Castañer, C. Rovira and O. Armet, *J. Org. Chem.*, 1986, **51**, 2472.

13 For other strongly (*E*)-selective reactions, see for example, H. Pommer and A. Nurenbach, *Angew. Chem., Int. Ed. Engl.*, 1977, **16**, 423.

14 H. O. House, *Modern Synthetic Reactions*, Benjamin, Menlo Park, CA, 1972, 2nd edn., 608.

15 (*a*) K. Wurst, O. Elsner and H. Schottenberger, *Synlett*, 1995, **1**, 833; (*b*) J.-G. Rodríguez, A. Oñate, R. M. Martin-Villamil and I. Fonseca, *J. Organomet. Chem.*, 1996, **513**, 71.
16 L. Fitjer and U. Quabeck, *Synth. Commun.*, 1985, **15**, 855.
17 Compound **1**. Anal. calcd. for $C_{48}H_8Cl_{28}$: C, 36.55; H, 0.51; Cl, 62.94; found C, 36.59; H, 0.62; Cl, 62.86%. MS m/z 1576 (M^+), 1506 ($M^+ - Cl_2$), IR(KBr): $\nu = 3025$, 1624, 1510, 1330, 1255, 960, 810, 800, 800 cm^{-1}. Compound **2**. Anal. calcd. for $C_{48}H_8Cl_{28}$: C, 36.55; H, 0.51; Cl, 62.94; found C, 37.20; H, 0.53; Cl, 62.27%. MS m/z 1576 (M^+), 1506 (M^+-Cl_2), IR(KBr): $\nu = 2920$, 1510, 1340, 1260, 960, 815, 780, 755 cm^{-1}. Compound **3**. Anal. calcd. for $C_{64}H_{40}Cl_{28}$: C, 41.92; H, 2.19; Cl, 54.13; O, 1,74 found C, 42,12; H: 2,18; Cl, 54.27%. MS m/z 1572 (M^+), 1762 ($M^+ - Cl_2$), I R(KBr): $\nu = 2918$, 1613, 1508, 1336, 1258, 1202, 967, 814, 734, 710, 665 cm^{-1}.Compound **6**. Anal. calcd. for $C_{48}H_{10}Cl_{28}$: C, 36.51; H, 0.64; Cl, 62.85; found C, 36.56; H, 0.64; Cl, 62.81%. MS m/z 1579 (M^+), 1544 (M^+-Cl) IR(KBr): $\nu = 3010$, 2920, 1635, 1530, 1510, 1410, 1360, 1335, 1290, 1135, 960, 865, 800, 710, 680, 640, 525, 505 cm^{-1}. ^{1}H-NMR(CCl_4.D_2O): $\delta = 7.25$(4 H, s), 7.00 (6 H, m) ppm.Compound **7**. Anal. calcd. for $C_{48}H_{10}Cl_{28}$: C, 36.50; H, 0.63; Cl, 62.86; found C, 36.63; H, 0.47; Cl, 62.50%. MS m/z 1579 (M^+), 1544 (M^+-Cl) IR(KBr): $\nu = 3010$, 2920, 1640, 1535, 1520, 1365, 1340, 1295, 1240, 1135, 960, 860, 808, 715, 680, 645, 530, 495 cm^{-1}. ^{1}H-NMR($CDCl_3$): $\delta = 7.67$(1 H, m), 7.53 (3 H, m), 7.12 (4H, m), 7.07 (2H,s) ppm.Compound **8** was not isolated in order to improve the yield of biradical **3**.
18 S. D. McGlynn, T. Azumi and M. Kinoshita, in *Molecular Spectroscopy of the Triplet State*, Prentice Hall, Englewood Cliffs, NJ, 1969.
19 (*a*) MOPAC 93.00: J. J. Stewart, M. J. S. Dewar, E. G. Zoebish and E. F. Healy, *J. Am. Chem. Soc.*, 1985, **107**, 3902; (*b*) M. J. S. Dewar and E. G. Zoebish, *J. Mol. Struct. (THEOCHEM)*, 1988, **180**, 1.
20 It is well known that molecules of the type Ar_3Z, in which three aryl groups (Ar) are bonded to a central Z atom, which have a sp^2 hybridization, are propeller-like in shape. A large number of theoretical studies on the stereochemistry of these propeller-like systems were performed by Mislow *et al.* (K. Mislow, *Acc. Chem. Res.*, 1976, **9**, 26). They demonstrated that in the most simple cases, where the three aryl groups (blades) are not substituted, or else symmetrically substituted around the C_{ar}–Z bond axis of each aryl group (*i.e.*, groups having local C_2 axes), the molecule has always at least two enantiomeric forms due to the intrinsic stereogenic center (helicity) of any propeller that provides chirality to the molecule. If the helix adopts a clockwise sense the enantiomer is called *Plus* (*P*) whereas if it adopts the opposite sense is called *Minus* (*M*). For diradicals **1** and **2** it is straightforward to predict the existence of three stereoisomers, one pair of enantiomers-namely (*P*,*P*)-, and (*M*,*M*)- and a *meso* form—namely (*M**,*P**)—due to the presence of two triphenylmethyl stereogenic centers. (N. Ventosa, D. Ruiz-Molina, J. Sedó, C. Rovira, X. Tomas, J.-J. André, A. Bieber and J. Veciana, *Chem. – Eur. J.*, 1999, **5**, 3533, J. Sedó, N. Ventosa, M. A. Molins, M. Pons, C. Rovira and J. Veciana, *J. Org. Chem.*, 2001, **66**, 1567; J. Sedó, N. Ventosa, M. A. Molins, M. Pons, C. Rovira and J. Veciana, *J. Org. Chem.*, 2001, **66**, 1579). Semi-empirical calculations were done by using initial input molecules with one of the triphenylmethyl units in the *M* form and the other in the *P* form. It was also confirmed that the use of another stereoisomer should not have any influence in the final minimized geometries of the bridges.
21 The upper limit of J^{intra} was determined from the experimental Weiss constant and Bleany-Bowers expression for a dimer using the approximation for the high temperature limit. See R. L. Carlin, *Magnetochemistry*, Springer, Berlin, 1986, p. 76. The lower limit of J^{intra} was deduced from the value of the ^{1}H hyperfine coupling constants. See L. Salem and C. Rowland, *Angew. Chem., Int. Ed. Engl.*, 1972, **11**, 92.
22 A.-C. Ribou, J.-P. Launay, K. Takahashi, T. Nihira, S. Tarutani and C. W. Spangler, *Inorg. Chem.*, 1994, **33**, 1325.
23 S. F. Nelsen, S. C. Blacstock and Y. Kim, *J. Am. Chem. Soc.*, 1987, **109**, 677.
24 It is well known that the unpaired electron of highly chlorinated triphenylmethyl radical and the negative charge of their corresponding carbanions are mainly localized on the α-carbons.[10] Therefore, the two alpha carbon atoms of **1** can be considered in a first approximation as the electron active sites for the electron transfer phenomena. It must also be noticed that as far as intramolecular electron transfer phenomena are concerned, the main distance to be considered is the through-bond distance between the two electron active sites; *i.e.*, the sum of bond lengths corresponding to the conjugated pathway. Such distance, measured from the two radical alpha carbon atoms is 24.1 Å for diradical **1**. However, as shown by EPR spectroscopy, the average distance between the two spins –the average interspin separation- differs from the nominal separation, either through-space or through-bond, between the two radical alpha carbon atoms obtained from the AM1 minimized structures, due to the different extent of the delocalization in both diradicals.
25 J. Bonvoisin, J. P. Launay, M. Van der Auweraer and F. C. De Schryver, *J. Phys. Chem.*, 1994, **98**, 5052.
26 The variation of the intervalence absorption band *versus* the average reduction state can be used to determine the comproportionation constant, K_c, according to eqn. (5), $K_c = [\mathbf{1}^{\bullet-}]^2/[\mathbf{1}][\mathbf{1}^{2-}]$ (6) In the present case, the resulting value of K_c is 7. This value is employed in eqn. (6) for the calculation of the corrected spectra.

$$\varepsilon(\mathbf{1}^{-\bullet}) = \left[\left(2+\sqrt{K_c}\right)/\sqrt{K_c}\right]\varepsilon(\mathbf{1}^{-\bullet})_{app} - \left(\varepsilon(\mathbf{1}^{\bullet\bullet}) + \varepsilon(\mathbf{1}^{2-})\right)/\sqrt{K_c} \quad (7)$$

where ($\mathbf{1}^{-\bullet}$) and ($\mathbf{1}^{-\bullet}$)$_{app}$ are the true and the apparent extinction coefficients of the radical anion $1^{-\bullet}$, and ($\mathbf{1}^{-\bullet}$) and ($\mathbf{1}^{-2}$) are the extinction coefficients of diradical **1** and dianion $\mathbf{1}^{-2}$, respectively. Band

deconvolution of the corrected spectrum of $1^{-\bullet}$ was performed, assuming Gaussian profiles, in order to separate the intervalence band from other nearby bands (see Fig. 4). This deconvolution procedure has already been described in ref. 21, and has the advantage of giving a better estimate of the Marcus λ value.
27 (*a*) J. Heinzer, *Mol. Phys.*, 1971, **22**, 167; (*b*) J. Heinzer, *Quantum Chem. Program Exchange*, 1972, N° 209.
28 O. Elsner, D. Ruiz-Molina, J. Vidal-Gancedo, C. Rovira and J. Veciana, *Chem. Commun.*, 1999, 579.
29 C. Patoux, C. Coudret, J. P. Launay, C. Joachim and A. Gourdon, *Inorg. Chem.*, 1997, **36**, 5037.
30 (*a*) M. J. Shephard, M. N. Paddon-Row and K. D. Jordan, *J. Am. Chem. Soc.*, 1994, **116**, 5328; (*b*) M. N. Paddon-Row, *Acc. Chem. Res.*, 1994, **27**, 18, and obtained again, references cited therein.
31 AM1 semiempirical calculations gave the same distances as for diradical **1** and from the EPR parameters the distance is also the same.

General Discussion

Professor Tao opened the discussion of Professor Rampi's paper: In your experiment, where thin molecular layers are sandwiched between two relatively large mercury droplets, how will the gate field penetrate into the molecular layers without getting screened by the mercury droplets?

Professor Rampi replied: The electrical equivalent circuit used for both junctions Hg–$SC_{10}COO$–//$Ru(NH_3)_6^{3+}$//–$OOCC_{10}S$–Hg and Hg–SAM–R//R–SAM–Hg is schematized in the figure below. The potentials of the mercury electrodes were controlled such that one (cathode) acted as an electron donor and the other (anode) as an electron acceptor. At the reference electrode there is no current flowing through, because the impedence is extremely high. We designated the cathode and anode as the drain and source electrodes, respectively. We measured the conductance through the junction as a function of the potentials of the drain and source electrodes with respect to the reference electrode (that is, V_{DG} and V_{SG}, respectively) and as a function of the potential between the source and drain (V_{DS}) using the electrolyte solution as a gate. We controlled the potential of the gate by controlling the potential applied to the reference electrode relative to the ground potential. Since the source electrode is at ground potential, this voltage is V_{SG}.

The redox sites are in contact with the aqueous medium as is indicated by the fact that the electrochemical potential at which the maximum value of the NDR curve is found is the $E^{\ominus}$ of the Ru(II)/Ru(III) couple: this results suggest that there is a layer of electrolyte solution between the two Hg drops and/or in the pockets of the SAM defects. The change of the electrodes potential, induce a change of the redox state, similarly to a standard CV experiment. The change of the oxidation state is in turn neutralized by the electrolyte ions that are present in a ratio of 100 to 1 with respect to the redox sites. On the basis of this ratio, the movements of the ions to neutralize a change of from Ru^{2+} to Ru^{3+} are of the order of micrometres.

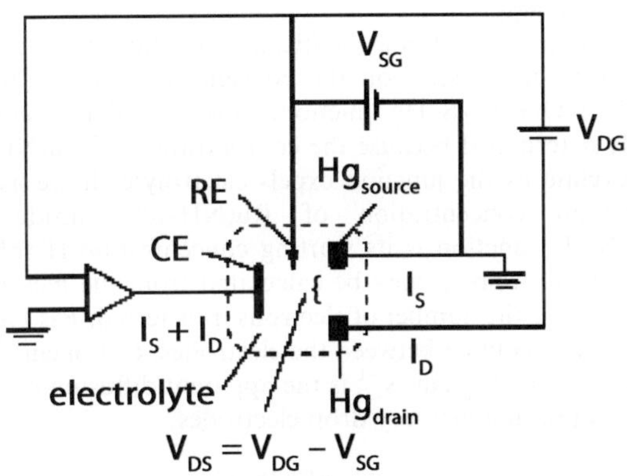

Professor Fermin asked: In your analysis of the current–voltage characteristics for the junction Hg–SAM//R//SAM–Hg, it is mentioned that the current response is related to a generation–diffusion–collection mechanism involving the $Ru(NH_3)_6^{3/2+}$ couple. I have several questions in relation to this point:

1. Can you estimate, from the current values and reasonable diffusion coefficients, the distance separating the electrodes?

2. Can you also comment on the value of the current plateau as a function of the distance separating the two electrodes? This could provide information comparable to the "approach curves" familiar from electrochemical scanning microscopy.

3. Several reports in the literature have shown that the charge termination of SAMs can have important effects on the electron transfer kinetics involving charged redox species.[1] In particular, it appears that carboxyl terminated SAMs can strongly block electron transfer to $Fe(CN)_6^{3-}$, but not

significantly to $Ru(NH_3)_6^{3+}$. Therefore, I could anticipate very different results to those shown in Fig. 6 of the paper in the presence of the former redox species. Have you investigated this system?

1 K. Takehara, H. Takemura and Y. Ide, *Electrochim. Acta*, 1994, **39**, 817–822.

Professor Rampi replied: We have made an estimation of the inter-electrode distance using the following approach.

The apparent diffusion coefficient, $D_{ap}/cm^2\ s^{-1}$, resulting from a combination of physical displacement and electron hopping is given by eqn. (1),

$$D_{ap} = D_{phys} + \frac{k_{ex}\delta^2 C}{6} \quad (1)$$

where $D_{phys}/cm^2\ s^{-1}$ is the diffusion coefficient for physical displacement of the redox molecules and the second term represents the diffusion coefficient due to electron hopping. In this second term, $k_{ex}/cm^3\ mol^{-1}\ s^{-1}$ is the bimolecular rate constant for electron self-exchange, C/M is the total concentration of redox species, and δ/cm is the center-to-center distance between redox centers at the time of electron transfer.

The experimental results indicate that immediately after the formation of the junction and during the first few potential cycles, charge transport across the junction probably occurs by a diffusion-controlled process, most plausibly by redox shuttling in which the $Ru(NH_3)_6^{3+}$ diffuses back and forth between the two electrodes.

After the junction has been subjected to repeated cycling, the results suggest that a strong electrostatic interaction exists between the COO^- groups and the $Ru(NH_3)_6^{3+}$ ions present in the gap between the electrodes and that such electrostatic interaction arises as a result of expulsion of electrolyte solution from the junction and a decrease in the interelectrode spacing. We propose that without sufficient electrolyte solution and space between the electrodes, the diffusion of the ruthenium centers from one electrode to the other becomes hindered.

Charge transport may thus no longer occur exclusively by diffusion, but rather may become competitive with electron hopping.

In principle, if electron hopping makes a significant contribution to the diffusion process, then D_{ap} should exhibit a linear dependence on the concentration of $Ru(NH_3)_6^{3+}$ inside the Hg–$SC_{10}COO^-//Ru(NH_3)_6^{3+}//{}^-OOCC_{10}S$–Hg junction. This dependence on concentration, unfortunately, cannot be readily determined because the concentration of $Ru(NH_3)_6^{3+}$ changes substantially during potential cycling as the junction expels electrolyte. If we assume, however, that at steady-state the minimum concentration of $Ru(NH_3)_6^{3+}$ inside the Hg–$SC_{10}COO$–// $Ru(NH_3)_6^{3+}//{}^-OOCC_{10}S$–Hg junction is its starting concentration (1 mM), the contribution of physical diffusion *vs.* electron hopping may be calculated from the measured value of $I_{lim}/C\ s^{-1}$ according to eqn. (2), where n is the number of electrons transferred, F is Faraday's constant (96490 C mol^{-1}), A/cm^2 is the area of contact between the electrodes, $C/mol\ cm^{-3}$ is the concentration of $Ru(NH_3)_6^{3+}$ inside the junction, $D_{ap}/cm^2\ s^{-1}$ is the apparent diffusion coefficient, and d/cm is the distance between the two opposing mercury drop electrodes.

$$I_{lim} = \frac{nFAD_{ap}C}{d} \quad (2)$$

In this approach, D_{ap} is approximated to be D_{phys} for a process dominated by physical diffusion, and $k_{ex}\delta^2 C/6$ for a process dominated by electron hopping. Assuming that $D_{ap} = D_{phys}$ and taking D_{phys} to be $7 \times 10^{-6}\ cm^2\ s^{-1}$, A to be $0.0079\ cm^2$, and I_{lim} to be *ca.* 90 μA (value extrapolated from Fig. 5a of the paper), d is calculated to be about 534 nm. If we now approximate D_{ap} to be equal to the second term of eqn. (1) and assume that $k_{ex} \sim 4 \times 10^3\ M^{-1}\ s^{-1}$ and $\delta \sim 6 \times 10^{-8}$ cm, substituting these values into eqn. (2) then gives d to be 4×10^{-7} nm. Both these distances are not reasonable numbers suggesting that in this very confined space the used parameters are inappropriate and difficult to evaluate. We nonetheless suggest that charge transport across the junction occurs primarily by physical diffusion. In support of this conclusion is the observation that the ratio of the collector to the generator currents never equals 1 for any junction that exhibits steady-state behavior.

The current plateau is determined by eqn. (2). A relationship between I_{lim} and the distance relays on constant values of C and D_{phys}. By applying an increasing number of redox cycles, and therefore by decreasing the inter-electrode separation, we believe that the redox sites are expelled from the gap, and therefore C will change. We expect also D_{phys} to change in a confined space. Consequently, comparison with the "approach curves" of ESM are very difficult.

The groups terminating the SAMs can have indeed a big effect of the electron transfer process. We have not yet studied any other group. We know from the literature and also form our experiments that CH_3 and COOH terminating SAMs block significantly the electron transfer to $Ru(NH_3)_6^{3+}$. In our system, the SAMs are functionalized by negatively charged COO^- with the aim of both trapping the Ru at the interface and to give access to the electrodes *via* the SAM defect created by the electrostatic repulsion of the COO^- groups.

Professor Calvo commented: In the Hg–SAM//R//SAM–Hg experiments one would expect the $Ru(NH_3)_6^{3+}$ ions to be electrostatically bound to terminal COO^- groups of the thiol layer. Then, in the limiting case would you not expect them to behave the same as in the Hg–SAM–R//R–SAM–Hg system?

Have you got any evidence for this?

Professor Rampi replied: In the Hg–SC_{10}COO–//$Ru(NH_3)_6^{3+}$//–$OOCC_{10}$S–Hg junction, on decreasing the interelectrode gap, the redox site movement is limited by the confined space. Under these experimental conditions, two mechanisms of charge transport can become competitive: the redox cycling and/or electron hopping between the redox center. Given the difficulty of evaluating the kinetic parameters in such a confined space, it is impossible to identify the charge transfer mechanism (see my previous answer) in the limiting case.

Dr Nichols remarked: As you change the potential the surface tension of the mercury junction electrode may change and this surface tension change could lead to thickness changes (particularly if water is entrapped in the junction).

Professor Rampi replied: Deformation of the SAMs is one of main issue in using junctions based on Hg electrodes, where the distance between the electrodes in not fixed as in junctions based on planar gold electrodes. The change in the distance between the electrodes under applied voltages is evident from the results and discussion concerning the junction Hg–SC_{10}COO–//$Ru(NH_3)_6^{3+}$// –$OOCC_{10}$S–Hg. Significantly, in this case, the COO^- terminating SAMs are disordered and can easily interpenetrate under electrostriction. Working with junctions carrying SAMs formed by CH_3 terminating alkyl chains, that are known to be tightly packed; some authors indicate that interpenetration can take place, in particular when short chains are used.[1,2]

1 K. Slowinski and M. Majda, *J. Electroanal. Chem*.., 2000, **491**, 139–147.
2 A. Salomon, D. Cahen, S. Lindsay, J. Tomfohr, V. B. Engelkes and C. D. Frisbie, *Adv. Mater*.., 2003, **15**, 188.

Dr Higgins opened the discussion of Professor Bryce's paper: In the longer (7 nm) molecule with three or more fluorenone groups, have I interpreted your CV correctly in that all three fluorenones are reduced at the same potential (*i.e.* no communication between fluorenones)?

Professor Bryce responded: Yes, no clearly-defined splitting of the redox peaks is observed for compound **17**.

Professor Rovira asked: Does the measured current in the STS experiments change with the molecular length?

Professor Bryce replied: It is too early to say. Data for compounds **15** and **17** are being obtained at present in Professor Ashwell's laboratory at Cranfield University.

Professor McCreery said: You commented that molecules containing several fluorenone units show one redox wave, implying weak coupling between fluorenone centers. In view of he highly conjugated "connectors" between fluorenones, do you find weak interaction to be surprising?

Professor Bryce answered: There are no precedents for the electrochemical behaviour of class of compounds with more than one fluorenone unit in the backbone, so we were not sure what to expect. It seems clear from our data that the electrochemically-generated radical anions are mainly localised on the ketone group of the fluorenone moieties.[1]

1 C. Wang, A. S. Batsanov and M. R. Bryce, *J. Org. Chem.*, DOI: 10.1021/jo051711o.

Dr Higgins said: Just a comment on the fluorenone radical anion sites not communicating significantly in the longer oligomers containing several fluorenone groups—in the conducting polymer literature, in conjugated polymers incorporating fluorene units, fluorenone impurities act as charge trapping sites—perhaps these observations are related.

Professor Bryce replied: Yes, we believe there is a link between these observations. In poly (fluorene)s the presence of a few fluorenone defect sites can drastically lower the photoluminescence and electroluminescence quantum yields due to charge trapping. For a discussion of this point see ref. 1.

1 X. Gong, P. K. Iyer, D. Moses, G. C. Bazan, A. J. Heeger and S. S. Xiao, *Adv. Funct. Mater.*, 2003, **13**, 325.

Dr Nichols remarked: If a three electron process operates then the peak separation in the CV should quantitatively agree with what we expect for a three electron process.

Professor Bryce responded: The electrochemistry of these aryleneethynylenes is not cleanly reversible. The behaviour is quasi-reversible, so the data cannot be analysed in this way.

Professor Fitzmaurice said: It is possible that the properties of the longer molecule will depend on its orientation, with respect to the electrode substrate, and that this may on average be different in solution compared to a gap.

Professor Bryce replied: Yes, there are likely to be differences. These are rigid rod molecules, so the molecular length (along the long axis) will not change, but the orientation could be different, *e.g.* upright in a gap with the sulfur attached to gold, and perhaps flat (parallel) to an electrode surface in the solution experiments. We don't know this from experimental data as yet.

Dr Nichols asked: You favour cyanoethyl protection for thiols, and you have pointed out problems with "Tour" molecular wires where nitro group reduction can occur. Can you expand on these comments?

Professor Bryce replied: For multi-step syntheses, *e.g.* Scheme 1 of the paper, we find that cyanoethylthio end groups lead to easier purification than for thioacetate analogues. We have not tried to deprotect a cyanoethylthio group in the presence of a nitro substituent, so we do not know if competing nitro group reduction or other side-reactions will occur.

Professor Fitzmaurice commented: A tripodal linker may be used to ensure a more strongly adsorbed species and also to ensure adsorption is normal to the substrate.

Dr Higgins asked Professor Bryce: Do your cyanoethyl-protected thiols adsorb directly at gold? If they only adsorb very slowly or not at all, this would be very useful in orthogonal protection, and subsequent controlled orientation, of dithiol-containing molecules between contacts.

Professor Bryce responded: Yes, they adsorb slowly onto gold without the need to add any deprotecting reagents. Orthogonal protecting groups are being explored in this context. For example, the selective deprotection of cyanoethylthio (using NaOEt in ethanol/THF) in the presence of trimethylsilylethylthio has been reported.[1]

1 G. V. Morales, P. Jiang, S. Yuan, Y. Lee, A. Sanchez, W. You and L. Yu, *J. Am. Chem. Soc.*, 2005, **127**, 10456.

Professor Bjørnholm commented: We experienced in our STM studies of single molecules in alkylthiol SAMs on gold that you do not need to deprotect a thiolacetate protected thiol. If you don't you get exactly the same contact to Au as if you do deprotect. The only difference is that insertion is slower.

Professor Bryce commented: The *I–V* curves for compound **15** are very similar with and without deprotecting reagents (ref. 45 of the paper). It appears, therefore, that cyanoethylthio behaves in a similar way to thioacetate in this regard.

Professor McCreery opened the discussion of Professor Barbero's paper: On HOPG, there is a significant difference between edge sites and basal sites regarding reactivity with diazonium ion. Would you expect a similar difference for end *vs.* wall reactivity of nanotubes?

Professor Barbero replied: I agree that edge sites and basal sites on carbon nanotubes should have very different reactivities. However, only if we consider the nanotubes as perfect cylinders of graphite sheets (basal sites) with open ends (edge sites), such a difference can be clearly detected. On the other hand, since it is known that carbon nanotubes present defects in the walls, with exposed edge sites,[1] it seem that there are enough reactive edge sites in the walls to allow for a more or less even distribution of functionalities along the nanotubes. Such distribution of functional groups could explain the strong effect of functionalization on interaction dependent properties (solubility, electrostatic interactions, luminescence[2] of objects with a high aspect ratio, like carbon nanotubes. If only the extremes are reacting, the interaction properties of the whole nanotube should be only slightly altered.

1. X. Feng, S. Irle, H. Witek, K. Morokuma, R. Vidic, E. Borguet, *J. Am. Chem. Soc.*, 2005, **12**, 10533.
2. Y. Lin, B. Zhou, R. B. Martin, K. B. Henbest, B. A. Harruff, J. E. Riggs, Z.-X. Guo, L. F. Allard, Y.-P. Sun, *J. Phys. Chem. B*, 2005, **109**, 14779.

Professor McCreery asked:
1. Compare the reactivity of the ends with the walls of carbon nanotubes with diazonium reagents.
2. Do you see a change in the tube properties (*e.g.* conductivity) upon modification?

Professor Barbero responded:
1. This answer is similar to my last one. In addition, the fact that the reactivity of multiwall carbon nanotubes, with several exposed edge sites, is similar to that of single wall carbon nanotubes, mostly closed, suggests that the reaction occurs at edge sites present at ends and/or defects.
2. The X-ray diffraction measurements suggest that the graphite sheets of the nanotubes are not significantly altered. The bulk properties (bulk conductivity, solubility, mechanical properties) are only slightly altered. The conductivity decrement upon modification is likely to be due to a diminished interaction between individual nanotubes. While it is possible for other factors, like changes in the graphite electron densities, to be responsible, the lack of a clear relationship between the electronic properties of the groups attached to the aryl ring (e.g. $-NO_2$) with the conductivity suggests that electronic effects have less influence.

Professor Naaman commented: Working with oxidized single wall carbon nanotubes we did observe a change in the Raman spectrum as a result of the reaction. Upon heating, the Raman spectrum returned to its original form.

Professor Barbero replied: The X-ray diffraction measurements (reported in the paper), conductivity measurements and preliminary Raman measurements suggests that covalent modification using aryl diazonium ions does not alter significantly the electronic properties of the carbon nanotubes. Only the characteristic vibrations of the attached groups can be detected.

This is in contrast with oxidative treatments where oxygen rich groups (carboxylic, quinone, lactone) which are directly attached to the graphite layer, are created. In that case, the electronic properties (like conductivity) are strongly affected. Upon heating, the oxygenated groups could be pyrolysed back into carbon, restoring the electronic properties of the carbon nanotubes.

Dr Nichols asked: Could wrapped CNT structures be delivered into nanoscale gaps by controlling the surface charge of the conducting polymer so they could be delivered into the gap, perhaps using zwitterionic conducting polymers?

Professor Barbero responded: This is an interesting idea: fine-tuning the charge distribution in the nanotubes using the conducting polymer's charge. The difficulty is that block conducting polymers are not easy, if possible, to synthesize. A simpler alternative would be to synthesize vinylic block copolymers, having differently charged segments, by living polymerization methods (*e.g.* ATRP,[1]). Some or all the blocks could have dangling conducting oligomer units to allow specific interaction with the nanotubes.

1. M. C. Iovu, M. Jeffries-EL, E. E. Sheina, J. R. Cooper, R. D. McCullough, *Polymer* 2005, **46**, 8582–8586.

Dr Higgins suggested: Polythiophenes might be an interesting alternative to PANI in these systems, since polythiophene–C60 and polythiophene–C nanotube systems are of great current interest in photovoltaics.

Professor Barbero replied: Polythiophenes are indeed a good alternative to polyanilines towards functionalization/solubilization of carbon nanotubes since it is easy to synthesize functionalized polythiophenes by homo- or copolymerisation of functionalized thiophenes. We have found that multiwall carbon nanotubes (MWNT) can be dispersed in water by wrapping with a poly(3,4-ethylenedioxythiophene)–poly(styrenesulfonate) (PEDOT-PSS) complex (see Fig. 1). Since we have built electrostatic layer-by-layer multilayers using PEDOT-PSS and poly(diallyl dimethylammonium chloride) (PDAMAC), it should be possible to build multilayers using the MWNT dispersed with PEDOT-PSS and cationic polyelectrolytes.

Professor Fitzmaurice opened the discussion of Professor Ulstrup's paper: Why is there a $\sim$50 mV shift in the cathodic and anodic scans in Fig. 7?

Dr Albrecht answered: This question may be related to two issues. One is the shift in peak positions between anodic and cathodic peak at the same tip/substrate bias, the other one the apparent asymmetry of the peak positions relative to zero V_{bias}, *i.e.* the negative intercept at the ordinate. The reason for the former is not clear, but may be instrumental. If the substrate potential is swept slowly enough, say on the order of 0.1 V s^{-1}, the peak separation in the anodic and cathodic scans approaches zero.

The second issue relates to the small overall shift towards negative peak potentials (with respect to E^0) as observed in Fig. 7C. Its absolute value is comparable to the uncertainty of the individual peak potentials. This is particularly the case for the data set presented in Fig. 7C as Pt was used as a (pseudo)reference electrode. The accuracy of the latter is only ±50 mV under these experimental conditions. Thus, the above-mentioned downshift of ($E_{\text{p}} - E^0$) *vs.* V_{bias} should be interpreted with care and we hesitate to draw further conclusions based on present experimental data.

Professor Bjørnholm asked: Can you observe the shift you describe in your 'elevator' mechanism between the oxidation and reduction process driven by molecular relaxation in, for example, the oxidation and reduction wave of a redox active SAM on an electrode?

Fig. 1 (A) MWNT dispersed using a solution of PEDOT-PSS (0.1% w/v) in water (A). Compare with the same MWNT suspended in pure water (B). The image was taken 48 h after immersing both vials in an ultrasonic bath for 15 min (vial B was shaken briefly before taking the image to reveal the nanotubes).

Professor Ulstrup replied: Direct observation of such an "on–off" mechanism would require either a very fast experiment (nanosecond–picosecond time range) or very slow vibrational relaxation. The shift is, however, reflected in the spectroscopic tunnelling feature, which emerges out of the view of the tunnelling process as a two-step interfacial electrochemical electron transfer pattern.

Professor Fermin commented: Scanning tunnelling spectroscopy (STS) of quantum dots at low temperature can be described within two limiting conditions, shell-*tunnelling* and shell-*filling* spectroscopy.[1] The former is commonly observed when the tip–dot tunnelling resistance is higher than the dot–substrate resistance, *i.e.* the electron occupation in the dot is not substantially affected. The peaks observed in the dI/dV spectra correspond to the electronic energy levels of the dot. In the shell-filling limit, the tip–dot resistance is smaller than the dot–substrate one and the energy levels are partially filled with electrons. Under these conditions, new features appear in the spectra due to electron–electron interactions in the highly confined media. The transition metal complexes studied in your paper describe resonant-tunnelling behaviour as a function of the substrate potential. Can you predict the conditions required, in terms of electronic structure of the molecule and charge transfer kinetics, for observing electron–electron interactions at the molecular level?

1 E. P. A. M. Bakkers, Z. Hens, A. Zunger, A. Franceschetti, L. P. Kouwenhoven, L. Gurevich and D. Vanmaekelbergh, *Nano Lett.*, 2001, **1**, 551–556.

Professor Ulstrup replied: "Shell-tunnelling" and "shell-filling" are expected when the tip–dot tunnelling resistance is higher than the dot–substrate resistance for negative bias voltage. The "quantum dot" in the reference alluded to are too large for discreteness effects in the electronic spectrum. The analogy to the transition metal complexes that we have used is therefore not very conspicuous.

The relevant charge states of the transition metal complexes used only differ by a single electronic charge. There is no charge accumulation as for the "quantum dots". Charge equilibration, for example between the substrate and the molecule can be expected if the electron transfer step between molecule and tip is rate determining, and *vice versa*. Single-electron transfer to or from the transition metal complex is also accompanied by electron relaxation, which is essentially a result of many-electron interaction effects.

Professor Calvo referred to Fig. 5a of Professor Ulstrup's paper: Laviron theory, which you are using to work out $k_{ET}{}^{\circ}$, does not predict the peak potential separation that can be seen in your results at low scan rate. I wonder if you have considered a kinetic hindrance due to the charge compensation by the counter ions. Have you looked for instance at the effect of ionic strength? Could such a kinetic hindrance explain the potential difference of the anodic and cathodic peak separation in Fig. 7B?

Professor Ulstrup replied: The surface monolayer is composed of quite highly charged $Os^{2+/3+}$ complex ions and almost certain to involve counterions in the packing. Counter ion reorganization will also accompany the redox process, either by additional reorganization free energy terms or in a separate kinetic step. Ionic pairing has in fact been documented by Faulkner and Forster[1] in the electrochemical surface behaviour of very similar complexes. This could perhaps be further illuminated by studies of the ionic strength effects which, however, we have not undertaken at present. It would entail extending the simplest Laviron form to include quadratic driving force terms and counterion reorganization terms.

1 R. J. Forster and L. R. Faulkner, *J. Am. Chem. Soc.* 1994, **116**, 5444–5452.

Professor Rampi asked: Osmium complexes are known to have a substantial reorganisation energy: what values of λ did you extract from your measurements?

Professor Ulstrup replied: The values are in the range 0.3–0.4 eV, but these are most likely associated solely with the solvent environment as there are only small differences between the Os-complexes in the two oxidation states.

Professor Rampi asked: A parallel between quantum dots and transition metal complexes is straightforward—how correct is this comparison and where does it fail?

Professor Ulstrup answered: If "quantum dot" is meant to allude to a metallic or semiconductor nanoparticle, then the parallel to transition metal complexes applies when the particle is small enough that the electronic spectrum of the particle is discrete. In this sense the particle then resembles a transition metal complex. The comparison fails for larger particles, say larger than about 5 nm where the particle spectrum approaches a continuous, band-like structure. Electronic transitions are here no longer of the discrete, "molecular-like" kind but collective plasmon or interband transitions of a different nature. Single-electron charging effects (Coulomb blockade) can, however, still be expected in "quantum dots" larger than the limiting size for electronic discreteness effects.

Professor Calvo made a general comment: During this Discussion the importance of metal–molecule contact in the molecular conductance has been raised. This is indeed an important contribution to the total molecular junction conductance.[1,2] So far the metal–alkanethiol and metal–pyridine bonds have been the contacts of choice. However, it has been recently reported that C–C[3] and C–Au[4,5] bonds may be obtained by electro-reduction of diazonium salts. For the latter, some infrared evidence of surface functional groups has been reported[4] but no conclusive proof of molecular monolayer has been presented. The ready formation of multilayers in the diazonium cation reduction reaction on carbon.[6]

We would like to disclose new experimental evidence in our laboratory of FTIR reflection absorption spectra of *p*-nitroaniline diazonium salt electroreduced on poly-crystalline Au at 0.1 *vs.* Ag/AgCl; 3 M NaCl that demonstrates the formation of a monolayer.

Fig. 2 shows the spectra for s- and p-polarized radiation at an angle of incidence of 80°. While the s-polarized spectrum is featured in the 600–2400 cm^{-1} range, p-polarized radiation shows characteristic bands due to quadrant C–C vibration of phenyl group at 1600 cm^{-1}, ν_a-NO_2 at 1530 cm^{-1} and ν_s-NO_2 at 1352 cm^{-1} and δ-NO_2 at 855 cm^{-1}. Surface selection rules in the IR reflection at metals determines that the molecule attached to the surface has a component of the dynamic transition moment perpendicular to the surface and the absence of s-polarized signals indicates that a monolayer has been formed. Furthermore the CN_2^+ vibration at 2280–2300 cm^{-1} found in the diazonium salts, are absent in the reflectance spectrum confirming the loss of N_2 upon electroreduction and formation of the aryl radical on the Au surface.

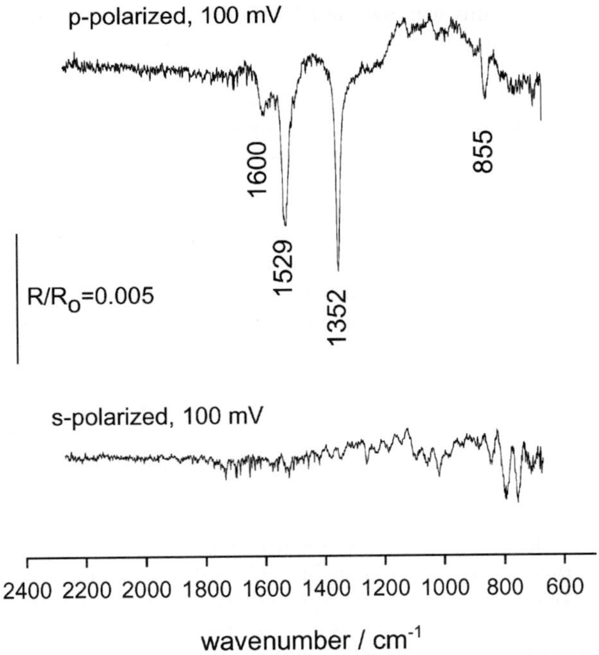

Fig. 2 Spectra for s- and p-polarized radiation at an angle of incidence of 80°.

1. J. R. Heath, M. A. Ratner, *Phys. Today*, 2003, 43–49.
2. R. L. McCreery, *Chem. Mater.*, 2004, **16**, 4477–4496.
3. M. Delamar, R. Hitmi, J. Pinson, J. M. Savéant, *J. Am. Chem. Soc.*, 1992, **114**, 5883–5884.
4. M. C. Bernard, A. Chaussé, E. Cabet-Deliry, M. M. Chehimi, J. Pinson, F. Podvorica, C. Vautrin-Ul, *Chem. Mater.*, 2003, **15**, 3450–3462.
5. A. Laforgue, T. Addou, D. Bélanger, *Langmuir*, 2005, **21**, 6855–6865.
6. J. K. Kariuki, M. T. McDermott, *Langmuir*, 2001, **17**, 5947–5951.

Professor McCreery commented: I am a strong proponent of diazonium chemistry for surface modification, and I agree that Professor Calvo's IR evidence provides good evidence for monolayer bonding to Au. However, be careful of multilayer formation, as diazonium chemistry is not self limiting, like Au–thiol SAM formation.

Professor McCreery then addressed Professor Ulstrup: Professor Rampi's experiment represents the limit of complete reorganisation and redox exchange. The other limit is coherent tunnelling with no reorganisation. What happens in the Albrecht–Ulstrup experiment if reorganisation is absent? Can tunnelling still occur through redox levels?

Professor Ulstrup replied: Tunnelling could still occur but only by tuning the redox level to values inside the energy window between the Fermi levels of the enclosing electrodes. This would give a rectangular, pulse-like dependence of the tunnelling current on the overpotential, and a step-like pattern for the bias voltage dependence. In this limit the tunnelling process would accord with a coherent tunnelling view rather than with the view of incoherent two-step tunnelling.

Professor Schiffrin asked: In the picture you have presented, at low temperatures, would the LUMO simply behave as an indentation to the barrier?

Professor Ulstrup replied: Yes, the LUMO behaves as an indentation in the barrier as long as neither the overpotential nor the bias voltage has taken the LUMO to the energy window between the two Fermi levels. Inside this window the LUMO opens a new electron transfer channel and assumes a different mechanistic role, also at low temperatures.

Dr Davis communicated: In section 4.4 the authors state that the bias voltage drops mostly between the substrate and redox site whilst there is strong electronic coupling to the probe. Can the authors explain the physical origin of this conclusion?

Professor Ulstrup communicated in response: In a simple view of the bias voltage drop as being linear in the tunnelling gap, and close to contact distance between the molecule and the tip, most of the drop would be between the substrate and the redox site. This is due to the structural asymmetry of the molecule. This seems to accord with the rate parameters extracted from the tunnelling spectra. Theoretical analysis of the potential distribution in the tunnelling gap is, however, highly needed.

Dr Davis communicated: Is Fig. 8 of the paper correct? It is annotated as compound **2** where it has already been stated that coupling to the substrate is limiting and varies with *substrate* potential (Fig. 7a of the paper)–why is there also tip potential dependence?

Professor Ulstrup communicated in reply: The reason why the conductivity is plotted *vs.* the tip potential is that it is actually the tip potential which is kept constant during a tip–substrate bias scan. (This is because the tip is grounded in the MI STM instrument). The substrate potential is thus swept ±50 mV with respect to E_t, and as the conductivity of the complex depends on the substrate potential, a modulation of the conductivity is observed. The latter is still evaluated at zero bias, so in principle $E_t = E_s$, but from the experimental point of view E_t remains constant.

Dr Davis communicated: Could the authors comment on the difference between the conclusions presented in this paper (weak substrate–adsorbate coupling) and Dr Chi's paper (strong substrate coupling) despite the fact that these are both ECSTM configurations.

Dr Albrecht communicated the answer: The conclusions in this paper are based on the view of strong substrate-adsorbate coupling in the sense that the system accords with the adiabatic limit of interfacial electron transfer. This is likely to induce transition of a number of electrons between substrate and tip in a single-two-step STM process. The system discussed in Dr Chi's paper corresponds to much weaker couplings and smaller interfacial rate constants. The number of electrons transferred is therefore smaller. Other features of the system in Dr Chi's paper are discussed in Professor Ulstrup's earlier answer to Professor Tao.

Professor Kornyshev made a general comment: The introductory lecture introduced us to how we can build circuits of single-molecular devices. Later we heard these talks about the performance of single-molecule constructions: in solid state environment or in electrochemistry. These are two different worlds, and electrochemistry provides us with a number of new options, which some of the talks today, Professor Ulstrup's, in particular have brilliantly demonstrated.

It is straightforward how to integrate solid state architectures in the circuits, but how to do it with electrochemical devices? Is it not the time to think about it? References to solid electrolytes or solid polymer electrolytes, or even ionic liquid are not yet substantiated by any serious experimental studies. I am not aware of any STM experiments on ionic solids or polymer electrolytes.

Dr Chi replied: These are inspirational comments for future investigations.

Professor Fitzmaurice commented: Nanofluidics are the way to integrate classical electrochemistry and classical solid-state chemistry.

Professor McCreery replied: In order to incorporate electrochemical devices into solid state electronics, it is possible to remove some of the traditional properties of a "wet" cell. What do you really need to do redox reactions? Electric field, mobile ions, reorganisation, but not necessarily solvent. There is a range of devices between a wet cell and a solid state device which may be useful commercially.

Dr Brust commented: A lot of the work we have seen here is driven by the availability of tools, such as STM, rather than by the quest to assemble computing devices from single molecules. I think these goals, *i.e.* understanding molecules and making computers, should be regarded as separate problems each having their merits. Molecular devices for signal processing may well be best made from networks of very many molecules that do not require a precise knowledge of what is happening on the level of each individual molecule.

Professor Kornyshev commented: The examples from Professors Lindsay and Fermín are quite convincing. One may speculate also on colloid electrolytes or electrolyte gels, intermediate in viscosity between solids and liquids. It seems, however, is that for the time being we still don't know how to do it. And the first one who finds the right system will make a revolution in this business.

Dr Geskin opened the discussion of Professor Weber's paper: Let there be a redox-active molecule in a junction, in equilibrium electrochemically controlled conditions. Supposing a redox potential translates, *via* Nernst equation, into defining the probability, over the statistical ensemble, for the active species to be in the reduced/oxidised state. It can be supposed that, for a given molecule, the same probability applies to its state as a function of time. In this respect, do you expect any predictable manifestations of noise related to the redox fluctuations of the molecule caught in the junction?

Professor Weber replied: The experiment is performed in vacuum, not in electrolyte; therefore it might be misleading to apply the Nernst equation.

If one considers two charge states of the molecule as a two-level system in contact with a charge reservoir, which can be identified by the electrodes, then there is indeed a Lorentz-oscillator. Its characteristic time scale is then given as the inverse of the exchange rate Γ. Within the standard coherent transport models, the same Γ is proportional to the conductance and is in our case of the order of 10^{-11} s. At lower frequencies, this is white (meaning frequency-independent) noise, whereas at GHz frequencies (beyond our experimental window) one would eventually see a Lorentz-like f^{-2} frequency dependence related to fluctuating charge states. This estimation is based on purely electronic considerations. If you consider a transition between different configurations associated with different charge states, the same Lorentz-like statistics apply, but on a much slower time scale, eventually within our experimental range. So far, we have not enough information to distinguish this scenario from environmental noise.

Professor Tao commented: For random stochastic processes, the time distribution should be very broad, from fast to slow. Is it possible that although the average time is fast there is some slow tail in the time distribution? That is: if some slow process happens, it will be observed. I guess that the question is: what is such probability.

Professor Lindsay asked: Fluctuations are not likely to be related to classical redox processes because (a) the system is in vacuum (b) the electron addition steps are at high bias and are sharp and not noisy. Could the steps be contact fluctuations? Simulations show a huge range of activation energies for thiol–gold on gold.

Professor Weber responded: Our picture is that the steps in the current voltage characteristics represent resonant tunnelling across "conducting" orbitals, which are entering the bias window when a certain threshold voltage is reached. In the purely electronic picture, this is unrelated to a classical redox process, because the charge on the molecule is not necessarily integer in non-equilibrium conditions. Possible configuration changes at the contacts are rather associated to our observation of discrete transitions from one stable contact to another that occur as sudden switching (sometimes back and forth) between two (bi-)stable I–V characteristics, which are in itself reproducible. One should have in mind that we slightly stretch the contact in order to get more stable data. This leads to a reduced free space for moving thiols. However, these are just likely interpretations and we can not unambiguously prove them.

Professor Bjørnholm asked: Could you explain how you envision the states that give rise to the steps you observe in your *I–V* curve?

Professor Weber replied: As a first approximation, we assume a fully electronic picture, where the current steps are related to resonant tunnelling. In the description of the asymmetric diode-like molecule, we make an attempt to describe the steps as level crossings between the two conjugated sub-systems of the molecules, which shift with respect to each other as a function of applied field. In the latter case, inelastic processes have to be assumed in order to obtain steps in the *I–V* curves.

Professor Naaman commented: The frequency of the noise is very low indicating nuclear motion and perhaps a rotation-like motion of the molecule. This motion will involve a change from *trans* to *gauche* configuration.

Professor Weber replied: The degrees of freedom you are referring to are actively moving at elevated temperature, but are frozen out at low temperature, where the presented data were taken. It is conceivable that the applied voltage or the current activate this rotation above a certain bias threshold. To investigate this further, we have designed molecules with controlled rotational degrees of freedom, which are currently under investigation.

Professor Schiffrin said: We have heard today that different groups have observed noise of different types in single molecule conductance measurements. I think that the physical origins of the noise is inelastic tunnelling that results in the excitation of vibrational/rotational modes. This is very well known, for instance, from the work of Ho[1] and from techniques such as inelastic tunnelling spectroscopy. Dr Haiss has shown that different conformers of a molecule can have very different conductivities.[2]

1 G. V. Nazin, S. W. Wu and W. Ho, *Proc. Natl. Acad. Sci., USA*, 2005, **105**, 8832.
2 W. Haiss, 2005, unpublished work.

Dr Nichols asked: You postulate that the "flicker" noise could arise from rotations about the ethynyl C≡C bond which is not hindered. But your "flicker" noise is a lot lower in frequency than this rotational frequency. You suggest that this could arise from "packing" around molecules. However, I thought that there were few molecules in the gap?

Professor Weber replied: Our observations indicate that a single molecule (or, at most, very few molecules) *bridge* the gap and form the electronic contact. However, the sample to sample fluctuations we observe are due to the poorly controlled local environment. This may consist of other molecules of the same species, or other residues (water, solvent *etc.*) and eventually slow down the free oscillations/vibrations of the bridging molecules.

Professor Tao asked: Can you cool down your sample even further with your setup?

Professor Weber responded: We are currently building a ^{3}He system to reach the sub-Kelvin range.

Professor Rampi said: The *I–V* curve measured for the symmetric molecules show an abrupt increase in the experimental and determined potential. How does this potential correlate with the energy level of the molecule?

Professor Weber replied: From quantum chemical computations we know for this type of molecules that the HOMO is the level closest to the electrochemical potential μ of electrodes.[1] If the energetic distance between HOMO and μ is ε, then the first peak will occur at $U = 2\varepsilon/e$.[2]

1 H. B. Weber, J. Reichert, F. Weigend, R. Ochs, D. Beckmann, M. Mayor, R. Ahlrichs and H. v. Löhneysen, *Chem. Phys.*, 2002, **281**, 113.
2 S. Datta, *Electronic Transport in Mesoscopic Systems*, Cambridge University Press, Cambridge, 1995.

Professor Bjørnholm asked: Is the distance between the first step in the *I–V* curve in the positive and negative direction a measure of the HOMO–LUMO gap?

Professor Weber answered: In a symmetric setup, the distance is given by 2ε (*cf.* last answer). Only in the case of strongly asymmetric contacts (like an STM tip above a molecule on a conducting surface) does the step distance represent the HOMO–LUMO gap.

Professor Lindsay commented: While the position of the first step may be very hard to predict, the separation between the first and subsequent, second, third...peaks should be reproducible and it should be possible to compare with theoretical models of level spacing in these systems.

Professor Weber agreed: Indeed, for the diode we observed that the distance between the first and the second peak (in $\mathrm{d}I/\mathrm{d}V$) was very similar for several samples, independent of the position of the first peak. We have not yet tried to assign this distance to theoretically determined orbitals, but it should be less affected by external charges then the first peak.

Professor Bryce opened the discussion of Professor Rovira's paper: It would be interesting to define the length of the bridge which will allow an interaction to be observed in your systems, by synthesising an analogue with a shorter phenyleneethynylene bridge.

Professor Rovira replied: Yes, it will be interesting. What we know is that whereas no electron transfer was observed in a compound bearing a long (~3.2 nm) phenyleneethynylene based bridge between the polychlorotriphenylmethyl radical and the polychlorotriphenylmethide anion[1] in other compound having a phenylenevinylene based bridge with the same length[2] electron transfer is clearly observed. This result confirms that phenylenevinylene based bridges are better to promote the electronic coupling between polychlorotriphenylmethyl radical (A) and the polychlorotriphenylmethyde anion (D).

1 O. Elsner, D. Ruiz-Molina, J. Vidal-Gancedo, C. Rovira, J. Veciana, *Nano Lett.*, 2001, **1**, 117.
2 J. Veciana *et al.*, unpublished results.

Professor Rampi asked: The electron transfer rate through polyphenyl chains depends on the twist angle between the aromatic rings. In your bridges you have alkyl chains that influence the twist angle. Did you observe any change in the electron transfer rate depending on the presence of the alkyl chain?

Professor Rovira replied: It is true that the alkyl chains present in the bridge of compound **3** can influence the twist angle with respect to the one in molecule **1** that do not have the substituents and as a consequence the electron transfer rate. Nevertheless the V_{ab} coupling parameter value (100 cm^{-1}) of $\mathbf{3}^{-\bullet}$ as well as the thermal energy barrier for the electron transfer process ($\Delta G^* = 5.9$ kcal mol^{-1}) obtained from EPR experiments, are very close to that of the topologically related mixed valence compound $\mathbf{1}^{\bullet-}$ ($V_{\mathrm{ab}} = 121$ cm^{-1}, $\Delta G^* = 5.9$ kcal mol^{-1}) which suggest that ring substituent effects on electron-transfer rates in this class of bridging molecules is very small.

Dr Higgins asked: Is it correct to say from your data for **1**(diradical) and $\mathbf{1}^{\bullet-}$ that there is better delocalisation in $\mathbf{1}^{\bullet-}$? (If so, why is this?) Presumably there isn't much communication between sites in these molecules because of the large steric interactions caused by the perchlorophenyl moieties?

Professor Rovira replied:
1: Obviously the presence of an additional electron in $\mathbf{1}^{\bullet-}$ with respect to diradical **1** should change the wave function due to electronic correlation. In order to know precisely the extension of

the delocalization in both compounds it will be necessary to perform *ab initio* calculations that are very time-consuming due to the large size of the molecules.

2: The high torsion caused by the steric interactions of perchlorophenyl moieties precludes in fact having a very large coupling between the two sites. For this reason $\mathbf{1}^{\bullet -}$ does not belong to the class III mixed valence system of the Robin and Day classification (see ref. 1) but to class II in which there is moderate electronic coupling between sites.

1 M. B. Robin and P. Day, *Adv. Inorg. Chem. Radiat. Chem.*, 1967, **10**, 247.

Professor Tao said: Professor Rovira measures electron transfer rate (from donor to acceptor groups) while Dr Haiss and Professor Weber measure electron transport current. Is there a connection between these two different measurements?

Professor Rovira replied: There should be a connection. If the same molecule that acts as a bridge between donor and acceptor groups is connected by two electrodes, conductivity should be observed.

Nevertheless, due to the huge difference in the local environment of the molecule, solution media in the case here presented and dry media in the other cases, the relationship between the two measurements is complex.

In fact as stated in the introduction of our paper, even though the conclusions from the solution studies do not necessarily correlate for the same molecules in other environment as is the solid state, in electrochemical experiments in which the donor or acceptor is replaced by a conducting solid also seek to reveal how the ET rate depends on the "bridge" structures (see ref. 7 of our paper). On the other hand, the values of the "decay factors" β obtained by measurements of current density through saturated and unsaturated molecules of different length, using junctions of the type Hg–SAM(1)/SAM(2)–Ag, are in good agreement with corresponding values obtained in solution (see ref. 8).

Professor Ulstrup replied: There is a connection but it is not so transparent. In the former case the electron is localized on the donor or acceptor group, stabilized by environmental nuclear relaxation. There is no such relaxation in the latter case. In addition both the tunnelling factors and the pre-factors in the rate constants or currents are different.

Professor Weber: There are substantial differences between donor–acceptor electron transfer and a steady state current in a molecular junction. In the first case, the electron number on the isolated molecule remains integer and constant during the single-charge transfer process. In a junction, the donation and acceptance of electrons is taken over by the electrodes as a function of the bias direction, resulting in a steady-state current. The difference becomes obvious in the classical example of the Aviram–Ratner thought experiment, where the current direction from the donor to the acceptor is counter intuitively the (blocking) reverse bias direction, whereas charge transport is favourable from the acceptor to the donor in the forward direction.

Professor Naaman commented: There is a basic difference between electron transfer and conduction. In the first the molecule is neutral; in the second the system includes an additional electron.

Professor Tao replied: Yes. In the two-step transfer theories, in which conduction arises when the electrons transfer from one electrode to a redox molecule, and then from the molecule to the second electrode, the two are different but related.

Professor Kuznetsov commented: In addition to what has been said by Professors Naaman and Rampi I would like to emphasize that we have to compare tunnel systems with reaction charge transfer systems under similar conditions. For example, if the tunnel barrier involves two redox groups, the rate constant for electron transfer between such groups in solution may be informative provided tunnelling is studied in the same medium. Similarly the electrochemical rate constants provide information about the rate of electron exchange between the electrodes and end groups of the bridge. In particular, equations for tunnel current through one redox group in an "*in situ*" system explicitly involve electrochemical rate constants.

Professor Ulstrup asked Professor Tao: Have you tried to convert the observed tunnelling currents for the two-step mechanism, first to a current density per molecule and then to a value of the interfacial rate constant? Could you comment on the expected or estimated values of such a rate constant?

Professor Tao replied: We have not. We are currently performing temperature dependence measurements which should allow us to determine whether it is a single-step or two-step mechanism.

Dr Davis communicated a question on Dr Haiss's paper: The data outlined in this manuscript is exceedingly interesting and I would like to congratulate the authors. The data presented in Fig. 4 is important but shows very considerable scatter—could the authors comment on this?

Dr Haiss communicated in response: The standard deviation of I_w at fixed I_0 in Fig. 4 is in the order of 10–15%, which is typical for this type of experiments (see Fig. 1c of our paper). This scatter may be caused by the various adsorption geometries of the sulphur contact at the tip and at the sample (*i.e.* threefold hollow site, bridge site, on top site, step site, kink site).

Professor Kornyshev opened the discussion of Dr Haiss's paper: I don't understand the resulting dependence of activation energy on the length of this molecule. In your brilliant experiments you have shown that the activation energy of conductivity is proportional to the length. To me that sounds like a requirement to align all the units in the chain in some ideal way for tunnelling. Then the activation energy will be sum of activation energy of each unit conformation.

Dr Haiss replied: In Fig. 3 of our paper we have shown that the activation energy scales linearly with the number of C–C bonds and have determined the activation energy per C–C bond (E_{C-C} = 39 meV). Since the C–C bonds are the conformer creating bonds and E_{C-C} is approximately the energy difference between a *trans* and a *gauche* conformation in an alkane chain, it was concluded that temperature induced changes of the conformer distribution could be responsible for the observed temperature-dependence of the single molecule conductivity.

If one assumed that the tunnelling probability between two neighbouring carbon atoms is increased by a certain amount upon a transition from *trans* to *gauche* conformation then one could express the total probability (P_{tot}) for electron tunnelling through an alkane chain as:

$$P_{tot} = P_t^k (P_t C)^{n-k}$$

where $P_t C = P_g$ is the probability of tunnelling through a *gauche* conformation, P_t is the probability of tunnelling through a *trans* conformation, n is the number of C–C bonds in the chain, k is the number of C–C bonds in *trans* conformation and $C = P_g/P_t > 1$ is a constant expressing the increased probability of tunnelling through a *gauche* conformer which is assumed to be independent of temperature. The normalised logarithmic dependence of P_{tot} on n would then read:

$$\ln(P_{tot}) = (n - k)\ln(C) + n\ln(P_t)$$

Since C and n are independent of temperature one can write:

$$\frac{\partial \ln(P_{tot})}{\partial(1/T)} = \ln(C)\frac{-\partial k}{\partial(1/T)}$$

But k can be expressed as $k = nP(T)$ where $P(T)$ is the thermodynamic probability that a C–C bond is in *trans* conformation and it follows:

$$\frac{\partial \ln(P_{tot})}{\partial(1/T)} = n\ln(C)\frac{\partial P(T)}{\partial(1/T)}$$

This result would explain the linear dependence of the activation energy on the chain length since P_{tot} is proportional to the tunnelling current (I_t) and $\partial \ln(P_{tot})/\partial(1/T)$ increases linearly with n.

Professor Lindsay said: The following aspects of your model seem unrealistic.
(a) with the contacts fixed, the end to end length of the molecule is not so free to fluctuate
(b) with through-bond tunnelling dominating, it doesn't seem right to use the 'free space' end-to-end length in your calculations.

Dr Haiss replied:
(a) Thiols attached to gold surfaces display mobility at room temperature and therefore transitions between different conformers are possible.
(b) I agree that more sophisticated theoretical methods should be used to describe the tunnelling through alkanedithiols (see also my reply to Professor Kornyshev). In our paper the emphasis is on the experimental part and the theory presented is very simple.

Professor Lindsay said: In addition to a small, and temperature dependent current, this new process you have discovered also has an anomalously small value of β. Could the two factors be related?

Dr Haiss responded: The experimentally determined conductivity of octanedithiol between Au electrodes is a factor 20[1] to 500[2] smaller than those predicted from first-principles theory. The reason for that is not clear yet but one should keep in mind that theoretical calculations are strongly dependent on contact geometry. Coadsorption may also change the conductivity significantly.

The β values determined in our experiments are dependent on temperature and chain length as shown in Fig. 3. For nonanedithiol at room temperature we calculate a β factor of ~0.7 per CH_2 group at 0.2 V which is slightly lower than the value determined in densely packed decanedithiol monolayers (0.79 $Å^{-1}$).[3]

1 X. D. Cui, A. Primak, X. Zarate, J. Tomfohr, O. F. Sankey, A. L. Moore, T. A. Moore, D. Gust, G. Harris, and S. M. Lindsay., *Science*, 2001, **294**(5542), 571–574.
2 C. C. Kaun and H. Guo, *Nano Lett.*, 2003, **3**(11), 1521–1525.
3 W. Y. Wang, T. Lee and M. A. Reed, *Phys. Rev. B*, 2003, **68**(3), 035416.

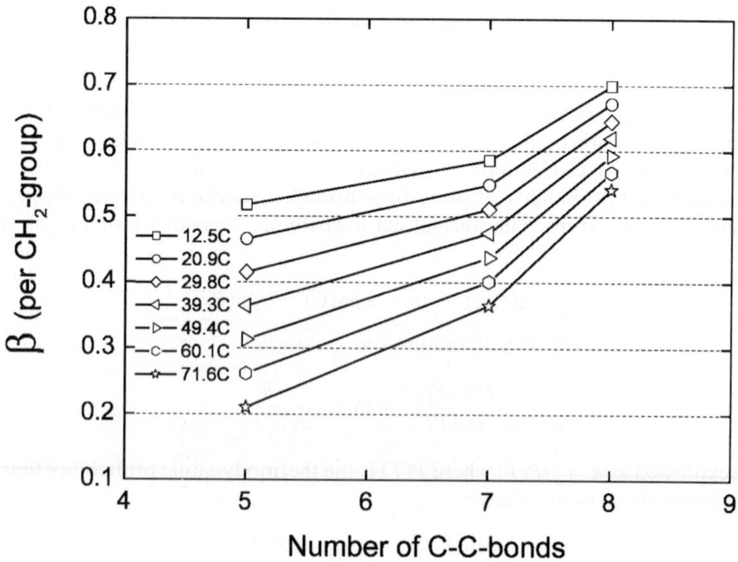

Fig. 3 Dependence of β values on temperature and chain length.

Professor Ulstrup commented: The model introduced in your paper does not of course represent details of thiolate linking of the alkanedithiolates to the enclosing gold electrodes. It does, however, illuminate strikingly the importance of configurational nuclear gating and the origin of observed temperature dependences as rooted in this effect and not directly in the electronic tunnelling phenomenon. Given suitable representations of the conformational dependence of the covalent electron exchange interactions and Au–S contact features, this observation carries over to through-bond tunnelling through the linked molecules.

Professor Larsson asked: Which conformation is the most efficient for tunnelling? Many years ago we calculated the transmission in alkane chains in different conformations.[1] Surprisingly we obtained smaller couplings $|H_{12}|^2$ the shorter the chain. The reason can be traced to how the orbitals overlap from one end of the molecule to the other. Although I agree with many aspects of your paper your result for coupling is contrary to ours. This is probably due to a too simplified tunnelling model. The Gamow model does not work very well in chemistry.

1 A. Broo and S. Larsson, *Chem. Phys.*, 1990, **148**, 103–115.

Dr Haiss replied: I agree that a very simple model has been used in our calculations. In a different approach, which has been briefly outlined in my response to Professor Kornyshev, one would have to assume that *gauche* conformations give rise to an increased tunnelling probability compared to the *trans* conformation. Otherwise our experimental data could not be explained by a temperature-induced change in the conformational distribution.

Dr Smith asked: If the temperature dependent tunnelling current is due to conformations with different lengths should it not be possible to relate current–distance plots to the temperature dependence? Would it be possible to observe different activation energies for different contact separations?

Dr Haiss replied: This is a nice idea but there are formidable experimental difficulties in carrying out the experiment suggested. Although the tip–substrate separation can be determined with some degree of accuracy, the absorption geometry is not necessarily determined by the tip–sample separation since the thiol contact at the tip may not necessary be located at the outermost apex of the tip.

Dr Davis communicated: At higher temperature one expects more molecular fluctuation and, despite the "greater population of the more conductive conformer" one might expect an increase in "blinking"/decrease in stability of contact. Did the authors make any such observations?

Dr Haiss communicated in response: There was no clear indication that the normalised noise in a molecular wire (*i.e.* $\Delta I/I$) increases upon an increase of temperature and the standard deviation of the observed current jump values was independent of temperature.

Dr Macdonald asked: Would it not be useful to repeat your beautiful temperature-dependent measurements on other molecules which have a much higher persistence length to confirm the role of conformation?

Dr Haiss responded: We have carried out experiments with conformationally rigid molecules where no temperature dependence of single molecule conductivity has been found.

Ms Solomon commented: Calculations have shown that relationship between thiol binding and conductivity is not always intuitive, for example those from Mark Ratner's group. Could it not be that changes in the thiol binding cause the effects shown and in fact may dominate over the changes to the molecule?

Dr Haiss replied: A temperature dependence of the thiol binding is rather unlikely to dominate the observed temperature dependence for three reasons:

The $E_a(n)$ plot, presented as Fig. 3 in our paper shows that the experimentally observed temperature dependence is decreasing linearly with the number of C–C bonds and an extrapolation of the experimental data to $n = 0$ results in $E_a \sim 0$. These observations suggest that the contacts do not contribute significantly.

Experiments by Wang *et al.*[1] have shown that no pronounced temperature dependence of the conductivity of densely packed SAM's of alkanedithiols can be observed.

We have carried out experiments with conformationally rigid molecules where no temperature dependence of single molecule conductivity has been found.

1 W. Y. Wang, T. Lee and M.A. Reed, *Phys. Rev. B*, 2003, **68**(3), 035416.

Dr Wandlowski commented: One possibility to explain the issue of adsorbate mobility on i_T–Δt characteristics might be to work with surfaces of different crystallographic orientations [Au(100)-(1 × 1), Au(110)-(1 × 1)]. This approach might help to distinguish between temperature induced conformation changes and adsorbate–gold mobility-related effects.

Dr Haiss replied: This is an interesting idea and certainly worth pursuing. Using different crystallographic orientations will also help to determine the influence of the adsorption site on single molecule conductivity. Concerning the mobility of the thiols one should keep in mind that there may be still a high degree of mobility at the tip, since it is difficult to control the crystallographic orientation of the tip surface.

Professor Lindsay commented: Your low currents and small β are exactly what we observe for measurements with a tethered gold nanoparticle. In our case, we account for the anomalously low current by poor contact between the nanoparticle and AFM. Could a similar process be operating here? A 'weak' contact might permit conformational functions to influence the contact.

Dr Haiss said: No, we do not think that there is a Au–S contact problem in our experiments. To find out if there is a contact problem in your outstanding AFM experiments[1] it may be worth recording current–force curves on a gold surface and compare these to the current–force curves observed in your nanoparticle experiments.

1. X. D. Cui, A. Primak, X. Zarate, J. Tomfohr, O. F. Sankey, A. L. Moore, T. A. Moore, D. Gust, G. Harris, and S. M. Lindsay., *Science*, 2001, **294**(5542), 571–574.

Professor Barbero said: To test the model where different conformers give different conductances, one of the hydrogens of a methylene group in the alkyl chain could be replaced by a methyl group. Therefore, the rotation of the C–C bond will get less free, rising the energy barrier and increasing the number of different conformer states.

Dr Haiss replied: This is a very good idea, which we will pursue.

Professor Kornyshev and **Professor Kuznetsov** communicated: Stimulated by Haiss's work we rationalized his idea of the effect of conformational fluctuation on tunnelling in a very simple model, which nonetheless gives quite instructive results in different regimes.[1] In that paper we present a model for a tunnelling conductance controlled by quasi-independent fluctuations of relative conformations between the nearest neighbour units of the chain molecule. The model leads to a simple analytical formula which in one regime reproduces the Haiss-type Arrhenius dependences for alkanedithiols with activation energy proportional to the length of the molecule, whereas in other regimes it shows the activationless behaviour corresponding to the tunnelling across more rigid molecules. The general formula covers the transition between the two limits, predicting some new dependences that could be interesting to verify experimentally.

1 A. A. Kornyshev and A. M. Kuznetsov, "Single molecule tunnelling conductance: the temperature and length dependences controlled by conformational fluctuations", *Chem. Phys.*, to be published in the N. Hush *festschrift*.

Self-assembled electrical circuits and their electronic properties

Miron Hazani, Dmitry Shvarts, Dana Peled, Victor Sidorov and Ron Naaman

Received 20th April 2005, Accepted 12th May 2005
First published as an Advance Article on the web 26th September 2005
DOI: 10.1039/b505583k

A straightforward method for the self assembly of single walled carbon nanotubes (SWNTs) between gold electrodes was developed. The technique utilizes the hybridization between short complementary DNA sequences located on metal contacts and SWNTs. The new technique enables simple production of hundreds of devices with high yields. The electrical characteristics are shown to depend strongly on the existence of the chemical binding groups at the contacts as well as along the tubes. This technique was used to drive the self assembly of SWNT-based field effect transistors (CNTFETs). In principle, the devices made by this method behave like those made using direct metal–carbon nanotube contacts. The inverse subthreshold slope of the CNTFETs depends on the source–drain voltage applied to the device, confirming that the conductance of CNTFETs is determined by the Schottky barriers at the interfaces between the CNTs and the gold electrodes.

Introduction

Single-walled carbon nanotubes (SWNTs) are considered as promising building blocks for nanoscale structures and hence the tremendous effort invested in trying to control the way the tubes are arranged between electrodes or to assemble them into predetermined structures. In order to achieve these goals methods have been developed for controlling the growth of the tubes,[1–6] and for achieving alignment by mechanical forces,[7,8] by electric[9–13] and magnetic[14,15] fields, and by self-assembly methods.[2,16,17,18]

For any practical application in electronic devices, the uniformity of the electronic properties of the SWNTs is of high importance, and therefore efforts are made to control the structure and properties of the SWNT themselves and of the SWNT -metal contacts.[19] However, if devices were to be made with multiple SWNT contacts, the electronic properties would be averaged and a rather uniform performance may be achieved, despite the non-uniformity amongst the different SWNTs themselves and the variability in the contacts.

Controlled self assembly of SWNTs was recently achieved by interphasing them with biological molecules.[20] This approach has the potential for driving self assembly of SWNT-based devices in light of the immense richness of biological recognition. Recently, the fabrication of a DNA-templated SWNT field effect transistor was reported.[21] In this work a long DNA molecule featuring RecA proteins was used as a scaffold onto which streptavidin-functionalized SWNTs were assembled, utilizing anti-RecA primary antibodies and biotinylated secondary antibodies. Electrical contact to the tubes was achieved by metallizing the scaffold DNA molecule. Despite the ability to fabricate SWNT based devices, many of the production methods developed so far involved multiple steps resulting in low yield.

Recently we developed a method in which the organization of SWNT between gold electrodes is driven by biochemical affinity.[22] The SWNTs are chemically modified with single strand DNA which is complementary in sequence to DNA capture sequences located on the electrodes. This approach enables the production and full physical characterization of hundreds of devices. Here we present new findings on the electrical properties of these devices and compare their performance to

Department of Chemical Physics, Weizmann Institute of Science, Rehovot, 76100 Israel

SWNT based devices fabricated using different methods. In addition CNTFETs were produced and characterized in order to establish what are the structural parameters that control their electronic properties.

Currently we have no way of accurately controlling the number of SWNTs that connect the two electrodes and the technique developed results in a wide distribution of the number of contacts between the electrodes. This distribution also leads to situations in which SWNT networks involving multiple junctions are formed between the electrodes.

Nanoscale study of conduction through SWNT networks revealed that the conductance does not decrease linearly with the distance from the electrodes but rather drops drastically at certain points in the network, coinciding with junctions between metallic and semi-conducting SWNTs. It was also shown that SWNT clusters with many paths leading to the electrode have higher conductances than those with fewer paths. This was reasoned by the higher probability of obtaining a metallic pathway to the electrode in clusters with many paths.[23]

Experimental section

The approach which we have realized, as depicted schematically in Fig. 1, consists of two steps. In the first step, a self assembled monolayer of single stranded DNA was adsorbed onto gold contacts by reaction with thiol-terminated oligonucleotides. All oligonucleotides used were high purity salt free grade from MWG, and used without further purification. For the formation of the self assembled monolayers, 26-mer poly-T oligonucleotides, modified with a 3′ C_3 propyl protected dithiol linker [–$(CH_2)_3$–S–S–$(CH_2)_3$–OH] were used. For the covalent modification of the oxidized SWNTs, 26-mer poly A oligonucleotides, modified with a 3′ C_3 primary amine linker [–$(CH_2)_3$–NH_2] were used.

In the next step, oxidized SWNTs modified with oligonucleotides of the complementary sequence were allowed to hybridize with the DNA located on the gold contacts. The devices bearing the gold electrodes were cleaned by successive sonication in acetone, methanol, and isopropanol (5 min each) followed by extensive washing in de-ionized water and drying. The devices were then placed in a UV-ozonator for 20 min followed by immersion in ethanol for another 20 min. The samples were then washed with ethanol, dried and immediately covered with a few drops from a solution of 1 μM 3′-thiolated oligonucleotides in 0.4 M phosphate buffer pH 7. After over-night incubation the samples were washed by successive immersions in 0.4 M phosphate buffer pH 7.

Modification of the oxidized SWNTs with 3′-amino modified oligonucleotides was carried out with slight modifications to the procedure described previously.[24] Oxidized SWNTs suspended in 0.1 M phosphate buffer pH 7 were reacted with 5 μM 3′ amino terminated oligonucleotides in the presence of 100 mM 1-ethyl-3-(3-dimethylaminopropyl)carbodiimide and 10 mM *N*-hydroxysuccinimide for 12 h at room temperature. The reaction mixture was filtered over a 0.2 μm polycarbonate membrane, washed with 100 ml of de-ionized water and resuspended in H_2O.

Devices were fabricated by thermally growing 100 nm SiO_2 on top of p-type boron-doped Si wafers, followed by deposition of hundreds of metal pads, consisting of 10 nm Ti adhesion layer, 30 nm Au and 200 nm Ni protection layer, by e-beam evaporation and standard lift-off process. The distance between the source and drain electrodes is between 1.5 to 2 μm, and the highly doped silicon wafer serves as a gate electrode. Two types of devices differing in the thickness of the oxide layer were fabricated. The first, having a 100 nm thick SiO_2 layer, was obtained by simply removing the Ni protection layer from the Au contacts by HCl wet etching. Another type of device was obtained by defining active areas between the source and drain electrodes by optical lithography, followed by selective dry etching of SiO_2 using a CHF_3-based RIE process. This process results in local openings in the thermally-grown SiO_2 layer at positions intended to contain SWNTs.

Deposition onto the electrodes was achieved by immersing the devices in a aqueous suspension of DNA modified SWNTs containing 0.1 M phosphate buffer pH 7, and 0.5 M NaCl, at room temperature for 12 h. The samples were then extensively washed by successive immersions in fresh hybridization buffer followed by washing with a solution of 0.1 M *N*-2-hydroxyethylpiperazine-*N*′-2-ethanesulfonic acid (Hepes) pH 7, 0.5 M NaCl. Finally, the samples were washed briefly with H_2O and dried by nitrogen stream. Arrays of hundreds of electrode devices bearing DNA were exposed to SWNTs modified with complementary or non-complementary DNA sequences, as described, and

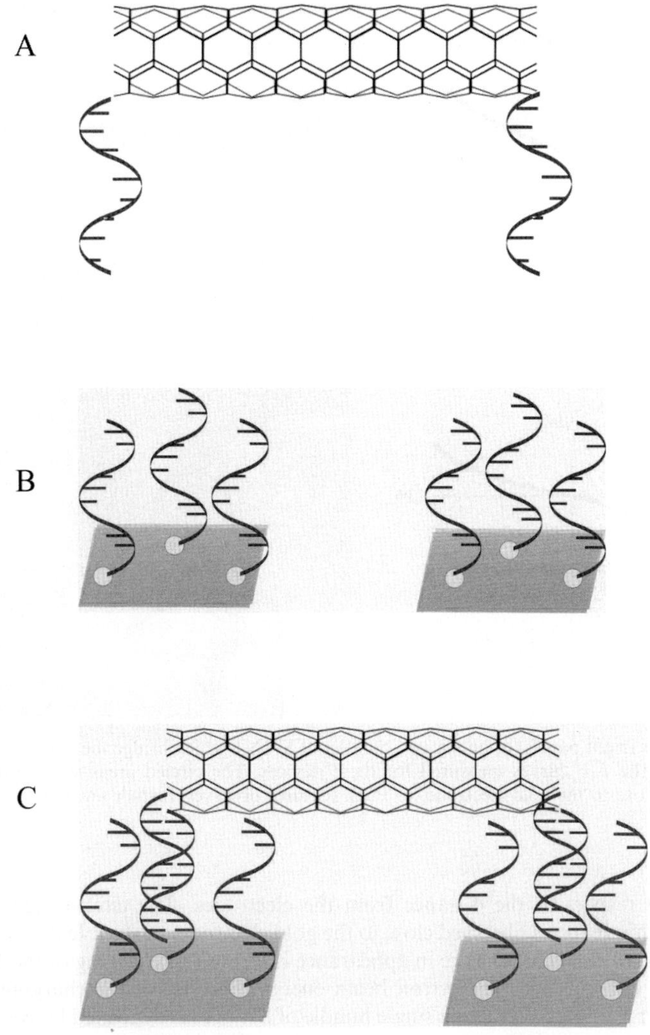

Fig. 1 Schematic representation of DNA-mediated deposition of a SWNT between two gold electrodes. (A) DNA modified SWNT is reacted with (B) gold electrodes bearing the complementary oligonucleotides (light-grey circles represent thiol groups at the 3′ of the capture sequences, used to assemble them on the gold electrodes). (C) Hybridization between the complementary strands results in bridging the two electrodes by the SWNT.

characterized by scanning electron microscopy (SEM) and room temperature current-voltage measurements.

Electrical measurements were carried out at ambient conditions using two Keithley Model 236 Source-Measure Units, which applied source–drain voltage V_{sd} and gate–drain voltage V_g and measured source–drain current I_{sd} and leakage current I_{leak}, respectively. The latter was negligible for all measurements.

Results

The ability to produce many devices and to characterize them using SEM and electrical measurements provides the ability for obtaining new insights on the relation between the geometry and the electrical properties of SWNT based devices. SEM images reveal a variation in the appearance of

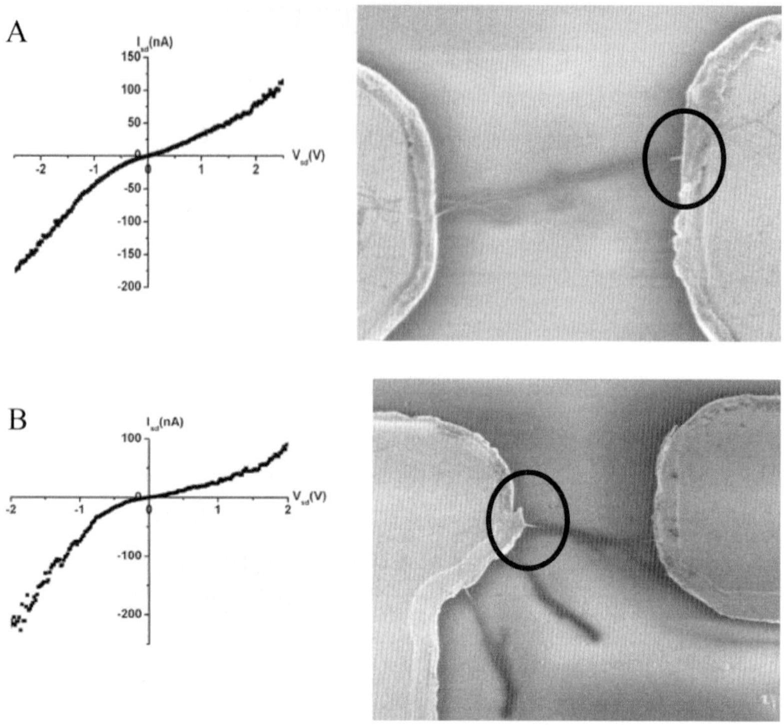

Fig. 2 SEM images (right panels) of one or two bundles of SWNT which bridge the gap between the electrodes. On the left panels the *I–V* curves measured for these devices. The circled areas show transitions from thin features observed close to the gold electrode, to thick features observed further away from the electrodes.

the SWNTs with respect to the distance from the electrodes. The circled areas in Fig. 2 show transitions from thin features observed close to the gold electrode, to thick features observed further away from the electrodes. This change in appearance is a result of charging of the DNA present on the SWNTs, and depends on the electron beam energy. Because of the charging it is difficult to determine whether a single SWNT or a single bundle of SWNTs is deposited between the electrodes. The fact that the charge builds up away from the electrodes and dissipates close to them already sheds some light on the nature of the contact between the gold electrodes and the SWNTs. The distance dependence indicates the non-Ohmic nature of the coupling between the SWNTs and the gold electrodes. In most cases individual SWNTs and bundles of up to ten SWNTs bridge the gap between the source and the drain electrodes. The average diameter and length of the individual SWNTs used as judged by atomic force, transmission, and scanning electron microscopy are 1.4 nm and 1.5 μm, respectively.

Fig. 2 shows the "ideal" situations in which one or two bundles of SWNT bridge the gap between the electrodes. From the SEM images it appears that in these cases the SWNT bundle originating from one electrode reaches the other and the conduction does not pass through a SWNT junction. The fact that in both cases shown, there is an additional bundle originating from one electrode and joining the other, may account for the slightly non-symmetric *I–V* curves. The *I–V* features observed can be modeled by assuming two leaky back to back diodes. Fig. 3 shows SWNT networks where the contact between the electrodes obviously involves SWNT junctions. The *I–V* curve in Fig. 3A is highly non linear and not symmetric in the range of −3 V to +3 V bias voltage, while the device shown in Fig. 3B, which involves a very large number of SWNTs, features a linear *I–V* curve with a current exceeding by an order of magnitude the current shown in Fig. 3A. When the contact between electrodes is formed by a SWNT network many conduction paths are established which may include junctions between various types of SWNTs. Multiple parallel

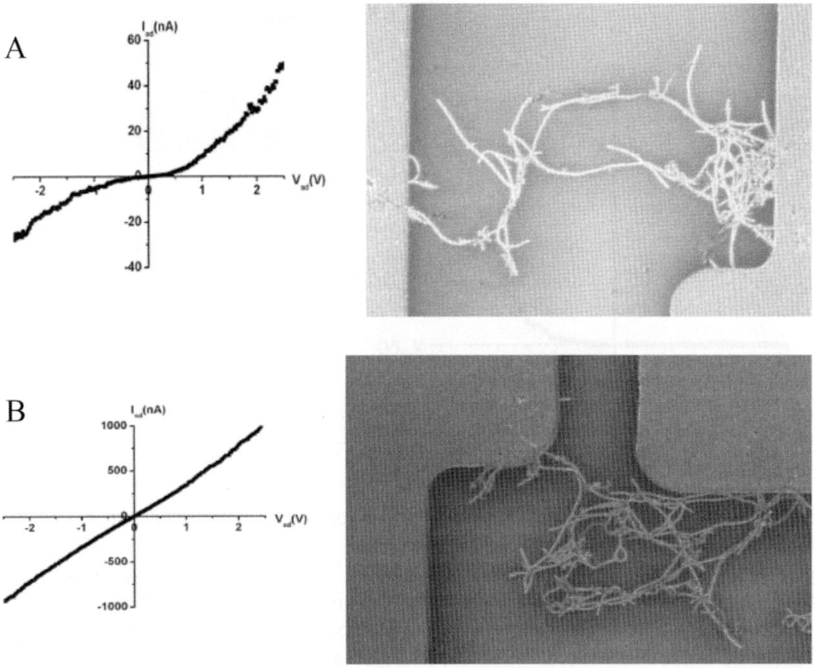

Fig. 3 SWNT networks where the contact between the electrodes obviously involves SWNT junctions. The *I–V* curve in (A) is highly non linear and not symmetric in the range of −3 V to +3 V bias voltage, while the device shown in (B), features a linear *I–V* curve with a current exceeding by an order of magnitude the current shown in (A).

conduction channels will obviously reduce the total resistance especially if metallic pathways are formed, however junctions formed between metallic and semi-conducting SWNTs may increase the resistance due to formation of Schottky barriers. In the case of conducting channels that involve metallic SWNT an Ohmic like *I–V* curve is observed. Fig. 4 shows a case where the contact between the electrodes involves a single junction between SWNTs (circled). In this case a diode-like behavior is observed. The *I–V* curve and the small magnitude of the current may be explained by the formation of a Schottky barrier at the contact point between a metallic and semi-conducting SWNT.

Fig. 5 demonstrates the effect of mild thermal treatment (200 °C for 30 min). The thermal treatment results in an increase of the current for a given bias voltage of up to two orders of magnitude. We believe that the increase in the current is due to the destruction of the DNA molecules binding the SWNT and the electrodes and the formation of a direct gold–SWNT contact. Importantly, despite the increase in the current the shape of the *I–V* curve remains the same.

Fig. 6 presents SEM images demonstrating the effect of high currents on the SWNT based devices. In practically all cases investigated by us, current induced damage did not occur at the SWNT/electrode interface, but rather along the SWNTs and especially at SWNT junctions. Current induced breakdown of the devices therefore depends not only on the number of SWNTs connecting the electrodes but more importantly on the manner in which the SWNTs bridge the gap, namely direct contact or through SWNT junctions. In the case of single SWNT connector, the current induced breakdown occurred at a current of about of few hundred nanoampere. When the devices where treated thermally (as described above) the breakdown occurred at current of up to 20 μA. This indicates that the thermal treatment actually annealed some defects on the SWNT causing it to be more uniform. Fig. 6C and 6D show a case where a network of multiple SWNTs connects the electrodes, a situation obviously involving multiple junctions, a few of which are expected to be highly resistive. It is striking, however, that the current-induced breakdown is highly localized and spatially distinct (red circles).

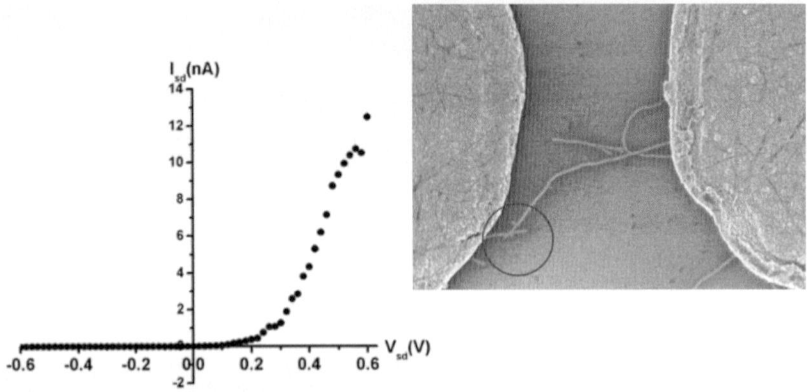

Fig. 4 A device where the contact between the electrodes involves a single junction between SWNTs (circled). The *I*–*V* curve is diode like.

For farther probing the electronic properties of the self-assembled devices, field effect transistors (FET) with back-gates were fabricated and investigated. Many research groups studying CNTFETs have reported on the hysteretic behavior of the *I*–*V* characteristics of their devices which is manifested in the shift of the threshold voltage.[25] This shift is attributed to a redistribution of charges in the vicinity of the nanotube caused by the applied gate voltage. In Fig. 7 we present the typical hysteresis curves measured for sweeps of the gate voltage in different ranges. The anti-clockwise direction of the I_{sd}–V_g hysteresis loop indicates that the injection of the charges into traps in and on the surface of the gate dielectric is the cause of the hysteresis. The threshold voltage clearly depends on the starting voltage of each sweep and is very close to the latter up to high values of about 100 V as shown in the insert to Fig. 7. The hysteresis observed in our measurements is thus much wider than observed by other groups[26,27] who found that water molecules bound to both silicon oxide surface and carbon nanotubes themselves play the role of the charge traps.[28] We should note here that in our devices DNA molecules are chemisorbed at defect sites along the SWNT. The adsorbed molecules can serve as additional surface charge traps, which increase the total density of traps and, consequently lead to a wider hysteresis loop.

Stability of the device current under constant voltage would be desirable for any practical application of the observed hysteresis in a memory device. Fig. 8A presents the hysteresis loops measured with different delays between the voltage setting and the current measurement. It is clearly

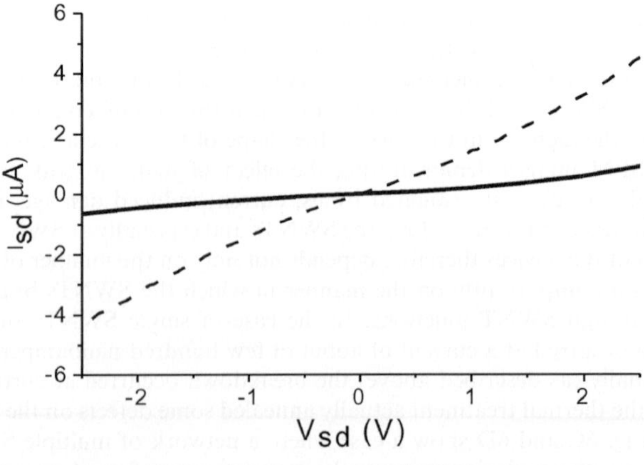

Fig. 5 The effect of mild thermal treatment (200 °C for 30 min) on the *I*–*V* curve. The current at a given voltage increases by a factor of about 20 (dashed line) following the treatment, however the shape of the curve does not change.

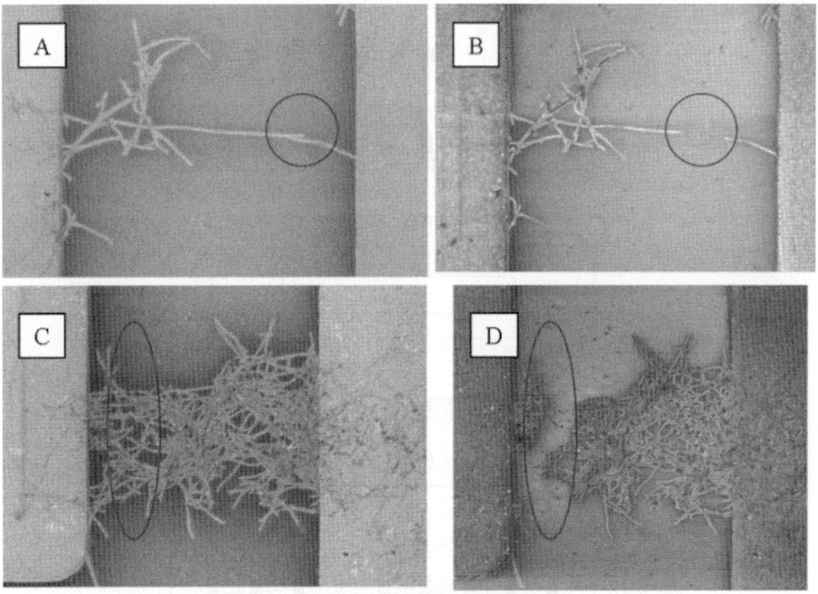

Fig. 6 SEM images demonstrating the effect of high currents on the SWNT based devices. The images of the devices before the high-current induced breakdown are shown on the left panels, while the same devices after the breakdown are shown on the right panels.

seen that for long enough delays the hysteresis almost disappears, *i.e.* the current relaxes to its equilibrium value after the charges are released from their traps. The similar effect is observed when the delays are made between the individual V_g points with no delay between the voltage setting and the current measurement (not shown). Fig. 8B shows several write-read cycles of our CNTFETs while switching the gate voltage between +20 V, 0 V and −20 V whereas Fig. 8C presents the variation of the device current on long-range scale while switching the gate voltage between +20 V and −20 V. These graphs confirm that the characteristic timescale of the trap discharge is in the range of hours, thus confining the action of the memory devices to this timescale. It is interesting to

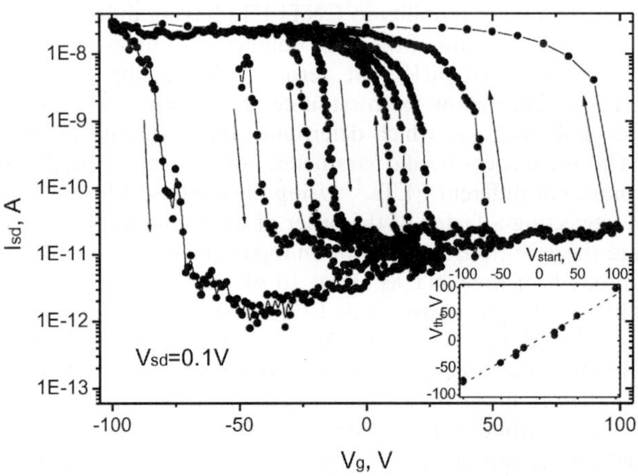

Fig. 7 Typical hysteresis curves measured for sweeps of the gate voltage in different ranges. The anticlockwise direction of the I_{sd}–V_g hysteresis loop indicates that the injection of the charges into traps in and on the surface of the gate dielectric is the cause of the hysteresis. In the insert: the dependence of the threshold voltage on the starting voltage of each sweep.

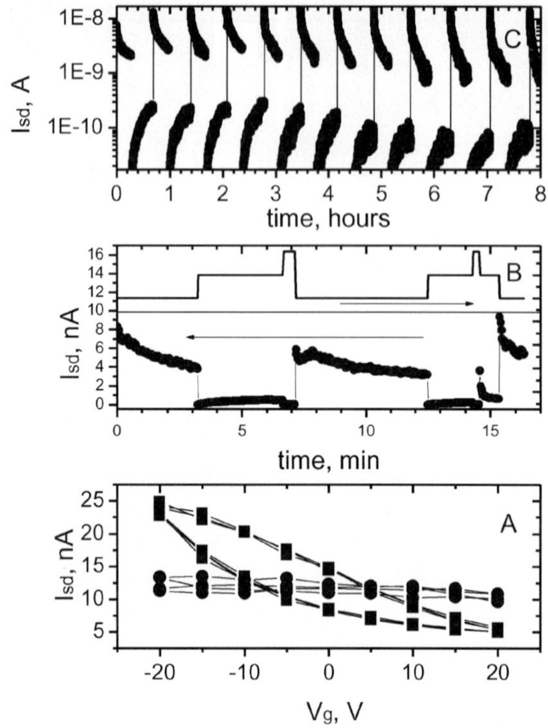

Fig. 8 (A) Hysteresis loops measured with different delays between the voltage setting and the current measurement. The squares correspond for measurements performed after a delay of 30 s, while the circles relate to measurements performed after a delay of 30 min. The source–drain potential is 0.1 V. B) Several write–read cycles while the gate voltage is switched between +20 V, 0 V and −20 V. V_{SD} is 0.1 V. C) The variation of the device current on long-range scale while switching the gate voltage between +20 V and −20 V. V_{SD} is 0.1 V.

note that the duration of the write pulse should also be long enough in order to produce measurable effects. After a 1 s pulse of the gate voltage to ± 20 V the I_{sd} remained the same as before the pulse (not shown).

In order to compare the carrier mobility in CNTFETs having different gate dielectric thickness we picked a few particular devices out of the large number of samples measured that exhibit similar differential conductance, $G = dI_{sd}/dV_{sd}$, at zero V_g. Extracting the derivative dG/dV_g and calculating the gate capacitance allow us to estimate an important technological characteristic of FETs, the field effect mobility, μ_{FE}, which determines the sensitivity of the device to variations of the gate voltage. This is a device-specific parameter, not a material-specific, and is usually used to compare the performance of different FETs.[25] Using the analytical formula for a cylinder over an infinite plane,[29] and neglecting a factor of the order of unity due to presence of the dielectric only between the CNT and the gate electrode but not on top of the nanotube,[30] allows us to estimate the capacitance per unit length of the CNT as about 10 pF m^{-1} for our 100 nm-thick gate-dielectric FETs and about 50 pF m^{-1} for the native-oxide FETs. The corresponding mobilities are, therefore, $\mu_{FE} \sim 25$ cm^2 V^{-1} s^{-1} and $\mu_{FE} \sim 35$ cm^2 V^{-1} s^{-1}, for thicker and thinner gate dielectrics, respectively. The field-effect mobility of our devices seems to depend only slightly on the dielectric thickness and is also slightly lower than values reported for back-gate CNTFET structures with SiO_2 gate dielectric.[31] We attribute both phenomena to the geometry of the devices, where CNT/Au interfaces are located on the top or on the edge of the metal contact pads. This leads to partial screening of the contact Schottky barriers, which determine the conduction in our devices, from the gate voltage.

This conclusion is further confirmed by the results obtained from measuring the inverse subthreshold slope $S = (d(\log I_{sd})/dV_g)^{-1}$. For channel-determined conduction, as in the case of

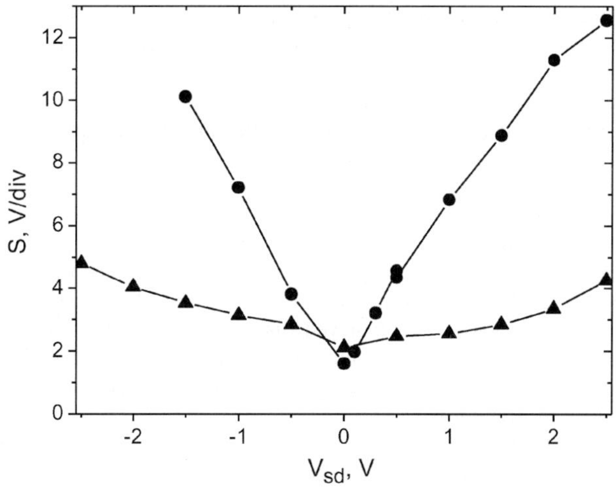

Fig. 9 Inverse subthreshold slope as a function of source–drain voltage. The circles correspond to the 100 nm-thick SiO_2 gate layer device; the triangles correspond to the native oxide device.

conventional long-channel MOSFETs, *S* depends linearly on the temperature but is independent of the source–drain voltage.[32] The temperature dependence of the inverse subthreshold slope of CNTFETs is reported to be very weak and is far from being linear, confirming the Schottky-barrier nature of the CNT conduction.[33,34] The variation of *S* with the source–drain voltage is shown on Fig. 9. It provides yet another indication that the conductance in CNTFETs is determined indeed by the interface Schottky barriers.

Discussion

Fabrication of devices from individual carbon nanotubes (SWNTs) involves technical difficulties stemming from uncontrolled chirality and length, and the tendency to bundle. Networks of SWNTs positioned between electrode pairs occupy more space yet are easier to obtain. Recent studies indicate that the electrical properties of devices fabricated from modified SWNTs are controlled by the Schottky barriers formed at the SWNT-electrode contacts.[35] The present study indicates that despite the structural non-uniformity of devices obtained, using the self-assembly method, the electrical properties of the devices are quite uniform. Typically for an electrical contact which is based on single (or few) SWNTs, the *I–V* curve is not linear and is characterized by an S-shape which may be the result of the Schottky barriers at the contacts. Based on the current measured at 1 V, the typical resistivity of the devices was found to be between 10^{-2} to 10^{-3} Ω cm, assuming that the diameter of the SWNT is 1 nm. Only when the number of SWNTs was very large, then linear *I–V* curve was obtained and the resistance decreased significantly.

It is interesting to realize that the DNA molecules do not contribute significantly to the shape of this barrier. This is proved by the fact that upon thermal treatment, although the resistance of the devices drops significantly, the shape of the *I–V* curve remains unchanged. This finding leads to the conclusion that the DNA molecules act as resistors, but do not change significantly the density of states of the SWNT and therefore do not affect the shape of the barrier at the metal–SWNT contact.

The devices could function for very long time as long as the current through them did not exceed the breakdown current. Despite the fact that the connection between the SWNT and the electrodes was done with a short double stranded DNA (typically 26 base-pair about 8.8 nm long), the DNA was not the limiting current component and devices were not burned at the SWNT-electrode contact. Since many DNA single strands are bound at the end of each SWNT, we estimate that up to ten DNA strands are involved in each contact made between the SWNT and the electrodes. Despite this parallel contact, the current density through each of the DNA contact is very high. This ability of the short double stranded DNA to carry relatively high current density, has been reported recently in several separate experiments.[36]

The thermal treatment caused a dramatic decrease in the resistivity. This can be caused by two effects, namely decomposition of the DNA at the contact, resulting in a better metal–SWNT contact, and annealing of defects on the SWNT. The fact that the breakdown current increases by two orders of magnitude, following the thermal treatment, is a strong indication that the second process is also taking place and that the uniformity of the SWNT improves upon thermal treatment.

The formation of a junction between SWNTs definitely causes an increase of the resistivity and increases the contribution of leakage through the substrate to the measured current. This may explain the diode like behavior shown in Fig. 4 and the current which is an order of magnitude smaller then the currents shown in Fig. 2. The higher resistance of such junctions also explains the current driven breakdown observed at these locations. Previous work on current saturation and electrical breakdown in multiwall carbon nanotubes showed that current induced breakdown resulted in substantial carbon loss.[37] It is believed that the breakdown process most likely involves excitation of high energy optical or zone boundary phonons leading to self heating and reaction with oxygen. The highly localized breakdown involving multiple tubes, shown in Fig. 6D, supports the role of local heating in breakdown.

The hystheresis effect in the *I–V* curve was demonstrated previously in devices fabricated from SWNTs and is attributed to charging of defect states in the vicinity of the nanotubes. In our devices, the effect of hystheresis is pronounced and the electric potential range in which hystheresis is observed is broader than reported before. However, the lifetime of the charge on the defects is shorter. These findings may be explained by assuming that the adsorbed DNA blocks the deep traps, yet increases the total number of traps. These effects may open the possibility to chemically modify and control the hystheresis properties.

Conclusions

The present work provides a method for producing SWNT based devices with high efficiency and proves that the transport properties of these devices are controlled by the chemical linker, namely the double stranded DNA. The ability to use biological active linkers opens the door towards interacting self-assembled devices with biological systems, enabling the fabrication of bio-sensors and bio-controlled electronic devices. We demonstrate the ability to produce CNTFETs with high yields, which in principle perform like those fabricated using direct metal–CNT contacts. We have shown that the inverse subthreshold slope of the CNTFETs depends on the source–drain voltage applied to the device, confirming that the conductance of the CNTFETs is determined by the Schottky barriers at the interfaces between the CNTs and the gold.

The results indicate that DNA-based electronics can be robust and the technology is reproducible. The ability to remove the biomolecules by heat treatment opens the possibility to use bio molecules in the self-assembly process and to end up with all inorganic system.

Acknowledgements

We deeply acknowledge Prof. Manfred Kappes and Dr Frank Hennrich for providing the oxidized carbon nanotubes. This work was partially supported by the Ministry of Science. MH acknowledges the support of the Ministry of Science through the Eshkol Fellowship.

References

1 V. Derycke, R. Martel, M. Radosavljevic, F. M. Ross and P. Avouris, *Nano Lett.*, 2002, **2**, 1043.
2 S. Orlanducci, V. Sessa, M. L. Terranova, M. Rossi and D. Manno, *Chem. Phys. Lett.*, 2003, **367**, 109.
3 S. M. Huang, X. Y. Cai and J. Liu, *J. Am. Chem. Soc.*, 2003, **125**, 5636.
4 L. Ki-Hong, C. Jeong-Min and W. Sigmund, *Appl. Phys. Lett.*, 2003, **82**, 448.
5 S. J. Wind, R. Martel and Ph. Avouris, *J. Vac. Sci. Technol. B*, 2002, **20**, 2745.
6 P. Mauron, C. Emmenegger, A. Zuttel, C. Nutzenadel, P. Sudan and L. Schlapbach, *Carbon*, 2002, **40**, 1339.
7 N. Bendiab, R. Almairac, J. L. Sauvajol, S. Rols and E. Elkaim, *J. Appl. Phys.*, 2003, **93**, 1769.
8 L. Jin, C. Bower and O. Zhou, *Appl. Phys. Lett.*, 1998, **73**, 1197.
9 Y. T. Jang, J. H. Ahn, B. K. Ju and Y. H. Lee, *Solid State Commun.*, 2003, **126**, 305.
10 T. Ono, E. Oesterschulze, G. Georgiev, A. Georgieva and R. Kassing, *Nanotechnology*, 2003, **14**, 37.

11 Y. Zhang, A. Chang, J. Cao, Q. Wang, W. Kim, Y. Li, N. Morris, E. Yenilmez, J. Kong and H. Dai, *Appl. Phys. Lett.*, 2001, **79**, 3155.
12 K. Yamamoto, S. Akita and Y. Nakayama, *J. Phys. D: Appl. Phys.*, 1998, **31**, L34.
13 R. Krupke, F. Hennrich, H. Weber, D. Beckmann, O. Hampe, S. Malik, M. Kappes and H. von Löhneysen, *Appl. Phys. A*, 2003, **76**, 397.
14 (*a*) J. E. Fischer, W. Zhou, J. Vavro, M. C. Llaguno, C. Guthy, R. Haggenmueller, M. J. Casavant, D. E. Walters and R. E. Smalley, *J. Appl. Phys.*, 2003, **93**, 2157; (*b*) B. W. Smith, Z. Benes, D. E. Luzzi, J. E. Fischer, D. A. Walters, M. J. Casavant, J. Schmidt and R. E. Smalley, *Appl. Phys. Lett.*, 2000, **77**, 663.
15 M. Fujiwara, E. Oki, M. Hamada, Y. Tanimoto, I. Mukouda and Y. Shimomura, *J. Phys. Chem. A*, 2001, **105**, 4383.
16 B. Kim and W. M. Sigmund, *Langmuir*, 2003, **19**, 4848.
17 X. Gao, T. o Hu, L. Liu and Z. Guo, *Chem. Phys. Lett.*, 2003, **370**, 661.
18 Z. Liu, Z. Shen, T. Zhu, S. Hou, L. Ying, Z. Shi and Z. Gu, *Langmuir*, 2000, **16**, 3569.
19 R. Krupke, F. Hennrich, H. von Löhneysen and M. M. Kappes, *Science*, 2003, **301**, 344.
20 H. Xin and A. T. Wooley, *J. Am. Chem. Soc.*, 2003, **125**, 8710.
21 K. Keren, R.S. Berman, E. Buchstab, U. Sivan and E. Braun, *Science*, 2003, **302**, 1380.
22 M. Hazani, F. Hennrich, M. Kappes, R. Naaman, D. Peled, V. Sidorov and D. Shvarts, *Chem. Phys. Lett.*, 2004, **391**, 389.
23 M. Stadermann, S. J. Papadakis, M. R. Falvo, J. Novak, E. Snow, Q. Fu, J. Liu, Y. Fridman, J. J. Boland, R. Superfine and S. Washburn, *Phys. Rev. B*, 2004, **69**, 201402.
24 M. Hazani, R. Naaman, F. Hennrich and M. M. Kappes, *Nano Lett.*, 2003, **3**, 153.
25 T. Durkop, B. M. Kim and M. S. Fuhrer, *J. Phys. Condens. Matter*, 2004, **16**, R553, and references therein.
26 M. S. Fuhrer, B. M. Kim, T. Durkop and T. Brintlinger, *Nano Lett.*, 2002, **2**, 755.
27 C. H. Lee, K. T. Kang, K. S. Park, M. S. Kim, H. S. Kim, H. G. Kim, J. E. Fischer and A. T. Johnson, *Jpn. J. Appl. Phys.*, 2003, **42**, 5392.
28 W. Kim, A. Javey, O. Vermesh, Q. Wang, Y. Li and H. Dai, *Nano Lett.*, 2003, **3**, 193.
29 See, for example, L. Yung-kuo, *Problems and Solutions on Electromagnetism*, World Scientific, Singapore, 1993.
30 T. Durkop, S. A. Getty, E. Cobas and M. S. Fuhrer, *Nano Lett.*, 2004, **4**, 35.
31 Ph. Avouris, R. Martel, V. Derycke and J. Appenzeller, *Phys. B*, 2002, **323**, 6.
32 H.-M. Mühlhoff and D. V. McCaughan, in *Device Physics*, ed. C. Hilsum, Elsevier, Amsterdam, 1993.
33 J. Appenzeller, J. Knoch, V. Derycke, R. Martel, S. Wind and Ph. Avouris, *Phys. Rev. Lett.*, 2002, **89**, 126801.
34 L. Marty, V. Bouchiat, C. Naud, M. Chaumont, T. Fournier and A. M. Bonnot, *Nano Lett.*, 2003, **3**, 1115.
35 M. Hazani, D. Shvarts, D. Peled, V. Sidorov and R. Naaman, *Appl. Phys. Lett.*, 2004, **85**, 5025.
36 (*a*) C. Nogues, S. R. Cohen, S. S. Daube and R. Naaman, *Phys. Chem. Chem. Phys.*, 2004, **6**, 4459; (*b*) S. M. Iqbal, G. Balasundaram, S. Ghosh, D. E. Bergstrom and R. Bashir, *Appl. Phys. Lett.*, 2005, **86**, 153901; (*c*) H. Cohen, C. Nogues, R. Naaman and D. Porath, *Proc. Natl. Acad. Sci. USA*, 2005, **102**, 11589.
37 P. G. Collins, M. Hersam, M. Arnold, R. Martel and Ph. Avouris, *Phys. Rev. Lett.*, 2001, **86**, 3128.

Strong electronic coupling between single C_{60} molecules and gold electrodes prepared by quench condensation at 4 K. A single molecule three terminal device study

Andrey V. Danilov,[ab] Sergey E. Kubatkin,[b] Sergey G. Kafanov[b] and Thomas Bjørnholm[ab]

Received 11th May 2005, Accepted 29th June 2005
First published as an Advance Article on the web 23rd September 2005
DOI: 10.1039/b506679b

We report the first measurements of single C_{60} molecules trapped in three terminal devices prepared by quench condensation of a gold source and drain electrode on top of an aluminium gate electrode covered with a thin oxide. Our experimental platform allows the source and drain electrodes to be fabricated on the gate oxide at low temperatures and in high vacuum. In a subsequent step, single molecules are evaporated *in situ* onto the surface and caught in the gap between a source and a drain electrode. This fabrication method ensures a clean contact between the molecule and the gold electrode due to the unbroken vacuum. Our measurements reveal a strong interaction between the C_{60} molecule and the gold electrodes resulting in the absence of the Coulomb blockade effects observed by others. In addition, we observe an insignificant gate dependence but a pronounced negative differential resistance (NDR) at bias voltages from 20–50 meV. The position of the peak in the NDR shows a pronounced and universal temperature dependence for all six devices included in the study. The results are related to previous measurements in such devices which focus on the detailed nature of the contact region between the molecule and the gold electrode.

Introduction

A number of recent experiments have revealed different facets of the properties of single molecules coupled to two[1,2] or three electrical terminals[3–6] both in the solid state[3–6] and solution based devices.[7,8] Contrary to *e.g.* standard electrochemical measurements performed on equilibrated molecules in electrolyte solution, the molecular scale proximity of two or three electrodes with a molecule fixed between them tends to dramatically change the properties of the molecule.[6]

The degree to which the electrode influences the electronic properties of the molecule is intimately connected to the detailed geometry of the molecule–metal contact. In one extreme case, characterized by a poor electronic coupling, one could experience Coulomb blockade behavior[3] due to the limited hybridization between the molecular states and the electrode surface states. In another extreme, characterized by good electronic coupling, one should expect strong hybridization resulting in ballistic transport through the molecule. These physically very different regimes may be separated by quite subtle changes in the detailed molecule electrode contact emphasizing the need to study well defined contact regions. One way to accomplish this is to form a chemical bond between the electrode and the molecule. Numerous examples of this strategy uses gold–thiol bonds as the covalent link (see *e.g.*, refs. 1, 2 and 7). The present paper focuses on a different strategy, which relies on *in situ* preparation of "clean" electrode–molecule interfaces in high vacuum and at low temperatures allowing simple physi- or chemisorption to the surface to be studied in a

[a] *Nano-Science Center, University of Copenhagen, Universitetsparken 5, DK-2100 Copenhagen, Denmark*
[b] *Department of Microtechnology and Nanoscience (MC2), Chalmers University of Technology, Fysikgrand 3, S-41296 Göteborg, Sweden*

controlled fashion. The first example of such studies involved an oligo phenylenevinylene (OPV) derivative which was physisorbed to the gold electrode surfaces leading to a small degree of hybridization and hence Coulomb blockade behaviour.[6] In this particular case, analysis of the experimental data shows that image charge effects localize the charges on the molecule.[6,9]

The present paper presents our first results on C_{60} measured in such devices, revealing very different behaviour compared to OPV and other reports on C_{60} in three terminal devices. We explained these data by a strong electronic coupling between the C_{60} molecule and the clean gold surface.

In the experiment by Park *et al.*[5] C_{60} was the first molecule to be used in a three terminal device consisting of a nanogap between gold electrodes on top of an electrostatic gate. Park *et al.* found nanomechanical oscillations of the fullerene molecule in the trapping potential of the metallic lead.[5] Charge transport through the fullerene molecule was of the Coulomb blockade type, as evidenced by a strong electrostatic gate modulation of the source–drain current. This first experiment has been successfully repeated by other groups with minor variations, for example, when the fullerenes, instead of being captured from solution while electromigrating, have been pre-deposited on gold before the formation of a break junction.[10] Recently, similar results have been reported in ref. 11, where the fullerene was affected both mechanically and electrostatically in a mechanical break-junction on a bending substrate. Again, all these experiments report the absence of current at low bias, and a gate modulation of the devices in agreement with Coulomb blockade physics.

A Coulomb blockade picture of the electron transport through a fullerene molecule in direct contact with a gold electrode contradicts, however, many theoretical and experimental results of surface physics and chemistry: The electron tunnelling from a cluster representing an STM tip through the fullerene molecule on gold (111) has been considered theoretically from first principles in ref. 12. Results of this paper are in good agreement with STM data,[13] where a finite zero-bias conductance above the fullerene have been measured, and it was found that the fullerene LUMO orbital is substantially broadened by interaction with gold, indicating strong coupling between the gold electrode and the fullerene molecule. An estimate for the low bias fullerene conductance made on the basis of theoretical calculations presented in ref. 12, is about 0.1 of quantum conductance (G_0). Recently, mechanically controlled break junction experiments in UHV[14] gave similar results, indicating that the conductance of a fullerene molecule on gold is about 0.1 G_0. It is worth noting that in these experiments, the fullerene was deposited by sublimation, and not from solution. This difference in sample fabrication apparently led to a better fullerene–gold contact and better fullerene transparency. Recent low temperature STM data taken in UHV[15] also do not support the Coulomb blockade scenario of transport through a fullerene molecule.

To summarize, the data taken in UHV environments show that fullerene molecules are strongly coupled to a clean gold surface, in agreement with first-principle calculations, whilst all experiments where the molecules were deposited from solution demonstrate a weak coupling resulting in Coulomb blockade behaviour. We suggest that this is a strong indication that immersing a sample into solution is a critical step where the gold surface is affected, influencing significantly the charge transport regime through the fullerene molecule. In this paper we describe fullerene-based devices prepared by quench condensation, as described in refs. 6 and 16, and argue that our results demonstrate a strong C_{60}-gold coupling.

Experimental

To fabricate the nanogaps at 4 K we deposited gold at an angle through a mask elevated above the aluminium oxide gate by placing an evaporator in the helium cryostat. The mask was Tailor-made by electron-beam lithography to allow accurate deposition of source and drain electrodes separated by a nanogap.[6,16] The tilt angle was changed until an appropriate gap was obtained (see ref. 16 for details). After initial gold deposition, a multigrain structure was often formed in the electrode gap (Fig. 1a) with a dI/dV plot typical for a single metallic grain situated between the source and drain electrodes (Fig. 1d). By measurements and analysis of such differential conductance plots we extracted the strength of the electrostatic coupling to the gate, and we were also able to determine how many grains were situated in the constriction. In a final step, the sample topology was reduced to a single gap geometry by a few extra gold depositions (Fig. 1b), as described in ref. 17. When a single gap topology was reached, the nanogap resistance was typically a few MΩ (Fig. 1e). By

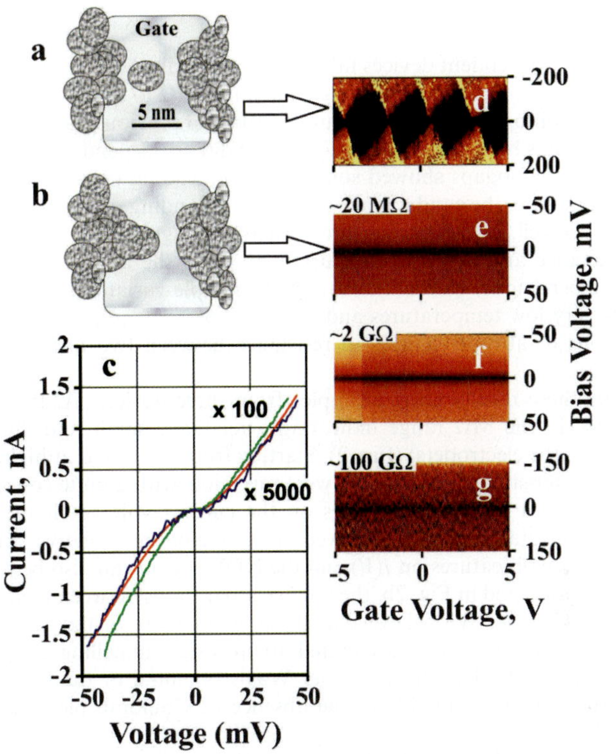

Fig. 1 Fabrication and transport properties of metallic nanogaps (T = 4.2 K). (a) Sample topology when a single grain is left between the source and drain electrodes. The corresponding differential conductance plot is shown in Fig. 1d. (b) Sample topology when the grain is attached to one of the leads; the absence of a gate dependence in differential conductance plots (Fig. 1e–g) confirms the single tunneling gap topology, free from parasitic tunneling through extra grains of gold. (c) $I(V)$ curves for the nanogaps before fullerene deposition. The red, green and blue curves correspond to plots of e, f and g, respectively. The green and blue curves are scaled 100- and 5000-times, respectively. (d) Differential conductance plot for a single gold grain between macroscopic electrodes, demonstrating a good coupling of electrostatic gate to the grain, $Cg/C_{\sum} \sim 0.2$. e. The same plot after deposition of additional 2 Å of gold, resulting in single gap topology with no gate modulation. (f) Differential conductance plot after annealing up to 100 K. The annealing resulted in the irreversible increase of the nanogap resistance, corresponding to a wider gap, while preserving perfect single-gap topology. (g) Differential conductance plot of a wider gap prepared by annealing up to 150 K. The noise on this plot is a noise from the current preamplifier (see also a blue curve from Fig. 1c).

annealing the substrate up to 100–150 K we were able to increase the gap resistance up to 2×10^{11} Ω which corresponds to a tunneling gap width of roughly 2 nm.[18] The transport properties of these nanogaps did not depend on the gate voltage as it should be for a single tunnelling gap (see examples in Fig. 1e, f, g). We also checked that the nanogaps are stable and can be reversibly thermocycled up to 70 K.

When the proper tunnelling gap was prepared, we deposited about 0.005–0.01 of a fullerene monolayer also at 4.2 K. At such a low surface coverage the probability for a fullerene molecule to land in the gap was very low; in fact we have only once observed a sharp nanogap conductance change during the fullerene deposition. In all other samples C_{60} was trapped when the sample was later annealed up to 30 K—the temperature which activates a surface diffusion of the fullerene molecule on aluminium oxide. At this temperature the current through the gap was monitored at the constant bias of 60 mV until a single conductance jump occurred. Immediately after such an event, assumed to be due to trapping of a single C_{60} molecule in the gap, we cooled the sample back to 4 K and studied the sample transport properties. All subsequent measurements were made at temperatures below 30 K.

Results

We have measured six independent devices fabricated as described above, by depositing fullerene in the nanogaps of different conductance, varying from a few MΩ to a few GΩ. None of these devices demonstrated the Coulomb blockade behaviour described in previous papers.[5,10,11] Instead, a finite zero-bias conductance has been measured, whose value did not depend on the gate voltage.[19] Since the initial width of our nanogaps showed some variation leading to low bias resistances covering several orders of magnitude, the conductance of the corresponding devices with fullerene trapped in the nanogap varied as well, also covering three orders of magnitude in conductance. Despite this variation in conductance and hence in the current through the device, all devices demonstrated qualitatively similar behaviour, with minor sample-to-sample variations. This is:

(1) Hysteresis at very low temperatures and

(2) Narrow regions of abrupt changes in resistance between two distinct values *e.g.* negative differential resistance.

We shall describe these results using examples from three devices; A, B, C. Device A showed a low bias resistance in the MΩ range indicating a relatively small tunnelling gap between the fullerene molecule and the electrode(s) (Fig. 2). Starting from 4.2 K and at higher temperatures, the current–voltage characteristics of this sample were reversible with a finite conductance at low bias. At higher bias voltages, a few small features on the current–voltage characteristic can be seen (each about 1–2% of total current), followed by a region of negative differential resistance (NDR) (Fig. 2a). The small features on $I(V)$ and the NDR region can also be seen in the derivative plot of the same data presented in Fig. 2b, the voltage separations between them varied from sample to sample from 5 to 11 meV. We believe that these features are of the same nature as the ones observed in ref. 5, where they were attributed to inelastic tunnelling processes involving the excitation of the fullerene oscillation in van der Waals potential of the metallic lead. Excitation of molecular vibrations is also possible in the absence of Coulomb blockade—see, for example refs. 20 and 21.

All samples demonstrated hysteretic behaviour at low enough temperatures. The type of hysteresis, however, varied from sample to sample. One type of hysteresis is shown in Fig. 2a for the device A (inset). The hysteresis of other samples with a larger initial gap was of a different type illustrated in Fig. 3 (device B, the data were taken with a voltage bias to ensure that the hysteresis is not a trivial circuit effect due to a combination of NDR and a finite bias resistor).

For all samples, the NDR region shifted to the higher biases at higher temperatures as seen for device A in Fig. 2 and for device C in Fig. 4 which presents dI/dV plots for this sample obtained at different temperatures. To quantify the temperature dependence we define V_C as a bias corresponding to dI/dV minimum.

The V_C is shown in Fig. 5 as a function of temperature for the four out of six devices that had regions with no hysteresis. Intriguingly, the temperature dependence of V_C, $V_C(T)$, seems to be universal: $V_C(T) = V_0 + \Delta V_C(T)$, where V_0 is a constant and $\Delta V_C(T)$ a universal function applicable for all samples and at all temperatures where the samples do not show hysteresis. This universality of the temperature dependence of V_C is very robust since the data shown in Fig. 5 belong to several samples prepared with different lithography masks patterned on different wafers and using gold from different suppliers.

We also note that there is no systematic correlation between V_0 and the device resistance shown in coloured numerals in Fig. 5 (the latter changes by several orders of magnitude). This experimental observation implies that the device current is not important for the processes leading to NDR, and the NDR is rather triggered by a certain bias voltage. The observation of hysteretic behaviour at very low temperature (below 4 K) indicates that there are two metastable states in a device at these low temperatures; one dominates when the energy of tunnelling electrons is below eV_C, and another one above the threshold energy. Finally we note that moderate sample annealing up to 20 K could change V_0 irreversibly.

Discussion

Voltage-induced switching between the two metastable positions, also leading to NDR, was also found in STM experiments on a pyrrolidine molecule.[22–24] The pyrrolidine molecule has two

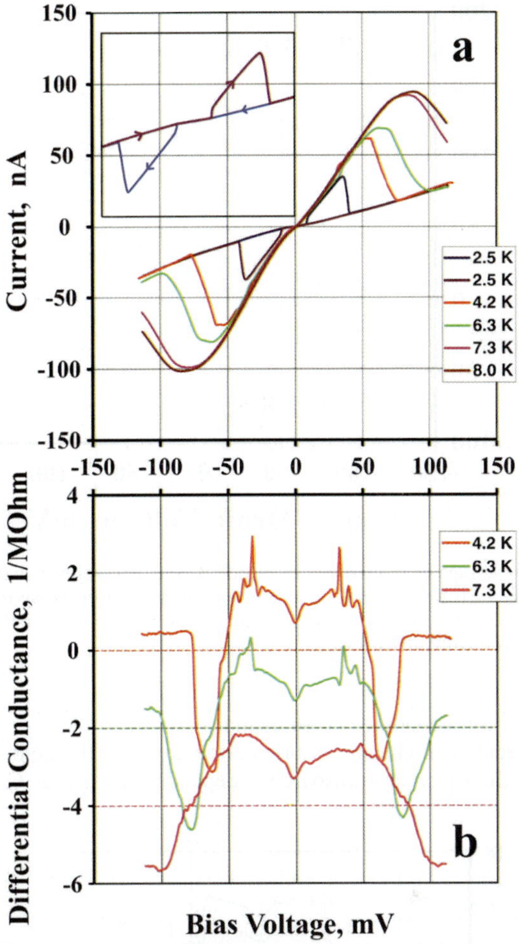

Fig. 2 Temperature dependence of transport properties of device A with C_{60} trapped in nanogap. (a) Set of current–voltage characteristics, measured at the indicated temperatures. At all temperatures, negative differential resistance is observed in a certain region of source–drain voltages; the voltage of the conductance drop shifts to higher bias with higher temperatures. At 2.5 K, the sample demonstrated hysteresis: the trace and retrace plots are schematically blown-up in the insert of this plot. Differential conductance of selected current–voltage characteristics from this plot is shown in Fig. 2b. (b) Differential conductance (dI/dV) plots of the same device (A) at 4.2 K, 6.3 K, and 7.3 K. Curves are shifted for clarity. Colour gridlines mark the corresponding zero-conductance level. At low bias, two features around ±5 mV can be seen; we attribute these to inelastic tunneling processes resulting in excitation of the fullerene motion in the potential resulting from the interaction with the metallic lead. Note, a few peaks preceding the dI/dV minima, we interpret these peaks as due to inelastic tunneling processes involving other fullerene vibrations.

conformations with different electron transmission. At low tip bias the molecule is in highly conductive conformation. At some critical bias the energy of tunnelling electrons exceeds the excitation energy of a C–H stretching vibration. This opens the possibility for the pyrrolidine molecule to change its conformation. The molecule then spends a fraction of time in a low conducting metastable state. This switching between two states with different conductance was observed on real time current traces taken above the critical bias for pyrrolidine. On the time-averaged $I(V)$ curve this process leads to an NDR as explained in ref. 24.

The fullerene molecule itself has no metastable conformations. But when placed on a gold surface, it does have a few known geometrical transitions which could play a similar role as the

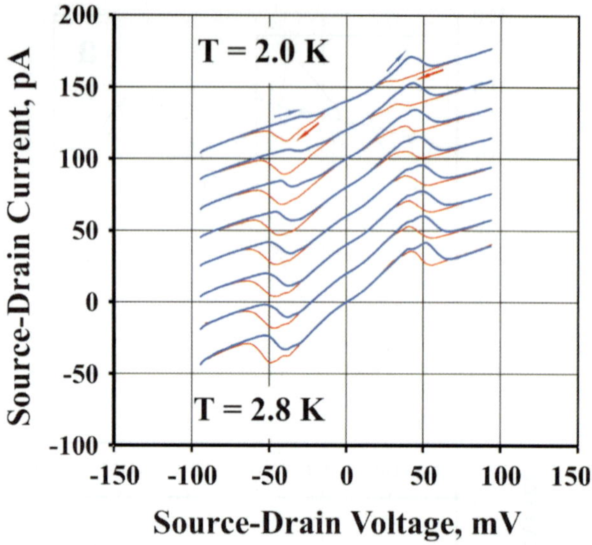

Fig. 3 Hysteretic current–voltage characteristics of device B. Current–voltage characteristics of device B are equidistantly spaced between 2.0 K and 2.8 K. The bias was swept from negative to positive voltages for blue traces and in the reverse direction for the red traces. The curves taken at 2.8 K are not displaced; the curves taken at lower temperatures are vertically offset for clarity.

conformational changes in pyrrolidine. These include:

1. The fullerene molecule can change its orientation towards the gold surface. STM data and calculations show that the fullerene transparency depends on whether C_{60} faces the gold with a

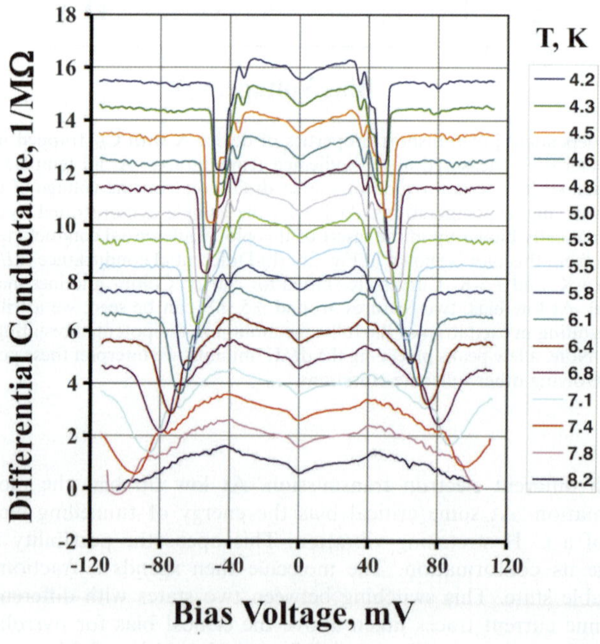

Fig. 4 Temperature dependence of differential conductance for device C. A set of dI/dV plots taken at different temperatures. The curves are offset vertically for clarity. The conductance plot for 8.2 K is not displaced. Bias voltages, corresponding to differential conductance minima (V_C) for this sample, are shown in red in Fig. 5.

23 J. Gaudioso and W. Ho, *Angew. Chem., Int. Ed.*, 2001, **40**, 4080.
24 W. Ho, *J. Chem. Phys.*, 2002, **117**, 11033.
25 J. K. Gimzevski, S. Modesti and R. R. Schlittler, *Phys. Rev. Lett.*, 1994, **72**, 1036.
26 M. Pedio, R. Felici, X. Torrelles, P. Rudolf, M. Capozi, J. Ruis and S. Ferrer, *Phys. Rev. Lett.*, 2000, **85**, 1040.
27 R. J. Baxter, P. Rudolf, G. Teobaldi and F. Zerbetto, *ChemPhysChem*, 2004, **5**, 245.
28 K. L. Ekinci and J. M. Valles, *Phys. Rev. B*, 1998, **58**, 7347.
29 S. Braig and K. Flensberg, *Phys. Rev. B*, 2003, **68**, 205324.
30 K. D. McCarthy, N. Prokof'ev and M. Tuominen, *Phys. Rev. B*, 2003, **67**, 245415.
31 Indeed, for a sample resistance 140 MΩ (Fig. 5 magenta) at $V_C = 50$ mV the tunnelling rate is about 2×10^9 electrons s^{-1}. Even for the lowest energy C_{60} vibration mode (5 meV) the relaxation time is a few picoseconds,[29] *i.e.* much shorter than the time between two tunnelling events. The switching starts when tunnelling electrons can excite, in a single shot, a high energy vibration mode (C–H stretching mode for pyrrolidine) with vibration quanta $\hbar w > \Delta E$. It is the decay of this excitation which flips the conformation. The tunnelling electron cannot switch the fullerene directly from the initial conformation to the final conformation because all the conformational changes listed above involve the fullerene displacement over a distance much higher than 3 pm—the amplitude of C_{60} zero-point fluctuations.[5] It means that there is no overlap between the wave functions corresponding to different conformations and the corresponding Franck–Condon factor is zero (see, for example, ref. 30). The process *must* be mediated by excitation of the intermediate state.
32 A. Danilov, S. Kubatkin, P. Hedegård, R. Shekter and T. Bjørnholm, in preparation.
33 Low—about 10^{-4} G_0—fullerene conductance was reported in a paper by C. Joachim, J. K. Gimzewski, R. R. Schlittler and C. Chavy, *Phys. Rev. Lett.*, 1995, **74**, 2102; this value of conductance, however, does not lead to the Coulomb blockade scenario of electron transport through fullerene.

Molecular three-terminal devices: fabrication and measurements

Herre S. J. van der Zant,*[a] Yann-Vai Kervennic,[a] Menno Poot,[a] Kevin O'Neill,[a] Zeger de Groot,[a] Jos M. Thijssen,[a] Hubert B. Heersche,[a] Nicolai Stuhr-Hansen,[b] Thomas Bjørnholm,[b] Daniel Vanmaekelbergh,[c] Cornelis A. van Walree[d] and Leonardus W. Jenneskens[d]

Received 3rd May 2005, Accepted 30th June 2005
First published as an Advance Article on the web 30th September 2005
DOI: 10.1039/b506240n

Incorporation of a third, gate electrode in the device geometry of molecular junctions is necessary to identify the transport mechanism. At present, the most popular technique to fabricate three-terminal molecular devices makes use of electromigration. Although it is a statistical process, we show that control over the gap resistance can be obtained. A detailed analysis of the current–voltage characteristics of gaps without molecules, however, shows that they reveal features that can mistakenly be attributed to molecular transport. This observation raises questions about which gaps with molecules can be disregarded and which not. We show that electrical characteristics can be controlled by the rational design of the molecular bridge and that vibrational modes probed by electrical transport are of potential interest as molecular fingerprints.

Introduction

Control of the molecular conductance by a third electrode is not only a necessary requirement for integration of molecular devices[1] in circuits, but also essential for a thorough study of transmolecular conduction. To date, three-terminal molecular devices have been fabricated by several approaches mostly based on lithographic techniques. These include electromigration[2] in which a metal wire is broken with an electric field, shadow evaporation[3] and electrochemical techniques.[4–7] In all these cases, nanoscale gaps are produced in metallic wires that are bridged by molecules, sitting on top of a gate electrode separated by an oxide layer. The gate electrode electrostatically modifies the potential of the molecular orbitals independently from the bias voltage. With the gate, the molecule can be oxidized or reduced and tuned between nonconducting and conducting states.[8–12]

The extent to which the orbital levels can be shifted up and down is controlled by the gate coupling parameter. This parameter is related to the gain of the device and in Coulomb theory it equals the ratio of the gate capacitance to the total capacitance of the islands. In an experiment, it should be close to unity in order to access different charge states. The geometry plays an important role in the gate coupling and one should take care that the electrodes themselves do not screen the gate potential. For example, when using electro-deposition to create nanogaps, three-dimensional growth usually occurs and structures may be created that are effective screens (see Fig. 1a).

[a] *Kavli Institute of Nanoscience, Delft University of Technology, PO Box 5046, 2600 GA, The Netherlands. E-mail: herre@qt.tn.tudelft.nl*
[b] *Nano-Science Center & Department of Chemistry, University of Copenhagen, Universitetsparken 5, DK-2100 Ø, Denmark*
[c] *Condensed Matter and Interfaces, Debye Institute, Utrecht University, Princetonplein 1, 3584 CC Utrecht, The Netherlands*
[d] *Organic Chemistry and Catalysis, Debye Institute, Utrecht University, Padualaan 8, 3584 CH Utrecht, The Netherlands*

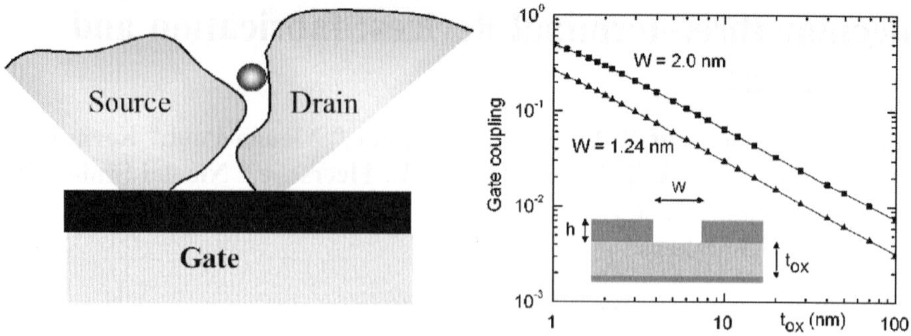

Fig. 1 Left: For the nano-object trapped in between two metallic electrodes fabricated for example by three-dimensional electro-deposition growth, the two leads can be effective screens against the gate. Right: gate coupling as a function of oxide thickness for two different gap sizes W. Calculation performed by solving Laplace's equation for the geometry shown in the inset with $\varepsilon_r = 9$ and $h = 0.6$ nm.

The molecule–electrode distance and the break-through voltage of the gate oxide are other important parameters. Currently, two gate materials are frequently used: heavily doped silicon substrates with thermally grown SiO_x on top and aluminium strips with a native oxide of only a few nanometers. In Fig. 1b, we plot the gate-coupling factor calculated from Laplace's equation for a 1.24 and 2 nm wide junction as a function of the gate oxide thickness. For large thicknesses the dependence is $\sim 1/t_{ox}$. For the aluminium gates with $\varepsilon_r = 9$, the gate coupling is about 0.1 so that with a typical break voltage of 4 V at low temperatures, the potential of the molecular levels can be shifted by ± 0.4 eV. For a silicon device ($\varepsilon_r \approx 3$) with an oxide thickness of 250 nm, extrapolation yields a gate coupling of about 10^{-3}. With a typical breakthrough voltage of 100 V, the range over which the potential on the molecule can be varied equals ± 0.1 eV. Here, we have not corrected for the different relative permeability of the two oxides. Calculations show that this is a small effect.

Wire fabrication

We have developed two techniques to create nanogaps on top of aluminium gates: electro-chemical etching and electromigration. The former method uses a combination of e-beam lithography and electrochemical etching. The approach allows the creation of gaps with subnanometer precision by *in situ* monitoring of the gap conductance during etching. Details of this technique are documented elsewhere.[7]

Electromigration is currently a widely adapted technique to fabricate molecular junctions and is reported to be a statistical method with poor control over the gap size.[2] We have extensively studied this electromigration method and found that some control can be obtained. We will now outline our method in more detail starting with device fabrication. In short, it consists of four lithographic steps. In the first step, contact pads and wide wires are defined by evaporation of 3 nm Ti, 35 nm Au and 15 nm AuPd. Secondly, the aluminium gate electrode is defined. The one-micron wide electrode is 50 nm thick with a sticking layer of 3 nm Ti. When taken out of the vacuum chamber of the evaporator, the aluminium oxidizes in air. The native oxide is typically 3 nm thick. The third step involves the definition of the thin and narrow gold wire that will be broken. A 10–12 nm thick and typically 100 nm wide gold wire without sticking layer is evaporated on top of the aluminium gate. In the last step, the thin wire is connected to the coarse pads by evaporation of 110 nm of AuPd.

Electromigration: Gap fabrication

The electrical circuit that is used to break the wire is shown in Fig. 2. It consists of a voltage source with a resistor in series, R_s. The current through the gold wire increases by ramping the voltage of the source; the current is determined by measuring the voltage across the series resistance. Electromigration-induced breaking[13] in gold is dominated by momentum transfer from charge carriers to the atoms, which is directed opposite to the electric field. The corresponding force is proportional to the current density and breaking occurs above a certain threshold value.

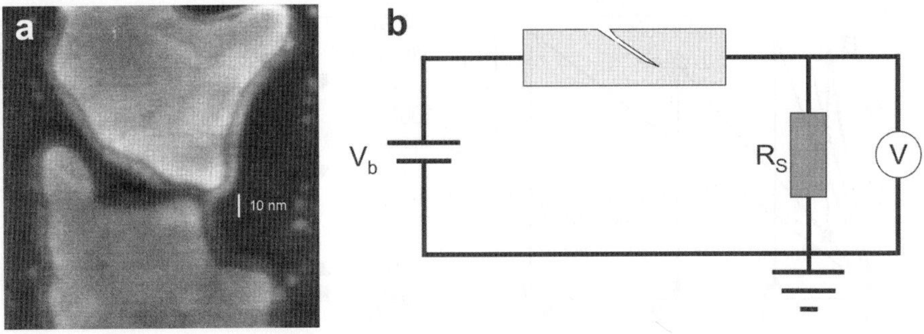

Fig. 2 Left: Scanning electron microscopy (SEM) picture of an electromigrated gold wire on top of an aluminium gate. The gap is smaller than a few nanometers. Right: Electrical circuit used to break the thin metallic wires.

Experiments performed at 4.2 K demonstrate that the ratio between the series resistance and the resistance of the gold wiring on the chip plays a crucial role (see Fig. 3). Wires with a resistance of 80 Ω break completely when using R_s = 1 kΩ. With R_s = 10 Ω, they break only partially as illustrated in the left hand side of Fig. 3. The wire first breaks to a plateau of 1 kΩ and only after increasing the voltage to 0.6 V, the wire breaks completely.

To investigate the dependence of the zero-bias gap resistance on R_s, 26 Au wires on the same chip have been broken with different R_s. The resistance of the gold wires before breaking was approximately 50 Ω. Six wires are broken with R_s = 10 Ω, six with R_s = 50 Ω, six with R_s = 100 Ω, four with R_s = 300 Ω and four with R_s = 500 Ω. Immediately after breaking the wire, the voltage has been swept to zero. Fig. 4 (left) shows the break curves for the five different values of R_s. Higher values of R_s yield higher break voltages, while the current at which breaking occurs remains approximately constant as expected for electromigration. The zero-bias resistance of each broken wire is plotted in the figure on the right hand side of Fig. 4. A clear dependence is observed: the higher the series resistance in the electrical circuit, the larger the gap resistance.

The difference in breaking can be understood as follows. The circuit with a high R_s compared to the bridge resistance is equivalent to that of a current source in which the current is kept constant. When increasing the bias voltage, the junction cross section decreases and the current density increases thereby accelerating the electromigration process. Before breaking most of the voltage is across R_s; during breaking this voltage is transferred to the junction. For low series resistances on the other hand, the voltage drop is mainly across the junction during the whole breaking process (connection to a voltage source)—this results in a less powerful breaking compared to the high-series-resistance case.

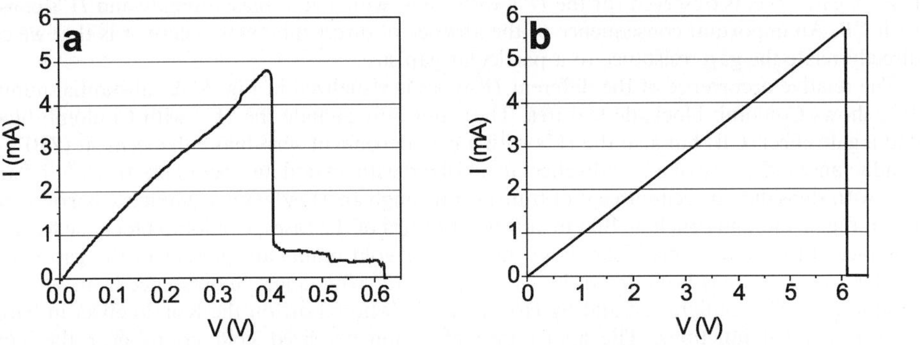

Fig. 3 Break curves for two different series resistances. The current, determined by measuring the voltage across R_s, as a function of the voltage that is ramped linearly in time. Left (a): Break curve for R_s = 10 Ω. The wire first breaks to a plateau of 1 kΩ before it breaks completely. Right (b): Break curve for R_s = 1 kΩ. The wire breaks completely.

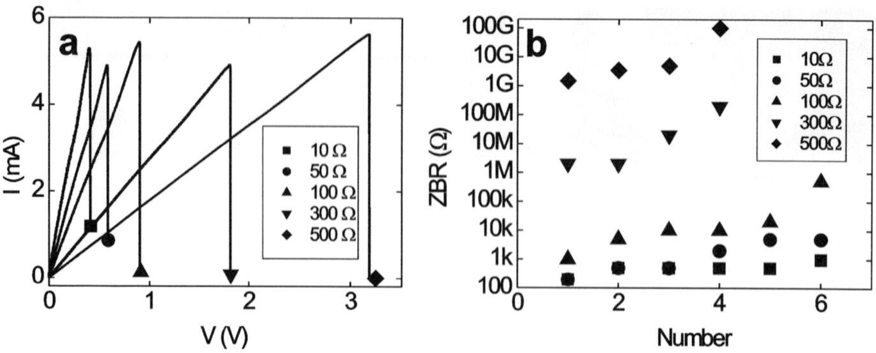

Fig. 4 Twenty six Au wires broken with five different R_s. The higher the series resistance, the higher the zero-bias resistance of the gap. Left (a): Representative break curves for different R_s, each represented with a different symbol. Right (b): Zero-bias resistance of each wire. Different symbols correspond to different R_s (see legend). The values within one set are ordered from left (low) to right (high).

A further improvement can be obtained by using an active feedback loop to monitor the resistance during breaking. While the electromigration process is inherently unpredictable, the catastrophic event that forms the void typically occurs in a timescale of ~100 μs—enough time to actively monitor the breaking process and adjust the voltage that is driving it. Different algorithms may be implemented to do this.[14,15] We ramp the voltage until a decrease in the absolute conductance is observed upon which the applied voltage is returned to zero and the loop is repeated until a desired resistance has been reached. Compared with simply ramping the voltage to a fixed limit, the yield of devices made with controlled electromigration is greatly increased from about 50% to more than 90%.

Current–voltage characteristics

To characterize the bare gold nanogaps, we have measured their current–voltage (*IV*) characteristics. A surprising result is that the *IV*'s curves generally do not follow the dependence expected from direct, through-space tunneling with its characteristic sinh(V) form.[18] We have collected the *IV*'s in seven different categories: (1) *IV*'s that do not show a current (<1 pA for $V = 100$ mV) indicating that the gap distance is a few nanometers or larger, (2) *IV*'s that are linear, (3) *IV*'s that show a suppression (ZBS) of the current at low bias[16] (see Fig. 5a), (4) smooth asymmetric *IV*'s (see Fig. 5b), (5) *IV*'s with steps (see Fig. 5c), (6) *IV*'s that are consistent with Coulomb blockade[17] (CB) (see Fig. 5d,e), and (7) *IV*'s with a zero-bias enhancement (ZBE) of the conductance around zero-bias. A gate effect is observed for the *IV*'s with steps, with a zero-bias anomaly and *IV*'s consistent with CB. An important consequence of the absence of direct tunneling behavior is that we cannot directly relate the gap resistance to a particular gap size.

The relative occurrence of the different *IV*-types is visualized in Fig. 6. A substantial number of *IV*'s shows Coulomb blockade features. These not only include the *IV*'s with Coulomb blockade and a gate effect (CB) but also the *IV*'s with a gate-dependent zero-bias enhancement (ZBE) of the conductance indicative of a Kondo effect in an island with an odd number of electrons,[9,10,12] and the *IV*'s with steps that are reminiscent of transport through an island with asymmetric barriers (Steps). From Fig. 6, one can conclude that in at least one third of the cases, Coulomb blockade features are observed. Most likely, small islands in the form of gold grains are present in the gap region. A similar conclusion has been drawn by Consalez *et al.*[19] who detect fluorescence from small grains produced by electromigration and by Houck *et al.*[15] who report on the Kondo effect in bare gold electro-migrated junctions. The appearance of nanometer-sized gold grains near the gap is a problem if one wants to distinguish transport through molecules from that through grains: the energies to add or withdraw an electron may be of the same order in both cases.

We have also investigated the breaking process with wires that contain a self-assembled layer of OPV-3 molecules ((*E*,*E*)-1,4-Bis[4-(acetylthio)styryl]benzene; see Fig. 6). Prior to breaking, the

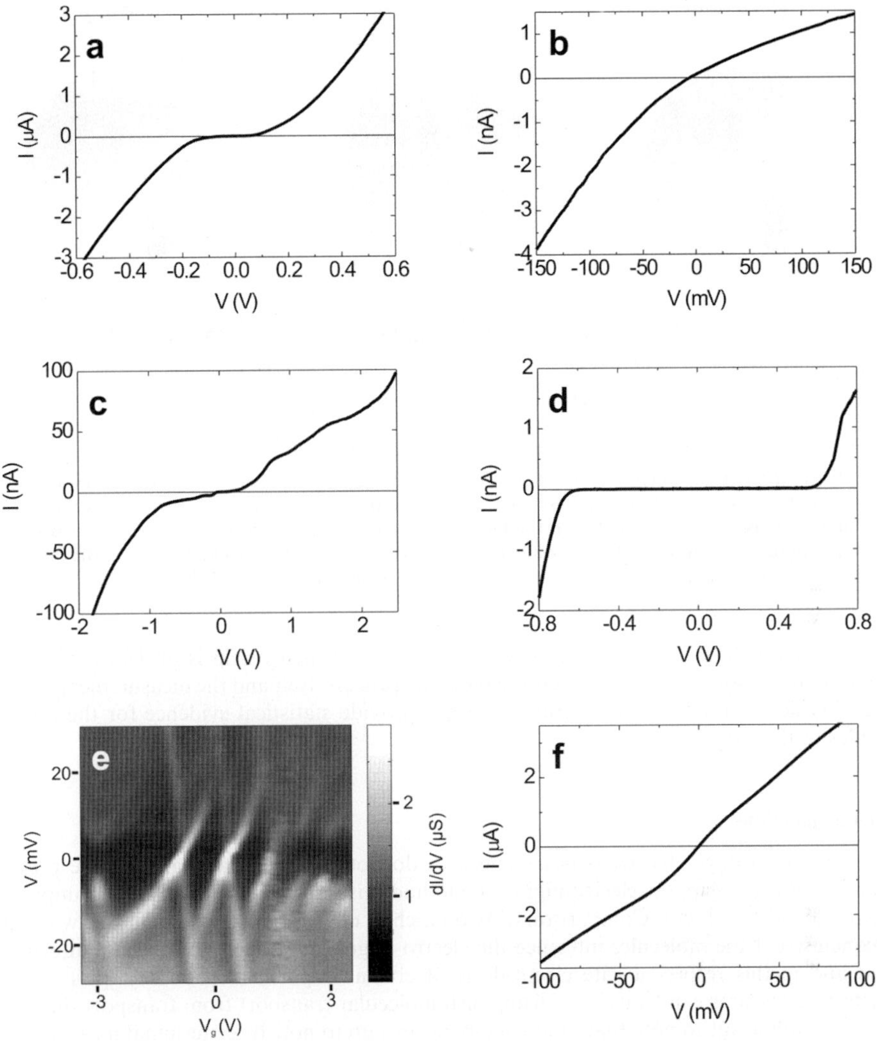

Fig. 5 Different types of current–voltage characteristics of bare gold nanogaps measured at 4.2 K: (a) *IV*'s with a zero-bias suppression (ZBS) of the current, (b) asymmetric, smooth *IV*'s, (c) *IV*'s with steps, (d) *IV*'s consistent with Coulomb blockade and (f) with a zero-bias enhancement of the conductance. In (e) a stability diagram of a sample with Coulomb blockade is shown. Plotted is $\mathrm{d}I/\mathrm{d}V$ as a function of gate and bias voltage. In the black regions, the current is blocked (Coulomb diamond). Each black diamond-like structure corresponds to a different charge state.

wires are dipped in a 1 mM solution of OPV-3 in chloroform. Upon binding to gold, the acetate group falls off and the sulfur atom ensures a good bonding. After 12 h, the sample with typically 24 individual wires is rinsed with chloroform and blow-dried. Then it is transferred to the measuring set-up and cooled to 4.2 K, where the wires are broken. As a control measurement, we have measured nanogaps that were only dipped in the solvent for 12 h. Fig. 6 summarizes the result. Samples are all from the same fabrication run; measurements on other batches—not included—show similar behavior.

In all three types of junctions, in about one third of the cases no current is detected (<1 pA for $V = 100$ mV). In about one third of the cases, Coulomb blockade features (Steps, ZBE and CB) are seen. Surprisingly, samples that have been exposed to chloroform show more *IV*'s with a ZBE, but less *IV*'s with pure CB features. Another significant difference is that the samples with OPV-3 show

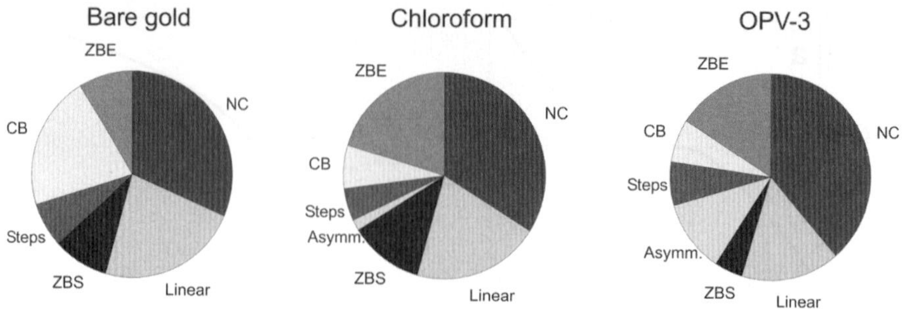

	Total	*NC*	*Linear*	*ZBS*	*Asymm.*	*Steps*	*CB*	*ZBE*
Bare gold	52	18	13	5	0	4	12	5
Chloroform	47	20	12	7	1	3	4	12
OPV-3	74	34	14	4	10	6	6	14

Fig. 6 Distribution of the different types of current–voltage characteristics measured on three different types of samples from the same fabrication run: bare gold wires, wires that have been exposed to the solvent for 12 h and wires that are exposed to the solvent with the OPV-3 molecules. NC stands for "no current", ZBS stands for "zero-bias suppression", Asymm. for "asymmetric *IV*", CB for "Coulomb blockade", and ZBE for "zero-bias enhancement". The table lists the number of junctions corresponding to each of the cases.

more asymmetric *IV*'s. These *IV*'s show no gate effect and their origin is unclear at the moment. From the comparison, we conclude that the breaking process itself and the measurement of current–voltage characteristics at low temperature do not provide statistical evidence for the presence of molecules in the gap.

Molecular junctions

As the current–voltage characteristics themselves do not provide clear signs for the presence of molecules inside the gap, the electromigration method raises questions about which samples can be disregarded and which not. Comparison of two batches, one with molecules and one without, may be inconclusive if the molecules influence the electro-migration process. Additional information is needed and in this respect a gate electrode is of crucial importance. The key question is what characteristic features can be used to distinguish molecular transport from transport through gold grains. It is important to note that at low temperatures up to now three-terminal measurements on molecular junctions have all shown Coulomb blockade features[8–12] including the Kondo effect.[9,10,12] Our measurements presented in the previous section, however, show that Coulomb blockade itself is not a distinctive feature. In the remainder of this paper, we present two measurements with different molecules that may provide information about the presence of molecules in the gap.

Molecular junctions with oligo(cyclohexylidene)

The sulfur end-functionalized oligo(cyclohexylidenes) molecules consist of cyclohexane rings[20] connected by double bonds. These are rigid molecules in which the sulfur-end atom in the cyclohexane rings binds strongly to gold atoms, leading to ordered self-assembled monolayers. Photo-induced charge transfer in Au/spacer/CdSe quantum dots assemblies[21] with the sulfur end-functionalized oligo(cyclohexylidenes) as molecular spacers has been measured at room temperature. Transport has been related to a favorable intramolecular interaction between the sulfur functionalities and the olefinic π orbitals, which are mediated by the σ C–C orbital system (through-bond interactions). Photo-electron spectroscopy[22] and photo-induced-electron transfer experiments[23,24] have confirmed the occurrence of transannular electronic interactions mediated *via* the σ–π–σ hydrocarbon skeleton.

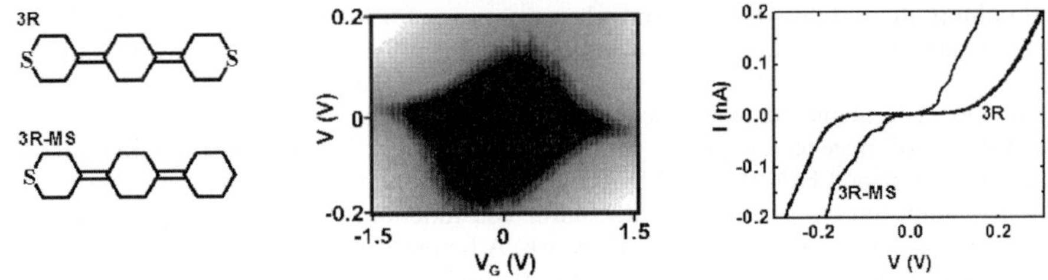

Fig. 7 Left: 3R molecules with two and one sulfur end groups. Middle: Stability diagram of a 3R sample measured at 8 K. Plotted is dI/dV as a function of bias and gate voltage. The conductance ranges from zero (dark) to 10 nS (light). Right: Two current–voltage characteristics measured at 10 K. The asymmetric junction with one sulfur end group (3R-MS) shows steps; the junction with the 3R molecule bonded to both electrodes by sulfur atoms exhibits a smooth curve.

Three-terminal devices were constructed with the technique that combines e-beam lithography and electrochemical etching.[7] Three different molecules were used: Molecules with two rings sulfur-functionalized at both ends, and three rings with either one sulfur or two sulfur termini. Here, we discuss results on junctions with the mono-sulfur three ring compound (3R-MS) and the three ring compound (3R) with two sulfur end atoms (see Fig. 7). It is important to note that junctions made with the same molecules share common features in their temperature dependence and gate effect.[25] No samples have been disregarded in contrast to the electromigration method. Furthermore, in the electrochemical method, the molecules are introduced after gap fabrication and the gap resistance before and after can be compared. In all cases a dramatic decrease of gap conductance has been observed. Control experiments show that the resistance decrease is due to the molecules.[7]

At room temperature, current–voltage (*IV*) characteristics of the molecular junctions are slightly nonlinear. The zero-bias room-temperature resistance varies from sample to sample between a few MΩ to 100 MΩ. One 3R-MS junction had a resistance of 500 MΩ. These differences may reflect variations in bonding strengths or in molecule–lead geometries, *i.e.*, in the exact chemisorption of the molecule on the gold. Another factor may be the number of molecules; we cannot be sure that only a single molecule is bound between the gold leads.

At low temperatures current–voltage characteristics are strongly nonlinear. Surprisingly, we consistently find that the *IV*'s from the junctions with the asymmetric, 3R-MS molecules (3 studied) show steps whereas the junction with the symmetric 3R (5 studied of which 3 had the gate connected) do not. Fig. 7 shows an example. For the mono-sulfur three ring compound 3R-MS, the sulfur-functionalized site is well coupled to the gold electrode while the other side is—most likely—only weakly coupled through a van der Waals interaction. The set-up resembles the one of a scanning tunneling microscopy (STM) experiment in which the two barriers (tip–object and object–substrate) are very asymmetric. In STM experiments steps in the *IV* are generally observed and identified as Coulomb staircases.[26] In our case, steps may also be due to vibrational modes of the molecules (see also discussion next section).

When changing the voltage on the gate, a strong influence on the current is observed for both types of molecular junctions. For the asymmetric junction, we observe shifts in the step position but switches are often observed making a detailed analysis over the full gate range (± 2V) difficult. In contrast, gate dependent measurements on the symmetric junctions are stable. In Fig. 7 we plot the numerically obtained derivative dI/dV as a function of gate and bias voltage. The resulting stability diagram comprises more than one hundred curves. In the black region in the middle, current is below the measuring accuracy of one pA. From the diamond-like structure in Fig. 7 the value of the gate-coupling—estimated as half the diamond height divided by its width—equals about 0.1 in agreement with the calculated value of Fig. 1. Two other samples show similar stability diagrams as the one in Fig. 7 with similar gate coupling values.

The differential conductance plot does not show the features that are commonly observed in weakly coupled molecular junctions.[8–12] To be more specific, a surprising feature of the data in Fig. 7 is the absence of degeneracy points along the gate axis for low bias.[17] The crossing of a degeneracy point corresponds to the appearance of a new charge state. For sharp levels, this crossing is a well-defined point in the stability diagram and when measuring for low bias the current as a function of gate, sharp current peaks occur. In our experiment on the other hand, we observe a steady increase of the current as the gate voltage is increased. Our data are therefore incompatible with the usual Coulomb blockade theory for weakly coupled systems.

The absence of degeneracy points in our junctions may be explained by the presence of substantially broadened HOMO and LUMO levels[27] due to a molecule–gold hybridization. In this respect, it is important to realize that the sulfur end-functionalities in the oligocyclohexylidenes interact effectively through the σ–π–σ framework. A favorable interaction between the sulfur lone pairs and the olefinic double bond exists.[22] The situation is different from that in aromatic thiols or sulfides, in which the sulfur lone pairs are more distant from the aromatic levels. The sulfur and aromatic moieties may therefore be effectively more decoupled leading to the formation of higher barriers at the molecule–lead contact and hence a weaker coupling.

Molecular junctions with OPV-3

With electromigration, junctions with the π-conjugated OPV-3 molecule are fabricated as outlined in the section on current–voltage characteristics. Thiol groups at both ends of this molecule (see Fig. 6) ensure a good bonding to the gold electrodes. In Fig. 8a, the differential conductance as a function of gate and bias voltage is shown (stability diagram). The plot shows one crossing point at a gate voltage of approximately −0.5 V associated with the separation of two charge states that differ by one electron. As the plot shows no indications of diamond edges that close again, the energy to add an additional electron (for both charge states) is well above 200 meV, the maximum bias applied in Fig. 8a.

The stability diagram shows a bright excitation running parallel to the diamond edges. It does not extend all the way to the diamond edge. Extrapolation to the diamond edge gives an energy of 50 meV for this excitation; its origin, however is unclear. The stability diagram also shows two black lines, symmetrically located around the $V = 0$ axis. Their color is black as they exhibit negative differential conductance as illustrated in Fig. 8b, in which a dI/dV curve taken at a gate voltage of −2 V is shown. On the left hand side of the plot the black lines appear at ±150 mV in the region where current is blocked and they are therefore due to an inelastic co-tunnel process.[28] In the case of an electronic excitation, cotunnel lines are expected to transfer into lines that run parallel to the diamond edges as they enter the region where current flows. In Fig. 8a this expected behavior is indicated by the dashed lines. Clearly, the data in Fig. 8 show a completely different pattern that does not follow straightforwardly from the usual Coulomb blockade theory.

Recently, Kushmerick *et al.*[29] have measured current–voltage characteristics of OPV-3 molecules in parallel in a crossed wire geometry. With inelastic tunneling spectroscopy (IETS), they have

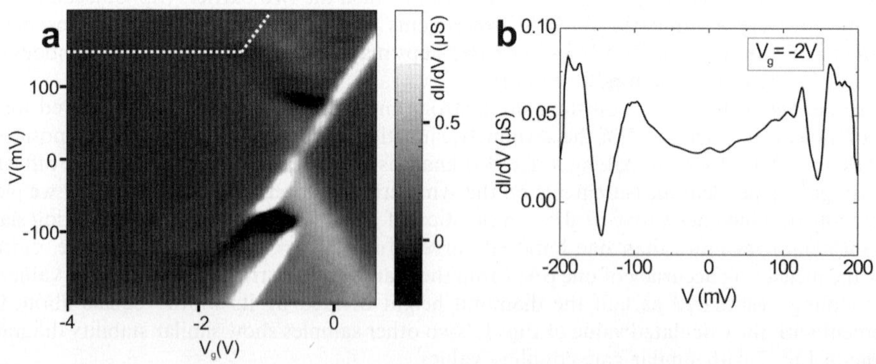

Fig. 8 (a) Large-bias stability diagram of a molecular junction with OPV-3 measured at 2 K. The plot shows one diamond crossing at a gate voltage of −0.5 V and at the left hand side two black horizontal lines whose position changes when reaching the Coulomb diamond edge. The dashed line indicates the expected behavior for an electronic excitation. (b) Differential conductance *vs.* bias voltage measured at a gate voltage of −2 V showing dips at approximately ±150 meV.

studied vibrational modes of OPV-3 molecules. Features in the second derivative are observed at 96, 134, 144, 174 and 195 mV. The mode at 144 mV is ascribed to a vibration of the benzene ring.

The question now arises: Can we assign the 150 meV feature we observe in the stability diagram to the same vibrational mode? This cannot easily be answered. Firstly, it should be emphasized that only one out of the 75 junctions with OPV-3 studied showed a stability diagram with the large addition energies and the 150 meV feature that are present in Fig. 8b. A second issue concerns the vibrational induced line shapes. In inelastic tunneling spectroscopy (IETS) one usually plots[29–32] the second derivative of the *IV* curves. Peaks then correspond to the location of the vibrational modes. It has theoretically been shown that also dips in the second derivative of the current–voltage characteristic may appear.[33] In Coulomb physics one expects a peak in the dI/dV as soon as a new electronic level enters the bias window. In the cotunnel regime, a step instead of a peak appears. For vibrational modes, the same general features are expected but electron–phonon coupling in movable dots is presently a field of intense study. For example, recent theoretical studies have shown that vibrational excitations may exhibit negative differential resistance when nonequilibrium processes associated with a strong electron–phonon coupling are present.[34] A more detailed understanding of the line shape is needed to be able to assign features in the current–voltage characteristics to vibrational modes of the molecule. Such studies should include the dependence of the vibrational energy on the gate voltage[32] and possibly on the charge state and the environment.

Conclusions

There are several approaches to contact single molecules. Each of them has its advantages and disadvantages. To determine the transport mechanism of molecular junctions, measurements with a gate provide crucial information. Planar technologies such as electromigration and electrochemical methods are usually adopted, but recently gates have also been incorporated in mechanical break junctions[35] and electrochemical gates have been used for single molecule transport at room temperature.[36]

With the planar techniques, it is very difficult to distinguish molecular transport from transport through small gold grains or other imperfections near the gap region. Although impressive improvement has been made in recent years, it is still not clear what the criteria are to determine molecular transport with a high degree of certainty. Inelastic-tunneling spectroscopy is an important tool that may be used as a molecular fingerprint, but a better understanding is necessary. A second strategy is to use molecule functionality. We have shown that the lead–molecule coupling can be used to make this distinction. Other routes include molecules with magnetic properties different from those expected for gold grains and light-sensitive molecules that switch conformation back and forth upon radiation with different wavelengths.

It is important to agree on which "molecular" fingerprints in the (I, V, V_G) characteristics must be observed before any conclusion can be made on the nature of the junction. Lastly, it would also be very helpful if measurements on molecular junctions fabricated with different techniques are compared with each other as well as with theoretical calculations.

Acknowledgements

We thank Edgar Osorio for discussions. Financial support is obtained from the Dutch Organization for Fundamental Research on Matter (FOM), the Danish Nanotechnology Program and by the EC through the FP6 program (contract no. FP6-2004-IST-003673). This publication reflects the views of the authors and not necessarily those of the EC. The Community is not liable for any use that may be made of the information contained herein. K. O'Neill acknowledges support of the EC under a Marie Curie Host for incoming International Fellowships.

References

1 C. Joachim, J. K. Gimzewski and A. Aviram, *Nature*, 2000, **408**, 541.
2 H. Park, A. K. L. Lim, A. P. Alivisatos, J. Park and P. L. McEuen, *Appl. Phys. Lett.*, 1999, **75**, 301.
3 (*a*) K. Liu, Ph. Avouris, J. Bucchignano, R. Martel and S. Sun, *Appl. Phys. Lett.*, 2002, **80**, 865; (*b*) M. S. M. Saifullah, T. Ondarçuhu, D. K. Koltsov, C. Joachim and M. E. Welland, *Nanotechnology*, 2002, **13**, 659.
4 C. Z. Li and N. J. Tao, *Appl. Phys. Lett.*, 1998, **72**, 894.

5 A. F. Morpurgo, C. M. Marcus and D. B. Robinson, *Appl. Phys. Lett.*, 1999, **74**, 2084.
6 Y. V. Kervennic, H. S. J. van der Zant, A. F. Morpurgo, L. Gurevich and L. P. Kouwenhoven, *Appl. Phys. Lett.*, 2002, **80**, 321.
7 Y. V. Kervennic, D. Vanmakelbergh, L. P. Kouwenhoven and H. S. J. van der Zant, *Appl. Phys. Lett.*, 2003, **83**, 3782.
8 H. Park, J. Park, A. K. L. Lim, E. H. Anderson, A. P. Allivisatos and P. L. McEuen, *Nature*, 2000, **407**, 57.
9 J. Park, A. N. Pasupathy, J. I. Goldsmith, C. Chang, Y. Yaish, J. R. Petta, M. Rinkoski, J. P. Sethna, H. D. Abruña, P. L. McEuen and D. C. Ralph, *Nature*, 2002, **417**, 722.
10 W. Liang, M. P. Shores, M. Bockrath, J. R. Long and H. Park, *Nature*, 2002, **417**, 725.
11 S. Kubatkin, A. Danilov, M. Hjort, J. Cornil, J.-L. A. L. Bredas, N. Stuhr-Hansen, P. Hedegård and T. Bjørnholm, *Nature*, 2003, **425**, 698.
12 L. H. Yu and D. Natelson, *Nano Lett.*, 2004, **4**, 79.
13 (*a*) P. S. Ho and T. Kwok, *Rep. Prog. Phys.*, 1989, **52**, 301; (*b*) M. Mahadevan and R. M. Bradley, *Phys. Rev. B*, 1999, **59**, 11037.
14 D. R. Strachan, D. E. Smith, D. E. Johnston, T.-H. Park, M. J. Therien, D. A. Bonneel and A. T. Johnson, *Appl. Phys. Lett.*, 2005, **86**, 43109.
15 A. A. Houck, J. Labaziewicz, E. K. Chan, J. A. Folk and I. L. Chang, *Kondo effect in electromigrated break junctions, Cond-mat*, 2004, 410752.
16 The *IV*'s with a linear behavior at high current/voltage are consistent with Environmental Coulomb Blockade (see *e.g.* K. Flensberg., S. M. Girvin, M. Jonson, D. R. Penn and M. D. Stiles, *Phys. Scr., T*, 1992, **42**, 189) but more measurements (temperature dependence) are needed to draw definite conclusions.
17 See *e.g.*, L. P. Kouwenhoven, D. G. Austing and S. Tarucha, *Rep. Prog. Phys.*, 2001, **64**, 701.
18 J. G. Simmons, *J. Appl. Phys.*, 1963, **34**, 1793.
19 J. Gonzales, T.-H. Lee, M. D. Barnes, Y. Antoku and R. M. Dickson, *Phys. Rev. Lett.*, 2004, **93**, 147402.
20 F. J. Hoogesteger, R. W. A. Havenith, J. W. Zwikker, L. W. Jenneskens, H. Kooijman, N. Veldman and A. L. Spek, *J. Org. Chem.*, 1995, **60**, 4375.
21 E. P. A. M. Bakkers, A. W. Marsman, L. W. Jenneskens and D. Vanmaekelbergh, *Angew. Chem., Int. Ed.*, 2000, **39**, 2297.
22 A. W. Marsman, R. W. A. Havenith, S. Bethke, L. W. Jenneskens, R. Gleiter, J. H. van Lenthe, M. Lutz and A. L. Spek, *J. Org. Chem.*, 2000, **65**, 4584.
23 F. J. Hoogesteger, C. A. van Walree, L. W. Jenneskens, M. R. Roest, J. W. Verhoeven, W. Schuddeboom, J. J. Piet and J. M. Warman, *Chem. – Eur. J.*, 2000, **6**, 2948.
24 W. D. Oosterbaan, C. Koper, T. W. Braam, F. J. Hoogesteger, J. J. Piet, B. A. J. Jansen, C. A. van Walree, H. J. van Ramesdonk, M. Goes, J. W. Verhoeven, W. Schuddeboom, J. M. Warman and L. W. Jenneskens, *J. Phys. Chem. A*, 2003, **107**, 3612.
25 A comparison between the two and three ring molecular junctions can be found in: Y.-V. Kervennic, J. M. Thijssen, D. Vanmaekelbergh, R. Dabirian, C. A. van Walree, L. W. Jenneskens and H. S. J. van der Zant, *J. Am. Chem. Soc.*, submitted.
26 See *e.g.*, E. S. Soldatov, V. V. Khanin, A. S. Trifonov, D. E. Presnov, S. A. Yakovenko, G. B. Khomutov, C. P. Gubin and V. V. Kolesov, *JETP Lett.*, 1996, **64**, 556.
27 The gate measurements suggest that the gold Fermi energy is approximately in the middle of the HOMO–LUMO gap. It is emphasized that in both cases only the tails of the levels are probed. The *IV* curves presented in this paper do not show this step, indicating that we only probe the rise of the current.
28 See *e.g.*, S. De Franceschi, S. Sasaki, J. M. Elzerman, W. G. van der Wiel, S. Taracha and L. P. Kouwenhoven, *Phys. Rev. Lett.*, 2001, **86**, 878.
29 J. G. Kusmerick, J. Lazorcik, C. H. Patterson, R. Shashidhar, D. S. Seferos and C. G. Bazan, *Nano Lett.*, 2004, **4**, 639.
30 W. Wang, T. Lee, I. Kretzschmar and M. A. Reed, *Nano Lett.*, 2004, **4**, 643.
31 R. H. M. Smit, Y. Noat, C. Untiedt, N. D. Lang, M. C. van Hemert and J. M. van Ruitenbeek, *Nature*, 2002, **419**, 906.
32 L. H. Yu, Z. K. Keane, J. W. Ciszek, L. Cheng, M. P. Stewart, J. M. Tour and D. Natelson, *Phys. Rev. Lett.*, 2004, **93**, 266802.
33 M. Galperin, M. A. Ratner and A. Nitzan, *J. Chem. Phys.*, 2004, **121**, 11965.
34 J. Koch and F. van Oppen, *Phys. Rev. Lett.*, 2005, **94**, 206804.
35 A. R. Champagne, A. N. Pasupathy and D. C. Ralph, *Nano Lett.*, 2005, **5**, 305.
36 B. Xu, X. Xia, X. Yang, L. Zang and N. Tao, *J. Am. Chem. Soc.*, 2005, **127**, 2386.

Polaronic semiconductor behavior of long-range charge transfer in DNA oligomers in solution: controlling barriers to long-distance radical cation migration in DNA with thymine analogs†

Abraham Joy,[a] Gozde Guler,[a] Shahadat Ahmed,[b] Larry W. McLaughlin[b] and Gary B. Schuster*[a]

Received 20th April 2005, Accepted 20th May 2005
First published as an Advance Article on the web 21st September 2005
DOI: 10.1039/b505550d

A series of anthraquinone-linked DNA oligonucleotides was prepared and the efficiency of long-distance radical cation migration was measured. In one set of oligonucleotides, two GG steps are separated by either a TATA or an ATAT bridge. In these two compounds, the efficiency of radical cation migration from GG to GG differs by more than an order of magnitude. Replacement of the thymines in the TATA or ATAT bridges with 3-methyl-2-pyridone (*t*, a thymine analog) results in the much more efficient radical cation migration across the bridge in both cases. This is attributed to a decrease in the oxidation potential of ***t*** to a value below that of A. In contrast, replacement of the thymines in the TATA or ATAT bridges with difluorotoluene (*f*, a thymine analog with high oxidation potential) does not measurably affect radical cation migration. These findings are readily accommodated by the phonon-assisted polaron-hopping mechanism for long-distance charge transfer in duplex DNA and indicate that DNA in solution behaves as a polaronic semiconductor.

Introduction

DNA in solution has been described as a "molecular wire".[1,2] One challenge in assessing this claim has been uncertainty in the precise definition of a molecular wire. At one extreme, the term is limited to those specific cases of a conductor having an electronic transmission of unity that exhibits the Landauer resistance of $h/2e^2$ (12.9 kΩ).[3] A more liberal characterization of molecular wires identifies them as having bridges of many bond lengths through which charge moves rapidly and efficiently.[4] However, the problematic nature of such a qualitative definition has been noted,[5] and this difficulty has plagued discussion of the properties of DNA.

It is apparent now that consideration of the electrical properties of DNA requires a division into at least two main groups: "dry" and wet DNA. A careful assessment of the available data led to the explicit conclusion that DNA in the dry state is an insulator.[6] In contrast, long-distance charge transport through DNA oligonucleotides in appropriate aqueous buffer solutions has been observed and confirmed in several laboratories.[7,8] These findings show that radical cations (holes) migrate in the DNA. However, these studies show that DNA oligomers in solution are not

[a] *School of Chemistry and Biochemistry, Georgia Institute of Technology, Atlanta, Georgia 30332, USA. E-mail: gary.schuster@cos.gatech.edu*
[b] *Department of Chemistry, Merkert Chemistry Center, Boston College, Chestnut Hill, Massachusetts, 02467*

† Electronic supplementary information (ESI) available: Autoradiogram from the irradiation of DNA duplexes (5) and (7). CD spectra of DNA duplexes used in this study. A description of a minor difference from previously published synthetic procedure. See http://www.rsc.org/suppdata/fd/b5/b505550d/

molecular wires or conductors in any sense. The behavior of DNA places it in the class of polaronic semiconductors. Over very short distances, holes may tunnel from guanine to guanine through bridges composed of A–T base pairs,[9,10] but over longer distances charge transfer occurs by the ion-gated phonon-assisted polaron-like hopping mechanism.[11–14]

We have suggested that radical cations in DNA are stabilized by delocalization over some number of contiguous purine bases (forming a polaron) and that migration occurs when thermal motions (phonons) of the DNA and its solvent and ionic environment provide sufficient energy to overcome activation barriers formed around interposing pyrimidines.[11,12,15,16] Thus, the rate of radical cation hopping is determined by the sequence of base pairs that controls both the extent of delocalization and the nature of the barrier. A recent report by Rösch and co-workers supports this view.[17] Using quantum mechanics/molecular dynamics modeling, they show that structural and environmental fluctuations facilitate the transfer of a hole from a guanine to an adenine. The relative energies of adenine and guanine radical cations are modulated by fluctuations of the electrostatic interaction between DNA and its electrolyte environment thereby forming charge transfer effective configurations[12] where a radical cation at an A has a lower energy than at a neighboring G. The formation of these configurations is thought to be the rate determining step that gates hole transfer from one polaron to the next. Recently, O'Neill and Barton[18] proposed an alternative to the ion-gated phonon-assisted polaron-hopping model that they identify as conformationally gated hopping through stacked domains. However, this proposal is part of the polaron hopping mechanism[19] and indistinguishable from it when intermediates and transitions states are considered properly and it is recalled that activation energy has both enthalpy and entropy as components.

A key postulate of the ion-gated phonon-assisted polaron hopping mechanism is the creation of barriers to charge transfer by pyrimidines, particularly thymine, between polarons. The oxidation potentials (E_{ox}) of C and T are both significantly greater than that of A, which is greater than for G.[20,21]A radical cation encountering an adjacent cytosine as part of a G–C base pair can migrate to the complementary guanine, but there is no especially low oxidation potential site for the radical cation when it encounters an adjacent thymine that is part of an A–T base pair. We examined the postulate that intervening thymines create barriers to charge migration by construction of DNA oligomers that contain the thymine mimics 3-methyl-2-pyridone[22] and difluorotoluene[23] (***t*** and ***f***, see Fig. 1). These unnatural nucleobases maintain the stacking characteristic of the A–T base pair but provide different electronic and solvent environments. Investigation of long-distance charge transfer in these modified oligomers confirms the central role that thymine plays in the creation of barriers to radical cation migration in duplex DNA. In addition, the findings indicate that radical cations migrate to the complementary adenine when a T–A barrier is encountered.

Experimental

The 2,4-difluorotoluene phosphoramidite was purchased from Glen Research, Sterling, VA and used as received. The protected, activated deoxyribose 3-methyl-2-pyridone nucleoside {2-[2-(4-nitrophenyl)ethyl]-5-[5′-*O*-(4,4′-dimethoxytrityl)]-1′,2′-dideoxy-3′-*O*-[(2-cyanoethoxy)diisopropylaminophosphino]-β-D-erythro-pentofuranosyl)-3-methylpyridine: ^{1}H NMR ($CDCl_3$): δ = 8.10 (2H, d, ArH), 7.94 (1H, s, ArH), 7.56–6.82 (16H, m, ArH), 5.08 (1H, m, $H_{1'}$), 4.60 (2H, t, CH_2), 4.58 (1H, m, $H_{3'}$), 4.22 (1H, s, $H_{4'}$), 3.78 (6H, s, OCH_3), 3.58–3.75 (4H, m, OCH_2CH_2CN), 3.24 (2H, d, $H_{5'}$), 3.20 (2H, t, CH_2), 2.14 (2H, m, $H_{2'}$), 2.00 (3H, s, CH_3), 1.21–1.23 (12 H, m, iPr)} was synthesized using the published procedure with minor modifications (see supplementary information).[22] The DNA oligomers were synthesized on a standard Applied Biosystems Expedite Nucleic Acid Synthesis system. A modified protocol was used for deprotecting the oligomers containing *p*-nitrophenethyl protected 3-methyl-2-pyridone. This protecting group was removed as the first step after sequence assembly by treating the solid support with a solution of 40% TEA/anhydrous pyridine for 2 h. The support was then treated with a solution of 0.5 M 1,8-diazabicyclo[5.4.0]undec-7-ene (DBU) in anhydrous pyridine for 48 h (the oligomer is still on the support after this step) and finally by washing the sample three times with acetonitrile. The wash was followed by ammonium hydroxide treatment in the conventional fashion to cleave the oligomer from the

DNA(1)	TTTC **GG$_1$** TATA **GG$_2$**CATC TTG*-5′
	AQ-5′AAAG CC ATAT CC GTAG AAC-3′
DNA(2)	TTTC **GG$_1$** ***t*A*t*A** **GG$_2$**CATC TTG*-5′
	AQ-5′AAAG CC ATAT CC GTAG AAC-3′
DNA(3)	TTTC **GG$_1$** ATAT **GG$_2$**CATC TTG*-5′
	AQ-5′AAAG CC TATA CC GTAG AAC-3′
DNA(4)	TTTC **GG$_1$** A***t***A***t*** **GG$_2$**CATC TTG*-5′
	AQ-5′AAAG CC TATA CC GTAG AAC-3′
DNA(5)	TTTA **GG$_1$** TATA **GG$_2$** TATA **GG$_3$** TATA **GG$_4$** TACG*-5′
	AQ-5′AAAT CC ATAT CC ATAT CC ATAT CC ATGC-3′
DNA(6)	TTTA **GG$_1$*t*A*t*A** **GG$_2$** TATA **GG$_3$** TATA **GG$_4$** TACG*-5′
	AQ-5′AAAT CC ATAT CC ATAT CC ATAT CC ATGC-3′
DNA(7)	TTTA **GG$_1$*f*A*f*A** **GG$_2$** TATA **GG$_3$** TATA **GG$_4$** TACG*-5′
	AQ-5′AAAT CC ATAT CC ATAT CC ATAT CC ATGC-3′
DNA(8)	AAAT **GG$_1$** ATAT **GG$_2$** ATAT **GG$_3$** ATAT **GG$_4$** ATCG*-5′
	AQ-5′TTTA CC TATA CC TATA CC TATA CC TAGC-3′
DNA(9)	AAAT **GG$_1$** A***t***A***t*** **GG$_2$** ATAT **GG$_3$** ATAT **GG$_4$** ATCG*-5′
	AQ-5′TTTA CC TATA CC TATA CC TATA CC TAGC-3′
DNA(10)	AAAT **GG$_1$** A***f***A***f*** **GG$_2$** ATAT **GG$_3$** ATAT **GG$_4$** ATCG*-5′
	AQ-5′TTTA CC TATA CC TATA CC TATA CC TAGC-3′

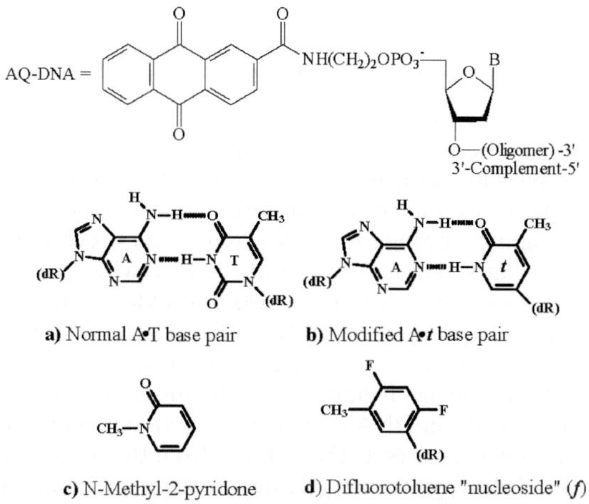

Fig. 1 Structures of DNA oligomers.

support and to deprotect the remaining groups. The oligomers were purified on a Hitachi preparative HPLC system equipped with a Dynamax C-18 reversed phase column. All DNA samples were analyzed by mass spectrometry and by UV melting and cooling curves. [γ-^{32}P] radioactive isotopes were purchased from Amersham Biosciences. T4 polynucleotide kinase was purchased from New England Biolabs.

The samples for irradiation were prepared by hybridizing a mixture of the unlabeled (5 μM) and radiolabeled (10 000 cpm) oligonucleotides with the complementary strands (5 μM) in 10 mM sodium phosphate buffer (pH 7). Samples were irradiated at 350 nm at room temperature in microcentrifuge tubes in a Rayonet photoreactor (Southern New England Ultraviolet Company, Barnsford, CT). The irradiation times, 4 min with four lamps for DNA(1–4), or 4 min with two lamps for DNA(5-8), were shown to be satisfactory for maintenance of single-hit conditions. After irradiation, the samples were precipitated with cold ethanol (100 μL) in the presence of glycogen (1.25 μL, 20 mg mL^{-1}), washed with 80% ethanol (2 × 100 μL), dried and treated with piperidine (50 μL, 1 M) at 90 °C for 30 min. After evaporation of the piperidine, lyophilization and dissolution in formamide-dye solution, the samples were run on a 20% 19 : 1 acrylamide : bis-acrylamide gel containing urea (7 M) at 70 W. The gels were dried and the cleavage sites were visualized by autoradiography. Quantification of the cleavage bands was performed on a Fuji phosphorimager.

Table 1

DNA	T_m/°C[a]	GG_1/GG_{tot}[b]	GG_2/GG_{tot}	GG_3/GG_{tot}	$(GG_1 + GG_2)/GG_3$
1	48	0.82	0.18		
2	42	0.48	0.52		
3	49	0.98	0.02		
4	42	0.33	0.67		
5	50	0.95	0.05	0.005	200
6	44	0.50	0.47	0.03	30
7	44	0.95	0.05	0.003	
8	48	0.96	0.04	<0.005[c]	>200
9	42	0.50	0.48	0.02	50
10	40	0.95	0.04	0.01	

[a] Melting temperature of the DNA duplex. [b] Fraction of reaction that occurs at GG_n, errors are typically 10% of the reported value. [c] Reaction at GG_3 is not significantly above background.

Results

(1) Radical cation migration across 3′–TATA–5′ and 3′–ATAT–5′ bridges

The DNA oligomers shown in Fig. 1 were prepared, purified and characterized by standard procedures. Each contains an anthraquinone group (AQ) linked covalently to a 5′-terminus and a ^{32}P-radiolabel (* in Fig. 1). We have shown previously that UV irradiation of AQ at 350 nm[24] results in one-electron oxidation of an adjacent nucleobase that injects a radical cation into the DNA.[25] The radical cation migrates through the DNA and reacts irreversibly with H_2O or O_2 at GG steps resulting in damage that is detected as strand cleavage when the irradiated samples are treated with piperidine.[19] These experiments were carried out under "single-hit" (low conversion) conditions. In this circumstance, on average each DNA oligomer reacts once or not at all, and the number of duplexes that contain two or more damaged bases is insignificant. Under these conditions the amount of strand cleavage is linked with the probability that a radical cation will encounter a particular GG step. Thus, in DNA oligomers that contain more than one GG step, the amount of strand cleavage usually decreases as the distance from the AQ to the GG step increases.

The relative amount of strand cleavage at each GG step in oligomers containing several GG steps is determined by the magnitudes of the rate constant for charge migration (k_{hop}) and for irreversible trapping of the radical cation (k_{trap}), the latter of which is largely independent of the identity of the bases around the GG step.[16] If hopping is much slower than trapping, all of the strand cleavage will occur at the GG step that is closest to the AQ. On the other hand, if hopping is much faster than trapping, then the amount of strand cleavage is determined by thermodynamic equilibration of the radical cation among all of the available GG steps, and the amount of reaction is independent of distance from the AQ. For cases where the hopping and trapping rates are comparable, the amount of strand cleavage will decrease approximately exponentially with distance from the AQ.[26]

The DNA duplexes shown in Fig. 1 were characterized spectroscopically and by analysis of their thermal stability. Each of these duplexes exhibits a circular dichroism (CD) spectrum that is characteristic of B-form DNA[27] (see electronic supplementary information (ESI)†). The melting temperatures (T_m) of these compounds decrease when ***t*** or ***f*** is substituted for T, see Table 1. For ***t***, this has been attributed to the absence of the O2 carbonyl group in the minor groove and the concomitant destabilization of the spine of hydration.[28] Similarly, the decrease in T_m observed when ***f*** replaces T is attributed to uncompensated desolvation of the complementary adenine during base pair formation.[29] X-ray crystallography has shown that the conformation of the A–T base pair is not affected by replacement of T with ***t***.[30] Similarly, NMR spectroscopy shows that the ***f***–A base pair very closely resembles the T–A base pair in the same context.[31] These findings indicate that substitution of ***t*** or ***f*** for T does not significantly alter the local or the global structure of the modified oligomers.

As expected, irradiation of a DNA(1) sample to low conversion and its subsequent treatment with piperidine leads to cleavage of the DNA strand at GG_1 and at GG_2. Gel electrophoresis and

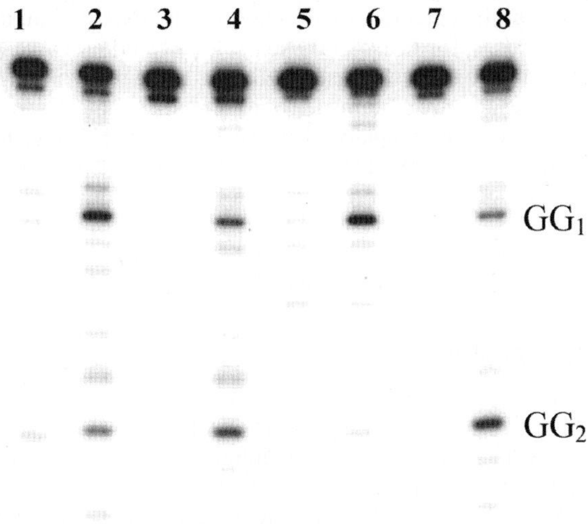

Fig. 2 Autoradiogram resulting from the irradiation of DNA(1) (lanes 1,2), DNA(2) (lanes 3,4), DNA(3) (lanes 5,6) and DNA(4) (lanes 7,8). In lanes 2,4,6,8, the samples were irradiated at 350 nm (pH 7, phosphate buffer) and worked up with hot piperidine (1 M). Lanes 1,3,5,7 are control experiments (no irradiation).

phosphorimagery of irradiated DNA(1) reveal that the amount of reaction at GG_1 is 4.5 times that at GG_2 (see Fig. 2 and Table 1). This result indicates that the 3′–TATA–5′ bridge presents a modest barrier to radical cation hopping. In contrast, for DNA(3), the 3′–ATAT–5′ segment introduces a more significant barrier to radical cation migration from GG_1 to GG_2. In this case, GG_1 reacts 50 times more often than GG_2. These results allow the comparison of polaron stabilization and barrier creation with computational predictions.

(2) The effect of replacing T with *t* on radical cation hopping

DNA(2) contains a 3′-***t***A***t***A-5′ bridge in place of the 3′–TATA–5′ bridge of DNA(1). The irradiation of DNA(2) leads to readily detected strand cleavage at both GG_1 and GG_2. However, unlike the 3′–TATA–5′ bridge of DNA(1), the radical cation readily crosses the 3′-***t***A***t***A-5′ bridge in DNA(2). Substitutions of T with ***t*** in this bridge causes the ratio of cleavage at GG_1 to GG_2 to go from 4.5 : 1 in DNA(1) to about 1 : 1 in DNA(2). The effect of substituting ***t*** for T in the bridge is even more striking when DNA(3) is compared with DNA(4). In this case the ratio of reactivity goes from 50 : 1 for DNA(3) to approximately 1 : 2 in DNA(4).

Because the absorption spectrum of ***t*** tails beyond 300 nm, we considered the possibility that the portion of the irradiating light directly absorbed by this chromophore could result in reaction of the DNA that leads to strand cleavage. However, control experiments showed that irradiation of the duplex oligomer that contains the modified ***t*** but lacks the AQ, [DNA(2)-AQ] results in no detectable strand cleavage. In addition, even extensive (10 fold excess) irradiation of DNA(2)-AQ caused no change in the T_m of the duplex or in its absorption spectrum in regions characteristic of ***t***.

Clearly, though structurally similar to T, the replacement of thymine with ***t*** has a profound effect on the ability of the radical cation to traverse bridges composed of A–T base pairs. In principle, this might be due to an enhancement of k_{hop} from GG step to GG step, or this substitution might modify k_{trap} by H_2O or O_2 at nearby guanines. We prepared and investigated compounds DNA (5, 6, 8, 9) to extend this analysis and differentiate between these two possibilities.

DNA(5) contains four GG steps separated by a 3′–TATA–5′ bridging sequence. Its irradiation leads to measurable amounts of strand cleavage at GG_1 through GG_3, the amount of reaction at GG_4 is too small to quantify reliably (autoradiograms of these gels are presented in the supplementary information). DNA(6) is identical to DNA(5) except that ***t*** has been substituted for T in the bridge between GG_1 and GG_2. If this substitution causes an increase in k_{hop} between

GG_1 and GG_2, this will result in a greater reaction at GG_3 in DNA(6) than in DNA(5). In contrast, if changing T to ***t*** increases k_{trap} for nearby GG steps, less strand cleavage will occur at GG_3. The experimental results (Table 1) support the first possibility; more reaction is seen at GG_3 in DNA(6) than in DNA(5). Similar results are obtained from comparison of DNA(8) with DNA(9); more strand cleavage results at GG_3 when ***t*** replaces T in the 3′–TATA–5′ bridge between GG_1 and GG_2.

Introduction of a radical cation into DNA results from the one-electron oxidation of a base. Because G and A have lower oxidation potentials than C and T,[32] it is expected that radical cations in duplex DNA will reside primarily on the purine partner of normal Watson–Crick base pairs. Despite its structural similarity to thymine, electronically ***t*** is not a pyrimidine. In particular, we surmised that the replacement of N1 of T with a carbon atom in ***t*** would lower its E_{ox} significantly.

(3) The E_{ox} of *N*-methyl-2-pyridone

Thymine, with an E_{ox} of *ca.* 2.1 V *vs.* NHE[21] is the most difficult to oxidize of the four common DNA nucleobases. The E_{ox} of *N*-methyl-2-pyridone (see Fig. 1), which has an electronic structure analogous to that of ***t***, was determined by cyclic voltammetry in an acetonitrile solution containing tetrabutylammonium tetrafluoroborate as the supporting electrolyte. The oxidation wave shows a peak potential at 1.68 V *vs.* NHE. Under similar conditions, we showed that 2′,3′,5′-*tris* (*t*-butyldimethylsilyloxy)guanosine[33] has a peak potential of 1.5 V. This compares well with the value obtained by Seidel and co-workers[21] for guanine (1.49 V). They determined the E_{ox} of adenine to be 1.96 V under these conditions. Thus, the E_{ox} of ***t*** is greater than guanine but is *ca.* 0.28 V less than that of adenine. Consequently, it may be anticipated that in an A–***t*** base pair the radical cation will reside on the ***t*** rather than on the A as it does in a normal A–T base pair. This matter is discussed further below.

(4) The effect of replacing T with *f* on radical cation hopping

It is clear that replacement of T with ***t*** in bridges between GG steps increases the charge transfer efficiency. As suggested above, this may be attributable to the reduced E_{ox} of ***t***. However, it may also be a consequence of disruption to the normal hydration of the duplex DNA.[34] To distinguish between these two possibilities we examined duplexes DNA(7) and DNA(10) in which the thymines in bridging base pairs between GG steps were replaced by difluorotoluene. Both ***t*** and ***f*** disrupt the hydration of the DNA, but the E_{ox} of ***f*** should be well above 2.37 V *vs.* NHE (the E_{ox} of toluene[35]), which is significantly greater than that of adenine.

In contrast to the results obtained by substitution of ***t*** for T in the bridge, substitution with ***f*** in DNA(7) gives strand cleavage results (Table 1) that are indistinguishable from those obtained from irradiation of DNA(5), which contains only normal nucleotides (see supplementary information). Similarly, irradiation of DNA(10), which also has ***f*** in place of T in the bridge, gives results that are the same as from irradiation of DNA(8), which has A–T base pairs in the bridge. On the basis of these findings, the effect of substitution of ***t*** for T on radical cation migration efficiency is attributed primarily to electronic perturbation of the charge transfer pathway and not a consequence of structural modification of the DNA or its solvent environment.

Discussion

(1) Stabilization of radical cations in DNA by polaronic delocalization

The results from irradiation of DNA(1) and DNA(3) demonstrate again that long-distance radical cation migration occurs in duplex DNA. Light absorbed by the AQ chromophore creates an electronically excited singlet state that intersystem crosses[36] rapidly to a triplet. This is a critical step. Excited singlet states are generally inefficient sensitizers of DNA oxidation because of rapid charge annihilation of the resulting radical ion pair,[37,38] which reduces the quantum efficiency and can complicate mechanistic interpretations.[39] Energy transfer from triplet AQ to ***t*** or ***f*** is energetically impossible, but the AQ triplet readily oxidizes an adjacent nucleobase to form the AQ radical anion and base radical cation.[24] In the oligonucleotides investigated here, three A–T base pairs follow the AQ. We have shown[40] that this arrangement promotes efficient reaction because the radical cation is rapidly delocalized. Delocalization of the radical cation further slows charge annihilation in the (overall) triplet radical ion pair and this allows consumption of the AQ radical anion by reduction of molecular oxygen to superoxide to proceed more efficiently.[24] After the AQ radical anion has been consumed, the base radical cation can exist in the DNA for hundreds of microseconds,[16,41] which permits the long distance migration of the radical cation to occur.[42]

Numerous investigations have shown that radical cations in DNA react predominantly at $(GG)_n$ sequences where $n = 2,3$.[7,8] In both DNA(1) and DNA(3), the radical cation encounters GG_1 after an identical four base pair sequence. As expected, strand cleavage is observed at GG_1. The second GG sequence in these duplexes (GG_2) is separated from GG_1 by bridges composed of A–T base pairs arranged in different sequences, and the efficiency of radical cation migration across the bridge depends dramatically on that detail. Simply put, the bridge is crossed about 10 times more efficiently in DNA(1) than it is in DNA(3). The phonon-assisted polaron-hopping model can explain this effect.

A polaron is a distortion in the DNA structure that is formed as a result of delocalization of the radical cation over nearby bases. We have suggested that polaronic delocalization occurs primarily among adjacent purines on one strand of the DNA duplex because that is adequate to explain radical cation migration data.[16] Thus, for DNA(1), delocalization of the polaron at GG_1 is restricted to the GG step (P_1 in Fig. 3) because it is surrounded on its 3′ and 5′-sides by pyrimidines. Similarly, polaron P_2 at GG_2 in DNA(1) is presumed to be delocalized over the 3′-AGG-5′ sequence. The more limited delocalization of P_1 results in a somewhat higher energy polaron than for P_2. A related circumstance is encountered in DNA(3). In this case, P_1 is assigned to a 3′-GGA-5′ polaron that is somewhat lower energy than P_2, which is delocalized only over the GG step. The nucleobases between the assigned polaronic sites create the activation barrier (‡) to charge migration from one site to the next. This barrier for both DNA(1) and DNA(3) is formed by a TAT sequence.

One-electron oxidation of DNA(1) or DNA(3) introduces a radical cation adjacent to the AQ group. From the structure of these duplexes, this radical cation must encounter GG_1 before it migrates as far as GG_2. The activation energy for migration from GG_1 to GG_2 ($\Delta G^{\ddagger}_{1,2}$) in DNA(1) is less than $\Delta G^{\ddagger}_{2,1}$ because P_2 is more stable than P_1, see Fig. 3. Consequently, for DNA(1) $\Delta\Delta G^{\ddagger}_{1\to 2}$ is negative. In contrast, for DNA(3) $\Delta\Delta G_{1\to 2}{}^{\ddagger}$ is positive and this difference accounts for the factor of ten difference in relative reactivity observed for the two GG steps of DNA(1) and DNA(3). Simply put, the radical cation has a harder time getting from GG_1 to GG_2 in DNA(3) than it does in DNA(1), so it is trapped more often at GG_1 in DNA(3). This analysis is consistent with calculations of the relative ionization potentials of three-base sequences where it is found that the identity of the base on the 3′ side has a greater effect on the ionization potential of G than the base on the 5′ side.[43,44]

(2) The pathway for radical cation migration

The substitution of ***t*** for T may effect both the extent of charge delocalization of the radical cation and the energy of the activation barrier encountered when it hops from one polaronic site to

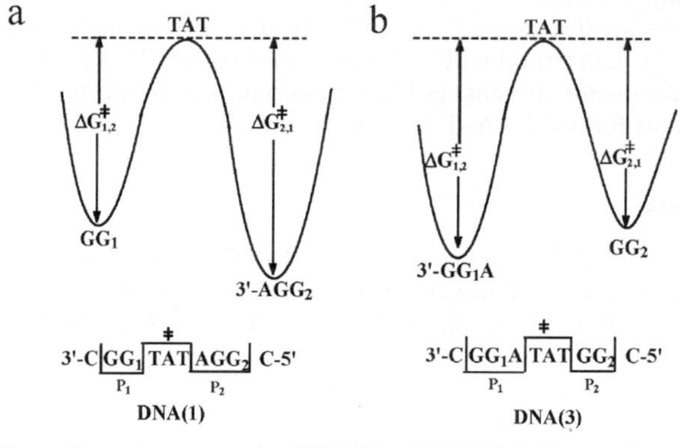

Fig. 3 Reaction coordinate diagrams representing DNA(1), a, DNA(3), b. For DNA(1), GG_1, the first GG step encountered by the radical cation after injection by the AQ, is somewhat higher in energy than GG_2 because of delocalization to adjacent adenine. For DNA(3), the situation is reversed. This differential stabilization affects the activation barriers for charge hopping $\Delta G^{\ddagger}$.

another. The results reported above show that the preferences for reaction at GG_1 in DNA(1) and DNA(3) are absent for DNA(2) and DNA(4), where the bridges contain no thymines. The experiments also show that this occurs because the magnitude of k_{hop} increases. The variation of k_{hop} could be due to a disruption of the hydration environment of the DNA in the vicinity of the modified base, but the experiments where *f* replaces T show that this is not an important factor. Moreover, in contrast to the factor of ten decrease observed between reaction at GG_1 and GG_2 when the 5′–ATAT–3′ sequence of DNA(1) is changed to 5′–TATA–3′ in DNA(3), reactions at GG_1 and GG_2 are of nearly equal efficiency for the 5′-A***t***A***t***-3′ and 5′-***t***A***t***A-3′ sequences of DNA(2) and DNA(4), respectively. These findings are attributed to the reduction of the oxidation potential of ***t*** compared to T.

The oxidation potential of ***t*** is much lower than T, greater than guanine, but less than that of adenine. Consequently, when a radical cation encounters a ***t***–A base pair, we conclude that it will reside primarily on the ***t*** rather than on the A as it does in a normal A–T base pair. Since the ***t***–A base pairs are part of the barrier to charge hopping between the GG-containing sites, the reduced oxidation potential of ***t*** relative to A lowers the activation energy for hopping so that k_{hop} is significantly greater than k_{trap}. As a result, the amount of reaction at the GG steps in DNA(2) and DNA(4) is determined by the relative energies of the polarons at GG_1 and GG_2, which are evidently nearly the same. These findings are fully consistent with the polaron-hopping model for long-distance charge migration in DNA. It has been concluded on the basis of calculation that unnatural nucleobases of low oxidation potential facilitate charge transfer.[45]

The substitution of *f* for T illuminates an additional aspect of the mechanism for long-distance radical cation migration in DNA. The oxidation potential of *f* is far greater than that of T. However, the experiments reported above show that there is no detectable effect on the relative reactivity at GG sites when *f* replaces T in the bridges of DNA(7) and DNA(10). This finding indicates that the radical cation in an A–T base pair resides primarily on the adenine whether or not that adenine is adjacent to a purine or in the complementary strand. Consequently, the increase in barrier height observed when the bridge separating GG sites is changed from AAAA to 5′–TATA–3′, for example,[46] can be attributed to decreased delocalization associated with reduced electronic overlap of adenines when they are in complementary strands.

Conclusions

We have shown that radical cations migrate across ATAT and TATA bridges in DNA(1) and DNA(3) with different efficiency. This is attributed primarily to the difference in stabilization energy of radical cations in CGGA and CGGT sequences. Substitution of ***t*** for thymine in these bridges lowers the barrier to radical cation migration. This result is ascribed to an electronic effect. The E_{ox} of ***t*** is lower than A, so in an A–***t*** base pair the radical cation will reside primarily on the ***t***. In contrast, substitution of *f* for A has no measurable effect on radical cation migration through ATAT or TATA bridges because the radical cation resides primarily on the A. These sequence and structural effects can be applied generally to DNA because the efficiency of radical cation hopping is predictable[16] and easily understood within the phonon-assisted polaron-hopping model.[11,19] DNA in solution is not a conductor. It behaves like a polaronic semiconductor with thermal barriers to radical cation transport formed by A–T base pairs.

Acknowledgements

We thank Dr Sriram Kanvah for the preparation of the DNA oligomers and Dr Mira Josowicz for assistance with the electrochemical measurements. This work was supported by the Vassar Woolley Foundation and by grants from the National Science Foundation to GBS and to LWM (MCB 0066776).

References

1 C. J. Murphy, M. R. Arkin, Y. Jenkins, N. D. Ghatlia, S. H. Bossman, N. J. Turro and J. K. Barton, *Science*, 1993, **262**, 1025.
2 N. J. Turro and J. K. Barton, *JBIC, J. Biol. Inorg. Chem.*, 1998, **3**, 201.

3 D. M. Adams, L. Brus, C. E. D. Chidsey, S. Creager, C. Creutz, C. R. Kagan, P. V. Kamat, M. Lieberman, S. Lindsay, R. A. Marcus, R. M. Metzger, M. E. Michel-Beyerle, J. R. Miller, M. D. Newton, D. R. Rolison, O. Sankey, K. S. Schanze, J. Yardley and X. Zhu, *J. Phys. Chem. B*, 2003, **107**, 6668.
4 E. A. Weiss, M. J. Ahrens, L. E. Sinks, A. V. Gusev, M. A. Ratner and M. R. Wasielewski, *J. Am. Chem. Soc.*, 2004, **126**, 5577.
5 R. G. Endres, D. L. Cox and R. R. P. Singh, *Rev. Mod. Phys.*, 2004, **76**, 195.
6 A. Filoramo, C. Dekker, U. Sivan, C. Schonenberger and M. E. Michel-Beyerle, *Phantoms Newsl.*, 2003, **4**.
7 G. B. Schuster, in *Long-Range Charge Transfer in DNA I*, Springer, Heidelberg, 2004.
8 G. B. Schuster, in *Long-Range Charge Transfer in DNA II*, Springer, Heidelberg, 2004.
9 J. Jortner, M. Bixon, A. A. Voityuk and N. Rosch, *J. Phys. Chem. A*, 2002, **106**, 7599.
10 F. D. Lewis, J. Liu, X. Zuo, R. T. Hayes and M. R. Wasielewski, *J. Am. Chem. Soc.*, 2003, **125**, 4850.
11 P. T. Henderson, D. Jones, G. Hampikian, Y. Kan and G. B. Schuster, *Proc. Natl. Acad. Sci. USA*, 1999, **96**, 8353.
12 R. N. Barnett, C. L. Cleveland, A. Joy, U. Landman and G. B. Schuster, *Science*, 2001, **294**, 567.
13 M. A. O'Neill, H.-C. Becker, C. Wan, J. K. Barton and A. H. Zewail, *Angew. Chem., Int. Ed.*, 2003, **42**, 5896.
14 A. Voityuk, K. Siriwong and N. Rosch, *Angew. Chem., Int. Ed.*, 2003, **43**, 624.
15 C.-S. Liu and G. B. Schuster, *J. Am. Chem. Soc.*, 2003, **125**, 6098.
16 C.-S. Liu, R. Hernandez and G. B. Schuster, *J. Am. Chem. Soc.*, 2004, **126**, 2877.
17 A. A. Voityuk, K. Siriwong and N. Rosch, *Angew. Chem., Int. Ed.*, 2004, **43**, 624.
18 M. A. O'Neill and J. K. Barton, *J. Am. Chem. Soc.*, 2004, **126**, 11471.
19 G. B. Schuster, *Acc. Chem. Res.*, 2000, **33**, 253.
20 S. Steenken and S. V. Jovanovic, *J. Am. Chem. Soc.*, 1997, **119**, 617.
21 C. A. M. Seidel, A. Schulz and M. H. M. Sauer, *J. Phys. Chem.*, 1996, **100**, 5541.
22 T. Lan and L. W. McLaughlin, *Bioorg. Chem.*, 2001, **29**, 198.
23 S. Moran, R. Ren, R.-F. I. Rumney and E. T. Kool, *J. Am. Chem. Soc.*, 1997, **119**, 2056.
24 B. A. Armitage, C. Yu, C. Devadoss and G. B. Schuster, *J. Am.Chem. Soc.*, 1994, **116**, 9847.
25 S. M. Gasper and G. B. Schuster, *J. Am. Chem. Soc.*, 1997, **119**, 12762.
26 M. Roginskaya, W. A. Bernhard and Y. Razkazovskiy, *J. Phys. Chem. B*, 2004, **108**, 2432.
27 D. M. Gray, S.-H. Hung and K. H. Johnson, *Methods Enzymol.*, 1995, **246**, 19.
28 T. Lan and L. W. McLaughlin, *J. Am. Chem. Soc.*, 2000, **122**, 6512.
29 B. A. Schweitzer and E. T. Kool, *J. Am. Chem. Soc.*, 1994, **117**, 1863.
30 K. K. Woods, T. Lan, L. W. McLaughlin and L. D. Williams, *Nucleic Acids Res.*, 2003, **31**, 1536.
31 K. M. Guckian, T. R. Krugh and E. T. Kool, *Nat. Struct. Biol.*, 1998, **5**, 954.
32 S. Steenken, *Chem. Rev.*, 1989, **89**, 503.
33 C. Sheu and C. S. Foote, *J. Org. Chem.*, 1995, **60**, 4498.
34 R. N. Barnett, C. L. Cleveland, U. Landman, E. Boone, S. Kanvah and G. B. Schuster, *J. Phys. Chem. B*, 2003, **107**, 3525.
35 E. Li, G. Shi, X. Hong and P. Wu, *J. Appl. Polym. Sci.*, 2004, **93**, 189.
36 J. Michl and V. Bonacic-Koutecky, *Electronic Aspects of Organic Photochemistry*, John Wiley & Sons, New York, 1990.
37 J. Reynisson, G. B. Schuster, S. B. Howerton, L. D. Williams, R. N. Barnett, C. C. L. U. Landman and N. Harrit, *J. Am. Chem. Soc.*, 2003, **125**, 2072.
38 C. Dohno, E. D. A. Stemp and J. K. Barton, *J. Am. Chem. Soc.*, 2003, **125**, 9586.
39 T. T. Williams, C. Dohno, E. D. Stemp and J. K. Barton, *J. Am. Chem. Soc.*, 2004, **126**, 8148.
40 L. Sanii and G. B. Schuster, *J. Am. Chem. Soc.*, 2000, **122**, 11545.
41 M. Misiaszek, C. Crean, A. Joffe, N. E. Geacintov and V. Shafirovich, *J. Biol. Chem.*, 2004, **279**, 32106.
42 T. Takada, K. Kawai, M. Fujiska and T. Majima, *Proc. Natl. Acad. Sci. USA*, 2004, **101**, 14002.
43 A. A. Voityuk, J. Jortner, M. Bixon and N. Rosch, *Chem. Phys. Lett.*, 2000, **324**, 430.
44 K. Senthilkumar, F. C. Grozema, C. Guerra Fensca, F. M. Bickelhaupt and L. D. A. Siebbeles, *J. Am. Chem. Soc.*, 2003, **125**, 13658.
45 A. A. Voityuk and N. Rösch, *J. Phys. Chem. B*, 2002, **106**, 3013.
46 B. Giese, M. Spichty and S. Wessely, *Pure Appl. Chem.*, 2001, **73**, 449.

Electrical characterization of self-assembled single- and double-stranded DNA monolayers using conductive AFM

Hezy Cohen,†[a] Claude Nogues,†[ab] Daniela Ullien,[a] Shirley Daube,[b] Ron Naaman[b] and Danny Porath*[a]

Received 31st May 2005, Accepted 13th June 2005
First published as an Advance Article on the web 14th October 2005
DOI: 10.1039/b507706k

We recently reported electrical transport measurements through double-stranded (ds)DNA molecules that are embedded in a self-assembled monolayer of single-stranded (ss)DNA and attached to a metal substrate and to a gold nanoparticle (GNP) on opposite ends. The measured current flowing through the dsDNA amounts to 220 nA at 2 V. In the present report we compare electrical transport through an ssDNA monolayer and dsDNA monolayers with and without upper thiol end-groups. The measurements are done with a conductive atomic force microscope (AFM) using various techniques. We find that the ssDNA monolayer is unable to transport current. The dsDNA monolayer without thiols in the upper end can transport low current on rare occasions and the dsDNA monolayer with thiols on both ends can transport significant current but with a much lower reliability and reproducibility than the GNP-connected dsDNA. These results reconfirm the ability of dsDNA to transport electrical current under the appropriate conditions, demonstrate the efficiency of an ssDNA monolayer as an insulating layer, and emphasize the crucial role of an efficient charge injection through covalent bonding for electrical transport in single dsDNA molecules.

Introduction

The issue of DNA conductivity has created much debate in recent years, with a wide variation in the reported results, depending on the sample preparation method and the measuring technique.[1–12] In measurements where long dsDNA is lying horizontally on a hard flat substrate, it has been suggested that the interactions between the DNA backbone and the underlying solid surface may interfere with the ability of the DNA molecules to conduct charge.[1,13] One possible way to minimize this type of interaction is to arrange the dsDNA molecules perpendicular to the surface within a closed packed organized layer so that on one end the molecule is attached to a metal substrate and on the opposite end to a gold nanoparticle (GNP) or to a metallized AFM tip, so that the molecules are free-standing.[14–16]

Another parameter that affects the charge transport through DNA is the contact between the molecule and the macroscopic world, *i.e.* the metal electrodes. It has been shown that covalent bonding of the molecule to the electrodes may increase the charge transport efficiency by up to three orders of magnitude and may therefore be advantageous for electrical transport measurements.[17]

To optimize the measurement configuration and overcome the above obstacles, a novel measurement configuration was designed, in which the dsDNA is attached through thiol linkers to a gold substrate and a GNP on opposite ends.[16] This configuration has some similarity to those

[a] *Physical Chemistry Department, The Hebrew University, Jerusalem 91904, Israel. E-mail: porath@chem.ch.huji.ac.il*
[b] *Department of Chemical Physics, The Weizmann Institute, 76100 Rehovot, Israel*

† These authors contributed equally to the reported work.

described by Xu *et al.*[18] and Cui *et al.*[19,20] The latter formed mixed monolayers of alkanethiol and alkanedithiol, to which they adsorbed gold nanoparticles. They demonstrated the ability to obtain reproducible current–voltage curves by contacting the gold nanoparticles with a conductive AFM tip, and to estimate the resistivity of a single octaneditiol molecule. They emphasize the importance of chemical bonds at both ends of the molecule to obtain effective charge transport between the single molecule and the gold contacts.

In our configuration, the molecule is free standing and embedded in a monolayer of ssDNA. This way the non-specific interaction of the molecule along its length is minimized and the charge injection to the molecule is optimized by the covalent gold–thiol bonding *on both ends* and by the efficient contact through the GNP. With this configuration currents higher than 220 nA at 2 V are measured.[15] In addition, using this preparation method a large number of "devices" can be produced on a small area.

Here we employ a more sophisticated AFM technique, using the 3D-mode,[21] that enables better control during the measurement and provides more comprehensive data. This method is used to compare electrical measurements on ssDNA monolayers with dsDNA with and without thiols in the upper ends.

Materials and methods

Sample preparation

ssDNA monolayer and GNP-connected dsDNA preparation. ssDNA molecules (5′-CAT TAA TGC TAT GCA GAA AAT CTT–AG-3′–$(CH_2)_3$–SH) were connected *via* a propyl thiol end group to a gold substrate, forming a packed monolayer. Complementary strands were connected *via* thiol groups to a 10 nm GNP. One or a few of the complementary strands are then hybridized with the surface adsorbed single-strands to form a dsDNA molecule(s) connecting the GNP and the surface.

Full details of the sample preparation were recently reported.[16] Briefly, the 3′-thiolated ssDNA is maintained protected in its oxidized form-$(CH_2)_3$-S-S-$(CH_2)_3$-OH until usage. Prior to adsorption on the gold surface or on a GNP, the protecting group is removed by reduction with tris (2-carboxyethyl) phosphine (TCEP). The reduced DNA oligomer is then separated from all residues using a column (BioSpin 6, BioRad), and is immediately pipetted onto a clean, annealed gold surface. For the ssDNA connected to GNPs, the reduced and purified thiolated complementary ssDNA (5′-CTA AGA TTT TCT GCA TAG CAT TAA–TG-3′–$(CH_2)_3$–SH) oligomer is stirred with the GNP suspension. The GNPs are rinsed in deionized water prior to the DNA adsorption to remove the excess surfactant that is adsorbed on the GNPs surface. During the preparation of complementary ssDNA–GNPs the concentration of oligomers was tuned to be several orders of magnitude larger than the GNPs concentration in order to achieve high ssDNA coverage on the GNP.

The resulting complementary ssDNA–GNP derivative is diluted with Tris/NaCl buffer solution and pipetted onto the ssDNA monolayer. Incubation of the sample for 2–12 h at 100% relative humidity causes the hybridization and the formation of dsDNA. The sample is rinsed with fresh buffer solution to remove non-specifically bound ssDNA–GNP. Just prior to AFM measurements the sample is rinsed with deionized water to remove excess salts. The above preparation procedure leads to samples where dsDNA molecules are covalently linked to both the substrate and the GNP. The sample configuration is schematically shown in Fig. 1(a). The GNP is lying on the ssDNA monolayer and the dsDNA is tilted relative to the substrate normal.

Depending on the monolayer density, up to ten parallel dsDNA molecules can fit underneath a GNP. The surface area of the GNP is 300 nm^2. If 25% of the GNP bottom is available for dsDNA attachment and the surface area per dsDNA is $\sim 2.5 \times 2.5$ nm^2, then there is room for ~ 10 molecules at most. This value is in accord with the ssDNA coverage density, $\sim 3 \times 10^{13}$ cm^2 (as found from radioactive labeling of the ssDNA).[16] The coverage of dsDNA is, however, likely to be much smaller than 100% so probably less then 10 molecules are connecting the GNP and the substrate.

dsDNA monolayer. Double-stranded DNA monolayer was prepared on a gold surface. Reduced and purified ssDNA molecules (5′-CAT TAA TGC TAT GCA GAA AAT CTT–AG-3′–$(CH_2)_3$–

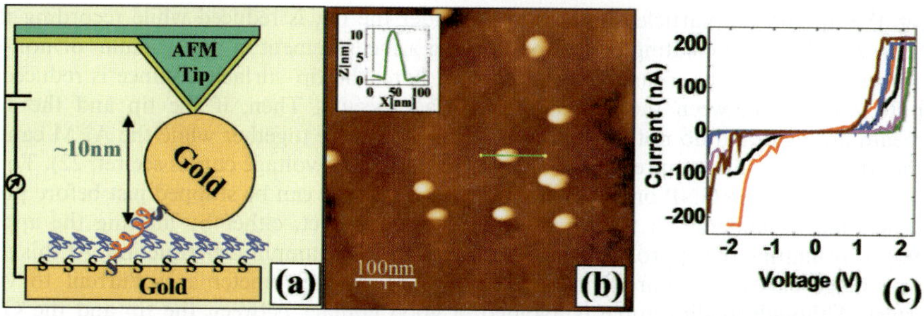

Fig. 1 (a) Schematic presentation of the GNP-connected dsDNA that is embedded in the ssDNA monolayer and the measurement configuration. (b) AFM topography image showing a top view of the GNPs connected to dsDNA molecules on the background of the ssDNA monolayer. Inset: A cross section along the marked line shows the surface contour of a GNP lying on the ssDNA monolayer and connected to a dsDNA molecule. (c) Collection of current–voltage curves measured through various dsDNA molecules connected on opposite ends to a gold substrate and a GNP using a metallized AFM tip. The curves were taken on different samples and in various occasions with different tips.

SH) with a propyl thiol end group and non-thiolated complementary ssDNA (5′-CTA AGA TTT TCT GCA TAG CAT TAA–TG-3′) were used. Both ssDNA oligomers were diluted to a final concentration of ∼10 μM each, with 25 mM Tris buffer in the presence of 0.4 M NaCl, hybridized in solution and incubated onto a gold surface as dsDNA. Incubation of the sample for 2–12 h at 100% relative humidity affords a monolayer of dsDNA. The sample is rinsed with a fresh buffer solution to remove non-specifically bound dsDNA. Just prior to the AFM measurements the sample is rinsed with deionized water to remove excess salts. The preparation procedure leads to samples where dsDNA molecules are covalently linked to the substrate without a thiol group at the upper end.

Dithiolated dsDNA monolayer. A dsDNA monolayer thiolated on the side connecting the molecule to the underlying gold surface and presenting a thiol functional group to be contacted by the AFM tip is formed. The reduced and purified thiolated ssDNA oligomer complementary to the above ssDNA monolayer (5′-CTA AGA TTT TCT GCA TAG CAT TAA–TG-3′–$(CH_2)_3$–SH) is diluted to 15 μM with 25 mM Tris buffer in the presence of 0.4 M NaCl and incubated onto the ssDNA monolayer. The incubation and rinsing were the same as for the thiolated dsDNA samples. The above preparation procedure leads to samples where dsDNA molecules are covalently linked to the substrate and with a thiol group at the upper end.

AFM measurements—topography. Images were obtained with a commercial AFM (Nanotec Electronica S.L. Madrid) in dynamic mode. Soft cantilevers (OMCL-RC800PSA, Olympus Optical Co., LTD.) of nominal force constant 0.4 and 0.8 N m^{-1}, resonance frequency ∼80 kHz and tip radius 15–20 nm were used (the tip radius was measured with a scanning electron microscope). For the electrical measurements the tips were covered with ∼3 nm Cr and ∼15–30 nm evaporated gold or directly sputtered gold/palladium. Consequently, the tip radius grows to 25–40 nm, the force constant is increased to 0.5–1.0 N m^{-1} and the resonance frequency drops to 60–75 kHz upon coverage. The cantilever is oscillated at a frequency close to its resonance frequency while the amplitude and the phase, relative to the oscillatory driving force, are measured through the tip deflection signal as detected by the photo-detector. The feedback is performed on the amplitude signal channel. The tip is typically a few to tens of nanometres above the surface and the oscillation amplitude for imaging is 5–50 nm.

AFM measurements—combined force–distance and current–voltage (*F–Z*/*I–V*). Current–voltage curves were measured between the metallized tip and the substrate under various configurations. The AFM software was especially modified in order to have good control over the applied force and the tip position relative to the sample while performing the electrical measurements. The distance between the tip and the particle, or the surface under the tip, is reduced while recording the tip deflection as the tip is oscillating. When performing a measurement in non-vacuum or non-liquid environment, as are all the measurements reported here, the tip–surface distance is reduced until "jump to contact" between the tip and the surface occurs. Then, if the tip and the surface underneath are rigid and do not get deformed then they move together while the AFM cantilever is further deflected (for a comprehensive analysis of deflection–voltage curves see ref. 22). To avoid tip-loading force on the GNP or the monolayer, the approach can be stopped just before pressing the surface that is underneath the tip by the jump to contact, either by limiting the approach distance or by limiting the approach using a feedback from the amplitude reduction. At this point a current–voltage curve is acquired (using Keithley 6517A electrometer as a current to voltage converter). Although loading force is avoided, a good contact between the tip and the GNP is formed, as evident from the hysteresis in the deflection–voltage and from the current–voltage curves. This probably occurs due to an upward movement of the GNP. The measurements were done at various humidity conditions ranging from ambient (RH ~50–85%) to N_2 atmosphere (RH ~15%) with no observable difference. Detailed description of these measurements appears in ref. 15.

AFM measurements—3D-mode. Conductivity measurements were carried out utilizing an AFM with a conductive tip in a 3D-mode operation of the instrument.[21] In this mode, the voltage is scanned while the deflection and the current are measured simultaneously, then the tip is approached stepwise towards the surface. After each step the voltage is scanned again and the deflection and current are recorded. The measurements are presented graphically such that the fast axis (*X*) is the voltage, the slow axis (*Y*) is the approach distance perpendicular to the sample towards its surface (30 nm displacement from top to bottom, starting at the set-point position above the surface). The magnitudes that are shown using the color scale of the image are the deflection and the current respectively. This mode of operation provides control over the tip–sample distance and enables one to check if tip bending occurs upon voltage application at various distances from the surface during the measurement. It also directly provides the correlation between the tip displacement and the possible rise of current flow between the tip and the metal substrate.

Experimental results. The surface morphology of an ssDNA monolayer with protruding dsDNA molecules connected to GNPs is characterized using AFM and shown in Fig. 1(b). The surface of the single-stranded monolayer appeared rather smooth (<1 nm roughness), although the "bulges" of the ssDNA molecules can be still observed. The 10 nm GNPs, which are connected to the dsDNA, appear very clear on the background of the monolayer. The measured height of the GNP above the monolayer surface is ~10 nm (inset in Fig. 1(b)). This means that the double strands formed do not protrude appreciably above the thickness of the monolayer, which is about ~4 nm.‡ Since the length of the dsDNA is about 9 nm, this implies that the dsDNA molecules are tilted relative to the surface normal, as expected for thiols (see also Fig. 1(a)).[23,24] When non-complementary ssDNA oligomers are attached to the GNPs the number of particles observed on the surface is two orders of magnitude smaller.[16]

Upon contacting various GNPs with a metallized AFM tip, in different samples and with different tips, a collection of S-shaped current–voltage curves are measured as shown in Fig. 1(c).[15] The current usually rises beyond a threshold voltage of ~1 V. The observed current goes up to 220 nA at 2 V (the saturation of the measuring system) although sometimes the current measured is lower. Note that the maximal Coulomb blockade signature of a completely isolated 10 nm GNP is ~0.2 eV and therefore its contribution to the shape of the current–voltage curve is minimal.

A more controlled and informative method for measuring the electronic properties is the 3D-mode procedure described above. It is applied in a similar way to an ssDNA monolayer, a dsDNA

‡ The ssDNA and dsDNA monolayers thickness were characterized by ellipsometry and found to be ~3–4 and 5–6 nm, respectively. The thickness measured by the AFM (by "shaving" a square on the surface) was 2–3 nm for the ssDNA monolayer and 3–4 nm for the dsDNA monolayer. The thickness difference originates from the inelastic interaction of the tip with the monolayer. The GNP coverage was measured using AFM and was found to be ~500 GNP μm^{-2} for complementary strands (dsDNA) and ~5 GNP μm^{-2} for non-complementary strands (same strand as the one connected to the surface).

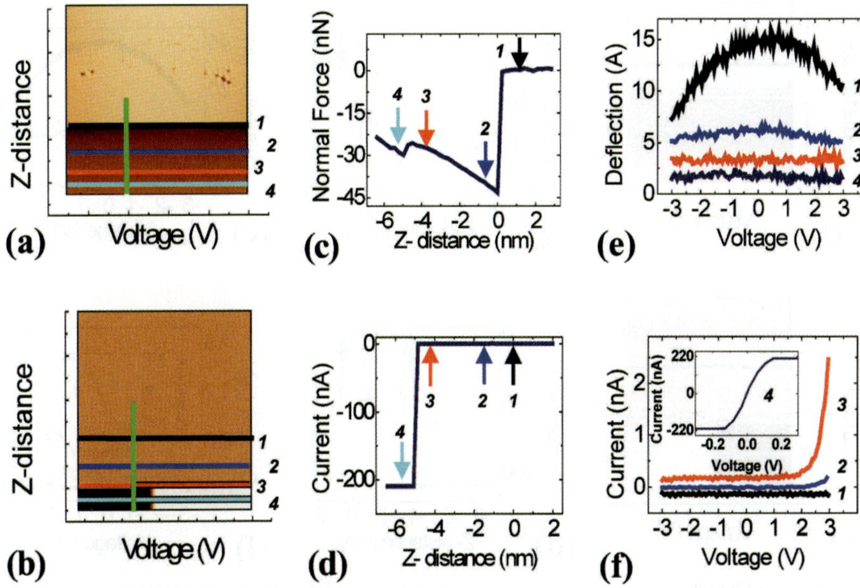

Fig. 2 3-D-mode AFM images and selected profiles from the images. (a) and (b) are deflection and current maps acquired simultaneously by displacing a metallic tip towards an ssDNA monolayer surface while scanning the tip–sample bias voltage (left to right −3 to +3 V) at each distance from the surface (top to bottom 30 nm displacement-part of it is shown in the image). (c) and (d) show corresponding top to bottom deflection–distance and current–distance profiles (right to left), along corresponding vertical lines in (a) and (b). (e) and (f) show corresponding deflection–voltage and current–voltage curves taken at the position of the corresponding colored lines in (a) and (b) and the colored arrows in (c) and (d). No current is measured upon establishment of contact between the tip and the monolayer (first jump to contact in (c)) as can be seen in (d). Only after a second jump to contact (∼4.8 nm), probably with the gold substrate, the current is rising to 220 nA.

monolayer, a dsDNA monolayer with thiols also on the upper end and as a control to bare gold. The results, presented in Figs. 2–6, are arranged respectively. Panels (a) and (b) show the tip-deflection and tip-sample current, respectively, as a function of the bias voltage applied between the tip and the gold substrate and as a function of the vertical movement of the tip towards the surface. In practice, each horizontal line in the image in panel (a) is a deflection–voltage curve and each line in (b) is a current–voltage curve measured simultaneously. The (a) and (b) images in Figs. 2–6 show only the informative part of the acquired images. Corresponding vertical lines in the images provide simultaneous deflection–distance, *e.g.* (c), and current–distance, *e.g.* (d). The correlation between the latter corresponding curves enables us to compare the ability of the various monolayers to transport current. The vertical tip displacement between the first jump to contact and the onset of the current is used to determine whether the layer transports current right after the jump to contact, where the tip establishes the first contact with the monolayer, after a short penetration or only after piercing deep through the monolayer down to the gold substrate. Corresponding horizontal lines that are measured at various tip displacements: before, right after the tip jumps to contact with the monolayer surface just before a second jump to contact, and after a good contact is established with the metal substrate, provide detailed comparison of the electronic properties of the monolayer and in particular the molecule under the tip apex. Deflection–voltage curves taken just before the tip jumps to contact demonstrate the tip bending upon voltage application. Corresponding examples of the two types of curves are shown in (e) and (f). The positions of the horizontal colored lines in (a) and (b) are marked by arrows with the same color in (c) and (d). The explicit curves are those presented in (e) and (f).

We now review the results obtained from these measurements on the various types of samples. Fig. 2 shows the results measured on the ssDNA monolayer. This is the layer that serves as a separating layer between the GNPs and the gold substrate and through which the dsDNA molecules

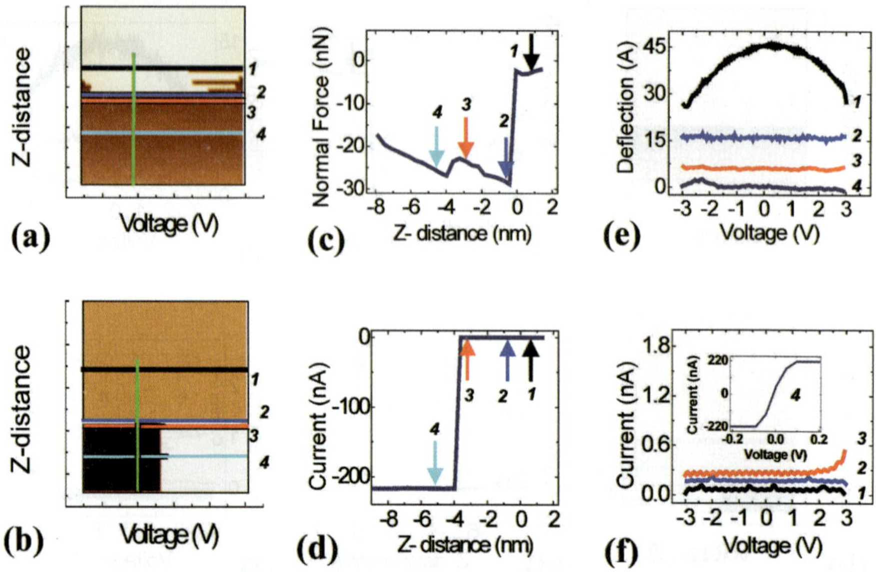

Fig. 3 3D-mode AFM images and selected profiles measured on a dsDNA monolayer. The panels are similar in arrangement to those in Fig. 2. The general behavior is similar to that of the ssDNA monolayer shown in Fig. 2. No current is measured until the second jump to contact that occurs, however, at a smaller deflection and penetration into the monolayer ($\sim$3.6 nm), probably due to the greater stiffness of the dsDNA with respect to the ssDNA monolayer.

are protruding in the measurements shown in Fig. 1 (an AFM image of this monolayer is shown in Fig. 1(b)). The measurements presented here prove that the monolayer is insulating and it prevents direct contact between the metal AFM tip (or the GNP) and the metal substrate. Panels 2(a) and 2(b) show data measured for this monolayer, taken in a position where no GNPs were present under

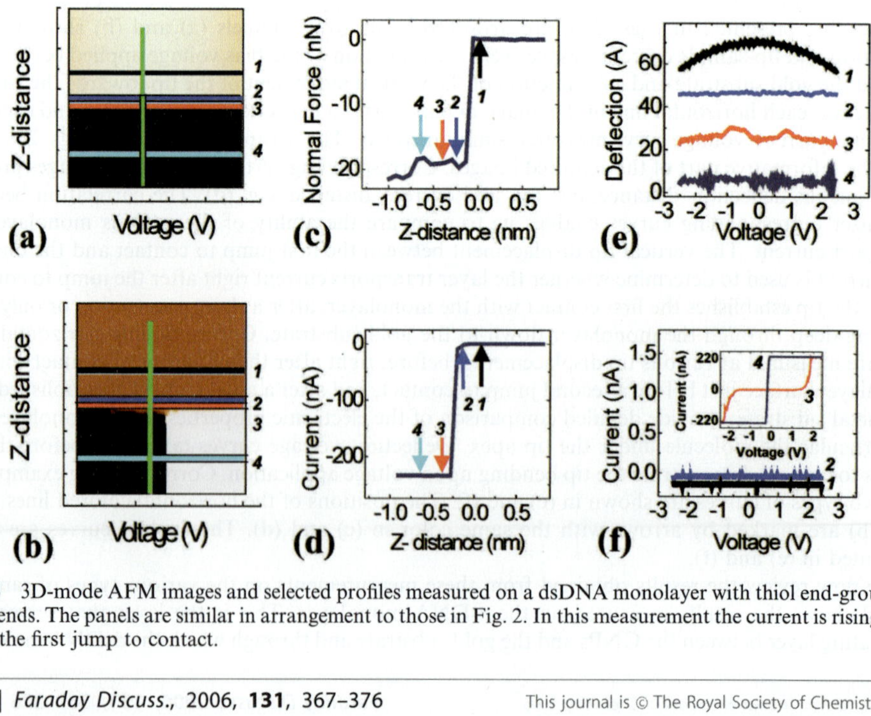

Fig. 4 3D-mode AFM images and selected profiles measured on a dsDNA monolayer with thiol end-groups on both ends. The panels are similar in arrangement to those in Fig. 2. In this measurement the current is rising right after the first jump to contact.

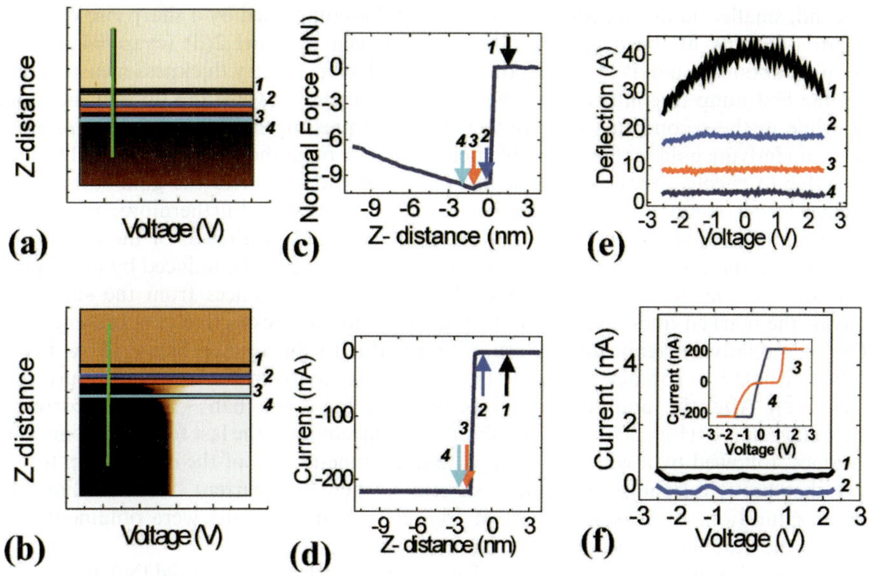

Fig. 5 3D-mode AFM images and selected profiles measured on a dsDNA monolayer with thiol end-groups on both ends in a different position from in Fig. 4. The panels are similar in arrangement to those in Fig. 2. In this measurement current is rising after the tip penetrates ~1.4 nm to the monolayer together with the second jump to contact.

the metal tip upon its approach to the monolayer. The curve in panel 2(c) shows the approach to the surface (right to left here and in the following figures) which is followed by a sharp jump to contact (immediately after the black–#1 arrow). No current onset is monitored in the simultaneous current–distance curve in panel 2(d). The tip is displaced ~4.8 nm downwards (blue–#2 and red–#3 arrows)

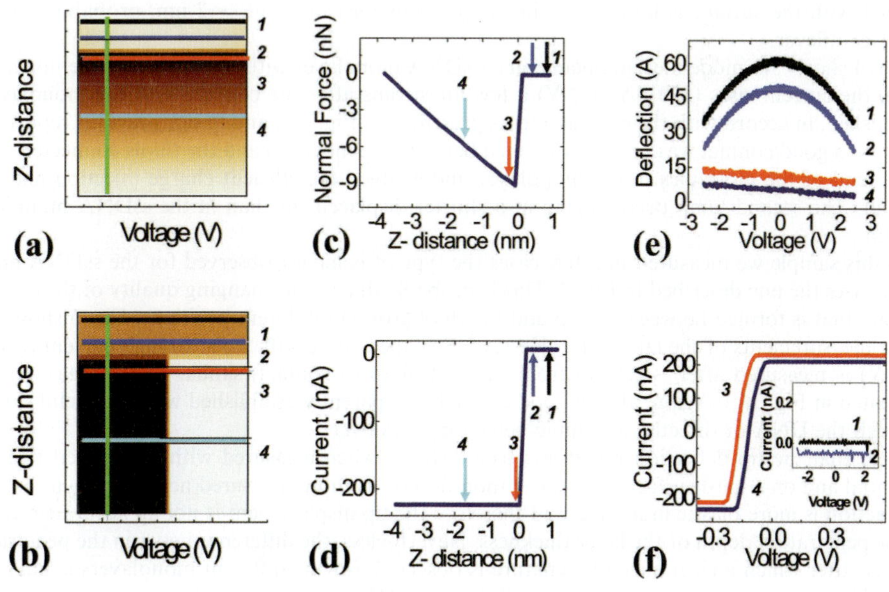

Fig. 6 3D-mode AFM images and selected profiles measured on bare gold. The panels are similar in arrangement to those in Fig. 2. In this measurement current is rising right after the jump to contact, as expected for a metal surface.

and a second, smaller, jump to contact is observed and accompanied by a sharp rise of the current (from 0 to −220 nA at the negative voltage range) seen in panel 2(d) (cyan–#4 arrow). This separation, of about 4.8 nm, is in rough consistency with ellipsometry thickness measurements. We interpret the first jump to contact as establishment of contact between the tip and the monolayer surface, while in the second jump to contact the tip is pressing hard on the ssDNA monolayer against the underlying gold surface, possibly penetrating through the soft monolayer. The measured current is due to effective tunneling through the pressed monolayer to the gold surface and then ohmic when good contact with the gold substrate is achieved. Furthermore, to support this interpretation it is important to verify that the tip is deflected as a result of the interaction with the DNA monolayer, and not due to the electrostatic field that may be induced by the applied bias. The deflection of the tip *versus* the applied bias at various distances from the surface (at the positions of the marked lines and arrows in Fig. 2(a) and 2(c), respectively) is presented in panel 2(e). The only relatively significant deflection that is due to the applied bias (~5 Å) is observed *before* the first jump to contact, when the tip is further away from the monolayer surface (upper most line in (e)). After the jump to contact, the tip deflection is less than ~2 Å. The corresponding current–voltage curves (Fig. 2(f)) show no observable current until the last few angstroms above the metal surface, followed by a wide gap (~4 V) and asymmetric rise of the current (up to ~2 nA). The tip is then brought right onto the metal substrate, where high current, >220 nA limited by the instrument saturation, is measured (inset of Fig. 2(f)). Similar results were obtained at various positions on the monolayer and on many samples.

Fig. 3 shows a 3D-mode measurement performed in a similar way on a dsDNA monolayer with no thiols on the upper end of the molecules. The panels present the same type of measurements as those described above. The overall response of the monolayer is similar to the one described in Fig. 2. The difference in electrical response between this dsDNA monolayer and the single dsDNA molecules connected to GNPs (see Fig. 1(c)) is, however, striking. For the non-thiolated dsDNA, no current is measured until the tip is displaced ~3.6 nm into the monolayer. Then a low current (<1 nA) is measured for the positive bias. In other positions on rare occasions some current was measured before the second jump to contact with the metal surface. Note that the monolayer thickness, as determined by ellipsometry, is ~6 nm, as expected for the higher persistence length of the dsDNA molecules. The tip displacement upon penetration into the layer until the second jump to contact occurs is shorter, ~3.6 nm, probably due to the stiffer nature of the dsDNA compared to the soft ssDNA monolayer. In this measurement the tip bending upon voltage application prior to contact with the surface is larger than in the previous measurement, ~2 nm, probably due to a softer cantilever.

Fig. 4 shows 3D-mode measurements on a dsDNA monolayer with thiols also on the upper end. Here, the current rises (220 nA at 2 V) a few angstroms after the contact with the monolayer is established, in accord with the measurements presented in Fig. 1(c) and in our previous report.[15] In this case a good contact was probably formed between the metal tip and the thiols connected to the upper end of the molecules in the monolayer and enabled the efficient charge injection and high current. The second jump occurs again at a shorter displacement than in the ssDNA monolayer, 0.8 nm.

In this sample we measured in a few cases the type of behavior observed for the ssDNA and in other cases the one described in Fig. 4. This is probably due to the changing quality of the covalent contact that is formed between the tip and the thiol group in different positions. Fig. 5 shows 3D-mode measurements of the latter case. Here, an S-shaped curve with a rise to high current (220 nA at 2 V) is measured after ~1.4 nm from the first jump to contact, similar to the measurements presented in Fig. 1(c). Again, after a good contact is apparently established with the metal surface, through the DNA or directly, an ohmic behavior takes over.

Note that the tip deflection represents true distance when measured with no applied electrical potential and on a hard surface. For softer monolayers, as those measured here, the tip–monolayer interaction is more plastic in nature, and therefore the tip displacement is not an accurate estimate of the penetration depth or the layer thickness. Nevertheless, the difference between the penetration depths, after which a clear current signature is observed, for the different monolayers is unambiguous. The tip penetrates through most or all the soft ssDNA monolayer before current is measured. The dsDNA monolayers are stiffer and react faster to the penetrating tip and therefore the second jump to contact appears after a shorter penetration. This distance was similar in many positions on both types of dsDNA monolayers measured and therefore the measurements on them can be safely compared.

The positive control measurement shown in Fig. 6 is done directly on a bare gold surface. An ohmic behavior is achieved right after the first jump to contact with 220 nA (limited by the preamplifier saturation) already at ~0.1 V.

Conclusions and discussion

Here we show that the dsDNA molecules are capable of transporting relatively high currents in two different configurations: (i) dsDNA connected to GNP and a metal substrate and embedded in an ssDNA monolayer and (ii) a monolayer of dsDNA with thiols on both ends. In both configurations the dsDNA does not undergo large deformation due to non-specific interaction with the surface along its length, and in both cases the charge injection is optimized by covalent bonding. Similar currents at comparable bias voltages were observed by Xu *et al.*,[18] who contacted shorter dsDNA molecules with thiols on both ends in solution using a scanning tunneling microscope.

In addition, we show that a well packed self-assembled monolayer, made of 26 base long single-stranded DNA, forms an insulating layer on gold surfaces. Electrical transport measurements with a conductive AFM tip indicate that no current (<10 pA) passes through the single-stranded monolayer until the tip is pressed hard onto and into the monolayer. Similarly, measurements on a dsDNA monolayer where the contact to the metal tip is physical and not covalent give, in most cases, no significant current until the tip penetrates through the monolayer and touches the substrate. These results emphasize the critical role of an efficient covalent bonding between the metal electrodes and the molecule. This understanding is further supported by the results measured on the similar monolayer, where the upper, 3′ end, of the molecule was connected to the tip *via* a thiol group. In this case high currents were indeed measured, although the contact to the AFM tip is less efficient than through a GNP.

It is important to realize that the dsDNA molecules are not behaving as a "conductor". Our data show that the molecule can generally be described as "a wide-band-gap" material, for which the conductivity can be quite efficient once the potential is raised above the band gap.

Acknowledgements

The authors would like to thank Igor Brodsky for assistance and fruitful discussions. The research was supported by the FIRST foundation, The Israeli Academy of Sciences and Humanities and by European grant for Future & Emerging Technologies (IST-2001-38951).

References

1 D. Porath, G. Cuniberti and R. Di Felice, *Top. Curr. Chem.*, 2004, **237**, 183.
2 C. Dekker and M. A. Ratner, *Phys. World*, 2001, **14**, 29.
3 R. G. Endres, D. L. Cox and R. R. P. Singh, *Rev. Mod. Phys.*, 2004, **76**, 195.
4 M. D. Ventra and M. Zwolak, in *DNA Electronics. Encyclopedia of Nanoscience and Nanotechnology*, ed. H. S. Nalwa, American Scientific Publishers, Stevenson Ranch, CA, 2003, p. 475.
5 E. Braun, Y. Eichen, U. Sivan and G. Ben-Yoseph, *Nature*, 1998, **391**, 775.
6 H. W. Fink and C. Schonenberger, *Nature*, 1999, **398**, 407.
7 D. Porath, A. Bezryadin, S. de Vries and C. Dekker, *Nature*, 2000, **403**, 635.
8 P. J. de Pablo, F. Moreno-Herrero, J. Colchero, J. Gómez Herrero, P. Herrero, A. M. Baró, P. Ordejón, J. M. Soler and E. Artacho, *Phys. Rev. Lett.*, 2000, **85**, 4992.
9 A. J. Storm, J. van Noort, S. de Vries and C. Dekker, *Appl. Phys. Lett.*, 2001, **79**, 3881.
10 H. Watanabe, C. Manabe, T. Shigematsu, K. Shimotani and M. Shimizu, *Appl. Phys. Lett.*, 2001, **79**, 2462.
11 Y. Zhang, R. H. Austin, J. Kraeft, E. C. Cox and N. P. Ong, *Phys. Rev. Lett.*, 2002, 89.
12 T. Shigematsu, K. Shimotani, C. Manabe, H. Watanabe and M. Shimizu, *J. Chem. Phys.*, 2003, **118**, 4245.
13 A. Y. Kasumov, D. V. Klinov, P. E. Roche, S. Gueron and H. Bouchiat, *Appl. Phys. Lett.*, 2004, **84**, 1007.
14 A. Csaki, R. Moller, W. Straube, J. M. Kohler and W. Fritzsche, *Nucleic Acids Res.*, 2001, **29**, e81.
15 H. Cohen, C. Nogues, R. Naaman and D. Porath, *Proc. Natl. Acad. Sci. USA*, 2005, **102**, 11589.
16 C. Nogues, S. R. Cohen, S. S. Daube and R. Naaman, *Phys. Chem. Chem. Phys.*, 2004, **6**, 4459.
17 A. Salomon, D. Cahen, S. Lindsay, J. Tomfohr, V. B. Engelkes and C. D. Frisbie, *Adv. Mater.*, 2003, **15**, 1881.
18 B. Q. Xu, P. M. Zhang, X. L. Li and N. J. Tao, *Nano Lett.*, 2004, **4**, 1105.

19 X. D. Cui, A. Primak, X. Zarate, J. Tomfohr, O. F. Sankey, A. L. Moore, T. A. Moore, D. Gust, G. Harris and S. M. Lindsay, *Science*, 2001, **294**, 571.
20 X. D. Cui, X. Zarate, J. Tomfohr, O. F. Sankey, A. Primak, A. L. Moore, T. A. Moore, D. Gust, G. Harris and S. M. Lindsay, *Nanotechnology*, 2002, **13**, 5.
21 C. Gomez-Navarro, A Gil, M Alvarez, P. J. De Pablo, F. Moreno-Herrero, I. Horcas, R. Fernandez-Sanchez, J. Colchero, J. Gomez-Herrero and A.M. Baro, *Nanotechnology*, 2002, **13**, 314.
22 B. Cappella and G. Dietler, *Surf. Sci. Rep.*, 1999, **34**, 1.
23 A. Haran, D. H. Waldeck, R. Naaman, E. Moons and D. Cahen, *Science*, 1994, **263**, 948.
24 S. O. Kelley, J. K. Barton, N. M. Jackson and M. G. Hill, *Bioconjugate Chem.*, 1997, **8**, 31.

The mechanism of long-range electron exchange in molecular-scale photonic wires

Anthony Harriman,*[a] Sarah A. Rostron,[a] Abderahim Khatyr[b] and Raymond Ziessel[b]

Received 25th April 2005, Accepted 27th May 2005
First published as an Advance Article on the web 23rd September 2005
DOI: 10.1039/b505817c

Several disparate systems have been examined as putative molecular-scale wires able to conduct charge over relatively long distances under illumination. The key feature of such systems concerns the level of electronic coupling between the terminals, which are themselves formed from photoactive metal poly(pyridine) complexes. In particular, a series of linear polynuclear ruthenium(II) tris(2,2′-bipyridine) complexes has been synthesized whereby individual chromophores are separated by 1,4-diethynylenebenzene subunits bearing dialkoxy groups for improved solubility. These arrays contain two, three, four or five metal centres. The compounds are reasonably soluble in polar organic solvents and possess optical absorption spectral properties that are dominated by transitions associated with the polytopic ligand. Weak luminescence is observed for each complex in deoxygenated acetonitrile at room temperature that appears to be characteristic of emission from a metal-to-ligand, charge-transfer triplet state. The emission lifetime is essentially independent of temperature, at least over a modest range. There is no indication for interaction between close-lying triplet states and no obvious sign of a low-energy π,π^* triplet associated with the polytopic ligand. The photophysical properties suggest, however, that the longer arrays are segmented due to intramolecular charge-transfer interactions. The ligands bind zinc(II) cations in solution and thereby affect the absorption and emission spectra.

1. Introduction

The future development of effective systems for use as molecular photonic devices is contingent on the availability of molecular materials able to conduct charge over reasonable distances and along preferred pathways.[1] Ideally, the mechanism for charge migration will involve through-bond electron transfer (or exchange) and will facilitate efficient transport, storage and retrieval under illumination with visible light.[2] It will be necessary to transfer such information over distances in excess of 100 Å with high efficiency and excellent directionality. In recent years, it has become clear that there are two major mechanisms that favour long-range information transfer at the molecular level; namely, electron tunneling[3] and charge hopping.[4] The prime difference between these two extreme mechanisms is illustrated in Fig. 1. Thus, electron tunneling occurs *via* superexchange interactions and makes use of virtual orbitals on the connecting bridge. On the other hand, charge hopping occurs over a series of short-range steps and involves redox states on the bridge as real intermediates. The design of the molecular system, especially the energy levels of the bridge, depends on the choice of mechanism. Although early indications suggested that charge hopping provides the more likely route to success there are strong suggestions[5] that electron tunneling might prove to be more effective in the long run.

[a] *Molecular Photonics Laboratory, School of Natural Sciences, University of Newcastle, Bedson Building, Newcastle upon Tyne, UK, NE1 7RU. E-mail: anthony.harriman@ncl.ac.uk; Fax: +44 191 222 8660*
[b] *Laboratoire de Chimie Moléculaire, Ecole Européenne de Chimie, Polymères et Matériaux, Université Louis Pasteur, 25 rue Becquerel, 67087, Strasbourg Cedex 02, France*

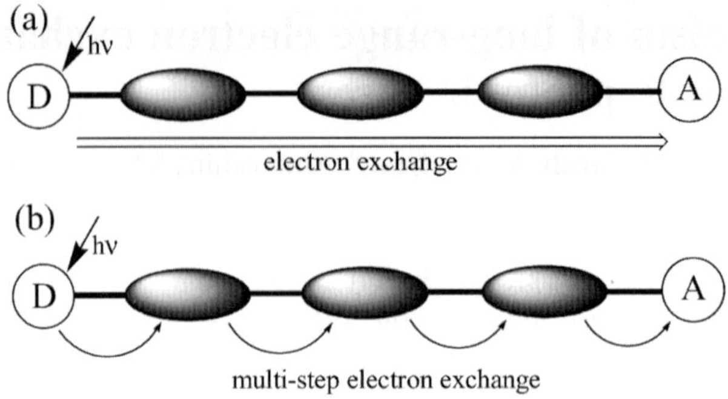

Fig. 1 Pictorial representation of long-range electron exchange occurring *via* (a) electron tunneling and (b) electron hopping processes from donor (D) to acceptor (A).

Controlling and optimising such systems requires detailed knowledge of the precise role of the bridge. This is particularly important with light-activated devices and it is essential that we obtain a proper understanding of the excited-state behaviour of all components in the system. Since there exists the possibility for π,π^*, $n–\pi^*$, intramolecular charge-transfer, metal-to-ligand and ligand-to-metal charge-transfer excited states,[6] of both singlet and triplet parentage, it is clear that the excited state manifold will be heavily congested. Add to this the realisation that the energy levels of π,π^* and charge-transfer states are strongly dependent on the degree of conjugation[7] and it becomes obvious that resolving the various excited-state identities will not be straightforward. In fact, there are several examples where the various excited states reside in thermal equilibrium at ambient temperature, thereby further complicating the situation.[6]

As the bridge becomes longer there is a greater propensity for its involvement as a real intermediate in the overall chemistry. It is possible that the mechanism will evolve with increasing length of the bridge, moving from electron tunneling to charge hopping as the bridge becomes more extended.[8] It is also important to consider the flexibility of the bridge under operating conditions since few molecular scaffolds are really rigid in solution at room temperature. Segments of the bridge might act together rather than blend into continuous states. Clearly, this is a rich and varied field that provides numerous opportunities for the design of unusual materials.

Ruthenium(II) poly(pyridine) complexes, and the corresponding osmium(II) and rhenium(I) analogues, have found prominent use in a wide variety of photo-activated molecular systems.[9–12] These materials often emit in fluid solution at ambient temperature and have been used as chemical sensors,[13] electron[14] or photon[15] donors, light harvesters[16] and electro-generated luminescent centres.[17] In many cases, the photophysical properties of the metal complexes are sensitive to temperature,[18] nuclearity[19] and the nature of the poly(pyridine) ligand.[20] Attaching aryl hydrocarbon residues[21–24] close to the metal centre provides a means by which to prolong the triplet lifetime of the complex whilst the addition of electron donating or withdrawing units opens up the possibility to involve the metal complex in intramolecular charge-transfer reactions.[25,26] Attaching conjugated substituents to the metal complex introduces the likelihood that ligand-localised excited states will figure in the triplet manifold.[27] Conjugated substituents also favour electron delocalisation at the triplet level.[28–30] As such, the photophysical properties of the metal complex become complicated and difficult to unravel. This is a particular problem for arrays, or networks, of luminescent metal complexes designed for use in molecular photonic devices[31] where it is important to retain the metal-to-ligand, charge-transfer state as the lowest-energy triplet in the system.

Here, we describe the luminescence properties of a series of linear arrays containing up to five ruthenium(II) tris(2,2′-bipyridine) complexes linked by π-conjugated bridges.[32] Such materials could be useful as luminescent wires for signalling purposes or as light-harvesting units. Prior work[5] has shown that this bridge facilitates electron exchange between terminal metal complexes whilst related research[33] has established that π-conjugated ligands of this generic type display interesting electro-generated luminescence properties. In such systems, it is important to identify the nature of the lowest-energy triplet excited state and to ascertain if individual chromophores operate separately or in a cooperative manner. Although there is considerable literature information pertaining to closely coupled binuclear[34] and trinuclear[35] metal complexes, little is known about the more extended systems. This situation is in marked contrast to the corresponding organic fragments where it is well known that the optical properties depend on the degree of oligomerisation.[36,37]

2. Objective

The aim of this study is to explore the excited state manifold for a series of oligomeric metal complexes of increasing nuclearity. These materials are designed to operate as prototypic molecular-scale wires that conduct electronic charge. The system comprises a series of 2,2′-bipyridine ligands separated by diethynylene-phenoxyl groups. The key issue is to resolve excited states localised on the metal complex and on the organic framework. It is recognised that the energy of these latter excited states will likely change on attachment of the metal cation. The emergence of next-generation photonic devices depends on our ability to resolve these issues.

3. Experimental

Raw materials were purchased from Aldrich Chemicals Co. and used as received. The free ligands were synthesized as described earlier[32,38] while the resultant metal complexes were synthesized by routine methods.[39] All metal complexes were prepared by gently heating the free ligand with the required amount of $[Ru(bipy)_2Cl_2]\cdot 2H_2O$ in ethanol. The course of reaction was followed by thin-layer chromatography and the target complexes were isolated as the hexafluorophosphate salts by column chromatography using alumina as support and a mixture of dichloromethane and methanol as the mobile phase. All complexes were unambiguously characterised by ^{1}H-NMR, FT-IR, FAB-MS or ES-MS, cyclic voltammetry and elemental analysis. Solvents were spectroscopic grade and distilled from suitable drying agents immediately before use. Full synthetic details will be published elsewhere. Structures of the compounds used throughout this study are shown at appropriate places in the text. The 200.1 (^{1}H) NMR spectra were recorded at room temperature on a Bruker AC 200 spectrometer, using perdeuterated acetone as internal standard. High-resolution mass spectral analyses (HRMS) were performed using a Mariner ESI-Tof instrument from Applied Bio-System/Perkin Elmer. Solvents were spectroscopic grade and were freshly distilled before use.

Absorption spectra were recorded with a Hitachi U3310 spectrophotometer while corrected emission spectra were recorded with a Hitachi F4500 spectrophotometer. All luminescence measurements were made using optically dilute solutions and were corrected for spectral imperfections of the instrument by reference to a standard lamp. Luminescence quantum yields for the various metal complexes were determined relative to ruthenium(II) tris(2,2′-bipyridine) in deoxygenated acetonitrile.[40] Fluorescence quantum yields were measured relative to 9,10-dimethylanthracene in cyclohexane[41] and corrections were made for changes in refractive index of the solvent. Time-resolved emission studies were made with a Spex Fluorolog tau-3 spectrophotometer. Temperature dependence studies were made with either an Oxford Instruments Optistat DN cryostat or a liquid circulating bath. In the latter case, the temperature was measured by a thermocouple maintained in direct contact with the cell. Deconvolution of the emission spectra into the minimum number of Gaussian components was made with PEAKFIT. The spectra were reconstituted using purpose-written software. Quantum mechanical calculations were made using GAUSSIAN-03 or TITAN, running on a fast PC.

4. Results and discussion

4.1 Synthesis

The series of soluble ligands was synthesized around a 1,4-diethynyl-2,5-diiododecyloxybenzene bridging unit.[32] The synthesis required a protection/deprotection methodology, as well as a chemioselective palladium-catalysed Sonogashira–Hagihara cross-coupling protocol to obtain the target multitopic ligands. Within this strategy,[38] the pivotal intermediates bearing one or two bipyridine and phenyl units are required and such entities have been isolated in excellent yield. The products are highly soluble and photostable. In each case, the final step involves a double cross-coupling reaction between the appropriate constituents, with the best preparative conditions

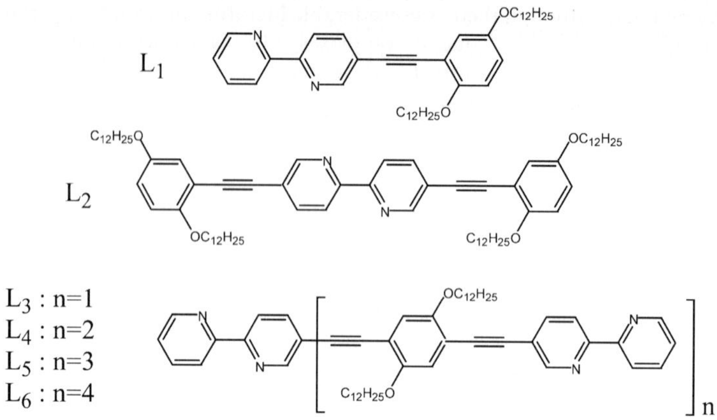

Fig. 2 Structural formulae for the free ligands used in this work.

involving [$Pd(PPh_3)_4$] (6 mol%) in *n*-propylamine at 70 °C. The main advantage of this methodology lies in its synthetic versatility and adaptability for creating multitopic metal-binding scaffolds with a potentially large variety of bridging units and phenyl substituents.

4.2 Photophysical properties of the free ligands

The structures are shown in Fig. 2. In order to provide for adequate solubility in organic solvents, the polytopic ligands were built around 1,4-dialkoxybenzene groups. These latter groups were intended to increase the solubility and also to isolate the individual bipyridine units, at least in an electronic sense. One way to assess the success of this strategy is to measure the photophysical properties of the ligands prior to attaching the metallo-fragments. Thus, each of the free ligands displays pronounced absorption in the near-UV region that can be assigned to a combination of π,π^* transitions.[38] The wavelength corresponding to the lowest-energy transition (λ_{max}) shows a modest dependence on the degree of oligomerisation (Table 1). Strong fluorescence is observed for each compound but the fluorescence maximum (λ_{flu}) shows a pronounced Stokes shift (Table 1) in *N*,*N*-dimethylformamide solution. The fluorescence quantum yields (Φ_F) are very high, approaching unity in some cases, but the emission lifetimes (τ_F) are quite short (Table 1). The radiative rate constants (k_{rad}) depend only on the average frequency of the fluorescence spectrum and are otherwise independent of the nature of the compound. This latter finding holds true in a range of solvents of different polarity. Despite the high fluorescence yields, weak phosphorescence could be detected in an ether : pentane : alcohol 5 : 5 : 2 glass at 77 K. The phosphorescence maxima (λ_{pho}) show a slight dependence on the molecular length (Table 1).

These results imply that the degree of electronic communication along the molecular axis is modest. At first sight, this seems surprising but it was found that the lowest-energy excited singlet state possesses considerable charge-transfer character. This effect was evidenced by the strong

Table 1 Photophysical properties recorded for the free ligands L_1–L_6 in *N*,*N*-dimethylformamide solution at room temperature. NB The phosphorescence spectra were recorded in an EPA glass at 77 K

Compound	λ_{max}/nm	λ_{flu}/nm	Φ_F	τ_F/ns	λ_{pho}/nm
L_1	347	477	0.64	4.1	545
L_2	355	504	0.19	2.0	540
L_3	388	448	0.88	1.5	547
L_4	399	478	0.44	1.6	545
L_5	397	478	0.37	1.6	545
L_6	401	480	0.35	1.7	545

dependence of the Stokes shift on the polarity of the surrounding solvent. For L_3, this dependence corresponds to a change in dipole moment upon excitation of 22.2 D. In contrast, the excited triplet state appears to retain similar polarity to that of the ground state. This means that the singlet–triplet energy gap depends markedly on solvent polarity. At room temperature, nonradiative deactivation of the excited singlet state is primarily *via* intersystem crossing to the triplet manifold. By correlating the derived rate constants with the measured energy gap, in accordance with classical Marcus–Jortner theory,[42] the reorganisation energy associated with intersystem crossing was calculated to be *ca.* 1500 cm^{-1}.

On the basis of quantum chemical calculations, it appears that the electron donor in these ligands is the dialkoxybenzene unit while the electron acceptor is the bipyridine group. The molecular structure favours short-range interactions and it is this realisation that inhibits the creation of a giant dipole along the molecular axis. Since increasing the degree of oligomerisation merely adds incremental numbers of charge-transfer units to the chain, there is only a weak correlation between excited state energy and the molecular length. The energy of the triplet state, in particular, is relatively insensitive to increasing degree of oligomerisation. This is good news for the future development of molecular-scale photoactive wires, especially those based on triplet states.

4.3 Attachment of zinc(II) cations

Each of the ligands L_1–L_6 was titrated with a stock solution of $Zn(ClO_4)_2$ in chloroform solution and the course of reaction was followed by UV–visible absorption and fluorescence spectroscopy. The presence of the cation caused significant changes in the UV–visible absorption spectrum (Fig. 3) and led to a serious extinction of fluorescence from the ligand (Fig. 4). It is clear that the cation coordinates to the vacant bipyridine sites available on the ligand. For both L_1 and L_2, coordination of zinc(II) could be analysed satisfactorily in terms of the stepwise formation of 1 : 2 and 1 : 3 (metal : ligand) complexes. The derived stability constants are collected in Table 2. More complicated binding behaviour is evident, however, for the higher-order ligands. Indeed, molecular mechanics studies indicated that these polytopic ligands facilitate formation of metallo-helicates, despite the bulky dialkoxybenzene units.

The effect of added zinc(II) cations on the absorption spectrum of the free ligand is to induce formation of a new absorption band at longer wavelength. This band is characteristic of the metal complex and is most likely due to an intramolecular charge-transfer transition. The fluorescence spectra show progressive loss of emission characteristic of the free ligand and the concomitant growth of a new fluorescence profile at long wavelength. Again, this new band is attributed to the metal complex and possesses the expected features of an intramolecular charge-transfer transition. Clearly, isosbestic points are not preserved throughout the titrations; indicating the involvement of

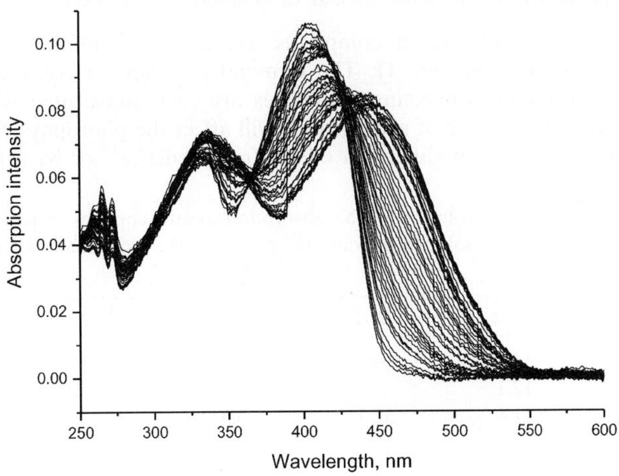

Fig. 3 Titration of L_5 with $Zn(ClO_4)_2$ in $CHCl_3$ as followed by UV–visible absorption spectroscopy.

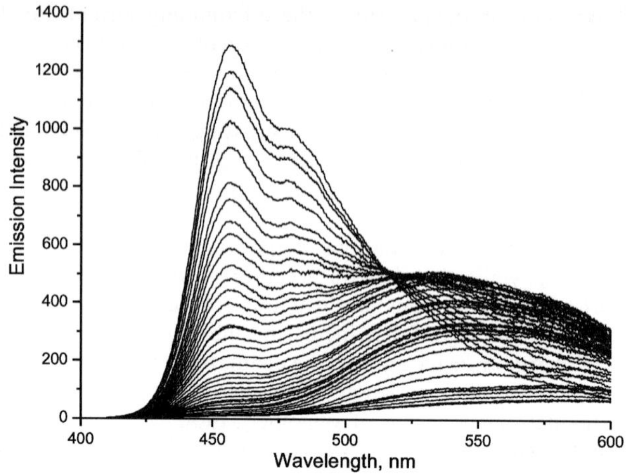

Fig. 4 Spectrofluorescence titration of L_5 with $Zn(ClO_4)_2$ in $CHCl_3$.

several species. Stability constants (β) were determined using SPECFIT to analyse the data collected over many different concentrations of ligand and cation. Comparable values were found from fitting the absorption and fluorescence data.

In fitting the titration data, we have assumed that the zinc(II) cation tries to maximise the coordination geometry to 6. We did not see clear evidence for formation of the 1 : 1 complex, except with L_2 and a large excess of cation. For ligands of intermediate length, namely L_3 and L_4, there is a propensity to wrap three ligands around two or three cations, each being coordinated to a bipyridine site. These are stable complexes. With L_5, both 3 : 3 and 4 : 3 complexes are formed, together with the 2 : 3 complex. However, the longest ligand, L_6, does not form the expected 5 : 3 helicate-type complex and prefers to assemble complexes with lower stoichiometry. This might be because of electrostatic repulsion or steric crowding.

The presence of the zinc(II) cation lowers the energy of the first-excited singlet state, which is of intramolecular charge-transfer character, by raising the reduction potential of the 2,2′-bipyridine ligand to a less negative value. There is no obvious dependence on chain length and each metal complex appears to act independently. Phosphorescence could not be detected at 77 K but this does not mean that the corresponding triplet excited states are not formed.

4.4 Photophysical properties of the mononuclear ruthenium(II) complexes

Apart from the dimer, the polynuclear complexes studied here contain two different types of substituted bipyridine ligand (Scheme 1): The terminal complex has only a single ethynylene attachment but the central or connecting complexes are each substituted with two ethynylene groups. It is likely that the number of such groups will affect the photophysical properties of the ruthenium(II) complex due to a modification of the triplet energy[37] or by a perturbation of the

Table 2 Stability constants derived from UV–visible absorption spectroscopy for attachment of zinc(II) cations to the various ligands in chloroform solution at room temperature. The stoichiometries refer to metal : ligand ratios

log β	1 : 2	1 : 3	2 : 3	2 : 2	3 : 3	4 : 3	5 : 3
L_1	10.9	15.3	—	—	—	—	—
L_2	9.8	12.1	—	—	—	—	—
L_3	—	11.8	32.3	23.8	—	—	—
L_4	—	11.9	16.6	—	21.3	—	—
L_5	—	—	25.7	—	31.7	37.8	—
L_6	—	13.9	19.0	—	33.0	7.1	—

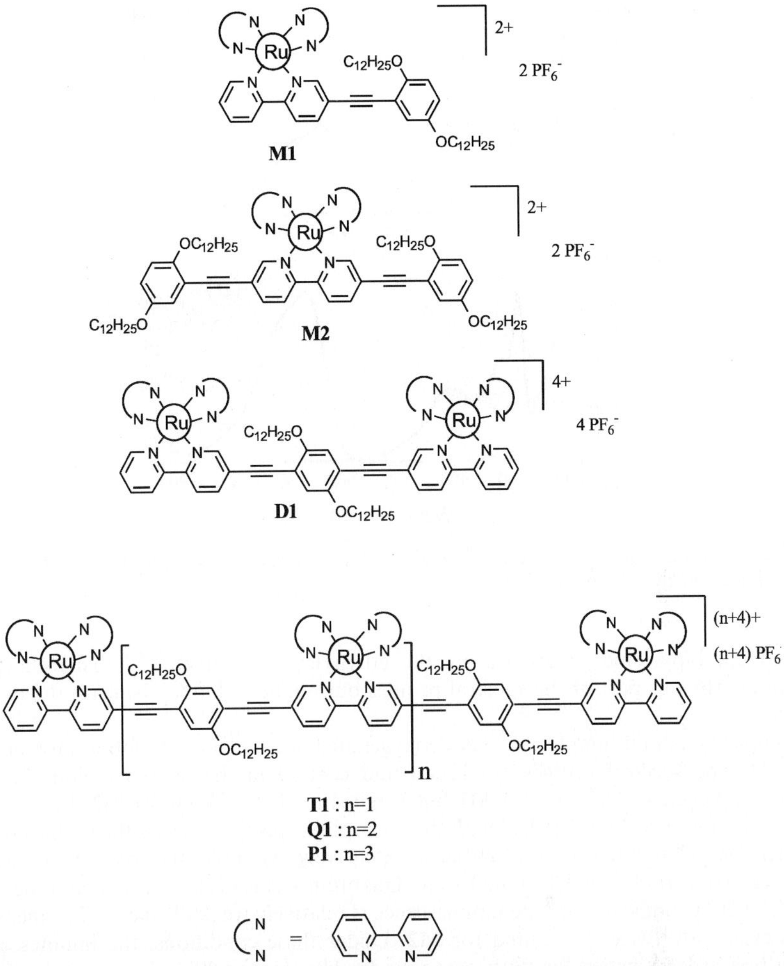

Scheme 1 Structures of the various ruthenium(II) poly(pyridine) complexes examined in this work.

extent of electron delocalisation at the triplet level.[28–30] In order to assess the importance of this structural change, the photophysical properties of the relevant metal complexes were evaluated in deoxygenated acetonitrile solution at room temperature. For **M1**, the absorption spectrum shows the expected metal-to-ligand, charge-transfer (MLCT) transitions centred around 460 nm (Fig. 5). There is a weak tail stretching beyond 600 nm that can be attributed to the spin-forbidden MLCT transitions. The parent 2,2′-bipyridine ligands show prominent absorption centred around 290 nm. There are two obvious absorption bands associated with the ethynylene-substituted ligand. One band appears around 320 nm whilst the second band is apparent between 350 and 420 nm. This latter band overlaps to some degree with the MLCT transition.

The absorption spectrum recorded for **M2** is qualitatively similar to that described above and, in particular, π,π* transitions associated with the parent ligands are seen clearly at 290 nm (Fig. 5). The disubstituted 2,2′-bipyridine ligand displays a relatively intense absorption band at 320 nm and a broad, structureless transition centred around 400 nm. The MLCT transition appears as a weak shoulder at 460 nm on the low-energy side of this latter band, although the spin-forbidden MLCT transition is well resolved. The ligand-centred (LC) absorption transition seen around 400 nm is similar to that found for the free ligand.[32] In the latter case, both absorption bands were assigned to π,π* transitions. It is possible that **M1** and **M2** also display an intramolecular charge-transfer transition in which the dialkoxybenzene unit acts as electron donor and the acceptor is the

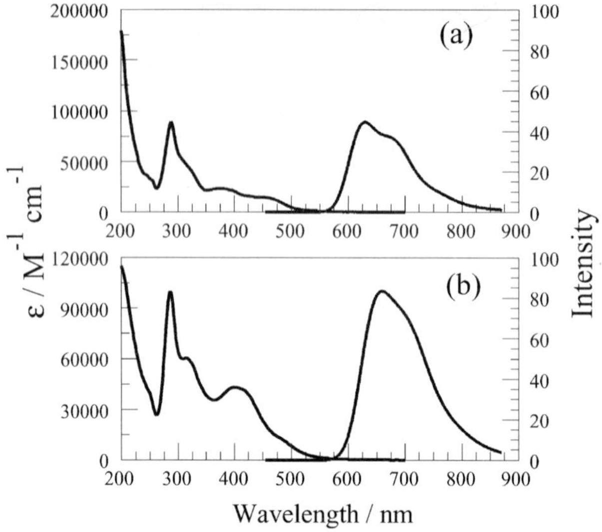

Fig. 5 Absorption and emission spectra recorded for the mononuclear complexes (a) **M1** and (b) **M2** in deoxygenated acetonitrile solution at 20 °C.

coordinated 2,2′-bipyridine group bearing the ethynylene substituent.[34,43] This charge-transfer band is not an obvious part of the spectral profiles but might be hidden beneath the more intense π,π* bands.

Both complexes exhibit luminescence in deoxygenated acetonitrile solution at ambient temperature (Fig. 5). The spectral profiles are broad and centred in the far-red region. The emission maximum (λ_{lum}) occurs at 628 nm for **M1** but is red-shifted to 653 nm for **M2**. In both cases, the corrected excitation spectrum matched with the absorption spectrum across the visible and near-UV regions. The emission intensity was reduced extensively by molecular oxygen, indicating that luminescence arises from a triplet-excited state. Quantum yields (Φ_{lum}), measured in deoxygenated acetonitrile at 20 °C, indicate that the luminescence is relatively weak (Table 3). The measured Φ_{lum} for **M1** is essentially twice that found for **M2**. Under these conditions, the luminescence signal decays *via* first-order kinetics for both compounds. The derived lifetimes (τ_{lum}) are collected in Table 3 and were found to be independent of laser intensity and monitoring wavelength. Again, it is notable that the triplet lifetime found for **M1** exceeds that recorded for **M2** under identical conditions.

For both complexes, the triplet lifetime was found to be essentially independent of temperature over the range 250 to 330 K (Fig. 6a). For **M1**, the variation in triplet lifetime was less than 10% whilst that for **M2** was even less pronounced. It was not possible to construct a meaningful Arrhenius-type plot for these data. The experimental results give no indication for electronic coupling between triplet levels, at least over this modest temperature range. It should be noted that

Table 3 Photophysical properties recorded for the various ruthenium(II) poly(pyridine) complexes in deoxygenated acetonitrile solution at 20 °C

Cmpd	λ_{lum}/nm	Φ_{lum}	τ_{lum}/ns	$10^{-4}k_{rad}$/s^{-1}	$10^{-4}k_{nr}$/s^{-1}	α/°C^{-1}	ν_0/cm^{-1}
M1	628	0.055	850	6.5	110	2.5	15 875
M2	653	0.026	370	7.0	265	2.2	15 270
D1	634	0.033	620	5.3	155	2.4	15 725
T1	653	0.022	400	5.5	245	2.0	15 275
Q1	660	0.018	370	4.9	265	1.9	15 115
P1	665	0.014	360	3.9	275	1.9	15 000

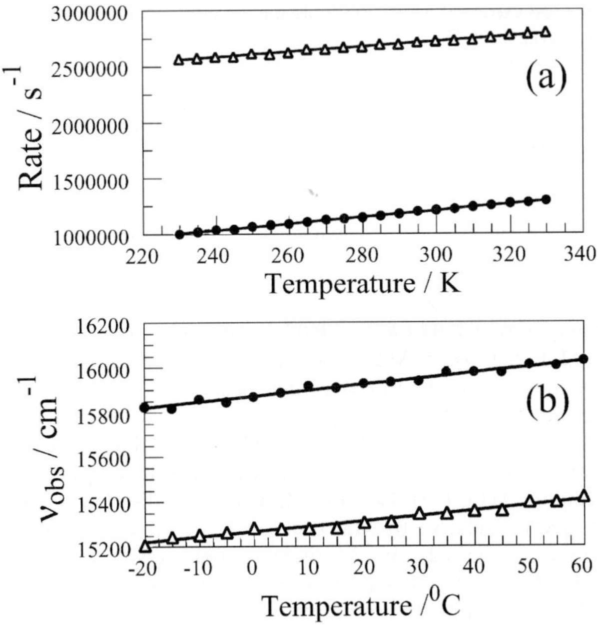

Fig. 6 Effect of temperature on (a) the reciprocal triplet lifetime and (b) the emission maximum for the mononuclear complexes **M1** (●) and **M2** (△) in deoxygenated acetonitrile solution.

Meyer *et al.*[44] have shown that the rate of nonradiative deactivation of certain MLCT triplet states should show a very weak temperature dependence, of the type described here.

The emission maxima are also affected by temperature (Fig. 6b), although again this is a rather small effect. For both complexes, the peak maximum, expressed in wavenumbers (ν_{OBS}), obeys the following relationship over the temperature range −20 to +60 °C.

$$\nu_{obs} = \nu_0 + \alpha T \tag{1}$$

Here, the temperature is expressed in °C and the derived value for the coefficient α is listed in Table 3. It is seen that **M1** is somewhat more sensitive to temperature variations but the emission maximum at 0 °C (ν_0) is lower for **M2**. This latter observation is in keeping with the room temperature results.

The photophysical properties derived for **M2** can be compared with those determined earlier by Schanze *et al.*[45] for a somewhat related mononuclear ruthenium(II) tris(2,2′-bipyridine) complex, **M3** (Fig. 7). It was shown that the absorption spectral profile recorded for **M3** contained important contributions from transitions associated with the parent ligand, together with LC and MLCT bands. Luminescence was observed from **M3** in deoxygenated acetonitrile, with a maximum at

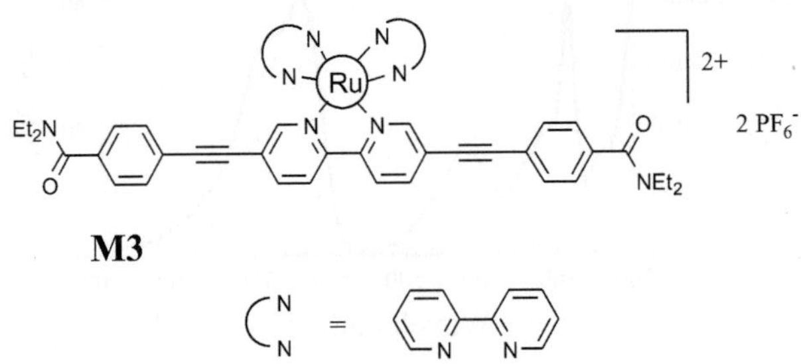

Fig. 7 Structure of the related mononuclear complex **M3** described in ref. 45.

Table 4 Parameters extracted from the emission spectral analyses carried out for spectra recorded in acetonitrile at 20 °C

Cmpd	E_{00}/cm^{-1}	S_M	$\hbar\omega_M$/cm^{-1}	$\Delta\nu_{1/2}$/cm^{-1}
M1	16 025	0.71	1,350	1,450
M2	15 420	0.64	1,290	1,560
D1	15 875	0.70	1,330	1,400
T1	15 410	0.65	1,300	1,620
Q1	15 250	0.63	1,280	1,650
P1	15 150	0.62	1,290	1,650

693 nm. The emission quantum yield ($\Phi_{lum} = 0.035$) and lifetime ($\tau_{lum} = 520$ ns) remain similar to those measured for **M2**. In the case of **M3**, it was concluded[45] that the emissive state was the lowest-energy MLCT triplet state formed upon charge injection from the metal centre to the substituted ligand. It was further considered that the promoted electron was extensively delocalised over the ligand,[46] in line with the behaviour noted for several different binuclear complexes.[34] Although mindful of the LC state associated with **M2**, it seems reasonable to suppose that emission from both **M1** and **M2** arises from the MLCT triplet state. Electron delocalisation is likely to be less extensive with these latter complexes, relative to **M3**, because of the absence of the amide groups. This difference might account for the red-shifted emission noted for **M3**.

Several other factors are in apparent support of the assignment of the emitting state as being the lowest-energy MLCT triplet. Thus, the radiative rate constants of 6.5×10^4 and 7.0×10^4 s^{-1}, respectively, calculated for **M1** and **M2** are in keeping with those recorded for other ruthenium(II) poly(pyridine) complexes.[45,46] The weak temperature dependence found over the ambient temperature range is also characteristic of these chromophores.[34] The corresponding rate constants for nonradiative deactivation of the lowest-energy triplet state (k_{NR}) are comparable to those found for certain closely-coupled binuclear ruthenium(II) tris(2,2'-bipyridine) complexes.[34] The difference in k_{NR} found between **M1** and **M2** is easily accounted for in terms of the energy-gap law.[47,48] Finally, fitting the room temperature emission spectra to a single-mode Franck–Condon analysis[49] gives parameters considered typical of MLCT luminescence. Thus, the luminescence spectra were analysed in terms of eqn. (2) where $I(\nu)$ is the normalised emission intensity at wavenumber ν (in cm^{-1}).

$$I(\nu) = \sum_{m=0}^{5}\left\{\left(\frac{E_{00} - m\hbar\omega_M}{E_{00}}\right)^3 \frac{(S_M)^m}{m!}\exp[-4\ln 2Q]\right\}$$
$$Q = \left(\frac{\nu - E_{00} + m\hbar\omega_M}{\Delta\nu_{1/2}}\right)^2 \tag{2}$$

Here, E_{00} is the energy of the zero–zero transition, $\hbar\omega_M$ is a weighted-averaged, medium-frequency vibrational mode coupled to the emitting state, m is the vibrational quantum number of this

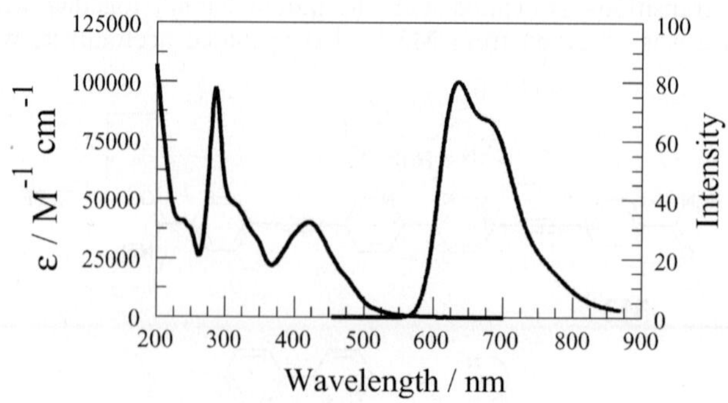

Fig. 8 Absorption and emission spectra recorded for the binuclear complex **D1** in deoxygenated acetonitrile solution at 20 °C.

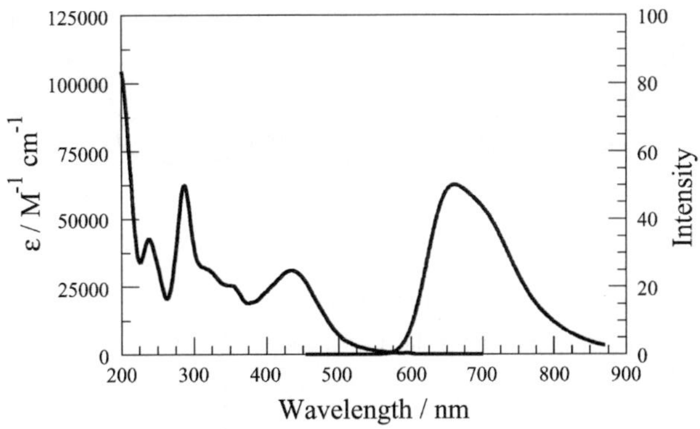

Fig. 9 Absorption and emission spectra recorded for the trinuclear complex **T1** in deoxygenated acetonitrile solution at 20 °C.

medium-frequency mode, S_M is the Huang–Rhys factor and $\Delta\nu_{1/2}$ is the half-width of the individual vibronic bands. The parameters extracted from this analysis are collected in Table 4. They appear to be as expected for emission from an MLCT triplet state possessing extended electron delocalisation.[28–30] On this basis, we believe that the luminecent state in both **M1** and **M2** is the lowest-energy MLCT triplet. This state is formed upon charge injection from the metal centre into the substituted 2,2′-bipyridine ligand. Presumably, the promoted electron is subject to delocalisation over part of the substituent.[46]

4.5 Photophysical properties of the ruthenium(II) dimer

The binuclear complex **D1** is built around the unit characterised for **M1**. The absorption spectrum recorded in acetonitrile solution shows the expected transitions associated with the parent ligand at 290 nm (Fig. 8). The polytopic ligand exhibits absorption peaks at 320 and 350 nm that appear to be similar to the higher-energy transition noted for the two mononuclear complexes. The lower-energy transition appears as a broad, structureless band centred at 430 nm. This latter band is somewhat red shifted with respect to the mononuclear complexes and serves to obscure the MLCT transitions associated with the metal complexes. In fact, the spin-allowed MLCT transition can be recognised only as a slight shoulder around 480 nm. The spin-forbidden MLCT transition is still apparent in the spectrum as a tail stretching beyond 600 nm. The absorption bands attributed to the polytopic ligand agree well with those characterised earlier for the free ligand,[32] except that the presence of the metal centres induces a modest red shift. As such, these absorption bands are probably due to π,π^* transitions.

Weak emission is observed with **D1** in deoxygenated acetonitrile solution at room temperature (Fig. 8). The luminescence profile is broad and centred around 634 nm, which is remarkably similar to that found for **M1** under these conditions. Good agreement was noted between the corrected excitation spectrum and the absorption spectrum over the range 300–600 nm. It is interesting to note that whereas the emission lifetime found for **D1** remains quite similar to that found for **M1** there is a more substantial decrease in the quantum yield (Table 3). This suggests that the radiative rate constant ($k_{rad} = 5.3 \times 10^4\ s^{-1}$) is lower for the dimer than for the corresponding monomer but it should be noted that recent work[34] has reported triplet quantum yields less than unity for closely coupled binuclear complexes. Again, the emission peak moves towards lower energy as the temperature increases while the emission lifetime is essentially independent of temperature over a modest range. Comparing the various parameters with those recorded for the mononuclear complexes (Tables 3 and 4) shows that the binuclear complex exhibits very similar behaviour to that displayed by **M1**. Apart from the slightly lower Φ_{lum}, there are no obvious effects due to the increase in nuclearity and it appears that the central dialkoxybenzene unit does not promote increased conjugation along the molecular axis in this case. There is no indication that the two metal complexes act cooperatively and it appears that the dialkoxybenzene subunit functions as an electronic insulator for the triplet state. In particular, it should be noted that the corresponding

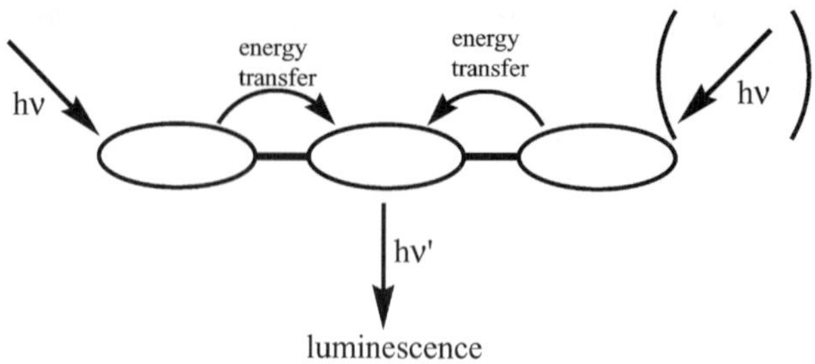

Fig. 10 Pictorial representation of intramolecular triplet energy transfer, followed by emission, proposed for the trinuclear complex **T1**.

binuclear complex lacking the central aryl unit exhibits a pronounced temperature effect.[34] Such behaviour is in marked contrast to that found for **D1**.

4.6 Photophysical properties of the trinuclear ruthenium(II) complex

The trinuclear complex **T1** comprises a central unit comparable to **M2** and terminal complexes reminiscent of **M1** (Fig. 1). The absorption spectrum recorded in acetonitrile is dominated by π,π^* transitions associated with the polytopic ligand (Fig. 9). Absorption due to the parent ligand is seen clearly at 290 nm while the polytopic ligand absorbs over the range 300–500 nm. Some fine structure is apparent in the near-UV and the absorption maximum occurs at 430 nm. However, the spin-allowed MLCT transition is not obvious in the spectrum. As above, the spin-forbidden MLCT band can be seen as a tail on the low-energy side of the π,π^* transitions. Relative to the free ligand,[32] the lowest-energy π,π^* transitions noted for **T1** are red shifted by about 20 nm but the spectral profile remains the same. The origin of this red shift can be traced to the increased planarity of the ligand upon coordination and to the reduction of the LUMO energy that accompanies formation of the metal complex.

Weak luminescence occurs for **T1** in deoxygenated acetonitrile at room temperature (Fig. 9). The emission maximum is centred at 653 nm, as is that found for **M2** under identical conditions. Good agreement was observed between the corrected excitation spectrum and the absorption spectrum over the range 250–600 nm. The emission quantum yield was relatively low but the lifetime remained similar to those found for the mononuclear complexes (Table 3). The decay kinetics remained strictly mono-exponential at all monitoring wavelengths while the lifetime was shortened in the presence of molecular oxygen. The emission maximum was found to be slightly dependent on

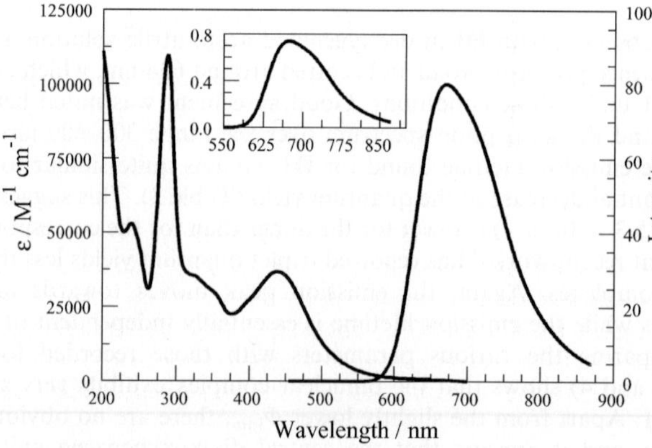

Fig. 11 Absorption and emission spectra recorded for the tetranuclear complex **Q1** in deoxygenated acetonitrile solution at 20 °C. The insert shows the emission spectrum recorded for **P1** under the same conditions.

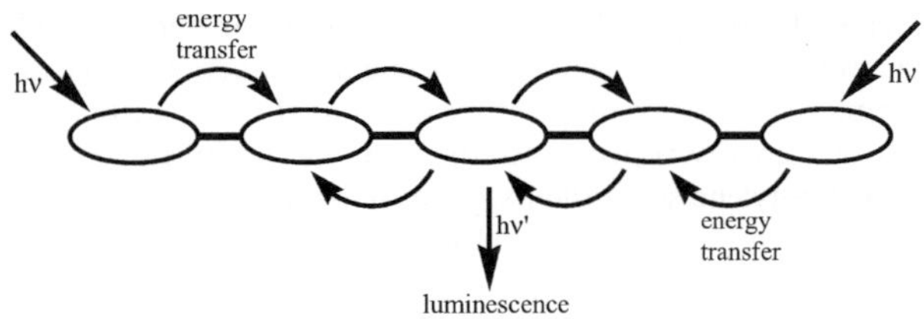

Fig. 12 Pictorial representation of intramolecular triplet energy transfer and the localisation effect proposed for the longer arrays.

temperature (Table 3), with the value recorded at 0 °C being somewhat lower than that found for **M2**. The radiative rate constant, calculated assuming the triplet state is formed in a quantum yield of unity, is slightly lower ($k_{rad} = 5.5 \times 10^4\ s^{-1}$) than those found for the other complexes.

The nonradiative decay rate constant was found to be essentially independent of temperature over the range 240–330 K. The derived value (Table 3) is comparable to that found for **M2** but it is significantly larger than the k_{NR} characterised for **M1**. Likewise, the various parameters extracted by fitting the emission spectrum to eqn. (2) are in keeping with luminescence arising solely from the central metal complex (Table 4). In fact, the experimental data suggest that the three metal complexes function independently. Since the two terminal complexes, these being akin to **M1**, possess somewhat higher triplet energies it is reasonable to suppose that they will transfer excitation energy to the central unit (Fig. 10). This latter unit emits with the characteristic features of **M2**. The absence of any apparent temperature effect on the triplet lifetime rules out the possibility of LC triplet states of comparable energy. Thus, as with **D1**, it appears that the trinuclear complex is segmented into individual complexes that do not communicate in an electronic fashion.

4.7 Photophysical properties of the longer arrays

The absorption spectrum recorded for the tetranuclear complex **Q1** shows the main features already recognised for **T1** (Fig. 11). The parent 2,2′-bipyridine units absorbs strongly at 290 nm and two distinctive absorption bands can be seen for the polytopic ligand. The lower energy LC absorption band is centred at 438 nm and completely obscures the MLCT transition. It is difficult to resolve the spin-forbidden MLCT band from the overlapping π,π* transitions. There is a small red shift relative to **T1** whilst the absorption peak is red shifted by about 30 nm with respect to that of the free ligand. The absorption spectrum recorded for the pentanuclear complex **P1** is very similar to that described for the tetranuclear complex except that the absorption peak is shifted to 445 nm. Again, the anticipated MLCT transitions cannot be resolved in the spectrum.

Luminescence could be detected from the tetranuclear complex in deoxygenated acetonitrile at room temperature (Fig. 11). The emission maximum is red shifted by about 7 nm compared to that observed for the trinuclear complex (Table 3). Both the emission quantum yield and the lifetime are reduced somewhat with respect to **T1**, probably because of the lower triplet energy. Again, the parameters extracted from the spectral analysis (Table 4) are in keeping with the luminescence arising from a triplet MLCT state. Very similar behaviour is found for the pentanuclear complex **P1**. In both cases, the emission is quenched by molecular oxygen and is hardly affected by modest changes in temperature. These latter studies provide no indication for the presence of a relatively low energy LC triplet state that could be populated by thermal equilibration from the lowest-energy triplet. Overall, the emission spectral properties remain similar to those recorded for the trinuclear complex. The various subunits do not appear to act cooperatively and the effective conjugation length remains comparable to that identified for **M2**. Again, the appearance is that the central dialkoxybenzene subunit acts to isolate a relatively small luminescent centre (Fig. 12). If so, we could suppose that energy transfer occurs from the terminals to the central units.

5 Concluding remarks

The most interesting feature to emerge from this study is that the lowest-energy triplet is associated with the MLCT state in all the compounds examined here. There is no obvious indication for a LC π,π* triplet state localised on the polytopic ligand, even for the pentanuclear species. Such behaviour might arise from the central dialkoxybenzene unit functioning as an insulator or because the pendant metal complexes interrupt the electronic conjugation of the polytopic ligand. It is likely that intramolecular triplet energy transfer occurs from the peripheral metal complexes to those complexes bearing two ethynylene substituents, because of the lower triplet energy. The segmented nature of these molecular wires might be a useful facet for constructing artificial light-harvesting arrays.

Acknowledgements

The Engineering and Physical Sciences Research Council (GR/S00088/01), the CNRS, the University of Newcastle and the Université Louis Pasteur de Strasbourg provided financial support for this research programme. We thank Johnson Matthey for the loan of precious metal salts.

References

1 S. I. Khondaker, Z. Yao, L. Cheng, J. C. Henderson, Y. X. Yao and J. M. Tour, *Appl. Phys. Lett.*, 2004, **85**, 645.
2 E. A Weiss, M. J. Ahrens, L. E. Sinks, A. V. Gusev, M. A. Ratner and M. R. Wasielewski, *J. Am. Chem. Soc.*, 2004, **126**, 5577.
3 J. P. Lin and D. N. Beratan, *J Phys. Chem. A*, 2004, **108**, 5655.
4 Y. A. Berlin, G. R. Hutchinson, P. Rempala, M. A. Ratner and J. Michl, *J. Phys. Chem. A*, 2003, **107**, 3970.
5 A. Harriman, A. Khatyr, R. Ziessel and A. C. Benniston, *Angew. Chem. Int. Ed.*, 2000, **39**, 4287.
6 A. Harriman, A. Khatyr and R. Ziessel, *Dalton Trans.*, 2003, 2061.
7 C. Y. Tan, M. R. Pinto and K. S. Schanze, *Chem. Commun.*, 2002, 446.
8 W. B. Davis, M. A. Ratner and M. R. Wasielewski, *J. Am. Chem. Soc.*, 2001, **123**, 7877.
9 A. Juris, V. Balzani, F. Barigelletti, S. Campagna, P. Belser and A. Von Zelewsky, *Coord. Chem. Rev.*, 1988, **84**, 85.
10 J.-P. Sauvage, J.-P. Collin, J.-C. Chambron, S. Guillerez, C. Coudret, V. Balzani, F. Barigelletti, L. De Cola and L. Flamigni, *Chem. Rev.*, 1994, **94**, 993.
11 A. Harriman and J.-P. Sauvage, *Chem. Soc. Rev.*, 1996, 41.
12 A. Harriman and R. Ziessel, *Chem. Commun.*, 1996, 1707.
13 A. Harriman, M. Hissler, P. Jost, G. Wippf and R. Ziessel, *J. Am. Chem. Soc.*, 1999, **121**, 14.
14 Y. Liu, A. De Nicola, O. Reiff, R. Ziessel and K. S. Schanze, *J. Phys. Chem. A*, 2003, **107**, 3476.
15 V. Grosshenny, A. Harriman and R. Ziessel, *Angew. Chem., Int. Ed. Engl.*, 1995, **34**, 1100.
16 C. Kim and H. Kim, *J. Organomet. Chem.*, 2003, **77**, 673.
17 G. B. Cunningham, Y. T. Li, S. X. Liu and K. S. Schanze, *J. Phys. Chem. B*, 2003, **107**, 12569.
18 F. Barigelletti, A. Juris, V. Balzani, P. Belser and A. Von Zelewsky, *J. Phys. Chem.*, 1986, **90**, 5190.
19 E. C. Constable, C. E. Housecroft, E. R. Schofield, S. Encinas, N. Armaroli, F. Barigelletti, L. Flamigni, E. Figgemeier and J. G. Vos, *Chem. Commun.*, 1999, 869.
20 M. Hissler, A. El-ghayoury, A. Harriman, R. Ziessel and R. Coord, *Chem. Rev.*, 1998, **180**, 1251.
21 W. E. Ford and M. A. J. Rodgers, *J. Phys. Chem.*, 1991, **96**, 2917.
22 G. J. Wilson, W. H. F. Sasse and A. W.-H. Mau, *Chem. Phys. Lett.*, 1996, **250**, 583.
23 J. A. Simons, S. L. Curry, R. H. Schmehl, T. R. Schatz, P. Piotrowiak, X. Jin and R. P. Thummel, *J. Am. Chem. Soc.*, 1997, **119**, 11012.
24 M. Hissler, A. Harriman, A. Khatyr and R. Ziessel, *Chem. – Eur. J.*, 1999, **5**, 3366.
25 A. D. Guerzo, S. Leroy, F. Fages and R. H. Schmehl, *Inorg. Chem.*, 2002, **41**, 359.
26 A. C. Benniston, A. Harriman, D. J. Lawrie and A. Mayeux, *Phys. Chem. Chem. Phys.*, 2004, **6**, 51.
27 K. A. Walters, L. L. Premvardhan, Y. Liu, L. A. Peteanu and K. S. Schanze, *Chem. Phys. Lett.*, 2001, **339**, 255.
28 A. C. Benniston, V. Goulle, A. Harriman, J.-M. Lehn and B. Marczinke, *J. Phys. Chem.*, 1994, **98**, 7798.
29 A. C. Benniston, V. Grosshenny, A. Harriman and R. Ziessel, *Angew. Chem., Int. Ed. Engl.*, 1994, **33**, 1884.
30 V. Grosshenny, A. Harriman, F. M. Romero and R. Ziessel, *J. Phys. Chem.*, 1996, **100**, 17472.
31 Z. Peng and L. Yu, *J. Am. Chem. Soc.*, 1996, **118**, 3777.
32 A. Khatyr and R. Ziessel, *J. Org. Chem.*, 2000, **65**, 7814.
33 K. D. Ley, Y. Li, J. V. Johnson, D. H. Powell and K. S. Schanze, *Chem. Commun.*, 1999, 1749.
34 A. C. Benniston, A. Harriman, F. Romero and R. Ziessel, *Dalton Trans.*, 2004, 1227.
35 P. J. Connors, D. Tzalis, A. L. Dunnick and Y. Tor, *Inorg. Chem.*, 1998, **37**, 1121.
36 H. Kukula, S. Veit and A. Godt, *Eur. J. Org. Chem.*, 1999, 277.
37 K. A. Walters, D. M. Dattelbaum, K. D. Ley, J. R. Schoonover, T. J. Meyer and K. S. Schanze, *Chem. Commun.*, 2001, 1934.
38 A. Khatyr and R. Ziessel, *Tetrahedron Lett.*, 2000, **41**, 3837.
39 A. Harriman, M. Hissler, A. Khatyr and R. Ziessel, *Eur. J. Inorg. Chem.*, 2003, 955.

40 J. N. Demas and G. A. Crosby, *J. Am. Chem. Soc.*, 1971, **93**, 2841.
41 S. L. Murov, I. Carmichael and G. L. Hug, *Handbook of Photochemistry*, Marcel Dekker, New York, 1993.
42 P. A. Van Hal, S. C. J. Meskers and R. A. J. Janssen, *Appl. Phys. A*, 2004, **79**, 41.
43 S. Leroy, T. Soujanya and F. Fages, *Tetrahedron Lett.*, 2001, **42**, 1665.
44 J. P. Claude and T. J. Meyer, *J. Phys. Chem.*, 1995, **99**, 51.
45 Y. Wang, S. Liu, M. R. Pinto, D. M. Dattelbaum, J. R. Schoonover and K. S. Schanze, *J. Phys. Chem. A*, 2001, **105**, 11118.
46 N. H. Damrauer, T. R. Boussie, M. Devenny and J. K. McCusker, *J. Am. Chem. Soc.*, 1997, **119**, 8253.
47 R. Englman and J. Jortner, *Mol. Phys.*, 1970, **18**, 145.
48 E. M. Kober, J. V. Caspar, R. S. Lumpkin and T. J. Meyer, *J. Phys. Chem.*, 1986, **90**, 3722.
49 Z. Muraza, A. P. Zipp, L. A. Worl, D. K. Graff, W. E. Jones Jr., W. D. Bates and T. J. Meyer, *J. Am. Chem. Soc.*, 1991, **113**, 5113.

General Discussion

Dr Schuster opened the discussion of Professor Naaman's paper: Is DNA an "active" electrical element or is its role to bind the carbon nanotube to the electrode?

Professor Naaman replied: The role of the DNA is to bind the nanotube to the substrate in a controlled way. However, as we show we can use the DNA also as an electrical element, when charge is tunnelled between the electrodes and the nanotubes through the short DNA oligomers. By thermal heating we can remove the DNA and thereby form a direct contact between the electrode and the nanotube.

Professor Kornyshev asked: If the key point is DNA "conductance", have you changed the length of DNA to see the change in the resistivity? If with single stranded DNA you get wrapping and winding of DNA over nanotubes, could you do the experiments working with DNA duplexes from the very beginning? Up to 100 base pairs they will definitely be "rigid rods", because the persistence length is about 120 base pairs. Then the complication you are talking about will be gone! But were the experiments performed in solution or do you dry it out? If it is dry, then what you are measuring is not DNA in the conventional sense.

Professor Naaman responded: In the current study the length of the DNA was not changed. The fact that we have selective recognition between the single strand attached to the carbon nanotube and the one adsorbed on the electrode[1] indicates that the single strand is not wrapped around the nanotubes. The assembling of the devices was done in solution. The electrical measurements were performed in ambient conditions.

1 See also: M. Hazani, F. Hennrich, M. Kappes, R. Naaman, D. Peled, V. Sidorov and D. Shvarts, *Chem. Phys. Lett.*, 2004, **391**, 389–392.

Professor Calvo asked: Can you comment on how the current magnitude and the *I–V* characteristics depend on the DNA base sequence?

Professor Naaman answered: From preliminary results we see that the "band gap" we observe depends on the number of G–C pairs in the double strand. As the number increases the gate becomes smaller.

Professor Tao asked: How many of your electrically connected SWNTs are metallic?

Professor Naaman replied: Due to the oxidation procedure that we apply for attaching the DNA to the SWNTs, they all become semiconducting as is evident by their Raman spectra.

Professor Barbero commented: As you mention, oxidation of carbon nanotubes leads to conductivity loss. There has been reports that DNA strands can "wrap" the carbon nanotube. Using long DNA chains, it could be possible to have α-strands for hybridization and the metallic character of the nanotube will be maintained.

Professor Naaman replied: One should not confuse metallic properties with conduction properties. Even the semiconductor nanotubes conduct once the potential exceeds the band gap. Actually to ensure uniform conductive properties we prefer not to have metallic nanotubes.

Dr Schuster said: DNA should be referred to as a wide band-gap semiconductor, which it is.

Professor Naaman agreed.

Dr Schuster then asked: Is it possible to distinguish charge transport by an ionic mechanism or by an electronic mechanism?

Professor Naaman replied: In the case of the carbon nanotubes, the conduction is always by electrons (either electrons or holes). The only check we did for probing the conduction through the

DNA that is attached to the nanotubes was to change the humidity. It has no effect on the conduction. Although it seems unlikely we cannot absolutely reject the ionic conduction mechanisms.

Dr Toppari asked: Why are the thiol–gold bonds not desorbed due to high applied negative biases (< -2 V) at the *IV* measurements? Are the DNAs sticking more strongly to the surface after drying or could it be the case that current is not running through the DNA and thus the thiol contact at all?

Professor Naaman answered: It is important to realize that the measurements are performed in ambient conditions. Hence the redox reactions that take place in solutions are not occurring at the same potentials *etc*. Indeed we do observe electrochemical desorption when the device is in solution, but never when it is in ambient conditions.

Professor Kornyshev asked: Do you see anything special in your conductivity at the melting point of DNA (≈ 50 °C)? Is there a decrease in conductivity?

Professor Naaman replied: We did not study the temperature dependence of the conductivity.

Mr Adisa asked: As the DNA is heated which leads to a loss of the current and then a subsequent increase in the current due to the contact produced between the Au electrodes, is a drop in current observed as the DNA is heated, *i.e.* is the current measured *in situ* and can this drop in current with temperature be related to the temperature of denaturation of the DNA?

Professor Naaman responded: The thermal annealing procedure was not performed while measuring the current. Hence we have no information on the thermally-induced changes due to denaturation of the DNA.

Professor Porath opened the discussion of Professor Bjørnholm's paper: Is it possible that the negative differential resistance (NDR) originates in the presence of two molecules between the electrodes and the mismatch in levels between them at the contact?

Professor Bjørnholm replied: The experimental observation that six different devices display the same behaviour strongly indicates that we are dealing with a well defined system. This together with the careful preparation, where C_{60} molecules are evaporated onto the cold surface one by one, and not in clusters, would disagree with the notion of two molecules.

Professor McCreery said: I am confused about the hysteresis shown in Fig. 2 of the paper. It shows higher current initially as you scan from positive from $V = 0$, then a decrease, then the same when you scan negative from $V = 0$. The system must be returning to its initial state at $V = 0$ in both cases. Is this correct? And if so, why?

Professor Bjørnholm replied: Fig. 3 of the paper and the inset in Fig. 2 show examples of the hysteresis observed below 2.8 K. I think the confusion arises because the graphs were not reproduced in color: Below T_c the system gradually changes from state A to state B above a critical voltage V_c. Once shifted to state B, the systems stays in state B until you reach a negative critical voltage, $-V_c$. In short the hysteresis loop is symmetric around $V = 0$.

Professor Remacle said: To explain the observed NDR, you assume that the free energies of the two conformers (C_{60} oriented with a pentagonal facet towards the surface or C_{60} oriented with a hexagonal facet) vary as a function of the applied field with opposite slopes and cross for a value of the applied voltage V_c (Fig. 1). Could you explain why it is so? Is it supported by simulations?

Professor Bjørnholm replied: So far it is a result of a fit to the data. We are currently working on a detailed model.

Professor Bryce asked: Have functionalised C_{60} derivatives been used in similar devices? For example, use of dimers of C_{60} could affect contact geometries and alter the possible motion of the fullerene in the gap.

Professor Bjørnholm replied: Paul McEuen and co-workers have studied dimerized C_{60} in similar devices and Karsten Flensberg's group have made model calculations showing how transport couples to contact geometries and so on.

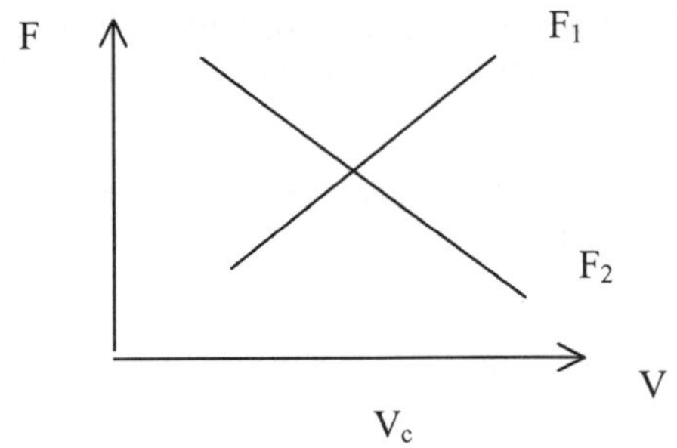

Fig. 1 Plot of free energies against applied voltage.

Professor Lindsay commented: Gold is covered rapidly with hydrocarbons in ambient conditions, so you will not obtain hybridization between gold and C_{60}. Do these type of experiments require either a clean surface in UHV, or strong direct chemisorption to displace adventitious contamination?

Professor Porath added:

1. A measurement in a "dirty" system where a junction is formed between the C_{60} and the gold (instead of hybridization at the π orbital with the gold) would cause Coulomb blockade to be observed and so confirm the observed results.[1,2]
2. Is there a direct tunnelling component between the electrodes in the calculation?

1 D. Porath, Y. Levi, M. Tarabiah and O. Millo, *Phys. Rev. B*, 1997, **56**, 9829.
2 D. Porath and O. Millo, *J. Appl. Phys.*, 1997, **81**, 2241.

Professor Bjørnholm replied:

1. Yes—that is my guess. But we have not done that "control" experiment.
2. No.

Dr Smith opened the discussion of Dr Van der Zant's paper: What is the effect of a magnetic field on the conductivity of the Mn single molecule magnet?

Dr Van der Zant replied: We have only performed preliminary measurements in a magnetic field. We have observed an enhancement of the current at negative bias; at positive bias the current level remains the same. The current increase may be related to the mixing of spin eigenstates allowing transitions that are inhibited due to spin selection rules in the absence of a field. Unfortunately, the angle of the external field with respect to the easy axis is unknown, so that a more quantitative analysis is difficult.

Professor Rovira asked: Have you made measurements on the temperature range where the hysteresis of spin can be observed? Do you think that is this interesting for the changes in conductivity that can be observed?

Dr Van der Zant replied: We measure our samples at 1.6 K well below the anisotropy barrier in the regime where measurements on crystals show magnetic hysteresis. We have not studied this in detail, but plan to perform more magnetic field dependent measurements in the near future.

Professor Fermín said: Results have been presented in the meeting concerning the electrical signature of the Mn_{12} complex in a three-terminal device. I would appreciate if the role of the spin states in the current signatures is further clarified.

Dr Van der Zant responded: A qualitative picture of our current understanding is as follows: Both the total spin of the molecule (S) and its projection (M) on the intrinsic anisotropy axis of the molecule are important. The spin-selection rules $|\Delta S|$, $|\Delta M| = \pm 1/2$ apply to adding or subtracting an electron. Spin states of the molecule that differ by more than 1/2 from the ground states are only accessible *via* sequences of tunnel processes, provided that these are energetically allowed. In this way, sequential tunnel processes lead to a non-equilibrium population of excited states that can only be depopulated slowly by a violation of the spin-selection rules induced by quantum tunneling. When such a sequence becomes energetically allowed, transport can be hindered or blocked, causing negative differential conductance or even a total suppression of the current up to the energy scale of the anisotropy barrier. It is important to note that this new blocking mechanism involves *non-degenerate* spin multiplets.

Dr Higgins asked Professor Bjørnholm and Dr Van der Zant: Do you just use the native oxide in your devices? Is it worth treating the oxide with HMDS or octyltrichlorosilane, for instance, to make it hydrophobic, to see what effect this has (particularly on the C_{60} device)?

Professor Bjørnholm replied: We fabricate the devices in UHV, so I would not expect an oxide on the gold electrodes. The Al_2O_3 present on the gate is a sufficient gate insulator and it would not be an advantage to add an organic molecule to that.

Dr Kubatkin commented: We keep track of electrical properties of our devices at three separate stages of device formation: during the nanogap fabrication, in the process of deposition of molecules and, of course, during the measurements of device with molecules. Therefore for all devices with fullerenes in metallic nanogap we have the reference data on the electron transport in the nanogap as it existed just before the fullerene capture. During this characterization we used the full range of source–drain biases and electrostatic gate voltages which were going to be applied for the measurements of devices with molecules. Examples of these reference data can be found in Fig. 1 of our paper, where featureless current–voltage characteristics, not affected by electrostatic gate are presented. If there was any reconstruction of gold, it would have been seen in these reference data. Therefore, we attribute the temperature dependent NDR effects to the presence of fullerene in the nanogap, and not to the reconstruction of gold.

Professor Tao commented: Your cyclic voltammogram shows no redox activity which led you to think C_{60} should insulating. But your STM image clearly shows that tunnelling current is possible through C_{60}.

Professor Ulstrup replied: The voltammogram would only show redox activity on electron transfer from the electrode to the C_{60}-molecule and trapping of the electron on the molecule. Tunnelling in the STM mode can also be by superexchange with no electron electron trapping and nuclear molecular relaxation on the molecule. The two observations are therefore not mutually exclusive.

Professor Kornyshev opened the discussion of Dr Schuster's paper: What is the nature of the sequence effect on conductivity? Is it like your polaron is pinned to (trapped on) a certain combination of base pairs?

Dr Schuster replied: In general, it can be said that thymines act as the barriers to CT. So a run of Ts would lower the efficiency of CT. The extent of this effect would depend upon the sequence of the DNA. The Ts cannot be looked upon in isolation—it has to be considered with respect to the rest of the sequence.

Professor Naaman asked: What about non-adiabatic interactions that may "smear" the polaron?

Dr Schuster replied: Those are in balance at some polaron length with the forces that result in its location. In both cases they are controlled by the sequence of DNA bases, which varies in normal DNA as does the number of bases included in the polaron.

Professor Bjørnholm said: You mentioned that one possibility to define the term 'wire' was that ballistic tunnelling took place. I find that an unnecessary restriction in the term. It would *e.g.* exclude the polaron transport along a conjugated polymer, which is one of the key examples of a change transport system: a molecular wire!

Professor Calvo asked: May I disagree with your definition of a molecular wire as only a ballistic conductor? For instance in a cell respiratory chain there is a good electrical conduction between the cytochrome *c* oxidase and the florin protein. This is not ballistic but the flow of charge is modulated as needed when you rest, move, dance, *etc*, and it can be fast. Moreover, it has been shown that enzymes can be wired by redox relays by electron hopping between adjacent redox centers in biosensors and biomolecular electronics.

Dr Schuster answered: The term ballistic tunnelling was used to illustrate the general understanding of the term "wire". This term also puts into perspective that charge transport in DNA is quite the opposite of ballistic tunnelling.

It's a definition, which is a way of deciding whether a particular phenomenon belongs in a particular class. I have suggested a very restrictive definition of the phrase molecular wire specifically to provide some clarity to a field that has lacked clear boundaries.

Professor Kornyshev asked: What is the contribution of internal degrees of freedom of DNA to the polaron-formation? How much important is their contribution relative to the obvious contribution of the polarizability of water molecules? As DNA has different flexibility in the AT-rich sections, will the contribution to the polaron shift also be sequence dependent?

Dr Schuster replied: The flexibility of DNA—or the internal degrees of freedom—allows the DNA to distort its structure so as to stabilize the radical cation. The degree of this distortion would depend on the sequence—how much energy is expended to distort the structure and how much stabilization is gained by doing so.

Professor Porath commented: The terms conductor, semiconductor, superconductor and so on should be used with care since defects are dominating the charge transport behaviour. Even a superconductor would not transport charges in 1D in the presence of a defect. The material is practically mediating between defects. The term conductor refers to non-zero slope in the *I–V* curve around zero bias voltage. This is not the case in molecular wires like DNA, where non-gap curves can be observed. They are not metallic conductors but they are still molecular wires.

The main questions are then: is it possible to transport charges between the electrodes through the molecule? How fast and by which mechanism?

Dr Schuster agreed.

Professor Larsson commented: Among the factors contributing to polaron formation are bond length changes as the hole moves from one base to the other. Those contributions are by themselves likely to cause localization, although inhomogeneities may be even more important. I would like to ask you whether there have been any experiments which show marked differences in conduction properties between uniform DNA and λ-DNA.

Dr Schuster replied: As far as I know, all determinations of the conduction properties have been done on uniform DNA and not on λ-DNA. These experiments include the solution phase donor–acceptor type experiments and also experiments wherein DNA has been anchored to electrodes and the conduction measured.

Dr Davis communicated a question for Professor Porath: Could the author comment on the statement that "sometimes the measured current is lower".

Professor Porath communicated in reply: The statement in the Experimental results section that "sometimes the measured current is lower" means that in some of the *I–V* curves the maximal observed current can be significantly lower than 220 nA. This is demonstrated in some of the curves shown in Fig. 1c.

Professor Naaman opened the discussion of Professor Porath's paper: It is important to realise that there is a band gap in the *I–V* curves which indicates that DNA is a semiconductor with a wide band-gap.

Professor Porath replied: In our measurements we observe a gap in most of the measured *I-V* curves. In some of our measurements, however, as in Tao's measurements,[1] we do not observe a gap in the *I–V* curves. The *I–V* measurements reflect the energetic properties of the DNA in the measured configuration and the mechanism of charge transport in that configuration. DNA material probably behaves as a "semiconductor like" material with a gap but its size can not be extracted accurately from our measurements.

1 B. Xu, P. Zhang, X. Li and N. Tao, *Nano Lett.*, 2004, **4**, 1105–1108.

Professor Ulstrup asked: You have alluded to the notion of "coherent" charge transport or "coherence" effects in the DNA-conductivity. Could you give a precise characterization of what is meant by these terms and the physical effects that they imply?

Professor Porath responded: Generally coherence refers to phase preservation and no thermalization. Calculations by Bixon and Jortner[1] showed that hole hopping (which does not preserve phase or coherence) can not lead to currents as high as we measure (even at high voltage) and therefore a faster conduction mechanism is needed in order to account for them. To identify the conduction mechanism we would probably need to measure length, sequence and mismatch dependence of the current flow through DNA.

1 M. Bixon and J. Jortner, "Incoherent charge hopping and conduction in DNA and long molecular chains", *Chem. Phys.*, in press.

Dr Nichols asked: In your Conclusions and discussion you state that "similar currents at comparable bias voltages were observed by Xu *et al.*[1]". Can you clarify this statement, since they used up to 14 base-pair long sequences while you use 26 base-pair long sequences. You see a band gap, while they do not. In these two respects the comparison is not obvious to me.

1 B. Xu, P. Zhang, X. Li and N. Tao, *Nano Lett.*, 2004, **4**, 1105–1108.

Professor Porath replied: Indeed Xu *et al.* measure 8–14 base-pair long molecules while we measure 26 base-pair long molecules. Xu *et al.* report currents of ~150 nA at ~1 V while we measure ~200 nA at 1–2 V. Although the length is not the same both measurements indicate that relatively high currents can flow through single dsDNA molecules. In most cases we do observe a gap and sometimes not. They do not observe a gap in most cases but not always. The differences in the gap appearence may be related to the specific differences in the measurement configuration, contacts and environment. The overall behaviour seems to agree in both measurements and in particular the ability of the molecules to transport relatively high currents.

Dr Nichols addressed Dr Schuster: Ref. 17 of your paper predicts very large fluctuations of the relative energies for hole transfer between groups of bases. These fluctuations are in the order of 0.5–1 eV or so. This is energetically much larger than the differences in oxidation potential between bases. Given this why is there any sequence dependence in hole transfer kinetics or conductance? Should the large fluctuations not simply wash out any base sequence dependence?

Dr Schuster replied This is precisely the mechanism we have called phonon-assisted polaron hopping. The thermal fluctuations (phonons) provide the energy required to overcome the activation barrier caused by different nucleobases having different ionization potentials. The height of the barrier is controlled by the sequence of bases. For example, ATTA between two GG steps creates a higher barrier than AAAA, and this preserves the sequence dependence of the kinetics of the hole transfer rate.

Dr Schuster then commented: I want to suggest that under the conditions of your experiment the motions of "wet" DNA are significantly reduced, which allows the formation of a band.

Professor Porath replied: This is possible. Our experiments do not provide information regarding the origin of possible band formation. Note, however, that Xu *et al.* measured high currents (~150 nA at ~1 V) in a wet environment. I would like to add that the attachment of the molecule to the metal leads is likely to induce Fermi level equalization, level reorganization and charge transfer between the electrodes and the molecule. This may affect the position and width of a possible band as well.

Professor Naaman commented: From the result presented it is clear that the DNA oligomer behaves as a broad band gap semiconductor. The bands that are formed are due to delocalized molecular orbitals on the oligomer. These states are broadened by the interaction with the electrodes and form bands and allow an efficient charge transfer across the oligomer. It is important to distinguish between the fact that DNA is not a conductor and the fact that once the bias potential is high enough so that the Fermi level equalized with the non-localized MO on the DNA, then the electron transmission can be coherent and very efficient.

Professor Calvo addressed Professor Porath and Dr Schuster: Can you comment on the possibility of proton conduction through phosphates along the chain assisting the electron transfer process through the bases?

Professor Porath replied: The current itself is not carried by protons in the outside sugar–phosphate backbone since in our previous publication[1] we have shown that currents that flow through DNA molecules decay upon stretching, which should not happen if it would be a backbone proton current. It is also inconsistent with our previous measurements at low temperatures[2] where proton current is unlikely. To check that experimentally and directly, nicks should be introduced to the DNA backbone. The high currents are likely to be carried by electrons or holes and not protons. Whether or not protons are indirectly assisting the conduction through the inner core of the molecules along the bases I do not know.

1 H. Cohen, C. Nogues, R. Naaman and D. Porath, *Proc. Natl. Acad. Sci., USA*, 2005, **102**, 11589.
2 D. Porath, A. Bezryadin, S. de Vries and C. Dekker, *Nature*, 2000, **403**, 635.

Professor Rampi opened the discussion of Professor Harriman's paper: The back electron transfer from ligand to metal is usually very fast—do you think that in an experiment for measuring the *I–V* curve when the molecular bridges are connecting the electrodes, the back electron transfer rate can be lowered?

Professor Harriman replied: Yes, you are correct in asserting that the rate of back charge transfer is fast in these metal complexes. This rate is easily tuned, however, by substitution of acetylenic groups into the coordinated ligand and/or attachment of aryl hydrocarbons. Lifetimes of around 150 ms can be reached rather easily at ambient temperature. The kind of applications we have in mind, namely photonic wires, would operate very well with such long lifetimes.

Professor Ulstrup asked: Can the reason for the slow fall-off with distance of the photo-induced processes be viewed in terms of a small energy gap between the photo-excited local state and LUMO's of intermediate bridge groups? This effect would give a low distance decay factor, for example in a superexchange view of the photo-induced process.

Professor Harriman replied: We have little doubt that the small attenuation factor arises because the energy gap between donor and bridge orbitals changes with increasing length of the bridge. Because we are looking at electron exchange, which involves simultaneous transfer of an electron and a positive hole, the important gap relates to the HOMO's not the LUMO's. A possibility is that the longer bridges introduce charge-transfer character into the excited state manifold and this affects the HOMO energies. It is clear from temperature dependence studies that the mechanism involves superexchange interactions.

Professor Lindsay said: Experiments on D–B–A systems are not easily transferable to metal–molecule–metal systems because β depends critically on the energy difference between the injected electrons and the molecular level mediating the transport.

Professor Bryce said: Length dependence of fluorescence quantum yield is reported in Table 1 of the paper. It is striking that L_3 is more fluorescent than all the other ligands. I would not have predicted that this specific ligand would be so different from the others in this regard. Do you have an explanation for this?

Professor Harriman replied: These free ligands have closely-spaced charge-transfer and π,π^* excited states and rather small variations in structure affect the relative contributions of the two species. This means that the fluorescence yields and lifetimes are very sensitive to changes in structure, geometry and solvent polarity. This sensitivity tends to plateau as the molecular length increases but is noticeable for the shorter analogues.

Professor Tao opened a general discussion, addressing Dr Schuster: Does your polaron-like model explain exponential dependence of the rate constant on the length? The simple hopping model tells us that it should be inversely proportional to the length.

Dr Schuster replied: It is in fact inversely proportional to the length. We do not think that there is an exponential dependence of the rate constant on the length.

Professor Kuznetsov commented: When one speaks about the bands for the electrons, it is assumed that there is a periodic structure. Electron motion in the conduction band is ballistic, no resistance. However, the atoms are not located all the time in periodic positions due to thermal fluctuations which produce phonons. The interaction of electrons with phonons produces the resistance. If the interaction is very strong, polaron formation takes place and the quasi-free motion may change for hopping. If a hopping mechanism is assumed for DNA, do you have any idea about the activation energy of hopping?

Dr Schuster replied: We have not measured the hopping rate, however our estimate and experiments of others puts it in the microsecond time scale, which is consistent with strong interactions leading to polarons.

Dr Brust asked: What are the effects of changing (i) the doping level (number of vacancies) and (ii) the type of counter ion on the conductance by the hole hopping mechanism?

Dr Schuster responded: Experimentally, we have seen that there is no significant effect of the identity of the counter ion (monovalent ions) on the CT process.

Professor Ulstrup said: There has been repeated reference to "charge transport". Can anything be said regarding whether this is electron or hole transfer in the present case?

Professor Porath replied: This is an excellent question. It seems to be a consensus that holes are carrying the charges in the solution donor–bridge–acceptor experiments. According to calculations by Jortner, holes cannot transport faster than 1 nA (actually way below) by hopping even at high voltage. It is therefore more likely that electrons carry the high currents in the charge transport experiments where high relative currents are measured. More experiments, demonstrating the sequence dependence in particular, are needed to clarify this point.

Dr Nichols and Dr Van Zalinge commented: It has been shown recently that both single and double stranded oligonucleotides in air and in a 10 mM sodium phosphate buffer solution ($pH = 6.8$), shows conductive behaviour.[1] Furthermore the length dependence of the single stranded oligonucleotides has been determined, shown in Fig. 2. These data show two remarkable features. The first is that the hybridisation of the single stranded to a double stranded nucleotide leads to a single molecule conductivity that is orders of magnitude higher. This can be seen by comparing the extrapolated decay lines at a length of 15 nucleotides. Furthermore, the decay constants of the single stranded oligonucleotides are (0.62 ± 0.05), (0.73 ± 0.11), (1.33 ± 0.06) and (0.34 ± 0.01) per nucleotide for oligoadenine, oligothymine, oligocytosine and oligoguanine, respectively.

1. H. van Zalinge, D. J. Schiffrin, A. D. Bates, W. Haiss, J. Ulstrup and R. J. Nichols, *ChemPhysChem*, 2005, accepted for publication.

Professor Kornyshev commented: I suggest that we end by discussing the biological significance of electron transport in the DNA.

Dr Schuster replied: Oxidative damage to genomic DNA is believed to be linked to aging and mutations, which could lead to cancer. Our research has led to an understanding of the factors that control charge transport in oligonucleotides in the test tube, but we are not yet in a position to state that we could apply these principles to genomic DNA. As you have suggested, it is important to apply our understanding to biological systems and we are starting out in that direction.

Professor Porath said: I am aware of experiments carried out in the Barton group that are aimed at addressing the issue of the biological relevance of charge transfer.[1,2] The charge transport experiments that we reported were performed on short sequences of artificially synthesized DNA and I can only speculate on their biological relevance.

1 A. K. Boal, E. Yavin, O. A. Lukianova, V. L. O'Shea, S. S. David and J. K. Barton, *Biochemistry*, 2005, **44**, 8397–8407.
2 E. Yavin, A. Boal, E. D. A. Stemp, E. M. Boon, A. L. Livingston, V. L. O'Shea, S. S. David and J. K. Barton, *Proc. Natl. Acad. Sci, USA*, 2005, **102**, 3546–3551.

Professor Kornyshev communicated in response to Professor Lindsay's Concluding Remarks: Analytical models with parameters should not replace simulations. They should help each other. In the analytical theory the point is not to fit parameters but to see the trends—what happens when you vary them. In ideal simulations you don't explain the nature but reproduce it.

Professor Lindsay communicated in reply: Yes, but quantum chemical simulations claim to provide absolute, quantitative results for transport. So it is important to put them to the test.

Professor Barbero communicated: In several papers presented at the discussion, including Professor Lindsay's Concluding Remarks, the question was raised about how adequate is the S–Au bond, produced by spontaneous adsorption of thiols on gold surfaces, to link metal contacts with electronic molecules. It is clear that such a bond is easy to form, electrochemically stable in a wide potential window and mobile (allowing for molecules to move on the surface until self-assembly occurs). However, the calculations presented by Professor Remacle suggest that charge is

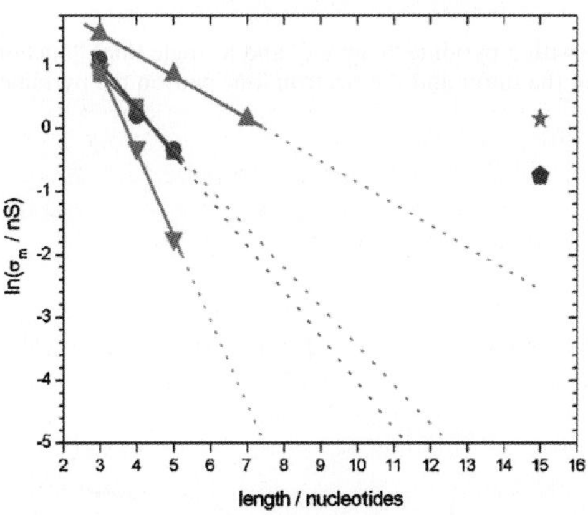

Fig. 2 Dependence of the single molecule conductance of the oligonucleotides on the number of bases, ■ oligo-dA ● oligo-dT, ▲ oligo-dG and ▼ oligo-dC. Also shown are the conductances of double stranded $(dA)_{15}$-$(dT)_{15}$ (⬟) and $(dG)_{15}$-$(dC)_{15}$ (★). The dotted lines are extrapolations of the length dependence in order to compare the shorter single stranded oligonucleotides to the longer double stranded compounds.

localized in the Au–S moiety, making a barrier to charge transfer. Moreover, the size of the sulfur atom makes overlap of non-shared electron orbitals with carbon atoms less effective than smaller atoms (N, O). Therefore, the sharing of electrons with conjugated moieties, like aromatic rings, is difficult, further creating barriers to electron movement. Professor Tao shows data of single molecule conductance suggesting that pyridine groups are better than sulfur atoms in contacting conjugated molecules with gold contacts. This is to be expected since the lone pair of nitrogen in pyridine is conjugated with the ring electron cloud. However, in the same paper it was observed that the molecule desorb at relatively low negative potentials. This is to be expected because the interaction is a weak complex instead of a covalent bond. A possible way to have both behaviours, strong binding and good electronic contact could be to produce molecules where the molecular wire (extended conjugation) is linked to a pyridine which have a thiol linked to the carbon nearest to the nitrogen (Fig. 3). The anchor group will maintain the molecule on the surface in a wide potential window while $\geq$N: of the pyridine will provide electronic contact. A more stable molecule could be designed by having two thiol anchors linked by alkyl chains to the pyridine ring (Fig. 4). The rotation of the C–C bonds in the alkyl chains would provide the flexibility needed for the interaction of the nitrogen lone pair with the gold surface. The effective contact of the pyridine with the surface gold could be easily checked *in situ* using SERS, a technique widely used to study pyridine adsorption. Although other contact groups could be used, this one has the advantage that either setting the gold to negative potentials or decreasing the pH should desorb the pyridine group, breaking the contact. However the anchor groups should maintain the molecule in place. In that way, the conduction mechanism could be tested and/or sensors fabricated.

Obviously, the same scheme could apply to gold surfaces and metal tips, allowing for break-junction or similar methods to be used.

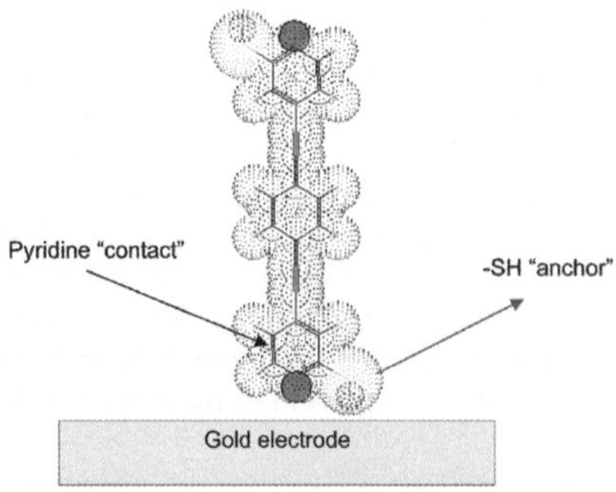

Fig. 3 Conjugated molecule with a pyridine "contact" and a single thiol "anchor" without alkyl spacer. Note the possible steric mismatch of the sulfur and the electron lone pair on the pyridine nitrogen related to the direct link of the sulfur to the ring.

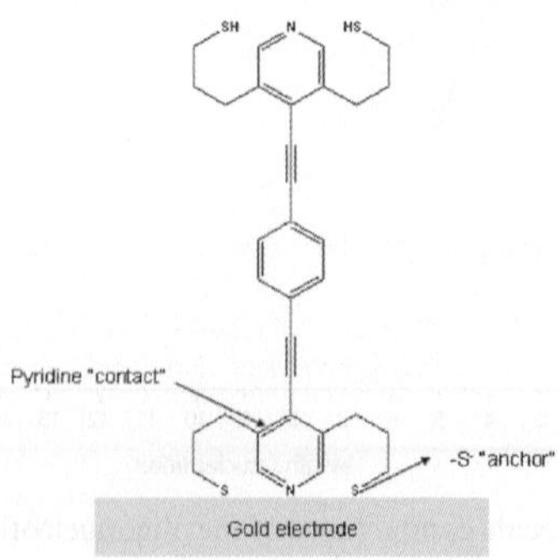

Fig. 4 Conjugated molecule with two thiol "anchors" and one pyridine "contact". The alkyl chain in the anchoring groups would allow for better interaction of the nitrogen lone electron pair with the gold surface.

Molecular wires and devices: Advances and issues

Stuart Lindsay*

Received 9th September 2005, Accepted 9th September 2005
First published as an Advance Article on the web 5th October 2005
DOI: 10.1039/b512828p

Substantial advances have been made in (a) our experimental and theoretical understanding of simple tunnel junctions based on molecules, (b) the role of contacts in tunneling and (c) experimental studies of redox-state gated transport. In addition, a new thermally-activated transport process has been discovered in a non-redox active system. Substantial challenges remain. The engineering of molecular contacts to technologically important electronic materials will stretch the ingenuity of synthetic chemists. The theory community must rise to the challenge of describing redox-mediated transport using first-principles, parameter free approaches. These advances will enable a new generation of devices that span the silicon (electronic) and carbon (bio-organic) worlds.

Introduction

The world of molecular electronics is, in some important ways, an unhappy one. Published experimental data have been all over the map. As just one example, DNA is reported to be an insulator,[1] semiconductor,[2] conductor[3] and even superconductor[4] (though the issue is now probably resolved[5]). This is a field where reasonable agreement between theory and experiment should have been achieved long ago, but with experimental disagreements like this, it is hardly surprising that this been a long time in coming.[6] Hyperbole about imminent molecular computing has been compounded by the effects of outright fraud,[7] leaving the field with little credibility in the eyes of many scientists. This meeting marked a milestone, in that broad quantitative agreement was reached by several laboratories about the magnitude of tunneling in simple systems.[8,9] Redox gated transport, first reported by Tao nearly a decade ago,[10] was the subject of reports by many of the discussants[11–16] and many aspects of the process are in qualitative agreement with the proposals of Kuznetsov and Ulstrup.[7] Interesting issues about contacts and environment remain to be clarified. Haiss and coworkers in the Liverpool group[8] reported a remarkable temperature dependence to 'tunneling' in alkanedithiols and subsequent discussion suggested that this is a new process existing alongside the simple tunneling that is already quite well described by theory.[18] These developments surely mark a turning point in this field. But difficult problems remain if molecular electronics is ever to make a real impact on technology. The key challenges are for the synthetic chemists in developing reliable contacts to materials that are more suitable than gold, and in finding ways to covalently attach devices at both sides of a junction. Theory of redox gated transport will present a major challenge, but it should become possible given the power of modern computers. We turn now to a more detailed review of these topics.

Simple tunneling processes in molecular junctions

Robust comparison with theoretical simulations, the only real measure of our degree of understanding, requires single molecule data. Reproducible single-molecule data requires both well-defined covalent contacts to molecules and statistical analysis of substantial data sets.[19] Tunnel transport in alkanethiols has now been studied with methods that incorporate these features by at least three groups.[19–21] Data taken in this way is in broad agreement with first-principles simulation

Biodesign Institute and Departments of Physics, Chemistry and Biochemisty, Arizona State University, Tempe AZ 85287, USA. E-mail: Stuart.Lindsay@asu.edu

Table 1 Examples of some types of response in redox-mediated charge transport. The letters refer to Fig. 2

Group	Molecule	Response	Ref.
Tao	Perylene derivative	c	11
Lindsay	Oligo-aniline	a	16
Rampi	Tethered Ru	a	13
Ulstrup	Os, Co complexes	a	12
Chi	Cu–azurin	a[a]	14
Haiss	Viologen	b	32
Wandlowski	Viologen	a and b	15

[a] Means intermediate response between a and b.

for a number of molecules as shown in Table 1 of a contribution elsewhere in this volume.[9] (An exception is benzenedithiol, which presents both experimental and theoretical challenges as described by Avik Ghosh and co-workers in Condensed Matter Preprint cond-mat/0505375.) To take one example, values of the conductance reported for a single octanedithiol attached to gold electrodes are approximately 20 nS,[9,21] 3 nS[9] and 1 nS.[19,20] This range of 20 : 1 may seem like a large disagreement, but measured on the scale of the disagreements prior to the introduction of these techniques,[6] it is not. Furthermore, this set of features has been observed in several laboratories, suggesting that they are not spurious, but rather correspond to discrete types of arrangement in the tunnel junction.

Haiss and coworkers have shown that the 1 nS feature is strongly temperature dependent (an observation confirmed in as yet unpublished work from other labs). The 20 nS and 3 nS features are not temperature dependent (also unpublished data). We had suggested that differences in contact geometry could account for these different sets of conductances, although simulations of the difference between top-site and hollow-site attachments did not produce much difference.[9] Recent (unpublished) simulations indicate that 'pyramidal' connection geometries (Fig. 1) might result in substantial reduction of the calculated 20 nS conductance, and so could account for the 3 nS feature. Thus there appear to be at least two distinct temperature independent processes that are described well by tunneling theory.

A new temperature-dependent transport process in alkanethiols

The 1 nS temperature dependent feature reported for octanedithiol by Haiss *et al.* in this discussion[8] has several things in common with the system used in our original work on this molecule:[19]

(a) The low bias conductance of octanedithiol attached to gold electrodes is 1 nS in both cases.

(b) The tunnel decay constant, β, is *ca.* 0.45 Å^{-1} in both cases (reports by Haiss *et al.*[8] and Cui *et al.*[22]). This contrasts with a value of *ca.* 0.8 Å^{-1} for the 20 nS[21] and 3 nS (unpublished) temperature-independent features.

(c) The current–voltage characteristic is linear over a $\pm$1 V range[8] in complete disagreement with simple tunneling theory, but as observed for the 1 nS process when a small gold particle was used as an attachment.[23]

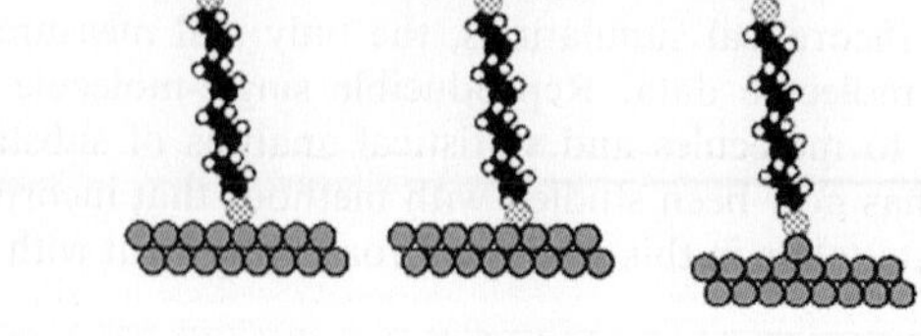

Fig. 1 Showing (a) on-top, (b) hollow site and (c) pyramidal connections of an octanedithiol molecule to gold electrodes. The precise connection geometry influences the tunnel transport significantly.

It is therefore tempting to suppose that that same process underlies the 1 nS feature in both cases. It is interesting to note that in the case of a gold nanoparticle contact[19,23] *no* current is expected in the low-bias blockaded region[23] but one flows anyway. The process may, or may not require the attachment of a gold cluster—this is not done deliberately in the work of Haiss *et al.*, but gold is undoubtedly picked up by thiolated molecules.[24] The temperature dependence suggests that a trap is involved, a factor that could also explain the anomalous value of β.[25] Further theoretical developments and experiments are eagerly awaited.

Resonant tunneling in vacuum

Simple tunneling theories suggest that when the bias applied to a junction aligns the Fermi level of the metal with a molecular orbital of the molecule, resonant tunneling should be manifested by abrupt steps in the current–voltage characteristic.[26] These have not, in general, been observed, presumably because environmentally relaxed and broadened electrochemical processes occur at a lower energy first (when the environment is not ultrahigh vacuum). Weber[27] reported data in this discussion that show these features clearly (and see also the reports by Bjørnholm's group[28]). The energies at which the steps are observed are difficult to account for, because polarization, contact dipoles, image charges and potential distributions present formidable theoretical challenges.[28]

Redox-state gated tunneling

Given the preceding remarks about the difficulty of locating molecular states in a vacuum experiment, one might think that conducting measurements in an electrolyte is nothing short of madness. But the great power of electrochemistry lies in the capability of *controlling the molecular potential* at the electrode with respect to the solution electrochemical potential precisely and reproducibly (though tethered molecular films present some challenges[29]). Thus a molecular electronic process may be quantitatively associated with an oxidation state of a molecule if the measurement is made under potential control. This was first demonstrated by the pioneering work of Tao, who showed that the STM contrast of a protoporphyrin molecule containing an Fe(II) was a maximum near the formal potential for the Fe(II) ⇔ Fe(III) redox process.[10] Work in this area has increased recently[16,30–33] and this discussion marked a flurry of new reports.[11–15] These are summarized in Table 1 and the types of response observed are illustrated in Fig. 2. We will discuss these responses in terms of electronic transitions, but at this early stage, it is important to note that chemical and structural changes could come into play. Thus, a molecule might undergo permanent structural or chemical changes that alter its electronic transmission, giving the appearance of a potential-dependent electronic phenomenon. One important test is to establish the reversibility of the process.[16,33,34]

Most of the responses show a characteristic maximum in the conductance when the surface potential is somewhat above the formal potential for the redox process, E_0 (Fig. 2a). This is in line with the predictions made by Kuznetsov and Ulstrup[17] for the case where the extra current is mediated by partial occupation of the redox state during transport. As the level fluctuates thermally between the Fermi energy on the left electrode ($E_F + V/2$) and the Fermi energy on the right

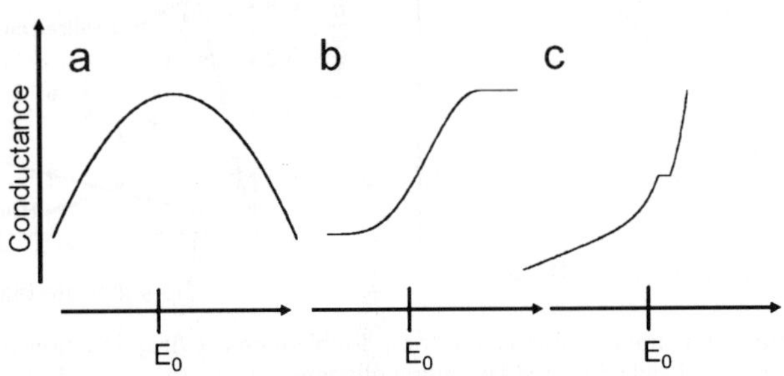

Fig. 2 Showing three types of dependence of the molecular conductance on surface potential. (a) Parabolic with a maximum a little above E_0, (b) sigmoidal with the half-maximum conductance occurring near E_0 and (c) a rapid rise in current with step features reminiscent of resonant tunneling.

electrode ($E_F - V/2$) it serves as a transient resting place for a partially-localized electron that completes its transition by a series of hops, from the left electrode to the molecule, and from the molecule to the right electrode. As the molecule is driven towards its stable (oxidized or reduced) end point above E_0, the molecular state relaxes below the two Fermi levels, cutting off the current channel and causing the fall-off in conductance as the potential is raised. V is the bias applied across the junction, and it is usually small. Note that the width of the response is controlled by thermal fluctuations, so that it should be some small multiple of 25 meV (as it generally is). As an aside, redox processes are conventionally discussed in terms of the (potential-dependent) fraction of the total population of molecules that has undergone an electron transfer process, but if the Kuznetsov–Ulstrup fluctuation-model is correct, then the individual molecules are highly dynamic, so that the change in population with potential occurs through fluctuations of *individual* molecules that move between oxidation states rapidly until the overpotential is sufficiently large to trap the population in its final redox state, well below the lower Fermi level.

Another type of response is sigmoidal, the conductance rising near E_0, and then saturating above it (Fig. 2b). This type of response might occur in a system with densely spaced unoccupied states so that the redox gating sweeps a 'conduction band' into the gap. This might be the case for long oligomers that form highly conductive oxidized states.[35] Wandlowski has shown a given molecule (a viologen derivative) can be switched from the parabolic response (when it is weakly coupled to one electrode) to the sigmoidal response (when it is strongly coupled to one electrode). The response reported for a perylene derivative[11] (Fig. 2c) is unusual, and reminiscent of a resonant tunneling process.

Contacts, environment and voltage-induced changes

Contacts

It is remarkable that such common behavior is found in such disparate experiments as those described above, but it is a testament to the power of electrochemistry. The potential at which a conduction maximum occurs is independent of the means used to contact the molecule because the potential is controlled by the double-layer, which is, in turn, established with respect to a non-polarizable reference electrode. This will not be the case for current–voltage characteristics, where the potential change at the molecule owing to the applied bias is critically-dependent on the microscopic details of the connections to the electrodes and the electrode geometry (this is, of course, precisely the problem that potential control circumvents in the low-bias experiments described above).

Fig. 3a shows *i–V* data for a double-mercury drop junction containing tethered Ru complexes.[13] Coupling to the electrodes is weak, with the molecules located in the middle of the junction by rather long tethers, with no direct covalent linkage across the junction. The current follows the

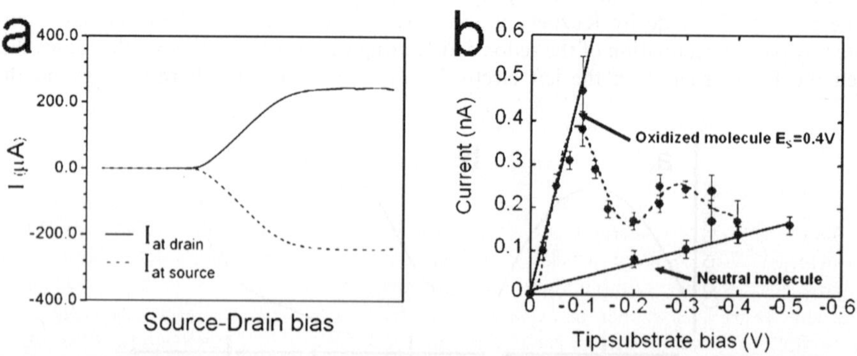

Fig. 3 Showing (a) current–voltage characteristic for a double-mercury drop junction containing redox active molecules[13] and (b) single molecule data for an aniline oligomer covalently tethered to gold electrodes.[16] Data are shown both for the molecule immersed in H_2SO_4 electrolyte with the substrate poised at the potential for maximum conductance (labeled 'oxidized molecule') and for the molecule immersed in toluene (labeled 'neutral molecule').

classical redox-mediated behavior. It rises exponentially with overpotential, saturating when limited by the maximum turn-over rate of the redox centers. This behavior should be contrasted with that of an oligoaniline molecule covalently tethered between gold electrodes and poised at a potential near its peak conductance[16] (Fig. 3b). As the bias is increased in a direction that opposes the surface double layer field (that maintains the molecule near the formal potential) the molecule is driven back towards its less conductive neutral form, resulting striking negative differential resistance. This behavior is accounted for semi-quantitatively using the measured dependence of conductance on surface potential.[16]

Environment

Fig. 3b also illustrates the critical role of environment in charge transport. Measurements made in toluene, lacking the electrolyte and protons that stabilize the oxidized state, show only a linear current–voltage behavior. The applied bias cannot drive changes in oxidation state.

The environment itself can modify the transport dramatically through specific electrochemical reactions. For example, negative differential resistance in ferrocenylundecyl monolayers was recently shown to be a consequence of an irreversible reaction with ambient oxygen.[34]

Voltage-induced changes

McCreery's Raman-scattering studies of molecules in optically-thin junctions[36] show that significant structural changes occur in molecular monolayers sandwiched in a junction when high (*ca.* 3 V) biases are applied. Changes may occur even at small applied bias. For example, we have reported current–voltage data for oligoaniline molecules, both as taken point-by-point using the method of repeated break-junctions, and as obtained by sweeping the bias with a single molecule trapped in the gap. The characteristics are similar, but an additional rising background current is visible in the data taken by sweeping the bias applied to a fixed gap. As pointed out by Haiss during these discussions, potential induced distortions in the gap geometry might be expected to alter the transport over and above any effects owing to molecules in the gap, and this could account for the somewhat different characteristics we obtain using the two methods.[16]

Challenges

Theory

The observation of redox gated charge transfer makes molecular electronics much more interesting: We now have a robust way to build non-linear devices. But understanding these processes at a fundamental level presents many theoretical challenges. A parametric theory of these phenomena is complex[17] as might be expected, and a full-blown first principles simulation might appear to be beyond the range of current techniques. However, much was learned about ion-gated polaron motion in DNA from a 2 ns molecular dynamics simulation combined with electronic structure calculations.[37] It would seem feasible therefore to capture the states of molecule, water and ions that generate the transition states for charge transfer, provided that the molecules are small enough so that nanosecond scale simulations are adequate.

Chemical synthesis

Only a small part of the discussion touched on the issue of contacts, but they are the key to many developments (*cf.* Fig. 3). Even the simple problem of contact to just one electrode needs much more work. How does one make stable attachments of molecules to materials such as silicon that form (easily hydrolyzed) oxides? The same is true for even more complex (and important) electronic materials like the optically-transparent ITO and FTO. Can one engineer contacts to anchor the molecule in a desired orientation with one, and only one type of desired contact bonding?

These are difficult problems, but even more challenging is the problem of putting molecules into *fixed* gaps of sort that might be used to make an integrated circuit. van der Zant[38] alluded to the problems of doing this in a laboratory research setting (much more forgiving than device manufacturing). Our experience with these junctions has not been encouraging, and we see little correlation between the state of the molecules put into fixed gaps (made by electromigration) and the measured characteristics.[39] One might think that distinctive signatures like the Kondo effect would be clear indicators of the presence of a molecule in the gap[40,41] yet Houck *et al.*[42] report observation of a Kondo-type response in up to 30% of bare gold junctions (*i.e.*, containing no molecules). The problem is that not only must the gap be just the right distance to be spanned by a molecule, but it must also accommodate the accompanying relaxation of the metal surface when covalent bonding occurs.

One remarkable recent development is the catalytic growth of alkenes directly onto a ruthenium surface in a process that should be capable of self-termination in a covalent bond at a second electrode.[43]

Other issues in synthesis include materials to replace liquid electrolytes to enable electrochemical gating in manufacturable devices, and materials for passivation of devices to avoid undesired reactions with, *e.g.*, oxygen and water vapor.

Future directions

Molecular electronics is often touted as a potentially revolutionary technology for the next generation of compact electronic devices. While, of course, it is already part of computer hardware (LCDs, OLEDs) the high density applications are fraught with pitfalls. Power is a major problem already, and conventional semiconductor platforms develop apace with feature sizes based on lithography that would have been regarded as impossible a few years ago. Yet there is an obvious and potentially revolutionary application in the interfacing of the carbon and silicon worlds. The completion of the Human Genome Project will drive the development of personalized medicine, with a need for complex *molecular* diagnostic devices that are widely available and cheap (the so-called 'doc-in-a-box'). These will require clever engineering of the electronic materials to molecule interface. Perhaps even more critically, the energy needs of the future might be met by photovoltaics and hydrogen generators based on cleverly engineered interfaces between electronic materials and molecules. These would be noble outcomes of the present research.

Acknowledgements

I thank Nongjian Tao, Rich Nichols, David Schiffrin, Wolfgang Haiss and Tom Wandlowski for useful discussions outside the meeting sessions. This work has been supported by a NIRT grant of the NSF (ECS 01101175) and the Biodesign Institute at Arizona State University.

References

1 D. D. Dunlap, R. Garcia, E. Schabtach and C. Bustamante, *Proc. Natl. Acad. Sci. USA*, 1993, **90**, 7652.
2 D. Porath, A. Bezryadin, S. de Vries and C. Dekkar, *Nature*, 2000, **403**, 635.
3 H.-W. Fink and C. Schoenberger, *Nature*, 1999, **398**, 407.
4 A. Y. Kasumov, M. Kociak, S. Guéron, B. Reulet, V. T. Volkov, D. V. Klinov and H. Bouchiat, *Science*, 2001, **291**, 280.
5 B. Xu, P. M. Zhang, X. L. Li and N. J. Tao, *Nano Lett.*, 2004, **4**, 1105.
6 A. Salomon, D. Cahen, S. Lindsay, J. Tomfohr, V. B. Engelkes and C. D. Frisbie, *Adv. Mater.*, 2003, **15**, 1881.
7 G. Brumfiel, *Nature*, 2002, **418**, 907.
8 W. Haiss, H. van Zalinge, D. Bethell, J. Ulstrup, D. J. Schiffrin and R. J. Nichols, *Faraday Discuss.*, 2005, **131**, DOI: 10.1039/b507520n.
9 J. He, O. Sankey, M. Lee, X. Li, N. J. Tao and S. M. Lindsay, *Faraday Discuss.*, 2005, **131**, DOI: 10.1039/b508434m.
10 N. Tao, *Phys. Rev. Lett.*, 1996, **76**, 4066.
11 X. Li, B. Xu, X. Xiao, X. Yang, L. Zang and N. J. Tao, *Faraday Discuss.*, 2005, **131**, DOI: 10.1039/b512828p.
12 T. Albrecht, K. Moth-Poulsen, J. B. Christensen, A. Guckian, T. Bjornholm, J. G. Vos and J. Ulstrup, *Faraday Discuss.*, 2005, **131**, DOI: 10.1039/b505451f.
13 E. Tran, M. Duati, G. M. Whitesides and M. A. Rampi, *Faraday Discuss.*, 2005, **131**, DOI: 10.1039/b505860k.
14 Q. Chi, J. Zhang, P. S. Jensen, H. E. M. Christensen and J. Ulstrup, *Faraday Discuss.*, 2005, **131**, DOI: 10.1039/b506136a.

15 Z. Li, B. Han, G. Meszaros, I. Pobelov, T. Wandlowski, A. Blaszczyk and M. Mayors, *Faraday Discuss.*, 2005, **131**, DOI: 10.1039/506623a.
16 F. Chen, J. He, C. Nuckolls, T. Roberts, J. Klare and S. M. Lindsay, *Nano Lett.*, 2005, **5**, 503.
17 A. M. Kuznetsov and J. Ulstrup, *Electron Transfer in Chemistry and Biology*, Wiley, Weinheim, 1999.
18 J. Tomfohr and O. F. Sankey, *Phys. Rev. B*, 2002, **65**, 245105.
19 X. D. Cui, A. Primak, X. Zarate, J. Tomfohr, O. F. Sankey, A. L. Moore, T. A. Moore, D. Gust, H. G. and S. M. Lindsay, *Science*, 2001, **294**, 571.
20 W. Haiss, R. J. Nichols, H. van Zalinge, S. J. Higgins, D. Bethell and D. J. Schiffrin, *Phys. Chem. Chem. Phys.*, 2004, **6**, 4330.
21 B. Xu and N. J. Tao, *Science*, 2003, **301**, 1221.
22 X. D. Cui, A. Primak, X. Zarate, J. Tomfohr, O. F. Sankey, A. L. Moore, T. A. Moore, D. Gust, L. A. Nagahara and S. M. Lindsay, *J. Phys. Chem B*, 2002, **106**, 8609.
23 J. Tomfohr, G. Ramachandran, O. F. Sankey and S. M. Lindsay, in *Introducing Molecular Electronics*, ed. G. Fagas and K. Richter, Springer, Berlin, 2005.
24 G. K. Ramachandran, T. J. Hopson, A. M. Rawlett, L. A. Nagahara, A. Primak and S. M. Lindsay, *Science*, 2003, **300**, 1413.
25 J. Halbritter, *J. Appl. Phys.*, 1985, **58**, 1320.
26 S. Datta, *Electronic Transport in Mesoscopic Systems*, Cambridge University Press, 1990.
27 R. Ochs, D. Secker, M. Elbing, M. Mayor and H. B. Weber, *Faraday Discuss.*, 2005, **131**, DOI: 10.1039/b506390f.
28 S. Kubatkin, A. Danilov, M. Hjort, J. Cornil, J.-L. Bredas, N. Stuhr-Hansen, P. Hedegard and T. Bjornholm, *Nature*, 2003, **425**, 698.
29 C. P. Smith and H. S. White, *Anal. Chem.*, 1992, **64**, 2398.
30 D. I. Gittins, D. Bethell, D. J. Schriffrin and R. J. Nichols, *Nature*, 2000, **408**, 67.
31 E. Tran, M. A. Rampi and G. M. Whitesides, *Angew. Chem. Int. Ed.*, 2004, **43**, 3835.
32 W. Haiss, H. van Zalinge, S. J. Higgins, D. Bethell, H. Hobenreich, D. J. Schiffrin and R. J. Nichols, *J. Am. Chem. Soc.*, 2003, **125**, 15294.
33 X. Xiao, L. Nagahara, A. Rawlett and N. J. Tao, *J. Am. Chem. Soc.*, 2005, **127**, 9235.
34 J. He and S. Lindsay, *J. Am. Chem. Soc.*, 2005, **127**, 11932.
35 J.-L. Bredas and G. B. Street, *Acc. Chem. Res.*, 1985, **18**, 309.
36 R. L. McCreery, U. Viswanathan, R. P. Kalakodimi and A. M. Nowak, *Faraday Discuss.*, 2005, **131**, DOI: 10.1039/b505684p.
37 R. N. Barnett, C. L. Cleveland, A. Joy, U. Landman and G. B. Schuster, *Science*, 2001, **294**, 567.
38 H. S. J. van der Zant, Y.-V. Kervennic, M. Poot and S. G. Kafanov, *Faraday Discuss.*, 2005, **131**, DOI: 10.1039/b506240n.
39 J. He, F. Chen, P. A. Liddell, J. Andréasson, S. D. Straight, D. Gust, T. A. Moore, A. L. Moore, J. Li, O. F. Sankey and S. M. Lindsay, *Nanotechnology*, 2005, **16**, 695.
40 J. Park, A. N. Pasupathy, J. I. Goldsmith, C. Chang, Y. Yaish, J. R. Petta, M. Rinkoski, J. P. Sethna, H. D. Abruna, P. L. McEuen and D. C. Ralph, *Nature*, 2002, **417**, 722.
41 W. Liang, M. P. Shores, M. Bockrath, J. R. Long and H. Park, *Nature*, 2002, **417**, 725.
42 A. A. Houck, J. Labaziewicz, E. K. Chan, J. A. Folk and I. L. Chuang, *Nano Lett.*, 2005, DOI: 10.1021/nl050899i.
43 G. S. Tulevski, M. B. Myers, M. S. Hybertsen, M. L. Steigerwald and C. Nuckolls, *Science*, 2005, **309**, 591.

Poster titles

Development of a novel test bed device for molecular electronics applications **Catherine E. Gardner**, **Mohamed A. Ghanem**, **James W. Wilson** and **David C. Smith**, *University of Southampton, UK*

Understanding STM of physisorbed aromatic molecules: from band structures to transmission probabilities and bias-dependent imaging **Victor M. Geskin**, **J. Cornil** and **R. Lazzaroni**, *Université de Mons-Hainaut, Belgium*

Electrochemical fabrication of PEDOT nanowires between Au electrodes on SiO_2 substrates **A. Das**, **C. H. Lei**, **M. Elliott** and **J. Emyr Macdonald**, *Cardiff University, UK*

ToF SIMS, PM-RAIRS and XPS studies of π-conjugated aromatic molecular wires grown on gold **Rosella Pinna**, **F. Jamme**, **F. J. M. Rutten**, **E. F. Smith**, **M. R. Willis**, **D. Briggs** and **M. R. S. McCoustra**, *University of Nottingham, UK*

Transistor effects in redox molecules at room temperature **Tim Albrecht**, **K. Moth-Poulsen**, **J. B. Christensen**, **A. Guckian**, **T. Bjørnholm**, **J. Ulstrup** and **J. G. Vos**, *Technical University of Denmark, Denmark* and *Dublin City University, Republic of Ireland*

Preparation of mixed Langmuir Blodgett films containing a molecular wire **A. Villares**, **S. Martín**, **M. Haro**, **B. Giner**, **P. J. Low** and **P. Cea**, *Universidad de Zaragoza, Spain* and *University of Durham, UK*

Construction of molecular wire with photo-switching functions on hydrogen-terminated Si(111) surface **T. Masuda**, **M. Irie** and **K. Uosaki**, *Hokkaido University, Japan*

Unidirectional two-step photoinduced energy transfer in metallo-cyclodextrin assemblies **J. A. Faiz**, **M. J. Pereira Silva**, **C. J. Milios**, **J. M. Haider**, **P. Contrera**, **R. M. Williams**, **L. DeCola** and **Z. Pikramenou**, *University of Birmingham, UK* and *University of Amsterdam, The Netherlands*

Supramolecular wires based on metallocyclodextrins: solution and surface assembly **E. T. Kefalas**, **P. Bertoncello**, **C. J. Milios**, **J. M. Haider**, **P. Unwin**, **R. Forster** and **Z. Pikramenou**, *University of Birmingham, UK*, *University of Warwick, UK* and *Dublin City University, Republic of Ireland*

Dielectrophoresis of nanoscale double stranded DNA and humidity effects on its electrical conductivity **S. Tuukkanen**, **A. Kuzyk**, **J. J. Toppari**, **V. P. Hytönen**, **T. Ihalainen** and **P. Törmä**, *University of Jyväskylä, Finland*

Molecular "wiring" enzymes to electrodes in LbL self-assembled enzyme multilayers with redox polyelectrolytes **Mario Tagliazucchi**, **Victoria Flexer**, **Claudia B. Danilowicz**, **Alejandro Wolosiuk** and **Ernesto J. Calvo**, *Universidad de Buenos Aires, Argentina*

Atomic and electronic structure and STM images of molecular adsorbates on silicon(100) surface **Ante Bilic**, **Jeffrey R. Reimers** and **Noel S. Hush**, *University of Sydney, Australia*

Electrical conduction through single molecules: the nature of self-assembled monolayers: the bonding of nitrogen to the gold surface in 1,10-phenanthroline **Peter Cafe**, **Ante Bilic**, **Palli Thordarson**, **Jeffrey R. Reimers**, **Maxwell J. Crossley** and **Noel S. Hush**, *University of Sydney, Australia*

Inelastic electron tunnelling spectroscopy: the molecular signature in single molecule conduction **G. C. Solomon**, **A. Gagliardi**, **A. Pecchia**, **T. Frauenheim**, **Aldo Di Carlo**, **Jeffrey R. Reimers** and **Noel S. Hush**, *University of Sydney, Australia*

Fabrication of nanoscale gaps for the electrical interrogation of molecules **P. Steinmann** and **J. M. R. Weaver**, *University of Glasgow, UK*

Biomolecular self assembly to create nanoscale metallic wires and structures **A. D. Bates**, **B. Callen**, **J. M. Cooper**, **R. Cosstick**, **D. R. S. Cumming**, **Andrew Glidle**, **L. Jager**, **J. Pearson**, **M. Proupín-Pérez** and **C. Xu**, *University of Glasgow, UK*

Nanowires and nanogaps for molecular electronics **Farouk Hadeed** and **C. Durkan**, *University of Cambridge, UK*

A spectroscopic view on interfacial electron transfer. *In situ* ATR-SEIRAS of viologen adlayers on Au electrodes **B. Han**, **Z. Li**, **Th. Wandlowski**, **A. Błaszczyk** and **M. Mayor**, *Forschungszentrum Jülich, Germany* and *Forschungszentrum Karlsruhe, Germany*

Molecules that mimic Schottky diodes **Geoffrey J. Ashwell**, **Anna Chwialkowski**, **Barbara Urasinska**, **Wayne D. Tyrrell**, **Abdul Mohib** and **Anne J. Whittam**, *Cranfield University, UK*

The electrical characterisation of single and double stranded DNA molecules **H. van Zalinge**, **D. J. Schiffrin**, **A. D. Bates**, **W. Haiss**, **J. Ulstrup** and **R. J. Nichols**, *University of Liverpool, UK*

Single molecule conductivity at electrochemical interfaces – an attempt to correlate adlayer structure and functionality **Zhihai Li**, **Ilya Pobelov**, **Bo Han**, **Marcel Mayor** and **Thomas Wandlowski**, *Forschungszentrum Jülich, Germany* and *Forschungszentrum Karlsruhe, Germany*

The Skinner Prize for the best poster was awarded to Jonathan Faiz from the University of Birmingham, UK, for his poster on unidirectional two-step photoinduced energy transfer in metallo-cyclodextrin assemblies

List of Participants

Mr A. Adisa, *University of Manchester, United Kingdom*
Dr T. Albrecht, *Technical University of Denmark, Denmark*
Professor G. Ashwell, *Cranfield University, United Kingdom*
Professor C. Barbero, *Universidad Nacional de Rio Cuarto, Argentina*
Dr C. Batchelor, *Royal Society of Chemistry, United Kingdom*
Dr J. Bell, *Royal Society of Chemistry, United Kingdom*
Dr A. Benniston, *University of Newcastle, United Kingdom*
Professor D. Bethell, *University of Liverpool, United Kingdom*
Professor T. Bjornholm, *University of Copenhagen, Denmark*
Dr A. Bronowska, *University of Leeds, United Kingdom*
Professor M. Bryce, *University of Durham, United Kingdom*
Mr P. Cafe, *University of Sydney, Australia*
Professor E. Calvo, *Universidad de Buenos Aires, Argentina*
Dr Q Chi, *Technical University of Denmark, Denmark*
Dr T. Cox, *QINETIQ, United Kingdom*
Dr H Crichton, *Royal Society of Chemistry, United Kingdom*
Dr D. Crouch, *University of Manchester, United Kingdom*
Mr A. Danilov, *Chalmers University of Technology, Sweden*
Dr A. Das, *Cardiff University, United Kingdom*
Dr J. Davis, *University of Oxford, United Kingdom*
Miss A. Demetriou, *University of Liverpool, United Kingdom*
Dr R. Dryfe, *University of Manchester, United Kingdom*
Mr D. Evans, *University of Leeds, United Kingdom*
Mr J. Faiz, *University of Birmingham, United Kingdom*
Professor D. Fermin, *University of Bern, Switzerland*
Professor D. Fitzmaurice, *University College Dublin, Republic of Ireland*
Dr C. Gardner, *University of Southampton, United Kingdom*
Dr V. Geskin, *Université de Mons-Hainaut, Belgium*
Miss M. Gilbert, *Royal Society of Chemistry, United Kingdom*
Dr D. Giovanelli, *Royal Society of Chemistry, United Kingdom*
Dr A. Glidle, *University of Glasgow, United Kingdom*
Mr F. Hadeed, *University of Cambridge, United Kingdom*
Dr W. Haiss, *University of Liverpool, United Kingdom*
Mr B. Han, *Forschungszentrum Jülich, Germany*
Professor A. Harriman, *University of Newcastle, United Kingdom*
Dr S. Harris, *University of Leeds, United Kingdom*
Professor J. Heath, *Caltech, USA*
Dr J. Herrmann, *CSIRO, Australia*
Dr S. Higgins, *University of Liverpool, United Kingdom*
Professor N Hush, *University of Sydney, Australia*
Dr M. Kalaji, *University of Wales, Bangor, United Kingdom*
Mr E. Kefalas, *University of Birmingham, United Kingdom*
Professor A. Kornyshev, *Imperial College London, United Kingdom*
Mr S. Kubatkin, *Chalmers University of Technology, Sweden*
Professor A. Kuznetsov, *A. N. Frumkin Institute of Electrochemistry, Russia*
Professor S. Larsson, *Chalmers University of Technology, Sweden*
Mr E. Leary, *University of Liverpool, United Kingdom*
Mr Z. Li, *Forschungszentrum Jülich, Germany*
Professor S. Lindsay, *Arizona State University, USA*
Mr E. Macdonald, *Cardiff University, United Kingdom*
Dr S. Makhseed, *Kuwait University, Kuwait*
Mr T. Masuda, *Hokkaido University, Japan*

Mr R. McCreery, *Ohio State University, USA*
Dr S. Mertens, *Technical University of Denmark, Denmark*
Mr A. Mohib, *Cranfield University, United Kingdom*
Professor R. Naaman, *Weizmann Institute of Science, Israel*
Dr I. Nandhakumar, *University of Southampton, United Kingdom*
Dr R. Nichols, *University of Liverpool, United Kingdom*
Dr K. O'Donoghue, *Ciba Specialty Chemicals, United Kingdom*
Miss H. Paloniemi, *University of Turku, Finland*
Miss R. Pinna, *University of Nottingham, United Kingdom*
Professor D. Porath, *The Hebrew University, Israel*
Miss K. Psychogyiopoulou, *University of Liverpool, United Kingdom*
Professor M. Rampi, *University of Ferrara, Italy*
Professor F. Remacle, *University of Liège, Belgium*
Professor C. Rovira, *Institut de Ciencia de Materials de Barcelona, Spain*
Professor D. Schiffrin, *University of Liverpool, United Kingdom*
Dr G. Schuster, *Georgia Institute of Technology, USA*
Miss I. Sendroiu, *University of Liverpool, United Kingdom*
Mr H. Simmonds, *Cardiff University, United Kingdom*
Dr D. Smith, *University of Southampton, United Kingdom*
Ms G. Solomon, *University of Sydney, Australia*
Dr P. Steinmann, *University of Glasgow, United Kingdom*
Professor N. Tao, *Arizona State University, USA*
Dr J. Toppari, *University of Jyväskylä, Finland*
Dr W. Tyrrell, *Cranfield University, United Kingdom*
Professor J. Ulstrup, *Technical University of Denmark, Denmark*
Dr H. Van Der Zant, *Delft University of Technology, The Netherlands*
Dr H. Van Zalinge, *University of Liverpool, United Kingdom*
Miss A. Villares, *Universidad de Zaragoza, Spain*
Dr T. Wandlowski, *Forschungszentrum Jülich, Germany*
Dr T. Wear, *Kodak, United Kingdom*
Professor J. Weaver, *University of Glasgow, United Kingdom*
Professor H. Weber, *Forschungszentrum Karlsruhe, Germany*
Mr W. Xi, *University of Oxford, United Kingdom*

Index of Contributors*

* The page numbers in **bold** type indicate papers submitted for discussions.